AF692703

Emerging Agricultural Engineering Sciences, Technologies, and Applications—Volume I

Emerging Agricultural Engineering Sciences, Technologies, and Applications—Volume I

Topic Editors

Muhammad Sultan
Yuguang Zhou
Redmond R. Shamshiri
Muhammad Imran

MDPI

Basel • Beijing • Wuhan • Barcelona • Belgrade • Novi Sad • Cluj • Manchester

Topic Editors

Muhammad Sultan
Bahauddin Zakariya University
Multan
Pakistan

Yuguang Zhou
China Agricultural University
Beijing
China

Redmond R. Shamshiri
Leibniz Institute for
Agricultural Engineering
and Bioeconomy
Potsdam
Germany

Muhammad Imran
Aston University
Birmingham
UK

Editorial Office
MDPI AG
Grosspeteranlage 5
4052 Basel, Switzerland

This is a reprint of the Topic, published open access by the journals *Agriculture* (2077-0472), *AgriEngineering* (2624-7402), *Energies* (1996-1073), *Sustainability* (2071-1050) and *Water* (2073-4441), freely accessible at: https://www.mdpi.com/topics/Agricultural_Engineering_Technologies.

For citation purposes, cite each article independently as indicated on the article page online and as indicated below:

Lastname, A.A.; Lastname, B.B. Article Title. *Journal Name* **Year**, *Volume Number*, Page Range.

Volume I
ISBN 978-3-7258-2711-4 (Hbk)
ISBN 978-3-7258-2712-1 (PDF)
https://doi.org/10.3390/books978-3-7258-2712-1

Set
ISBN 978-3-7258-2709-1 (Hbk)
ISBN 978-3-7258-2710-7 (PDF)

© 2024 by the authors. Articles in this book are Open Access and distributed under the Creative Commons Attribution (CC BY) license. The book as a whole is distributed by MDPI under the terms and conditions of the Creative Commons Attribution-NonCommercial-NoDerivs (CC BY-NC-ND) license (https://creativecommons.org/licenses/by-nc-nd/4.0/).

Contents

About the Editors

Muhammad Sultan

Dr. Muhammad Sultan is an Associate Professor of Agricultural Engineering at Bahauddin Zakariya University, Pakistan. He earned his B.Sc. and M.Sc. degrees in Agricultural Engineering with distinction from the University of Agriculture Faisalabad, Pakistan. Dr. Sultan completed his Ph.D. and Postdoctoral Research in Energy & Environmental Engineering at Kyushu University, Japan. He furthered his expertise with Postdoctoral Research in Mechatronic Systems Engineering at Simon Fraser University, Canada. Dr. Sultan is a Research Fellow at the University of South Africa and INTI International University (Malaysia). His previous experience includes posts at WPI-I2CNER—International Institute for Carbon-Neutral Energy Research, Japan, and CORE, Canada. Recognized as one of the World's Top 2% Scientists by Stanford University/Elsevier, Dr. Sultan is also ranked as the leading scientist in Pakistan in the field of Agricultural Engineering according to the AD Scientific Index Ranking. He has also been honored with several awards in recognition of his outstanding research contributions. Currently, Dr. Sultan is engaged in several local- and foreign-funded research projects. He has authored more than 200 journal articles, alongside 15 books, 25 book chapters, and 110 conference articles. He serves as an editor for 20 renowned journals and a reviewer for more than 100 journals. His research focuses on advancing state-of-the-art energy-efficient systems for agricultural applications—innovative solutions that optimize energy use, reduce environmental impact, and improve sustainability in agricultural practices.

Yuguang Zhou

Dr. Yuguang Zhou is currently an Associate Professor at the Department of Agricultural Engineering, College of Engineering, at China Agricultural University, secretary of Key Technologies for Cleaner Utilization of Renewable Energy by the Ministry of Agriculture and Rural Affairs and National International Joint Research Center for Bioenergy Science and Technology, Executive Secretary of the ISO/TC255 (Biogas) Secretariate, Chinese expert on ISO/TC285 (clean biomass stoves international standards technical committee), expert on EPD (Environmental Product Declaration, member of the Chinese Society of Agricultural Engineering, and a member of the Beijing Society of Agricultural Engineering. Dr. Zhou's research focus is related to agro-biomass feedstock utilization via engineering measures.

Redmond R. Shamshiri

Dr. Redmond R. Shamshiri is a senior scientist at the Leibniz-Institut für Agrartechnik und Bioökonomie, working on digitizing agriculture to enhance food security. He holds a Ph.D. in Mechanical Engineering from the Technische Universität Berlin, a Ph.D. in Agricultural Automation from Universiti Putra Malaysia, and an M.Sc. and a Doctor of Engineering in Biosystems from the University of Florida. His research interests include the autonomous navigation of agricultural mobile robots, sensor fusion for collision avoidance, teleoperation, wireless systems, and the optimization of controlled environments. With more than 150 publications, Dr. Shamshiri has been recognized among the top 2% of scientists worldwide in the past three years.

Muhammad Imran

Dr. Muhammad Imran is a prominent academic and researcher in the field of thermal energy systems, currently serving as a Reader in Energy Systems at Aston University, Birmingham, UK. His work focuses on pioneering innovations in energy conversion, thermal energy systems, and waste heat recovery. Since 2012, Dr. Imran has driven advancements in thermal energy systems design, optimization, and system integration, contributing significantly to both the theoretical and practical dimensions of thermal energy systems. His expertise spans hybrid energy systems—including geothermal, solar, wind, and biomass—along with combined heat and power systems, thermal energy management, district heating and cooling networks, advanced heat pumps, and thermal energy storage. Dr. Imran's extensive research contributions include both numerical modeling and experimental validations, resulting in over 180 publications in high-impact journals and conference proceedings. As a Fellow of the Higher Education Academy (FHEA), Dr. Imran is recognized for his commitment to both research and education. He has successfully guided research projects and mentored students, establishing himself as a leader in sustainable energy research. Through his work, Dr. Muhammad Imran continues to advance the future of energy technology, supporting the engineering sector's shift toward greater efficiency and sustainability.

AgriEngineering

Editorial

Emerging Agricultural Engineering Sciences, Technologies, and Applications

Muhammad Sultan

Department of Agricultural Engineering, Bahauddin Zakariya University, Multan 60800, Pakistan; muhammadsultan@bzu.edu.pk

1. Introduction

The closing Editorial of this comprehensive special collection presents the journey from this project's inception to the publication of around five dozen outstanding studies that have been a testament to the dedication, innovation, and collective wisdom of the global agricultural engineering community.

Agricultural engineering plays a vital role in the modernization and revitalization of agriculture [1]. It encompasses various fields such as agricultural production design and management, agricultural mechanization, water/soil engineering, land use optimization, and agricultural resource recycling [1,2]. The scope of agricultural engineering also includes the integration of agricultural engineering technologies, development of agricultural engineering patterns, and the promotion of science and technology innovation in the agricultural sector [2]. Overall, agricultural engineering has a broad scope that encompasses various aspects of agricultural development and innovation, e.g., agricultural greenhouses [3–8], agricultural storage systems [9–12], irrigation and drainage [13–15], agricultural energy systems [16–18], agricultural fuels [19,20], fam machinery [21–23], etc.

Over the past year, the field of agricultural engineering has witnessed remarkable advancements. The published articles in this collection span a broad spectrum, covering themes from precision farming and remote sensing to sustainable practices, robotics, and the integration of cutting-edge technologies. These contributions collectively weave a narrative of progress, highlighting the dynamic nature of agricultural engineering in addressing contemporary challenges.

Agricultural engineering is inherently interdisciplinary, requiring a holistic understanding of engineering and technology for various agricultural applications, as shown in Figure 1. This collection has successfully filled crucial knowledge gaps by delving into diverse areas. The pursuit of knowledge has been relentless, and our contributors have shed light on unexplored facets of soil science, crop management, and post-harvest activities.

Citation: Sultan, M. Emerging Agricultural Engineering Sciences, Technologies, and Applications. *AgriEngineering* **2024**, *6*, 2057–2066. https://doi.org/10.3390/agriengineering6030120

Received: 7 April 2024
Accepted: 30 April 2024
Published: 3 July 2024

Copyright: © 2024 by the author. Licensee MDPI, Basel, Switzerland. This article is an open access article distributed under the terms and conditions of the Creative Commons Attribution (CC BY) license (https://creativecommons.org/licenses/by/4.0/).

Figure 1. An overview of modern research areas in the field of agricultural engineering.

2. List of Contributions

In total, 190 manuscripts were submitted for consideration in this Topic Collection, and all of them were subject to rigorous review processes by their respective journals. Just under 31% of the submitted papers were accepted for publication and inclusion in this Topic Collection. The overall scientific contributions to this Topic Collection are listed below:

1. Peralta Manjarrez, R.M.; Delgado Martínez, R.; Benavides Mendoza, A.; Juárez Maldonado, A.; Cabrera De la Fuente, M. Calcium, Potassium, and Magnesium Affect the Nutritional Value of Tomato Grafted Fruits Grown in a Nutrient Film Technique System. *Agriculture* **2023**, *13*, 2189. https://doi.org/10.3390/agriculture13122189.
2. Mardiah, M.; Samadhi, T.W.; Wulandari, W.; Aqsha, A.; Situmorang, Y.A.; Indarto, A. Recent Progress on Catalytic of Rosin Esterification Using Different Agents of Reactant. *AgriEngineering* **2023**, *5*, 2155–2169. https://doi.org/10.3390/agriengineering5040132.
3. Abhiram, G.; Diraj, R.; Eeswaran, R. Optimization of Black Tea Drying Temperature in an Endless Chain Pressure (ECP) Dryer. *AgriEngineering* **2023**, *5*, 1989–1999. https://doi.org/10.3390/agriengineering5040122

4. Linets, G.I.; Bazhenov, A.V.; Malygin, S.V.; Grivennaya, N.V.; Melnikov, S.V.; Goncharov, V.D. Evaluation of the Accuracy of the Remote Determination of the Brewster Angle When Measuring Physicochemical Parameters of Soil. *AgriEngineering* **2023**, *5*, 1893–1908. https://doi.org/10.3390/agriengineering5040116
5. Chaomuang, N.; Nuangjamnong, T.; Rakmae, S. Performance Evaluation of a Wet Medium Made of Mangosteen Peels for a Direct Evaporative Cooling System. *AgriEngineering* **2023**, *5*, 1865–1878. https://doi.org/10.3390/agriengineering5040114
6. Siriani, A.L.R.; Miranda, I.B.d.C.; Mehdizadeh, S.A.; Pereira, D.F. Chicken Tracking and Individual Bird Activity Monitoring Using the BoT-SORT Algorithm. *AgriEngineering* **2023**, *5*, 1677–1693. https://doi.org/10.3390/agriengineering5040104
7. Ghanim, A.A.J.; Anjum, M.N.; Rasool, G.; Saifullah; Irfan, M.; Alyami, M.; Rahman, S.; Niazi, U.M. Analyzing Extreme Temperature Patterns in Subtropical Highlands Climates: Implications for Disaster Risk Reduction Strategies. *Sustainability* **2023**, *15*, 12753. https://doi.org/10.3390/su151712753
8. Oh, B.-W.; Seo, H.-J.; Seo, I.-H. Ventilation Operating Standard for Improving Internal Environment in Pig House Grafting Working Conditions Using CFD. *AgriEngineering* **2023**, *5*, 1378–1394. https://doi.org/10.3390/agriengineering5030086
9. Raza, Q.-U.-A.; Bashir, M.A.; Rehim, A.; Geng, Y.; Raza, H.M.A.; Hussain, S.; Ahmad, I.; Wasif, M. Identifying the Role of Biostimulants in Turnip (*Brassica rapa* L.) Production Compared with Chemical Fertilization. *Sustainability* **2023**, *15*, 11851. https://doi.org/10.3390/su151511851
10. Kaskous, S.; Pfaffl, M.W. Milking Machine Settings and Liner Design Are Important to Improve Milking Efficiency and Lactating Animal Welfare—Technical Note. *AgriEngineering* **2023**, *5*, 1314–1326. https://doi.org/10.3390/agriengineering5030083
11. Xiao, Y.; Sun, C.; Wang, D.; Li, H.; Guo, W. Analysis of Hotspots in Subsurface Drip Irrigation Research Using CiteSpace. *Agriculture* **2023**, *13*, 1463. https://doi.org/10.3390/agriculture13071463
12. Schmidt, D.; Butturi, M.A.; Sellitto, M.A. Opportunities of Digital Transformation in Post-Harvest Activities: A Single Case Study of an Engineering Solutions Provider. *AgriEngineering* **2023**, *5*, 1226–1242. https://doi.org/10.3390/agriengineering5030078
13. Ashraf, H.; Qamar, S.; Riaz, N.; Shamshiri, R.R.; Sultan, M.; Khalid, B.; Ibrahim, S.M.; Imran, M.; Khan, M.U. Spatiotemporal Estimation of Reference Evapotranspiration for Agricultural Applications in Punjab, Pakistan. *Agriculture* **2023**, *13*, 1388. https://doi.org/10.3390/agriculture13071388
14. Alves, M.d.F.A.; Pandorfi, H.; Montenegro, A.A.d.A.; Silva, R.A.B.d.; Gomes, N.F.; Santana, T.C.; Almeida, G.L.P.d.; Marinho, G.T.B.; Silva, M.V.d.; Silva, W.A.d. Evaluation of Body Surface Temperature in Pigs Using Geostatistics. *AgriEngineering* **2023**, *5*, 1090–1103. https://doi.org/10.3390/agriengineering5020069
15. Aon, M.; Aslam, Z.; Hussain, S.; Bashir, M.A.; Shaaban, M.; Masood, S.; Iqbal, S.; Khalid, M.; Rehim, A.; Mosa, W.F.A.; et al. Wheat Straw Biochar Produced at a Low Temperature Enhanced Maize Growth and Yield by Influencing Soil Properties of *Typic calciargid*. *Sustainability* **2023**, *15*, 9488. https://doi.org/10.3390/su15129488
16. Akinhanmi, F.O.; Ayanda, O.I.; Ahuekwe, E.F.; Dedeke, G.A. Mitigating the Impacts of the COVID-19 Pandemic on Crop Farming: A Nanotechnological Approach. *Agriculture* **2023**, *13*, 1144. https://doi.org/10.3390/agriculture13061144
17. Jiang, Y.; Xuan, K.; Gao, C.; Liu, Y.; Zhao, Y.; Deng, H.; Li, X.; Liu, J. Investigating the Potential of Cosmic-Ray Neutron Sensing for Estimating Soil Water Content in Farmland and Mountainous Areas. *Water* **2023**, *15*, 1500. https://doi.org/10.3390/w15081500
18. Neethirajan, S. The Significance and Ethics of Digital Livestock Farming. *AgriEngineering* **2023**, *5*, 488–505. https://doi.org/10.3390/agriengineering5010032

19. Darvishi, Y.; Hassan-Beygi, S.R.; Massah, J.; Gancarz, M.; Bieszczad, A.; Karami, H. Determining the Influence of a Magnetic Field on the Vibration and Fuel Consumption of a Heavy Diesel Engine. *Sustainability* **2023**, *15*, 4088. https://doi.org/10.3390/su15054088
20. Neupane, J.; Maja, J.M.; Miller, G.; Marshall, M.; Cutulle, M.; Greene, J.; Luo, J.; Barnes, E. The Next Generation of Cotton Defoliation Sprayer. *AgriEngineering* **2023**, *5*, 441–459. https://doi.org/10.3390/agriengineering5010029
21. Md-Tahir, H.; Zhang, J.; Zhou, Y.; Sultan, M.; Ahmad, F.; Du, J.; Ullah, A.; Hussain, Z.; Xia, J. Engineering Design, Kinematic and Dynamic Analysis of High Lugs Rigid Driving Wheel, a Traction Device for Conventional Agricultural Wheeled Tractors. *Agriculture* **2023**, *13*, 493. https://doi.org/10.3390/agriculture13020493
22. Deeken, H.F.; Lengling, A.; Krommweh, M.S.; Büscher, W. Improvement of Piglet Rearing's Energy Efficiency and Sustainability Using Air-to-Air Heat Exchangers—A Two-Year Case Study. *Energies* **2023**, *16*, 1799. https://doi.org/10.3390/en16041799
23. Holtorf, L.; Titov, I.; Daschner, F.; Gerken, M. UAV-Based Wireless Data Collection from Underground Sensor Nodes for Precision Agriculture. *AgriEngineering* **2023**, *5*, 338–354. https://doi.org/10.3390/agriengineering5010022
24. Vahdanjoo, M.; Gislum, R.; Sørensen, C.A.G. Operational, Economic, and Environmental Assessment of an Agricultural Robot in Seeding and Weeding Operations. *AgriEngineering* **2023**, *5*, 299–324. https://doi.org/10.3390/agriengineering5010020
25. Wang, S.; Zhao, X.; Liu, W.; Du, J.; Zhao, D.; Yu, Z. Impact-Type Sunflower Yield Sensor Signal Denoising Method Based on CEEMD-WTD. *Agriculture* **2023**, *13*, 166. https://doi.org/10.3390/agriculture13010166
26. Park, S.H.; Koutsospyros, A.; Moon, D.H. Optimization of a High-Pressure Soil Washing System for Emergency Recovery of Heavy Metal-Contaminated Soil. *Agriculture* **2022**, *12*, 2054. https://doi.org/10.3390/agriculture12122054
27. Ismail, M.; Ahmed, E.; Peng, G.; Xu, R.; Sultan, M.; Khan, F.U.; Aleem, M. Evaluating the Impact of Climate Change on the Stream Flow in Soan River Basin (Pakistan). *Water* **2022**, *14*, 3695. https://doi.org/10.3390/w14223695
28. Xin, Y.; Gao, L.; Hu, W.; Gao, Q.; Yang, B.; Zhou, J.; Xu, C. Genome-Wide Association Study Based on Plant Height and Drought-Tolerance Indices Reveals Two Candidate Drought-Tolerance Genes in Sweet Sorghum. *Sustainability* **2022**, *14*, 14339. https://doi.org/10.3390/su142114339
29. Ma, X.; Gong, Q.; Wang, Q.; Xu, D.; Zhou, Y.; Chen, G.; Cao, X.; Wang, L. Design of an Air Suction Wheel-Hole Single Seed Drill for a Wheat Plot Dibbler. *Agriculture* **2022**, *12*, 1735. https://doi.org/10.3390/agriculture12101735
30. Cui, N.; Pozzobon, V. Food-Grade Cultivation of *Saccharomyces cerevisiae* from Potato Waste. *AgriEngineering* **2022**, *4*, 951–968. https://doi.org/10.3390/agriengineering4040061
31. Wu, Y.; Cao, Y.; Zhai, Z. Early Detection of Bacterial Blight in Hyperspectral Images Based on Random Forest and Adaptive Coherence Estimator. *Sustainability* **2022**, *14*, 13168. https://doi.org/10.3390/su142013168
32. Sakar, E.H.; Khtira, A.; Aalam, Z.; Zeroual, A.; Gagour, J.; Gharby, S. Variations in Physicochemical Characteristics of Olive Oil (cv 'Moroccan Picholine') According to Extraction Technology as Revealed by Multivariate Analysis. *AgriEngineering* **2022**, *4*, 922–938. https://doi.org/10.3390/agriengineering4040059
33. Nguyen, T.-H.; Nguyen, T.-N.; Ngo, B.-V. A VGG-19 Model with Transfer Learning and Image Segmentation for Classification of Tomato Leaf Disease. *AgriEngineering* **2022**, *4*, 871–887. https://doi.org/10.3390/agriengineering4040056
34. Bilal, M.; Sultan, M.; Majeed, F.; Farooq, M.; Sajjad, U.; Ibrahim, S.M.; Khan, M.U.; Azizi, S.; Javaid, M.Y.; Ahmad, R. Investigating Adsorption-Based Atmospheric Water Harvesting Potential for Pakistan. *Sustainability* **2022**, *14*, 12582. https://doi.org/10.3390/su141912582

35. Mu, G.; Wang, W.; Zhang, T.; Hu, L.; Zheng, W.; Zhang, W. Design and Experiment with a Double-Roller Sweet Potato Vine Harvester. *Agriculture* **2022**, *12*, 1559. https://doi.org/10.3390/agriculture12101559
36. Chen, W.; Wang, G.; Hu, L.; Yuan, J.; Wu, W.; Bao, G.; Yin, Z. Research on the Control Strategy of Leafy Vegetable Harvester Travel Speed Automatic Control System. *AgriEngineering* **2022**, *4*, 801–825. https://doi.org/10.3390/agriengineering4040052
37. Rasheed, M.W.; Tang, J.; Sarwar, A.; Shah, S.; Saddique, N.; Khan, M.U.; Imran Khan, M.; Nawaz, S.; Shamshiri, R.R.; Aziz, M.; et al. Soil Moisture Measuring Techniques and Factors Affecting the Moisture Dynamics: A Comprehensive Review. *Sustainability* **2022**, *14*, 11538. https://doi.org/10.3390/su141811538
38. Vrochidou, E.; Tsakalidou, V.N.; Kalathas, I.; Gkrimpizis, T.; Pachidis, T.; Kaburlasos, V.G. An Overview of End Effectors in Agricultural Robotic Harvesting Systems. *Agriculture* **2022**, *12*, 1240. https://doi.org/10.3390/agriculture12081240
39. Cheng, S.; Xing, Z.; Tian, C.; Li, S.; Tian, J.; Liu, Q.; Hu, Y.; Guo, B.; Hu, Q.; Wei, H.; et al. Effects of Controlled Release Urea Formula and Conventional Urea Ratio on Grain Yield and Nitrogen Use Efficiency of Direct-Seeded Rice. *Agriculture* **2022**, *12*, 1230. https://doi.org/10.3390/agriculture12081230
40. Rasheed, A.; Lee, J.W.; Kim, H.T.; Lee, H.W. Study on Heating and Cooling Performance of Air-to-Water Heat Pump System for Protected Horticulture. *Energies* **2022**, *15*, 5467. https://doi.org/10.3390/en15155467
41. Grimberg, R.; Teitel, M.; Ozer, S.; Levi, A.; Levy, A. Estimation of Greenhouse Tomato Foliage Temperature Using DNN and ML Models. *Agriculture* **2022**, *12*, 1034. https://doi.org/10.3390/agriculture12071034
42. Mohd Ali, M.; Hashim, N.; Abd Aziz, S.; Lasekan, O. Characterisation of Pineapple Cultivars under Different Storage Conditions Using Infrared Thermal Imaging Coupled with Machine Learning Algorithms. *Agriculture* **2022**, *12*, 1013. https://doi.org/10.3390/agriculture12071013
43. Nadeem, M.U.; Anjum, M.N.; Afzal, A.; Azam, M.; Hussain, F.; Usman, M.; Javaid, M.M.; Mukhtar, M.A.; Majeed, F. Assessment of Multi-Satellite Precipitation Products over the Himalayan Mountains of Pakistan, South Asia. *Sustainability* **2022**, *14*, 8490. https://doi.org/10.3390/su14148490
44. Zakir-Hassan, G.; Punthakey, J.F.; Shabir, G.; Yasmeen, F.; Sultan, M.; Ashraf, H.; Sohoo, I.; Majeed, F. Physicochemical Investigation of Rainfall for Managed Aquifer Recharge in Punjab (Pakistan). *Water* **2022**, *14*, 2155. https://doi.org/10.3390/w14142155
45. Molaei, F.; Ghatrehsamani, S. Kinematic-Based Multi-Objective Design Optimization of a Grapevine Pruning Robotic Manipulator. *AgriEngineering* **2022**, *4*, 606–625. https://doi.org/10.3390/agriengineering4030040
46. Wu, K.; Zhang, M.; Wang, G.; Chen, X.; Wu, J. A Continuous Single-Layer Discrete Tiling System for Online Detection of Corn Impurities and Breakage Rates. *Agriculture* **2022**, *12*, 948. https://doi.org/10.3390/agriculture12070948
47. Zhu, Q.; Zhang, H.; Zhu, Z.; Gao, Y.; Chen, L. Structural Design and Simulation of Pneumatic Conveying Line for a Paddy Side-Deep Fertilisation System. *Agriculture* **2022**, *12*, 867. https://doi.org/10.3390/agriculture12060867
48. Madrid, B.; Zhang, H.; Miles, C.A.; Kraft, M.; Griffin-LaHue, D.; Wasko DeVetter, L. Humic and Acetic Acids Have the Potential to Enhance Deterioration of Select Plastic Soil-Biodegradable Mulches in a Mediterranean Climate. *Agriculture* **2022**, *12*, 865. https://doi.org/10.3390/agriculture12060865
49. Li, B.; Liu, B.; Li, S.; Liu, H. An Improved EfficientNet for Rice Germ Integrity Classification and Recognition. *Agriculture* **2022**, *12*, 863. https://doi.org/10.3390/agriculture12060863
50. Petrakis, T.; Kavga, A.; Thomopoulos, V.; Argiriou, A.A. Neural Network Model for Greenhouse Microclimate Predictions. *Agriculture* **2022**, *12*, 780. https://doi.org/10.3390/agriculture12060780

51. Wang, H.; Ling, G.; Wang, W.; Hu, X.; Gao, X. A Prediction Model of Labyrinth Emitter Service Duration (ESD) under Low-Quality (Sand-Laden Water) Irrigation. *Water* **2022**, *14*, 1690. https://doi.org/10.3390/w14111690
52. Cheng, Q.; Wang, J.; Liu, K.; Chao, J.; Liu, D. Design of Rice Straw Fiber Crusher and Evaluation of Fiber Quality. *Agriculture* **2022**, *12*, 729. https://doi.org/10.3390/agriculture12050729
53. Fang, W.; Wang, X.; Han, D.; Chen, X. Review of Material Parameter Calibration Method. *Agriculture* **2022**, *12*, 706. https://doi.org/10.3390/agriculture12050706
54. Zhou, Y.; Ma, W.; Liu, F.; Wang, J. Simulation and Evaluation of Hydrothermal Conditions in Crop Growth Period: A Case Study of Highland Barley in the Qinghai-Tibet Plateau. *Sustainability* **2022**, *14*, 5932. https://doi.org/10.3390/su14105932
55. Lisičar Vukušić, J.; Millenautzki, T.; Barbe, S. Bridging the Implementation Gap between Pomace Waste and Large-Scale Baker's Yeast Production. *AgriEngineering* **2023**, *5*, 2238–2252. https://doi.org/10.3390/agriengineering5040137
56. Pereira, J.S.; Magalhães, R.R.; Santos, F.L.; de Andrade, E.T.; Marques, L.S. Modeling of Coffee Fruit: An Approach to Simulate the Effects of Compression. *AgriEngineering* **2023**, *5*, 2303–2313. https://doi.org/10.3390/agriengineering5040141
57. Monteleone, S.; Alves de Moraes, E.; Protil, R.M.; Tondato de Faria, B.; Maia, F.R. Proposal of a model of irrigation Operations Management for exploring the factors that can affect the adoption of Precision Agriculture in the context of Agriculture 4.0. *Agriculture* ***2024***, *14*, 134; https://doi.org/10.3390/agriculture14010134.
58. Yu, L.; Wang, H.; Ren, A.; Han, F.; Jia, F.; Hou, H.; Liu, Y. The Influencing Factors Analysis of Aquaculture Mechanization Development in Liaoning, China. *AgriEngineering* ***2024***, *6*, 34–51; https://doi.org/10.3390/agriengineering6010003.
59. Ajayi-Banji, I.; Monono, E.; Teboh, J.; Yuja, S.; Hellevang, K. Post-Harvest Management of Immature (Green and Semi-Green) Soybeans: Effect of Drying and Storage Conditions (Temperature, Light, and Aeration) on Color and Oil Quality. *AgriEngineering* **2024**, *6*, 135–154; https://doi.org/10.3390/agriengineering6010009.

3. Innovative Farm Machinery and Design

1. Engineering Design for Tractors: Contribution 21 focused on the engineering design and analysis of a high lugs rigid driving wheel for conventional agricultural tractors. It not only presented technical specifications but discussed the potential for scalability and adoption in diverse agricultural settings. This article extended beyond technical innovation, considering the practical implications of adopting new technologies in farming.
2. Advanced Harvester Design: Contribution 35 introduced a double-roller sweet potato vine harvester, not only showcasing its technical advancements but evaluating its energy efficiency and labor requirements. This article extended beyond technological novelty, providing insights into the economic and labor implications of adopting new harvesting technologies.
3. Robotic Pruning for Grapevines: Contribution 45 presented a kinematic-based multi-objective design optimization of a grapevine pruning robotic manipulator, and it not only discussed the technical specifications but explored the adaptability of robotic technologies to different grape varieties and vineyard settings. This article contributed to technical innovation as well as to the practical considerations of adopting robotics in viticulture.
4. Pneumatic Conveying Line Design: Contribution 47 explored the structural design and simulation of a pneumatic conveying line for a paddy side-deep fertilization system, which not only provided insights into material handling but discussed the potential for automation in paddy cultivation. This article extended beyond structural design, considering the broader implications of adopting automated material handling systems in agriculture.

5. Rice Straw Fiber Crusher: Contribution 52 discussed the design of a rice straw fiber crusher and evaluated fiber quality. This not only presented technical specifications but explored the potential applications of rice straw fiber in diverse industries. This article extended beyond equipment design, considering the broader implications of valorizing agricultural residues.

4. Innovations in Crop Management

1. Nutritional Enhancement: Contribution 1 explored the intricate relationship between calcium, potassium, and magnesium, revealing their nuanced impact on the nutritional value of grafted tomatoes. This in-depth understanding opens avenues for precise nutrient management in horticultural practices.
2. Drought-Tolerance Genes: Contribution 28 presented a groundbreaking genome-wide association study identifying not only candidate drought-tolerance genes in sweet sorghum but also uncovering the complex genetic networks that govern resilience to water stress. This nuanced insight provides a roadmap for targeted breeding programs.
3. Precision Agriculture Technologies: Contribution 11 not only identified hotspots in subsurface drip irrigation research but went further to analyze the evolving trends and methodologies within these hotspots. By utilizing advanced techniques such as CiteSpace, this article provides a meta-analysis that informs researchers of the most critical areas for future exploration.
4. Energy Efficiency in Pig Farming: Contribution 22 delved into the realm of energy efficiency, offering a detailed two-year case study on improving piglet rearing energy efficiency and sustainability using air-to-air heat exchangers. This comprehensive analysis extends beyond energy savings, exploring the broader implications on animal welfare and farm economics.
5. Crop Yield Sensor Technology: Contribution 25 not only introduced an impact-type sunflower yield sensor signal denoising method but also validated its effectiveness through extensive field trials. This article not only contributes a novel technological advancement but also provides practical insights into its real-world application.

5. Technological Solutions

1. Precision Agriculture Robotics: Contributions 23 and 24 offered more than a demonstration of potential; they presented a comprehensive analysis of the operational, economic, and environmental implications of UAV-based wireless data collection and agricultural robots in precision farming operations. These findings bridge the gap between technological promise and on-field practicality.
2. Innovative Harvesting Techniques: Contribution 46 not only introduced a continuous single-layer discrete tiling system for online detection of corn impurities and breakage rates but also explored the economic implications of adopting this type of technology. This article provides a holistic view, considering both technological and economic aspects of innovation.
3. Climate Change Resilience: Contribution 16 addressed the mitigation of COVID-19 impacts on crop farming through nanotechnological approaches, diving deep into the ethical considerations and societal implications of employing nanotechnology in agriculture. These findings extend beyond technical solutions, prompting reflections on the ethical dimensions of agricultural innovation during global crises.
4. Adsorption-Based Water Harvesting: Contribution 34 investigated the adsorption-based atmospheric water harvesting potential for Pakistan. This article provides a nuanced perspective, considering not only technological feasibility but also its alignment with local socio-economic conditions.
5. Digital Livestock Farming: Contribution 18 discussed the significance and ethics of digital livestock farming. It not only examines the technological aspects but also delves into the ethical considerations of employing digital technologies in animal

husbandry. These findings contributed to technological discourse as well as broader conversations on ethical practices in agriculture.

6. Sustainability and Climate Resilience

1. Multi-Satellite Precipitation Products: Contribution 43 assessed multi-satellite precipitation products over the Himalayan Mountains for sustainable water management. It not only evaluates the technical aspects but also discusses the policy implications of utilizing satellite data for water resource management. This article extends beyond technical validation, contributing to the discourse on policy-relevant applications of remote sensing in agriculture.
2. Impact of Climate Change: Contribution 27 evaluated the impact of climate change on stream flow in Soan River Basin (Pakistan), not only providing insights into hydrological changes but also exploring the socio-economic impacts on local communities. This article bridges the gap between climate science and social resilience, offering a holistic understanding of the effects of climate change.
3. Humic Acids in Agriculture: Contribution 48 investigated how humic and acetic acids can enhance the deterioration of plastic soil-biodegradable mulches and delved into the long-term environmental implications of their use. This article contributes not only to immediate agricultural practices but also to broader conversations on the sustainability of soil management strategies.
4. Greenhouse Microclimate Modeling: Contribution 50 presented a neural network model for greenhouse microclimate predictions. It demonstrated the accuracy of predictions and discussed the interpretability of machine learning models in practical farming scenarios. This article contributes not only to the technical discourse but also to the practical considerations of implementing machine learning in agriculture.
5. Precision Irrigation: Contribution 51 developed a prediction model for labyrinth emitter service duration (ESD) under low-quality (sand-laden water) irrigation. It not only provides a technical solution but examines the economic feasibility and practical considerations of adopting precision irrigation in challenging water conditions. The article contributes not only to the technological discourse but also to the practical considerations of precision agriculture.

7. Future Directions and Conclusion

As we celebrate the achievements of this collection, it is imperative to chart the course for future research endeavors in agricultural engineering. Some key areas that demand attention include the following:

- Interdisciplinary Collaboration: Encourage deeper collaboration between agriculturalists, engineers, data scientists, and environmental scientists to address complex agricultural challenges comprehensively.
- Climate-Resilient Farming: Investigate strategies for enhancing climate resilience in agriculture, considering the unpredictable impacts of climate change on crop yields and water availability.
- Digital Transformation: Explore the full potential of digital technologies, including artificial intelligence, machine learning, and the Internet of Things, in revolutionizing various aspects of agricultural practices.
- Sustainable Precision Farming: Continue the exploration of sustainable precision farming technologies to optimize resource use, reduce environmental impact, and enhance overall agricultural productivity.
- Innovations in Post-Harvest Activities: Focus on developing innovative solutions for post-harvest activities, storage, and transportation to minimize food loss and ensure food security.

In conclusion, the "Emerging Agricultural Engineering Sciences, Technologies, and Applications" collection has been a collaborative endeavor that has pushed the boundaries of knowledge in our field. I extend my heartfelt gratitude to all contributors, reviewers, and MDPI for making this venture a success. May the insights gained from this collection propel us toward a future where agricultural engineering continues to be at the forefront of sustainable, efficient, and resilient food production.

Acknowledgments: As the Topic Editor of the Topic Collection "Emerging Agricultural Engineering Sciences, Technologies, and Applications", Muhammad Sultan would like to express deep appreciation to all authors whose valuable work was published in this topic and thus contributed to the success of this collection. In addition, Muhammad Sultan acknowledges all kinds of support from Bahauddin Zakariya University, Multan-Pakistan.

Conflicts of Interest: The author declares no conflicts of interest.

References

1. Sultan, M.; Shamshiri, R.; Ahamed, M.; Farooq, M. *Advances in Agricultural Engineering Technologies and Application*; Multidisciplinary Digital Publishing Institute: Basel, Switzerland, 2023; Volume I, ISBN 978-3-0365-9375-3.
2. Sultan, M.; Shamshiri, R.; Ahamed, M.; Farooq, M. *Advances in Agricultural Engineering Technologies and Application*; Multidisciplinary Digital Publishing Institute: Basel, Switzerland, 2023; Volume II, ISBN 978-3-0365-9377-7.
3. Bhat, S.A.; Huang, N.-F.; Hussain, I.; Bibi, F.; Sajjad, U.; Sultan, M.; Alsubaie, A.S.; Mahmoud, K.H. On the Classification of a Greenhouse Environment for a Rose Crop Based on AI-Based Surrogate Models. *Sustainability* **2021**, *13*, 2166. [CrossRef]
4. Ashraf, H.; Sultan, M.; Shamshiri, R.R.; Abbas, F.; Farooq, M.; Sajjad, U.; Md-Tahir, H.; Mahmood, M.H.; Ahmad, F.; Taseer, Y.R.; et al. Dynamic Evaluation of Desiccant Dehumidification Evaporative Cooling Options for Greenhouse Air-Conditioning Application in Multan (Pakistan). *Energies* **2021**, *14*, 1097. [CrossRef]
5. Shamshiri, R.R.; Bojic, I.; van Henten, E.; Balasundram, S.K.; Dworak, V.; Sultan, M.; Weltzien, C. Model-Based Evaluation of Greenhouse Microclimate Using IoT-Sensor Data Fusion for Energy Efficient Crop Production. *J. Clean. Prod.* **2020**, *263*, 121303. [CrossRef]
6. Rezvani, S.M.; Abyaneh, H.Z.; Shamshiri, R.R.; Balasundram, S.K.; Dworak, V.; Goodarzi, M.; Sultan, M.; Mahns, B. IoT-Based Sensor Data Fusion for Determining Optimality Degrees of Microclimate Parameters in Commercial Greenhouse Production of Tomato. *Sensors* **2020**, *20*, 6474. [CrossRef] [PubMed]
7. Sultan, M.; Miyazaki, T.; Saha, B.B.; Koyama, S. Steady-State Investigation of Water Vapor Adsorption for Thermally Driven Adsorption Based Greenhouse Air-Conditioning System. *Renew. Energy* **2016**, *86*, 785–795. [CrossRef]
8. Amani, M.; Foroushani, S.; Sultan, M.; Bahrami, M. Comprehensive Review on Dehumidification Strategies for Agricultural Greenhouse Applications. *Appl. Therm. Eng.* **2020**, *181*, 115979. [CrossRef]
9. Raza, H.M.U.; Sultan, M.; Bahrami, M.; Khan, A.A. Experimental Investigation of Evaporative Cooling Systems for Agricultural Storage and Livestock Air-Conditioning in Pakistan. *Build. Simul.* **2020**, *14*, 617–631. [CrossRef]
10. Hanif, S.; Sultan, M.; Miyazaki, T.; Koyama, S. Investigation of Energy-Efficient Solid Desiccant System for the Drying of Wheat Grains. *Int. J. Agric. Biol. Eng.* **2019**, *12*, 221–228. [CrossRef]
11. Mahmood, M.H.; Sultan, M.; Miyazaki, T. Solid Desiccant Dehumidification-Based Air-Conditioning System for Agricultural Storage Application: Theory and Experiments. *Proc. Inst. Mech. Eng. Part A J. Power Energy* **2020**, *234*, 534–547. [CrossRef]
12. Hussain, G.; Aleem, M.; Sultan, M.; Sajjad, U.; Ibrahim, S.M.; Shamshiri, R.R.; Farooq, M.; Usman Khan, M.; Bilal, M. Evaluating Evaporative Cooling Assisted Solid Desiccant Dehumidification System for Agricultural Storage Application. *Sustainability* **2022**, *14*, 1479. [CrossRef]
13. Aziz, M.; Khan, M.; Anjum, N.; Sultan, M.; Shamshiri, R.R.; Ibrahim, S.M.; Balasundram, S.K.; Aleem, M. Scientific Irrigation Scheduling for Sustainable Production in Olive Groves. *Agriculture* **2022**, *12*, 564. [CrossRef]
14. Ashraf, H.; Qamar, S.; Riaz, N.; Shamshiri, R.R.; Sultan, M.; Khalid, B.; Ibrahim, S.M.; Imran, M.; Khan, M.U. Spatiotemporal Estimation of Reference Evapotranspiration for Agricultural Applications in Punjab, Pakistan. *Agriculture* **2023**, *13*, 1388. [CrossRef]
15. Mujtaba, A.; Nabi, G.; Masood, M.; Iqbal, M.; Asfahan, H.M.; Sultan, M.; Majeed, F.; Hensel, O.; Nasirahmadi, A. Impact of Cropping Pattern and Climatic Parameters in Lower Chenab Canal System—Case Study from Punjab Pakistan. *Agriculture* **2022**, *12*, 708. [CrossRef]
16. Sultan, M.; Mahmood, M.H.; Ahamed, M.S.; Shamshiri, R.R.; Shahzad, M.W. Energy Systems and Applications in Agriculture. *Energies* **2022**, *15*, 9132. [CrossRef]
17. Ahamed, M.S.; Sultan, M.; Shamshiri, R.R.; Rahman, M.M.; Aleem, M.; Balasundram, S.K. Present Status and Challenges of Fodder Production in Controlled Environments: A Review. *Smart Agric. Technol.* **2023**, *3*, 100080. [CrossRef]
18. Majeed, Y.; Khan, M.U.; Waseem, M.; Zahid, U.; Mahmood, F.; Majeed, F.; Sultan, M.; Raza, A. Renewable Energy as an Alternative Source for Energy Management in Agriculture. *Energy Rep.* **2023**, *10*, 344–359. [CrossRef]
19. Ahmad, R.; Zhou, Y.; Zhao, N.; Pemberton-Pigott, C.; Annegarn, H.J.; Sultan, M.; Dong, R.; Ju, X. Impacts of Fuel Feeding Methods on the Thermal and Emission Performance of Modern Coal Burning Stoves. *Int. J. Agric. Biol. Eng.* **2019**, *12*, 160–167. [CrossRef]

20. Ahmad, R.; Ilyas, H.N.; Li, B.; Sultan, M.; Amjad, M.; Aleem, M.; Abbas, A.; Imran, M.A.; Riaz, F. Current Challenges and Future Prospect of Biomass Cooking and Heating Stoves in Asian Countries. *Front. Energy Res.* **2022**, *10*, 880064. [CrossRef]
21. Md-Tahir, H.; Zhang, J.; Zhou, Y.; Sultan, M.; Ahmad, F.; Du, J.; Ullah, A.; Hussain, Z.; Xia, J. Engineering Design, Kinematic and Dynamic Analysis of High Lugs Rigid Driving Wheel, a Traction Device for Conventional Agricultural Wheeled Tractors. *Agriculture* **2023**, *13*, 493. [CrossRef]
22. Miah, M.S.; Rahman, M.M.; Hoque, M.A.; Ibrahim, S.M.; Sultan, M.; Shamshiri, R.R.; Ucgul, M.; Hasan, M.; Barna, T.N. Design and Evaluation of a Power Tiller Vegetable Seedling Transplanter with Dibbler and Furrow Type. *Heliyon* **2023**, *9*, e17827. [CrossRef] [PubMed]
23. Md-Tahir, H.; Zhang, J.; Xia, J.; Zhou, Y.; Zhou, H.; Du, J.; Sultan, M.; Mamona, H. Experimental Investigation of Traction Power Transfer Indices of Farm-Tractors for Efficient Energy Utilization in Soil Tillage and Cultivation Operations. *Agronomy* **2021**, *11*, 168. [CrossRef]

Disclaimer/Publisher's Note: The statements, opinions and data contained in all publications are solely those of the individual author(s) and contributor(s) and not of MDPI and/or the editor(s). MDPI and/or the editor(s) disclaim responsibility for any injury to people or property resulting from any ideas, methods, instructions or products referred to in the content.

agriculture

MDPI

Article

Proposal of a Model of Irrigation Operations Management for Exploring the Factors That Can Affect the Adoption of Precision Agriculture in the Context of Agriculture 4.0

Sergio Monteleone [1,*], Edmilson Alves de Moraes [1], Roberto Max Protil [2], Brenno Tondato de Faria [3] and Rodrigo Filev Maia [4]

[1] School of Business Administration, Centro Universitário FEI, São Paulo 01525-000, SP, Brazil; edmilson@fei.edu.br

[2] Department of Agricultural Economics, Universidade Federal de Viçosa (UFV), Viçosa 36570-900, MG, Brazil; roberto.protil@ufv.br

[3] School of Electrical Engineering, Centro Universitário FEI, São Bernardo do Campo 09850-901, SP, Brazil; betondato@fei.edu.br

[4] Centre of Regional and Rural Futures, Deakin University, Hanwood 2680, Australia; r.filevmaia@deakin.edu.au

* Correspondence: smontelefei@fei.edu.br

Abstract: Agriculture is undergoing a profound change related to Agriculture 4.0 development and Precision Agriculture adoption, which is occurring at a slower pace than expected despite the abundant literature on the factors explaining this adoption. This work explores the factors related to agricultural Operations Management, farmer behavior, and the farmer mental model, topics little explored in the literature, by applying the Theory of Planned Behavior. Considering the exploratory nature of this work, an exploratory multi-method is applied, consisting of expert interviews, case studies, and modeling. This study's contributions are a list of factors that can affect this adoption, which complements previous studies, theoretical propositions on the relationships between these factors and this adoption, and a model of irrigation Operations Management built based on these factors and these propositions. This model provides a theoretical framework to study the identified factors, the relationships between them, the theoretical propositions, and the adoption of Precision Agriculture. Furthermore, the results of case studies allow us to explore the relationships between adoption, educational level, and training. The identified factors and the model contribute to broadening the understanding of Precision Agriculture adoption, adding Operations Management and the farmer mental model to previous studies. A future research agenda is formulated to direct future studies.

Keywords: precision agriculture; adoption; sensing technologies; irrigation; Agriculture 4.0; operations management; farmer mental model; center pivot; model

Citation: Monteleone, S.; Alves de Moraes, E.; Protil, R.M.; Faria, B.T.d.; Maia, R.F. Proposal of a Model of Irrigation Operations Management for Exploring the Factors That Can Affect the Adoption of Precision Agriculture in the Context of Agriculture 4.0. *Agriculture* **2024**, *14*, 134. https://doi.org/10.3390/agriculture14010134

Academic Editors: Redmond R. Shamshiri, Muhammad Imran, Muhammad Sultan and Yuguang Zhou

Received: 5 October 2023
Revised: 20 December 2023
Accepted: 22 December 2023
Published: 16 January 2024

Copyright: © 2024 by the authors. Licensee MDPI, Basel, Switzerland. This article is an open access article distributed under the terms and conditions of the Creative Commons Attribution (CC BY) license (https://creativecommons.org/licenses/by/4.0/).

1. Introduction

Agriculture is currently undergoing a profound change [1] due to a new technological revolution [2,3] related to the development of Agriculture 4.0 and the diffusion of technologies, such as the Internet of Things (IoT), big data, artificial intelligence, and cloud computing [4,5]. Agriculture is also experiencing a technological transformation related to the adoption of Precision Agriculture (PA), which, after 20 years of research, is not consolidated [6]. Despite the expected economic gains [7], PA adoption is often perceived as not occurring immediately [8], and it is progressing at a slower speed than predicted [9] due to several factors; among them, the lack of farmer knowledge about the use of PA technologies plays an important role in business outcomes [10]. There is a lack of consensus on whether PA improves farm profitability, in addition to the little attention paid to perceptions of PA profitability from the farmer's perspective [11].

Despite the abundant literature on the factors affecting PA adoption [12–14], factors related to Operations Management in agriculture [15] and farmer behavior [16,17], specifically their beliefs that are the basis of the farmer mental model [18,19], are little explored. In most studies, the factors are organized in models based on the theory applied in the research, like the Technology Acceptance Model [17] and Utility Maximization [20]. Theories for studying human behavior, such as the Theory of Planned Behavior (TPB) [21], were applied in rare cases in previous studies [22]. The TPB provides a solid conceptual framework useful for analyzing farmer behavior [23–25].

Operations Management (OM), applied in an agriculture context, as well as in manufacturing and services organizations, is the function associated with the design and management of the transformation processes of inputs into outputs [26] and supports designing, planning, scheduling, and executing operations involving humans and machines [27,28]. However, the factors affecting PA adoption related to the modeling of irrigation OM are scarce in the literature [15].

There is a limited availability of publications that delve into the farmer mental model within studies focusing on the adoption of Precision Agriculture (PA) [18,19]. Applying the concept of the mental model allows for an understanding of farmer decision making, linking perceptions, attitudes, and beliefs with behavior [29]. Farmers do not only make decisions based purely on economically rational motivations but also consider social and intrinsic objectives [30].

In the context of this ongoing technological revolution, Agriculture 4.0 has been developed through the evolution of PA [6,31]. Technologies included in the concept of Agriculture 4.0 allow us to improve planning and control in agricultural production [32], allowing us to follow plant and yield development progress through information collected by sensors [33] and to access a large amount of site-specific data to make more accurate decisions in farm management [34]. Most of the research carried out on Agriculture 4.0 is more theoretical [31,35]. Moreover, overemphasis is placed on high-tech solutions to the detriment of the inclusion and exclusion effects of Agriculture 4.0 technologies [36] and the social impacts of these new technologies [37,38], as well as to the factors related to OM, farmer behavior, and the farmer mental model, which can all affect PA adoption.

Therefore, in this work, we aim to contribute to filling the identified gaps, answering the following research questions: (1) What are the factors related to irrigation OM that can affect the adoption of PA in the context of Agriculture 4.0? (2) What are the factors related to the farmer's behavior, focusing on the mental model, that can affect the adoption of PA in the context of Agriculture 4.0? (3) What are the factors related to the modeling of irrigation operations, with a focus on OM and the farmer mental model?

To answer these questions, in this study, we aim to achieve the following research objectives: (1) exploring the factors related to irrigation OM that can affect the studied adoption; (2) exploring the factors related to farmer behavior, focusing on the mental model, that can affect the studied adoption; and (3) modeling the irrigation operations, with a focus on OM and the farmer mental model, to understand and formalize the relationships between adoption and the identified factors.

This work is characterized as exploratory since little research has been conducted on the topics being studied, an insufficient number of studies exists to document the research problems, and the variables and theory base are unknown. Exploratory study can help to develop theoretical propositions and frameworks [39,40]. This research adopts an exploratory multi-method made up of expert interviews, case studies, and the modeling of irrigation operations [39,41–43].

The empirical results of 27 expert interviews and two case studies on farms were used to develop the irrigation OM model [41]. For modeling, the Integration Definition for Function Modeling (IDEF0) technique was applied, which offers a standard and widely used approach to develop the conceptual models of systems [44], like farms; to represent products, controls, restrictions, and resource information in the same diagram [45]; and to identify system components, data requirements, and information flows [46].

Irrigation operations are the object of study in this research, according to the international project "Smart Water Management Platform" (SWAMP), in which context this study is carried out. The SWAMP project aims to bring the IoT concept to Precision Irrigation and to develop irrigation systems to optimize water use [47].

As for the scope of the research, in line with the SWAMP project, the farmer or farm manager is the central actor in the decision-making process on a farm, depending on its business and organizational structure [48]. Furthermore, in the literature and in the results of expert interviews and case studies, common factors between adoption and decision making in irrigation OM can be identified. According to the research objectives, the behavior "decision making in agricultural OM" is outside the scope of this work.

Considering the research gaps identified and the research objectives, the empirical results of this work contribute to theoretical progression in the literature on the adoption of PA, adding a set of factors related to OM and farmer behavior, including this mental model. The organization of these factors in a framework represents the irrigation OM and relationships between the factors in a structured way, contributing to an explanation of the adoption. The empirical results suggest a change in the farmer mental model in the adoption of OM models inspired by the industry. The results of the case studies also contribute to exploring the relationship between the adoption and some factors, such as educational level and training.

In addition to theoretical contributions, this research also provides a practical contribution. The proposed model can be used for the development of an Irrigation Management System (IMS), integrating the agronomic management of a crop and the irrigation OM.

A future research agenda is formulated, based on the results achieved, the theoretical propositions generated from the results of expert interviews and case studies, and the scope of the work.

This work is structured with the Materials and Methods in Section 2. Section 2.1 introduces the literature review; Section 2.2 describes the exploratory multi-method. The results of the study are displayed in Section 3. A discussion of the findings is provided in Section 4, and the limitations and the research agenda are covered in Section 5. At the end, Section 6 draws the conclusions.

2. Materials and Methods

This section presents the literature analyzed and the method applied in this work.

2.1. Literature Review

In this section, previous studies on Precision Agriculture adoption, Agriculture 4.0, Operations Management, farmer behavior, and Theory of Planned Behavior are analyzed to identify the theoretical concepts useful for studying the adoption and for identifying the topics that can be considered factors that can affect PA adoption (theoretical basis) and for collecting the factors that can affect the adoption cited in these studies (empirical basis). The theoretical concepts, topics, and factors analyzed in this section constitute the theoretical and empirical bases in conducting expert interviews, case studies, and building the model of irrigation OM. The definitions of Precision Agriculture and Agriculture 4.0 used in this work are also reported.

2.1.1. Precision Agriculture, Precision Irrigation, and Factors Affecting the Adoption

In the literature, there was no consensus on the definition of PA before 2019, when the International Society for Precision Agriculture [49] proposed the official definition that this work uses. This lack of a shared definition has made studying adoption difficult. In addition, the kind of technologies [50], like yield mapping, variable rate application, and remote sensing [7,51], affects the adoption.

The PA concept encompasses Precision Irrigation (PI), which is included in the PA technologies that can optimize crop responses for applied water units [16,52]. PA technologies for irrigation enable farmers to collect site-specific data that provide valuable quantitative information for effective irrigation management [53]. PI, despite being recent and still poorly studied, has great potential to contribute to improving irrigation efficiency, thus increasing economic returns with a reduction in energy use [54]. The irrigation systems used can be classified as sprinkler (e.g., center pivot), surface, and drip systems [55]. In the SWAMP project, center pivot and drip systems were studied [47].

Regardless of the irrigation system used, it is particularly important to accurately estimate the irrigation water requirement. According to the research carried out in the SWAMP project, this study applies the Food and Agriculture Organization version of the Penman Monteith model [56,57] and the approach based on the water balance [58,59] to estimate the irrigation water requirement.

The extensive number of studies on PA adoption direct the attention of researchers towards several factors, such as characteristics of farmers (e.g., age, educational level, computer use, training), farm characteristics (e.g., size, location), benefits (e.g., reduced water use), and barriers (e.g., software cost, funding sources) [13,14,60,61]. The factors related to values and beliefs determining the farmer behavior [29] and the factors associated with the agricultural OM [15] have been little studied. Most works apply the survey method [17], while qualitative methods are little used [62–65]. Regarding Precision Irrigation, its adoption is also influenced by factors, like soil quality, water cost, energy cost, crop type [16,66].

2.1.2. Agriculture 4.0 and Factors That Can Affect the Adoption

Agriculture 4.0 has a vague and unclear definition. A commonly accepted definition does not exist, and it is not entirely clear where the term Agriculture 4.0 originated [35,67]. Some researchers consider Agriculture 4.0 to be an evolution of PA [6,31,68]. According to them and to the study of Precision Irrigation adoption in pilot farms in the SWMP project, this research uses the following definition of Agriculture 4.0:

> *"Agriculture 4.0 is a management strategy, evolution of precision agriculture, realized through the automated collection, integration and analysis of temporal, spatial and individual data, collected by IoT sensing technologies and farm resources, making in this way possible the generation of knowledge, to support the design of applications for the farmer decision making process in irrigation operations management".* [69] (p. 21)

Most studies on Agriculture 4.0 are theoretical, but empirical works are increasing [70,71]. Research dealing with OM is still rare [32,72].

Agriculture 4.0 has the potential to bring numerous benefits, like water and energy cost decreases [73]. Through the automatic collection, integration, and analysis of data from different sources, Agriculture 4.0 allows us to generate knowledge and support the farmer in decision making, potentially increasing farm profitability [4]. Information gathered by sensors can be utilized to follow the plant and yield development progress [33].

Factors that can affect the adoption related to OM and farmer behavior include planning and control in agricultural production, degree of mobility of the production facilities [32], skills to operate equipment, limited knowledge, and capacity to adopt and to understand the technology [37].

2.1.3. Operations Management in Agriculture and in Irrigation

Operations Management is the management of a systemic transformation process to convert a set of inputs into outputs. These inputs include labor, equipment, raw materials, and information, while the outputs are goods and services. OM serves as a continuous improvement process to enhance quality, productivity, and customer satisfaction [74]. Its purpose is to plan and control the operations necessary to produce the goods and services demanded by the customer [28]. Production Planning and Control (PPC) is one of the core functions of OM. It aids in resource planning and utilization to ensure maximum productivity, quality, and competitiveness [75].

In agriculture, OM is responsible for designing, planning, scheduling, and executing operations involving humans and machines. An operation (e.g., seeding, fertilization, and irrigation) links resources and materials processed and produced [28,32,76]. The operations are performed with men and machines [77,78]. Resources include land, working hours, machinery, and equipment [76,79]. A management model for agricultural machinery operations, which is based on four management phases (planning, scheduling, operating, and controlling), was proposed by [27] (p. 71).

The vision of OM in agriculture has evolved since the 1950s: mechanization (1950s–1970s), work organization (1970s–1990s), information and communication technologies, automation (1990s–today), and robotics (in the future). In this evolution, the notions of planning and scheduling play a central role, leading to the need for advanced management, data acquisition, and analysis tools for making decisions [28].

Farmer decisions include decisions related to development stages (such as pesticide and irrigation applications); decisions related to planting (like choice of crop type, planting date); harvest and post-harvest decisions (such as harvest date) [80]. Decision making depends on uncontrollable factors (such as weather conditions) and factors that can be modified (like irrigation infrastructure) [76]. Decisions are influenced by farm planning, which involves seasonal determinations, such as types of crops to cultivate, equipment and workforce requirements, irrigation policies [81], operational production practices, and level of production costs [82]. In addition, the farmer must consider the timeliness of some operations and that some operations and decisions are interconnected [79]. Furthermore, many operations need to be performed in a specific order, such as field preparation [83].

Regarding OM in irrigation, two decisions take priority in planning: definitions of the amount of water to apply and of the irrigation time. These decisions are affected by factors like climate and crop development stages [55]. To decide the optimal level of irrigation for each field, at the beginning of each irrigation period, farmers need to consider other factors: water availability, irrigation costs, crop size, management zones, crops planted, and soil moisture [84]. Energy is currently one of the main production resources in irrigated agriculture, and its cost must always be considered in irrigation management [54]. The cost to irrigate a field is determined by the amount of water pumped and the cost to apply it. The factors that producers can influence include irrigation planning, application efficiency, pumping pressure required, type of energy used, distance from the water source, and unit cost of energy. The cost to pump water for irrigation depends on the type of energy used, like electricity and diesel fuel [85].

2.1.4. Farmer Behaviour, Theory of Planned Behaviour, and Farmer Mental Model Theories Used to Explain the Adoption of Innovations

A range of purely economic-based models have been developed to predict farmer decision making [86], which can be modeled in terms of individual action to maximize profit [87] or expected utility [88]. There is also extensive rural sociological literature in which the farmer attaches significant importance to aspects of lifestyle and personal considerations, which interact with management decisions. This literature suggests that purely economic models cannot capture the full complexity of farmer motivation and behavior [87], as adoption is based on farmer perceptions [89].

Four theories can be considered references to develop models to analyze the farmer decisions for innovation adoption and to establish the basis for choosing explanatory factors [90]: Utility Maximization [88], Theory of Reasoned Action and Theory of Planned Behavior [21], and Technology Acceptance Model [91].

As a result of comparing these theories and considering that the TPB provides a solid conceptual framework [92] to structure the research, this work applies the TPB. This choice is also justified by the attention to human behavior when the person has partial control over the decision making [93], characteristic of the PA context in which access to information is a barrier for the adoption [16,94]. New PA technologies are often produced through developers, and there is a large knowledge gap between developers and users [95]. Furthermore, studies based on TPB analyze farmer behaviors in a deeper and more complete way, thanks to including the Subjective Norm and Perceived Behavioral Control constructs [96].

Theory of Planned Behaviour

The TPB is a well-known and frequently applied framework to predict and explain human behavior [97] and to also study farmer behavior, like the switch from traditional to pressurized irrigation [98]. Intentions to perform behaviors can be predicted from attitude towards the behavior (ATB), Subjective Norm (SN), and Perceived Behavioral Control (PBC). The more favorable the ATB and SN are, and the higher the PBC, the stronger the intention to perform the behavior must be [21].

In TPB research, it is important to clearly define the behavior under study in terms of target to which it is directed, action performed, context in which it occurs, and relevant period [97].

The TPB is open to the inclusion of additional predictors, if they can be shown to capture a significant proportion of the variance in intention or behavior [21], and to the inclusion of background factors (such as age, educational level, experience) that can influence people's beliefs [97].

The standard research methods developed over the years for the use of TPB are largely quantitative. Furthermore, the TPB can be used as a framework to guide the questions in qualitative research [99], as in this work.

Farmer Beliefs and Farmer Mental Model

The constructs ATB, SN, and PBC are related to beliefs about the behavior [21], which represent the information people have about a behavior, its consequences, the expectations of other people, and the impediments. By measuring these beliefs, the TPB has the potential to create the links between beliefs and behavior of an adopter of PA technologies [12].

The farmer beliefs, relative to the adoption of PA, are the basis of this mental model, built by the human mind through perception, experience, attitudes, knowledge, and the comprehension on the behavior [29]. The mental model is also relevant in decision making in agriculture. The use of farmer mental model has the advantage of a certain element of adaptation, and the drawback is the lack of precision, particularly in areas where the farmer is inexperienced. The use of computerized models allows the farmer to base their decisions on a more comprehensive set of information, gaining more precision [28]. The literature on factors affecting farmer decisions also considers conservatism, which is included in the beliefs underlying the farmer mental model [92].

This work considers, through expert interviews and case studies, the factors that can affect the adoption of innovations mentioned in this section, using the TPB as framework and focusing on farmer beliefs and his mental model, to broaden understanding of the determining factors of the PA adoption. The SN and factors related to PBC, which involve actors operating in farm ecosystem, are classified outside the scope of this study.

2.2. *Method*

Considering the exploratory nature of the research questions and the gaps in the theoretical and empirical basis on the factors that can affect the PA adoption, and on Agriculture 4.0, an exploratory multi-method was adopted, shown in Figure 1, consisting of interviews with experts, case studies, and the modeling of irrigation OM [39,42,43,46].

Multiple research fields were combined to triangulate the results of each method, to provide a more complete and deeper understanding of the factors that can affect the adoption of PA [43,100].

In this methodology, the researcher adopts multiple methods of data collection and data analysis. Data are collected in expert interviews and case studies to generate ideas and theoretical propositions and then to build a theoretical framework (model of irrigation OM) [39]. Interviews with experts [43] and case studies [42,101] in a pilot of the SWAMP project and on a farm in Brazil, in the State of Bahia, were conducted to identify factors that can affect the adoption and to formulate theoretical propositions for building a model of irrigation OM with the IDEF0 technique [41]. The objective of this approach was to induce emerging insights from empirical research [102], to develop theory [40,42,101]. Expert interviews and case studies are especially useful when the researcher does not know the variables to examine [39], for new areas of research or areas for which existing theory seems inadequate [101], like the explored adoption. Furthermore, as highlighted by [103], talking to experts in the exploratory phase of research is a more efficient method of data collection than, for example, quantitative surveys. Main characteristics are described below for each method.

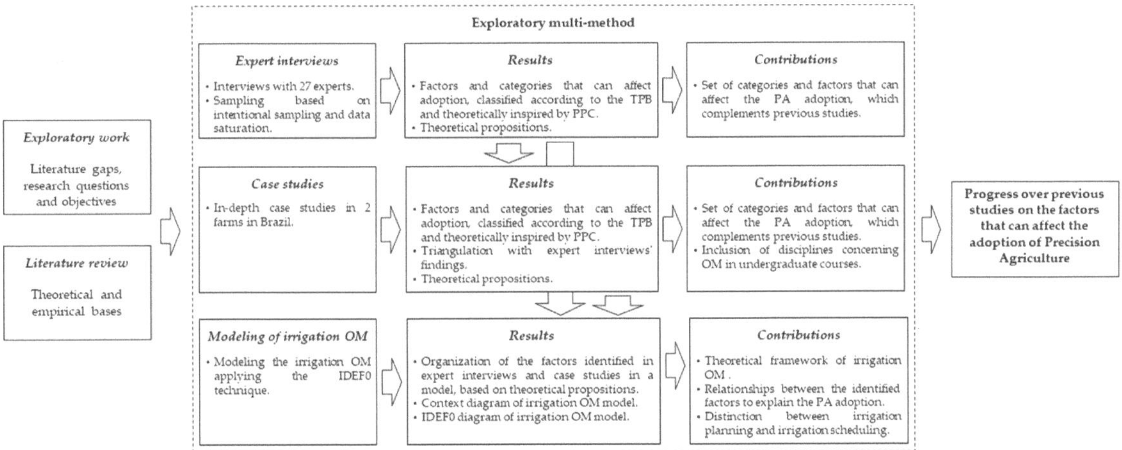

Figure 1. Exploratory multi-method.

2.2.1. Interviews with Experts

The expert interview is a qualitative method designed to draw on expert knowledge. According to the research objective, the researcher decides who to interview as an expert in their own field of knowledge [104]. In addition to exploring knowledge, an expert interview can be used to provide guidance on a recent or unclear thematic field and to generate theoretical propositions [103].

Due to the interdisciplinary nature of this project, 27 experts were selected for their expertise in agribusiness management, Precision Agriculture, irrigation, water resources management, and Operations Management. The protocol and questionnaire used in the interviews, illustrated in Supplementary Material A, were based on the precepts of the TPB [21] and focused on irrigation OM [105]. Additional questions were asked only if a participant response to the initial question did not cover certain topics of interest [106]. To reduce response bias, the definition of Agriculture 4.0 was not included in the questionnaire and was not provided to respondents [43,107].

To establish how many interviews to conduct, there are several factors affecting the sample size: population heterogeneity, budget, available resources, scope of the study, experience of the interviewer, availability of participants [108]. According to [106], in this work, intentional sampling was applied, with the aim of reaching data saturation (data saturation is the point in data collection and analysis when new information produces little or no change in the codes [106]) in relation to the codes (codes are theory or model building blocks on the basis on which the researcher arguments rest, according to the research question [106]) identified from data collected in the interviews and then in relation to the identified factors. So, the number of experts interviewed was 27, because when the 27th interview was completed, no new information was observed in the data. As, in this research, probabilistic sampling was not applied, the sample of experts interviewed is not representative. Furthermore, the codes ("deconstruction" of experts' knowledge) and the theoretically inspired perspective ("reconstruction" of experts' knowledge) were identified based on the knowledge of the literature on PA, Agriculture 4.0, OM, and farmer behavior [109]. The result was a review of the texts and terminology of the interviewees and the consequent creation of codes. This led to a "reconstruction of experts' knowledge" and to the representation of empirical results [104].

As a result of the evolution of the COVID-19 pandemic, it was not possible to access the university to use software to analyze the data collected in the interviews.

2.2.2. Case Studies

The researcher chooses single cases when the purpose is exploration and in-depth study [42], while he chooses multiple cases when the logic of replication in more cases is pursued rather than sampling [40,101]. Therefore, two in-depth case studies were conducted, considering the exploratory purpose for theory development and the possibility of carrying out in-depth studies.

In line with the objective of disseminating the SWAMP project results, case studies were conducted on two farms in Brazil: the first in the MATOPIBA pilot, and the second on a farm not involved in the project. To preserve and guarantee confidentiality, this company was named "Farm Bahia". The two case studies provided homogeneous characteristics in terms of interest in participating in research projects, farm size (industrial farm), location (State of Bahia), climatic region, irrigation system (center pivot), organizational structure (farm manager), and farmer characteristics (i.e., age, educational level, innovation capacity). Therefore, according to [40], these studies presented profiles to predict similar results.

The protocol for the case studies and the questionnaire with the objective of the interview, the context of which the research is part, and the proposed questions are shown in Supplementary Material B. The questions, based on the TPB precepts and focused on the irrigation OM, were the same as in the questionnaire used for the interviews with experts. Instead of the phrase "Agriculture 4.0", questions referred to sensors and drones, to be equipment the farmer could identify as representative of innovative technologies for irrigation, in addition to being equipment included in the context of Agriculture 4.0.

The data collection techniques applied were as follows: interviews, photos, farm reports, farm contracts, consultants' reports, and IMS database. The interviews were conducted remotely, and it was not possible to visit the farms due to the COVID-19 pandemic. To analyze the data collected in the case studies, the process of coding the data collected in the interviews with experts described in Section 3.1 is applied [63,64].

2.2.3. Modeling Techniques and IDEF0

In the literature, there are different techniques for modeling the structure, behavior, and organization of a company or information system, such as IDEF0 and Unified Modeling Language [44,45]. Considering that the IDEF0 is simple, it allows for the development of conceptual models of business models, and it is a standard governed by rules and conventions, representing data and resources in the same diagram. This work applied the IDEF0 technique to model the irrigation OM, also in accordance with the exploratory purpose of the question and research objectives. Furthermore, in the OM literature, the IDEF0 technique was applied by several authors to model PPC in companies in the manufacturing [46,110–112] and agricultural sectors [45].

An IDEF0 model involves a hierarchy of diagrams whose central components are boxes (representing a function: activity, action, process, operation) and arrows, representing an interface. An interface can be an input, entering the left box, transformed by the function and leaving the right box as output; a control, entering the top of the box and determining the function (i.e., procedure, command, worker experience); a mechanism, a tool or resource performing the function. Each box can be broken down to a lower level of detail, forming a hierarchy of information that is summarized in a tree of nodes [41].

3. Results

This section presents the results of the interviews with the experts, the case studies in the MATOPIBA pilot and on "Farm Bahia", and the model of irrigation OM. The data collected in the expert interviews and case studies, associated with the factors found in the theoretical and empirical bases of the literature, were used to identify the categories and factors that can affect the explored adoption. These data collected were coded based on the TPB constructs in scope (Attitude and Perceived Behavioral Control), on the categories and factors resulting from the literature review [21,101,109]. The coding process to identify the categories "Operations Management models inspired by the industry" and "Resources", and the factors "operations planning and control models", "resources (hardware—agricultural equipment, sensors)", is presented in Table 1.

Table 1. Example of coding process.

Theoretically Inspired Perspective: TPB, Production Planning and Control		**Deconstruction of the Knowledge**	**Reconstruction of the Knowledge**
TPB Constructs	**Literature**	**Interviews—Excerpts with Similar Topics**	**Categories**
Perceived Behavioral Control (impediments and obstacles, available resources and opportunities)	In agriculture OM is responsible for designing, planning, scheduling, and executing operations involving humans and machines [28]. Production Planning and Control (PPC) is one of the core functions of OM [75]. Agriculture 4.0 allows to improve planning and control in agricultural production [32]. Agriculture 4.0 allows to follow plant and yield development progress through information collected by sensors [33].	*"There are classic industry planning tools the farmer could use, such as PERT, CPM, and ABC costing system"*, expert 1. *"The adoption of Agriculture 4.0 allows the farmer to optimize the use of resources, the use of water. The adoption of sensors improves irrigation and equipment planning so that irrigation is more effective"*, expert 5.	Operations Management models inspired by the industry, Resources **Factors** Operations planning and control models, resources (hardware—agricultural equipment, sensors)

Responses from interviews and case study data were categorized based on TPB constructs and then analyzed to identify excerpts with similar topics from the interviews ("deconstruction" of the knowledge of experts and farm managers). Based on the factors identified in the literature on PA, Agriculture 4.0, farmer behavior, and OM, excerpts with similar topics were grouped into codes to define categories and factors ("reconstruction" of knowledge). The empirical results were framed by a theoretically inspired perspective [104], guided by the TPB and the introduction of inspired models from the industry, like the PPC.

The codes were organized into categories (e.g., "performance measures", "data access") and factors. The proposal to introduce industry-inspired models in irrigation was based on the factors related to OM analyzed in Section 3.1 from interviewed experts. This process was recursive, while in the analysis, it was necessary to return to a previous stage to verify the adequacy of generalizations [104]. In the analysis of data collected, a check was carried out to identify categories and factors found in other answers [40].

3.1. Results of Expert Interviews: Factors and Categories That Can Affect the Adoption

Between April and August 2021, 27 interviews with experts were carried out. Supplementary Material C summarizes the experts' profiles. The choice criteria were based on the interviewees' expertise in agribusiness management, PA, irrigation, water resource management, and OM. Experts with experience as researchers and in water resource management were included, considering the importance of this topic for this work. Initially, four experts were selected who illuminated the research questions. To select the other experts, the "snowball" sampling technique was adopted [113].

All interviews were conducted remotely because of the COVID-19 pandemic. A pilot interview was organized to test the questionnaire [102]. The questions were unambiguously formulated, and clarifying questions were asked to complete the expert answers [18]. All interviews were recorded and transcribed in full. Confidential and sensitive references or data were omitted or substituted to ensure the privacy and anonymity of the interviewee [106].

The answers of the experts interviewed were categorized and classified to produce the transcripts, with the results shown in Supplementary Material D. The factors and categories are shown in Table 2. The factors and categories are classified according to the TPB constructs and based on previous studies on PA, Agriculture 4.0, OM, and farmer behavior.

Table 2. Results of expert interviews: factors and categories that can affect the adoption.

TPB Constructs	Factors	Categories
Attitude	Crop water requirement, water use, energy use, cost of water use, cost of energy use, variable rate water use, production yield, profit, revenue, production cost.	Quantitative performance measures
	Management benefits, improvements in irrigation planning, improvements in irrigation control.	Qualitative performance measures
	Access to agronomic data, access to operational data.	Access to data
Perceived Behavioral Control	Technical training, managerial training, mental model, irrigation management, data-based management, farm management as a business.	Changes for the farmer
	Farm management, Operations Management, workforce qualification.	Changes for the farm
	Operations planning and control models, collaborative management models.	Operations Management models inspired by the industry
	Resources (inputs, hardware, people), inputs (water sources, water, energy), hardware (agricultural machinery, agricultural equipment, weather stations, soil probes, drones, satellites, sensors, Irrigation Management System, people (farmers, farm managers, workers, consultants).	Resources
Antecedent Factors	Age, educational level, income, experience in agriculture, familiarity with technologies, family of farmers, conservatism, managerial training, absorptive capacity, risk propensity, innovation capacity.	Farmer characteristics
	Farm size—family farm/industrial farm, farm location, crop type, cooperativism, production volume, product profitability.	Farm characteristics
	Technology type, equipment type, price, complexity.	Technology characteristics

Regarding the analysis of the interviews, if the quotes led to the identification of benefits and positive results expected following the adoption, the factors were classified in the attitude construct. If the quotes led to identifying obstacles, resources, and opportunities related to the adoption, the factors were classified in the PBC construct. The factors and categories identified in the responses to additional questions were included in the most proper factors and categories, based on the coding process. Factors unrelated to OM and relative to stakeholders in the farm ecosystem were classified as out of scope.

The benefits of the studied adoption can be measured through performance measures related to attitude, identifying quantitative or qualitative benefits that can be achieved with adoption. Quantitative performance measures are benefits to which a measurement can be associated concerning planning and the control of irrigation operations. All the experts pointed out that the adoption involves a cost reduction: irrigation water use and energy use reduction. Experts 11, 14, 16, 23, and 27 added that the adoption can lead to a rational use of water in terms of irrigation in the right amount, closer to the crop water requirement, at the right time and in the right place. The reduction in water use, and the relative cost, has an impact on energy savings, according to experts 2, 8, and 11: *"The farmer will have a water savings because he will be able to define the exact amount of water for that exact area, and with that he will be able to save energy which is one of the big problems in irrigated agriculture"*, expert 11. Experts 23 and 25 emphasized that the reduction in water use, and the relative cost, affects an increase in production yield due to irrigation efficiency. Therefore, saving water and energy has a positive impact on the farm economic results, as suggested by experts 6, 11, and 16.

Regarding qualitative performance measures, the interviews allowed us to identify benefits related to irrigation management, as stated by expert 3 (management benefits compared to increased productivity) and expert 26 (changing from planning based on experience to planning based on data).

The explored adoption can bring benefits thanks to access to more organized data, promoting better operation planning and control, as reported by experts 16, 26, and 27. These benefits could be related to access to agronomic data on crops, soil, and climate, possibly due to the adoption of sensing technologies, according to experts 11 and 27. These data can be used to calculate the crop water requirement and irrigation water requirement. The farmer also has the possibility to access operational data related to irrigation operations, such as water availability, water flow, cost of water use, water access contract, and energy supply contract, according to experts 1, 17, 18, and 27.

Factors concerning the changes required for the adoption are classified into two categories: "Changes for the farmer" and "Changes for the farm".

As for the changes for the farmer, they must be trained to adopt PA technologies, as reported by experts 15 and 16. Expert 16 states that: *"The farmer must be trained and prepared in terms of knowledge and skills to use these technologies. If he doesn't understand it well, he most likely won't adopt correctly"*.

Training includes technical training and management training, as reported by expert 13. Technical training is necessary for irrigation, to understand what irrigation is, what irrigated agriculture is, and that the agricultural field has variability, according to expert 17. Managerial training concerns specific skills for data analysis, planning, and management of new technologies, as reported by expert 6, who stated: *"It is necessary to change the farmer managerial training and the workforce qualification, because the adoption of Agriculture 4.0 involves the introduction of a production system that demands a set of specific skills, of data analysis, of handling machinery to operate according to the planning"*. Experts 1, 10, 13, and 24 show that this change must overcome some challenges in terms of farmer perception, low levels of education, and training, including the areas of operation engineering and management. Expert 1 emphasized that: *"Planning is not easy for the farmer, because he was not trained for it, he understands little about operation engineering, and often those who advise the farmer do not have this training either"*.

These challenges are linked to the need to change the farmer mental model, according to experts 2 and 20. Expert 14 stated that: *"The main barrier is cultural, in addition to the need to be trained to assimilate these technologies. And it depends on the farmer age, who mainly relies on his experience"*. This mindset shift involves irrigation management, data-driven management, and managing the farm as a business. The farmer needs to change their belief that irrigating is different from applying water and irrigated agriculture is different from rainfed agriculture plus water, as highlighted by experts 4, 11, and 17. The farmer also needs to change his mental model to move from intuition and experience-based management to data-based management, which brings benefits to the planning of operations, as reported by experts 14, 15, and 26. Expert 21 stated that: *"The culture of many farmers is based on family and community knowledge, transmitted from father to son, rather than adoption of techniques from scientific knowledge. There is a gap between the supplier of knowledge, which leads to training and the ability to understand, and the farmer in the adoption of irrigated agriculture techniques"*.

The shift to data-based management must also make farmers aware that water is a scarce resource. The adoption of sensing technologies can enable this change on water scarcity, which can be measured with a rational use of water, as said by experts 10, 19, 22, and 27.

Regarding the management of the farm as a business, the farmer has great difficulty in management, as reported by expert 12. Experts 5 and 23 added that the farm management must also include irrigation planning and control: *"Irrigation control is something that should always be considered in a planning process, which is part of managing a property"*.

In addition to the changes for the farmer, the analysis of the interviews enables one to identify the factors classified in the category "Changes for the farm". For the studied adoption, farm management results as the central change, as reported by experts 14 and 10, which must be built on the basic functions of administration (planning, implementation, and control). The changes for the farm also involve the Operations Management, according to expert 6: *"The OM has developed with a short delay in Brazil. These are aspects of agricultural production that have not evolved at the same speed as the agronomic aspects"*. Expert 1 highlighted that the OM is an area undervalued by the farmer because he does not have much knowledge about it. Expert 20 added to this with changes in organizational characteristics, such as control mechanisms and performance indicators, while expert 6 also emphasized the characteristics of integration between different systems within a farm to collect and analyze production data. Another factor classified in the changes for the farm is the qualification of the workforce, which is considered one of the cornerstones to overcome for adoption, according to experts 4, 6, 9, 12, and 19.

It is important to highlight that these changes related to the adoption studied must be gradual, and the adoption of Agriculture 4.0 is not a single path, as stated by experts 2, 4, 7, 12, and 26. Expert 12 said that *"Agriculture 4.0 has several stages and the farmer can move from the initial one, which do not involve robotization. The farmer can adopt with a more gradual process in which the cost-benefit analysis is easily identified"*.

Concerning the category "Operations Management models inspired by the industry", experts 1 and 6 inspired the adoption of planning and control models of operations, which consist of planning, scheduling, and control of irrigation management. Expert 1 pointed out that these models are not yet applied in agriculture and suggested industry models such as PPC: *"There are planning tools, classic in the industry, such as PPC, for the planning of operations: the farmer would need to have process mapping, production flow, activities, times, costs of machines, labor, demand forecast. With this information, the farmer could create a proper decision making support system"*.

Another opportunity available to the farmer, following the adoption, is constituted by collaborative management models, as reported by experts 16, 22, 25, and 27. Expert 16 suggested the use of a process of negotiated allocation of water: *"The water resources management agency can talk to the farmers to check who will plant, who will not, who will irrigate and who will not, to distribute the available water in the most critical period and prevent the farmer from planting"*. Expert 25 inspired the adoption of the Just-in-Time (JIT) management model [114] to control the use of motors and pumps in irrigation systems for monitoring water abstraction in farms. The adoption of these collaborative models enables them to overcome some challenges related to irrigation planning and control in the case of shared management of water resources, ensuring the availability of water throughout a season for the farmer.

The analysis of the responses relative to the PBC also identifies the category "Resources", which includes inputs, hardware, and people. These resources are necessary for the management of information concerning the planning and control of operations, according to experts 1 and 27. The adoption of hardware makes it easier to achieve benefits for irrigation planning, as in the case of sensors: *"If the farmer works with sensors, he has more accurate data on water withdrawal and applied for irrigation. He can plan the irrigation system so that irrigation is more effective"*, expert 5.

Regarding resources, experts 2, 3, 8, and 24 highlighted the role of consultants in irrigation planning. Expert 3 emphasized that: *"Companies that provide services on irrigation planning obtain data from equipment, sensors and have a team that processes this data and makes recommendations to the farmer"*. These services are predominantly contracted by large producers, who normally have a skilled workforce, as reported by expert 2.

Based on the quotes from the experts, the explored adoption can also be affected by the farmer characteristics, farm characteristics, and technology characteristics, as highlighted by expert 1: *"The application of the technologies included in the concept of Agriculture 4.0 will have different speeds, according to the regions of the country, the farm and farmer characteristics, the product types. In some plants and some farms, Agriculture 4.0 should arrive faster"*; expert 5 shared this view.

Concerning the study of the benefits of the adoption, and according to the importance of the attitude construct for the benefits associated with the adoption and perceived by the farmer [86,115], it is important to emphasize the farmer characteristic "conservatism", as highlighted by expert 3: *"The rural producer does not believe in things, he wants to see things"*. Expert 30 added: *"For the farmer to trust an innovation, he needs to see where it worked, it was tested, there are results and how he can adopt it"*.

As reported by expert 1, the adoption can also be affected by the farm characteristics. Cooperativism is a factor that can help in the adoption and use of models for irrigation OM, such as hiring consultant services, according to experts 3 and 24.

3.2. Results of Case Studies: Factors That Can Affect the Adoption

To conduct the case studies, the MATOPIBA pilot and "Farm Bahia" were selected due to their qualities in innovative irrigation systems. Theoretical sampling was adopted [101] due to the characteristics of the farms linked to the theory and the literature addressed in this work, such as exploring the adoption in farm size—industrial farm and in center pivot irrigation system. The choice of "Farm Bahia", familiar with the participation in research projects, was also consistent with the objective of the SWAMP project to replicate the results in terms of models and solutions.

The research protocol was designed with the questions to be applied in the interviews and the specific data to be collected [21,101,105]. Interviews with the farm manager of the MATOPIBA pilot were conducted between January 2020 and December 2021 and with the "Farm Bahia" manager between June 2021 and December 2021. The farm was the unit of analysis, according to the SWAMP project objective and the research questions [40].

The factors that can affect the adoption resulting from the cases, shown in Table 3, were classified according to the TPB constructs and the theoretical and empirical bases, and were triangulated with the findings of expert interviews [40,43].

Table 3. Results of case studies: factors that can affect the adoption.

Categories	Factors	Farm Bahia	MATOPIBA Pilot
Farmer characteristics	Educational level	Degree	Degree
	Age	30–35 years	30–35 years
Farm characteristics	Farm size	9800 hectares	915 hectares
	Crop types	Soybean and corn	Soybean, corn, sorghum, and cotton
Performance measures	Quantitative Qualitative	Used Not adopted	Used Not adopted
Access to data	Access to agronomic data	Weather station, IMS	Weather station, IMS
	Access to operational data	Contract for access to water, contract for energy supply, IMS	Contract for access to water, contract for energy supply, IMS
Changes for the farmer	Technical training	Skilled farm manager	Skilled farm manager
	Managerial training	Access to the market	Access to the market
	Mental model	Minimization of the cost	Minimization of the cost
	Cost of water use	Not yet charged	Not yet charged
	Data-based management	Also experience	Also experience
	Farm management as a business	Used	Used
Changes for the farm	Farm management	Industrial farm	Industrial farm
	Operations Management	Undervalued area, untrained farm manager	Undervalued area, untrained farm manager
	Qualification of workforce	Low educational level	Low educational level
OM models inspired by the industry	Models of planning and control of operations	Undervalued area, untrained farm manager	Undervalued area, untrained farm manager

Table 3. *Cont.*

Categories	Factors	Farm Bahia	MATOPIBA Pilot
Resources	Water source	River	River
	Energy	Electric pumps and motors	Electric pumps and motors
	Agricultural machinery	17 center pivots	7 center pivots, reservoir
	Weather station	Located at the farm	Located at the farm
	Communication system	Internet—4G	Internet—4G
	Soil probe	Not adopted	Research project
	Satellite	Not adopted	Not adopted
	Irrigation system management	Used	Used
	Workforce	25 agricultural workers	16 agricultural workers
	Consultants	Support for planning, control and contract management	Support for contract management

Furthermore, a "within-case analysis" (factors in columns of Table 3) and a "cross-case pattern search" (patterns between the two cases identified considering the factors in each row and described in the text illustrating the table content) [101] were conducted.

3.2.1. Within-Case Analysis: "Farm Bahia"

"Farm Bahia" is managed by a farm manager who graduated in Agricultural Engineering. The authors' perception is that the manager's conservatism is low, and he has a propensity to risk in relation to the adoption of innovations. Quantitative measures related to the use of water and electrical energy are used on the farm, such as crop water requirement, water use, energy use, cost of water use, and cost of energy use, in addition to economic measures.

Regarding the access to agronomic data, a weather station supplies meteorological data. Crop, soil, and irrigation data are accessible through an IMS, which also provides operational data, such as production yield, revenue, and cost per center pivot. The electrical energy supply contract makes data on operating time and the cost of energy use available.

As for the changes for the farmer, the farm manager has technical and managerial skills since he has a degree in Agricultural Engineering. Management is based on data the and scientific method (e.g., water balance to estimate the irrigation water requirement), although experience is a relevant factor in irrigation planning and control. The skills related to OM are limited due to the low valuation of this area. The possibility of access to the market does not pose challenges. He is aware that the main changes are managerial, related to the need for training to use new technologies, to read the data correctly, and to manage irrigation planning and control. His mental model is aimed at minimizing energy use cost, due to the lack of a definition for water use cost.

Concerning the farm changes, the case study presents two changes. The first involves the OM, since this is an area undervalued and the manager is not skilled. Workforce qualification results in another key change due to the low educational level of the farm workers. These challenges can be overcome considering the size of the farm, its possibilities of accessing the market, farmer low conservatism, risk propensity, and capacity for innovation. In relation to the models of OM inspired by the industry, the case shows a gap in the adoption due to little importance of this area and low managerial training.

The farm manager uses the IMS for irrigation planning. The estimation of the crop water requirement, the planning of irrigation water requirement, and of irrigation time are conducted by consultants using the water balance approach. The nightly, daytime, and peak operating times of the center pivots are defined based on the energy supply contract. The farm manager planning objective is to minimize the cost of electrical energy use to take advantage of the nightly operating time. As for irrigation control, he manually analyzes soil moisture based on experience. Center pivot control operations are also performed manually because of the bad quality of electrical energy.

Regarding the resources, the contract for accessing water is formalized, but the cost of using irrigation water and the time and volume of abstraction are not defined yet.

3.2.2. Within-Case Analysis: Farm of the MATOPIBA Pilot

The MATOPIBA pilot farm is directed by a manager who graduated in Agricultural Engineering, with low conservatism and propensity to risk in relation to adoption innovations. Quantitative measures related to the use of water and electrical energy are used. The farm manager uses a weather station and an IMS to access agronomic data, while he uses the water access contract and the electrical energy supply contract for access to operational data.

As for the technical and managerial skills, for the MATOPIBA pilot, manager experience is still a crucial factor in irrigation planning and control. The changes for the farmer, mainly managerial, do not represent major challenges, due to the possibility of access to the market. Regarding technical training, the central challenge is where to irrigate based on soil probe data and, therefore, implementing the variable rate irrigation. His mental model is aimed at minimizing the cost of electrical energy use due to the lack of water use cost in the current contract. Regarding the farm changes, the case presents two changes related to OM and the qualification of the workforce. The MATOPIBA pilot manager administers irrigation planning with the support of the IMS. The irrigation water requirement is estimated weekly using a water balance approach. Irrigation water requirement and irrigation time are planned based on the cost of electrical energy use, which prioritizes nightly operating time. Nightly irrigation is more economical, as the cost of electrical energy use is ten-times lower than during the day. As for irrigation control, center pivot operations are performed manually based on experience. Regardless of the contract for access to water, the cost of water use, the time, and the volume of abstraction are not defined yet.

3.2.3. Cross-Case Pattern Search

The two farms present homogeneous factors in terms of farmer characteristics, farm characteristics (mainly farm size—industrial farm, farm location), quantitative and qualitative performance measures, access to agronomic data, access to operational data, changes for the farmer and for the farm, models of OM inspired by the industry, and resources. The IMS used by the two farms is provided by the same supplier. On both farms, the contract for access to water is formalized, but the cost of water use, the time, and the volume of abstraction are not defined yet.

In relation to the models of OM inspired by the industry, in the MATOPIBA pilot, the irrigation planning is administered by the farm manager, while in "Farm Bahia", the estimation of the crop water requirement, the planning of irrigation water requirement, and the irrigation time are conducted by consultants. The two farms have some of the heterogeneous characteristics of the irrigation system. In the MATOPIBA pilot, two center pivots are powered by electric pumps, and the others are supplied through a reservoir, which enables one to reduce evaporation during water distribution. In "Farm Bahia", all 17 center pivots are supplied directly from a river. As for sensing technologies, the MATOPIBA pilot is carrying out a project for the adoption of soil probes and performing variable rate irrigation. The farm manager is assisted by agronomic experts in planning irrigation water requirements. He also hires management consultants to control the electrical energy use and for some issues inherent to the supplier agency (i.e., high fixed monthly cost, low quality of electrical energy). The manager of "Farm Bahia" hires management consultant services for the renewal of the water access contract and technical consultants to remotely control the electrical energy use on a weekly basis.

3.3. Results of the Expert Interviews and Case Studies: Theoretical Propositions

The results of the expert interviews and case studies allow us to generate ideas and formulate theoretical propositions [39,40] to explain the relationships between the adoption and the identified factors. The theoretical propositions, which provide a solid foundation for modeling irrigation OM and to direct future research, are illustrated below in the order of categories and factors shown in Tables 2 and 3.

Theoretical proposition 1: According to experts 2, 3, 6, 8, 11, 14, 16, 17, 23, 24, 25, 26, and 27 and the two farm managers, the adoption of PA in the context of Agriculture 4.0 allows for achieving advantages and benefits that can be measured through quantitative performance measures. The adoption involves cost reductions in irrigation, water, and energy use; the reduction in water use, and the relative cost, also affects an increase in production yield due to irrigation efficiency; saving water and electrical energy has a positive impact on the economic results of the farm. Theoretical proposition 1a: Based on interviews with experts 3, 20, 26, and 27, the adoption allows us to obtain advantages and benefits that can be measured through qualitative performance measures.

Theoretical proposition 2: As stated by expert 16, the explored adoption can bring benefits thanks to the access to more organized data. Theoretical proposition 2a: According to experts 16, 26, and 27, access to more organized data allows for better operation planning and control. Theoretical proposition 2b: Based on interviews with experts 11 and 27, the benefits of the adoption may also be related to the access to agronomic data on the three pillars of crop, soil, and climate. Theoretical proposition 2c: As stated by experts 11 and 27, these data can be used to estimate the crop water requirement and irrigation water requirement. Theoretical proposition 2d: According to experts 1, 17, 18, and 27, access to agronomic data is provided by the adoption of sensing technologies. Theoretical proposition 2e: Based on interviews with experts 11, 17, 18, 24, and 27 and the two farm managers, the adoption of PA in the context of Agriculture 4.0 makes it possible to obtain benefits thanks to the access to operational data concerning irrigation operations. Theoretical proposition 2f: According to both farm managers, the IMS provides operational data. Theoretical proposition 2g: Based on interviews with both farm managers, the electrical energy supply contract provides data on operating time and the cost of energy use in different periods; the operating times of the center pivot are defined based on the electrical energy supply contract. Theoretical proposition 2h: As stated by all the farm managers, the contract for access to water provides data on the cost of water use, abstraction time, and volume.

Theoretical proposition 3: The explored adoption requires farmer changes. Theoretical proposition 3a: According to experts 1, 5, 6, 7, 10, 11, 13, 15, 16, 17, and 24 and the two farm managers, to adopt these technologies, the farmer must be trained. Training includes technical and management training; managerial training comprises skills for data analysis, planning, and the management of new technologies. Theoretical proposition 3b: Based on interviews with experts 2, 4, 11, 14, 17, and 20, the adoption requires changes in the farmer mental model, which involve irrigation management due to the need to change the belief that irrigating is different from applying water, and irrigated agriculture is different from rainfed agriculture plus water. Theoretical proposition 3c: As stated by experts 14, 15, 19, 20, 21, 22, 26, and 27, changes in the farmer mental model involve data-driven management, as moving from experience-based management to data-based management provides benefits for operations planning. Theoretical proposition 3d. According to experts 10, 19, 22, and 27, changes in the mental model include the farmer awareness that water is a scarce resource. The adoption of sensing technologies can make this change possible. The farmer needs to understand that the agricultural field has variability. Theoretical proposition 3e: Based on interviews with experts 5, 12, 23, and 24, changes in the mental model involve the management of the farm as a business. The farm management must include irrigation planning and control; the farmer often does not know the production cost and does not understand their costs and income. Theoretical proposition 3f: As stated by expert 18 and the two farm managers, to overcome the challenges related to changes, the farmer can access technology, finance, and educational markets.

Theoretical proposition 4: The explored adoption requires farm changes. Theoretical proposition 4a: Based on interviews with experts 6, 10, 14, 17, and 20, farm management must be based on the basic functions of administration. Theoretical proposition 4b: According to experts 6 and 20 and the two farm managers, changes are also necessary in organizational characteristics in integration between different farm systems. Theoretical proposition 4c: As stated by experts 1 and 6 and the two farm managers, changes include OM, which has not evolved at the same speed as agronomic management. Theoretical proposition 4d: Based on interviews with experts 4, 6, 9, 12, and 19 and all farm managers, the changes involve the workforce qualification, which is one of the main barriers to overcome for the adoption.

Theoretical proposition 5: As stated by experts 2, 4, 7, 12, and 26, the changes about the adoption studied are gradual. Theoretical proposition 6: According to experts 1, 3, 4, and 6, for the adoption, the farmer makes available models of planning and control of irrigation operations, inspired by models of the industry, such as PPC. Theoretical proposition 7: According to experts 16, 22, 25, and 27, for the adoption, the farmer has the possibility to use collaborative management models, which allow one to overcome some challenges related to irrigation planning and control in the case of shared management of water resources.

Theoretical proposition 8: Based on interviews with experts 1 and 27 and both farm managers, for the explored adoption, resources to manage information concerning OM must be available to the farmer. Theoretical proposition 8a: As stated by experts 2, 8, 9, 14, 17, and 18 and both farm managers, the resources include water and energy inputs. Theoretical proposition 8b: According to experts 2 and 5 and the farm managers, the resources include hardware, such as sensors, drones, and satellites. Theoretical proposition 8c: Based on interviews with experts 2, 3, 5, 8, and 24 and the two farm managers, the adoption of resources like sensors allows us to obtain benefits for operations planning. Theoretical proposition 8d: As stated by experts 2, 3, 8, 23, and 24 and the farm managers, the resources include people.

The explored adoption can be affected by the farmer characteristics, according to experts 1, 3, 4, 5, 8, 9, 10, 13, 14, 16, 20, 21, 30, and 32 (theoretical proposition 9), by the farm characteristics, as stated by experts 1, 2, 3, 5, 9, 16, 17, 23, 24, and 25 (theoretical proposition 10), and by the technology characteristics, based on interviews with experts 1, 5, 6, 8, 20, and 25 (theoretical proposition 11).

Propositions 1, 1a, 2, 2a, 2b, 2c, 2d, 2e, 2f, 2g, 2g, 2h, 4d, 6, 8, 8a, 8b, 8c, 8d, and 10 support the proposal of the irrigation OM model presented in Section 3.4. The factors educational level, training, access to data, and resources (people—consultants) are discussed in Section 4.3. Other factors, out of the scope of this paper, will be analyzed in the future research agenda.

3.4. Model of Irrigation OM

A model of irrigation OM is built, based on theoretical propositions, factors identified in theoretical and empirical bases, and factors identified in expert interviews and case studies. This model, developed with the IDEF0 technique, is inspired by the PPC model, according to theoretical proposition 6. This irrigation OM model supports the development of theory on the adoption of PA in the context of Agriculture 4.0. The model of irrigation Operations Management (context diagram), shown in Figure 2, defines the scope of the model and describes the modeling of the top-level function.

The context diagram is broken down into four functions, creating child diagrams and the model of irrigation OM (IDEF0 diagram), as shown in Figure 3: irrigation planning, irrigation scheduling, irrigation execution, and irrigation control. Each of the sub-functions may be decomposed in lower-level child diagrams. The description of the IDEF0 diagram shows the identified factors and the pertinent theoretical propositions.

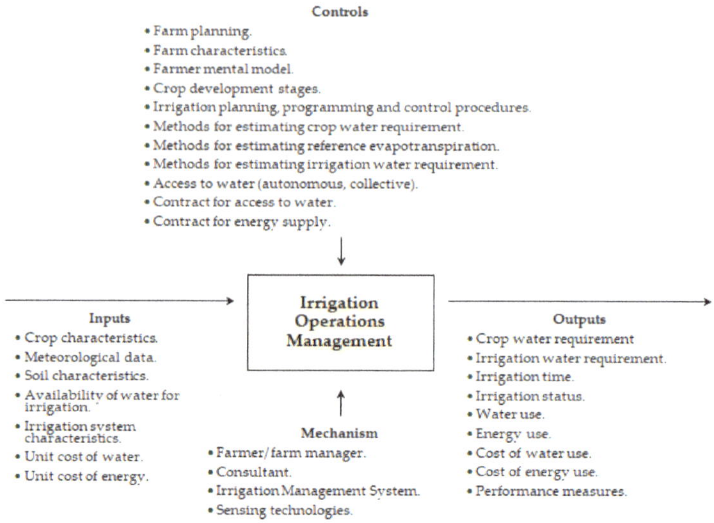

Figure 2. Model of irrigation OM (context diagram).

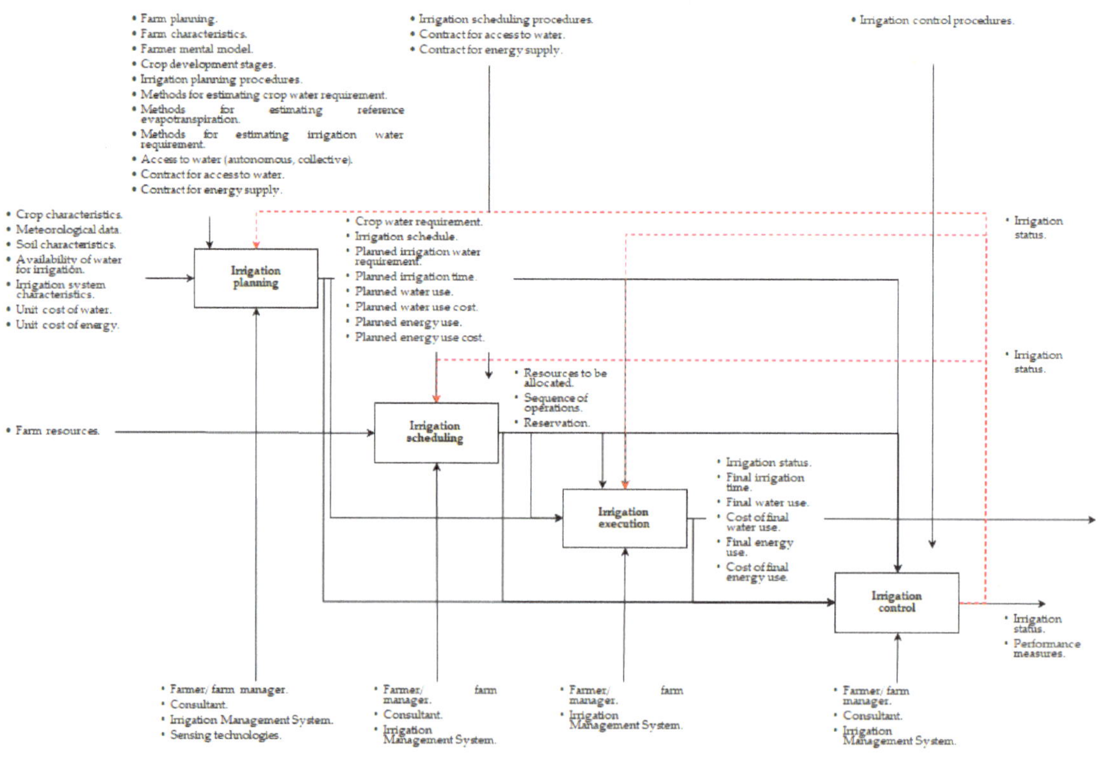

Figure 3. Model of irrigation Operations Management (IDEF0 diagram).

Irrigation planning emphasizes the dimension "planned" to the output factors: irrigation water requirement, irrigation time, water and energy use, and water and energy use cost. Irrigation planning can be referred to as a timescale of the harvest or within the harvest, according to experts 16 and 20.

To plan the crop water requirement, data from three pillars are essential: crop, soil, and climate (factors related to theoretical propositions 2b and 2c), as reported by experts 11, 15, 16, and 23. The irrigation schedule establishes irrigation days based on irrigation planning procedures, such as minimizing the cost of energy use, taking advantage of irrigation at night time, given the energy supply contract and the relative unit cost of energy at different times during the day, as reported by the two farm managers and expert 2. The factors mentioned are addressed in theoretical propositions 2e and 2g.

Irrigation time can be planned in relation to nightly, daytime, and peak operating times (factors related to theoretical propositions 2e and 2g), according to the two farm managers and experts 10, 14, and 25. Water use, water use cost, energy use, and energy use cost can be planned based on different operating times. Energy use and energy use cost can be planned depending on the irrigation equipment like pumps and motors, as reported by the two farm managers (factors related to theoretical propositions 2e, 2g, 2h, 8a, and 8b). The pumps and motors can be powered by diesel oil, renewables, or electrical energy, according to [116] and expert 27. Irrigation planning can be conducted at the level of the farm and field.

The controls facilitating the execution of the function consist of the same factors as in the context diagram, except for irrigation scheduling and control procedures, based on farm manager interviews and experts 1, 11, 15, 16, 17, and 23. The contract for access to water is essential to know the availability of water for irrigation, as reported by experts 16, 25, 26, and 27 (factors addressed in theoretical propositions 2b, 2c, 2g, 2h, and 10).

Water can be unlimited or limited, according to experts 14, 16, 25, 26, and 27. Water availability for irrigation is also influenced by collective access to water resources, which poses challenges for irrigation planning based on factors, such as water demand, water supply, available water volume, farm location, priority in access, and the economic sector (human activity, animal breeding, irrigation, industry) in the case of conflicts over water use, according to experts 9, 14, 15, 17, 18, 24, 26, and 27 (factors related to theoretical propositions 2h and 10).

The energy supply contract can contribute to establishing the planning procedures since the cost of electrical energy can vary based on the hours during the day and can impact the definition of irrigation time, as reported by the MATOPIBA pilot farm manager: "*Nightly irrigation is what makes the business viable*". The manager of "Farm Bahia" said the following: "*The priority is to take advantage of the time reserved for irrigation, which is 9 h a night. But irrigation is carried out during the day too, almost all pivots are off by 11:00*". Experts 2 and 14 agreed with this (factors addressed in theoretical propositions 2a, 2e and 2g).

Farm planning involves distinct factors affecting seasonal decisions the farmer must make, like crops to cultivate [81,117]. Expert 1 highlighted the strategic nature of this planning, which involves long-term decisions (factor related to theoretical proposition 6). Farm planning also involves economic factors the farmer needs to consider. Experts 3, 4, 16, 24, 25, and 26 cite input price and availability, production cost, product prices, and market trends. As highlighted by the results of the two case studies, the cost of the electrical energy plays a fundamental role because it can influence the use of electrical energy and the planning of the operating time (factors mentioned in theoretical propositions 2e and 2g).

Experts 1, 9, 15, 20, 22, 24, 25, 26, and 27 cite another characteristic of agriculture, which is the uncertainty related to climatic factors, such as precipitation, water availability, and air temperature (related to theoretical propositions 2b). Expert 18 added the volatility of rainfall and drought due to the extension of Brazilian territory, while expert 26 stated that the explored adoption can contribute to reduce the risk related to these factors: "*The more the farmer invests in Precision Agriculture in the context of Agriculture 4.0, the greater the information base and knowledge of these factors he has. Therefore, the greater the control of his business*" (factors related to theoretical proposition 2b). Experts 13 and 31 added that the availability of more information leads to reducing the risks in agricultural production, for example, with climate change, which has increasingly prolonged effects. According to expert 34, the risk, linked, for example, to a possible period of drought, which may be critical based on the type of crop that the farmer plans to cultivate, can be reduced if information is made available to the farmer in real time. Expert 10 highlighted that: *"In the manufacturing industry the production process is somehow dominated, particularly inside a factory or warehouse, it is a well-known process. In agriculture, it rains, it rains more, it rains less, there it's hail, there is a problem with an epidemic of foot and mouth disease, fruit flies, rust in soybeans. This is why the planning and control of agricultural operations are very critical and challenging. The adoption of innovative technologies and models of OM poses a challenge to the farmer aimed at reducing the risks of seasonality on production management"*. Expert 25 added that due to climate changes, the statistics used for planning increase their uncertainty and, for this reason, the adoption of PA would provide access to more reliable and complete data and, therefore, it could reduce the risk in irrigation OM.

Irrigation planning processes the following inputs: crop characteristics, meteorological data, soil characteristics [58,118,119], the availability of water for irrigation, as reported by experts 17 and 27, the characteristics of the irrigation system (like minimum depth of water to apply, water flow, irrigated area), the unit cost of water, and the unit cost of energy according to the two farm managers (factors covered by the theoretical propositions 2b, 2e, 2g, and 2h).

The mechanisms used in the planning of irrigation are considered in terms of the farmer or farm manager, the consultant, the IMS, and the sensing technologies [118,120] (factors related to theoretical propositions 8, 8b, 8d). The farmer oversees irrigation management. Farms such as the MATOPIBA pilot and "Farm Bahia" use an IMS to assist the manager in planning. Sensing technologies support planning thanks to the collection of field data and access to more correct data, as reported by expert 5 (factors related to theoretical propositions 2, 2b, 8b, and 8c). Experts 2, 3, 10, and 23 highlighted the role of the consultant (addressed in theoretical proposition 8d) as an opportunity to support the farmer.

Key outputs of irrigation scheduling are the allocation of agricultural resources: people, machines [77], irrigation technique [121–123], irrigation resources, and water sources, according to the two farm managers (factors related to theoretical propositions 8 and 10).

In the case of shared water sources, such as a collective irrigation canal or watershed, the access to water and its use depend on the water access contract, as highlighted by experts 2, 8, 14, 16, 17, 25, 26, and 27. Expert 9 added that farmer location also affects irrigation scheduling, for example, in an irrigated perimeter. The contract can regulate the use of resources for measuring the water use, which farmers must install to measure the water flow, according to expert 2 (factors analyzed in theoretical propositions 1, 2h, and 10).

Irrigation scheduling produces a reservation and an operation sequence to execute. An example is described in the case of "Farm Bahia", in which the operation *"execution of irrigation with the application of irrigation water requirement equal to 9.4 mm in 50% of the irrigated area"*, using the resource *"center pivot 4"*, is scheduled on *"day 1"* (factors related to theoretical propositions 2, 2e, and 6). Irrigation scheduling procedures involves priority rules such as the minimization of the cost of electrical energy use relative to the use of water. This rule defines the priority for executing the irrigation in the night period in the two case studies (factors addressed in Theoretical propositions 1, 2g, and 6).

According to the proposal to adopt a management model inspired by the industry (factor related to theoretical proposition 6), irrigation scheduling can be considered planning at the operational level, as reported by expert 1: *"The farmer has a planning that is more strategic, long, and medium term. And then there is a more operational planning, for defining the planning of the machines, irrigation, spraying, harvesting, post-harvesting"*.

The execution of irrigation affects the performance of the scheduled operations [124]. Scheduled resources are expected to be activated automatically in the context of Agriculture 4.0, as stated by expert 26. Experts 1 and 22 highlighted the benefits that automation can bring in relation to switching on and off the irrigation equipment (factors concerning theoretical propositions 1 and 8).

However, experts 1 and 2 added a challenge related to the quality of electrical energy (factor addressed in theoretical proposition 2g) for performing irrigation with an automated system: *"The farmer cannot install an automation system to turn the pump on and off and the control systems because the energy quality will burn this equipment more quickly, due to the overload"*, expert 2. Expert 17 also cited the challenge of Internet coverage on the farm location (factor related to theoretical proposition 10) to enable the execution of irrigation with an automated system, while expert 18 highlighted the challenge of training the workforce (factor addressed in theoretical proposition 4d) to execute a system based on automation. The challenges of electrical energy quality and Internet coverage are also confirmed by the manager of "Farm Bahia": *"Despite having installed 17 center pivots, our system is all manual, because the electrical energy quality is very bad. The less the system is automated, the lower the risk that equipment can burn out. If all the center pivots are turned on at the same time, the power drops, so each center pivot has a different starting time by one minute"*.

Outputs of the execution of irrigation result in status and some performance indicators: final irrigation time, final water use, final energy use, final water use cost, and final energy use cost, based on irrigation planning and scheduling outputs. The explored adoption allows us to achieve measurable benefits, according to experts 2, 11, 14, and 25 (factors analyzed in theoretical propositions 1, 2, and 2a).

The control of irrigation receives other function outputs to monitor the irrigation system and its resources [122,125]. In the case of irrigation equipment, for example, this function controls the input application, as reported by experts 2, 15, and 16 (factors related to theoretical propositions 2e, 6, 8, and 8b). Expert 2 highlighted that: *"The great advantage of adoption is that the farmer controls the application of water, energy, fertilizers. So, when he gains precision in applications, he controls the level of energy use, which has an extremely high cost"*. Experts 9, 16, and 23 added other benefits, such as localized water application, application uniformity, equipment efficiency, and pressure control (factors analyzed in theoretical propositions 1, 2, and 2a). This control helps the farmer in relation to the risks of insufficient and excessive irrigation [118] (factors related to theoretical propositions 1 and 2a).

Through resources such as sensing technologies, the three agronomic pillars are also monitored, according to experts 2, 14, 15, 16, and 23. Expert 23 stated: *"It is important to evaluate the result of the irrigation by looking at the plant, using satellite and drone. And controlling how the plant is responding"* (factors dealt with in theoretical propositions 1, 2b, 2f, 8b, and 8c). Therefore, controlling irrigation operations through sensing technologies can affect the adoption of PA. Expert 23 emphasized the large volume of data available and accessible thanks to these technologies, whose utilization can improve the visual inspection. The adoption also helps the control of irrigation machines, according to experts 20 and 25 (factors related to theoretical propositions 1, 2, 2d, 8b, and 8c).

Irrigation control also provides information for the next harvest, as stated by expert 35: "The farmer will use the information from the current or next harvest for the next one. How did the model respond? Did the gain of efficiency meet the expectations?" (factors analyzed in theoretical propositions 2 and 6).

The results of irrigation control are the irrigation status and some other performance measures, such as production efficiency, as well as the measures related to water and energy use [126] (factors addressed in theoretical proposition 1). In the two case studies, the IMS produces performance measures by center pivots, such as production, revenue, and cost. Resource control is also emphasized by experts 2 and 4 as an advantage of adoption (factors related to theoretical propositions 1, 2, 2a, 2f and 8b). As in the industry, the farms can adopt measures to monitor and review performance [127], with a focus on OM, benefits highlighted by experts 10 and 27 (factors concerning theoretical propositions 1, 1a, and 6).

Irrigation control procedures refer to information to perform irrigation operations (factors related to theoretical proposition 6). Regarding the electrical energy use, the operating time of a center pivot must be planned and executed to avoid periods of higher costs of energy [128]. According to expert 8, in collective water use, the farmer receives a volume of water for a specific time, which must be controlled during the execution.

4. Discussion

In this section, the results and their contributions are discussed to answer the research questions and verify that the research objectives are achieved, beginning with the factors that can affect the explored adoption. Then, the irrigation OM model and the relationships between some factors identified in the case studies are discussed.

4.1. Factors That Can Affect the Explored Adoption

This work contributes to the current knowledge on PA adoption with a set of categories and factors that can affect the adoption, which complements previous studies [12–14,60] and allows us to overcome the identified gaps. The performance measures related to attitude are grouped into quantitative and qualitative benefits that can be achieved with the adoption. Quantitative performance measurements are integrated with crop water requirement, variable rate water use, production yield, profit, revenue, and production cost. The category "Access to data" is divided into access to agronomic data and access to operational data to differentiate the data sources (agronomic pillars and irrigation operations). The category "Changes" is organized into changes for the farmer and changes for the farm to highlight the challenges and barriers that the farmer and the farm must face and overcome to successfully adopt.

Empirical results add to [69] the following factors: technical training, managerial training, mental model, irrigation management, data-based management, and farm management as a business. Regarding the farmer mental model, the main barrier to the adoption is cultural, which depends mainly on farmer age and experience. According to experts 8, 9, and 20, younger farmers are more likely to be successful in adoption due to their familiarity with innovative technologies, which involve experience and knowledge of computer science and mathematics. For older farmers and those with low educational levels, adoption can represent a cultural shock, which is also linked to their conservatism. The case studies highlighted other significant beliefs, since the mental model of the two farm managers is aimed at minimizing the energy use cost due to the lack of definition of water use cost. The cost of water use is not a priority factor in irrigation OM of the two farms evaluated, even though the two farm managers are young and graduates. At the same time, the farmer mindset is involved in other factors, such the scarcity of water, the need to change the belief that irrigating is different from applying water and to move from experience-based management to data-based management, which the farmer can adopt through sensing technologies and OM models inspired by the industry. As for the changes for the farm, the empirical results emphasize that the OM has not evolved at the same speed as agronomic management in Brazil. OM is undervalued by the farmer due to little knowledge and skills in the OM subject. The previous category "Operations planning and control" is more generically named "Operations Management models inspired by the industry", thanks to the contribution of several experts. This category introduces a new factor: collaborative management models.

The identified categories and factors, shown in Tables 2 and 3, like electrical energy supply contract, and planned irrigation water requirement, can influence the perception of the farmer, affecting the intention of adoption. According to [129,130], the greater the benefits perceived from sensing technology adoption (like decrease in use of water and relative cost), the more likely their intention to adopt it. In addition, the less the farmer perceives the obstacles, such as training in irrigation OM, and the more resources he believes are available, such as training, the greater the perceived control and the more likely the intention of adopting.

The identified factors also provide progress in knowledge as they allow us to overcome the gaps in the literature related to the factors concerning irrigation OM and farmer behavior, themes little explored to explain PA adoption. These factors contribute to explain the potential of PA for irrigation operations, considering the applications available, adopting the proposed model for irrigation planning, scheduling, executing, and controlling, such as planning water use and energy use costs, measuring final irrigation time, and water use.

This work contributes to previous studies, complementing the factors specific to the studied domain, relevant for a comprehensive understanding of the adoption explored [21]. Examples of these factors are as follows: agronomic data access and operational data access (water access contract, electrical energy supply contract).

4.2. Model of Irrigation OM

As PPC is one of the core functions of OM, which aids in planning and utilizing resources [75], and as experts 1 and 6 suggested the adoption of planning and control models of operations such as PPC, these theoretical perspectives inspired the building of the model of irrigation OM, through the foundation of theoretical proposition 6.

The organization of the factors in a model contributes to representing the irrigation OM and the relationships between the factors in a structured way. Considering the gap in the literature relative to the factors concerning the OM, this model contributes to the understanding of PA adoption by theoretical propositions about the relationships between these factors to explain the adoption. These relationships are represented in the model, which is broken down into four functions in Figure 3.

This model complements the irrigation OM model shown in Figures 4 and 5 [69] with the factors presented in Sections 3.1 and 3.2. The context diagram of irrigation OM, shown in Figure 2, highlights the complementing factors. In relation to the controls, the following factors are added in this work: farm characteristics, farmer mental model, access to water (autonomous, collective), contract for access to water, and contract for energy supply. In relation to the inputs, the factor "characteristics of the irrigation system" is integrated, while regarding the mechanisms, the factors "consultant" and "Irrigation Management System" are added. As for the outputs, irrigation status, water and energy use, cost of water, and energy use complement the previous diagrams published in [69]. These factors are significant for the completeness of the proposed model and for the progress of knowledge on PA adoption.

In the model, irrigation planning differs from scheduling, based on inspiration from the planning and control models of operations (factor related to theoretical proposition 6) and from experts 1 and 6. The model contributes to the research of [28], in which irrigation is not the focus.

4.3. Results of Case Studies: Relationships between Some Factors in the Model

The case study findings complement the categories and factors in the literature that can affect the adoption of PA, allowing one to explore the relationships between factors. In this section, the relationships between the adoption and the factors "educational level", "training", "access to data" and "resources (people—consultants)" are discussed.

The experts interviewed highlight the importance and role of "training" and "consultants" factors. Farmer training has a crucial role in spreading PA technologies [61] and in the explored adoption, according to theoretical proposition 3a. The farmer can also hire consultant services as an opportunity to support his OM, according to theoretical proposition 8d.

Using the factors "educational level", "training", "access to data", and "resources (people—consultants)" in relation to the adoption (factors addressed in theoretical propositions 2, 3a, 8d, and 9), the empirical results allow us to study the relationships between these factors in the MATOPIBA pilot and "Farm Bahia".

Regarding technical training, the two managers have skills and knowledge of irrigation and PA technologies, as they graduated in Agronomic Engineering and take part in training courses. However, at "Farm Bahia", the planning of irrigation water requirement is carried out with the service of consultants, who define the crop water requirement and the planned irrigation water requirement. In the MATOPIBA pilot, the farm manager plans these two factors. Therefore, the manager of "Farm Bahia" proved less knowledgeable about agronomic data. The data collection to conduct the case study was completed thanks to the support of the consultants, while the MATOPIBA pilot manager had complete knowledge on the agronomic data.

As reported by expert 2 and observed in "Farm Bahia", the outsourcing of irrigation operations can lead to a loss of control by the farmer. Still, expert 3 highlighted that outsourcing is growing significantly in agriculture. Therefore, the farmer can lose control over some operations and the related data [131,132].

The two farm managers have low managerial skills in irrigation OM, since they did not take courses in this area at the universities where they studied. Furthermore, on the two farms, mapping of irrigation OM is not available, and irrigation planning and control models are not adopted. Access to data is hampered by the lack of integration between agronomic databases in the IMS and operational databases, in particular the cost of water use and cost of electrical energy use. Both farm managers are supported by consultants to manage the relationship with the companies that supply the two inputs. Furthermore, the two managers do not differentiate between irrigation planning and scheduling terms and operations.

Concerning the educational market for training future farmers in Brazil, it is important to note that in relation to the National Curriculum Guidelines for the undergraduate course in Agronomic Engineering or Agronomy, a discipline of agricultural OM is not included in the core of essential professional content. This core can be complemented by specific content inserted in the pedagogical project of a course [133]. As for the Agricultural Economics community, a discipline concerning the irrigation OM can be included in the curricula of undergraduate courses.

The experts interviewed also highlight that the irrigation OM is a gap in the training and knowledge of farmers in Brazil, as reported by experts 1, 6, and 11.

Therefore, the results of the case studies allow us to identify two training models: in relation to technical training, "Farm Bahia" adopts an outsourcing model through the support of consultants, while the MATOPIBA pilot manager administers all irrigation operations. As for managerial training, the two farm managers have low training in irrigation OM. These preliminary findings must be analyzed in relation to the need for changes for both the farmer and farm in order to adopt PA, with the objective of exploring the farmer empowerment model and the farm organizational model to support the adoption. However, these empirical results cannot be generalized, due to the scope and the limitations of this work. Future research is proposed to explore the farmer training model and the farm organizational model for the explored adoption.

5. Limitations and Research Agenda

Based on the results achieved and discussed, the theoretical propositions, the limitations relative to the scope of work, and the data collected in the interviews with the experts and the two farm managers, a future research agenda is formulated (shown in Table 4). For each study, the level influencing the farmer decision-making process on the adoption is indicated: micro—farm, medium—farm ecosystem, macro—watershed or irrigated perimeter.

Future research 1: The irrigation OM model shown in Figure 3 is proposed as a theoretical framework to direct future studies to investigate the relationships between the factors involved in the explored adoption. According to [28], in this model, "Irrigation planning" plays a central role in OM. "Irrigation planning" may be decomposed in lower-level child diagrams based on the IDEF0 technique [41], which can be used in combination with simulation models for the support it provides in model documentation and data collection [134]. Therefore, future research is proposed to study the relationships between the factors related to the decomposition of "Irrigation planning" and the adoption of sensing technologies, such as satellites and drones, using simulation techniques.

Table 4. Future research agenda.

Components	Factors	Decision Making Level	Future Research
Model of irrigation OM	Factors related to the decomposition of the "Irrigation planning" and the adoption of sensing technologies	Micro	Future research 1. Study of the relationships between the factors related to the decomposition of the "Irrigation planning" and the adoption of sensing technologies, using simulation techniques.
Unit of analysis	Farm size—industrial farm, educational level, training, access to data, resources (people—consultants)	Micro	Future research 2. Study of the training model and organizational model of industrial farms for the explored adoption with focus on OM.
	Farm size—family farm	Micro	Future research 3. Study of the relationship between the adoption and the factor "farm size—family farm" in the irrigation OM model.
	Factors related to other agricultural operations	Micro, medium, macro	Future research 4. Study of the explored adoption with focus on extending the irrigation OM model to other agricultural operations.
Constructs of TPB	Factors related to the Subjective Norm construct	Medium	Future research 5. Exploring the relevant factors for the explored adoption, in relation to the farm ecosystem.
Behavior	Factors related to decision making in irrigation OM	Micro, medium	Future research 6. Study of the explored adoption with focus on decision making in OM of irrigation.
	Factors related to the diffusion of PA in the context of Agriculture 4.0	Macro	Future research 7. Study of the factors affecting the diffusion of PA in the context of Agriculture 4.0 with focus on farmer behavior and irrigation OM.
Determining factors of adoption	Factors related to collaborative management models inspired by the industry	Macro	Future research 8. Study of the application of the irrigation OM model to the management of watersheds and irrigated perimeters, using collaborative management models.

Future research 2: Regarding the unit of analysis and the factor "farm size—industrial farm" (addressed in theoretical proposition 10), this work involved the MATOPIBA pilot and "Farm Bahia". According to [42], two in-depth case studies were conducted. This choice is a limitation of this research in terms of the replication of the results [42,101], especially concerning the relationships between the adoption and the factors: educational level, training, access to data, and resources (people—consultants). Considering the factor "farm size" of the farms involved in the cases, future research is suggested to study whether the results can be replicated and generalized and to investigate the farmer training model and the organizational model of industrial farms for adoption in the state of Bahia and in other Brazilian states. Considering that training and educational level are key factors in empirical findings, this research analyzes how the results will be communicated to farmers and how the extent of changes in farm management will be communicated to scientists.

Future research 3: According to previous studies on PA adoption [135] and as reported by experts 1 and 8, farm characteristics, such as farm size (family farm), affect technology adoption. Future research is suggested to study the interconnection between the adoption and the size of the family farm in the model of irrigation OM.

Future research 4: The case studies and the expert interviews highlight the connections between irrigation and other agricultural operations. Some experts, like expert 15, emphasize that the adoption for irrigation should be seen as complementary and integrated with other agricultural operations.

The literature on irrigation also underlines the integration between agricultural operations. For example, center pivots provide an excellent vehicle to apply some fertilizers to meet the exact crop requirements [116]. The farmer must consider that some operations and decisions are interlinked. For example, irrigation for competing crops is an important agricultural planning issue when water is scarce [79]. Furthermore, each operation has several agronomic conditions that must be satisfied in the field before the operation can be performed efficiently and effectively. The assessment of these conditions is included in the field preparation operation [83].

Therefore, with inspiration from [27] (p. 71), this work proposes an integrated model of agricultural Operations Management, shown in the graphical abstract, that expands the field of study of irrigation, involving the factors addressed in theoretical propositions 4c and 6. This integrated management model offers a holistic view of the agricultural OM based on the three agronomic pillars of a farm production system. This model includes the corn development stages [136] and emphasizes the integration between the agronomic pillars, the irrigation operations, and the other agricultural operations, as highlighted by experts 2, 11, 12, 15, and 16. Future research is proposed to study the explored adoption with a focus on extending the model of irrigation OM to other agricultural operations.

Future research 5: By exploring the factors relevant for the adoption of PA in the context of Agriculture 4.0, the TPB construct "Subjective Norm" was considered out of scope. Future research is suggested to include the "Subjective Norm" and to investigate the relationships between the farmer and the actors operating on the farm ecosystem, including extension professionals [34], governments, and technology companies [36]. This proposal is focused on the analysis of the factors out of the scope of this research (related to theoretical propositions 3f, 9, 10, and 11) and their impacts on the adoption.

Future research 6: In the literature [28,137,138] and in the results of expert interviews and case studies, common factors can be found between adoption and decision making in irrigation OM. For example, the relationship between the planned irrigation water requirement and water use can affect decision making in irrigation management. Another example is cited by expert 5: "*The farmer can make a more accurate decision regarding irrigation time and water flow of the center pivot, because he can work with sensors that provide precision in information. He can plan the operations so that irrigation is more effective*". Since the farm managers in the two case studies do not manage the irrigation planning operations, including the planned cost of water use and the planned cost of electrical energy use, except that at farm planning level, following the explored adoption, the proposed model can be useful to support decision making. Future research is suggested to study the adoption with a focus on decision making in irrigation OM.

Future research 7: Regarding the behavior "diffusion" (i.e., the process by which an innovation is disseminated among several individuals [139]), it was considered out of scope in exploring the relevant factors for the adoption studied, such as extension professionals who can help effectively facilitate the diffusion of PA technologies and bridge the gap in PA information dissemination channels [34]. Future research is needed to investigate the factors that can affect the diffusion of PA in the context of Agriculture 4.0, with a focus on irrigation OM and farmer behavior. According to [36], considering the overemphasis placed on high-tech solutions in the literature, special attention is needed for the inclusion and exclusion effects of the new technologies. The research should consider at what scale to spread innovations to investigate whether this would create more losers than winners.

Future research 8: Regarding the opportunities for the farmer following the explored adoption related to irrigation planning and control, experts 16, 22, 25, and 27 mention the collaborative management models (factor related to theoretical proposition 7). The adoption of these models allows us to overcome some challenges inherent in irrigation OM in the case of shared management of water resources, such as ensuring the availability of water during the harvest. The suggestion of experts 1 and 6 of adopting a management model inspired by the industry (factor addressed in theoretical proposition 6) for irrigation leads to the consideration of other models, such as Sales and Operations Planning (S&OP). S&OP is the long-term planning of sales-related production levels within the framework of a manufacturing planning and control system. The main objective is to balance aggregate supply and aggregate demand, through updates of the annual business plan [140,141]. The inspiration for the S&OP model suggests the idea of exploring the possibility of overcoming the challenges related to poor and ineffective communication between the demand and supply of water in a watershed. This idea involves factors that can be useful to overcome the challenge of balancing water supply (shared water managed by the water resource management agency) and water demand (water users, including farmers).

Furthermore, to help effectively facilitate the diffusion of PA technologies and bridge the gap in ineffective communication, extension professionals can play a key role [34]. Future research is proposed to study the application of the irrigation OM model to the management of watersheds and irrigated perimeters, using collaborative management models inspired by the industry.

6. Conclusions

The research objectives in this work were formulated to answer the research questions and overcome the gaps in the literature about the factors that can affect the adoption of PA in the context of Agriculture 4.0.

The identified categories and factors in this study contribute to broadening the understanding of the determinants of PA adoption in the 4.0 era, adding OM and the farmer mental model to the literature. The findings show the importance of sociological and psychological factors in the research on PA adoption. The central role of farmer behavior and perceptions expands the sensing technology applications in the domain of PA, according to the farmer needs and his mental model, in terms of beliefs about the perceived benefits and disadvantages in irrigation OM.

The categories and factors concerning the irrigation OM also suggest the establishment of operational solutions, thanks to the proposal of adopting the irrigation OM model. The irrigation OM model can be applied as a theoretical framework to direct future research to investigate the relationships between the factors identified in the explored adoption. This model provides applications, enabling one to manage decision making in irrigation OM.

The empirical results emphasize the role of farmer and workforce training, and the related gaps reported by the experts interviewed and the case studies, to success in the adoption. Therefore, this work suggests the inclusion of a discipline related to irrigation OM in courses of Agronomic Engineering and Agricultural Economics for training future farmers for the revolution in progress.

The results of the case studies allow one to study the relationship between the adoption and the factors "educational level", "training", "access to data", and "resources (people—consultants)". These findings highlight the adoption of an outsourcing model related to technical training in "Farm Bahia", with the support of consultants.

In addition to the theoretical contributions, this work provides some practical contributions. The model of irrigation OM can be useful to support the farmer decision making in managing operations and for the development of an IMS integrating agronomic crop management and irrigation OM.

Considering the exploratory nature of this work, a future research agenda is formulated, addressing the factors and components considered out of scope and including an integrated model of agricultural OM based on the three agronomic pillars.

Supplementary Materials: The following supporting information can be downloaded at https://www.mdpi.com/article/10.3390/agriculture14010134/s1. Supporting information can be found in Supplementary Material A, Supplementary Material B, Supplementary Material C, and Supplementary Material D.

Author Contributions: Conceptualization: S.M. and E.A.d.M.; methodology: S.M.; validation: S.M. and E.A.d.M.; formal analysis: S.M.; investigation: S.M. and B.T.d.F.; resources: S.M. and E.A.d.M.; data curation: S.M., E.A.d.M. and B.T.d.F.; writing—original draft preparation: S.M.; writing—review and editing: all authors; supervision: E.A.d.M. and R.M.P.; project administration: E.A.d.M. and R.F.M.; funding acquisition: R.F.M. All authors have read and agreed to the published version of the manuscript.

Funding: This research was jointly funded by the European Commission in Europe and by MCTIC/RNP in Brazil, under the call EUB-02-2017 IoT Pilots.

Institutional Review Board Statement: Ethical approval was not required for this study, which does not involve humans or animals.

Data Availability Statement: The data generated or analyzed during this study are included in this article and in Supplementary Materials A, B, and C. Supplementary Material D presents a selection of quotes from expert interviews.

Acknowledgments: Authors would like to thank the experts interviewed, the farm managers involved in the case studies, the Everardo Chartuni Mantovani for the contact with "Farm Bahia", and its owner, André Torre Neto, for his support with the MATOPIBA pilot.

Conflicts of Interest: The authors declare no conflict of interest.

References

1. Andujar, D. Back to the Future: What is Trending on Precision Agriculture? *Agronomy* **2023**, *13*, 2069. [CrossRef]
2. Rose, D.C.; Chilvers, J. Agriculture 4.0: Broadening Responsible Innovation in an Era of Smart Farming. *Front. Sustain. Food Syst.* **2018**, *2*, 87. [CrossRef]
3. Bernhardt, H.; Bozkurt, M.; Brunsch, R.; Colangelo, E.; Herrmann, A.; Horstmann, J.; Kraft, M.; Marquering, J.; Steckel, T.; Tapken, H.; et al. Challenges for Agriculture through Industry 4.0. *Agronomy* **2021**, *11*, 1935. [CrossRef]
4. Scuderi, A.; La Via, G.; Timpanaro, G.; Sturiale, L. The Digital Applications of "Agriculture 4.0": Strategic Opportunity for the Development of the Italian Citrus Chain. *Agriculture* **2022**, *12*, 400. [CrossRef]
5. Bolfe, É.L.; de Jorge, L.A.C.; Sanches, I.D.; Júnior, A.L.; da Costa, C.C.; de Victoria, D.C.; Inamasu, R.Y.; Grego, C.R.; Ferreira, V.R.; Ramirez, A.R. Precision and Digital Agriculture: Adoption of Technologies and Perception of Brazilian Farmers. *Agriculture* **2020**, *10*, 653. [CrossRef]
6. Weltzien, C. Digital Agriculture-or Why Agriculture 4.0 Still Offers Only Modest Returns. *Landtechnik* **2016**, *71*, 66–68. [CrossRef]
7. Barnes, A.; Soto, I.; Eory, V.; Beck, B.; Balafoutis, A.; Sánchez, B.; Vangeyte, J.; Fountas, S.; van der Wal, T.; Gómez-Barbero, M. Exploring the Adoption of Precision Agricultural Technologies: A Cross Regional Study of EU Farmers. *Land Use Policy* **2019**, *80*, 163–174. [CrossRef]
8. Rodrigues, G.C. Precision Agriculture: Strategies and Technology Adoption. *Agriculture* **2022**, *12*, 1474. [CrossRef]
9. Lowenberg-Deboer, J.; Erickson, B. Setting the Record Straight on Precision Agriculture Adoption. *Agron. J.* **2019**, *111*, 1552–1569. [CrossRef]
10. Cisternas, I.; Velásquez, I.; Caro, A.; Rodríguez, A. Systematic Literature Review of Implementations of Precision Agriculture. *Comput. Electron. Agric.* **2020**, *176*, 105626. [CrossRef]
11. Wang, T.; Jin, H.; Sieverding, H.; Kumar, S.; Miap, Y.; Rao, X.; Obembe, O.; Nafchi, A.; Redfearn, D.; Cheye, S. Understanding Farmer Perceptions of Precision Agriculture Profitability in the U.S. Midwest. In Proceedings of the Agricultural & Applied Economics Association Annual Meeting, Anaheim, CA, USA, 31 July 2022; pp. 2–37.
12. Pathak, H.S.; Brown, P.; Best, T. A Systematic Literature Review of the Factors Affecting the Precision Agriculture Adoption Process. *Precis. Agric.* **2019**, *20*, 1292–1316. [CrossRef]
13. Karunathilake, E.M.B.M.; Le, A.T.; Heo, S.; Chung, Y.S.; Mansoor, S. The Path to Smart Farming: Innovations and Opportunities in Precision Agriculture. *Agriculture* **2023**, *13*, 1593. [CrossRef]
14. Vrchota, J.; Pech, M.; Švepešová, I. Precision Agriculture Technologies for Crop and Livestock Production in the Czech Republic. *Agriculture* **2022**, *12*, 1080. [CrossRef]
15. Robertson, M.J.; Llewellyn, R.S.; Mandel, R.; Lawes, R.; Bramley, R.G.V.; Swift, L.; Metz, N.; O'Callaghan, C. Adoption of Variable Rate Fertiliser Application in the Australian Grains Industry: Status, Issues and Prospects. *Precis. Agric.* **2012**, *13*, 181–199. [CrossRef]
16. Galioto, F.; Raggi, M.; Viaggi, D. Assessing the Potential Economic Viability of Precision Irrigation: A Theoretical Analysis and Pilot Empirical Evaluation. *Water* **2017**, *9*, 990. [CrossRef]
17. Pierpaoli, E.; Carli, G.; Pignatti, E.; Canavari, M. Drivers of Precision Agriculture Technologies Adoption: A Literature Review. *Procedia Technol.* **2013**, *8*, 61–69. [CrossRef]
18. Ayerdi Gotor, A.; Marraccini, E.; Leclercq, C.; Scheurer, O. Precision Farming Uses Typology in Arable Crop-Oriented Farms in Northern France. *Precis. Agric.* **2020**, *21*, 131–146. [CrossRef]
19. Barnes, A.; De Soto, I.; Eory, V.; Beck, B.; Balafoutis, A.; Sánchez, B.; Vangeyte, J.; Fountas, S.; van der Wal, T.; Gómez-Barbero, M. Influencing Factors and Incentives on the Intention to Adopt Precision Agricultural Technologies within Arable Farming Systems. *Environ. Sci. Policy* **2018**, *93*, 66–74. [CrossRef]
20. D'Antoni, J.M.; Mishra, A.K.; Joo, H. Farmers' Perception of Precision Technology: The Case of Autosteer Adoption by Cotton Farmers. *Comput. Electron. Agric.* **2012**, *87*, 121–128. [CrossRef]
21. Ajzen, I. The Theory of Planned Behavior. *Organ. Behav. Hum. Decis. Process.* **1991**, *50*, 179–211. [CrossRef]
22. Reichardt, M.; Jurgens, C.; Kloble, U.; Huter, J.; Moser, K. Dissemination of Precision Farming in Germany: And Training Activities. *Precis. Agric.* **2009**, *10*, 525–545. [CrossRef]
23. Sutherland, L. "Effectively Organic": Environmental Gains on Conventional Farms through the Market? *Land Use Policy* **2011**, *28*, 815–824. [CrossRef]
24. Xu, Y.; Lyu, J.; Xue, Y.; Liu, H. Intentions of Farmers to Renew Productive Agricultural Service Contracts Using the Theory of Planned Behavior: An Empirical Study in Northeastern China. *Agriculture* **2022**, *12*, 1471. [CrossRef]

25. Lee, C.S.; Chen, Y.C.; Tsui, P.L.; Chiang, M.C. Using the Theory of Planned Behavior to Examine the Sustainable Extension of Rural Food Preparation Techniques. *Agriculture* **2023**, *13*, 1012. [CrossRef]
26. Chopra, S.; Lovejoy, W.; Yano, C. Five Decades of Operations Management and the Prospects Ahead. *Manag. Sci.* **2004**, *50*, 8–14. [CrossRef]
27. Bochtis, D.; Sørensen, C.G.C.; Busato, P. Advances in Agricultural Machinery Management: A Review. *Biosyst. Eng.* **2014**, *126*, 69–81. [CrossRef]
28. Bochtis, D.; Sorensen, C.A.G.; Kateris, D. *Operations Management in Agriculture*, 1st ed.; Academic Press: Cambridge, MA, USA; Elsevier: Amsterdam, The Netherlands, 2018.
29. Prager, K.; Curfs, M. Using Mental Models to Understand Soil Management. *Soil Use Manag.* **2016**, *32*, 36–44. [CrossRef]
30. Gasson, R. Goals and Values of Farmers. *J. Agric. Econ.* **1973**, *24*, 521–542. [CrossRef]
31. Zambon, I.; Cecchini, M.; Egidi, G.; Saporito, M.G.; Colantoni, A. Revolution 4.0: Industry vs. Agriculture in a Future Development for SMEs. *Processes* **2019**, *7*, 36. [CrossRef]
32. Braun, A.T.; Colangelo, E.; Steckel, T. Farming in the Era of Industrie 4.0. *Procedia CIRP* **2018**, *72*, 979–984. [CrossRef]
33. Mukherjee, S.; Baral, M.M.; Chittipaka, V.; Srivastava, S.C.; Pal, S.K. Discussing the Impact of Industry 4.0 in Agriculture Supply Chain. In *Lecture Notes in Mechanical Engineering*; Springer: Singapore, 2021; pp. 301–307.
34. Lee, C.L.; Strong, R.; Briers, G.; Murphrey, T.; Rajan, N.; Rampold, S. A Correlational Study of Two U.S. State Extension Professionals' Behavioral Intentions to Improve Sustainable Food Chains through Precision Farming Practices. *Foods* **2023**, *12*, 2208. [CrossRef] [PubMed]
35. Bertoglio, R.; Corbo, C.; Renga, F.M.; Matteucci, M. The Digital Agricultural Revolution: A Bibliometric Analysis Literature Review. *IEEE Access* **2021**, *9*, 134762–134782. [CrossRef]
36. Klerkx, L.; Rose, D. Dealing with the Game-Changing Technologies of Agriculture 4.0: How Do We Manage Diversity and Responsibility in Food System Transition Pathways? *Glob. Food Sec.* **2020**, *24*, 100347. [CrossRef]
37. Barrett, H.; Rose, D.C. Perceptions of the Fourth Agricultural Revolution: What's In, What's Out, and What Consequences are Anticipated? *Sociol. Rural.* **2020**, *62*, 162–189. [CrossRef]
38. Rose, D.C.; Wheeler, R.; Winter, M.; Lobley, M.; Chivers, C.A. Agriculture 4.0: Making It Work for People, Production, and the Planet. *Land Use Policy* **2021**, *100*, 104933. [CrossRef]
39. Creswell, J.W.; Creswell, J.D. *Research Design: Qualitative, Quantitative, and Mixed Methods Approaches*, 4th ed.; Sage Publications: Thousand Oaks, CA, USA, 2014; ISBN 9781452226095.
40. Yin, R.K. *Case Study Research: Design and Methods*, 4th ed.; SAGE Publications: Thousand Oaks, CA, USA, 2009.
41. Kusiak, A.; Nick Larson, T.; Wang, J. (Ray) Reengineering of Design and Manufacturing Processes. *Comput. Ind. Eng.* **1994**, *26*, 521–536. [CrossRef]
42. Voss, C.; Tsikriktsis, N.; Frohlich, M. Case Research in Operations Management. *Int. J. Oper. Prod. Manag.* **2002**, *22*, 195–219. [CrossRef]
43. Flick, U. *An Introduction to Qualitative Research*, 4th ed.; SAGE Publications Limited: London, UK, 2009.
44. Kim, C.-H.; Weston, R.H.; Hodgson, A.; Lee, K.H. The Complementary Use of IDEF and UML Modelling Approaches. *Comput. Ind.* **2003**, *50*, 35–56. [CrossRef]
45. Abt, V.; Vigier, F.; Schneider, M. Enterprise Business Modelling Languages Applied to Farm Enterprise: A Case Study for IDEF0, GRAI Grid, and AMS Languages. In *Advances in Modeling Agricultural Systems*; Papajorgji, P.J., Ardalos, P.M., Eds.; Springer: Boston, MA, USA; New York, NY, USA, 2009; Volume 25, pp. 167–191.
46. Venkateswaran, J.; Son, Y.J. Hybrid System Dynamic-Discrete Event Simulation-Based Architecture for Hierarchical Production Planning. *Int. J. Prod. Res.* **2005**, *43*, 4397–4429. [CrossRef]
47. Kamienski, C.; Soininen, J.P.; Taumberger, M.; Dantas, R.; Toscano, A.; Cinotti, T.S.; Maia, R.F.; Neto, A.T. Smart Water Management Platform: IoT-Based Precision Irrigation for Agriculture. *Sensors* **2019**, *19*, 276. [CrossRef]
48. Silva, C.F.; de Moraes, M.A.F.; Molin, P. Adoption and Use of Precision Agriculture Technologies in the Sugarcane Industry of São Paulo State, Brazil. *Precis. Agric.* **2011**, *12*, 67–81. [CrossRef]
49. International Society for Precision Agriculture-ISPA Precision Ag Definition. Available online: https://www.ispag.org/about/definition (accessed on 8 May 2020).
50. Groher, T.; Heitkämper, K.; Walter, A.; Liebisch, F.; Umstätter, C. Status Quo of Adoption of Precision Agriculture Enabling Technologies in Swiss Plant Production. *Precis. Agric.* **2020**, *21*, 1327–1350. [CrossRef]
51. Mulla, D.J. Twenty Five Years of Remote Sensing in Precision Agriculture: Key Advances and Remaining Knowledge Gaps. *Biosyst. Eng.* **2013**, *114*, 358–371. [CrossRef]
52. Evans, R.G.; LaRue, J.; Stone, K.C.; King, B.A. Adoption of Site-Specific Variable Rate Sprinkler Irrigation Systems. *Irrig. Sci.* **2013**, *31*, 871–887. [CrossRef]
53. Haghverdi, A.; Leib, B.; Washington-Allen, R.; Wright, W.C.; Ghodsi, S.; Grant, T.; Zheng, M.; Vanchiasong, P. Studying Crop Yield Response to Supplemental Irrigation and the Spatial Heterogeneity of Soil Physical Attributes in a Humid Region. *Agriculture* **2019**, *9*, 43. [CrossRef]
54. Souza, S.A.; Rodrigues, L.N. Increased Profitability and Energy Savings Potential with the Use of Precision Irrigation. *Agric. Water Manag.* **2022**, *270*, 107730. [CrossRef]

55. Koech, R.; Langat, P. Improving Irrigation Water Use Efficiency: A Review of Advances, Challenges and Opportunities in the Australian Context. *Water* **2018**, *10*, 1771. [CrossRef]
56. Allen, R.G.; Pereira, L.S.; Raes, D.; Smith, M. Crop Evapotranspiration-Guidelines for Computing Crop Water Requirements-FAO Irrigation and Drainage Paper 56. *FAO Rome* **1998**, *300*, D05109.
57. Pereira, L.S.; Allen, R.G.; Smith, M.; Raes, D. Crop Evapotranspiration Estimation with FAO56: Past and Future. *Agric. Water Manag.* **2015**, *147*, 4–20. [CrossRef]
58. Brouwer, C.; Heibloem, M. *Irrigation Water Management: Irrigation Water Needs*; Training Manuals on Irrigation; FAO: Rome, Italy, 1986.
59. Brouwer, C.; Prins, K.; Heibloem, M. *Irrigation Water Management: Irrigation Scheduling*; Training Manuals on Irrigation; FAO: Rome, Italy, 1989.
60. Ofori, E.; Griffin, T.; Yeager, E. Duration Analyses of Precision Agriculture Technology Adoption: What's Influencing Farmers' Time-to-Adoption Decisions? *Agric. Financ. Rev.* **2020**, *80*, 647–664. [CrossRef]
61. Cutini, M.; Bisaglia, C.; Brambilla, M.; Bragaglio, A.; Pallottino, F.; Assirelli, A.; Romano, E.; Montaghi, A.; Leo, E.; Pezzola, M.; et al. A Co-Simulation Virtual Reality Machinery Simulator for Advanced Precision Agriculture Applications. *Agriculture* **2023**, *13*, 1603. [CrossRef]
62. Batte, T.; Arnholt, M.W. Precision Farming Adoption and Use in Ohio: Case Studies of Six Leading-Edge Adopters. *Comput. Electron. Agric.* **2003**, *38*, 125–139. [CrossRef]
63. Jarvis, D.H.; Wachowiak, M.P.; Walters, D.F.; Kovacs, J.M. Adoption of Web-Based Spatial Tools by Agricultural Producers: Conversations with Seven Northeastern Ontario Farmers Using the Geo Visage Decision Support System. *Agriculture* **2017**, *7*, 69. [CrossRef]
64. Hundal, G.S.; Laux, C.M.; Buckmaster, D.; Sutton, M.J.; Langemeier, M. Exploring Barriers to the Adoption of Internet of Things-Based Precision Agriculture Practices. *Agriculture* **2023**, *13*, 163. [CrossRef]
65. Rupnik, R.; Vavpotič, D.; Jaklič, J.; Kuhar, A.; Plavši´c, M.P.; Žvanut, B.; Rodrigues, G.C. A Reference Standard Process Model for Agriculture to Facilitate Efficient Implementation and Adoption of Precision Agriculture. *Agriculture* **2021**, *11*, 1257. [CrossRef]
66. Moreno, G.; Sunding, D.L. Joint Estimation of Technology Adoption and Land Allocation with Implications for the Design of Conservation Policy. *Am. J. Agric. Econ.* **2005**, *87*, 1009–1019. [CrossRef]
67. da Silveira, F.; Lermen, F.H.; Amaral, F.G. An Overview of Agriculture 4.0 Development: Systematic Review of Descriptions, Technologies, Barriers, Advantages, and Disadvantages. *Comput. Electron. Agric.* **2021**, *189*, 106405. [CrossRef]
68. Symeonaki, E.; Arvanitis, K.; Piromalis, D. A Context-Aware Middleware Cloud Approach for Integrating Precision Farming Facilities into the IoT toward Agriculture 4.0. *Appl. Sci.* **2020**, *10*, 813. [CrossRef]
69. Monteleone, S.; De Moraes, E.A.; de Tondato Faria, B.; Aquino Junior, P.T.; Maia, R.F.; Neto, A.T.; Toscano, A. Exploring the Adoption of Precision Agriculture for Irrigation in the Context of Agriculture 4.0: The Key Role of Internet of Things. *Sensors* **2020**, *20*, 7091. [CrossRef]
70. Erdoğan, M. Assessing Farmers' Perception to Agriculture 4.0 Technologies: A New Interval-Valued Spherical Fuzzy Sets Based Approach. *Int. J. Intell. Syst.* **2022**, *37*, 1751–1801. [CrossRef]
71. Latino, M.E.; Menegoli, M.; Lazoi, M.; Corallo, A. Voluntary Traceability in Food Supply Chain: A Framework Leading Its Implementation in Agriculture 4.0. *Technol. Forecast. Soc. Chang.* **2022**, *178*, 121564. [CrossRef]
72. Filip, M.; Zoubek, T.; Bumbalek, R.; Cerny, P.; Batista, C.E.; Olsan, P.; Bartos, P.; Kriz, P.; Xiao, M.; Dolan, A.; et al. Advanced Computational Methods for Agriculture Machinery Movement Optimization with Applications in Sugarcane Production. *Agriculture* **2020**, *10*, 434. [CrossRef]
73. Maffezzoli, F.; Ardolino, M.; Bacchetti, A.; Perona, M.; Renga, F. Agriculture 4.0: A Systematic Literature Review on the Paradigm, Technologies and Benefits. *Futures* **2022**, *142*, 102998. [CrossRef]
74. Bayraktar, E.; Tatoglu, E.; Jothishankar, M.C.; Teresa, T. Evolution of Operations Management: Past, Present and Future. *Manag. Res. News* **2007**, *30*, 843–871. [CrossRef]
75. Gunasekaran, A.; Ngai, E.W.T. The Future of Operations Management: An Outlook and Analysis. *Int. J. Prod. Econ.* **2012**, *135*, 687–701. [CrossRef]
76. Recio, B.; Rubio, F.; Criado, J.A. A Decision Support System for Farm Planning Using AgriSupport II. *Decis. Support Syst.* **2003**, *36*, 189–203. [CrossRef]
77. Van Elderen, E. Scheduling Farm Operations. *Eur. J. Oper. Res.* **1980**, *4*, 19–23. [CrossRef]
78. Wijngaard, P.J.M. A Heuristic for Scheduling Problems, Especially for Scheduling Farm Operations. *Eur. J. Oper. Res.* **1988**, *37*, 127–135. [CrossRef]
79. Lowe, T.J.; Preckel, P.V. Decision Technologies for Agribusiness Problems: A Brief Review of Selected Literature and a Call for Research. *Manuf. Serv. Oper. Manag.* **2004**, *6*, 201–208. [CrossRef]
80. Hu, Q.; Zillig, L.M.P.; Lynne, G.D.; Tomkins, A.J.; Waltman, W.J.; Hayes, M.J.; Hubbard, K.G.; Artikov, I.; Hoffmann, S.J.; Wilhite, D.A. Understanding Farmer's Forecast Use from Their Beliefs, Values, Social Norms, and Perceived Obstacles. *J. Appl. Meteorol. Climatol.* **2006**, *45*, 1190–1201. [CrossRef]
81. Glen, J.J. Mathematical Models in Farm Planning: A Survey. *Oper. Res.* **1987**, *35*, 641–666. [CrossRef]
82. Osaki, M.; Batalha, M.O. Optimization Model of Agricultural Production System in Grain Farms under Risk, in Sorriso, Brazil. *Agric. Syst.* **2014**, *127*, 178–188. [CrossRef]

83. Edwards, G.; Sørensen, C.G.; Bochtis, D.D.; Munkholm, L.J. Optimised Schedules for Sequential Agricultural Operations Using a Tabu Search Method. *Comput. Electron. Agric.* **2015**, *117*, 102–113. [CrossRef]
84. Cid-Garcia, N.M.; Bravo-Lozano, A.G.; Rios-Solis, Y.A. A Crop Planning and Real-Time Irrigation Method Based on Site-Specific Management Zones and Linear Programming. *Comput. Electron. Agric.* **2014**, *107*, 20–28. [CrossRef]
85. Martin, D.; Kranz, W.; Smith, T.; Irmak, S.; Burr, C.; Yoder, R. *Center Pivot Irrigation Handbook*; University of Nebraska-Lincoln: Lincoln, NE, USA, 2017.
86. Edwards-Jones, G. Modelling Farmer Decision-Making: Concepts, Progress and Challenges. *Anim. Sci.* **2006**, *82*, 783–790. [CrossRef]
87. Austin, E.J.; Willock, J.; Deary, I.J.; Gibson, G.J.; Dent, J.B.; Edwards-Jones, G.; Morgan, O.; Grieve, R.; Sutherland, A. Empirical Models of Farmer Behaviour Using Psychological, Social and Economic Variables. Part II: Nonlinear and Expert Modelling. *Agric. Syst.* **1998**, *58*, 225–241. [CrossRef]
88. Weersink, A.; Fulton, M. Limits to Profit Maximization as a Guide to Behavior Change. *Appl. Econ. Perspect. Policy* **2020**, *42*, 67–79. [CrossRef]
89. Pannell, D.J.A.; Marshall, G.R.B.; Barr, N.C.; Curtis, A.D.; Vanclay, F.E.; Wilkinson, R.C. Understanding and Promoting Adoption of Conservation Practices by Rural Landholders. *Aust. J. Exp. Agric.* **2006**, *46*, 1407–1424. [CrossRef]
90. de Oca Munguia, O.M.; Llewellyn, R. The Adopters versus the Technology: Which Matters More When Predicting or Explaining Adoption? *Appl. Econ. Perspect. Policy* **2020**, *42*, 80–91. [CrossRef]
91. Davis, F. Perceived Usefulness, Perceived Ease of Use, and User Acceptance of Information Technology. *MIS Q.* **1989**, *13*, 319–340. [CrossRef]
92. Burton, R.J.F. Reconceptualising the "behavioural Approach" in Agricultural Studies: A Socio-Psychological Perspective. *J. Rural. Stud.* **2004**, *20*, 359–371. [CrossRef]
93. Bianchi, C.G.; de Figueiredo, J.C.B. Characteristics of Brazilian Scientific Research on Diffusion of Innovations in Business Administration. *J. Adm. Innov. Rev. Adm. E Inovação-RAI* **2017**, *14*, 311–320. [CrossRef]
94. Reichardt, M.; Jurgens, C. Adoption and Future Perspective of Precision Farming in Germany: Results of Several Surveys among Different Agricultural Target Groups. *Precis. Agric.* **2009**, *10*, 73–94. [CrossRef]
95. Lamb, D.W.; Frazier, P.; Adams, P. Improving Pathways to Adoption: Putting the Right P' s in Precision Agriculture. *Comput. Electron. Agric.* **2008**, *61*, 4–9. [CrossRef]
96. Armitage, C.J.; Conner, M. Efficacy of the Theory of Planned Behaviour: A Meta-Analytic Review. *Br. J. Soc. Psychol.* **2001**, *40*, 471–499. [CrossRef]
97. Sok, J.; Borges, J.R.; Schmidt, P.; Ajzen, I. Farmer Behaviour as Reasoned Action: A Critical Review of Research with the Theory of Planned Behaviour. *J. Agric. Econ.* **2020**, *72*, 388–412. [CrossRef]
98. Lanza Castillo, G.M.; Engler, A.; Wollni, M. Planned Behavior and Social Capital: Understanding Farmers' Behavior toward Pressurized Irrigation Technologies. *Agric. Water Manag.* **2021**, *243*, 106524. [CrossRef]
99. Ajzen, I. Frequently Asked Questions. Available online: https://people.umass.edu/aizen/faq.html (accessed on 6 January 2021).
100. Johnson, R.B.; Onwuegbuzie, A.J.; Johnson, R.B.; Onwuegbuzie, A.J. Toward a Definition of Mixed Methods Research. *Educ. Res.* **2007**, *33*, 14–26. [CrossRef]
101. Eisenhardt, K.M. Building Theories from Case Study Research. *Acad. Manag. Rev.* **1989**, *14*, 532–550. [CrossRef]
102. Flynn, B.B.; Sakakibara, S.; Schroeder, R.G.; Bates, K.A.; Flynn, E.J. Empirical Research Methods in Operations Management. *J. Oper. Manag.* **1990**, *9*, 250–284. [CrossRef]
103. Bogner, A.; Littig, B.; Menz, W. *Interviewing Experts*, 1st ed.; Palgrave MacMillan: London, UK, 2009; ISBN 9788578110796.
104. Meuser, M.; Nagel, U. The Expert Interview and Changes in Knowledge Production. In *Interviewing Experts*; Palgrave Macmillan: London, UK, 2009; pp. 17–42.
105. Jumman, A. Using System Dynamics to Explore the Poor Uptake of Irrigation Scheduling Technologies in a Commercial Sugarcane Community in South Africa. Ph.D. Thesis, University of KwaZulu-Natal, Pietermaritzburg, South Africa, 2016.
106. Guest, G.; Bunce, A.; Johnson, L. How Many Interviews Are Enough?: An Experiment with Data Saturation and Variability. *Field Methods* **2006**, *18*, 59–82. [CrossRef]
107. Paulhus, D.L. Measurement and Control of Response Bias. In *Measures of Personality and Social Psychological Attitudes*; Robinson, J.P., Shaver, P.R., Wrightsman, L.S., Eds.; Academic Press, Inc.: New York, NY, USA, 1991; Volume 1, pp. 17–59.
108. Mason, M. Sample Size and Saturation in PhD Studies Using Qualitative Interviews. *Forum Qual. Sozialforschung* **2010**, *11*. [CrossRef]
109. Miles, M.B.; Huberman, A.M. *Qualitative Data Analysis: An Expanded Sourcebook*, 2nd ed.; SAGE Publications: Thousand Oaks, CA, USA, 1994.
110. Mackulak, G.T. High Level Planning and Control: An IDEF0 Analysis for Airframe Manufacture. *J. Manuf. Syst.* **1984**, *3*, 121–133. [CrossRef]
111. Ciurana, J.; Garcia-Romeu, M.L.; Ferrer, I.; Casadesús, M. A Model for Integrating Process Planning and Production Planning and Control in Machining Processes. *Robot. Comput. Integr. Manuf.* **2008**, *24*, 532–544. [CrossRef]
112. Hassan Zadeh, A.; Afshari, H.; Ramazani Khorshid-Doust, R. Integration of Process Planning and Production Planning and Control in Cellular Manufacturing. *Prod. Plan. Control* **2014**, *25*, 840–857. [CrossRef]
113. Patton, M.Q. *Qualitative Research & Evaluation Methods*, 3rd ed.; Sage Publications: Thousand Oaks, CA, USA, 2002.

114. Kannan, V.R.; Tan, K.C. Just in Time, Total Quality Management, and Supply Chain Management: Understanding Their Linkages and Impact on Business Performance. *Omega* **2005**, *33*, 153–162. [CrossRef]
115. Reimer, A.P.; Thompson, A.W.; Prokopy, L.S. The Multi-Dimensional Nature of Environmental Attitudes among Farmers in Indiana: Implications for Conservation Adoption. *Agric. Hum. Values* **2012**, *29*, 29–40. [CrossRef]
116. Evans, R.G. Center Pivot Irrigation. Available online: https://www.ars.usda.gov/ (accessed on 23 December 2021).
117. García-Vila, M.; Fereres, E. Combining the Simulation Crop Model AquaCrop with an Economic Model for the Optimization of Irrigation Management at Farm Level. *Eur. J. Agron.* **2012**, *36*, 21–31. [CrossRef]
118. Jabro, J.D.; Stevens, W.B.; Iversen, W.M.; Allen, B.L.; Sainju, U.M. Irrigation Scheduling Based on Wireless Sensors Output and Soil-Water Characteristic Curve in Two Soils. *Sensors* **2020**, *20*, 1336. [CrossRef]
119. Savva, A.P.; Frenken, K. *Crop Water Requirements and Irrigation Scheduling*; Irrigation Manual; Springer: New York, NY, USA, 2002.
120. Dhungel, R.; Allen, R.G.; Trezza, R.; Robison, C.W. Comparison of Latent Heat Flux Using Aerodynamic Methods and Using the Penman–Monteith Method with Satellite-Based Surface Energy Balance. *Remote Sens.* **2014**, *6*, 8844–8877. [CrossRef]
121. Buchleiter, B.G.W.; Heermann, D.F. Using Computers to Manage Irrigation Systems. *J. Water Resour. Plan. Manag.* **1986**, *112*, 354–365. [CrossRef]
122. Keswani, B.; Mohapatra, A.G.; Keswani, P.; Khanna, A.; Gupta, D.; Rodrigues, J.J.P.C. Improving Weather Dependent Zone Specific Irrigation Control Scheme in IoT and Big Data Enabled Self Driven Precision Agriculture Mechanism. *Enterp. Inf. Syst.* **2020**, *14*, 1494–1515. [CrossRef]
123. Ojha, T.; Misra, S.; Raghuwanshi, N.S. Wireless Sensor Networks for Agriculture: The State-of-the-Art in Practice and Future Challenges. *Comput. Electron. Agric.* **2015**, *118*, 66–84. [CrossRef]
124. Pinedo, M.L. Chapter 1-Introduction. In *Scheduling: Theory, Algorithms and Systems*; Springer Nature: New York, NY, USA, 2012; pp. 1–676.
125. Liu, Z.; Xu, Q. An Automatic Irrigation Control System for Soilless Culture of Lettuce. *Water* **2018**, *10*, 1692. [CrossRef]
126. Córcoles, J.; Tarjuelo, J.; Moreno, M.; Ortega, J.; Juan, J. DE Evaluation of Irrigation Systems by Using Benchmarking Techniques. In Proceedings of the XVIIth World Congress of the International Commission of Agricultural and Biosystems Engineering (CIGR), Québec City, QC, Canada, 13 June 2010; Canadian Society for Bioengineering: Ottawa, ON, Canada, 2010.
127. Dora, M.; Lambrecht, E.; Gellynck, X.; Van Goubergen, D. Lean Manufacturing to Lean Agriculture: It's about Time. In Proceedings of the 2015 Industrial and Systems Engineering Research Conference, Nashville, TN, USA, 30 May 2015; Cetinkaya, S., Ryan, J.K., Eds.; Institute of Industrial and Systems Engineers (IISE): Peachtree Corners, GA, USA, 2015; pp. 633–642.
128. Moreno, M.; Medina, D.; Ortega, J.F.; Tarjuelo, J.M. Optimal Design of Center Pivot Systems with Water Supplied from Wells. *Agric. Water Manag.* **2012**, *107*, 112–121. [CrossRef]
129. Thompson, N.M.; Bir, C.; Widmar, D.A.; Mintert, J.R. Farmer Perceptions of Precision Agriculture Technology Benefits. *J. Agric. Appl. Econ.* **2019**, *51*, 142–163. [CrossRef]
130. Walton, J.C.; Roberts, R.K.; Lambert, D.M.; Larson, J.A.; English, B.C.; Larkin, S.L.; Martin, S.W.; Marra, M.C.; Paxton, K.W.; Reeves, J.M. Grid Soil Sampling Adoption and Abandonment in Cotton Production. *Precis. Agric.* **2010**, *11*, 135–147. [CrossRef]
131. Lankford, W.M.; Parsa, F. Outsourcing: A Primer. *Manag. Decis.* **1999**, *37*, 310–316. [CrossRef]
132. Lau, K.H.; Zhang, J. Drivers and Obstacles of Outsourcing Practices in China. *Int. J. Phys. Distrib. Logist. Manag.* **2006**, *36*, 776–792. [CrossRef]
133. Ministry of Education National Curriculum Guidelines for the Undergraduate Course in Agronomic Engineering or Agronomy, Resolution No. 1 of 2 February 2006. Available online: http://portal.mec.gov.br (accessed on 25 April 2023).
134. Perera, T.; Liyanage, K. Methodology for Rapid Identification and Collection of Input Data in the Simulation of Manufacturing Systems. *Simul. Pract. Theory* **2000**, *7*, 645–656. [CrossRef]
135. Paustian, M.; Theuvsen, L. Adoption of Precision Agriculture Technologies by German Crop Farmers. *Precis. Agric.* **2017**, *18*, 701–716. [CrossRef]
136. Alam, M.R.; Nakasathien, S.; Molla, M.S.H.; Islam, M.A.; Maniruzzaman, M.; Ali, M.A.; Sarobol, E.; Vichukit, V.; Hassan, M.M.; Dessoky, E.S.; et al. Kernel Water Relations and Kernel Filling Traits in Maize (*Zea mays* L.) Are Influenced by Water-Deficit Condition in a Tropical Environment. *Front. Plant Sci.* **2021**, *12*, 717178. [CrossRef] [PubMed]
137. Willock, J.; Deary, I.J.; Mcgregor, M.M.; Sutherland, A.; Edwards-Jones, G.; Morgan, O.; Dent, B.; Grieve, R.; Gibson, G.; Austin, E. Farmers' Attitudes, Objectives, Behaviors, and Personality Traits: The Edinburgh Study of Decision Making on Farms. *J. Vocat. Behav.* **1999**, *54*, 5–36. [CrossRef]
138. Zhai, Z.; Martínez, J.F.; Beltran, V.; Martínez, N.L. Decision Support Systems for Agriculture 4.0: Survey and Challenges. *Comput. Electron. Agric.* **2020**, *170*, 105256. [CrossRef]
139. Fisher, D.K.; Norvell, J.; Sonka, S.; Nelson, M.J. Understanding Technology Adoption through System Dynamics Modeling: Implications for Agribusiness Management. *Int. Food Agribus. Manag. Rev.* **2000**, *3*, 281–296. [CrossRef]

140. Olhager, J.; Rudberg, M.; Wikner, J. Long-Term Capacity Management: Linking the Perspectives from Manufacturing Strategy and Sales and Operations Planning. *Int. J. Prod. Econ.* **2001**, *69*, 215–225. [CrossRef]
141. Thomé, A.M.T.; Scavarda, L.F.; Fernandez, N.S.; Scavarda, A.J. Sales and Operations Planning: A Research Synthesis. *Int. J. Prod. Econ.* **2012**, *138*, 1–13. [CrossRef]

Disclaimer/Publisher's Note: The statements, opinions and data contained in all publications are solely those of the individual author(s) and contributor(s) and not of MDPI and/or the editor(s). MDPI and/or the editor(s) disclaim responsibility for any injury to people or property resulting from any ideas, methods, instructions or products referred to in the content.

AgriEngineering

Article

Post-Harvest Management of Immature (Green and Semi-Green) Soybeans: Effect of Drying and Storage Conditions (Temperature, Light, and Aeration) on Color and Oil Quality

Ibukunoluwa Ajayi-Banji [1], Ewumbua Monono [1], Jasper Teboh [2], Szilvia Yuja [2] and Kenneth Hellevang [1,*]

[1] Department of Agricultural and Biosystems Engineering, North Dakota State University, Fargo, ND 58102, USA; ibukunoluwa.ajayiban@ndsu.edu (I.A.-B.); ewumbua.monono@ndsu.edu (E.M.)
[2] Carrington Research Extension Center, North Dakota State University, Carrington, ND 58421, USA; jasper.teboh@gmail.com (J.T.); szilvia.zilahisebess@ndsu.edu (S.Y.)
* Correspondence: kenneth.hellevang@ndsu.edu

Citation: Ajayi-Banji, I.; Monono, E.; Teboh, J.; Yuja, S.; Hellevang, K. Post-Harvest Management of Immature (Green and Semi-Green) Soybeans: Effect of Drying and Storage Conditions (Temperature, Light, and Aeration) on Color and Oil Quality. *AgriEngineering* **2024**, *6*, 135–154. https://doi.org/10.3390/agriengineering6010009

Academic Editors: Yuguang Zhou, Redmond R. Shamshiri, Muhammad Sultan and Muhammad Imran

Received: 2 November 2023
Revised: 29 December 2023
Accepted: 9 January 2024
Published: 15 January 2024

Copyright: © 2024 by the authors. Licensee MDPI, Basel, Switzerland. This article is an open access article distributed under the terms and conditions of the Creative Commons Attribution (CC BY) license (https://creativecommons.org/licenses/by/4.0/).

Abstract: Soybean downgrading due to immature (green and semi-green) color at harvest, caused by frost conditions, poses a significant loss to producers and processors. After harvest, drying and storage are important for preserving the quality of the harvested produce. This study investigated the impact of drying on color change in harvested immature soybeans and the effect of the soybean moisture content, storage environment (temperature, light, and aeration), and storage period on color change and oil quality of immature soybeans. Soybeans were harvested at three different maturity stages: R6 (green) and R7 (semi-green) in pods and R8 (fully matured) in seed. The soybeans in pods were dried, shelled, and conditioned to moisture contents of 12% and 17% (wet basis) prior to storage in 12 storage chamber (box) environments. The chambers were built to have four environments of "light" and "no light" with and without aeration and were stored at temperatures of either 4 °C or 23.5 °C for 24 weeks. Samples were taken every 2 weeks for 2 months and then bimonthly in storage. Soybean color change during drying and their chlorophyll, color, peroxide value (PV), and free fatty acid (FFA) status in storage were determined. Visual observation showed that R6 (green) soybean color faded after 48 h drying, which was supported with a colorimeter reading as the "a" value increased from −8.89 to −3.83 and −8.89 to −1.71 with 37 °C and 27 °C drying temperatures, respectively. The ANOVA analysis showed that light had the greatest contribution (~81%) to the color change compared to the other three storage environment factors of temperature (~9.1%), aeration (~8%), and moisture content (~1.5%) with <10% separate effects. During storage, the R6 green and R7 semi-green soybean color continued to fade with color a-values that exceeded the initial values of the R8 matured (control) by 353% and 350%, respectively, by the end of the storage period. Low amounts of peroxide and free fatty acids (FFA) were recorded throughout the storage period. Only the FFA of 17% M.C. soybeans stored at 23.5 °C exceeded acceptable limits at the end of the storage period. Exposing immature (green and semi-green) soybeans to light resulted in the fading of the green color. Seed producers in regions prone to frost can extend harvest time by allowing immature soybeans to field-dry.

Keywords: immature soybean; green and semi-green; drying; storage conditions; light; aeration; color; chlorophyll; oil quality

1. Introduction

Revenue generation from the United States (U.S.) soybean [*Glycine* max (L.) Merr] significantly surpassed other US agricultural export commodities with a value of USD 34.3 billion dollars in the year 2022 [1]. The global demand for this oilseed is based on its versatility as a plant-based protein source for humans and livestock, vegetable oil for food, cosmetics, and the growing biodiesel market.

The U.S. soybean market has been dominant due to its grain quality standards. According to the U.S. Federal Grain Inspection Service (FGIS), market and trade assessment of soybeans is based on the test weight, heat damage, total damage, foreign material, splits, and soybeans of other colors (SBOC) [2]. SBOC include seeds with black or bicolored coats and those with green color both in the cross-section and in the outer coat, which do not meet the requirement of mature yellow soybean color [3]. Environmental conditions such as early frost, cold, and freezing temperatures during late seed maturation result in immature (green and semi-green) seeds at harvest. This happens when soybeans are at the R6 (green) and R7 (semi-green) maturity stages before full maturity is attained at R8 (fully matured). Some studies indicate minimal effects on seed quality provided the soybean plant is at the R7 maturity stage prior to the frost [4,5]. However, when the soybean plant is at the R6 maturity stage, frost has been reported to be detrimental to the seed quality in terms of color and composition; the extent of seed quality damage is highly dependent on the severity of the freeze [6]. The seed weight, moisture content, protein and oil concentration, chlorophyll content, vigor, and germination are also affected [7].

While environmental conditions play a huge role in the outcome of seed color at maturity, seed genetics also affect soybean color at maturity. Recently, a new rule was issued by the United States Department of Agriculture (USDA) officially removing SBOC as an official grain grading factor [8]. This new change was permitted due to genetic factors causing some varieties of soybeans to be characterized with some amount of green and other colors of the soybean seed coat. Despite this new rule, soybeans with green color in the seed coat and the cross-section, caused by a frost, exceeding 10% of total shipment are still undesirable at the point of sale leading to dockage. The dockage of the green soybean seeds is higher when the green discoloration is observed both on the outer seed coat and the inner cut cross-section cotyledon of the seeds [9]. In this case, the seeds are graded as "damaged kernels or immature" by the Canadian Grain Commission [10]. In a scenario where the green color is only superficial, they are labeled as "lacking natural good color" but with very little dockage [11] and it is expected that the green color will disappear in storage [12].

Some of the techniques for color measurement include colorimetrics, the Munsell color notation system, and reflectance spectrophotometry [13]. The color of food materials is denoted by using color models. Common color models used in industries are RGB, XYZ, and L a b. The L a b model separates the component of color into a three-dimensional scale [14–16]. It has become the most widely accepted and used coloring model because of the uniformity in the scale of colors. The green color of immature seeds indicates the presence of chlorophyll. The presence of chlorophyll in soybeans increases oil processing costs due to the need for refinement of the green oil. Chlorophyll in the extracted oil makes it susceptible to oxidation and rancidity, thereby reducing the shelf life [17].

Post-harvest management of soybeans is dependent upon the duration and conditions of storage [18]. The factors that affect the storability of soybeans include temperature, seed moisture content, relative humidity, and length of storage [19]. Storage of seeds at high moisture content and higher temperatures leads to faster deterioration. Free fatty acid and peroxide values are the parameters used to evaluate oil quality. Oil deterioration is caused by oxidation, hydrolysis, and other factors. Factors such as light and heat can contribute to the oxidative reaction that speeds up the oil deterioration process [20].

While the goal of storage is to preserve the quality of soybeans in storage, preliminary studies in the United States and Canada have claimed that green soybean or canola seeds will change color in storage during the first five months [9,21]. Some other studies have stated that there is little to no improvement in the seed color, though there might be a slight lightening of the seed coat after some period in storage [22–24]. The effect of some of the storage conditions like temperature, moisture content, storage media, and storage period on the color and oil quality of matured grains, such as mature soybean and pinto beans, have also been reported [25–27].

Nevertheless, there are still limited scientific data published that assesses the effect of drying and storage conditions on the color and quality changes of immature R6 and R7 maturity-stage soybeans. Since soybean is a major US crop, there is a need to investigate drying and storage conditions that will facilitate the fading of the green color in immature (green and semi-green) soybeans. Therefore, in this study, the physical appearance, primarily the (surface) color, of green and semi-green soybeans over a period of six months at different drying and storage conditions (temperature, moisture content, light, and aeration) was evaluated. Furthermore, the effect of these storage conditions on oil quality, including free fatty acid and peroxide content, was examined. Research findings will help growers improve the market value of their produce and provide processors with a better quality of grain for oil processing.

2. Materials and Methods

2.1. Chemicals and Reagents

To conduct laboratory tests on oil quality, acetic acid (Glacial) was purchased from EMD Chemicals (Gibbstown, NJ, USA), while chloroform was procured from VWR International (Radnor, PA, USA). Potassium iodide (KI) was also acquired from VWR Chemicals (Solon, OH, USA) and soluble starch was purchased from Sigma Chemical Co. (St. Louis, MO, USA). In addition, sodium thiosulphate ($Na_2S_2O_3$) 0.01 N standardized solution was procured from Alfa Aesar (Ward Mill, MA, USA) and sodium hydroxide (NaOH) 0.01 N standardized solution was acquired from Acros Organics (Fairlawn, NJ, USA). Other reagents such as phenolphthalein powder and 99% Iso-propyl alcohol (IPA) were purchased from J.T. Baker (Center Valley, PA, USA) and Fisher Scientific (Rochester, NY, USA), respectively.

2.2. Soybean Samples Collection and Preparation

The soybean seed "Proseed 30-20" variety was grown at Carrington Research Extension Center (Carrington, ND, USA). The soybeans were planted on 20 May 2018 and grew under normal weather conditions (average of 68° F) during the summer months of June, July, and August until harvest.

2.2.1. Harvesting and Pod Removal

The soybeans were harvested at three different maturity stages: R6 (green pods with full green seeds), R7 (beginning of maturity: semi-green or yellow pods), and R8 (fully matured; 95% of mature pods). The R8 was the control. Harvesting was conducted on 31 August, 5 September, and 27 September 2018 for the R6, R7, and R8 seeds, respectively.

Due to the high moisture content (greater than 45% wet basis (w.b.) of the soybean seeds) at the R6 and R7 maturity stages, the plot combine harvester was not used. A forage harvester was used to cut the soybean plants at the base of the stem. The cut plants were temporarily stored in black plastic polythene bags in the cold room at 4 °C until the pods were manually removed. The pod removal took approximately two weeks to complete. Figure 1 shows the different stages of the pod removal, drying, and cleaning process after the plants were harvested. Pod removal from the stems was performed prior to drying of the pods. The R8 were fully matured seeds and harvested with the aid of a plot combine harvester. The R8 seeds were stored in airtight plastic pails under the same temperature conditions as the dried and shelled R6 and R7 seeds.

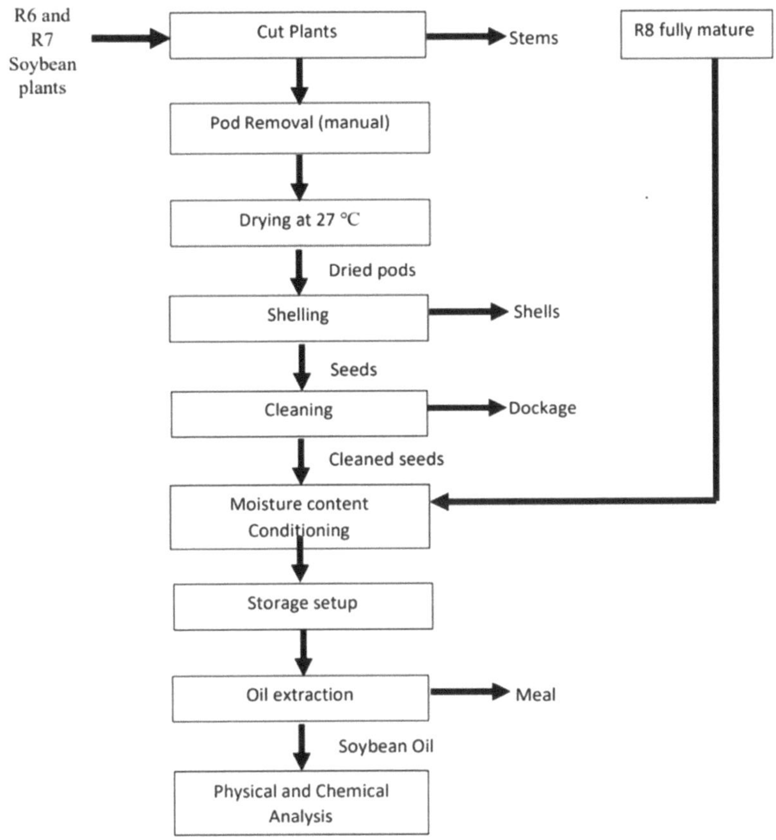

Figure 1. Preprocessing of chopped soybean (at R6 and R7 maturity stage) plants and fully matured (R8) seeds.

2.2.2. Drying, Shelling, and Cleaning

After manually removing the pods from the R6 and R7 plants, pods were dried in a drying cabinet at 27 ± 2 °C. The soybean pods were dried in batches for a period of 48–96 h to approximately 10% w.b. moisture. The pods were spread out in flat meshed trays inside the drying cabinet. The trays were 7.4 cm × 5.3 cm wide × 0.38 cm deep. A shallow layer of soybean pods was placed in each tray. After drying, pod shelling of R6 and R7 was conducted with the aid of a small bundle thresher model SBT (ALMACO, Nevada, IA, USA) at a threshing rate of approximately 55 kg h^{-1}. The threshed seeds were cleaned with the aid of an aspirator (KICE Metal Products Co. Inc., Park City, KS, USA) to remove foreign materials. The cleaned seeds were then stored in five-gallon airtight plastic pails and placed in the refrigerator prior to the seed conditioning and experimental setup. Fully mature soybeans at R8 were dry when harvested and cleaned prior to receipt, hence they were only conditioned before storage.

2.2.3. Moisture Content Determination

Due to the high initial moisture content of the soybean seeds in the R6 and R7 pods (greater than 45% w.b.), moisture content in the seeds was analyzed with a convection drying oven (Binder, model FD115-UL; Holzgerlingen, Germany) according to the AACC method No. 44-15.02 (1999) [28]. The moisture content was analyzed in triplicate. For each replicate, 15 g of whole soybean seeds was measured in triplicate in drying dishes and placed in an oven. The oven was set at 103 °C for approximately 72 h until no change in weight was recorded. Afterward, the samples were removed from the oven, cooled in the desiccator, and moisture in the seeds was calculated using Equation (1).

$$\%\ \text{Moisture content} = \frac{\text{Initial weight of sample} - \text{Final weight of sample}}{\text{Initial weight of sample (w.b.)}} \times 100 \tag{1}$$

The moisture contents of the dried soybean seeds were analyzed using the GAC 2100 capacitance-based moisture analyzer (Dickey John, Auburn, IL, USA).

2.2.4. Moisture Content Conditioning

The moisture contents of the cleaned seeds (R6 and R7) were adjusted to 12% and 17%, respectively. R8 seeds were received at a harvest moisture content of 12%; therefore, a proportion of the seeds was conditioned to 17%. Given the initial weight and moisture content of the cleaned seeds and the desired moisture of 12% and 17% for conditioning, the quantity of water added to the threshed seeds was calculated as shown in Equation (2) below.

$$\text{Mw} = \text{Mc}\left(\frac{\text{Wf} - \text{Wc}}{100\% - \text{Wf}}\right) \tag{2}$$

where Mw = mass of water added (g), Mc = initial weight of cleaned seeds (g), Wc = initial moisture content of cleaned seeds (%), and Wf = final desired moisture content (%).

Once the mass of water needed to achieve the desired moisture content (12% and 17%) was calculated from the above equation, the seeds were placed in a 27-gallon large tote on a weighing balance and the quantity of distilled water equal to the calculated Mw value was sprayed on the R6 seeds using a spraying bottle. The water was evenly distributed throughout the seeds by mixing thoroughly in the sealed container. Afterward, the tote was placed in the cold room to allow for proper equilibration of the samples for 48 h [29]. The same procedure was conducted for the conditioning of the R7 and R8 seeds. After conditioning, 300 g of seeds were weighed and placed in airtight plastic Ziploc freezer bags for non-aerated samples and mesh bags for aerated samples.

2.3. Experimental Setup

2.3.1. Storage Units Description

Eight small storage boxes of dimensions 61.0 × 58.4 × 57.1 cm and four large storage boxes of dimensions 121.9 × 61.0 × 61.0 cm were designed and made. The storage boxes were designed to have no aeration and no light (NANL), no aeration and light (NAL), aeration and no light (ANL), and aeration and light (AL). Henceforth, storage box conditions will be referred to as NANL, NAL, ANL, and AL unless otherwise stated. Figure 2; Figure 3 show a schematic diagram of the small and large boxes, respectively. The large boxes were designed to have NANL and NAL while the smaller boxes had ANL and AL. The boxes were made of plywood. Three tight mesh Rubbermaid wire shelves (121.9 × 40.6 cm) were placed in each of the boxes evenly spaced from each other. The conditioned soybean samples in meshed bags were placed on the wire shelves in the small boxes with ANL and AL conditions. The meshed bags facilitated the movement of air from the fan through the seeds during storage. Meanwhile, conditioned samples in Ziploc transparent freezer bags were stored in the large boxes under NANL and NAL.

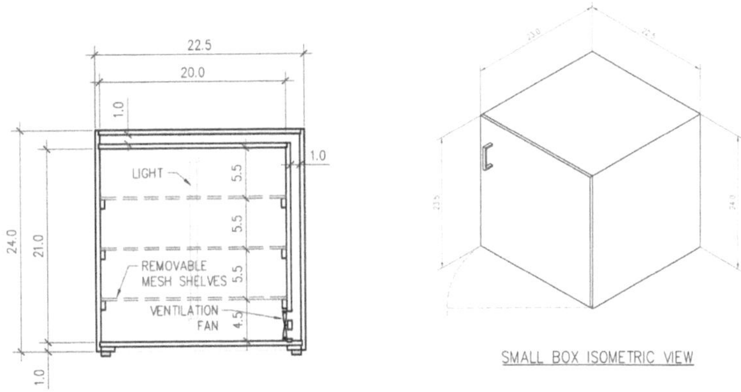

(**a**) (**b**)

Figure 2. Fabricated small storage box/chamber. (**a**) Front view of the inside of the box. (**b**) Exterior view of the assembled box with outer dimensions. All dimensions in cm.

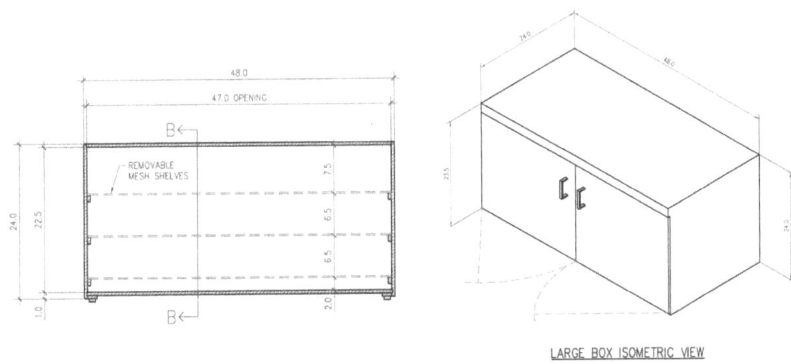

(**a**) (**b**)

Figure 3. Fabricated large storage box/chamber. (**a**) Front of the inside view of the box. (**b**) Complete view of the assembled box with outer dimensions (cm). "B" indicates the front view.

Aeration in the small boxes was achieved by a 12 V DC Dayton square 0.82 m^3/min axial fan (6 cm wide, 6 cm high) procured from Grainger. The small and large boxes with a light had one and two 45.7 cm full spectrum fluorescence bulbs (15 W), respectively. Lighted boxes were lined with commercial-grade heavy-duty aluminum foil to allow for proper reflection of light on the conditioned soybean samples. The boxes were designed to be airtight and stored at two different temperatures of 4 °C (cold room) and 23.5 °C–(room temperature).

2.3.2. Temperature Monitoring and Seed Sampling

Digi-Sense Mini Plug Type "T" thermocouples were attached to the boxes to monitor the temperature inside the boxes over the 6 month storage period. Temperature data were collected using a digi-sense temperature meter at the time of sampling. Analysis was performed every two weeks for the first two months and then monthly throughout the storage period. Likewise, moisture content monitoring and evaluation was conducted after each sampling date to ensure seed samples remained within the targeted storage moisture.

2.3.3. Seed Color Analysis

Color analysis of dried and stored seed samples was assessed with a portable tristimulus Konica Minolta Colorimeter (Minolta CR-300 Chroma meter, Ramsey, NJ, USA) that measures color readings in L, a, and b values. The instrument was calibrated with white calibration tiles before analysis. About 30 g of soybean seeds from each stored sample were placed in polystyrene sterilized Petri dishes (60 × 15 mm) purchased from VWR (Chicago, IL, USA). Samples from the cold room (4 °C) were allowed to warm up for 30 min before data collection. The filled Petri dishes were placed on the colorimeter lens and color readings of the samples were recorded in L, a, and b values. The values of lightness L (100—white, 0—black), a (−green to +red), and b (−blue to +yellow) were determined according to Hunter L a b color scale (2° standard observer, D65 Illuminant) and measurement was conducted in triplicate for each sample [30]. The percentage increase in color change was calculated using Equation (3).

$$\%\ \text{Increase} = \frac{\text{Final color ``a'' value} - \text{Initial color ``a'' value}}{\text{Initial color ``a'' value}} \times 100\% \quad (3)$$

2.3.4. Oil Extraction and Quality Analysis

After storage, an oilseed screw press (Komet, Germany) was used for expelling the oil from the soybean seed samples at a press head temperature of 62 ± 1 °C. The crude oil was then centrifuged using a Jouan CR 412 centrifuge set at 25 °C for 7 min at a speed of 4500 rpm. After centrifugation, the clear supernatant was poured into 50 mL plastic vials for oil quality analysis. The oil was analyzed for chlorophyll, peroxide, and free fatty acid values.

The number of chlorophyll pigments in the crude soybean oil was determined according to AOCS recommended practice Cc13i-96. Peroxide value analysis was conducted according to an adapted AOCS Cd 8-53 official method [31] for small sample sizes. The method involved using 0.001 N sodium thiosulfate solution standardized solution. The iodometric procedure was performed using a 0.5 g sample oil mass, with all reagents at 10% of the amounts recommended for the standard procedure. Oil samples were titrated with a 0.001 N sodium thiosulfate solution into 25-mL Erlenmeyer flasks. The titrant amount was recorded and the peroxide value (PV) was calculated using Equation (4).

$$\text{PV} = \frac{(\text{S} - \text{B}) \times \text{N} \times 1000}{\text{Sample oil mass (g)}} \quad (4)$$

where S = titration of the sample (mL), B = titration of the blank (mL), and N = normality of titrant. The free fatty acid content in the extracted oil samples was determined using the AOCS Ca 5a-40 official modified method [32].

2.4. Experimental Design and Data Analysis

A multilevel factorial experiment design was developed with six factors at various levels as shown in Table 1. Analysis of variance (ANOVA) was used to determine the effects of the maturity stage, storage conditions (moisture content, temperature, lighting, and aeration), and storage period on the seed color and they were statistically analyzed using the general linear model of Minitab 2020 (State College, PA, USA) and SAS 9.4 software. All factor types were fixed. A value of $p \leq 0.05$ was used to indicate significant differences among the main effects. Tukey's method was used to compare differences between the main effect means and interactions using the least square determination (LSD). Data presented in the graphs are in terms of mean values with their respective standard deviations.

Table 1. Factor information table.

Factor	Type	Levels	Values
Maturity Stage	Fixed	3	G, SG, Y
Moisture Content (%w.b.)	Fixed	2	12, 17
Temperature (°C)	Fixed	2	4, 25
Lighting	Fixed	2	L, NL
Aeration	Fixed	2	A, NA
Storage Period (weeks)	Fixed	5	0, 4, 8, 16, 24

3. Results and Discussion

3.1. Effect of Drying on (R6 and R7) Seed Color a-Value

Overall, the Hunter color a-value was noted to have significantly changed during the drying of the R6 and R7 maturity soybeans. Drying was important in reducing the seed moisture contents to an acceptable level before storage commenced. Interestingly, color changes in the seeds also occurred to some degree during the drying process. Since the Hunter color a-value shows the intensity of greenness of a sample, it was used primarily in measuring the color changes observed in the dried seeds. A higher negative Hunter color a-value indicates a higher intensity of the green color while a positive value illustrates the fading away of the green color. The color a-value due to the ease of measurement is referred to as a suitable means for grading purposes in soybean seed classification [13]. Figure 4 shows the color change in the R6 soybeans before and after drying.

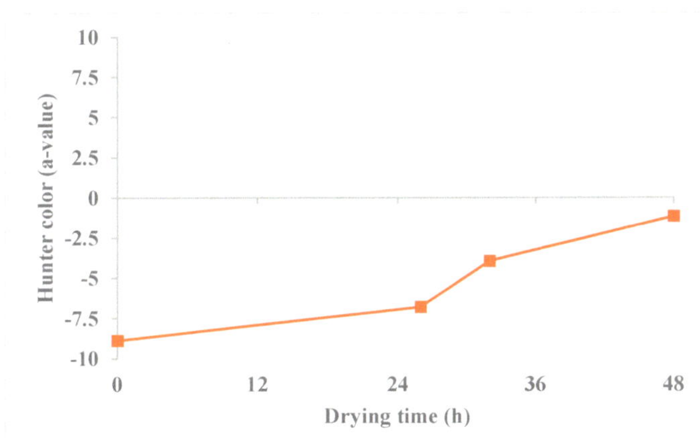

Figure 4. Color change in dried soybeans at the R6 maturity stage during drying for a period up to 48 h at drying temperature of 27 ± 5 °C.

An increase in the color a-value of more than 70% was recorded at the end of the drying period from −8.89 to −1.71 for the R6 soybeans. Consequently, the soybean seeds were characterized with a bright yellowish-green color after being subjected to the drying process, indicating the fading away of the green tone as shown in Figure 5. Ref. [13] reported similar yellowing and an increase in color a-value when Brazilian soybean seeds at R6 underwent drying at 25 °C for 10 days. The change in color can likely be attributed to the degradation in the chlorophyll content of the green seeds during the low-temperature drying process [13,17,33].

(**a**) (**b**)

Figure 5. R6 maturity green soybeans (**a**) before drying and (**b**) after 48 h drying.

3.2. Effect of Storage Conditions on the Seed Color a-Value

There was still a substantial amount of green tone to the immature soybeans after drying (Figure 5b). It was therefore important to investigate if the color will continue to change during storage and ascertain the optimal conditions that facilitate the complete degradation of chlorophyll in the beans after drying. Therefore, an ANOVA in Table 2 was developed to determine the main and interaction effects on the Hunter color a-value of the stored seeds. Interestingly, all the main factors were significant including all two-way and some three-way interactions at p-value < 0.05. In addition, the main effects (factors) contributed to more than 83% of the source of variation (SOV) observed in the color and less than 17% of the variation was due to the two-way and three-way interactions of these factors. Approximately 65% of the SOV of the color change from the main factors was attributed to differences in the maturity stage and storage time, of which the storage time alone contributed to approximately 38% of the variation in color. This clearly shows that a significant amount of color change will occur over time in storage irrespective of the other factors. The SOV of approximately 27% from the maturity stage alone can be ascribed to the inherent differences of the seeds at the three different maturity stages prior to storage. The other four main factors (moisture content, temperature, lighting, and aeration) can be termed as the storage environment and contribute to approximately 18% of the variation in color.

Table 2. Analysis of variance for Hunter color a-values showing main effects and two-way and three-way interactions for stored soybeans.

Source		DF	Adj SS		Adj MS	F-Value	p-Value
Maturity Stage		2	1784.36		892.178	7495.77	0.000
Moisture Content		1	17.30		17.298	145.33	0.000
Temperature	Storage environments	1	105.57	83 %	105.570	886.96	0.000
Lighting		1	947.05		947.055	7956.83	0.000
Aeration		1	92.06		92.063	773.49	0.000
Storage Time		4	2458.29		614.573	5163.43	0.000
Maturity Stage × Moisture Content		2	5.60		2.801	23.53	0.000
Maturity Stage × Temperature		2	8.95		4.475	37.60	0.000
Maturity Stage × Lighting		2	77.66		38.828	326.22	0.000
Maturity Stage × Aeration		2	2.03		1.017	8.54	0.000
Maturity Stage × Storage Time		8	122.81		15.351	128.97	0.000
Moisture Content × Temperature		1	0.45		0.446	3.75	0.053
Moisture Content × Lighting		1	2.28		2.278	19.14	0.000
Moisture Content × Aeration		1	18.05		18.050	151.65	0.000
Moisture Content × Storage Time		4	13.31		3.329	27.97	0.000
Temperature × Lighting		1	26.99		26.989	226.76	0.000
Temperature × Aeration		1	2.57		2.570	21.60	0.000
Temperature × Storage Time		4	43.18		10.794	90.69	0.000

Table 2. *Cont.*

Source	DF	Adj SS	Adj MS	F-Value	*p*-Value
Lighting × Aeration	1	23.83	23.835	200.25	0.000
Lighting × Storage Time	4	528.34	132.085	1109.73	0.000
Aeration × Storage Time	4	24.42	6.106	51.30	0.000
Maturity Stage × Moisture Content × Temperature	2	0.01	0.007	0.06	0.942
Maturity Stage × Moisture Content × Lighting	2	0.13	0.066	0.56	0.573
Maturity Stage × Moisture Content × Aeration	2	0.83	0.416	3.49	0.031
Maturity Stage × Moisture Content × Storage Time	8	1.59	0.199	1.67	0.102
Maturity Stage × Temperature × Lighting	2	3.89	1.945	16.34	0.000
Maturity Stage × Temperature × Aeration	2	0.08	0.041	0.35	0.708
Maturity Stage × Temperature × Storage Time	8	4.80	0.600	5.04	0.000
Maturity Stage × Lighting × Aeration	2	2.32	1.162	9.76	0.000
Maturity Stage × Lighting × Storage Time	8	48.65	6.082	51.10	0.000
Maturity Stage × Aeration × Storage Time	8	1.15	0.144	1.21	0.290
Moisture Content × Temperature × Lighting	1	1.22	1.215	10.21	0.001
Moisture Content × Temperature × Aeration	1	1.65	1.647	13.84	0.000
Moisture Content × Temperature × Storage Time	4	4.70	1.174	9.86	0.000
Moisture Content × Lighting × Aeration	1	0.18	0.181	1.52	0.218
Moisture Content × Lighting × Storage Time	4	1.54	0.386	3.24	0.012
Moisture Content × Aeration × Storage Time	4	6.18	1.545	12.98	0.000
Temperature × Lighting × Aeration	1	4.14	4.140	34.79	0.000
Temperature × Lighting × Storage Time	4	13.06	3.264	27.42	0.000
Temperature × Aeration × Storage Time	4	19.20	4.800	40.33	0.000
Lighting × Aeration × Storage Time	4	6.44	1.610	13.53	0.000
Error	599	71.30	0.119		
Lack-of-Fit	119	56.13	0.472	14.93	0.000
Pure Error	480	15.17	0.032		
Total	719	6498.18			

A closer look at these storage environment factors revealed that lighting conditions (factor) contributed the most to the SOV of the analyzed data. Of the remaining total 18% SOV of all storage environment factors, light alone contributed 14.6% (~81%) with less than 4% for the other factors (temperature contributing 1.6% and moisture content and aeration less than 1.8%). The differences in moisture content and aeration had a very small effect on the color change observed. In addition, the two-way and three-way interactions that were not significant in the ANOVA had either aeration or moisture content. Interactions that contributed the most SOV in the two-way and three-way interactions had either temperature, lighting, maturity stage, and storage time conditions present. Most of the two-way interactions that contributed to at least 5% SOV observed in color either had temperature, lighting, maturity stage, and storage time. In the three-way interactions, the only factors that contributed to this amount of SOV was the maturity stage, lighting, and storage time. Based on this finding, visualization plots were used to look in-depth at the effect of these interactions on the Hunter color a-value. A summary of the information table is provided in Table 3.

Table 3. Summary table of the analysis of variance for Hunter color a-values showing the main effects and two-way and three-way interactions for stored soybeans.

Source	DF	Adj SS	Adj MS	F-Value	*p*-Value
Maturity	2	1784.36	892.178	7495.77	0.000
Moisture Content	1	17.30	17.298	145.33	0.000
Temperature	1	105.57	105.570	886.96	0.000
Lighting	1	947.055	947.055	7956.83	0.000
Aeration	1	92.06	92.063	773.49	0.000
Storage Time	4	2458.29	614.573	5163.43	0.000
Maturity Stage × ighting	2	77.66	38.828	326.22	0.000
Maturity Stage × Storage Time	8	122.81	15.351	128.97	0.000
Temperature × Lighting	1	26.99	26.989	226.76	0.000
Temperature × Storage Time	4	43.18	10.794	90.69	0.000
Lighting × Storage Time	4	528.34	132.085	1109.73	0.000
Maturity Stage × Lighting × Storage Time	8	48.65	6.082	51.10	0.000

In Figure 6, three visualization plots were generated that show the three-way interactions comprising either temperature, light, maturity stage, and storage time significant at a *p*-value < 0.05. The visualization plots and charts average out the effect of moisture content and aeration. As illustrated earlier, moisture content and aeration had a minimal effect on the color change. Therefore, the focus was only on the effect of temperature and lighting over time across the different maturity stages on the Hunter color a-value of stored soybeans. The interaction conditions that give rise to the maximum change in a-values with reference to lighting and temperature conditions were observed.

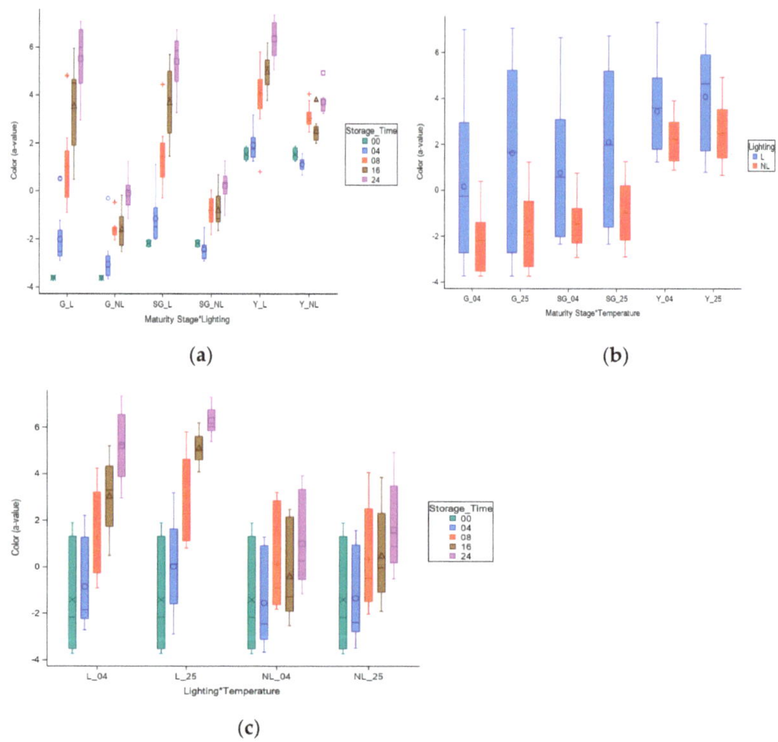

Figure 6. Three-way interaction effects plots of Hunter color a-values of each interaction. (**a**) Maturity stage, lighting, and storage time. (**b**) Maturity stage, temperature, and lighting. (**c**) Lighting, temperature, and storage time. N.B: * is used to denote interaction between factors.

In Figure 6a, the three-way interaction plot differs across the maturity stage and light combinations. The G_L (R6 Maturity Stage_Light) and SG_L (R7 Maturity Stage_Light) have very high color "a" values that increased steadily throughout the storage period when compared to the initial value of the R8 (fully matured) control. However, despite the increase in the a value for the Y_L-control (R8 Maturity Stage_Light), the profile looked different. The different maturity stages without light (NL) conditions for the same interaction plot showed a completely different pattern. The color "a" value does not increase steadily with a more compact profile for the G_NL (R6 Maturity Stage_No Light) and SG_NL (R7 Maturity Stage_No Light) throughout storage. For the Y_NL (R8 Maturity Stage_No Light) profile, this also behaved somewhat differently and these varying profiles are the outcome of this interaction. The effect of light on color change in R6 and R7 seeds is best visualized in the three-way interaction plots with a corresponding increase in the color a-value.

A similar pattern Is observed in the three-way maturity stage and temperature interaction combinations (Figure 6b) with higher bars in terms of color a-value for samples under light conditions irrespective of the storage temperature. On the contrary, the compact bars corresponded to lower color a-values for samples without light (NL) conditions, further showing the effect of minimal color change in the R6 and R7 maturity stage soybeans as previously observed in Figure 6a. In addition, samples stored at room temperature seemed to show slightly higher color change in comparison to the ones in the cold room irrespective of the maturity stage.

The last three-way interaction combination for lighting and temperature plot (Figure 6c) further emphasized the impact of light on the increase in color "a" value irrespective of the temperature condition in storage. The highest color a-value is recorded for samples under light conditions, again irrespective of the temperature of storage, with lower values observed for samples stored under no light conditions. These findings seemed to reveal the impact of light conditions on the increase in color a-value, which resulted in the fading away of the R6 green and R7 semi-green color of soybeans in storage. According to a University of Missouri study, sunlight and natural metabolic process were responsible for the decomposition of chlorophyll during soybean seed maturation [34]. Artificial lighting such as full spectrum fluorescent tubes has been shown to mimic and effectively simulate sunlight (natural light) because it characterizes the total visible spectrum [35]. A study reported the use of LED light to improve the quality of tomatoes [36]. Hence, in our study, the color change observed in the samples stored under florescence light can be attributed to similar results observed under field drying conditions because of sunlight.

Based on the outcome of the ANOVA analysis, to further appreciate the effect of light on the stored seeds, color a-value data were averaged out over moisture content, aeration, and temperature in Figure 7 to show the impact of the storage condition (specifically light) and storage period irrespective of other storage factors on the color change in the stored seeds. It was observed that generally, for all storage conditions without light or with light, an increase in the color a-value of stored soybeans was reported throughout the storage period.

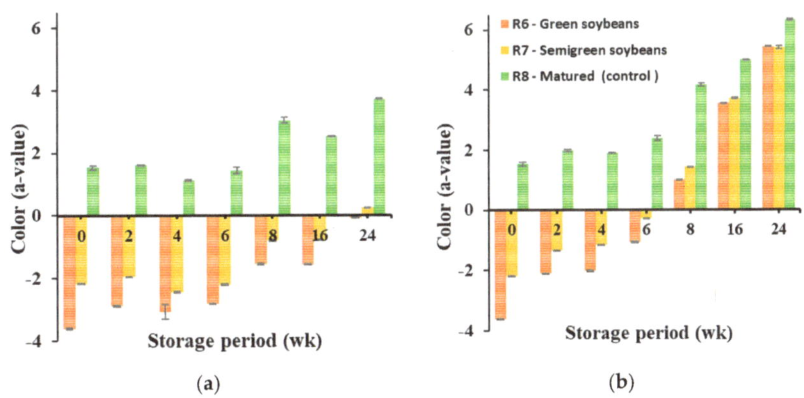

Figure 7. Color change (a-value) of soybeans under storage conditions of (**a**) without light and (**b**) with light.

For R6 (green) and R7 (semi-green) soybeans stored in boxes/chambers without light conditions (Figure 7a), only about a 63% increase in the initial color a-value of the fully matured R8 soybeans (control) was achieved at the end of the 24th week of storage. The color change was not close to that of the initial R8 (matured) control color a-value of +1.54. According to [24], a preliminary study finding showed that green soybeans showed very slight color change during six months storage in a dark cabinet. The effect of light on the development of green plants via photosynthesis has been known to be very important in their physiological maturity. Color change in terms of fading was also observed in the R8 control (fully matured) under this condition. As the length of the storage period increased, despite the absence of light, an increase in the color a-value of the R8 matured (control) recorded was about 141% from initial value of +1.54 to +3.72 by week 24 (Figure 7a).

In comparison, for soybeans stored under light conditions in Figure 7b, irrespective of the maturity stage, significant color change (an increase in the color a-value) was also marked by visible physical observation. This was evident via the fading away of the green tone color of the R6 and R7 soybeans shown in Figure 8 at the end of the 24th week of storage. R6 and R7 soybeans stored under light conditions by the 8th week of storage showed an increase from −3.61 to +1.02 and −2.18 to +1.44 in color a-values from week 0, respectively. This was about a 67% and 94% increase in color a-value with respect to the initial R8 (control) value of +1.54. By the end of the 24th week of storage, the increase in color a-values for R6 and R7 soybeans under this light condition had exceeded 353% and 350% with respect to the initial R8 (control) a-value (+1.54). At the end of the storage period, the maximum color a-values recorded in line with this percentage increase in color change are +5.44 and +5.40 for R6 green and R7 semi-green soybeans, respectively. Likewise, as observed for R8 soybeans stored without light, an increase in color a-value of R8 (control) was also logged under light conditions. In addition, the color a-value of R8 control under this condition exceeded the value of seeds without light by about 59% (a difference of +2.62) at the termination of the storage exercise.

Figure 8. Pictures of soybeans in storage for R6 (green seeds), R7 (semi-green seeds), and R8 (fully matured; control) before storage (week 0) and after twenty-four weeks of storage. (**a**) R6 soybeans at week 0, (**b**) R7 soybeans at week 0, (**c**) R8 soybeans at week 0, (**d**) R6 at week 24, (**e**) R7 at week 24, and (**f**) R8 at week 24.

Figure 8 shows the images of R6 (green seeds), R7 (semi-green seeds), and R8 (fully matured; control) maturity stage soybeans before storage (after drying-week 0) and after 24 weeks of storage. The images highlight the influence of lighting conditions (without or with light) on color change for all three maturity stages including the fully matured soybeans discussed in Figure 7.

3.3. *Chlorophyll Content*

The previous subsection clearly shows the visual color of the stored soybean changed externally, especially for samples stored under light conditions. The presence of green soybean seeds in a soybean lot that exceeds 10% are heavily discounted at the elevator. This aesthetic is caused by chlorophyll pigmentation which is also present in the oil. The green tone appears in oil extracted from the seeds, increasing its refining cost. It was then important to see if this change in the color of stored soybeans was reflected in the oil. Chlorophyll content degradation during soybean maturation is used as a predictor of color change in soybean seeds during physiological maturity prior to harvest. Figure 9 shows the chlorophyll content of the oil extracted from R6, R7, and R8 (control) soybeans after drying (prior to storage) at week 0 and after 24 weeks of storage under light conditions. Extracted oil from R6 (green) and R7 (semi-green) seeds irrespective of moisture content after drying exhibited deep green pigmentation shown in Figure 10.

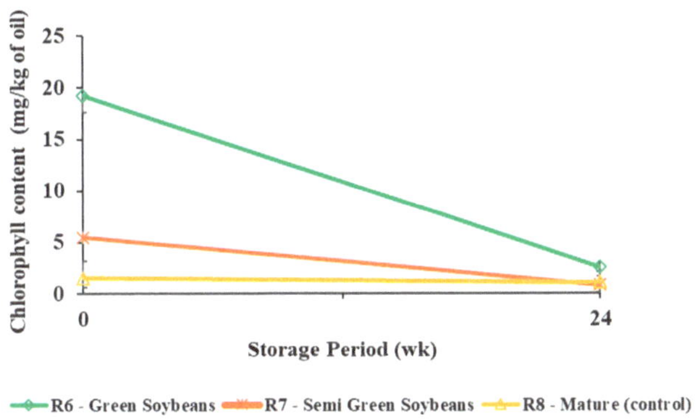

Figure 9. Chlorophyll content of R6, R7, and R8 soybean seeds after drying (before storage at week 0) and after storage under light conditions at week 24. The lines provide an estimate of degradation during storage after initial drying.

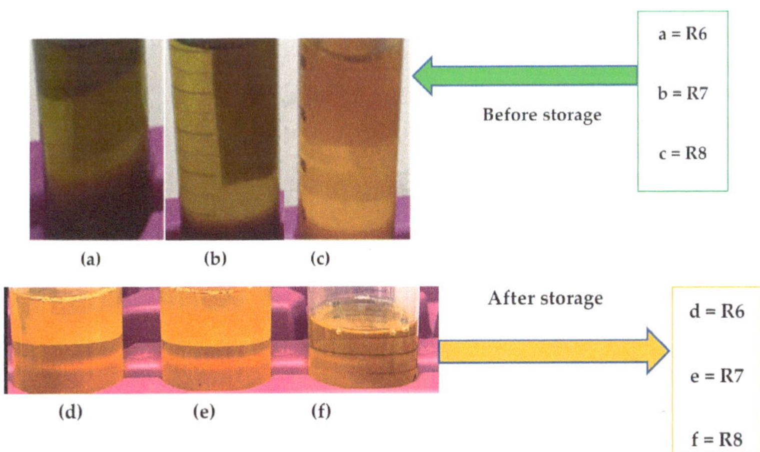

Figure 10. Color of extracted oil from R6 (green), R7 (semi-green), and R8 (matured) soybeans stored under light conditions. (**a**) R6 at week 0, (**b**) R7 at week 0, (**c**) R8 at week 0, (**d**) R6 at week 24, (**e**) R7 at week 24, and (**f**) R8 at week 24.

At the start of storage, the chlorophyll content in R6 and R7 samples were approximately 20 mg/kg and 6 mg/kg, respectively, while low levels of about 1.59 mg/kg of oil was recorded for the R8 (matured) control. These high chlorophyll contents recorded for the R6 and R7 at week 0 were characterized by a deep green tone to oil extracted from the seeds as shown in Figure 10. However, a deep yellow coloration existed in oil from R8 soybeans at week 0 due to the low amount of chlorophyll present in the fully matured seeds. At the end of the 24th week of storage, the chlorophyll content of the oil from R6 soybean stored under light conditions had dropped to 2.57 mg/kg. Likewise, the chlorophyll content of R7 seeds at the end of the 24th week of storage was similar to that of the R8 (matured control). The chlorophyll content of the R8 matured seeds remained relatively the same between weeks 0 and 24 in storage.

At the end of the storage period, oil extracted from R6 and R7, under this same condition, showed the most evident fading away of green color in stored soybeans as explained under the color change in the stored seeds section above. It is remarkably interesting to see the fading away of the green coloration in the R6 and R7 maturity soybeans for samples stored under light at the end of the experiment. These findings support the fading away of the green tone of R6 and R7 soybeans observed after 24 weeks of storage for soybeans stored under light. This also shows a relationship between the color change, chlorophyll content, and oil color of stored seeds.

3.4. Peroxide Value

The peroxide value is one of the most important indicators of primary oxidation in oils, especially during storage. Peroxidation occurs in the presence of oxygen and a favorable temperature, leading to the formation of hydroperoxides that contribute to rancidity in oils. Considering the early maturity stages at which soybeans were harvested in this study, it was important to determine the initial condition of the oil prior to storage and to monitor the changes in the oil under the different storage conditions. Oil processors refine oil prior to commercialization. Limits have been established to determine the acceptable peroxide limits in oils. According to the International Codex Alimentarius Commission [37], the maximum acceptable limit of peroxide value in virgin and cold-pressed crude oil is 15 meq of active oxygen/kg of oil. Figure 11 shows the peroxide value changes at the two storage temperatures under lighting conditions (with and without light). Initial peroxide values before storage were slightly higher in oil extracted from R6 soybeans compared to R7 and the R8 control at both moisture contents of 12% and 17%.

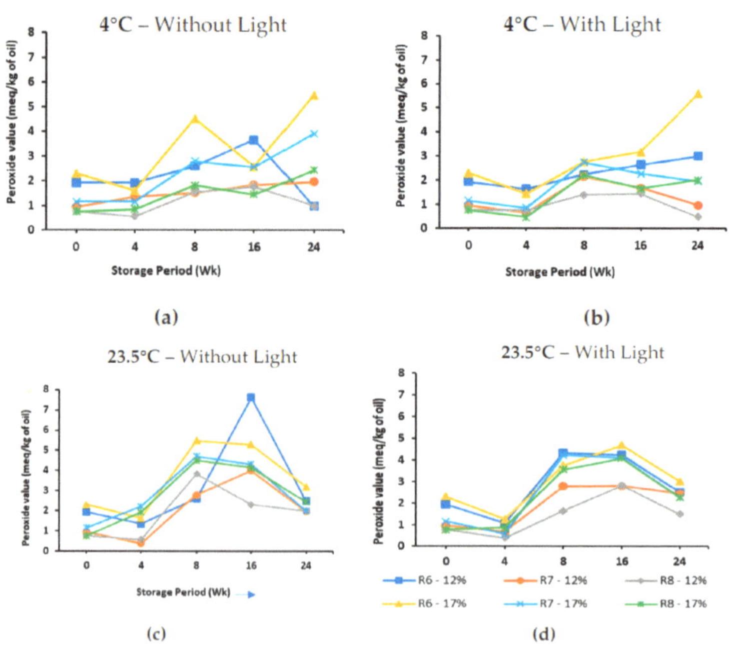

Figure 11. Peroxide value (meq/kg) for stored soybeans under lighting conditions at storage temperatures of 4 °C and 23.5 °C (**a**) without light and (**b**) with light at 4 °C (**c**) and without light and (**d**) with light at 23.5 °C.

Rapeseed oil oxidation was promoted in the presence of light, resulting in a high number of peroxides [38]. This was not the case in our findings as our soybean seeds were the primary recipient of lighting factor. The oil was extracted from the soybean seeds after each sampling date for immediate analysis. For samples stored without light and under light conditions at 4 °C, 17% moisture soybeans at R6 showed an increase in peroxide value of 5.48 meq/kg and 5.59 meq/kg of oil at the end of the 24th week of storage. A slight increase in peroxide value was also observed for R6 soybeans stored at 12% moisture content at the 16th week of storage at both temperatures. It appears that the presence of chlorophyll in R6 and R7 maturity stage soybeans served as a photosynthesizing agent that activated the increase in the peroxide value when compared to the R8 control with an initial value of 0.77 meq/kg throughout the storage period [38].

For R6 soybean samples at 17% M.C. (Figure 11a) and R6 at 12% M.C. (Figure 11c) stored under no light conditions, the moisture content, maturity stage, and storage temperature affected the peroxide value (PV). At 4 °C, between weeks 8 and 24, R6 17% M.C. seeds had higher PV compared to the R8 control. This could be due to the higher moisture content and initial values of peroxide.

At a warm temperature (23.5 °C), between weeks 4 and 16, a rise in PV was recorded for R6 and R7 soybeans including the R8 (control) irrespective of the moisture content. The initial increase could arise from the storage temperature and presence of chlorophyll compound, resulting in the formation of hydroperoxides causing lipid oxidation during storage [39]. The deterioration in oil quality was faster for R6 at both 12 and 17%, as reflected by the higher PV values at this temperature. After 16 weeks of storage, PV decreased. Since hydroperoxides are unstable, they break down into secondary oxidation products. The decrease observed after the 16th week of storage can be attributed to the rapid decomposition of the hydroperoxides to form volatile organic compounds [40].

R6 maturity stage soybeans at 12% moisture content had a maximum peroxide value of about 7.63 meq/kg of oil and 17% soybeans had a maximum of 5.49 meq/kg of oil under no light conditions. Interestingly, the peroxide value of R6 and R7 maturity soybeans stored under this same condition with light revealed lower values that could probably be attributed to the fading away of the green and semi green tone of the color of R6 and R7 soybeans. Due to the fading observed in the color of soybeans stored under light conditions, a reduction in the chlorophyll content was observed during storage till the end of the experiment as discussed in Section 3.3 above. The lower PV values recorded may be due to the lower chlorophyll content in the extracted oil at a storage temperature of 23.5 °C. In Figure 11d, the same trend of reduction in the PV was observed between week 16 and week 24 for samples stored at warm temperature (23.5 °C). The decrease showed the formation of secondary oxidation products after the 16th week of storage. In summary, the effect of the storage temperature, lighting condition, and maturity stage of soybean affected the chlorophyll content of the oils and contributed to the changes in the peroxide value. The results from this study indicated that the peroxide value of oil from stored seeds irrespective of the maturity stage and storage temperature were below the acceptable limit (<15 meq of active oxygen/kg of oil) for moderate oxidation at the end of the sixth month of storage.

3.5. Free Fatty Acid Value

Free fatty acids (FFA) are the direct product of hydrolysis due to the enzyme lipase under high temperature and high moisture from stored grains. The presence of free fatty acids in oil promotes oil oxidation through formation of hydroperoxides, hence causing off flavor and rancidity. The free fatty acid content has been used as a measure of the quality of stored grain. The US National Oilseed Processors Association (NOPA) grading and trading standards recommends a maximum acceptable level of 0.75% for good quality crude soybean oil. Figure 12 shows the changes in FFA values of soybeans stored at 4 and 23.5 °C with and without light.

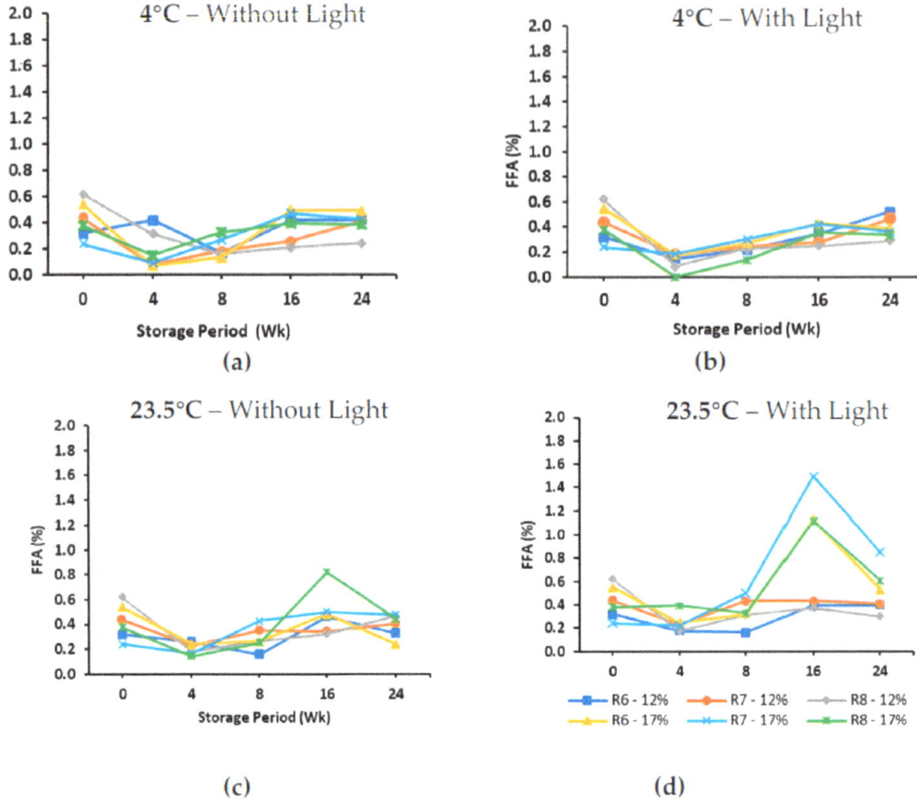

Figure 12. Percentage of free fatty acid (FFA) for stored soybeans at (**a**) 4 °C—without light, (**b**) 4 °C—with light, (**c**) 23.5 °C—without light, and (**d**) with light at 23.5 °C.

In general, very low levels of FFA were recorded for soybeans throughout the storage period at 4 °C. The values were within the acceptable limit irrespective of the maturity stage and moisture content of the soybeans. In addition, an effect of lighting conditions (with or without light) was not observed on the FFA content. Many studies have reported that FFA development is retarded at storage temperatures that are below 10 °C and with soybeans below 17% moisture content in stored grains over a six-month period [41–43]. However, there was a slight increase observed in FFA content for samples stored at a warm temperature of 23.5 °C. For samples under light conditions and at 17% moisture, it was observed that FFA content increased to 1.13%, 1.50%, and 1.12% after 16 weeks of storage for R6, R7, and R8 (control) soybeans. These values all exceeded the acceptable limit and the increase in FFA values can be likely due to the high moisture content of the seeds and temperature of storage.

Higher values of FFA content in immature soybeans at R6 and R7 maturity stages have been reported in the literature, which are due to the presence of chlorophyll. The low levels recorded in this study could be due to the degradation of chlorophyll in stored seeds especially for the samples under light conditions prior to oil extraction and oil analysis. Overall, the higher moisture content of stored seeds at warmer temperatures was more responsible for the rapid increase in FFA content. A study reported higher oil quality degradation in oil samples exposed to cold fluorescent light [44], which was not the case for our study.

4. Conclusions

The results of this study showed that drying affected the color change in the seeds before storage. Also, light and storage time affected the color of stored seeds. As a result of light conditions and 24 weeks of storage, an increase in color a-value was observed compared to the beginning of the storage period. This increase was very high and was observed after eight weeks of storage, showing the fading away of the green color in the seeds. The results also indicate that moisture content and storage temperature affected the increase in peroxide value. It might be helpful for farmers to leave their immature soybeans to field-dry. This can help improve the color of the soybeans at harvest.

Author Contributions: I.A.-B.: Methodology, Formal analysis, Resources and Material Preparation, Data Curation, Writing—Original Draft, and Writing—Review and Editing. K.H.: Conceptualization, Methodology, Funding acquisition, Supervision, Project administration, Reviewing, and Discussion. E.M.: Conceptualization, Methodology, Project administration, Supervision, and Writing—Review and Editing. J.T.: Methodology, Resources, and Writing—Review and Editing. S.Y.: Methodology and Resources. All authors have read and agreed to the published version of the manuscript.

Funding: This research was funded by orth Dakota Soybean Council, North Dakota Agricultural Experimental Station, USDA-NIFA Hatch Multi-State Project Number ND01491.

Data Availability Statement: Data used are contained within the article.

Acknowledgments: The authors would like to acknowledge the North Dakota Soybean Council for the funding support. Special appreciation goes to North Dakota State University for the facility made available to carry out the research work.

Conflicts of Interest: The authors declare no conflicts of interest.

References

1. USDA-FAS. United States Department of Agriculture—Foreign Agricultural Service 2022 Agricultural Export Yearbook. 2022. Available online: https://www.fas.usda.gov/sites/default/files/2023-05/2022-Yearbook.pdf (accessed on 6 January 2022).
2. Hammond, E.G.; Johnson, L.A.; Murphy, P.A. SOYBEAN | Grading and Marketing. In *Encyclopedia of Grain Science*; Wrigley, C., Ed.; Elsevier: Oxford, UK, 2004; pp. 155–159.
3. USDA-AMS-FGIS. United States Department of Agriculture-Agricultural Marketing Service-Federal Grain Inspection Service. 2023. Available online: https://www.ams.usda.gov/content/usda-seeks-comments-proposed-amendments-us-standards-soybeans-remove-soybeans-other-colors (accessed on 5 November 2023).
4. Bohner, H.; Tenuta, A. *What Happens to Soybeans When They Get Frosted*; Field Crop News: OMAFRA: Stratford, ON, Canada, 2013. Available online: https://fieldcropnews.com/2013/09/what-happens-to-soybeans-when-they-get-frosted/ (accessed on 15 June 2019).
5. Hurburgh, C.; Vittetoe, R.; Anderson, M. *Frost Damage to Soybeans*; IOWA State University Extension and Outreach, IOWA: Ames, IA, USA, 2019. Available online: https://crops.extension.iastate.edu/blog/charles-r-hurburgh-meaghan-anderson-rebecca-vittetoe/frost-damage-soybeans-0 (accessed on 6 September 2019).
6. Butzen, S. Managing delayed or frost-damaged soybeans. In *Soybean Reproductive Growth Stages*; Corteva: Johnston, IA, USA, 2019.
7. Judd, R.; TeKrony, D.; Egli, D.; White, G. Effect of Freezing Temperatures during Soybean Seed Maturation on Seed Quality 1. *Agron. J.* **1982**, *74*, 645–650. [CrossRef]
8. USDA-AMS. USDA Announces Amendments to the U.S. Standard Soybeans. 2023. Available online: https://www.ams.usda.gov/content/usda-announces-amendments-us-standards-soybeans (accessed on 20 October 2023).
9. Barker, B. *Soybean Green Seed and Pulse Seed Quality*; Saskatchewan Pulse Growers: Saskatoon, SK, Canada, 2017.
10. CGC. Soybeans: Grading Factors Official Grading Guide. 2021. Available online: https://www.grainscanada.gc.ca/en/grain-quality/official-grain-grading-guide/20-soybeans/grading-factors.html (accessed on 11 November 2022).
11. Dawson, A. Same Sample Different Grades for Green Seed Soybeans; 2018. Available online: https://www.manitobacooperator.ca/crops/shop-around-green-seeded-soybeans-to-get-your-best-grade/ (accessed on 4 September 2019).
12. Lange, D. *Advice on Green-Seeded Soybeans*; 2018.
13. Sinnecker, P.; Gomes, M.S.O.; Arêas, J.A.G.; Lanfer-Marquez, U.M. Relationship between color (instrumental and visual) and chlorophyll contents in soybean seeds during ripening. *J. Agric. Food Chem.* **2002**, *50*, 3961–3966. [CrossRef] [PubMed]

14. Leon, K.; Mery, D.; Pedreschi, F.; Leon, J. Color measurement in L* a* b* units from RGB digital images. *Food Res. Int.* **2006**, *39*, 1084–1091. [CrossRef]
15. Wu, D.; Sun, D.-W. Colour measurements by computer vision for food quality control—A review. *Trends Food Sci. Technol.* **2013**, *29*, 5–20. [CrossRef]
16. Francis, F.J. Color measurement and interpretation. In *Instrumental Methods for Quality Assurance in Foods*; Routledge: London, UK, 2017; pp. 189–209.
17. Cenkowski, S.; Sokhansanj, S.; Sosulski, F. The effect of drying temperature on green color and chlorophyll content of canola seed. *Can. Inst. Food Sci. Technol. J.* **1989**, *22*, 383–386. [CrossRef]
18. Alencar, E.R.d.; Faroni, L.R.D.; Peternelli, L.A.; da Silva, M.T.C.; Costa, A.R. Influence of soybean storage conditions on crude oil quality. *Rev. Bras. Eng. Agrícola Ambient.* **2010**, *14*, 303–308. [CrossRef]
19. Delouche, J.C. Physiological Changes during Storage That Affect Soybean Seed Quality; 2021. Available online: https://scholarsjunction.msstate.edu/cgi/viewcontent.cgi?article=1122&context=seedtechpapers (accessed on 5 November 2023).
20. Conceição, M.M.; Fernandes, V.J.; Araújo, A.S.; Farias, M.F.; Santos, I.M.; Souza, A.G. Thermal and oxidative degradation of castor oil biodiesel. *Energy Fuels* **2007**, *21*, 1522–1527. [CrossRef]
21. Carr, R.; Applewhite, T. Oilseed harvesting, storage and transportation. In *Proceedings of the World Conference on Oilseed Technology and Utilization*; AOCS Press: Urbana, IL, USA, 1993; pp. 118–125.
22. Anonymous. Use Caution When Drying Soybeans. In *No-Till Farmer*; 2009. Available online: https://www.no-tillfarmer.com/articles/919-use-caution-when-drying-soybeans (accessed on 16 July 2019).
23. OMAFRA. Green Soybeans—Can They Be Fed? 2021. Available online: https://omafra.gov.on.ca/english/live-stock/swine/facts/info_green_soybeans_fed_html (accessed on 10 January 2020).
24. Wilcke, B. Post-Harvest Management of Immature Corn and Soybeans. In *Proceedings of the Integrated Crop Management Conference*; Iowa State University: Ames, IA, USA, 1995.
25. Alencar, E.R.d.; Faroni, L.R.; Lacerda Filho, A.F.; Peternelli, L.A.; Costa, A.R. Quality of soy bean grains stored under different conditions. *Rev. Bras. Eng. Agrícola Ambient.* **2009**, *13*, 606–613. [CrossRef]
26. Hellevang, K. Post Harvest Management to Maintain Bean Quality. 2001. Available online: http://www.northarvestbean.org (accessed on 4 December 2019).
27. Hou, H.J.; Chang, K.C. Storage conditions affect soybean color, chemical composition and tofu qualities. *J. Food Process. Preserv.* **2004**, *28*, 473–488. [CrossRef]
28. International, A. Method 44-15.02: Moisture–air-oven methods. In *Approved Methods of Analysis*; Cereals and Grains Association: Eagan, MN, USA, 1999.
29. Deshpande, S.; Bal, S.; Ojha, T. Physical properties of soybean. *J. Agric. Eng. Res.* **1993**, *56*, 89–98. [CrossRef]
30. Kong, F.; Chang, S.; Liu, Z.; Wilson, L. Changes of soybean quality during storage as related to soymilk and tofu making. *J. Food Sci.* **2008**, *73*, S134–S144. [CrossRef] [PubMed]
31. Crowe, T.D.; White, P.J. Adaptation of the AOCS official method for measuring hydroperoxides from small-scale oil samples. *J. Am. Oil Chem. Soc.* **2001**, *78*, 1267–1269. [CrossRef]
32. Rukunudin, I.; White, P.; Bern, C.; Bailey, T. A modified method for determining free fatty acids from small soybean oil sample sizes. *J. Am. Oil Chem. Soc.* **1998**, *75*, 563–568. [CrossRef]
33. Salunkhe, D.K.; Desai, B.B. *Postharvest Biotechnology of Oilseeds*; CRC-Press: Boca Raton, FL, USA, 1986.
34. Wiebold, B. Soybean plants killed before maturity possess grain that remains green. In *Integrated Pest and Crop Management*; University of Missouri: Columbia, MO, USA, 2009.
35. Darko, E.; Heydarizadeh, P.; Schoefs, B.; Sabzalian, M.R. Photosynthesis under artificial light: The shift in primary and secondary metabolism. *Philos. Trans. R. Soc. B Biol. Sci.* **2014**, *369*, 20130243. [CrossRef] [PubMed]
36. Xu, H.-L.; Xu, Q.; Li, F.; Feng, Y.; Qin, F.; Fang, W. Applications of xerophytophysiology in plant production—LED blue light as a stimulus improved the tomato crop. *Sci. Hortic.* **2012**, *148*, 190–196. [CrossRef]
37. Alimentarius, C. Codex standard for named vegetable oils. *Codex Stan* **1999**, *210*, 1–13.
38. Li, X.; Yang, R.; Lv, C.; Chen, L.; Zhang, L.; Ding, X.; Zhang, W.; Zhang, Q.; Hu, C.; Li, P. Effect of chlorophyll on lipid oxidation of rapeseed oil. *Eur. J. Lipid Sci. Technol.* **2019**, *121*, 1800078. [CrossRef]
39. Choe, E.; Min, D.B. Mechanisms and factors for edible oil oxidation. *Compr. Rev. Food Sci. Food Saf.* **2006**, *5*, 169–186. [CrossRef]
40. Kaleem, A.; Aziz, S.; Iqtedar, M. Investigating changes and effect of peroxide values in cooking oils subject to light and heat. *FUUAST J. Biol.* **2015**, *5*, 191–196.
41. McNeal, X. *Conditioning and Storage of Soybeans*; University of Arkansas: Fayetteville, AR, USA, 1966.
42. Rani, P.; Chelladurai, V.; Jayas, D.; White, N.; Kavitha-Abirami, C. Storage studies on pinto beans under different moisture contents and temperature regimes. *J. Stored Prod. Res.* **2013**, *52*, 78–85. [CrossRef]

43. White, G.; Loewer, O.; Ross, I.; Egli, D. Storage characteristics of soybeans dried with heated air. *Trans. ASAE* **1976**, *19*, 306–310. [CrossRef]
44. Pignitter, M.; Stolze, K.; Gartner, S.; Dumhart, B.; Stoll, C.; Steiger, G.; Kraemer, K.; Somoza, V. Cold fluorescent light as major inducer of lipid oxidation in soybean oil stored at household conditions for eight weeks. *J. Agric. Food Chem.* **2014**, *62*, 2297–2305. [CrossRef] [PubMed]

Disclaimer/Publisher's Note: The statements, opinions and data contained in all publications are solely those of the individual author(s) and contributor(s) and not of MDPI and/or the editor(s). MDPI and/or the editor(s) disclaim responsibility for any injury to people or property resulting from any ideas, methods, instructions or products referred to in the content.

AgriEngineering

Article

The Influencing Factors Analysis of Aquaculture Mechanization Development in Liaoning, China

Lixingbo Yu [1,2,†], Haiheng Wang [1,2,†], Anqi Ren [1,2], Fengfan Han [1,2], Fei Jia [1,2], Haochen Hou [1,2,*] and Ying Liu [1,3]

1 Key Laboratory of Environment Controlled Aquaculture, Ministry of Education, Dalian Ocean University, Dalian 116023, China; ylixingbo@163.com (L.Y.); dream_heng@126.com (H.W.); raq1054778288@163.com (A.R.); hhanfengfan@163.com (F.H.); jiafei225@163.com (F.J.); liuyingzju@zju.edu.cn (Y.L.)
2 College of Marine Technology and Environment, Dalian Ocean University, 52 Heishijiao Street, Dalian 116023, China
3 College of Biosystems Engineering and Food Science, Zhejiang University, Hangzhou 310058, China
* Correspondence: houhaochen@dlou.edu.cn
† These authors contributed equally to this work and should be considered co-first authors.

Abstract: Promoting the mechanization of aquaculture is one of the most important supporting measures to ensure the high-quality development of the aquaculture industry in China. In order to solve the problems of predominantly manual work and to decrease the costs of aquaculture, the influencing factors of China's aquaculture mechanization were systematically analyzed. The triple bottom theory was selected, and three aspects were identified, including environmental, economic, and social aspects. Through the literature review, the Delphi method, and the analytic hierarchy process, the comprehensive evaluation indicator system, including 18 influencing factors, was proposed. Moreover, the fuzzy comprehensive evaluation method was combined with the model to solve the evaluation results. A case study in Liaoning Province was offered and, according to the analysis results, the economic aspect at the first level was the most critical factor; the financial subsidy for the purchase of aquaculture machinery, the energy consumption of the machinery and equipment, and the promotion and use of aquaculture technology were the most important factors and had the greatest impact on the development of aquaculture mechanization in China. The effective implementation paths and countermeasures were proposed, such as the promotion of mechanized equipment and the enhancement of the machinery purchase subsidies, in order to provide an important decision-making basis for the improvement of the level of aquaculture mechanization.

Keywords: aquaculture mechanization; influencing factors; low-carbon development; smart aquaculture

Citation: Yu, L.; Wang, H.; Ren, A.; Han, F.; Jia, F.; Hou, H.; Liu, Y. The Influencing Factors Analysis of Aquaculture Mechanization Development in Liaoning, China. *AgriEngineering* **2024**, *6*, 34–51. https://doi.org/10.3390/agriengineering6010003

Academic Editors: Muhammad Sultan, Yuguang Zhou, Redmond R. Shamshiri and Muhammad Imran

Received: 5 October 2023
Revised: 15 December 2023
Accepted: 27 December 2023
Published: 8 January 2024

Copyright: © 2024 by the authors. Licensee MDPI, Basel, Switzerland. This article is an open access article distributed under the terms and conditions of the Creative Commons Attribution (CC BY) license (https://creativecommons.org/licenses/by/4.0/).

1. Introduction

China is a major aquaculture country in the world, with aquaculture production accounting for two-thirds of the world's total aquaculture output. Since 2002, it has maintained a position as a major exporter of aquatic products [1]. In 2020, China's aquaculture output was about 52.24 million tons, accounting for approximately 46.4% of the world's aquaculture production; the per capita share of aquatic products was 46.39 kg [2]. Based on the 2022 China Fisheries Statistical Yearbook, the national freshwater aquaculture production in 2021 was 31.83 million tons, accounting for 59% of the total output [3]. In 2020, the production of marine aquaculture accounted for approximately 41% of the total production of aquatic products in China [4]. As one of the important supply channels, aquaculture can provide an important guarantee for China's food security [5].

Mechanization is an important component, support, and symbol of aquaculture modernization [6]. Since 1960, a variety of agricultural machinery has been extensively utilized in overseas nations. For instance, pneumatic feeders have been employed in the United

States and Denmark. Countries such as Sweden and Denmark used narrow gauge transport feeding vehicles for feeding. In Japan, water wheel aerators were used to aerate fish ponds, rail electric vehicles were used to transport fish ponds, and fish were classified and transferred through fish lifters and fish separators [7]. However, the mechanization of China's aquaculture industry was still in its early stages. With ongoing advancements in science and technology, automated and information-based equipment, such as that for water quality monitoring, feed distribution, harvesting, etc., has gained widespread recognition internationally [8]. Compared with other countries, despite the fact that China has a relatively high aquaculture production, its low production efficiency has led to a low level of aquaculture machinery and equipment [9]. The mechanization level of Chinese aquaculture started relatively late compared with other food production industries, which was a weak link in the application and development of mechanized agriculture [10]. Since 1970, fishery machinery and equipment has been used in China. In the early 1990s, the aerators and feeders were developed and popularized and were gradually applied in practical production. But it did not fully cover all the areas of aquaculture in China [11]. The major portion of freshwater aquaculture production in India was occupied by carp species such as *Labeo rohita*, *Catla catla* and *Cirrihinus mrigala*, accounting for approximately 87% of the overall freshwater yield [12]. Given their significant commercial value, it is essential to transition from small-scale farms to larger aquaculture enterprises. Therefore, it is necessary to invest in a mechanized system for the entire production process to ensure sustainable industrial development [13]. The marine fishery sector was highly mechanized; "Industrialization and mechanization were prominent features of the marine fishery department in the context of "big agriculture" [14]. Conventional aquaculture encompasses integrated agriculture–aquaculture systems (IAAS), utilizing on-farm or local agriculture by-products, manures, and vegetation; integrated peri-urban aquaculture systems (IPAS) using domestic sewage and by-products from the local agri-industry; and integrated fisheries–aquaculture systems (IFAS) with carnivorous fish fed with trash and low-value fish [15]. By the beginning of the 21st century, aerators and feeding machines had been widely applied. China's aquaculture industry had basically achieved the mechanization of its main operations. However, the degree of mechanization in the dredging, harvesting, grading, waste collection, and treatment of aquaculture was generally minimal. Some aquaculture equipment, such as the aquaculture tail water treatment equipment, fish pond dredging machines, harvesting devices, etc., was not widely applied [16]. Traditional extensive agriculture was still the main mode of production for most farmers. The mechanization and standardization levels of aquaculture were comparatively lower [17]. Therefore, promoting the transformation and upgrading the industry towards green and highly efficient production, accelerating the pace of modernization construction, and meeting the green and high-efficiency development of the aquaculture industry can be achieved by promoting the development of aquaculture mechanization in China [18]. The "Opinions on Accelerating the Development of Aquaculture Mechanization" issued by the Ministry of Agriculture and Rural Affairs in 2020 has set a clear target of reaching a mechanization level of over 50% in the aquaculture industry by 2025 [19]. However, the mechanization level of aquaculture in China was only about 32% in 2020, which was far from the goal of 50% in 2025 [20]. There were some issues in the development of aquaculture mechanization in China, such as relatively weak basic research and backward facilities and equipment. It was driven by the financial subsidies for the purchase of aquaculture machinery, the energy consumption of the machinery and equipment, and the promotion and utilization of aquaculture technology. At present, the lack of research on the implementation path and the decision-making recommendations for aquaculture mechanization was not conducive to the rapid development of aquaculture mechanization in China.

The Delphi method was a method of simulating a market that utilizes its knowledge, experience, and ability in analysis and judgment to verify problems [21]. Lucia et al. [22] employed the Delphi method to establish priorities for potential welfare issues associated with Europe's aquaculture industry, focusing specifically on Atlantic salmon and rainbow trout. The severity, duration, and prevalence of each issue were rated by experts, who also put forward interventions to address the most significant concerns. Anna et al. [23] employed the Delphi method to leverage the pan-European expertise to identify the key factors in site selection for European native oyster (*Ostrea edulis*) habitat restoration projects. This study established a strong foundation for future site selection efforts to build upon. The questionnaire survey was a widely used method in domestic and international social research. This controlled measurement was used by researchers to measure the problems being studied and to collect reliable data [24].

The analytic hierarchy process (AHP) was used to decompose a complex problem into various components and to group those factors according to the dominant relationship and thereby create a systematic hierarchical framework. This methodology was first introduced by American operations researcher Professor T. L. Saaty in the 1970s. By means of a pairwise comparison analysis, the relative importance of each factor in the hierarchy was determined; then, the overall ranking of the relative importance of the decision-making factors' relative importance was established by synthesizing the individuals' evaluations [25]. Francisco et al. [26] employed AHP technology to construct a multi-criteria decision-making problem, in which they assigned weights and rankings to factors such as mechanical properties, material quality characteristics, and manufacturing considerations. Du et al. [27] proposed a method that utilized the AHP to determine the importance of each evaluation indicator. Combined with a remanufacturing example of a heavy horizontal lathe, the proposed decision-making method was verified and analyzed. According to Du et al. [28], the weight of each indicator in the remanufacturing process of the waste planner was determined using the AHP method, which validated the feasibility and effectiveness of this approach. The combination of the AHP and fuzzy comprehensive evaluation (FCE) methods in this study resulted in more scientific and credible evaluation results. Therefore, they more objectively determine the factors and indicator systems that affect the current development status of aquaculture mechanization in China.

FCE was a scientific and reasonable evaluation method to quantify qualitative indicators; it was first put forward by Zadeh [29]. Its basic theory was to determine the value of the components of the evaluation indicator through the membership matrix and factor weight. It enabled a comprehensive assessment of factors that were challenging to analyze quantitatively and produced clear results and strong system suitability, and it could better solve various uncertainty problems [30]. Shi et al. [31] adopted membership degree conversion in the FCE method. The redundant data in the target classification indicator memberships were eliminated. The target membership degree of effective value was extracted. The new algorithm was applied to the FCE of the sustainable development of the equipment manufacturing industry. Zhao et al. [32] employed the FCE method to estimate the quality level of the dispensing process. The dispensing indicator and quality level mapping model based on a support vector machine was established. This allowed the online intelligent evaluation of drug dispensing quality. The experimental findings indicated the effectiveness of the method.

Therefore, in this paper, the influencing factors of aquaculture mechanization development were systematically analyzed based on the triple bottom theory. Through the literature review, the Delphi method and AHP were used to analyze and determine the environmental, economic, and social influencing factor indicator system, which including 18 indicators, such as financial subsidies for the purchase of aquaculture machinery, the energy consumption of machinery and equipment application, and the promotion and application of aquaculture technology. Subsequently, the FCE method was employed to develop the assessment framework, with Liaoning Province serving as a case study for the evaluation analysis. The qualitative and quantitative indicators were scientifically and rationally evaluated quantitatively to obtain various indicators, and the results were analyzed to propose effective implementation paths and countermeasures. In this way, the effective implementation path and countermeasures were proposed, and they will provide an important decision-making basis for the promotion of the high-efficiency mechanization of aquaculture in China, improving production efficiency, reducing labor intensity, reducing disaster losses, and improving the aquaculture mechanization level.

This paper's study considers the following aspects, which are also given in Figure 1.

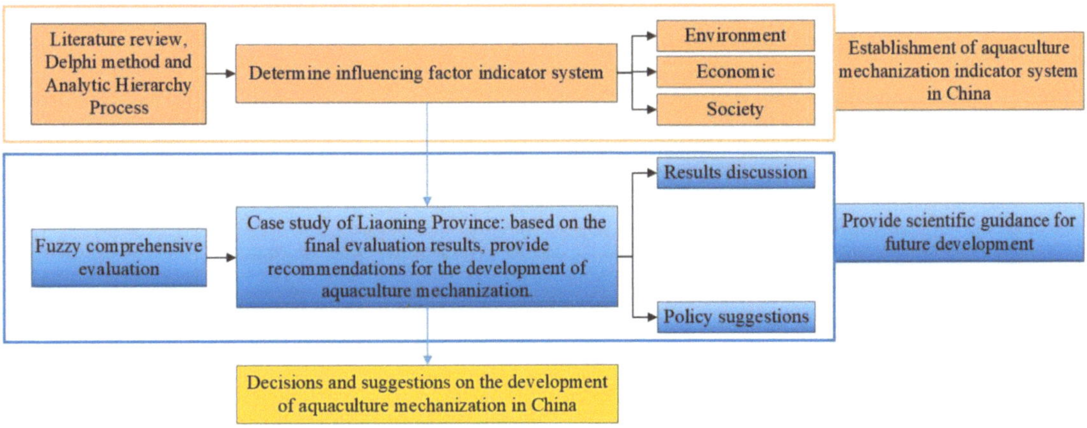

Figure 1. Overall flow chart of aquaculture mechanization development.

(1) Determination of the influencing factors evaluation indicator system. Through the method of the literature review, an analysis of the factors impacting the trajectory of aquaculture mechanization in China was conducted. This study adopts the model of expert scoring and the survey questionnaire, using experts to determine and screen indicators that can effectively reflect the current status of China's aquaculture mechanization. Based on the triple bottom theory, a research framework for an evaluation indicator system was ultimately derived using the Delphi method and AHP; the framework consists of 3 tiers and 18 influential factors.

(2) Evaluation method and case study. In this research, the FCE method was applied to construct the evaluation model. A scientific and rational quantitative evaluation of the qualitative and quantitative indicators was carried out. The analysis results of each indicator were obtained. Liaoning Province, China, was selected as the case study to carry out the evaluation and analysis. Discussion and policy recommendations. According to the final evaluation results, effective implementation paths and decision-making suggestions for the development of aquaculture mechanization were put forward, and beneficial reference and insights for China's aquaculture industry were provided.

2. Materials and Methods

2.1. Indicator Construction

Through the literature search, the Delphi method, and the consultation and interviews with experts, the obstacles and driving factors in the development direction of aquaculture mechanization in China were summarized, the important influencing factors were identified, and 33 influencing factors were retrieved. Then, through initial communication with 10 experts and scholars in the field of aquaculture, they were categorized based on the three levels of environment, economy, and society, resulting in the elimination of factors and the consolidation of 33 influencing factors into 25 factors. Then, a second interview was conducted for the "problem factors", and through this cycle, until the experts reached a consensus, they did not revise their views to obtain the final result. If the opinions of each expert were inconsistent, the Delphi questionnaire had to be modified until the questionnaire results converged. Finally, a questionnaire regarding the influencing factors of public participation in the development of aquaculture mechanization was formed; it included 3 levels and an indicator system with 18 influencing factors. A total of 1000 valid questionnaires were obtained through the online survey of "Sojump", and the effective rate reached 100%.

Evaluation Indicator System

Through the literature search, Delphi method, and AHP, the influencing factors of the aquaculture mechanization development in China were summarized. Finally, a comprehensive evaluation indicator system for the development of aquaculture mechanization, consisting of 3 levels and 18 influencing factors, was established, as depicted in Table 1.

Table 1. The description of influencing indicator system of mechanization development in China.

First-Level Indicators	Second-Level Indicators	Implication
A Environment	A1 Realize the sustainability of aquaculture	The key to harmony between man and nature was to achieve green development, low-carbon development, and sustainable development of aquaculture.
	A2 Utilization rate of aquaculture resources	Utilization of aquaculture resources was the maximum recycling.
	A3 Energy consumption of machinery and equipment	Mechanized and automated farming would produce a large amount of carbon emissions, and the increase in energy consumption would cause certain environmental pollution.
	A4 Energy consumption type of machinery and equipment	There are many types of machinery, which can be classified according to their scope, functions, and functions.
	A5 Manufacture mechanical equipment	Equipment will produce certain pollution and energy consumption during processing.
	A6 Purification needs of aquaculture tail water	The pollution problem of aquatic industry was most prominent in tail water discharge, which seriously restricts the harmonious development of ecology and green development.
B Economy	B1 Aquaculture cost management	Traditional aquaculture is characterized by high labor intensity, low efficiency, and high cost.
	B2 Aquaculture scale (small but scattered)	Traditional aquaculture is small in scale and scattered in the culture, which leads to low output.
	B3 Machinery purchase financial subsidy	Farmers purchased fishing machines through government subsidies to improve production efficiency.
	B4 Aquaculture advanced technology and equipment	Aquaculture mechanization technology played an important role in fishery economy and ecology.
	B5 Aquaculture machinery research investment	Increase the investment in scientific research of aquaculture facilities and equipment to promote the high-quality and efficient development of agricultural mechanization.
	B6 Government policy to benefit farmers	The government preferential agriculture policy was a convenient policy to promote the development of aquaculture mechanization.
C Society	C1 Aquaculture technology promotion and use	We should strengthen training, update farming concepts, strengthen technical management, and encourage the application of new technologies.
	C2 Aquaculture innovative talent demand	Those with innovative spirit, innovative ability, and practical ability are those who discover new problems, solve new problems, invent new things, and open up innovative fields.
	C3 Practitioners' policy awareness	Most of the farmers' cultural quality was not high, and the ability to accept new things, new technologies, and new achievements was quite different from their own needs.
	C4 Intensive farming level	Intensive farming was the maximization of the comprehensive benefits of production increase, energy conservation, and emissions reduction. Guiding farmers to develop intensive farming was an important way to achieve economies of scale.
	C5 Aquaculture labor resources	In the main production process, traditional aquaculture still needs to be managed by experienced workers, which leads to a labor force shortage.
	C6 Practitioners' technical training	We will increase the technical training of aquaculture practitioners, standardize the behavior of aquaculture farmers, improve the quality of aquatic products, and promote the green development of aquatic products.

2.2. *Analytic Hierarchy Process*

The analytic hierarchy process (AHP), as a systematic analysis method that combines quantitative and qualitative analysis, was suitable for the calculation of the factor weights. Its steps mainly include the establishment of a hierarchical organizational structure, the construction of a judgment matrix, the comparison of the importance of the factors, and the checking of the consistency. Ren et al. [33] employed the AHP to determine the weights of the indicators and the FCE method to evaluate the importance of each indicator, and they established a green supplier assessment indicator system for sea cucumber processing enterprises, verifying the effectiveness and practicality of this method. Hou et al. [34] developed a synthetic assessment model for the evaluation of the cleaner production indicator system of the sea cucumber aquaculture industry using the Delphi method, AHP, and FCE method. The pairwise comparisons matrix and membership degrees of the indicators were determined, and in order to reduce subjectivity in the subsequent quantitative calculation process, the AHP and the FCE method were also utilized. Finally, expert interviews were carried out to mitigate the subjective nature of pairwise comparison matrices and membership levels in future research. Chen et al. [35] proposed a safety performance measurement approach utilizing safety performance indicators to oversee the safety management process of civil aviation units using the AHP and the Delphi method. This method was successfully validated using actual data from a regional civil aviation bureau in China, and it could quantitatively reflect the safety management status of the bureau in a timely manner. Therefore, the construction of the model is realistic and effective.

2.2.1. Establish a Hierarchical Organizational Structure

An independent hierarchical organization structure that can describe the relationship of the contents of the system was established according to the requirements of the research questions.

2.2.2. Construct Judgment Matrix

Nine scales ranging from 1 to 9, along with their inverses, were employed to assess and balance the relative significance of each evaluation factor within the influencing factor set.

2.2.3. Consistency Test of Judgment Matrix

There were two parts to the consistency test.

(1) Hierarchical single sorting and consistency test

To assess the discrepancy of the error of the eigenvalue (λ_{max}) and the incompatibility-induced error in W caused by matrix incompatibility, the consistency of the judgment matrix was evaluated through calculation. The formula for determining the consistency indicator is as follows:

$$C.I. = \frac{\lambda_{max} - n}{n - 1} \tag{1}$$

Type, $C.I.$ was the consistency indicator, λ_{max} was the maximum value of judgment matrix, and n was for judging the matrix order. $C.I. = 0$ means that the judgment matrix is completely consistent. The larger the $C.I.$, the more serious the inconsistency of the judgment matrix.

The random consistency ratio calculation formula can be expressed as:

$$C.R. = \frac{C.I.}{R.I.} \tag{2}$$

Type, $C.R.$ was the random consistency ratio, $C.I.$ was the consistency indicator, and $R.I.$ was the average random consistency indicator.

When $C.R. < 0.1$, the judgment matrix was considered to meet the consistency requirements. Otherwise, the judgment matrix should be modified. Its consistency result is made to meet $C.R. < 0.1$ to ensure satisfactory consistency.

(2) Overall hierarchical ordering and consistency test

The overall hierarchical ordering and consistency test is based on the results of the single sorting and consistency test. The consistency ratio calculation formula of the total hierarchical ranking can be expressed as:

$$C.R. = \frac{a_1 C.I._1 + a_2 C.I._2 + \cdots + a_m C.I._m}{a_1 R.I._1 + a_2 R.I._2 + \cdots + a_m R.I._m} = \frac{C.I.}{R.I.} \tag{3}$$

When $C.R. < 0.1$, it is considered that the total ordering of the hierarchy passes the consistency test, otherwise the element values of the judgment matrix need to be readjusted. At this point, the final decision is made based on the overall ranking of the lowest layer.

In order to determine the difference in comparison with the previous $C.R.$, another formula for the overall hierarchical consistency ratio is used in this study; it is expressed as:

$$C.R.H = \frac{C.I.H}{R.I.H} \tag{4}$$

where the $C.I.H$ was the overall hierarchy consistency indicator, the $R.I.H$ was the overall stochastic value, and the $C.R.H$ was the overall hierarchical consistency ratio.

2.3. *Fuzzy Comprehensive Evaluation Method*

The FCE was a scientific and reasonable evaluation method to quantify the qualitative indicators. The FCE method has been validated as a scientific and rational assessment approach for quantifying qualitative metrics; it utilizes the concept of fuzzy relationship synthesis and can comprehensively and quantitatively evaluate factors that were difficult to quantitatively analyze. Its characteristic feature is its clear and systematic results, which provide better methods for solving various uncertainty problems. In 1965, Professor Chard et al. [36] proposed the concept of fuzzy set theory to express the uncertainty in the assessment process. Through the utilization of the FCE method, the factors that were difficult to analyze quantitatively were comprehensively and quantitatively evaluated, and various uncertainty problems could be better solved. With this method, the qualitative indicators can be quantified effectively, and the quantitative evaluation results can be obtained. The specific calculation steps are as follows.

2.3.1. Determine the Factor Universe of the Evaluated Content

Firstly, identify the factor universe of the assessed content. Assuming that the evaluated object contains P factors (or factors), the indicator set of the assessed object can be represented as:

$$U = \{u_1, u_2, u_3, \ldots, u_p\} \tag{5}$$

Type, p was the number of representative evaluation indicators.

2.3.2. Determine the Domain of Comment Level

This step was to determine the comment level domain of the FCE model. The comment level domain is a collection of comment levels, and each level can correspond to a fuzzy subset. Assuming that the comment level had an M level, the comment level universe of the evaluated object can be expressed as:

$$L = \{u_1, u_2, u_3, \ldots, L_m\} \tag{6}$$

Type, m represents the number of comment grades.

2.3.3. Constructing Judgment Matrix Based on Fuzzy Relation

After constructing fuzzy subsets, quantify the assessed content, determine the membership degree of each level of fuzzy subsets for the individual factor, and present the fuzzy relation matrix. The general expression of the fuzzy relation matrix is:

$$R = \begin{pmatrix} r_{11} & \cdots & r_{1m} \\ \vdots & \ddots & \vdots \\ r_{p1} & \cdots & r_{pm} \end{pmatrix}_{p \times m} \quad (7)$$

Type, r_{pm} represents the p indicator pair number m of the degree of membership of the level.

r_{pm} was calculated by membership function, and the membership degree of the indicators in the different reference values was calculated by the Delphi method. By issuing questionnaires, relevant experts were invited to score the inclusion of each secondary indicator relative to the benchmark value of the indicator. Finally, the membership degree of each secondary indicator was determined, and the fuzzy relation matrix was formed through sorting, summarizing, and normalizing the calculation.

2.3.4. Fuzzy Comprehensive Evaluation

After acquiring the fuzzy relation judgment matrix, multiply the matrix by the weight results of each first-level indicator and second-level indicator to obtain the FCE vector P. This process can be succinctly described as:

$$P = W \times R = (p_1, p_2, p_3, \ldots, p_n), \sum\nolimits_{i=1P_i}^{n} = 1 \quad (8)$$

Type, R is the fuzzy relation judgment matrix; W is the indicator weight.

This step was used to deal with each secondary indicator in the indicator system, and the calculation step of the first level was consistent with the secondary indicator. In the calculation process of the first-level indicator, the calculation results of the second-level indicator can be calculated and processed according to the known conditions.

2.3.5. Interpretation of Evaluation Results

The principle of the maximum membership degree principle was widely employed in the interpretation of the FCE results. However, this principle can only be applied to the qualitative analysis of the results, and it also cannot compare the differences between different factors for specific quantitative evaluation and analysis. Fuzzy vector singularization can be used to effectively solve the abovementioned problems with the results interpretation and the calculation method. The specific method was to multiply the membership score of the corresponding grade in fuzzy evaluation vector P by the normalized weight assignment in order to obtain the unified evaluation result.

Finally, through the comparison and analysis of the standardized results of the FCE vectors, the evaluation results of the various indicators and the influencing factors of aquaculture mechanization development were obtained; according to the evaluation results, the concrete improvement measures were put forward.

3. Results

3.1. *Evaluation Indicator of Influencing Factors of Aquaculture Mechanization*

3.1.1. Indicators Survey Results

The influencing factors of aquaculture mechanization were hierarchical and dynamic. They were affected and restricted by many factors, and finally, 18 effective factors were obtained. The environmental level was set as A: A1 realizes the sustainability of aquaculture (46.13%); A2 is the utilization rate of the aquaculture resources (43.77%); A3 is the energy consumption of the machinery and equipment (73.13%); A4 is the energy consumption type of the machinery and equipment (58.72%); A5 is the manufacture of the mechanical equipment (60.81%); and A6 is the purification needs of the aquaculture tail water (41.15%). On the economic level, it was set as B: B1 is the aquaculture cost management (47.28%); B2 is the aquaculture scale (small but scattered) (44.69%); B3 is the machinery purchase financial subsidy (70.64%); B4 is the advanced aquaculture technology and equipment (43.38%); B5 is the aquaculture machinery research investment (35.65%); and B6 is the government's policy to benefit farmers (64.61%). At the social level, it was set as C: C1 is the aquaculture technology promotion and use (68.15%); C2 is the aquaculture innovative talent demand (55.96%); C3 is the practitioners' policy awareness (38.27%); C4 is the intensive farming level (53.47%); C5 is the aquaculture labor resources (51.64%); and C6 is the practitioners' technical training (51.25%). The image sets are shown in Figure 2.

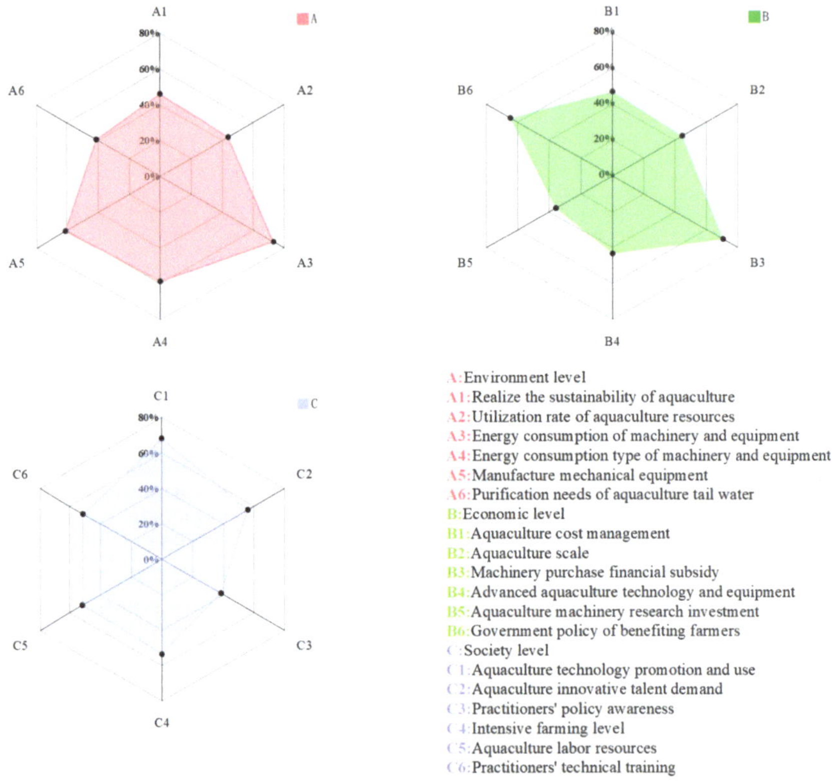

Figure 2. Composite structure diagram of influencing factors.

3.1.2. Weight Calculation and Consistency Test

The weight results of each indicator were calculated, and the judgment matrix of the importance relationship among the indicators was constructed according to the questionnaire results; the normalized results of the indicators are shown in Tables 2–5. The hierarchical single sorting and consistency test results of the first-level indicators and the second-level indicators are also reported in Tables 2–5.

Table 2. Judgment matrix and normalized results of first-level indicators.

Indicators	Environment	Economy	Society	Normalized Result
Environment	1	1/3	3	0.2604
Economy	3	1	5	0.6333
Society	1/3	1/5	1	0.1061
λ_{max} = 3.0387 $R.I.$ = 0.58 $C.I.$ = 0.019 $C.R.$ = 0.033 < 0.1				

Table 3. Judgment matrix and normalized results of second-level indicators (environment).

Indicators	A_1	A_2	A_3	A_4	A_5	A_6	Normalized Result
A1	1	2	1/6	1/4	1/4	3	0.08
A2	1/2	1	1/7	1/4	1/5	2	0.05
A3	6	7	1	4	4	7	0.45
A4	4	4	1/4	1	1/2	5	0.17
A5	4	5	1/4	2	1	5	0.21
A6	1/3	1/2	1/7	1/5	1/5	1	0.04
λ_{max}= 6.3531 $R.I.$= 1.24 $C.I.$= 0.071 $C.R.$ = 0.057 < 0.1							

Table 4. Judgment matrix and normalized results of second-level indicators (economy).

Indicators	B1	B2	B3	B4	B5	B6	Normalized Result
B1	1	2	1/6	2	4	1/5	0.12
B2	1/2	1	1/6	1	3	1/5	0.08
B3	6	6	1	6	7	3	0.49
B4	1/2	1	1/6	1	3	1/6	0.08
B5	1/4	1/3	1/7	1/3	1	1/6	0.04
B6	5	5	1/3	1/6	1/6	1	0.18
λ_{max}= 6.1014 $R.I.$= 1.24 $C.I.$= 0.020 $C.R.$ = 0.016 < 0.1							

Table 5. Judgment matrix and normalized results of second-level (society).

Indicators	C1	C2	C3	C4	C5	C6	Normalized Result
C1	1	4	6	4	5	5	0.45
C2	1/4	1	5	2	2	3	0.19
C3	1/6	1/5	1	1/4	1/4	1/4	0.04
C4	1/4	1/2	4	1	1	2	0.12
C5	1/5	1/2	4	1	1	1	0.11
C6	1/5	1/3	4	1/2	1	1	0.09
λ_{max}= 6.2747 $R.I.$= 1.24 $C.I.$= 0.055 $C.R.$ = 0.044 < 0.1							

Then, the consistency of the judgment matrix results was checked. According to Formulas (1) and (2), the consistency of each matrix was acceptable. This means that it was the effective weight value of each evaluation factor W in the judgment matrix. Therefore, the weight value of the first-level indicators and the second-level indicators of W can be expressed as:

$$W_T = (0.26\ 0.63\ 0.11) \tag{9}$$

$$W_{T1} = (0.08\ 0.05\ 0.45\ 0.17\ 0.21\ 0.04) \tag{10}$$

$$W_{T2} = (0.12\ 0.08\ 0.49\ 0.08\ 0.04\ 0.18) \tag{11}$$

$$W_{T3} = (0.45\ 0.19\ 0.04\ 0.12\ 0.11\ 0.09) \tag{12}$$

Moreover, according to Formulas (3) and (4), the overall hierarchical ordering and the consistency test results were performed and can be expressed as:

$$C.R.H = \frac{0.26 * 0.071 + 0.63 * 0.020 + 0.11 * 0.055}{0.26 * 1.24 + 0.63 * 1.24 + 0.11 * 1.24} = 0.029 \quad (13)$$

where the $C.R.H < 0.1$; the overall hierarchical consistency was acceptable. The weights of other the indicators were calculated according to the above methods, and all the outcomes passed the consistency assessment. Taking into account the aforementioned analysis and computation, the final evaluation indicator system of the influencing factors of Chinese aquaculture mechanization development included 3 first-class indicators and 18 second-class indicators. The reference values of each indicator and the calculation results of each indicator weight are shown in Table 6.

Table 6. Evaluation indicator system of influencing factors of aquaculture mechanization development in China.

Primary Indicator	Weight Value	Serial Number	Secondary Indicator	Weight Value
Environment	0.26	1	Realization of the sustainability of aquaculture	0.08
		2	Utilization rate of aquaculture resources	0.05
		3	Energy consumption of machinery and equipment	0.45
		4	Energy consumption type of machinery and equipment	0.17
		5	Manufacture of mechanical equipment	0.21
		6	Purification needs of aquaculture tail water	0.04
Economy	0.63	7	Aquaculture cost management	0.12
		8	Aquaculture scale (small but scattered)	0.08
		9	Machinery purchase financial subsidy	0.49
		10	Aquaculture advanced technology and equipment	0.08
		11	Aquaculture machinery research investment	0.04
		12	Government policy to benefit farmers	0.18
Society	0.11	13	Aquaculture technology promotion and use	0.45
		14	Aquaculture innovative talent demand	0.19
		15	Practitioners' policy awareness	0.04
		16	Intensive farming level	0.12
		17	Aquaculture labor resources	0.11
		18	Practitioners' technical training	0.09

3.2. *Case Analysis—Taking Liaoning Province as an Example*

Liaoning Province is one of the major provinces in China for aquaculture and is also a major province for the manufacturing of aquaculture machinery. The aquaculture machinery is developed, and the supporting facilities are complete. The overall mechanization level of aquaculture in the province has reached over 50%. There are many large aquaculture enterprises and aquaculture machinery manufacturing enterprises in the province. Therefore, based on the availability of the data, Liaoning Province is taken as an example to evaluate the importance of the influencing factors of aquaculture mechanization development.

Calculation of Fuzzy Comprehensive Evaluation Method

The questionnaire survey method was used to continue to invite the members of the expert group to rate the membership degree of each secondary indicator in the third-level comment grade. All the experts' scores were standardized to ascertain the overall level of association with each sub-indicator in the three comment grades. The normalized results of the secondary indicator membership degree are shown in Table 7.

Table 7. Normalized results of secondary indicator membership degree of influencing factors of aquaculture mechanization development in China.

Secondary Indicator	Membership Normalization Result		
	L1	L2	L3
Realization of the sustainability of aquaculture	0.3	0.6	0.1
Utilization rate of aquaculture resources	0.3	0.6	0.1
Energy consumption of machinery and equipment	0	1	0
Energy consumption type of machinery and equipment	0	1	0
Manufacture of mechanical equipment	0	0.8	0.2
Purification needs of aquaculture tail water	0	0.8	0.2
Aquaculture cost management	0	0.8	0.2
Aquaculture scale (small but scattered)	0	1	0
Machinery purchase financial subsidy	1	0	0
Aquaculture advanced technology and equipment	0	0.8	0.2
Aquaculture machinery research investment	0	0	1
Government policy to benefit farmers	0.2	0.8	0
Aquaculture technology promotion and use	0	1	0
Aquaculture innovative talent demand	0	1	0
Practitioners' policy awareness	0	0.6	0.4
Intensive farming level	0	0.8	0.2
Aquaculture labor resources	0	0.8	0.2
Practitioners' technical training	0.2	0.8	0

According to the results in Table 4 and Formula (5), the fuzzy relation judgment matrix of the first-class indicator can be obtained. Using the example of the first-level indicator of the economic impact factors, the fuzzy relational judgment matrix can be represented as:

$$R_E = \begin{bmatrix} 0.3 & 0.6 & 0.1 \\ 0.3 & 0.6 & 0.1 \\ 0 & 1 & 0 \\ 0 & 1 & 0 \\ 0 & 0.8 & 0.2 \\ 0 & 0.8 & 0.2 \end{bmatrix} \quad (14)$$

Multiply the obtained fuzzy relation judgment matrix of each first-level indicator by the indicator weight. Taking the first-level indicator of the economic factors as a case in point, the weights of its three second-level indicators can be expressed as vectors:

$$W_E = (0.076\ 0.051\ 0.45\ 0.17\ 0.21\ 0.037) \quad (15)$$

According to calculation Formula (6), the FCE vector of the first-level indicator of the economic level can be obtained. Taking the economic level indicator P_E as an example, its FCE vector is:

$$P_E = (0.038\ 0.90\ 0.062) \quad (16)$$

The FCE results of each first-class indicator of the influencing factors of aquaculture mechanization development in China are presented in Table 8.

Table 8. FCE results of first-level indicators.

Primary Indicator	Fuzzy Comprehensive Evaluation Vector Result
Environment	(0.038 0.90 0.062)
Economy	(0.53 0.39 0.08)
Society	(0.018 0.921 0.061)

According to the results in Table 5, Formula (6) is applied to calculate the fuzzy comprehensive evaluation vector P of the influencing factors of aquaculture mechanization development in China, which are:

$$P = (0.35\ 0.58\ 0.07) \tag{17}$$

Assign $q_1 = 3$, $q_2 = 2$, and $q_3 = 1$ to the weights of the three grades. Through the analysis of the assignment, it was concluded that the resulting interval of the fuzzy vector uniformization was between [1–3]. According to this interval, the single score range of the third-level evaluation results was evenly divided, and the allocation and scoring range for the third-level evaluation results are presented in Table 9.

Table 9. Table of fuzzy vector interval between evaluation grade, grade assignment, and uniformization result.

Primary Indicator	Fuzzy Comprehensive Evaluation Vector Result
Environment	(0.038 0.90 0.062)
Economy	(0.53 0.39 0.08)
Society	(0.018 0.921 0.061)

According to the normalization calculation of the FCE vector, the comprehensive evaluation outcomes of the factors influencing the development of aquaculture mechanization in China are as outlined:

$$\mathcal{V} = 2.28 \tag{18}$$

Based on the findings presented in Table 5, the FCE vector of the first-class indicator of the influencing factors of aquaculture mechanization in China was calculated in a single way. At the same time, the comparison in Table 3 determined the evaluation grade of the aquaculture mechanization factors of the first-level indicator, and the calculation results and evaluation results are shown in Table 10.

Table 10. Comparison table of single-value results and evaluation grades of first-level indicators.

Primary Indicator	Calculation Result	Comment Grade
Environmental level	1.98	L2
Economic level	2.45	L1
Social level	1.96	L2

According to the analysis results, the importance of the impacts among the different levels was economic, environmental, and social. In China, the top three influencing factors of aquaculture mechanization were A3: energy consumption of machinery and equipment; B3: financial subsidy for machinery purchase; and C1: promotion and use of aquaculture technology. The evaluation results of this study were largely in line with the existing production practices of the Chinese aquaculture industry, demonstrating the scientific validity of the evaluation system and the influencing factor methods, which were feasible and operable. Therefore, the financial subsidy for the purchase of aquaculture machinery, the energy consumption of the machinery and equipment, and the promotion and use of aquaculture technology were the effective improvement measures for the development of aquaculture mechanization in Liaoning Province at this stage.

4. Discussion

4.1. Domestic and Foreign Aquaculture Financial Policies

At present, the focus of the financial policies related to aquaculture in developed countries has changed from monoculture to diversification. Financial support, technical support, and talent support have been realized, and the scientific development of aquaculture mechanization has been given significant attention. Financial subsidies for aquaculture facilities are highly valued by some developed countries and regional governments. In addition to equipment subsidies for farmers, insurance support, administrative expenses subsidies, and credit support are provided in Japan [37]. Similarly, in the United States, financial subsidies and tax incentives have been provided, and the development and promotion of high-performance marine aquaculture equipment has been supported, such as in the research and development of large-scale storm-resistant cage equipment and deep-sea facilities [38], etc. Zambia has tried out and popularized the inclusive business model. The opportunities for investment, micro-credit, training, and the export of fishery facilities and equipment were increased. In addition, farmers who were interested in aquaculture were given technical training in aquaculture methods and provided with incentives or necessary support [39].

In conclusion, a variety of targeted improvement measures have been proposed for the mechanization of aquaculture, but more horizontal strategies are needed. These strategies will include policies, financing, training, infrastructure, and scientific elements to solve the difficulties faced by fishery farmers. In recent years, Liaoning Province has lacked funds and supporting policies for new machines and tools. In terms of investment in aquaculture technology promotion, about CNY 83.03 million was invested in Liaoning Province in 2020; thus, it ranked 17th in China. The infrastructure of the promotion system was relatively backward. The original value of the experimental equipment was only CNY 61.93 million, which was far lower than that of Jiangsu, Zhejiang, Guangdong and other strong marine fishery provinces [40]. In Liaoning Province, the number of subsidies for new equipment should be increased to guide farmers in using mechanical equipment such as oxygenation, feeding, catching, dredging, and water purification. Preferential policies such as interest discounts, tax reductions, and tax exemptions should be applied to financial institutions supporting the development of fishery mechanization in order to convince financial institutions to issue loans for fishery equipment [41]. At the same time, incentive measures should be formulated by the government; financial institutions that support the development of fishing machines should be rewarded; and modern publicity methods should be used to do a good job in the interpretation of fishing machine purchase subsidy policies and public opinion guidance [42]. For example, professionals should be organized to do a good job in policy consultation, in the application for operation procedures, in the steps of the purchase subsidy policy, etc.

4.2. Innovative Research and Development of Aquaculture Machinery and Equipment

In addition, in order to establish a precise facility and equipment system, such as fry loading and unloading and weighing and placing systems, measures to increase the research and development of new fishing equipment in Liaoning Province should be implemented, and a batch of advanced and applicable mechanical equipment should be developed. At the same time, the ability to fight against disaster risks should be strengthened, and the problems of fry weighing and the loading and unloading of vehicles in the process of breeding and releasing should be solved to achieve a labor-saving, accurate, safe, and efficient mechanization era [43]. By 2025, China's aquaculture mechanization level needs to reach more than 50%. It is essential to enhance investment in the facilities and equipment for critical areas such as breeding and seedling cultivation, epidemic prevention and treatment, planting and harvesting, and tail water treatment [44]. For example, the research and development of new equipment, such as the precise management and control of pond culture, efficient catching, light and simplified seedling planting and harvesting in raft culture, the counting and sorting of seedlings, and the harmless treatment of sick and dead aquatic products, should be focused on. Complete sets of technical facilities and equipment, such as large-scale intelligent aquaculture in the deep sea, deep-water anti-wind and wave cage aquaculture, factory aquaculture, container aquaculture and pond engineering circulating water aquaculture, should be improved [45]. At present, the scientific and technological level of Liaoning fishery production has been improved, and the cooperation between enterprises and scientific research institutions has brought new impetus to the research and development of new aquaculture equipment [46].

In a word, a marine fishery research base and innovation platform in Liaoning Province should be established. Resources in the field of fishery scientific research, such as marine universities and fishery scientific research institutions, should be integrated so that they can fully cooperate, give full play to their own advantages, and create a fishery science and technology innovation center with domestic and even international influence. Liaoning aquaculture industry will further accelerate the digital transformation of aquaculture modes such as factories, net cages, ponds, rice, and fisheries and promote the application of technologies such as the online monitoring of water quality, intelligent oxygenation, precise feeding, and the intelligent treatment of circulating water [47]. In the future, aquaculture facilities and equipment will transition towards mechanized research and comprehensive application and intelligent, information-based, and efficient aquaculture equipment, and the development of green, efficient, accurate, and intelligent fisheries will become a new development trend.

4.3. Financial Subsidies for the Purchase of Agricultural Machinery

Agricultural mechanization is the key to the comprehensive transition from traditional agriculture to modern agriculture. In order to promote the development of agricultural mechanization, China clearly stipulates that the government should arrange special funds separately to provide subsidies to farmers and agricultural production and operation organizations for purchasing advanced and applicable agricultural machinery supported and promoted by the state. The best cotton subsidy in Xinjiang is the mechanical subsidy. Xinjiang cotton has a large scale and high degree of mechanization, and the efficient utilization of mechanized facilities has a significant impact on cotton production. Therefore, the government established a special fund to strengthen the research and development of machinery, to support farmers in purchasing mechanical equipment, and to accelerate the process of modernization construction [48]. Hence, the implementation of financial subsidies for the purchase of agricultural machinery has significantly increased the investment in agricultural machinery. The substitution of agricultural machinery for labor will also drive enhancements in agricultural production efficiency and greatly stimulate the enthusiasm of farmers to purchase advanced and applicable agricultural machinery. This will improve the level of agricultural machinery equipment in China, drive the advancement of agricultural mechanization, liberate rural labor, promote grain production and farmers' income, and strengthen the foundation of agricultural mechanization; significant progress has been made in advancing the growth of the agricultural machinery sector and other aspects. The enhancement of agricultural productivity and income is intricately linked to the mechanization of agriculture. The Indian government has considered net agricultural income, household income, and household consumption, and has offered timely subsidies for machinery purchase subsidies to enable small farmers to access agricultural credit. The utilization of machinery has resulted in a 31% increase in agricultural net income, a 19% increase in household income, and a 19% increase in food consumption [49]. The Russian government has driven local farmers' ability to purchase advanced agricultural machinery through customized financial subsidy plans, increasing their willingness to use and purchase agricultural machinery [50]. Given the significant positive impact of agricultural mechanization, it is necessary to deepen the development of low-cost and suitable agricultural machinery for small farmers.

5. Conclusions

According to the characteristics of aquaculture mechanization in China, the evaluation indicator system of 18 influencing factors at 3 levels was screened out by a literature search, the Delphi method, and the AHP in this study. Through the FCE method, the research found that the economic level was the most critical factor in the development of mechanization in aquaculture. The financial subsidy for the purchase of aquaculture machinery, the energy consumption of machinery and equipment, and the promotion and use of aquaculture technology were the deepest factors and had the greatest impact on the development of aquaculture mechanization in China. In future, the emphasis will be on promoting mechanized equipment, such as high-density recirculating aquaculture for pond farming, factory circulating water aquaculture, deep-sea anti-wind and wave cages, etc.; this equipment occupies a small area, has high unit output, and is environmentally friendly. The technology of fishery equipment and aquaculture efficiency will be further improved, and the effective collection of aquaculture tail water, solid waste, and water resources recycling will be further strengthened to achieve an ecological, mechanized, and intelligent sustainable aquaculture mode.

Author Contributions: Investigation, formal analysis, writing—review and editing, L.Y.; investigation, writing—original Draft, H.W.; investigation, formal analysis, A.R.; investigation, formal analysis, F.H.; investigation, formal analysis, F.J.; writing—review and editing, funding acquisition, H.H.; supervision, Y.L. All authors have read and agreed to the published version of the manuscript.

Funding: This study was supported by the project of the Youth Foundation of the Department of Education of Liaoning Province [LJKQZ2021123], the earmarked fund for the Modern Agro-industry Technology Research System [CARS-49], and the Innovation and Entrepreneurship Project for High-Level Talents of Dalian [2020RQ109].

Institutional Review Board Statement: Not applicable.

Informed Consent Statement: Informed consent was obtained from all subjects involved in the study.

Data Availability Statement: Data are contained within the article.

Acknowledgments: We thank the editors and reviewers for their contributions to the publication of this study.

Conflicts of Interest: The authors declare that they have no known competing financial interests or personal relationships that could have appeared to influence the work reported in this paper; the authors declare the following financial interests/personal relationships which may be considered as potential competing interests; the researcher claims no conflicts of interests.

References

1. Li, X.; Li, J.; Wang, Y.; Fu, L.; Fu, Y.; Li, B.; Jiao, B. Aquaculture industry in China: Current state, challenges, and outlook. *Rev. Fish. Sci.* **2011**, *19*, 187–200. [CrossRef]
2. Newton, R.; Zhang, W.; Xian, Z.; McAdam, B.; Little, D.C. Intensification, regulation and diversification: The changing face of inland aquaculture in China. *Ambio* **2021**, *50*, 1739–1756. [CrossRef]
3. FAO. *World Fisheries and Aquaculture Status: Sustainable Development in Action*; FAO: Rome, Italy, 2020. [CrossRef]
4. Liu, Z.; Li, S.; Sun, L.; Guo, Y.; Zhao, W.; Xu, L.; Wu, F.; Yu, X.; Wang, D. *China Fishery Statistical Yearbook*; China Agriculture Press: Beijing, China, 2011–2021.
5. Fei, R.; Wang, H.; Wen, Z.; Yuan, Z.; Yuan, K.; Chunga, J. Tracking factor substitution and the rebound effect of China's agricultural energy consumption: A new research perspective from asymmetric response. *Energy* **2021**, *216*, 119307. [CrossRef]
6. Cao, T.; Zhao, X.; Yang, Y.; Zhu, C.; Xu, Z. Adaptive Recognition of Bioacoustic Signals in Smart Aquaculture Engineering Based on r-Sigmoid and Higher-Order Cumulants. *Sensors* **2022**, *22*, 2277. [CrossRef] [PubMed]
7. Xu, H.; Zhang, J.H.; Ding, J.L. The review of the research progress of fishery equipment and engineering technology at home and abroad. *Fish Mod.* **2010**, *37*, 1–8.
8. Wei, Y.; Wei, Q.; An, D. Intelligent monitoring and control technologies of open sea cage culture: A review. *Comput. Electron. Agric.* **2020**, *169*, 105119. [CrossRef]
9. Prem, R.; Tewari, V.K. Development of human-powered fish feeding machine for freshwater aquaculture farms of developing countries. *Aquac. Eng.* **2020**, *88*, 102028. [CrossRef]
10. Huang, Y.X.; Xu, H.; Ding, J.L. Situation of China land-based aquaculture engineering equipment and suggestion for its development. *Guizhou Agric. Sci.* **2016**, *44*, 87–91.
11. Wang, P.; Ji, J.; Zhang, Y. Aquaculture extension system in China: Development, challenges, and prospects. *Aquac. Rep.* **2020**, *17*, 100339. [CrossRef]
12. ICLARM. *Genetic Improvement of Carp Species in Asia: Final Report to the Asian Development Bank*; Regional Technical Assistance No. 5711; International Centre for Living Aquatic Resources Management: Penang, Malaysia, 2001.
13. Sharma, K.K.; Bikash, S.; Chand, S. A mechanized harvest system for freshwater fishpond. *Aquac. Eng.* **2009**, *41*, 147–151. [CrossRef]

14. Liu, S.; Tian, X.; Xiong, Y.; Zhang, Y.; Tanikawa, H. Challenges towards carbon dioxide emissions peak under in-depth socioeconomic transition in China: Insights from shanghai. *J. Clean. Prod.* **2020**, *247*, 119083. [CrossRef]
15. Peter, E. Aquaculture environment interactions: Past, present and likely future trends. *Aquaculture* **2015**, *447*, 2–14.
16. Simbeye, D.S.; Yang, S.F. Water quality monitoring and control for aquaculture based on wireless sensor networks. *J. Netw.* **2014**, *9*, 840. [CrossRef]
17. Li, D.L.; Liu, C. Recent advances and future outlook for artificial intelligence in aquaculture. *Smart Agric.* **2020**, *2*, 1.
18. Hu, F.; Zhong, H.; Wu, C.; Wang, S.; Guo, Z.; Tao, M.; Zhang, C.; Gong, D.; Gao, X.; Tang, C.; et al. Development of fisheries in China. *Reprod. Breed.* **2021**, *1*, 64–79. [CrossRef]
19. Wang, C.; Li, Z.; Wang, T.; Xu, X.; Zhang, X.; Li, D. Intelligent fish farm—The future of aquaculture. *Aquac. Int.* **2021**, *29*, 2681–2711. [CrossRef]
20. Naylor, R.L.; Hardy, R.W.; Buschmann, A.H.; Bush, S.R.; Cao, L.; Klinger, D.H.; Little, D.C.; Lubchenco, J.; Shumway, S.E.; Troell, M. A 20-year retrospective review of global aquaculture. *Nature* **2021**, *591*, 551–563. [CrossRef]
21. Landeta, J. Current validity of the Delphi method in social sciences. *Technol. Forecast. Soc.* **2006**, *73*, 467–482. [CrossRef]
22. Lucia, B.D.V.; Hans, S.; Laura, S. Prioritization of fish welfare issues in European salmonid aquaculture using the Delphi method. *Aquaculture* **2023**, *572*, 739557. [CrossRef]
23. Anna, H.; Kruno, B.; Tom, C.; Collins, K.; da Costa, F.; Debney, A.; van Duren, L.; Elzinga, J.; Fariñas-Franco, J.M.; Gamble, C.; et al. Site selection for European native oyster (*Ostrea edulis*) habitat restoration projects: An expert-derived consensus. *Aquat. Conserv. Mar. Freshw. Ecosyst.* **2023**, *33*, 721–736. [CrossRef]
24. Olson, K. An examination of questionnaire evaluation by expert reviewers. *Field Methods* **2010**, *22*, 295–318. [CrossRef]
25. Alphonce, C.B. Application of the analytic hierarchy process in agriculture in developing countries. *Agric. Syst.* **1997**, *53*, 97–112. [CrossRef]
26. Francisco, K.; Concepcion, R.; Relano, R.J.; Enriquez, M.L.; Baun, J.J.; Mayol, A.P.; Española, J.; Vicerra, R.R.; Bandala, A.; Dadios, E.; et al. Analytical Hierarchical Process-based Material Selection for Trailer Body Frame of an Underground Imaging System. In Proceedings of the 2021 IEEE 13th International Conference on Humanoid, Nanotechnology, Information Technology, Communication and Control, Environment, and Management (HNICEM), Manila, Philippines, 28–30 November 2021; pp. 1–6.
27. Du, Y.; Zheng, Y.; Wu, G.; Tang, Y. Decision-making method of heavy-duty machine tool remanufacturing based on AHP-entropy weight and extension theory. *J. Clean. Prod.* **2020**, *252*, 119607. [CrossRef]
28. Du, Y.; Cao, H.; Liu, F.; Li, C.; Chen, X. An integrated method for evaluating the remanufacturability of used machine tool. *J. Clean. Prod.* **2012**, *20*, 82–91. [CrossRef]
29. Zadeh, K.N.; Anderson, S.W. Sensitivity Analysis in Optimal Simultaneous Power Interchange. *IEEE Trans. Power Appar. Syst.* **1978**, *97*, 2405–2415. [CrossRef]
30. Wang, X.; Zhang, H.; Li, M.; Li, Y. Research on Interference Optimization Mechanism of Wireless Routing Signal Transmission Based on Fuzzy Comprehensive Evaluation Method. *Int. J. Online Eng.* **2017**, *13*, 136–148. [CrossRef]
31. Shi, C.; Zhang, K.; Ruan, J. Fuzzy comprehensive evaluation on the sustainable development of equipment manufacturing industry based on improved algorithm. In Proceedings of the 2009 Third International Symposium on Intelligent Information Technology Application Workshops, Nanchang, China, 21–22 November 2009; pp. 173–177.
32. Zhao, L.; Cheng, X.; Yao, Y. Online Intelligent Evaluation of Dispensing Quality Based on Entropy Weight Fuzzy Comprehensive Evaluation Method and Machine Learning. In Proceedings of the 2020 International Conference on Sensing, Measurement & Data Analytics in the Era of Artificial Intelligence (ICSMD), Xi'an, China, 15–17 October 2020; pp. 491–495.
33. Ren, A.Q.; Zhao, X.T.; Liu, Q.; Yu, L.; Han, F.; Jia, F.; Hou, H.; Liu, Y. Green Supplier Assessment and Selection for Sea Cucumber (*Apostichopus japonicus*) Processing Enterprise: Case Study in China. *Sustainability* **2023**, *15*, 15368. [CrossRef]
34. Hou, H.C.; Shao, S.; Zhang, Y.; Sun, D.; Yang, Q.; Qin, C.; Sun, X. Cleaner Production assessment for sea cucumber aquaculture: Methodology and case studies in Dalian, China. *Clean Technol. Environ. Policy* **2019**, *21*, 1751–1763. [CrossRef]
35. Chen, W.S.; Li, J. Safety performance monitoring and measurement of civil aviation unit. *J. Air Transp. Manag.* **2016**, *57*, 228–233. [CrossRef]
36. Zhang, K.; Si, B.F. The Application of Fuzzy Comprehensive Evaluation Method Based on AHP on the Distribution of Rail Transport. *Appl. Mech. Mater.* **2014**, *2916*, 1379–1382. [CrossRef]
37. Engle, C.R.; Stone, N.M. Competitiveness of US aquaculture within the current US regulatory framework. *Aquac. Econ. Manag.* **2013**, *17*, 251–280. [CrossRef]
38. Avadí, A.; Cole, S.M.; Kruijssen, F.; Dabat, M.-H.; Mungule, C.M. How to enhance the sustainability and inclusiveness of smallholder aquaculture production systems in Zambia. *Aquaculture* **2022**, *547*, 737494. [CrossRef]
39. Adekola, K.A.; Alabadan, B.A.; Akinyemi, T.A. China Agricultural mechanization development experience for developing countries. *Int. J. Agric. Innov. Res.* **2014**, *3*, 655–658.
40. Guan, H.; Sun, Z.; Wang, J. Decoupling Analysis of Net Carbon Emissions and Economic Growth of Marine Aquaculture. *Sustainability* **2022**, *14*, 5886. [CrossRef]
41. Spiertz, J.H.J. Nitrogen, sustainable agriculture and food security. A review. *Agron. Sustain. Dev.* **2010**, *30*, 43–55. [CrossRef]
42. Tuo, G. How to promote agricultural insurance in China. *China Agric. Econ. Rev.* **2016**, *8*, 194–205. [CrossRef]
43. Huang, J.; Yang, G. Understanding recent challenges and new food policy in China. *Glob. Food Secur.* **2017**, *12*, 119–126. [CrossRef]

44. Wang, G.; Hu, L.; Long, W.; Jiang, L. Evolution of Intelligent Feeding System for Aquaculture: A Review. In Proceedings of the 2021 3rd International Conference on Artificial Intelligence and Advanced Manufacture (AIAM), Manchester, UK, 23–25 October 2021; pp. 392–397.
45. Lee, P.G. Process control and artificial intelligence software for aquaculture. *Aquac. Eng.* **2000**, *23*, 13–36. [CrossRef]
46. Ji, J.; Guo, X.; Zhang, Y. The study of symbiotic relationships between the economic and the ecological system of China's mariculture industry—An empirical analysis of 10 coastal regions with Lokta–Volterra model. *Reg. Stud. Mar. Sci.* **2021**, *48*, 102051. [CrossRef]
47. Jothiswaran, V.V.; Velumani, T.; Jayaraman, R. Application of artificial intelligence in fisheries and aquaculture. *Biot. Res. Today* **2022**, *2*, 499–502.
48. Wang, X.Y.; Tan, Y.W.; Li, Y.H. The best choice of cotton subsidy modes based on principal component regression in Xin Jiang, China. In *Intelligent Decision Making System*; World Scientific: Singapore, 2010; pp. 544–549. [CrossRef]
49. Jena, P.R.; Tanti, P.C. Effect of farm machinery adoption on household income and food security: Evidence from a nationwide household survey in India. *Front. Sustain. Food Syst.* **2023**, *7*, 922038. [CrossRef]
50. Boiarskii, B.; Hasegawa, H.; Lioude, A.; Kolesnikova, E.; Sinegovskaia, V. Current Situation and Perspectives for Soybean Production in Amur Region, Russian Federation. *AMA Agric. Mech. Asia Afr. Lat. Am.* **2020**, *51*, 33–38.

Disclaimer/Publisher's Note: The statements, opinions and data contained in all publications are solely those of the individual author(s) and contributor(s) and not of MDPI and/or the editor(s). MDPI and/or the editor(s) disclaim responsibility for any injury to people or property resulting from any ideas, methods, instructions or products referred to in the content.

AgriEngineering

MDPI

Article

Modeling of Coffee Fruit: An Approach to Simulate the Effects of Compression

Janielle Souza Pereira *, Ricardo Rodrigues Magalhães, Fábio Lúcio Santos, Ednilton Tavares de Andrade and Leomar Santos Marques

School of Engineering, Federal University of Lavras, Lavras 37200-900, MG, Brazil; ricardorm@ufla.br (R.R.M.); fabio.santos@ufla.br (F.L.S.); ednilton@ufla.br (E.T.d.A.); leomar.marques@engenharia.ufjf.br (L.S.M.)
* Correspondence: janielle.engenheira@gmail.com

Abstract: The flavor, aroma, and color of coffee can be changed due to mechanical damage, reducing its quality. To measure the mechanical behavior of the fruit, compression tests can be performed at different stages of ripeness. In this study, we analyzed the deformation, strain energy, and von Mises stress of coffee fruits at mature, semi-mature, and immature stages under compression forces. Compression in three directions (x, y, and z) was simulated on coffee fruit models using the finite element method. A compression support was applied in the opposite direction to the force application axis. Numerical simulations of the compression process allowed us to verify that the more mature the fruit, greater the associated mean deformation (2.20 mm mm^{-1}, 0.78 mm mm^{-1}, and 0.88 mm mm^{-1}), the lower the mean strain energy (0.07 mJ, 0.21 mJ, and 0.34 mJ), and the lower the mean equivalent von Mises stress (0.25 MPa, 1.03 MPa, and 1.25 MPa), corresponding to ripe, semi-ripe, and immature fruits, respectively. These analyses not only save time and professional resources but also offer insights into how strain energy and von Mises stress affect fruits at different maturation stages. This information can guide machine adjustments to reduce coffee harvesting damages.

Keywords: deformation; finite element method; mechanical damage; mechanical properties; numerical simulation

Citation: Pereira, J.S.; Magalhães, R.R.; Santos, F.L.; de Andrade, E.T.; Marques, L.S. Modeling of Coffee Fruit: An Approach to Simulate the Effects of Compression. *AgriEngineering* **2023**, *5*, 2303–2313. https://doi.org/10.3390/agriengineering5040141

Academic Editors: Muhammad Sultan, Yuguang Zhou, Redmond R. Shamshiri and Muhammad Imran

Received: 14 August 2023
Revised: 16 November 2023
Accepted: 27 November 2023
Published: 1 December 2023

Copyright: © 2023 by the authors. Licensee MDPI, Basel, Switzerland. This article is an open access article distributed under the terms and conditions of the Creative Commons Attribution (CC BY) license (https://creativecommons.org/licenses/by/4.0/).

1. Introduction

In the production process of agricultural products, the loss rates accumulate throughout the entire production chain, from the harvest to the consumer's table. This scenario makes it essential to use techniques and technologies that promote a reduction in such losses [1]. Additionally, to provide conditions that favor the maintenance of product quality, one must know the physical and mechanical properties of the material being worked on for proper use of processing techniques and equipment [2,3].

Coffee is one of the products that has an increase in value and export volume due to better quality. However, its flavor, aroma, and color can be changed due to the occurrence of mechanical damage, which affects the desired attributes [4]. In this context, there has been a strengthening of the industry standards for the production of high-quality coffee, as well as of the institutions that evaluate its excellence and qualify specialty coffees. This influences the sales price of the lots and can favor the reputation of the countries and regions that produce this coffee [5,6].

Arabica coffee (*Coffea arabica* L.) belongs to the order Gentianales, family Rubiaceae, and genus *Coffea*. It is a shrub native to the region of southwestern Ethiopia, southeastern Sudan, and northern Kenya. It has also been cultivated in other locations, such as in Brazil, where there are approximately 132 cultivars registered with the responsible agencies [7,8].

The coffee plant, a shrub of continuous growth that reaches between 2 and 4 m in height, has a cylindrical, yellowish-white stem, with hardwood, dark green leaves, that are oval, opposite, and paired, as well as white aromatic flowers [7]. Its phenological cycle takes two years to complete. The formation of vegetative branches (plagiotropic or transverse), from the orthotropic branches (orthogonal or perpendicular to the soil), containing nodes with auxiliary buds, occurs in the first year. In the second year, flowering begins, followed by the formation of pellets, grain expansion, fruit germination, and maturation [9].

Among the coffee producing countries, Brazil produces the largest amount, (approximately 50 million bags in 2022), which contributes to the gross domestic product (GDP) of the country. The Brazilian coffee production accounts for 30% of global production [10–12].

The species *Coffea arabica* was first introduced in Brazilian soil in the state of Pará, in approximately 1727. It spread throughout the states of Maranhão, Bahia, Rio de Janeiro, São Paulo, Minas Gerais, Paraná, and Espírito Santo, occupying a relevant position in the worldwide supply in the twentieth century [13].

The country is also a significant consumer, emphasizing the product's importance for both export and domestic consumption [14]. Mechanical harvesting, while efficient, can cause branch breakage and fruit damage, necessitating the selection of appropriate harvesting technology [15]. This choice should consider various factors, including the tree's properties and the fruit's characteristics [15].

During the fruit ripening process, there is a reduction in chlorophyll and an increase in anthocyanins and carotenoids. This, combined with the action of ethylene and the increase in the activity of the enzymes responsible for the softening of the cell wall, transforms the hard, green, and rigid fruit into soft fruit [9]. The degree of maturation, together with the size of the beans, which are thinner than Robusta coffee, can reduce harvest efficiency and change the quality of the Arabica coffee beverage products, which tend to have a low caffeine content and favorable aroma and flavor [16,17].

However, if coffee is to continue to contribute effectively to Brazil's economic development, attention must be paid to the damage caused to this fruit by the mechanical, semi-mechanized and mechanized processes to which it is subjected from harvest to final destination [18]. Ref. [4] points out that during the harvesting process, mechanical damage can occur to the fruit, affecting the quality of the product (coffee) and, thus, jeopardizing the value and volume of commercialization. Therefore, there is a need to look for effective methods to optimize mechanical processing design and minimize these injuries, such as computer modeling and simulation [19].

Ref. [19] found that one of the ways to change the mechanical processes of coffee processing that can induce injuries to the fruits is the study and evaluation of scenarios. These scenarios involve damage prediction of the fruits via computer simulation, mechanical behavior during the storage, and harvesting processes for different stages of maturation. In this case, the use of known mechanical properties, such as elastic modulus, Poisson coefficient, and density is essential to predict the deformation, energy distribution, and equivalent stress of the material after the application of compression loads [20].

Ref. [21] highlighted that one of the ways to measure the mechanical response of materials is through compression testing and that parameters, such as moisture content and force application region, can affect the mechanical behavior of the fruit.

The Hertz theory is essential for understanding interactions between solids in contact. It establishes that, when compression occurs between two solid bodies, they initially touch at a single point in the contact region, resulting in elastic deformation in the vicinity of this primary contact point, particularly under light loads. The elastic properties and material geometries play a critical role in determining the pressure distribution in this contact region. This detailed understanding of pressure distribution is of utmost importance, particularly in applications, such as load measurement, where optimizing measurement precision and reliability is essential [22].

Ref. [19] highlighted that fruit cells in such a contact area are compressed, while the adjacent cells are subjected to stress, shearing, or bending loads. Mechanical response may be related to the internal structure of the fruit and the solid biomaterials of tissues and cellular components. Thus, it is important to analyze these responses in fruits with different maturation stages.

In modern industrial production, empirical design solutions, traditionally based on trial and error and heavily reliant on a designer's experience, have been both time-consuming and expensive. However, the recent surge in technological advancements has shifted the industry towards computational modeling as a more efficient approach to address these challenges [23].

The origins of modeling and simulation can be traced back to the 1950s in the United States; they were initially developed for military applications. Over time, its utility expanded, serving as a robust tool for decision making by facilitating the evaluation and analysis of real systems via computational models [23].

As suggested by Refs. [24,25], models can be conceptualized as a set of equations formulated to predict a process's behavior through an approximate representation. The complexity of these models often varies based on their intended application. Emphasizing the multidisciplinary nature of computational modeling, Ref. [26] described it as a technique that offers simplified representations of the system under study, employing tools to craft microsystems. Furthermore, Ref. [27] elaborated on the application of computational techniques and mathematical models, emphasizing their role in deciphering complex phenomena and simulating problem solutions.

One of the primary advantages of simulation, when paired with modeling, is its ability to emulate real-world scenarios without incurring the high costs associated with physical experiments. This method is versatile, facilitating system improvements, concept testing before actual implementation, and the optimization of processes and resources [23,25]. As Ref. [24] underscored, a model's intrinsic value lies in its capacity to probe a system's responses under diverse conditions without any physical alterations.

The modeling process entails problem definition, solution determination, model construction, and result analysis. Computer-aided design (CAD) tools, during model creation, offer a precise and reliable means to simulate the envisaged model numerically [26,28]. CAD employs the finite element method (FEM), a numerical technique rooted in mathematical formulas. Originally introduced in the 1960s for aeronautical applications, FEM has since found widespread use across various engineering disciplines [28].

FEM, or finite element analysis (FEA), is a computational strategy that approximates problem solutions by dissecting both partial and ordinary differential equations [29]. With the evolution of computational power, FEM has become indispensable, enabling predictions of the mechanical behavior of a diverse range of materials, encompassing organic, nonmetallic, and metallic entities, like vegetables, fruits, and agricultural machinery [30]. According to Ref. [31], FEM has been instrumental in simulating the ripening process of *Coffea arabica* L. var. Colombia, providing insights into its mechanical and geometrical characteristics.

When analytical methods become overly intricate, consuming significant time and professional resources, FEM-based analysis emerges as the preferred choice [28]. FEM is considered a system discretization, which involves segmenting the problem domain into small, geometrical, and simple regions [29]. This discretization results in a mesh, characterized by nodes that interconnect basic elements, whether linear, triangular, or rectangular. These foundational elements facilitate FEM analyses, supported by commercial software like Ansys, Nastran, Procal3D, Rocky DEM, and SolidWorks, which predict system behavior [28,29].

In this context, this study aimed to analyze deformation, strain energy, and von Mises stress of coffee fruit using a computer simulation. Mature, semi-ripe, and immature maturation stages of coffee fruits are analyzed after being compressed by collapse forces. The relevance of modeling in obtaining data can help producers to reduce the compression damage caused during harvesting and storage processes of coffee fruits at different stages of maturation.

2. Materials and Methods

In SolidWorks®, geometric models of coffee fruits at the mature, semi-mature, and immature maturation stages were developed (Figure 1) for subsequent simulation in Ansys®

version 14.5. The model was based on images of *Coffea arabica* L. variety Colombia fruits, with dimensions reflecting the average provided by Ref. [32]. Coffee fruits were modeled with a circumference of approximately 0.013 m at its smallest (x) and 0.017 m at its largest (z). The peduncle was created by moving a scan of a 2D profile along a 3D sketch path.

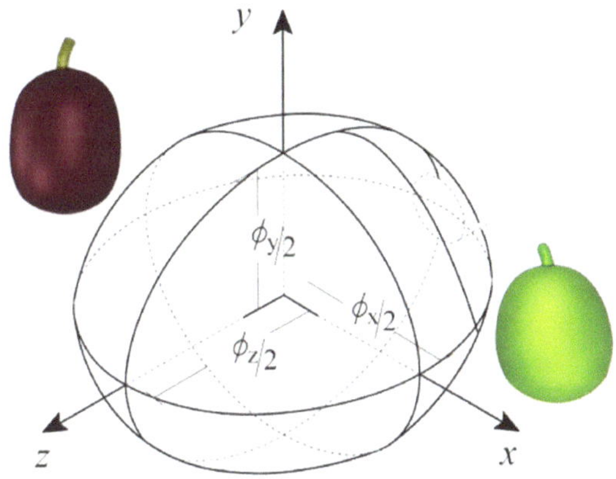

Figure 1. 3D model of coffee fruits with dimensions on axes x (φx), y (φy), and z (φz).

To check the reliability of the numerical analysis, a standard mesh convergence study was performed using the standard mode of the Ansys® software itself, with the use of meshes with different levels of refinement until the best one was found to represent the geometry of the coffee fruit with a greater number of elements and an element quality metric close to 1, a value that, according to Ref. [33], represents a perfect cube for a 3D element. For this purpose, a mesh with 3 mm elements was obtained initially (Figure 2a), and the final size was 0.7 mm (Figure 2b). The load intensity and its location (x, y, and z) were determined, in addition to the compression points, which were at the side of the model opposite to the applied forces. Thus, finite element models were generated with the tetrahedral mesh that best represented the fruit geometry, containing 6725, 13,393, and 9516 second-order elements for the mature, semi-mature, and immature maturation stages, respectively.

In the search to determine the response of the evaluated structure, numerical simulations were performed in Ansys®, by static structural analysis, compressing the fruit along three directions (x-, y-, and z-axes) to establish its deformation (ε), strain energy (μ), and equivalent von Mises stress ($\sigma_{vonMises}$), according to Equations (1)–(3), respectively. Equation (1) is as follows:

$$\{\varepsilon\} = \{\varepsilon xx\ \varepsilon yy\ \varepsilon zz\ \varepsilon xy\ \varepsilon xz\ \varepsilon yz\} \tag{1}$$

where $\varepsilon xx = \partial u/\partial x$, $\varepsilon yy = \partial s/\partial y$, $\varepsilon zz = \partial w/\partial z$, $\varepsilon xy = \partial u/\partial y + \partial s/\partial x$, $\varepsilon xz = \partial u/\partial z + \partial w/\partial x$, and $\varepsilon yz = \partial s/\partial z + \partial w/\partial y$. Equation (2) is as follows:

$$\mu = \int_0^{\varepsilon} \sigma \cdot d\varepsilon \tag{2}$$

where $\sigma = \frac{F}{A}$ and $d\varepsilon = \frac{\partial x}{T}$. Equation (3) is as follows:

$$\sigma_{vonMises} = \sqrt{\frac{[(\sigma 1 - \sigma 2)^{2} + (\sigma 2 - \sigma 3)^{2} + (\sigma 1 - \sigma 3)^{2}]}{2}} \tag{3}$$

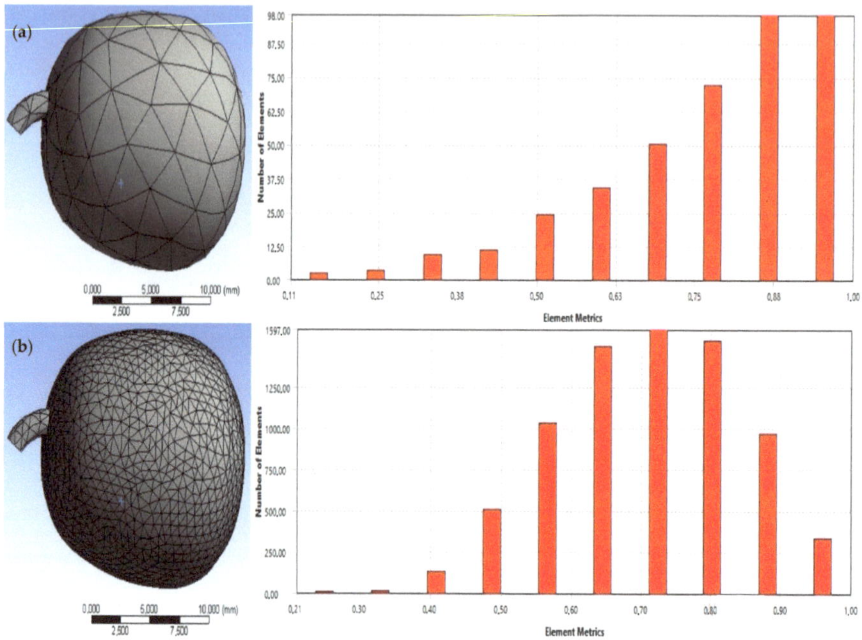

Figure 2. Second-order tetrahedral mesh refinement applied to coffee fruits with element sizes of 3 mm (**a**) and 0.7 mm (**b**).

In the proposed simulations, it was decided to position a compression support in the opposite direction to the applied force, which is distributed throughout the body, as shown in Figure 3.

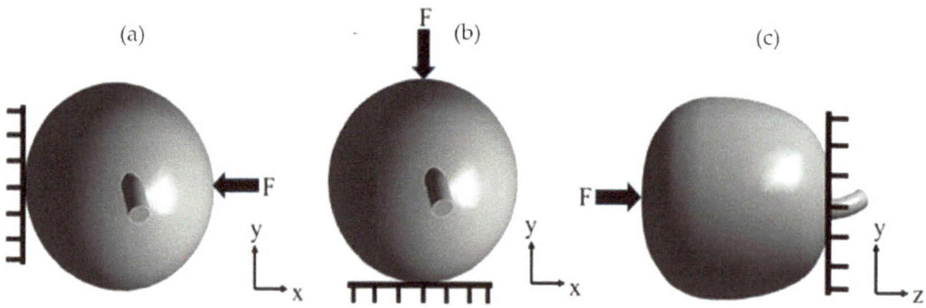

Figure 3. Loads applied to coffee fruit planes in the x-axis (**a**), y-axis (**b**), and z-axis (**c**).

To perform the simulations, known values of properties, including the collapse force (F), elastic or Young's modulus (E), Poisson coefficient (μ), and density determined in other studies (Table 1), were adopted, and the material was considered to be an elastic, homogeneous, and isotropic material for simplifications, as stated by Ref. [18].

Collapsing force represents the load acting during the compression of the material that reaches a maximum value upon reaching the greatest resistance of the fruit peel; then, its intensity is reduced upon reaching the pulp, which is milder and is reduced to a minimum after the breakage of the fruit. Young's modulus, which expresses the rigidity of the material, consists of the ratio between the applied stress and the strain felt by the material, while the Poisson coefficient measures the transverse strain related to the longitudinal direction of the force application [34,35].

Table 1. Mechanical properties of coffee fruit in the mature, semi-ripe, and immature stages of maturation.

Property	Fruit			Source
	Mature	**Semi-Mature**	**Immature**	
Collapse force (N)	50	195.00	231.80	[34]
Density (kg m^{-3})	1020	1087	1130	[18]
Young's modulus (Pa)	2,930,000	19,570,000	15,820,000	
Poisson coefficient (dimensionless)	0.27	0.25	0.24	

From the input data and the configuration of the scenarios to be studied, which related the applied compression and the maturation stages of the fruits, it was possible to perform computer simulations to determine the following parameters: total deformation, strain energy distribution, and equivalent von Mises stress. In this context, from the results of the simulations, the distributions of the fruit strain, the stress concentration points, and the strain energy involved in the process were analyzed.

3. Results and Discussion

Numerical simulations of the ripe coffee fruit, applying forces in three directions (x, y, and z) were performed. Clamping support was used in the opposite direction to the load, and deformation results were obtained (Figure 4).

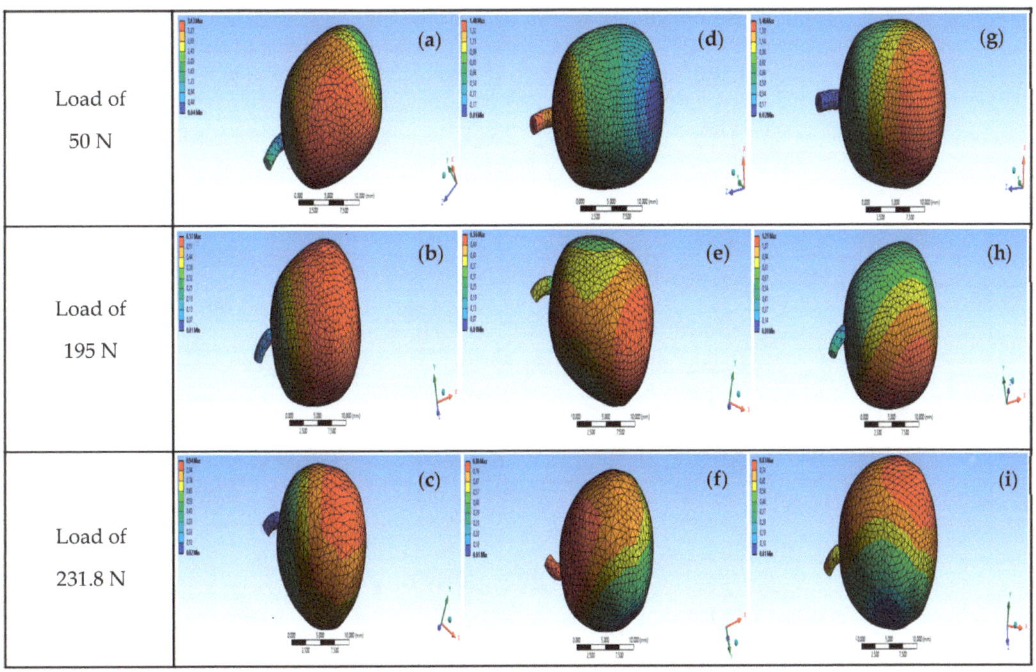

Figure 4. Total deformation (mm mm^{-1}) of ripe, semi-ripe, and immature coffee fruits when subjected to loads of 50 N, 195 N, and 231.8 N in x (**a–c**), y (**d–f**), and z (**g–i**), directions, respectively.

It was observed that the maximum deformation values of the ripe fruit were 3.63 mm mm^{-1}, 1.49 mm mm^{-1}, and 1.47 mm mm^{-1}, while those of the semi-ripe fruit were approximately 0.57 mm mm^{-1}, 0.55 mm mm^{-1}, and 1.21 mm mm^{-1}, and those of the immature fruit were 0.94 mm mm^{-1}, 0.86 mm mm^{-1}, and 0.84 mm mm^{-1}, corresponding to the application of loading in the x, y and z directions, respectively.

These values were applied opposite to the direction of the applied loading, to simulate crushing of the fruit, which can generate perceptible deformation in the sides of the fruit, corroborating the study by Ref. [36], who affirm the occurrence of transverse expansion in elastic material after compression, coinciding with Hertz theory, which states that deformation remains in the vicinity of the body after compression.

The results obtained suggest that, except for the semi-ripe fruit, the smallest deformations were obtained by compressing the fruit in the direction of the z-axis, corroborating the study by Ref. [37], who found lower firmness in the fruit peel when compressed in the direction of this same axis. Thus, there is a need to seek mechanisms that allow the storage of fruits with compression occurring in the direction of the z-axis, as this will be the best way to avoid deformation damage.

Notably, the largest deformation was found in the ripe fruit, with a tendency for less mature fruit to present smaller deformations (64% and 59%, respectively) more than in semi-ripe and immature fruit. This behavior was predictable because, according to Ref. [32], the more mature the outer layer of the fruit, the less the resistance; according to Ref. [38], during maturation, the cells grow and the cell wall thickness decreases, reducing this mechanical property; according to [39], a higher water content is present in ripe fruit, compared to the semi-ripe and immature maturation stages, making the fruit more flexible and resulting in a lower force required for collapse. Thus, the deformation of more mature fruit will be greater. This lower collapse force and greater deformation in wetter fruit is similar to the findings of Ref. [40].

When analyzing the strain energy distribution, it was observed that in the ripe fruit, the maximum values were 0.06 mJ, 0.05 mJ, and 0.11 mJ, corresponding to the application of force in the x, y, and z directions, respectively, while in the semi-ripe fruit, they were 0.31 mJ, 0.20 mJ, and 0.12 mJ, corresponding to the application of force in the x, y, and z directions, respectively; in the immature fruit, the maximum results were 0.47 mJ, 0.25 mJ, and 0.31 mJ, corresponding to the application of force in the x, y, and z directions (Figure 5), where the compression support was fixed.

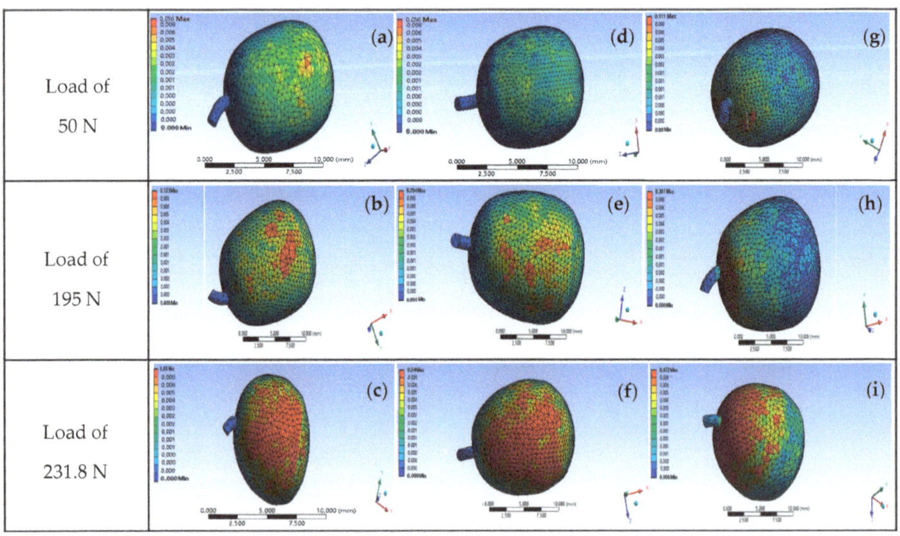

Figure 5. Strain energy distribution (mJ) in the ripe coffee fruit when subjected to loads of 50 N, 195 N, and 231.80 N in x (**a–c**), y (**d–f**). and z (**g–i**), directions, respectively, and compressed in the opposite load direction.

For the immature fruit, the values of the strain energy distribution exceeded those found in the ripe and semi-ripe fruit, being around 78.6% and 38.8% higher, respectively, which occurred due to the greater resistance to deformation found in this stage of maturation.

These conditions highlight the need for caution during the unloading process of the fruits from harvesters to the trucks transporting the coffee fruits, especially when their maturation stage is more advanced and their mechanical resistance is lower. This care is essential because unloading at a higher acceleration could result in greater damage to the coffee, as there will be more force applied to the fruits, given that force is the product of mass and acceleration.

It was also possible to observe that for the mature stage, when compressing the coffee fruit in the direction of the peduncle (z-axis), the higher the strain energy applied, the greater the distance was between the points of force application and the compression support in this direction of application.

This result corroborates with the study performed by Ref. [41] which used strawberries, in which the longitudinal compression of the fruit required a greater force and consequently higher strain energy, compared to the transverse compression.

Regarding the von Mises stress results, the maximum values in the ripe fruit for the application of force in the x, y, and z directions were 0.18 MPa, 0.18 MPa. and 0.39 MPa, respectively (Figure 6). These results also show that the highest von Mises stress value was located at the end of the fruit (close to the peduncle), representing a value more than 116% greater than the deformation on the x- and y-axes. This means that there was a greater concentration of stresses in this area and, consequently, a greater likelihood of collapse. Therefore, direct contact of the fruit detachment mechanism with this area during harvesting should be avoided.

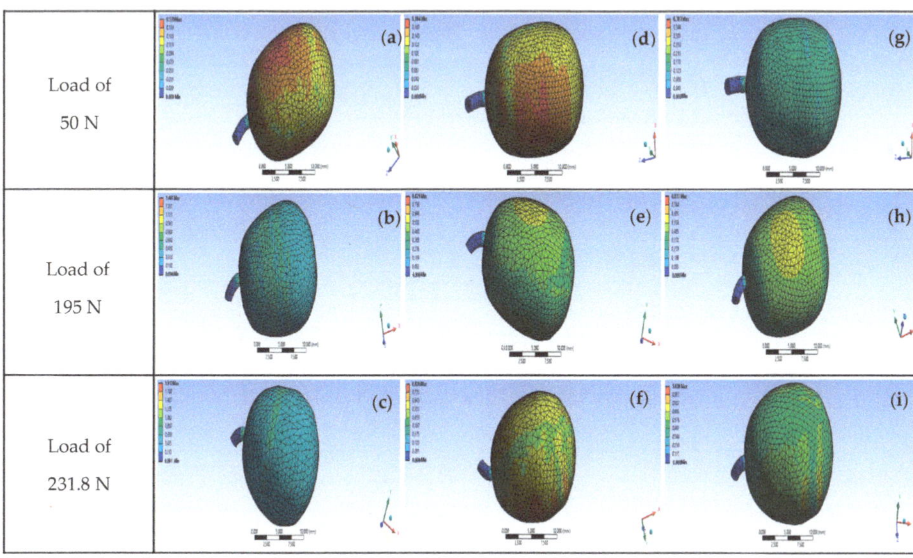

Figure 6. Von Mises equivalent stress distribution (MPa) in the ripe coffee fruit when subjected to loads of 50 N, 195 N, and 231.80 N in x (**a–c**), y (**d–f**), and z (**g–i**) directions, respectively, and compressed in the opposite load direction.

When analyzing the von Mises stress for the semi-ripe fruit, it was observed that the maximum values obtained when applying force along the x-, y-, and z-axes were approximately 1.44 MPa, 0.83 MPa, and 0.83 MPa, respectively. These values were 75.7% higher than those of ripe fruits due to the greater internal resistance to the application of external force (compression) inherent to this maturation stage; it can be inferred that during harvest by harvester stems, these find less resistance in mature fruits, making such fruits more susceptible to damage.

By analyzing von Mises stress for the immature fruit, it was observed that, concomitant with the ripe and semi-ripe fruit, the maximum values for the application of force in the x, y, and z directions were 1.91 MPa, 0.83 MPa, and 1.04 MPa, respectively. These values were higher where the compression support was fixed, indicating a higher concentration of this property in these locations. This makes them points with a higher probability of collapse. It was also evident that the highest stresses occurred at this stage of maturation. On average, these stresses were 80.1% and 18.2% higher than in ripe and semi-ripe fruit, respectively. These findings corroborate the results of Ref. [42], who also found maximum values for immature fruit.

Based on the two studies, it is possible to observe that compressive stress was higher in the green maturation condition compared to the mature maturation condition. These results are in line with the literature and support the conclusion of Ref. [42], which suggests that selective harvesting using mechanical vibrations and the brake force on the vibration rollers is feasible, as long as an appropriate excitation frequency is chosen, related to the modal parameters of the system to prevent damage to coffee fruits due to compression.

4. Conclusions

The numerical simulation of the coffee fruit compression process made it possible to verify that the more mature the fruit is, the greater the mean deformation (2.20 mm mm^{-1}, 0.78 mm mm^{-1}, and 0.88 mm mm^{-1}, corresponding to ripe, semi-ripe, and immature fruit, respectively), the lower the mean strain energy (0.07 mJ, 0.21, mJ and 0.34 mJ, corresponding to mature, semi-ripe, and immature fruit, respectively) and the lower the mean equivalent von Mises stress (0.25 MPa, 1.03 MPa, and 1.25 MPa, corresponding to ripe, semi-ripe, and immature fruit, respectively). These properties can guide adjustments in the harvest and storage systems to minimize losses due to mechanical damage by compression.

The computer simulation model proposed for the coffee fruit compression process proved to be efficient in predicting deformation, von Mises equivalent stress, and deformation energy, obtaining results that corroborate previous studies.

The performance of this type of analysis, besides reducing the time and professional resources used to solve problems, shows itself as an alternative to help improve the mechanized harvesting techniques, by indicating how the strain energy and the von Mises stress impact the fruit in the various stages of maturity. These data can provide more adequate indicators for machine regulation, which would minimize the injuries caused during harvesting.

For future works, it is proposed to analyze damages during the processing of the fruit in the drying and roasting stages, etc. and how it can be related with other variables, such as moisture, the temperature to which it is subjected during toasting. This can help to improve the technological processes and the quality control of the fruit that is processed in each of the stages of maturation.

Author Contributions: Conceptualization, J.S.P., R.R.M. and E.T.d.A.; Data curation, J.S.P.; Formal analysis, J.S.P. and R.R.M.; Investigation, J.S.P., R.R.M., E.T.d.A., F.L.S. and L.S.M.; Methodology, J.S.P., R.R.M. and E.T.d.A.; Project administration, R.R.M.; Supervision, R.R.M.; Visualization, J.S.P., R.R.M., E.T.d.A., F.L.S. and L.S.M.; Writing—original draft, J.S.P., R.R.M., E.T.d.A., F.L.S. and L.S.M.; Writing—review and editing, J.S.P., R.R.M., E.T.d.A., F.L.S. and L.S.M. All authors have read and agreed to the published version of the manuscript.

Funding: This research was funded by Coordenação de Aperfeiçoamento de Pessoal de Nível Superior—CAPES, grant number 001.

Data Availability Statement: No new data were created or analyzed in this study. Data sharing is not applicable to this article.

Acknowledgments: The authors greatly acknowledge CAPES, CNPq, and Fapemig for providing funding for this research.

Conflicts of Interest: The authors declare no conflict of interest.

References

1. Rosa, C.I.L.F.; Moribe, A.M.; Yamamoto, L.Y.; Sperandio, D. Pós-colheita e comercialização. In *Hortaliças-Fruto [Online]*; Brandão Filho, J.U.T., Freitas, P.S.L., Berian, L.O.S., Goto, R., Eds.; EDUEM: Maringá, Brazil, 2018; pp. 489–526. [CrossRef]
2. Munyendo, L.; Njoroge, D.; Hitzmann, B. The potential of spectroscopic techniques in coffee analysis—A review. *Processes* **2022**, *10*, 71. [CrossRef]
3. Oliveira, V.C.; Mendes, F.Q. Técnicas de preservação pós-colheita de frutas e hortaliças: Uma revisão narrativa. In *Ciência e Tecnologia de Alimentos: Pesquisa e Práticas Contemporâneas*, 2nd ed.; Cordeiro, C.A.M., da Silva, E.M., Barreto, N.S.E., Eds.; AREPB: Brasília, Brasil, 2021; p. 279. [CrossRef]
4. Abreu, G.F.; Pereira, C.C.; Malta, M.R.; Clemente, A.C.S.; Coelho, L.F.S.; Rosa, S.D.V.F. Alterações na coloração de grãos de café em função das operações pós-colheita. *Coffee Sci.* **2015**, *10*, 429–436.
5. Bastian, F.; Hutabarat, O.S.; Dirpan, A.; Nainu, F.; Harapan, H.; Emran, T.B.; Simal-Gandara, J. From plantation to cup: Changes in bioactive compounds during coffee processing. *Foods* **2021**, *10*, 2827. [CrossRef]
6. Gumecindo-Alejo, A.L.; Sánchez-Landero, L.A.; Ortiz-Ceballos, G.C.; Cerdán-Cabrera, C.R.; Alvarado-Castillo, G. Factors related to coffee quality, based on the "Cup of Excellence" contest in Mexico. *Coffee Sci.* **2021**, *16*, e161887. [CrossRef]
7. Carvalho, C.N.; Oliveira, Y.R.; Silva, P.H.; Abreu, M.C. *Coffea arabica* L.: Potencialidades e ações medicinais. *Revinter* **2018**, *11*, 87–105. [CrossRef]
8. Terceiro, M.G.; Meirelles, F.C.; Cavalcante, A.G.; Mingotte, F.L.; Lemos, L.B. Caracteres morfológicos de cafeeiro de porte baixo no primeiro ano de formação em Jaboticabal-SP. *Rev. Cienc. Agrar.* **2019**, *62*, 1–7. [CrossRef]
9. Negreiros, V.M.V.; Nascentes, R.F.; Nascentes, M.C.B. Efeitos da aplicação de regulador vegetal na maturação e na qualidade de bebida dos frutos do café. *Rev. AgroFIB* **2019**, *1*, 13–24.
10. CNA—Confederação da Agricultura e Pecuária do Brasil. PIB do Agronegócio 2022. 2022. Available online: https://cnabrasil.org.br/publicacoes/apos-alcancar-patamar-recorde-em-2021-pib-do-agronegocio-recua-4-22-em-2022 (accessed on 26 September 2023).
11. CONAB—Companhia Nacional de Abastecimento. Safra de Café deve Encerrar o Ciclo de 2022 com uma Produção de 50,92 Milhões de Sacas. 2022. Available online: https://www.conab.gov.br/ultimas-noticias/4857-safra-de-cafe-deve-encerrar-o-ciclo-de-2022-com-uma-producao-de-50-92-milhoes-de-sacas (accessed on 26 September 2023).
12. ICO—International Coffee Organization. Coffee Report and Outlook. 2023. Available online: https://icocoffee.org/documents/cy2022-23/Coffee_Report_and_Outlook_April_2023_-_ICO.pdf (accessed on 26 September 2023).
13. Queiroga, V.P.; Gomes, J.P.; Melo, B.A.; Albuquerque, E.M.B. *Cultivo do Café (Coffea arabica L.) Orgânico Sombreado para Produção de Grãos de alta Qualidade*, 1st ed.; Editora Científica Digital: São Paulo, Brazil, 2021; pp. 507–524.
14. ABIC—Associação Brasileira da Indústria de Café. Indicadores da Indústria de Café, 2022. 2022. Available online: https://estatisticas.abic.com.br/estatisticas/indicadores-da-industria/indicadores-da-industria-de-cafe-2022/ (accessed on 26 October 2023).
15. Afsah-Hejri, L.; Homayouni, T.; Toudeshki, A.; Ehsani, R.; Ferguson, L.; Castro-García, S. Mechanical harvesting of selected temperate and tropical fruit and nut trees. *Hortic. Rev.* **2022**, *49*, 171–242. [CrossRef]
16. Alves, G.S.P.; Coelho, A.P.; Lemos, L.B. Crescimento vegetativo, produtividade e qualidade pós-colheita de cultivares de café de porte baixo em região de baixa altitude. *Rev. Agroecossist.* **2021**, *13*, 63–83. [CrossRef]
17. Graciano, P.D.; Assis, G.A.; Siquieroli, A.C.S.; Ferreira Junior, L.D.; Fernandes, M.I.S.; Paiva, C.R. Estádios de maturação de cultivares de *Coffea arabica* L. em monte Carmelo-MG e suas características sensoriais. *Ciência Agrícola* **2019**, *17*, 7–14. [CrossRef]
18. Coelho, A.L.; Santos, F.L.; Pinot, F.A.C.; Queiroz, D.M. Determinação das propriedades geométricas, físicas e mecânicas do sistema fruto-pedúnculo-ramo do cafeeiro. *Rev. Bras. Eng. Agric. Ambient.* **2015**, *19*, 286–292. [CrossRef]
19. Li, E.; Maio, F.; Andrews, J. Mechanical models of compression and impact on fresh fruits. Comprehensive. *Rev. Food Sci. Food Saf.* **2017**, *16*, 1296–1312. [CrossRef] [PubMed]
20. Gazzola, J.; Fabbro, I.M.D.; Biasi, A.R.; Cardoso, K.C.; Rodrigues, S.; Enes, A.M. Propriedades mecânicas dos materiais biológicos. In *Ciências Agrárias: O Avanço da Ciência no Brasil*, 2nd ed.; Melo, J.O.F., Ed.; Editora Científica Digital: São Paulo, Brazil, 2021; pp. 507–524. [CrossRef]
21. Freccia, C.S.; Ramos, A.P.; Peres, L.G.; Vitto, D.C.; Carlos, E.B.; Cardoso, L.S.; Palhano, W.C.; Schimidt-Bellini, J.; Seibert, E. Aplicação de danos mecânicos e seus efeitos na qualidade pós-colheita de frutas cítricas. *Rev. Técnico Científica—IFSC* **2012**, *3*. Available online: https://periodicos.ifsc.edu.br/index.php/rtc/article/view/522 (accessed on 3 May 2023).
22. Melo Neto, J.V.; Silva, G.C. Avaliação experimental para detecção da elipsoide de Hertz em um alojamento de célula de carga para pontes de pesagem. *J. Ind. Prod. Eng.* **2019**, *2*, 39–54.
23. Andrade, L.O.; Silva Júnior, J.F.; Rodrigues, T.V.; Carvalho, G.D.G. Modelagem e simulação no estudo de melhoria da capacidade produtiva em uma empresa de serviços de lavanderia na cidade de Passos. *Rev. Tecnol. Apl.* **2020**, *9*, 38–62. [CrossRef]
24. Franco, I.C. Modelagem e Simulação: Processos Químicos. Notas de aula inovadoras. *J. Eng. Exact. Sci.* **2021**, *7*, 1–50. [CrossRef]
25. Santos, A.A.; Fagundes, C.H.; Ferreira Junior, J.S.; Rodrigues, T.V.; Sanjulião, L.K.A.F.; Borges, V.J.P. Modelagem e simulação do processo de atendimento cardiológico em um posto de saúde visando menor tempo de espera do paciente. *Braz. J. Bus.* **2019**, *1*, 719–734. Available online: https://ojs.brazilianjournals.com.br/ojs/index.php/BJB/article/view/2568/2578 (accessed on 26 September 2023).

26. Rodrigues, G.O.; Simonetto, E.O.; Barros, R.S.; Dalmolin, L.C.; Modro, N.R. Uso da modelagem computacional para comparativo entre adição de diferentes percentuais de biodiesel no transporte de resíduos sólidos urbanos. *Rev. CEPE* **2019**, *49*, 80–98. [CrossRef]
27. Batista, R.C.; Gomes, M.N. Modelagem computacional aplicada como uma tecnologia no currículo de engenharia mecânica. *Rev. Mundi Eng. Tecnol. E Gestão* **2020**, *5*, 238–301. [CrossRef]
28. Vieira, D.F.; Coelho, R.C.; Dias, M.J.; Neto, A.M.S.; Fontes, W.A. Elementos Finitos para o dimensionamento de uma viga perfil "I" utilizada em um pórtico móvel. *Rev. Process. Químicos* **2019**, *13*, 31–44. [CrossRef]
29. Carneiro, R.S.; Carneiro, R.S.; Silva, K.F. Método dos elementos finitos: Comparativo entre solução exata e soluções aproximadas pelo método de Galerkin em estruturas unidimensionais. *ReviSeM* **2020**, *2*, 407–428. [CrossRef]
30. Souza, V.H.S.; Santos, A.A.R.; Costa, A.L.G.; Santos, F.L.; Magalhães, R.R. Evaluation of the interaction between a harvester rod and a coffee branch based on finite element analysis. *Comput. Electron. Agric.* **2018**, *150*, 476–483. [CrossRef]
31. Tinoco, H.A.; Peña, F.M. Mechanical and geometrical characterization of fruits *Coffea arabica* L. var. Colombia to simulate the ripening process by finite element analysis. *Eng. Agric. Environ. Food* **2019**, *12*, 367–377. [CrossRef]
32. Tinoco, H.A.; Ocampo, D.A.; Peña, F.M.; Sanz-Uribe, J.R. Finite element modal analysis of the fruit-peduncle of *Coffea arabica* L. var. Colombia estimating its geometrical and mechanical properties. *Comput. Electron. Agric.* **2014**, *108*, 17–27. [CrossRef]
33. Kohnke, P. (Ed.) *ANSYS: Theory Reference*, 11th ed.; SAS IP, Inc.: Canonsburg, PA, USA, 1999.
34. Couto, S.M.; Batista, C.S.; Devilla, I.A.; Paim, V.T. Características de frutos de café sob compressão. *Rev. Bras. Eng. Agric. Ambient.* **2002**, *6*, 117–122. [CrossRef]
35. Silva, E.P.; da Silva, F.M.; de Andrade, E.T.; Magalhães, R.R. Structural static and modal frequency simulations in a coffee harvester's chassis. *Rev. Bras. Eng. Agric. Ambient.* **2020**, *7*, 511–515. [CrossRef]
36. Pallottino, F.; Costa, C.; Menesatti, P.; Moresi, M. Assessment of the mechanical properties of Tarocco orange fruit under parallel plate compression. *J. Food Eng.* **2011**, *103*, 308–316. [CrossRef]
37. Vieira, D.A.P.; Caliari, M.; Souza, E.R.B.; Soares Júnior, M.S. Mechanical resistance, biometric and physicochemical characteristics of tomato cultivars for industrial processing. *Food Sci. Technol.* **2019**, *39*, 363–370. [CrossRef]
38. Figueiredo Neto, A.; Almeida, F.A.C.; Olivier, N.C.; Dantas, A.C.S.; Irmão, M.A.S. Mechanical behavior of pumpkin fruits subjected to compression during maturation. *Pesqui Agropecu Trop.* **2013**, *43*, 223–231. [CrossRef]
39. Chakespari, A.G.; Rajabipour, A.; Mobli, H. Strength Behavior Study of Apples (cv. *Shafi Abadi & Golab Kohanz*) under Compression Loading. *Mod. Appl. Sci.* **2010**, *4*, 173–182.
40. Martins, B.A.; Ferraz, A.C.; Schmidt, F.L. Physical characteristics of baru tree fruit aimed at kernel extraction. *Semin. Cienc. Agrar.* **2017**, *38*, 1865–1874. [CrossRef]
41. Linhares, J.A.; Castillo, B.; Londoño, M.T. Characterization of the mechanical properties of the sweet passion fruit (*Passiflora ligularis* Juss.). *Agron. Colomb.* **2013**, *31*, 208–214.
42. Santos, F.L.; Queiroz, D.M.; Valente, D.S.M.; Coelho, A.L.F. Simulation of the dynamic behavior of the coffee fruit-stem system using finite element method. *Acta Sci. Technol.* **2015**, *37*, 11–17. [CrossRef]

Disclaimer/Publisher's Note: The statements, opinions and data contained in all publications are solely those of the individual author(s) and contributor(s) and not of MDPI and/or the editor(s). MDPI and/or the editor(s) disclaim responsibility for any injury to people or property resulting from any ideas, methods, instructions or products referred to in the content.

AgriEngineering

Review

Bridging the Implementation Gap between Pomace Waste and Large-Scale Baker's Yeast Production

Josipa Lisičar Vukušić, Thomas Millenautzki and Stéphan Barbe *

Faculty of Applied Natural Sciences, Technische Hochschule Köln, Campus Leverkusen, Campuspl. 1, 51379 Leverkusen, Germany
* Correspondence: stephan.barbe@th-koeln.de

Abstract: The objectives set in the European Green Deal constitute the starting point of this review, which then focuses on the current implementation gap between agro-industrial wastes as resources for large-scale bioprocesses (e.g., baker's yeast, bioethanol, citric acid, and amino acids). This review highlights the current lack of sustainability of the post-harvest processing of grapes and apples. In light of the European Green Deal, industrial biotechnology often lacks sustainability as well. We reviewed the recent progress reported in the literature to enhance the valorization of grape and apple pomace and the current failure to implement this research in technical processes. Nevertheless, selected recent papers show new perspectives to bridge this gap by establishing close collaborations between academic teams and industrial partners. As a final outcome, for the first time, we drew a circular flow diagram that connects agriculture post-harvest transformation with the industrial biotechnology and other industries through the substantial valorization of apple and grape pomace into renewable energy (solid biofuels) and sugar extracts as feedstock for large-scale bioprocesses (production of baker's yeast industry, citric acid, bioethanol and amino acids). Finally, we discussed the requirements needed to achieve the successful bridging of the implementation gap between academic research and industrial innovation.

Keywords: European Green Deal; apple pomace; grape pomace; industrial biotechnology; feedstock; agro-industrial wastes; sustainability; circular economy

Citation: Lisičar Vukušić, J.; Millenautzki, T.; Barbe, S. Bridging the Implementation Gap between Pomace Waste and Large-Scale Baker's Yeast Production. *AgriEngineering* **2023**, *5*, 2238–2252. https://doi.org/10.3390/agriengineering5040137

Academic Editors: Yuguang Zhou, Redmond R. Shamshiri, Muhammad Sultan and Muhammad Imran

Received: 25 October 2023
Revised: 20 November 2023
Accepted: 27 November 2023
Published: 1 December 2023

Copyright: © 2023 by the authors. Licensee MDPI, Basel, Switzerland. This article is an open access article distributed under the terms and conditions of the Creative Commons Attribution (CC BY) license (https://creativecommons.org/licenses/by/4.0/).

1. Introduction

In December 2019, the European Union (EU) introduced the European Green Deal (EGD) as a paradigm shift in environmental and climate policy [1]. It represents the EU's most serious action to reach climate neutrality by 2050. Pressure is currently increasing to reduce the abstraction of water and resources from our ecosystem and the release of emissions in to the environment. In this context, Europe aims to become the leader in clean products and technologies, as well as to ensure a just and inclusive transition [2]. The EGD's main objective is achieving net carbon neutrality within the European Union by 2050 and disassociating the economic growth from resource consumption. This deal refers to a general policy strategy, outlining the ambitions and goals in different policy sectors. The EGD must be understood as an approach to solve the existing implementation gap between scientific progress and European actions. In order for this new policy to be timely implemented, existing regulations and standards from the following eight areas are now subject to profound revisions and new policies are already in place. The EGD focuses on these eight areas [3,4]:

(1) Climate neutral Europe; introducing the European Climate Law with the goal of reducing the greenhouse gas emissions;
(2) Clean energy transition by securing an affordable energy supply, developing a new energy market and prioritizing energy efficiency;
(3) New industrial strategies to promote a sustainable and circular economic model;

(4) Energy and resource efficiency in construction and renovating;
(5) Transition towards sustainable and smart transportation;
(6) Reaching a pollution-free environment with zero toxicity;
(7) "Farm to fork strategy"—environmentally sustainable food system;
(8) Conserving and re-establishing ecosystems and biodiversity.

Overall, the EGD represents a set of targets, intentions, and objectives tackling Europe's many environmental and climate-related challenges that will be implemented over the next ten years. The COVID-19 crisis has affected the EGD by drawing resources and attention away from environmental crises to some extent. However, the commitment to the EGD and the green transformation remains high [3].

The EGD's research area "Farm to fork strategy" aims to create guidelines for an environmentally sustainable food system. Among these goals is a target to reduce food waste [5]. It represents a critical issue, due to its large quantity, therefore creating insecurity in food supply chain. There is contamination of water sources, air, and soil occurring due to practices such as direct dumping on landfills as well as burning. Improper handling, technological issues, and consumer behavior result in approximately one-third of the world's total food production turning into waste. It is estimated that approximately 1300 million tons of food is wasted annually worldwide due to the processing methods, inadequate storage handling, harvesting, climate influence and consumer behavior [6]. The largest share of approximately 50% refers to fruit and vegetable processing waste. This waste is usually disposed without being previously processed, which results in the environmental pollution. Nevertheless, numerous research topics covered the analyses of these wastes and the results indicate they are a good source of fiber, minerals, vitamins and bioactive compounds [6].

The present contribution focuses on lack of sustainability development in the large-scale process, such as industrial baker's yeast production and depicts recent advances achieved in sustainable valorization of waste originating from agricultural activities such as apple juice and wine production. This review is aiming to draft a circular economy approach and providing a way to achieve sustainable connections between the different industries.

2. Recent Advances in Sustainable Valorization of Apple and Grape Pomace as Problematic Agro-Industrial Wastes

In the last decade, efforts have been made trying to valorize agro-industrial wastes. This review focuses on grape pomace (by-product in winemaking), apple pomace (by-product in the production of apple juice) and the research made in the frame of their sustainable development.

Grape (*Vitis vinifera*) is one of the world's most widely cultivated fruit [7]. In the year 2021, over 70 million metric tons (MMT) were produced [8]. Approximately 50% of grapes are directed into winemaking, while the remaining 50% are being used as fresh and dried grapes [7]. During grape processing, in order to produce wine, large amounts of resources are utilized; not only water but also organic and inorganic fertilizers. In order to produce 1 L of wine, 1.3–1.5 kg of waste is generated (largely wastewater). Still, winemaking is seen as an eco-friendly process [9]. Grape pomace (GP) is the primary by-product in winemaking, approximately 20–25% of the total processed grape weight [10], which would amount to approximately 7 million tons. Its proximate chemical composition is presented in Table 1. The discharge of GP leads to environmental issues such as ground and surface water pollution, proliferation of organisms which can spread diseases, the depletion of oxygen in soil and groundwater, which may have an impact on local wildlife [11]. For a long time, GP was considered as waste/by-product. Traditionally, it has been utilized in the production of various distillates, used as fertilizer, as supplement in the animal nutrition or compost [7]. However, large quantities of GP, as a fertilizer with high organic content, disposed in landfills can exhibit harmful effects on biodegradation due to low pH and the presence of antibacterial substances (polyphenols). Although GP is rich in proteins it is reported that most of the animals are not able to successfully digest it and satisfy its dietary needs. Utilization of GP as a composting material showed non profitable and economic usage due to an absence of essential nutrients [11]. Nevertheless, the wine industry is definitely engaged in its responsibility to the environment, which is supported by increasing research found in the scientific publications [9,12–20]. Different utilizations and purposes of GP have been proposed in recent decades and can be observed in Table 2.

Table 1. Proximate chemical composition of grape pomace, apple pomace, cane and beet molasses based on dry weight (dry mass (DM), total dietary fiber (TDF), total polyphenolic content (TPC)); * adapted from [21], ** adapted from [22], *** adapted from [23].

Component	Grape Pomace (g/100 g DM) [11]	Apple Pomace (g/100 g DM) *	Cane Molasses (g/100 g DM) **	Beet Molasses (g/100 g DM) **	Vinasse (g/100 g DM) ***
Ash	1.73–9.10	0.54–6.58	13.14–20.99	8.79–25.02	27.02
Protein	3.57–14.17	3.21–6.12	2.86–11.99	14.47–21.10	21.71
Fat	1.14–13.90	1.30–4.21	–	–	–
TDF	17.28–88.70	5.07–55.15	–	–	–
TPC	0.28–8.70	–	–	–	–
Sucrose	–	4.10–6.26	50.48–86.67	62.88–89.38	–
Fructose	0.38–8.91	21.48–43.50	2.96–18.39	0.01–1.18	–
Glucose	0.21–26.34	21.05–21.26	1.67–15.54	0.03–2.65	–
Pectin	–	3.78–16.48	–	–	–

Apples (*Malus domestica* Borkh.) are the 4th most grown fruit worldwide (after oranges, bananas and grapes) [24]. In the year 2021, over 90 million tons were produced [8]. Fresh consumption accounts for 70–75% of the apples. From 5 to 30% of the world's total production is directed into production of various value-added products such as juice, wine, jams and dried product. However, the most necessitated apple product remains apple juice, which requires approximately 65% of the total amount of processed apple. In the production of apple juice, approximately 75% of apple fresh weight is extracted as juice, while approximately 25% is considered as food waste, i.e., apple pomace (AP) [24,25]. It is estimated that approximately 4 MMT/year of AP is generated. Although AP has been extensively researched for various purposes (Table 2), it is still an undervalued by-product used as animal feed or field fertilizer. However, AP's protein content is extremely low (Table 1) and therefore represents a poor nutritional supplement for animals [26]. AP is subjected to a fast fermentation by microbes, due to the AP's high content of water (>70%), sugars and organic acids. Therefore, direct disposal to the soil as field fertilizer with high organic content may cause the contamination of soil and water [27].

The direct disposal of apple and grape pomace as food waste represents a substantial loss of valuable biomass, which could be further processed and bio-converted into different value-added products [26]. However, the production of only one product from processing the large amount of these wastes is not economically profitable. Therefore, the efforts need to be made in exploring the additional applications as well as alternative products [28].

3. The Lack of Sustainability in Large-Scale Bioprocesses: Baker's Yeast Production as a Representative Example

This paragraph briefly explains some of the most important large-scale bioprocesses and their importance in the market. These are the production of bioethanol, citric acid, amino acids and baker's yeast, which require large amounts of resources. The microbial growth involved in these highly efficient processes require considerable quantities of organic carbon, which is usually supplied as sucrose, maltodextrin or glucose. The production of ethanol as a biofuel is associated with the substantial utilization of natural resources, including water, soil erosion and the required arable land for sugar beet and sugarcane cultivation (Figure 1a) [29]. Nowadays, bioethanol is mostly produced from grain starch or sugar crops (sugarcane, sweet sorghum) in the USA, China and Brazil. As an example, bioethanol in Brazil is mainly produced with sugarcane. The main production technology is very mature and includes pretreatment (cleaning and cutting) of sugarcane, extraction and concentration of sugarcane juice, ethanol fermentation, ethanol distillation and dehydration. Based on the environmental life cycle impact and energy balance assessment, from 1 t of sugarcane, 86.5 L of fuel ethanol is produced, 84 MJ of electricity, 150 kg of sucrose and other valuable by-products, which in return has capacity advantage and economic benefits [30]. With over 30 billion L produced in 2019, Brazilian bioethanol stands out as one of the most prominent biofuels produced on a large scale [31].

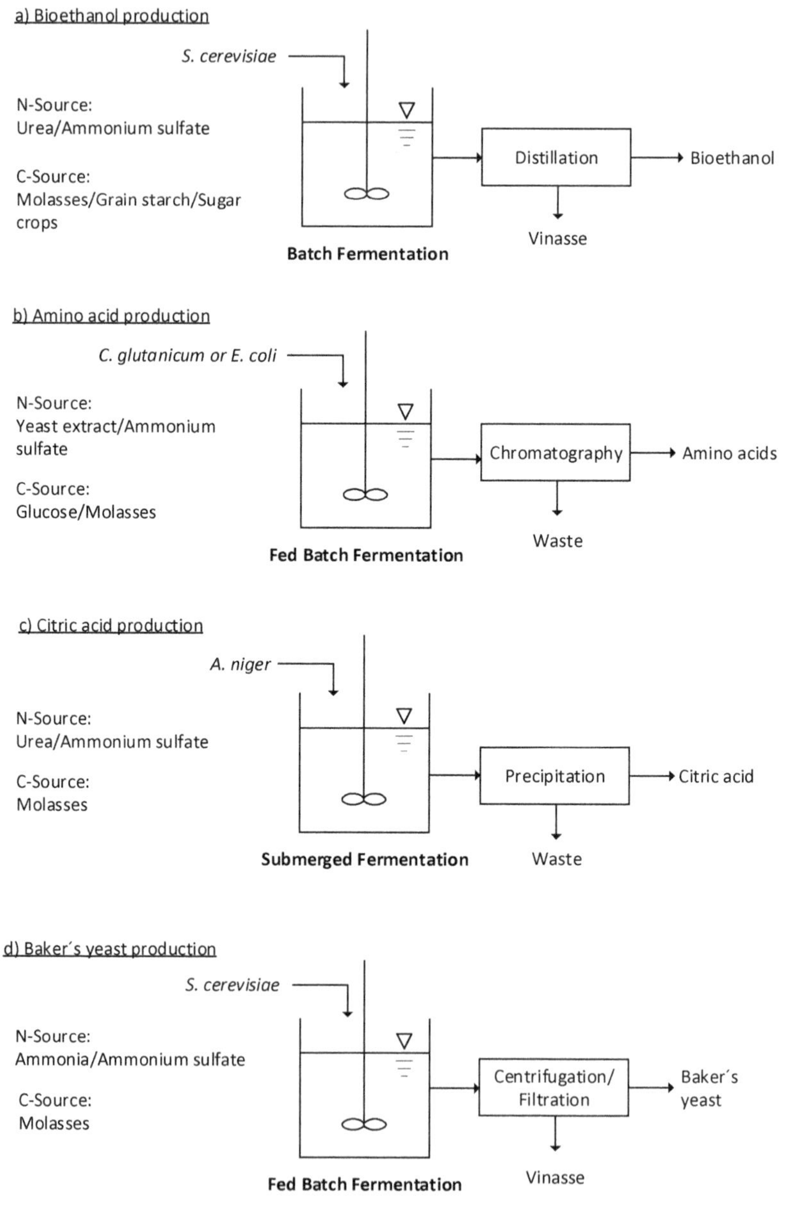

Figure 1. Flow diagrams for the production of bioethanol, amino acids, citric acid, and baker's yeast.

One of the most important organic acids widely used in pharmaceutical food, beverage, detergents, and cosmetics is citric acid. Industrially, it is produced by filamentous fungus *Aspergillus niger* on a medium containing molasses without any additional ammonium salts (Figure 1c) [32]. To reduce the cost of production, efforts have been made in utilizing the agricultural waste and by-products in the production of citric acid. This includes the following: coffee husk, rice bran, wheat bran, carrot waste, cassava bagasse, banana peel, vegetable wastes, sugarcane bagasse, tapioca, whey, rice straw, coconut husk, brewery wastes, rotting fruits, corn cob, orange peel, kiwifruit peel, pineapple peel, grape and apple pomace. The production of citric acid by fermentation is steadily growing with an annual rate of 5%. It is estimated that the global production is approximately 736,000 tones/year, with the market reaching over 3 billion USD by 2023 [32].

One of the major fields of industrial biotechnology and large-scale processes is industrial amino acid production (Figure 1b). They are widely used in seasoning and other food use, but also as ingredients in animal nutrients, pharmaceuticals, and cosmetics. The first amino acid to be commercialized was monosodium glutamate. A major revolution was the discovery of glutamate fermentation by *Corynebacterium glutamicu*, proving it is possible to produce amino acids in large amounts at low prices. At the beginning, glucose, fructose, and sucrose are used as a carbon source. However, after the discovery of an alternative method of biotin limitation, a more economically desirable source of molasses was used. Nowadays, the amino acid industrial fermentation sector is huge, so more than 5 million metric tons/year are produced worldwide [33]. Global demand is expected to grow with an annual growth rate of 5.6% in the period of 2017–2022 and will reach USD 25.6 billion by 2022 [34].

Table 2. Different purposes for grape pomace and apple pomace applications.

Waste	Application	Literature
Grape pomace	As source of bioactive compounds (phenolics) used in pharmaceutical, cosmetic and food industries	[7,35–39]
	As source of dietary fiber (as dietary supplement, dough improver, alternative source of antioxidants and dietary fiber for yogurt, for the fortification of meat and fish products)	[7,11,36,40–42]
	In the production of bioethanol	[39,43–45]
	In the production of baker's yeast	[10,46]
	In the production of lactic acid	[47,48]
Apple pomace	Green extraction of bioactive compounds:	[49–53]
	- extraction of pectin	[6,25–28,49,50,54–60]
	- extraction of polyphenols	[6,49,58]
	As solid biofuel	[25,27,28]
	As source of dietary fiber (used in baked food products)	[61–63]
	As functional ingredient in food products:	[6]
	- bakery products	[6,26,58,60,64,65]
	- meat products	[6,50,54,64,66–68]
	- dairy products	[6,24,61,69]
	In the production of citric acid	[6,26,70–74]
	In the production of baker's yeast	[25,75,76]

Baker's yeast production is an industrial process (Figure 1d), which has been both theoretically and technically very well described in the literature [77–80]. Industrial baker's yeast production involves multiple stages with variations in generation numbers, aeration levels and bioreactor types. As main carbon source, molasses is used and it is usually diluted, clarified, and sterilized before micro and macroelements, along with vitamins, are added. The initial cultivation of the pure culture starts in small flasks (less than 5 L) and is followed by two larger-scale stages (1140–26,500 L), lasting 13 to 24 h each. The main biomass production occurs in the so-called final trade cultivation, with large vessels (37,900 L to over 283,900 L) being carefully aerated. To prevent the Crabtree effect, molasses is added in a controlled manner to maintain low sugar concentrations (fed-batch). Molasses is usually supplemented with ammonium salts, phosphoric acid, and magnesium salts. The vitamin solution is dosed into the bioreactor. The final trade cultivation lasts 11 to 15 h, yielding 15,000 to 100,000 kg of compressed yeast per batch. Afterward, separation occurs in centrifugal separators and rotary vacuum filters in order to concentrate the yeast. The resulting cake is mixed with water, oil, and emulsifiers, extruded into blocks, wrapped, and cooled to below 8 °C [81,82]. Baker's yeast sector represents a huge market with annual production of 2.3 MMT worldwide [83]. Considering the large amount of molasses and other resources, such as water and electric energy, that needs to be directed into this production [23] and the waste it is generating (the production of 1 metric ton of baker's yeast generates approximately 60–130 metric tons of wastewater [84]), the producers need to start thinking about integrating the economical circularity. Although the process has been dramatically optimized over recent decades, mostly in the direction of producing high amounts of yeast biomass as fast as possible and at the lowest cost [77,81], more efforts are needed in developing new strategies in order to contribute to the goals of EGD. Nevertheless, research has been conducted regarding the sustainability improvement in the baker's yeast production and is focused on exploring alternative raw materials [10,25,46,75,76,85–87] and wastewater treatment [23,84,88,89]. However, very little of these outputs was implemented in the real industrial surroundings.

After mentioning some of the most important large-scale processes and the size of their markets, efforts need to be made in exploring the sustainability aspect of these productions. In this regard, the three primary products mentioned in Figures 1 and 2 (apple, grape and sugar beet) are taken as a starting point for the following discussion about the lack of sustainability and the links with the large-scale bioprocesses. Sugar beet is the main raw material involved in the European sugar production. The associated transformation is well optimized leading to the production of sugar (and specialty sugars), molasses and sugar beet pomace. The latter is further transformed into pellets for animal feeding. Because of its high sugar, protein and amino acid contents (Table 1), molasses is the most widely used raw material for industrial bioproductions. It was considered as a waste product of the sugar industry for a long time. Nowadays, it is termed as a by-product due to its low price compared to the other sugar sources and the presence of minerals, organic and inorganic compounds. Molasses is used in the production of alcohol, organic acid and single cell proteins [90]. From 2016 to 2018, sugar production worldwide increased steadily. Brazil, India and Thailand were the most important producers of sugarcane (33, 28 and 12 million tons of sugarcane were produced, respectively). The European Union is the world's largest producer of sugar beets (17 million tons). Since 2006, sugar beet production in Italy has dropped drastically, from 8.6% to 3.9% of the European Union (EU) production due to the EU reform, which liberalized the sugar market in all of Europe [91]. The volatile molasses market [81], descending trend of sugar market prices as well as increasing energy prices are an immediate threat to the survival of some sugar industry [92]. All large-scale productions associated with molasses are currently facing problems in terms of raw material supply and, consequently, the demand for alternative feedstocks is increasing.

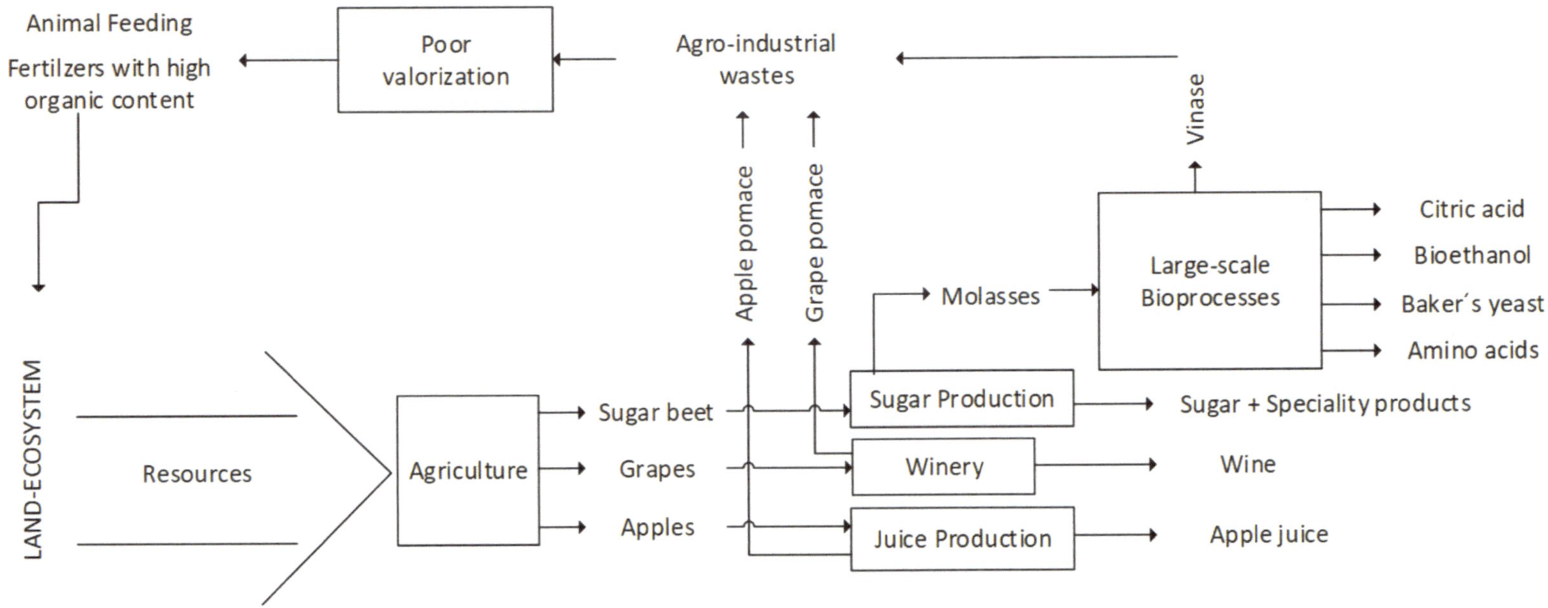

Figure 2. The implementation gap between agro-industrial wastes (fruit pomace/sugar beet vinasse) and large-scale bioprocesses. Apple juice production and wine making still work in a linear fashion while industrial bio-productions often lack adequate sustainability.

The production of apple juice and winemaking are still considered as linear economy systems, although a lot of research has been carried out to enhance the valorization of the corresponding by-product/waste (Table 2). Apple and grape pomaces are both highly available streams exhibiting high sugar contents (Table 1). They undergo rapid fermentation and cannot be stored for long. It is preferable to stabilize and process these fractions as soon as possible. As presented in Table 1, the compositions of apple and grape pomaces are comparable with beet and cane molasses making those very suitable raw materials for the preparation of alternative feedstocks. The efficient aqueous extraction of sugar from fruit pomace is not very challenging but effort is still needed for the proper assessment the resulting sugar extracts in a real industrial surrounding and by considering the other key aspects like logistics, quality standards, available infrastructures and resources. The large-scale bioprocesses mentioned in Figure 1 are highly optimized high-yield processes. For example, state-of-the-art industrial baker's yeast production requires the precise dosage of sugar during the fed-batch fermentation to avoid the Crabtree effect as well as the control of the fermentative capacity and the protein/dry matter content of the produced yeast. The challenges rely on the sustainable and cost-efficient production of sugar extract that can be easily transported and has a long storage capacity. The final step is to define to which extent the generated sugar extract can be used to substitute molasses without negatively affecting the production costs and the product quality. Figure 2 makes this implementation gap between large-scale bioprocesses and agro-industrial wastes (fruit pomace/sugar beet vinasse) visible. This in turn requires a mature technology, considerable changes in the infrastructure involved in bioprocesses as well, mentality, company's philosophy and marketing strategies [81]. The collaboration between academic teams and industrial partners plays an important role for the transition from research to industrial innovation and the efficient coordination between actors from different sectors with different backgrounds and interests. We believe that interdisciplinary academic actors are in charge of this task and their effort should focus on leading transdisciplinary research activities that embrace differences and generate global solution considering environmental, economic and ethical aspects.

At this stage, it seems necessary to discuss the lack of sustainability of both, primary and secondary production as the main problem in the framework of the implementation of the EGD. Due to the release of large amounts of vinasse, bioethanol and baker's yeast productions are currently quite problematic. The utilization of sugar beet vinasse requires the addition of co-substrates for the efficient production biogas production by means of anaerobic digestion [93]. It is common practice to concentrate vinasse, as a source of potassium and nitrogen, to generate a fertilizer of appreciable quality but requiring careful and precise dosage to avoid soil contamination [94]. This is usually achieved in energy demanding processes involving evaporation and centrifugation. The reuse of such condensates within the production process is limited and the produced wastewater is usually directed to a dedicated treatment station and finally released into the ecosystem.

4. Centralized Valorization Platforms as Well as a Close Collaboration between Academia and Industry Are Crucial to Bridge the Gap between Agro-Industrial Wastes and Large-Scale Bio-Industries

We now turn our attention to two distinguished recent works exploring the valorization of apple and grape pomace to generate alternative substrates for baker's yeast production [10,25]. In the mentioned research, apple pomace (obtained after extraction of apple juice) and grape pomace (by-product in wine production) were further processed in order to gain valuable by-products such as solid biofuels and sugar extracts as feedstocks for large-scale bioprocesses. The corresponding characteristics of the solid biofuels (heat release and mechanical stability) were very comparable to values reported for conventional wood pellets. Especially, the grape pomace fractionation process is characterized by a strong energy output that allow for covering the energy demand of the process and create an additional renewable energy surplus of 3 MJ/kg of processed grapes (dry mass). The efficiency of the mechanical dewatering of the treated pomace plays a key role in this technology by reducing the energy demand for drying.

Following a circular economy approach, this aspect makes this valorization process particularly interesting because it may supply other processes or citizens with heat. The feedstocks gained from these processes are the most interesting product, which were subsequently assessed for the production of baker's yeast under real industrial conditions. For this purpose, molasses was partly substituted with these extracts and the baker's yeast cultivations were performed in pilot bioreactor. The production procedure was optimized to reproduce the large-scale production conditions occurring in bubble columns. The results of the produced baker's yeast analyses indicate that the high quality yeast was produced by mixing molasses and sugar extract originating from pomace processing. It is worth mentioning that the particularity of these works relies in the close collaboration between academic teams and an industrial partner, who has a real technical, scientific and economic interest in the outcomes of the engaged research activities.

As a major outcome, this review combines these recent advances and proposes a way to transform largescale bioprocess by integrating them into a circular economy approach and gaining valuable feedstock from the processing of fruit pomace to achieve zero liquid discharge. In other works, the applicability of comparable sugar extracts originating from grape pomace transformation for the production of bioethanol was also reported in the literature [95]. Such feedstocks might also be appropriate as potential substrates for other large-scale bio-productions such as citric acid or amino acids but this has to be assessed and confirmed since these processes involve different microorganisms and fermentation conditions. Figure 3 shows a model to connect three fundamental sectors of the European economy namely, land/agriculture, post-harvest transformation and industrial biotechnology. During apple juice and winemaking transformations, the waste is converted into a valuable feedstock for the baker's yeast production or other large-scale bioprocesses. Renewable energy (solid biofuel) can be used in the energy balance of these transformations and heat surplus can used by other industries or citizens [10,25]. We believe that centralized valorization platforms are key enablers to bridge such implementation gaps by overwhelming the valorization chain and ensuring the efficient zero liquid discharge production of high value product such as feedstock for large-scale bioprocesses, renewable energy (solid biofuel) and pectin from apple pomace. Pectin is a polysaccharide originating from the cell wall of plants with excellent gelling properties. It is widely used in the food industry and in the pharmaceutical industry as well.

On the one hand, the organic content of fruit pomace (e.g., sugar, fibers, and polymers) constitutes the high valorization potential of this agro-industrial waste; on the other hand, it is exactly the biodegradation of this fraction that makes fruit pomace becoming a problematic fertilizer leading to soil contamination. In this regard, the main role of the centralized valorization platforms relies in the extraction, fractionation and conditioning of the organic fraction from these wastes to generate high value products and intermediates. Such a platform could simply consist of unit operations (solid–liquid extraction, membrane-based separation processes, ion-exchangers, mechanical dewatering, drying/evaporation) that could be combined in a customized way depending on the type of agro-industrial feed.

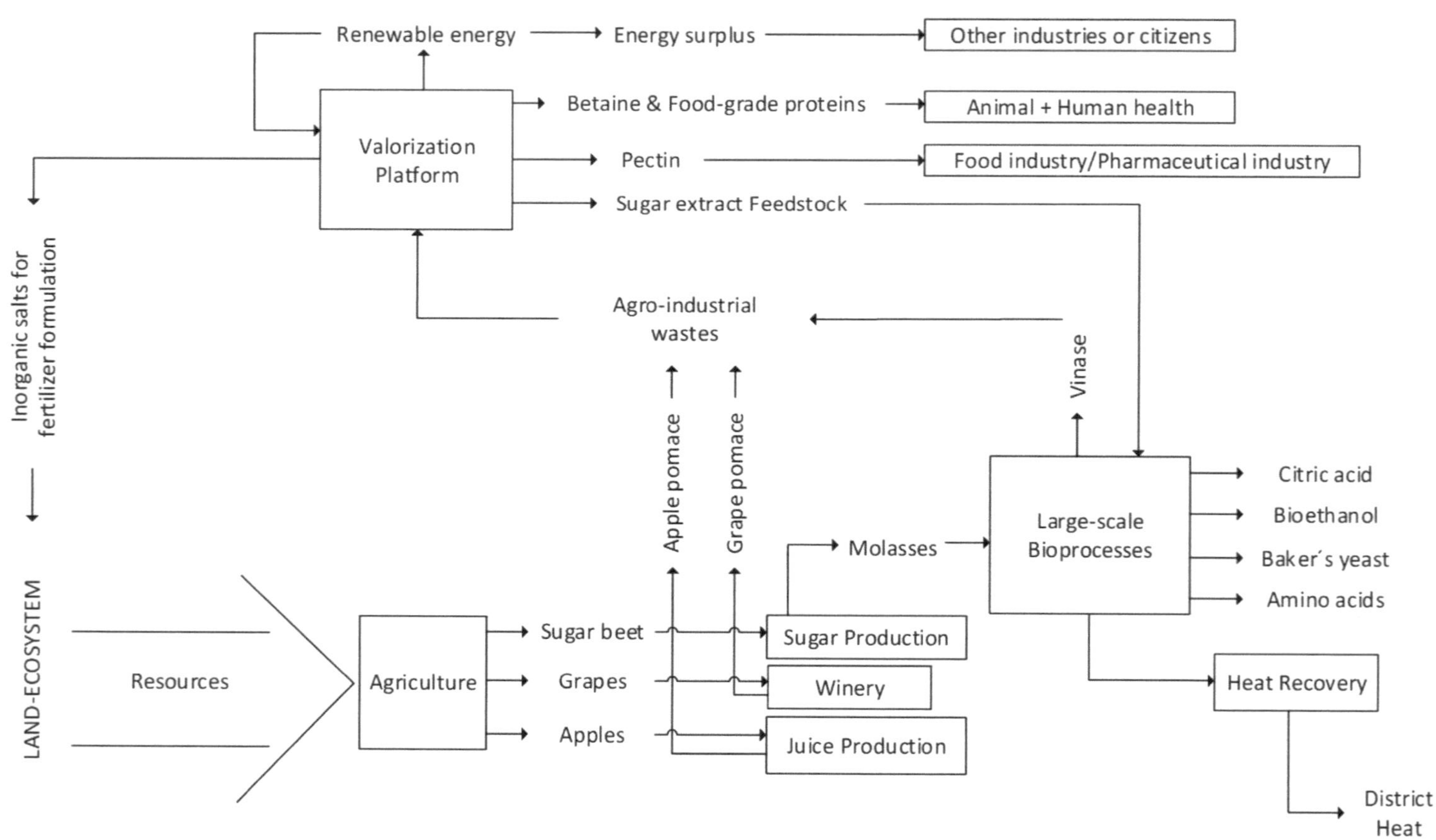

Figure 3. A circular flow diagram to bridge the implementation gap between agro-industrial wastes and large-scale bioprocesses. It connects agriculture and post-harvest transformation with the industrial biotechnology and other industries through the substantial valorization of vinasse, apple pomace and grape pomace into renewable energy (solid biofuels), betaine, food-grade proteins and sugar extracts as feedstock for large-scale bioproductions.

Aiming to integrate secondary product vinasse as an important member of the circular link with the valorization of grape and apple pomaces, we recently published interesting advances that demonstrate the enhanced valorization of sugar beet vinasse produced during baker's yeast production [23,96,97]. As presented in Table 1, the chemical composition of vinasse is quite simple when compared to molasses. It mainly contains minerals especially potassium and organic nitrogen (amino acids, peptides and proteins). In contrast to sugar cane vinasse, sugar beet vinasse contains an appreciable amount of betaine (approx. 22.4% w/w based on dry mass) [production [23]]. Natural betaine is a key osmolyte with applications in animal and human health [23]. In these works, we approached the valorization of sugar beet vinasse by separating the organic fraction from the inorganic fraction. Food grade proteins and betaine are recovered from the organic fraction and salts especially potassium are recovered from the inorganic fraction that can be used as formulation ingredient for fertilizers. Updating our circular workflow by implementing these advances within the above mentioned valorization platforms (Figure 3) allows for converting vinasse from a problematic waste to valuable products. Energy-efficient water reuse can be dramatically improved by substituting evaporation by state-of-the-art membrane processes such as ultrafiltration combined with electrodialysis and hybrid reverse osmosis leading to the energy-efficient dewatering of the different fractions and the generation of clean water, which can be further distributed to surrounding citizens [98–100]. Finally, the installation of state-of-the-art heat pumps allows for supplying residential neighborhood with 4th-generation district heat through the processing of the low-grade heat generated during aerobic fermentation operations [96].

5. Conclusions

In this work, we proposed a circular flow diagram that connects agriculture and post-harvest transformation with the industrial biotechnology and other industries through the substantial valorization of sugar beet vinasse as well as apple and grape pomaces into natural betaine, food-grade proteins, minerals, renewable energy (solid biofuels) and sugar extracts as feedstock for baker's yeast and bioethanol production. These feedstocks might also be appropriate for other large-scale bioprocesses such as citric acid or amino acid production but this has to be further assessed and confirmed. In this context, the transition from a linear to circular supply chain is feasible but requires:

- Openness and mentality changes, especially in companies involved in the bioproduction of one main product. New expertise and financial risks are often associated with the extension of a product portfolio. This move is often uncomfortable for small- and middle-sized companies, which then become new players in existing or novel markets. Centralized valorization platforms can play a crucial role in overwhelming the valorization chain and accelerating the transition towards a circular economy.
- Adequate policies to ensure the correct balance between the pressure to enforce, engage and consolidate the transition and the support to encourage companies to take risks and develop new market opportunities.
- Academic research plays a vital role for the establishment and the acceleration of transition towards a sustainable circular economy. However, the conversion of research outputs into innovation necessitates a close and trustful collaboration between academic and industrial partners. Prior to the project start, academic researchers should have a deep understanding of the workflow of the companies involved in the research with a special focus on the current constraints associated with the market dynamics, and the product quality standards.

Author Contributions: All authors contributed to this review paper. Conceptualization, S.B.; researching the literature, J.L.V. and S.B.; writing this paper, J.L.V., T.M. and S.B.; reviewing and editing this paper S.B.; funding acquisition, S.B. All authors have read and agreed to the published version of the manuscript.

Funding: This research was funded by the European Union's Horizon 2020 research and innovation program under grant agreement No. 958454—"intelWATT" Project.

Conflicts of Interest: The authors declare no conflict of interest. The funders had no role in the design of the study; in the collection, analyses, or interpretation of data; in the writing of the manuscript; or in the decision to publish the results.

References

1. Schunz, S. The 'European Green Deal'—A Paradigm Shift? Transformations in the European Union's Sustainability Meta-Discourse. *Polit. Res. Exch.* **2022**, *4*, 2085121. [CrossRef]
2. Kougias, I.; Taylor, N.; Kakoulaki, G.; Jäger-Waldau, A. The Role of Photovoltaics for the European Green Deal and the Recovery Plan. *Renew. Sustain. Energy Rev.* **2021**, *144*, 111017. [CrossRef]
3. Gheuens, J. The European Green Deal. *Mak. Eur. Green. Deal. Work.* **2023**, 15–28. [CrossRef]
4. European Commission. The European Green Deal. Available online: https://commission.europa.eu/strategy-and-policy/priorities-2019-2024/european-green-deal_en (accessed on 29 November 2023).
5. European Commission. Farm to Fork Strategy. Available online: https://food.ec.europa.eu/horizontal-topics/farm-fork-strategy_en (accessed on 29 November 2023).
6. Asif, M.; Javaid, T.; Razzaq, Z.U.; Khan, M.K.I.; Maan, A.A.; Yousaf, S.; Usman, A.; Shahid, S. Sustainable Utilization of Apple Pomace and Its Emerging Potential for Development of Functional Foods. *Environ. Sci. Pollut. Res.* **2023**, 1–19. [CrossRef] [PubMed]
7. Bordiga, M.; Travaglia, F.; Locatelli, M. Valorisation of Grape Pomace: An Approach That Is Increasingly Reaching Its Maturity—A Review. *Int. J. Food Sci. Technol.* **2019**, *54*, 933–942. [CrossRef]
8. FAO Global Fruit Production in 2019, by Selected Variety (in Million Metric Tons). Available online: https://www.statista.com/statistics/262266/global-production-of-fresh-fruit/ (accessed on 29 November 2023).
9. Maicas, S.; Mateo, J.J. Sustainability of Wine Production. *Sustainability* **2020**, *12*, 559. [CrossRef]
10. Lisičar Vukušić, J.; Millenautzki, T.; Reichert, L.; Saaid, A.M.; Müller, L.; Clavijo, L.; Hof, J.; Mösche, M.; Barbe, S. Conversion of Problematic Winery Waste into Valuable Substrate for Baker's Yeast Production and Solid Biofuel: A Circular Economy Approach. *Food Technol. Biotechnol.* **2023**, *61*, 1–19. [CrossRef]
11. Antonić, B.; Jančíková, S.; Dordević, D.; Tremlová, B. Grape Pomace Valorization: A Systematic Review and Meta-Analysis. *Foods* **2020**, *9*, 1627. [CrossRef]
12. Gómez-Brandón, M.; Lores, M.; Insam, H.; Domínguez, J. Strategies for Recycling and Valorization of Grape Marc. *Crit. Rev. Biotechnol.* **2019**, *39*, 437–450. [CrossRef]
13. Soceanu, A.; Dobrinas, S.; Sirbu, A.; Manea, N.; Popescu, V. Economic Aspects of Waste Recovery in the Wine Industry. A Multidisciplinary Approach. *Sci. Total Environ.* **2021**, *759*, 143543. [CrossRef]
14. Ardente, F.; Beccali, G.; Cellura, M.; Marvuglia, A. POEMS: A Case Study of an Italian Wine-Producing Firm. *Environ. Manag.* **2006**, *38*, 350–364. [CrossRef] [PubMed]
15. Kariyapperuma, N.; Collins, E. Family Logics and Environmental Sustainability: A Study of the New Zealand Wine Industry. *Bus. Strategy Environ.* **2021**, *30*, 3626–3650. [CrossRef]
16. Gilinsky, A.; Newton, S.K.; Vega, R.F. Sustainability in the Global Wine Industry: Concepts and Cases. *Agric. Agric. Sci. Procedia* **2016**, *8*, 37–49. [CrossRef]
17. Baiano, A. An Overview on Sustainability in the Wine Production Chain. *Beverages* **2021**, *7*, 15. [CrossRef]
18. Golicic, S.L.; Flint, D.J.; Signori, P. Building Business Sustainability through Resilience in the Wine Industry. *Int. J. Wine Bus. Res.* **2017**, *29*, 74–97. [CrossRef]
19. Flores, S.S. What Is Sustainability in the Wine World? A Cross-Country Analysis of Wine Sustainability Frameworks. *J. Clean. Prod.* **2018**, *172*, 2301–2312. [CrossRef]
20. Bandinelli, R.; Acuti, D.; Fani, V.; Bindi, B.; Aiello, G. Environmental Practices in the Wine Industry: An Overview of the Italian Market. *Br. Food J.* **2020**, *122*, 1625–1646. [CrossRef]
21. Bhushan, S.; Kalia, K.; Sharma, M.; Singh, B.; Ahuja, P.S. Processing of Apple Pomace for Bioactive Molecules. *Crit. Rev. Biotechnol.* **2008**, *28*, 285–296. [CrossRef]
22. Palmonari, A.; Cavallini, D.; Sniffen, C.J.; Fernandes, L.; Holder, P.; Fagioli, L.; Fusaro, I.; Biagi, G.; Formigoni, A.; Mammi, L. Short Communication: Characterization of Molasses Chemical Composition. *J. Dairy Sci.* **2020**, *103*, 6244–6249. [CrossRef]
23. Lisičar Vukušić, J.; Millenautzki, T.; Sedaghati, M.; Schallenberg, M.; Müller, P.; Hof, J.; Mösche, M.; Barbe, S. Fractionation of Baker's Yeast Vinasse via Ultrafiltration: Assessment of Feasibility. *Int. J. Food Sci. Technol.* **2019**, *54*, 1794–1803. [CrossRef]
24. Lyu, F.; Luiz, S.F.; Azeredo, D.R.P.; Cruz, A.G.; Ajlouni, S.; Ranadheera, C.S. Apple Pomace as a Functional and Healthy Ingredient in Food Products: A Review. *Processes* **2020**, *8*, 319. [CrossRef]
25. Lisičar Vukušić, J.; Millenautzki, T.; Cieplik, R.; Obst, V.; Saaid, A.M.; Clavijo, L.; Zlatanovic, S.; Hof, J.; Mösche, M.; Barbe, S. Reshaping Apple Juice Production into a Zero Discharge Biorefinery Process. *Waste Biomass Valorization* **2021**, *12*, 3617–3627. [CrossRef]

26. Gorawala, P.; Mandhatri, S. Apple Pomace: Source of Value Added Products. In *Agricultural Research Updates*; Prathames, G., Mandhatri, S., Eds.; Nova Science Publishers, Inc.: Hauppauge, NY, USA, 2014; Volume 8, pp. 167–203, ISBN 978-1-63321-485-9.
27. Gołębiewska, E.; Kalinowska, M.; Yildiz, G. Sustainable Use of Apple Pomace (AP) in Different Industrial Sectors. *Materials* **2022**, *15*, 1788. [CrossRef] [PubMed]
28. Shalini, R.; Gupta, D.K. Utilization of Pomace from Apple Processing Industries: A Review. *J. Food Sci. Technol.* **2010**, *47*, 365–371. [CrossRef] [PubMed]
29. Pereira, C.L.F.; Ortega, E. Sustainability Assessment of Large-Scale Ethanol Production from Sugarcane. *J. Clean. Prod.* **2010**, *18*, 77–82. [CrossRef]
30. Wu, B.; Wang, Y.-W.; Dai, Y.-H.; Song, C.; Zhu, Q.-L.; Qin, H.; Tan, F.-R.; Chen, H.-C.; Dai, L.-C.; Hu, G.-Q.; et al. Current Status and Future Prospective of Bio-Ethanol Industry in China. *Renew. Sustain. Energy Rev.* **2021**, *145*, 111079. [CrossRef]
31. Karp, S.G.; Medina, J.D.C.; Letti, L.A.J.; Woiciechowski, A.L.; De Carvalho, J.C.; Schmitt, C.C.; De Oliveira Penha, R.; Kumlehn, G.S.; Soccol, C.R. Bioeconomy and Biofuels: The Case of Sugarcane Ethanol in Brazil. *Biofuels Bioprod. Biorefining* **2021**, *15*, 899–912. [CrossRef]
32. Behera, B.C.; Mishra, R.; Mohapatra, S. Microbial Citric Acid: Production, Properties, Application, and Future Perspectives. *Food Front.* **2021**, *2*, 62–76. [CrossRef]
33. Yokota, A.; Ikeda, M. (Eds.) *Amino Acid Fermentation*; Advances in Biochemical Engineering/Biotechnology; Springer: Tokyo, Japan, 2017; Volume 159, ISBN 978-4-431-56518-5.
34. Wendisch, V.F. Metabolic Engineering Advances and Prospects for Amino Acid Production. *Metab. Eng.* **2020**, *58*, 17–34. [CrossRef]
35. Ferreira, S.M.; Santos, L. A Potential Valorization Strategy of Wine Industry By-Products and Their Application in Cosmetics—Case Study: Grape Pomace and Grapeseed. *Molecules* **2022**, *27*, 969. [CrossRef]
36. Fontana, A.R.; Antoniolli, A.; Bottini, R. Grape Pomace as a Sustainable Source of Bioactive Compounds: Extraction, Characterization, and Biotechnological Applications of Phenolics. *J. Agric. Food Chem.* **2013**, *61*, 8987–9003. [CrossRef] [PubMed]
37. Ferri, M.; Vannini, M.; Ehrnell, M.; Eliasson, L.; Xanthakis, E.; Monari, S.; Sisti, L.; Marchese, P.; Celli, A.; Tassoni, A. From Winery Waste to Bioactive Compounds and New Polymeric Biocomposites: A Contribution to the Circular Economy Concept. *J. Adv. Res.* **2020**, *24*, 1–11. [CrossRef] [PubMed]
38. Ioannou, L.A.; Puma, G.L.; Fatta-Kassinos, D. Treatment of Winery Wastewater by Physicochemical, Biological and Advanced Processes: A Review. *J. Hazard. Mater.* **2015**, *286*, 343–368. [CrossRef] [PubMed]
39. Corbin, K.R.; Hsieh, Y.S.Y.; Betts, N.S.; Byrt, C.S.; Henderson, M.; Stork, J.; DeBolt, S.; Fincher, G.B.; Burton, R.A. Grape Marc as a Source of Carbohydrates for Bioethanol: Chemical Composition, Pre-Treatment and Saccharification. *Bioresour. Technol.* **2015**, *193*, 76–83. [CrossRef]
40. Valiente, C.; Arrigoni, E.; Esteban, R.M.; Amado, R. Grape Pomace as a Potential Food Fiber. *J. Food Sci.* **1995**, *60*, 818–820. [CrossRef]
41. Iuga, M.; Mironeasa, S. Potential of Grape Byproducts as Functional Ingredients in Baked Goods and Pasta. *Compr. Rev. Food Sci. Food Saf.* **2020**, *19*, 2473–2505. [CrossRef]
42. García-Lomillo, J.; González-SanJosé, M.L. Applications of Wine Pomace in the Food Industry: Approaches and Functions. *Compr. Rev. Food Sci. Food Saf.* **2017**, *16*, 3–22. [CrossRef]
43. Energy, C.; Technologies, P. (Eds.) *Food Waste to Green Fuel: Trend & Development*; Springer: Berlin/Heidelberg, Germany, 2022; ISBN 978-981-19081-2-5.
44. Rodríguez, L.A.; Toro, M.E.; Vazquez, F.; Correa-Daneri, M.L.; Gouiric, S.C.; Vallejo, M.D. Bioethanol Production from Grape and Sugar Beet Pomaces by Solid-State Fermentation. *Int. J. Hydrogen Energy* **2010**, *35*, 5914–5917. [CrossRef]
45. Sirohi, R.; Tarafdar, A.; Singh, S.; Negi, T.; Gaur, V.K.; Gnansounou, E.; Bharathiraja, B. Green Processing and Biotechnological Potential of Grape Pomace: Current Trends and Opportunities for Sustainable Biorefinery. *Bioresour. Technol.* **2020**, *314*, 123771. [CrossRef]
46. Lo Curto, R.B.; Tripodo, M.M. Yeast Production from Virgin Grape Marc. *Bioresour. Technol.* **2001**, *78*, 5–9. [CrossRef]
47. Rivera, O.M.P.; Moldes, A.B.; Torrado, A.M.; Domínguez, J.M. Lactic Acid and Biosurfactants Production from Hydrolyzed Distilled Grape Marc. *Process Biochem.* **2007**, *42*, 1010–1020. [CrossRef]
48. Campanella, D.; Rizzello, C.G.; Fasciano, C.; Gambacorta, G.; Pinto, D.; Marzani, B.; Scarano, N.; De Angelis, M.; Gobbetti, M. Exploitation of Grape Marc as Functional Substrate for Lactic Acid Bacteria and Bifidobacteria Growth and Enhanced Antioxidant Activity. *Food Microbiol.* **2017**, *65*, 25–35. [CrossRef] [PubMed]
49. Putra, N.R.; Rizkiyah, D.N.; Abdul Aziz, A.H.; Che Yunus, M.A.; Veza, I.; Harny, I.; Tirta, A. Waste to Wealth of Apple Pomace Valorization by Past and Current Extraction Processes: A Review. *Sustainability* **2023**, *15*, 830. [CrossRef]
50. Costa, J.M.; Ampese, L.C.; Ziero, H.D.D.; Sganzerla, W.G.; Forster-Carneiro, T. Apple Pomace Biorefinery: Integrated Approaches for the Production of Bioenergy, Biochemicals, and Value-Added Products—An Updated Review. *J. Environ. Chem. Eng.* **2022**, *10*, 108358. [CrossRef]
51. Rashid, R.; Wani, S.M.; Manzoor, S.; Masoodi, F.A.; Dar, M.M. Green Extraction of Bioactive Compounds from Apple Pomace by Ultrasound Assisted Natural Deep Eutectic Solvent Extraction: Optimisation, Comparison and Bioactivity. *Food Chem.* **2023**, *398*, 133871. [CrossRef] [PubMed]

52. Zhang, Z.; Poojary, M.M.; Choudhary, A.; Rai, D.K.; Tiwari, B.K. Comparison of Selected Clean and Green Extraction Technologies for Biomolecules from Apple Pomace. *Electrophoresis* **2018**, *39*, 1934–1945. [CrossRef]
53. Wijngaard, H.; Brunton, N. The Optimization of Extraction of Antioxidants from Apple Pomace by Pressurized Liquids. *J. Agric. Food Chem.* **2009**, *57*, 10625–10631. [CrossRef]
54. Sobczak, P.; Nadulski, R.; Kobus, Z.; Zawiślak, K. Technology for Apple Pomace Utilization within a Sustainable Development Policy Framework. *Sustainability* **2022**, *14*, 5470. [CrossRef]
55. de Oliveira, T.C.G.; Sganzerla, W.G.; Ampese, L.C.; Sforça, B.P.; Goldbeck, R.; Forster-Carneiro, T. Sustainable Valorization of Apple Waste in a Biorefinery: A Bibliometric Analysis. *Biofuels Bioprod. Biorefining* **2022**, *16*, 891–919. [CrossRef]
56. Pyanikova, E.A.; Kovaleva, A.E.; Ovchinnikova, E.V.; Afanasieva, L.A. Researching the Possibility of Using Recycled Apple Raw Materials to Create Functional Food Products. *IOP Conf. Ser. Earth Environ. Sci.* **2021**, *677*, 6–11. [CrossRef]
57. Kodagoda, K.; Marapana, R. Utilization of Fruit Processing By-Products for Industrial Applications: A Review. *Int. J. Food Sci. Nutr.* **2017**, *2*, 24–30.
58. Guardia, L.; Suárez, L.; Querejeta, N.; Rodríguez Madrera, R.; Suárez, B.; Centeno, T.A. Apple Waste: A Sustainable Source of Carbon Materials and Valuable Compounds. *ACS Sustain. Chem. Eng.* **2019**, *7*, 17335–17343. [CrossRef]
59. Reis, S.F.; Rai, D.K.; Abu-Ghannam, N. Apple Pomace as a Potential Ingredient for the Development of New Functional Foods. *Int. J. Food Sci. Technol.* **2014**, *49*, 1743–1750. [CrossRef]
60. Sudha, M.L. *Apple Pomace (By-Product of Fruit Juice Industry) as a Flour Fortification Strategy*; Elsevier Inc.: Amsterdam, The Netherlands, 2011; ISBN 978-0-12-380886-8.
61. Wang, H.J.; Thomas, R.L. Direct Use of Apple Pomace in Bakery Products. *J. Food Sci.* **1989**, *54*, 618–620. [CrossRef]
62. Masoodi, F.A.; Sharma, B.; Chauhan, G.S. Use of Apple Pomace as a Source of Dietary Fiber in Cakes. *Plant Foods Hum. Nutr.* **2002**, *57*, 121–128. [CrossRef] [PubMed]
63. Sudha, M.L.; Baskaran, V.; Leelavathi, K. Apple Pomace as a Source of Dietary Fiber and Polyphenols and Its Effect on the Rheological Characteristics and Cake Making. *Food Chem.* **2007**, *104*, 686–692. [CrossRef]
64. Muntean, M.V.; Fărcaş, A.C.; Medeleanu, M.; Salanţă, L.C.; Borşa, A. A Sustainable Approach for the Development of Innovative Products from Fruit and Vegetable By-Products. *Sustainability* **2022**, *14*, 10862. [CrossRef]
65. Wang, X.; Kristo, E.; LaPointe, G. The Effect of Apple Pomace on the Texture, Rheology and Microstructure of Set Type Yogurt. *Food Hydrocoll.* **2019**, *91*, 83–91. [CrossRef]
66. Younis, K.; Ahmad, S. Waste Utilization of Apple Pomace as a Source of Functional Ingredient in Buffalo Meat Sausage. *Cogent Food Agric.* **2015**, *1*, 1119397. [CrossRef]
67. Lantto, R.; Plathin, P.; Niemistö, M.; Buchert, J.; Autio, K. Effects of Transglutaminase, Tyrosinase and Freeze-Dried Apple Pomace Powder on Gel Forming and Structure of Pork Meat. *LWT* **2006**, *39*, 1117–1124. [CrossRef]
68. Rather, S.A.; Akhter, R.; Masoodi, F.A.; Gani, A.; Wani, S.M. Utilization of Apple Pomace Powder as a Fat Replacer in Goshtaba: A Traditional Meat Product of Jammu and Kashmir, India. *J. Food Meas. Charact.* **2015**, *9*, 389–399. [CrossRef]
69. Issar, K.; Sharma, P.C.; Gupta, A. Utilization of Apple Pomace in the Preparation of Fiber-Enriched Acidophilus. *J. Food Process. Preserv.* **2017**, *41*, e13098. [CrossRef]
70. Dhillon, G.S.; Brar, S.K.; Verma, M.; Tyagi, R.D. Utilization of Different Agro-Industrial Wastes for Sustainable Bioproduction of Citric Acid by Aspergillus Niger. *Biochem. Eng. J.* **2011**, *54*, 83–92. [CrossRef]
71. Dhillon, G.S.; Brar, S.K.; Verma, M.; Tyagi, R.D. Apple Pomace Ultrafiltration Sludge—A Novel Substrate for Fungal Bioproduction of Citric Acid: Optimisation Studies. *Food Chem.* **2011**, *128*, 864–871. [CrossRef]
72. Hang, Y.D.; Woodams, E.E. Solid State Fermentation of Apple Pomace for Citric Acid Production. *MIRCEN J. Appl. Microbiol. Biotechnol.* **1986**, *2*, 283–287. [CrossRef]
73. Hoseyini, M.; Asefi, N.; Mozaffari, M. Production of Citric Acid from Apple Pomace by Using Surface Culture Method. *Agric. J.* **2011**, *6*, 226–230. [CrossRef]
74. Kumar, D.; Verma, R.; Bhalla, T.C. Citric Acid Production by Aspergillus Niger van. Tieghem MTCC 281 Using Waste Apple Pomace as a Substrate. *J. Food Sci. Technol.* **2010**, *47*, 458–460. [CrossRef]
75. Bhushan, S.; Joshi, V.K. Baker's Yeast Production under Fed Batch Culture from Apple Pomace. *J. Sci. Ind. Res.* **2006**, *65*, 72–76.
76. Joshi, V.K.; Bhushan, S. Apple Pomace Utilization for the Production of Baker's Yeast: Effect of Substrate Concentrations and Growth Stimulators. *Indian J. Biotechnol.* **2003**, *2*, 220–226.
77. Reed, G.; Nagodawithana, T.W. *Baker's Yeast Production*; Reed, G., Ed.; Springer: Dordrecht, The Netherlands, 2018.
78. Trivedi, N.B.; Jacobson, G.K.; Tesch, W.; Friend, J.P. Industrial Production of Bakers' Yeast. *Crit. Rev. Biotechnol.* **1986**, *4*, 75–109. [CrossRef]
79. Flynn, G.; Adams, M.R. *An Industrial Profile of Yeast Production*; Tropical Products Institute: London, UK, 1981.
80. Grba, S. Proizvodnja Pekarskog Kvasca. In *Kvasci u Biotehnološkoj Proizvodnji*; Ranić, I., Ed.; Plejada: Zagreb, Croatia, 2010; pp. 265–300.
81. Lisičar, J.; Scheper, T.; Barbe, S. Turning Industrial Baker's Yeast Manufacture into a Powerful Zero Discharge Multipurpose Bioprocess. *Ind. Biotechnol.* **2017**, *13*, 184–191. [CrossRef]
82. Lisicar, J.; Scheper, T.; Barbe, S. Industrial Baker's Yeast Fermentation: From Manufacture to Integrated Sustainability. *J. Biotechnol.* **2017**, *256*, S23–S24. [CrossRef]

83. Mamun-Or-Rashid, A.N.M.; Lucy, T.T.; Md Pramanik, K. Isolation, Identification, Optimization of Baker's Yeast from Natural Sources, Scale-Up Production Using Molasses as a Cheap Carbohydrate Source, and Evaluation for Bread Production. *Appl. Microbiol.* **2022**, *2*, 516–533. [CrossRef]
84. Igwegbe, C.A.; Obiora-Okafo, I.A.; Iwuozor, K.O.; Ghosh, S.; Kurniawan, S.B.; Rangabhashiyam, S.; Kanaoujiya, R.; Ighalo, J.O. Treatment Technologies for Bakers' Yeast Production Wastewater. *Environ. Sci. Pollut. Res.* **2022**, *29*, 11004–11026. [CrossRef]
85. Yusof, A.H.; Dailin, D.J.; Low, L.Z.M.I.; Abg Zaidel, D.N.; El Enshasy, H. Potential Application of Pineapple Waste as a Fermentation Substrate in Yeast Production. *Int. J. Sci. Technol. Res.* **2020**, *9*, 1933–1937.
86. Nancib, N.; Nancib, A.; Boudrant, J. Use of Waste Date Products in the Fermentative Formation of Baker's Yeast Biomass by *Saccharomyces cerevisiae*. *Bioresour. Technol.* **1997**, *60*, 67–71. [CrossRef]
87. Diboune, N.; Nancib, A.; Nancib, N.; Aníbal, J.; Boudrant, J. Utilization of Prickly Pear Waste for Baker's Yeast Production. *Biotechnol. Appl. Biochem.* **2019**, *66*, 744–754. [CrossRef]
88. Gengec, E.; Kobya, M.; Demirbas, E.; Akyol, A.; Oktor, K. Optimization of Baker's Yeast Wastewater Using Response Surface Methodology by Electrocoagulation. *Desalination* **2012**, *286*, 200–209. [CrossRef]
89. Xu, Y.; Cao, W.; Cui, J.; Shen, F.; Luo, J.; Wan, Y. Developing a Sustainable Process for the Cleaner Production of Baker's Yeast: An Approach towards Waste Management by an Integrated Fermentation and Membrane Separation Process. *J. Environ. Manag.* **2022**, *323*, 116197. [CrossRef]
90. Angumeenal, A.R.; Venkappayya, D. An Overview of Citric Acid Production. *LWT* **2013**, *50*, 367–370. [CrossRef]
91. Mordenti, A.L.; Giaretta, E.; Campidonico, L.; Parazza, P.; Formigoni, A. A Review Regarding the Use of Molasses in Animal Nutrition. *Animals* **2021**, *11*, 115. [CrossRef] [PubMed]
92. Dogbe, E.S.; Mandegari, M.; Görgens, J.F. Revitalizing the Sugarcane Industry by Adding Value to A-Molasses in Biorefineries. *Biofuels Bioprod. Biorefining* **2020**, *14*, 1089–1140. [CrossRef]
93. Moraes, B.S.; Triolo, J.M.; Lecona, V.P.; Zaiat, M.; Sommer, S.G. Biogas Production within the Bioethanol Production Chain: Use of Co-Substrates for Anaerobic Digestion of Sugar Beet Vinasse. *Bioresour. Technol.* **2015**, *190*, 227–234. [CrossRef] [PubMed]
94. Carpanez, T.G.; Moreira, V.R.; Assis, I.R.; Amaral, M.C.S. Sugarcane Vinasse as Organo-Mineral Fertilizers Feedstock: Opportunities and Environmental Risks. *Sci. Total Environ.* **2022**, *832*, 154998. [CrossRef] [PubMed]
95. Lisičar Vukušić, J.; Millenautzki, T.; Saaid, A.M.; Reichert, L.; Barbe, S. New Feedstocks for Bioethanol Production. In *Book of Proceedings of STEPsCON 2018*; TH Köln: Leverkusen, Germany, 2018.
96. Lisičar Vukušić, J.; Kneer, A.; Mösche, M.; Barbe, S. Turning Industrial Aerobic Fermentation Plants into Thermal Power Stations. *Int. J. Energy Res.* **2019**, *43*, 544–551. [CrossRef]
97. Lisičar, J.; Sedaghati, M.; Hof, J.; Mösche, M.; Barbe, S. Full Mass Balance Analysis During Industrial Baker's Yeast Fermentation Shows New Perspectives for Biomolecule Recovery. *Ind. Biotechnol.* **2019**, *15*, 303–310. [CrossRef]
98. Cai, B. Study of Membrane Separation Processes of a Betaine-like Medicinal Solution. *Desalination* **2006**, *191*, 432–437. [CrossRef]
99. Silva, S.C.D.; Moravia, M.C.S.A.; Couto, C.F. Combined Process of Ultrafiltration and Nanofiltration for Vinasse Treatment with and without Pre-Coagulation. *J. Water Process Eng.* **2020**, *36*, 101326. [CrossRef]
100. Hosseinipour, E.; Karimi, S.; Barbe, S.; Park, K.; Davies, P.A. Hybrid Semi-Batch/Batch Reverse Osmosis (HSBRO) for Use in Zero Liquid Discharge (ZLD) Applications. *Desalination* **2022**, *544*, 116126. [CrossRef]

Disclaimer/Publisher's Note: The statements, opinions and data contained in all publications are solely those of the individual author(s) and contributor(s) and not of MDPI and/or the editor(s). MDPI and/or the editor(s) disclaim responsibility for any injury to people or property resulting from any ideas, methods, instructions or products referred to in the content.

agriculture

Article

Calcium, Potassium, and Magnesium Affect the Nutritional Value of Tomato Grafted Fruits Grown in a Nutrient Film Technique System

Rocío Maricela Peralta Manjarrez [1], Rafael Delgado Martínez [2], Adalberto Benavides Mendoza [3], Antonio Juárez Maldonado [4] and Marcelino Cabrera De la Fuente [3,*]

1 CONAHCYT Posdoctoral, Departament of Horticulture, Universidad Autónoma Agraria Antonio Narro, Saltillo 25315, Coahuila, Mexico; rocio.peralta@uaaan.edu.mx
2 Faculty of Engineering and Sciences, Universidad Autónoma de Tamaulipas, Victoria City 87149, Tamaulipas, Mexico; rdelgado@docentes.uat.edu.mx
3 Departament of Horticulture, Universidad Autónoma Agraria Antonio Narro, Saltillo 25315, Coahuila, Mexico; adalberto.benavides@uaaan.edu.mx
4 Departament of Botany, Universidad Autónoma Agraria Antonio Narro, Saltillo 25315, Coahuila, Mexico; antonio.juarez@uaaan.edu.mx
* Correspondence: marcelino.cabrera@uaaan.edu.mx

Abstract: *Solanum lycopersicum* is a vegetable with a high mineral, nutraceutical, and vitamin content. It is a basic ingredient in the human diet, and its use is diverse within the kitchen. Grafting and Ca, K, and Mg applications can be used to increase production and raise the mineral contents of tomato fruit. An experiment was established with tomato plants using the "CID F1" variety with the "FORTAMINO" rootstock, established in the NFT system, determining the influence on the agronomic yield and mineral composition of the fruit. Grafted and non-grafted plants were considered, with five concentrations (0-0-0, 9-0-0, 0-12-0, 0-0-9, and 9-12-9 mEq L^{-1}) of Ca, K, and Mg, respectively. A highly significant difference was obtained in the grafted plants on high plants, number of leaves, number of fruits, polar diameters of fruits, equatorial diameters of fruits, and weight of fruits, with an increase in variables, FW 19% and NF 18%, and an improvement for the mineral composition in elements such as Ca 10%, P 1%, Mn 6%, Zn 7%, Cu 6%, Fe 64%, K 21%, and Mg 19%. The 9-12-9 meq formula improved Ca 6%, P 4%, Mn 12%, Zn 14%, Cu 8%, Fe 74%, and Mg 25%. The graft and the addition of calcium, potassium, and magnesium increased the mineral content in tomato fruits and improved the agronomic performance of the plants.

Keywords: graft; mineral composition; productivity; *Solanum lycopersicum*

Citation: Peralta Manjarrez, R.M.; Delgado Martínez, R.; Benavides Mendoza, A.; Juárez Maldonado, A.; Cabrera De la Fuente, M. Calcium, Potassium, and Magnesium Affect the Nutritional Value of Tomato Grafted Fruits Grown in a Nutrient Film Technique System. *Agriculture* **2023**, *13*, 2189. https://doi.org/10.3390/agriculture13122189

Academic Editor: Athanasios Koukounaras

Received: 16 October 2023
Revised: 17 November 2023
Accepted: 19 November 2023
Published: 23 November 2023

Copyright: © 2023 by the authors. Licensee MDPI, Basel, Switzerland. This article is an open access article distributed under the terms and conditions of the Creative Commons Attribution (CC BY) license (https://creativecommons.org/licenses/by/4.0/).

1. Introduction

The tomato (*Solanum lycopersicum* L.) is an important source of bioactive compounds beneficial to human health, and 180 million tons are produced worldwide, of which 20% are used for industrial transformation [1]. This vegetable also participates in providing humans with a wide variety of nutrients with potentially beneficial effects. Its consumption favors the intake of nutraceuticals due to their nutritional and pharmaceutical properties, given their richness in lycopene and biologically active metabolites that are beneficial to human health [2]. This vegetable increases antioxidant status and is associated with high mineral uptake; low mineral intake can induce inflammatory processes and affect defense systems [3].

Grafting confers resistance to root problems in plants and tolerance to abiotic stresses, which is why the use of alternative techniques has been adopted in tomato production that will guarantee vegetative and root development for efficient nutrient absorption and make the productive system more efficient [4–7]. Grafted tomatoes have been shown to modify the general mineral composition of the plant and improve yield and biomass accumulation [4,8,9]. Furthermore, nutrient uptake in horticultural crops is enhanced by the selection of suitable rootstocks, which play a vital role in manipulating the nutritional status of shoots by directly affecting ion uptake and transport [10].

Calcium, potassium, and magnesium are essential elements that, When added supplementally, intervene in the formation, production, and quality of tomatoes' fruits [9,11,12]. Ca modifies the total photochemical content and the concentration of lycopene in tomato fruits [11], as it is one of the essential elements for plant growth and fruit development [13]. The nutraceutical and mineral quality of the fruits is increased by the addition of potassium. It is a demanded element during the formation of tomato fruits [14]; the potassium-to-calcium ratio increases the quality attributes. Potassium participates in photosynthesis, respiration, and the activation of enzymes, and it has a significant influence on both the growth and the quality of fruits and vegetables [9,14]. Mg is involved in the photosynthetic reaction; it is a basic component of chlorophyll and activates enzymes that intervene in the synthesis of nucleic acids. The addition of this element contributes to the content of sugars and vitamin C, producing high-quality plants since it is an auxiliary in the metabolism of phosphates, plant respiration, and the activation of various enzyme systems involved in energy metabolism [12,15].

Therefore, grafting efficient nutrient absorption maximizes fertilization efficiency, increases plant vigor, ensures better yield and increases fruit quality [4,6]. The use of the grafting technique in tomato cultivation with the supplementation of Ca, K, and Mg is a technical–scientific novelty; it is considered a viable alternative to increase the quality of the crop and production. Its implementation, according to the formula recommended in the fruiting stage (9-12-9), could generate an improvement in the growth and development of plants.

Based on this, the objective of the present work was to evaluate the effect of grafting and the applications of Ca, K, and Mg to increase production and increase the mineral content of tomato fruit.

2. Materials and Methods

The research was carried out in a medium-technology micro-tunnel greenhouse at the horticulture department of the Antonio Narro Autonomous Agrarian University, Saltillo, Coahuila, Mexico. Latitude is 25°21′23.4″ and longitude 101°02′10.6″, with an altitude of 1760 m above sea level, where the average temperature is 25 °C.

2.1. Vegetal Material Used and Growing Conditions

Seeds of "Cid F1", from Harris Moran® produce an indeterminate tomato with uniform fruit size and shape, thick walls that provide excellent firmness, intense red fruits with long shelf life and high yield and these seeds work well in hydroponic systems. The rootstock Fortamino tomato was used, which is a tomato with a very high vigor and resistance to *Fusarium oxysporum* f.sp. *lycopersici*, races 0.1 and 2, and it is recommended for saline soils (Enza Zaden®, Enza Zaden®, Culiacan, Sinaloa, Mexico). Sowing was performed in a germinating tray of polystyrene (cavity measures: width 25 mm, length 25 mm, and depth 70 mm). Inside the greenhouse, there was an average irradiance of 187.47 W/m^2, the average temperatures recorded were 25 to 35 °C, and the average relative humidity inside ranged between 40 and 60%. The scion was sown on 19 February 2021 and the rootstock was sown 7 days later. This difference in sowing times was because the rootstock has greater vegetative vigor compared to the scion; with this, it was sought that both the stems of the scion and the rootstock had a similar diameter, which benefits the union of both plant structures. The grafted plants were placed in 250 mL Styrofoam containers. The substrate for filling the sowing trays was a mixture of peat moss with perlite at a 70:30 ratio.

2.2. Graft

The grafting process was carried out on 23 March 2021, when the plants had a 3 mm stem diameter, approximately, which was obtained 32 days after sowing of the scion. The splicing technique was used, which consisted of a cut at a 45° angle in the two plants to form one, and a silicone clip was applied to the union site. Subsequently, the grafted plants were kept for 15 days in an acclimatization chamber; they were in dark conditions for a period of 3 days, 5 with 50% shade and the rest of the days without shade. The relative humidity was 95%, and the temperature ranged between 25 and 35 °C for 10 days to favor the grafting process. Subsequently, the silicone clip was completely removed since the union between the scion and the rootstock had been produced and healed. The Steiner nutrient solution was used in mEq L^{-1} (chemical composition: NO_3^-, $H_2PO_4^-$, K^+, Ca^{++}, Mg^{++}, NH_4^+, and $SO_4^=$) in different concentrations: approximately 25% at the beginning of vegetative growth (3, 0.6, 1, 3, 1, 0.5, 1.6 mEq L^{-1}, respectively); 50% vegetative growth at the beginning of flowering (8, 1, 4, 7, 3, 1, 4 mEq L^{-1}, respectively), 75% fruit set (12, 1.2, 6.5, 9, 3.5, 1, 6 mEq L^{-1}, respectively), and 100% ripening and harvest (13, 1.2, 7.5, 9, 4, 1, 7 mEq L^{-1},respectively). With a pH of 6–6.5 and EC from 750 to 1500 mg L^{-1}, additional treatments were supplementary foliar applications of Ca, K, and Mg (9-12-9 mEq) to grafted and non-grafted plants.

2.3. Nutrient Film Technique (NFT)

The NFT system (Figure 1) used consisted of 6 PVC pipes 15.24 cm in diameter, with 16 holes 6 cm wide and 30 cm between holes. The distance between tubes was 60 cm, 68 L plastic boxes were used to collect the nutrient solution, and 25-watt submersible electric pumps for fish tanks with a circulation capacity of 1500 L per hour were used for recirculation of the nutrient solution. In a closed system, it is essential to maintain adequate EC during the growing season, renew the nutrient solution every time it reaches a value of 3.5 dS·m^{-1}, replace the transpired water with tap water, compensate for the transpired water with standard nutrient solution, and add all the nutrients according to the estimated consumption.

Figure 1. Plants established in greenhouse conditions.

2.4. Transplant to the NFT System

The transplant was carried out on 20 April (28 days after grafting), when the clips for the attachment had already been removed, with bare roots by completely removing the substrate from the root and immersing it in a 3% hydrogen peroxide solution as a preventive treatment against pathogenic microorganisms. Hydroponic baskets of three inches were used to adjust the plant inside the basket. A sponge was used to cover the root from direct sunlight, leaving the root free to have contact with the nutrient solution.

2.5. Treatments Application

The fertilizers that were added were the commercial products AMIFOL K®, HUMISOIL MG 14®, and HUMISOIL CA 16® (Tradecorp, Zapopan, Jalisco, Mexico). Supplementary applications were applied by spraying direct spraying on the fruit. Two factors were considered: factor 1 consisted of grafting with two levels (with and without grafting);

factor 2: five levels corresponding to plants treated with different doses of Ca, K, and Mg (0-0-0, 9-0-0, 0-12-0, 0-0-9, 9-12-9), with a total of 10 treatments with 10 replicates. The treatments were evaluated using a 2 × 5 factorial arrangement in a completely randomized design with 10 experimental units per treatment, with a total of 100 plants.

2.6. *Analysis of Crop Parameters*

The productive cycle of the plants lasted until the development of the fifth floral cluster, when the plants emitted said cluster, and the apex was cut to stop it. At 110 days after transplanting, plant height (PH) was measured. This parameter was evaluated with a TOOLCRAFT® model TC0234 (Toolcraft, Nuevo Leon, Mexico) measuring tape with a three-meter capacity, measuring from the base of the stem to the apex of the stem. From the transplant, the number of leaves (NL) was counted every 15 days until the end of the growth cycle, and 10 repetitions per treatment were evaluated (n = 10). The fruits were harvested 82 days after transplanting, and 5 harvests were carried out in total, based on the USDA classification, which indicates that the fruits should have 90% of the surface of a uniform red color. The number of fruits (NF) was counted and weighed on a portable scale (OHAUS®, model H-8109, OHAUS Latinoamierica, Mexico DF, Mexico) to estimate the total yield. A total of 6 fruits were selected per treatment at the time of harvest. The polar diameter (PD) and equatorial diameter (ED) of each fruit were measured with a digital caliper.

2.7. *Mineral Content of the Fruit*

The content of macromineral elements was carried out with a nitric acid digestion technique: 0.5 g of dry sample was weighed and placed in a 100 mL beaker, then 30 mL of concentrated HNO_3 were added and boiled for 3 h while maintaining the initial volume of 30 mL. The digested samples were filtered with a No. 42 Whatman® filter paper (Merck KGaA, Darmstadt, Alemania) and made up to 50 mL with deionized water. The resulting solution was analyzed with a PerkinElmer® spectrometer (Proinstra S.A., Quito, Ecuador), model PinAAcle 900H, at the UAT. The quantification of minerals was carried out for N, K, Ca, Mg, Fe, Cu, Zn, and Mn. The quantification of phosphorus was carried out with the technique of "Phosphorus in plant by UV-Visible spectrophotometry" by using 2 mL of heptamolybdate–vanadate and the amount of deionized water required to make 10 mL. Shake and let it rest for half an hour. Read absorbance of each of the solutions at 470 nm in a UV–visible spectrophotometer.

2.8. *Statistic Analysis*

For data processing, an analysis of variance and a comparison of means were performed using the LSD test ($p \leq 0.05$). Infostat® version 2017 software was used.

3. Results

3.1. *Crop Parameters for Grafts*

With the grafting factor, a yield of up to 5.2 kg per plant was achieved, which means an increase of 30% compared to the control, while an increase in the agronomic variables evaluated was observed in PH, NL, NF, PD, ED, and FW, which showed significant differences at early stages compared to the non-graft factor. The implementation of grafted plants showed an increase in fruit weight of 19%, number of fruits by 18%, and fruit firmness by 4%, considering that the modalities of Graft/No Graft did not show statistically significant differences in the concentration of pH (potential hydrogen) and TSS (total soluble solids).

3.2. *Crop Parameters for Doses*

Fertilized treatments did not show an increase in PH, NL, NF, PD, ED, and FW (Table 1), but in the interaction of the factors, there was an influence of the variable of ED increasing with the treatments 9-0-0, 0-12-0, and 0-0-9, up to 4%, which favors the application of the doses. The 9-0-0 dose participates in the 13% increase with respect to the other doses.

With all the doses, the fruit weight was increased compared to the absolute control, up to 6% for the 0-12-0, 0-0-9, and 9-12-9 doses and 5% for the dose 9-0-0.

Table 1. Effect of the application of calcium, potassium, and magnesium under agronomic parameters and commercial quality in tomato fruits.

	Treatment	PH	NL	NF	PD	ED	FW
Factor 1: Graft	Grafted	179.68 a	25.04 a	32.72 a	87.61 a	55.51 a	158.1 a
	Ungrafted	157.60 b	22.60 b	29.76 b	81.61 b	48.93 b	134.22 b
Factor 2: Fertilizer dose Ca-K-Mg	0-0-0	167.50 a	23.80 a	31.50 a	84.56 a	51.79 b	139.54 b
	9-0-0	158.50 a	23.90 a	31.90 a	85.15 a	53.37 a	146.95 a
	0-12-0	160.60 a	23.70 a	32.00 a	84.15 a	52.08 b	148.51 a
	0-0-9	169.70 a	23.90 a	31.40 a	84.29 a	51.98 b	148.30 a
	9-12-9	168.9 a	23.80 a	31.90 a	84.66 a	51.88 b	147.28 a
Interaction							
Grafted	0-0-0	178.00 a	22.60 b	33.60 a	87.00 b	54.72 b	152.94 b
Grafted	9-0-0	179.80 a	25.00 a	33.80 a	89.08 a	57.98 a	160.10 a
Grafted	0-12-0	179.80 a	24.80 a	33.60 a	86.44 b	55.28 b	159.60 a
Grafted	0-0-9	180.00 a	25.40 a	33.40 a	87.16 b	55.20 b	159.58 a
Grafted	9-12-9	180.80 a	25.00 a	34.20 a	87.88 a	54.38 b	158.48 a
Ungrafted	0-0-0	157.00 b	22.60 b	29.40 b	82.12 c	48.86 c	126.74 e
Ungrafted	9-0-0	157.20 b	22.80 b	30.00 b	81.22 c	48.76 c	133.80 d
Ungrafted	0-12-0	157.40 b	22.60 b	30.40 b	81.86 c	48.88 c	137.54 c
Ungrafted	0-0-9	159.40 b	22.40 b	29.40 b	81.42 c	48.78 c	136.92 c
Ungrafted	9-12-9	157.00 b	22.60 b	29.60 b	81.44 c	49.38 c	136.08 c
	CV (%)	1.88	4.18	5.32	1.76	2.6	1.5

PH (cm) = plant height, NL = number of leaves, NF: number of fruits, PD (mm): polar diameter of fruit, ED (mm): equatorial diameter of fruits, FW (g) = weight of fruits/plant. Means with the same letter are not significantly different ($p \leq 0.05$), ($p > 0.05$, Fisher's protected LSD test), CV (%) = coefficient of variation.

3.3. Crop Parameters for Interactions

Fertilizer dose × graft interaction had an increase in all doses where a grafting plant was used with respect to plant height, number of leaves, number of fruits, and fruit weight, highlighting an improvement for the dose where potassium fertilization was not used. The fruits' polar and equatorial diameters were positively influenced by the treatments with a supplementary application of calcium, and all the doses where at least one fertilizer applied is used in grafted plants increase the weight of the fruit by up to 17%, as can be seen in Table 1.

3.4. Mineral Profile of the Fruit of Grafted Plants

The implementation of grafted plants highly significantly increased the mineral contents of fresh fruits in Ca with an increase of 10%, P 1%, Mn 6%, Zn 7%, Cu 6%, Fe 64%, K 21%, and Mg 19% (Table 2).

3.5. Results of the Mineral Profile of the Fruit for the Different Doses

Table 2 shows that the dose of 9-12-9 milliequivalents statistically increased the percentage of minerals, such as Ca 6%, P 4%, Mn 12%, Zn 14%, Cu 8%, Fe 74%, and Mg 25%.

The interaction with grafted plants and Mg positively influenced the mineral content of the tomato fruits, although if other elements are added to this dose, as in the dose of 9-12-9 milliequivalents of Ca, K, and Mg, it gives as a result an increase in the mineral composition of the fruits of the grafted plants.

Table 2. Effect the application of calcium, potassium, and magnesium under mineral profile in tomato fruits.

Treatment		N	P	K	Ca	Mg	Fe	Cu	Zn	Mn
		%	mg/kg							
Factor 1: Graft	Grafted	2.43 a	352.00 a	3359.00 a	166.07 a	123.60 a	49.15 a	1.97 a	6.92 a	4.50 a
	Ungrafted	2.32 b	348.00 b	2937.00 b	151.20 b	104.30 b	29.85 b	1.85 b	6.45 b	4.24 b
Significance		0.0187	0.0001	0.00001	0.0001	0.0001	0.0001	0.0001	0.0001	0.0001
Factor 2: Fertilizer dose Ca-K-Mg	0-0-0	2.23 c	345.00 c	3092.00 a	155.37 b	102.80 b	34.10 c	1.84 c	6.41 b	4.20 c
	9-0-0	2.34 b	345.00 c	3216.00 a	155.55 b	105.10 b	34.30 c	1.90 b	6.64 b	4.38 b
	0-12-0	2.48 a	345.00 c	3113.00 a	153.00 b	103.80 b	34.17 c	1.86 b	6.52 b	4.29 b
	0-0-9	2.30 c	352.00 b	3120.00 a	164.17 a	129.10 a	35.19 b	1.93 a	6.42 b	4.27 b
	9-12-9	2.52 a	359.00 a	3198.00 a	164.29 a	129.00 a	59.74 a	1.99 a	7.46 a	4.73 a
Significance		0.0019	0.0001	0.3615	0.0001	0.0001	0.0001	0.041	0.0001	0.0001
Interaction										
Grafted	0-0-0	2.19 c	348.00 b	3306.00 a	158.61 b	103.60 b	35.45 c	1.86 c	6.56 b	4.21 c
Grafted	9-0-0	2.44 b	347.00 b	3446.00 a	159.26 b	104.50 b	35.64 c	1.93 b	6.76 b	4.41 b
Grafted	0-12-0	2.55 a	348.00 b	3330.00 a	156.26 b	103.60 b	36.46 c	1.89 c	6.44 b	4.25 b
Grafted	0-0-9	2.44 b	358.00 a	3311.00 a	178.09 a	154.00 a	40.66 b	2.02 a	6.51 b	4.34 b
Grafted	9-12-9	2.55 a	360.00 a	3402.00 a	178.15 a	152.50 a	89.52 a	2.13 a	8.34 a	5.29 a
Ungrafted	0-0-0	2.27 b	343.00 c	2878.00 b	152.14 c	102.20 b	29.74 d	1.83 c	6.25 c	4.18 c
Ungrafted	9-0-0	2.24 b	344.00 b	2987.00 b	151.84 c	105.70 b	29.96 d	1.88 c	6.52 b	4.34 b
Ungrafted	0-12-0	2.40 a	347.00 b	2897.00 b	151.35 c	104.00 b	29.85 d	1.83 c	6.44 b	4.32 b
Ungrafted	0-0-9	2.17 c	346.00 b	2929.00 b	150.25 c	104.20 b	29.72 d	1.84 c	6.33 c	4.21 c
Ungrafted	9-12-9	2.49 a	358.00 a	2995.00 b	150.42 c	105.50 b	29.96 d	1.85 c	6.57 b	4.16 c
Significance		0.1547	0.0013	0.9886	0.0001	0.0001	0.0001	0.0129	0.0001	0.0001
	CV (%)	6.17	0.8	4.7	1.56	4.1	0.66	4.13	3.78	3.1

N: nitrogen, P: phosphorus, K: potassium, Ca: calcium, Mg: magnesium, Fe: iron, Cu: copper, Zn: zinc, and Mn: manganese. Means with the same letter per column and within each statistical partition: graft, dose, and interaction are not statistically different ($p \leq 0.05$), LSD Fisher test. CV (%) = coefficient of variation.

4. Discussion

Grafting is a widely used technology in modern agriculture, and it plays an important role in the propagation of vegetative growth, renewing varieties, improving the absorption of nutrients, regulating the accumulation of ions, and improving the resistance of plants [16]. Regarding the factors evaluated, the results obtained in this study (Table 1) showed an increase in each of the variables that differentiate the two factors of grafted tomato plants versus non-grafted tomato plants, as well as in plant height, number of leaves, number of fruits, fruits' weight, and equatorial and polar diameters of the fruit, because a grafted plant is more efficient in the absorption and transport of water and ions from the root to the other organs, increasing the accumulation of biomass [16]. Treviño López [17] found that the effect of grafting has a positive effect on production variables, such as an increase in fruit diameter and weight compared to plants without grafting. Similar results to those of a research work where they reported an increase in fruit length, number of fruits, and yield per plant, the highest values were obtained in fruits of grafted plants [4]. The vegetative growth of grafted plants is largely manifested by the number of leaves. In a study carried out with cucumbers, differences between treatments were observed in favor of the grafted, and the trend was that its value increased by 27% with the graft [18]. The results regarding the number of leaves concur with those obtained by García-Ávila, where it was reported a greater specific leaf area (SLA) in grafted pumpkin plants than in non-grafted ones [10]. Plant height is an important variable, although this effect varies according to the cultivar studied [16]. The plant height variable in this research increased up to 14% compared to non-grafted plants; therefore, statistically significant differences were observed. A recent work evaluated plant height in grafted plants, and an increase was observed compared to non-grafted plants, where this variable is reflected as a plant with greater vigor that may be feasible for the translocation of elements through the plant's pathways [5]. The size of the fruits of the grafted plants increased, and this can occur depending on the rootstock used [19]. Similar results were presented by Grimaldo Juárez [20], where the average fruit size was greater in grafted plants, compared to the normal condition, with an increase of 17.7%, respectively. A study carried out on bell pepper shows that the Terrano rootstock increased yield by 50% [7]. The accumulation of K in grafted plants is attributed to physical characteristics of the root system, such as abundant lateral and vertical root development, exploration area, and, consequently, greater absorption of water and minerals [21]. A study carried out showed that there is an increase in the mineral content of grafted plants because of a higher concentration of the elements and accumulation of biomass [8]. Regarding the fruit weight variable, it is necessary to highlight that all the applications of the treatments where at least one dose of fertilizers was supplied increased fruits' weight compared to the control. As shown in Table 2, regarding the treatments, the best ones were where the interaction of each of the doses with the grafted plants was obtained, similar to that reported by Pérez-Grajales [22] where different treatments were tested, and all the treatments where the graft with the tomato variety CID was used improved the agronomic variables, increased yield, total biomass accumulation, and mineral composition. The two factors used show that the grafts have statistically significant differences in the mineral composition of tomato fruits due to the nutrient assimilation capacity provided by this technique [6]. When there is an adequate supply of potassium, the fruit tends to store more of one mineral than another due to the competition that exists between minerals. In conditions of potassium excess in the soil, the plant increases its consumption, making it able to also modify the absorption and availability of other minerals, such as calcium and magnesium, which can favor the fruits [14]. For the potassium content in the fruits, there were no statistically significant differences. The grafts increased the values of the agronomic variables, with higher values in variables such as leaf area. It may be said that the leaf area and height of the plant, which resulted in an increase of 0.96% in this variable, is reflected in a yield of 38% compared to the control [17]. A study shows that tomato fruits from grafted plants with Multifort/CID modified the mineral composition and improved yield [8]. Although the best dose was the supplementary addition of 9-12-9, with an increase in Ca, P, Mn, Zn, Cu, Fe, K, and Mg due to the interaction Ca-K-Mg (adsorption–desorption), these nutrients are considered poorly mobile. K is required in large quantities by crops (similar to N), while Ca and Mg are required in smaller quantities (similar to P and S) [10], which were made available for the fruits in this research. Therefore, the best interaction regarding the agronomic variables was obtained using the 9-0-0 dose where only Ca was applied in the grafted plants, similar to that reported by Rodríguez-Mendoza [23], who obtained favorable results in calcium applications in agronomic variables.

5. Conclusions

The grafting factor directly affects the agronomic variables of the tomato plant, where it mainly increases the diameter and weight of the fruit, which increases the total yield. Fruit quality variables and fruit mineral profile were also positively affected by the grafting factor. Ca-K-Mg applications affect production variables by increasing the number of leaves, number of fruits, polar diameter of the fruits, and weight. In fruit quality, grafting with the recommended dose (9-12-9) increases fruit firmness and fruit TSS; it also participates in the two doses where calcium is used: 9-12-9 and 9-0-0. In the mineral profile of the fruits, the grafted plants participate directly in the increase and stability of the fruits. Production increased with the interaction of factors in fruit weight, number of fruits, and polar and equatorial diameters. Due to the results obtained in this work, we recommend the use of grafts and the foliar application of Ca-K-Mg, which could be used in tomato production to increase productivity and the mineral profile of the fruit.

Author Contributions: Conceptualization, M.C.D.l.F. and R.M.P.M.; methodology, R.D.M.; software, A.J.M.; validation, A.J.M. and R.M.P.M.; formal analysis, M.C.D.l.F.; investigation, A.B.M.; resources, R.D.M.; data curation, M.C.D.l.F.; writing—original draft preparation, R.M.P.M.; writing—review and editing, M.C.D.l.F.; visualization, A.B.M.; supervision, M.C.D.l.F.; project administration,

M.C.D.l.F.; funding acquisition, M.C.D.l.F. All authors have read and agreed to the published version of the manuscript.

Funding: This research was funded by University Autonomus Agrarian Antonio Narro, grant number 3146.

Institutional Review Board Statement: Not applicable.

Data Availability Statement: The data presented in this study are available in article.

Acknowledgments: To the supporting research staff at Universidad Autónoma Agraria Antonio Narro and the post-doctorate scholarship by Mexico from CONAHCyT.

Conflicts of Interest: The authors declare no conflict of interest.

References

1. Méndez-Carmona, J.Y.; Ramírez-Guzmán, N.; Sandoval-Cortes, J.; Ascacio-Valdés, J.A.; Boone-Villa, V.D.; Govea-Salas, M.; Aguilar, C.N. Valorization of Tomato Fruit Processing Residues. In *Agricultural Waste: Environmental Impact, Useful Metabolites and Energy Production*; Ramawat, K., Mérillon, J.M., Arora, J., Eds.; Sustainable Development and Biodiversity; Springer: Singapore, 2023; Volume 31, pp. 215–243. [CrossRef]
2. González-García, Y.; López-Vargas, E.R.; Pérez-Álvarez, M.; Cadenas-Pliego, G.; Benavides-Mendoza, A.; Valdés-Reyna, J.; Pérez-Labrada, F.; Juárez-Maldonado, A. Seed priming with carbon nanomaterials improves the bioactive compounds of tomato plants under saline stress. *Plants* **2022**, *11*, 1984. [CrossRef] [PubMed]
3. Flores, I.R.; Vásquez-Murrieta, M.S.; Franco-Hernández, M.O.; Márquez-Herrera, C.E.; Ponce-Mendoza, A.; del Socorro López-Cortéz, M. Bioactive compounds in tomato (*Solanum lycopersicum*) variety saladette and their relationship with soil mineral content. *Food Chem.* **2021**, *344*, 128608. [CrossRef] [PubMed]
4. Colla, G.; Roupahel, Y.; Cardarelli, M.; Rea, E. Effect of salinity on yield, fruit quality, leaf gas exchange, and mineral composition of grafted watermelon plants. *HortScience* **2006**, *41*, 622–627. [CrossRef]
5. Hernández-González, Z.; Sahagún-Castellanos, J.; Espinosa-Robles, P.; Colinas-León, M.T.; Rodríguez-Pérez, J.E. Efecto del patrón en el rendimiento y tamaño de fruto en pepino injertado. *Rev. Fitotec. Mex.* **2014**, *37*, 41. [CrossRef]
6. Schwarz, D.; Rouphael, Y.; Colla, G.; Venema, J.H. Grafting as a tool to improve tolerance of vegetables to abiotic stresses: Thermal stress, water stress and organic pollutants. *Sci. Hortic.* **2010**, *127*, 162–171. [CrossRef]
7. Peralta-Manjarrez, R.M.; Benavides-mendoza, A.; Ramírez-Godina, F. Micromorphology of cucumber obtained by grafting and developed into two fertilization systems. *Rev. Mex. Cienc. Agrícolas* **2016**, *15*, 118–123.
8. Velasco-Alvarado, M.D.J.; Castro-Brindis, R.; Castillo-González, A.M.; Avitia-García, E.; Sahagún-Castellanos, J.; Lobato-Ortiz, R. Composición mineral, biomasa y rendimiento en tomate (*Solanum lycopersicum* L.) injertado. *Interciencia* **2016**, *41*, 703–708.
9. Vasileva, V.H.; Dinev, N.S. Mineral content and quality parameters of tomato fruits as affected by different potassium ferti-lization treatments and cultivar specifics. *Indian J. Agric. Res.* **2020**, *55*, 169–174. [CrossRef]
10. Ding, X.; He, L.; Li, R.; Qian, T.; Zhang, H.; Jin, H.; Cui, J.; Wang, H.; Zhou, Q.; Zou, J.; et al. Zero discharge of nutrient solution to the environment in a soilless greenhouse cucumber production systems. *Plants* **2022**, *11*, 2252. [CrossRef]
11. Mazumder, M.N.N.; Misran, A.; Ding, P.; Wahab, P.E.M.; Mohamad, A. Preharvest foliar spray of calcium chloride on growth, yield, quality, and shelf life extension of different lowland tomato varieties in Malaysia. *Horticulturae* **2021**, *7*, 466. [CrossRef]
12. He, H.; Jin, X.; Ma, H.; Deng, Y.; Huang, J.; Yin, L. Changes of plant biomass partitioning, tissue nutrients and carbohydrates status in magnesium-deficient banana seedlings and remedy potential by foliar application of magnesium. *Sci. Hortic.* **2020**, *268*, 109377. [CrossRef]
13. Ahmed, M.Z.; Hussain, T.; Gulzar, S.; Adnan, M.Y.; Khan, M.A. Calcium improves the leaf physiology of salt treated Limonium stocksii: A floriculture crop. *Sci. Hortic.* **2021**, *285*, 110190. [CrossRef]
14. Larbi, A.; Kchaou, H.; Gaaliche, B.; Gargouri, K.; Boulal, H.; Morales, F. Supplementary potassium and calcium improves salt tolerance in olive plants. *Sci. Hortic.* **2020**, *260*, 108912. [CrossRef]
15. Chen, Z.C.; Peng, W.T.; Li, J.; Liao, H. Functional dissection and transport mechanism of magnesium in plants. *Semin. Cell Dev. Biol.* **2018**, *74*, 142–152. [CrossRef] [PubMed]
16. Nawaz, M.A.; Imtiaz, M.; Kong, Q.; Cheng, F.; Ahmed, W.; Huang, Y.; Bie, Z. Grafting: A Technique to Modify Ion Accumulation in Horticultural Crops. *Front. Plant Sci.* **2016**, *7*, 1457. [CrossRef] [PubMed]
17. Treviño López, E.A.; Sandoval-Rangel, A.; Benavides Mendoza, A.; Benavides Mendoza, A.; Ortega Ortiz, H.; Cadenas Pliego, G.; Cabrera de la Fuente, M. Nanopartículas de selenio absorbidas en hidrogeles de quitosán-polivinil alcohol en la producción de pepino injertado. *Rev. Mex. Cienc. Agrícolas* **2021**, *26*, 159–169. [CrossRef]
18. García-Ávila CD, J.; Castillo-González, A.M.; Avitia-García, E.; Colinas-León MT, B.; Trejo-Téllez, L.I.; Vargas-Madriz, H. Magnesio y su relación con la calidad de *Lilium* cv. Casablanca. *Rev. Mex. Cienc. Agrícolas* **2017**, *6*, 265–276. [CrossRef]
19. Sánchez, E.; Torres, A.; Flores, M.; Córdova, F.; Preciado, P.; Quiroz, C. Use of rootstocks on the yield, fruit quality and resistance to Phytophthora capsici Leonian in bell peppers. *Nova Sci.* **2015**, *7*, 227–244.

20. Grimaldo Juárez, O.; Suárez Hernández, Á.M.; Vargas-Hernández, E.A.; Carrazco Peña, L.D.; Morales Zamorano, L.A. Concentración de nutrientes en hoja y calidad de pepino en plantas injertadas bajo condiciones salinas. *Idesia* **2020**, *38*, 41–48. [CrossRef]
21. An, S.; Bae, J.H.; Kim, H.C.; Kwack, Y. Production of grafted vegetable seedlings in the republic of korea: Achievements, challenges and perspectives. *Hortic. Sci. Technol.* **2021**, *39*, 547–559. [CrossRef]
22. Pérez-Grajales, M.; Pérez-Reyes, T.Q.; Cruz-Álvarez, Ó.; Castro-Brindis, R.; Martínez-Damián, M.T. Compatibilidad del portainjerto CM-334 y su respuesta sobre el rendimiento, calidad fisicoquímica y contenido de capsaicinoides en frutos de Capsicum pubescens. *Inf. Técnica Económica Agrar. Rev.* **2021**, *117*, 332–346. [CrossRef]
23. Rodríguez-Mendoza, M.N.; Baca-Castillo, G.; García-Cué, J.L.; Urrieta-Velázquez, J.A. Aclareo de frutos y aspersiones foliares de calcio y miel de abeja sobre la calidad de tomate tipo costilla. *Rev. Fitotec. Meicana* **2015**, *38*, 197. [CrossRef]

Disclaimer/Publisher's Note: The statements, opinions and data contained in all publications are solely those of the individual author(s) and contributor(s) and not of MDPI and/or the editor(s). MDPI and/or the editor(s) disclaim responsibility for any injury to people or property resulting from any ideas, methods, instructions or products referred to in the content.

AgriEngineering

Review

Recent Progress on Catalytic of Rosin Esterification Using Different Agents of Reactant

Mardiah Mardiah [1,2], Tjokorde Walmiki Samadhi [1], Winny Wulandari [1], Aqsha Aqsha [1,3], Yohanes Andre Situmorang [1,3] and Antonius Indarto [3,*]

1 Department of Chemical Engineering, Institut Teknologi Bandung, Faculty of Industrial Technology, Bandung 40132, Indonesia; mardiah@ft.unmul.ac.id (M.M.); twsamadhi@che.itb.ac.id (T.W.S.); winnywulandari@itb.ac.id (W.W.); aqsha@itb.ac.id (A.A.); yohanes.andrest@gmail.com (Y.A.S.)
2 Chemical Engineering Program, Engineering Faculty, Universitas Mulawarman, Samarinda 75119, Indonesia
3 Department of Bioenergy Engineering and Chemurgy, Institut Teknologi Bandung, Bandung 45363, Indonesia
* Correspondence: antonius.indarto@itb.ac.id

Abstract: Gum rosin is an important agricultural commodity which is widely used as a raw material for various industries. However, gum rosin has low stability, crystallizes easily, and tends to oxidize. This is due to carboxyl groups and conjugated double bonds in gum rosin's structure. Therefore, to reduce these weaknesses, it is necessary to modify the rosin compound to achieve better stability via the esterification process. This paper surveys esterification agents such as glycerol, pentaerythritol, methanol, ethylene glycol, polyethylene glycol (PEG), allyl group, and starch Rosin ester. The product is used in the manufacture of pressure-sensitive adhesives, drug delivery, solder flux for electronic devices, as a plasticizer, and as a coating agent in fertilizers. In general, the esterification reaction between alcohols and carboxylic acids is very slow without a catalyst. Heterogeneous catalysts have the advantage of controlling size, structure, spatial distribution, surface composition, thermal-chemical stability, and selectivity. Among the catalysts for gum rosin esterification are ZSM-5, Fe_3O_4, ZnO, Calcium, TiO_2, Kaolin, and Al_2O_3, among others. Different catalysts and esterification agents can produce various physical and chemical properties of rosin ester and will result in specific rosin ester products, such as glycerol ester, pentaerythritol ester, methyl ester, glycol ester, allyl ester, and acid starch-based rosin.

Keywords: catalyst; esterification; ester; rosin

Citation: Mardiah, M.; Samadhi, T.W.; Wulandari, W.; Aqsha, A.; Situmorang, Y.A.; Indarto, A. Recent Progress on Catalytic of Rosin Esterification Using Different Agents of Reactant. *AgriEngineering* **2023**, *5*, 2155–2169. https://doi.org/10.3390/agriengineering5040132

Academic Editor: Muhammad Sultan

Received: 17 September 2023
Revised: 19 October 2023
Accepted: 30 October 2023
Published: 9 November 2023

Copyright: © 2023 by the authors. Licensee MDPI, Basel, Switzerland. This article is an open access article distributed under the terms and conditions of the Creative Commons Attribution (CC BY) license (https://creativecommons.org/licenses/by/4.0/).

1. Introduction

The production of sustainable materials is a highly significant problem that future generations must address. The supply of fossil fuels for energy production and plastic manufacturing is limited. It will be reduced during the next century. Environmental concerns combined with the decrease of oil reserves have sparked a surge in interest in developing green materials generated from renewable natural resources. Rosin, also known as colophony, is a natural substance obtained from pine trees that is biodegradable and abundant with a wide range of applications [1]. Indonesia, China, and Brazil are the main producers of pine sap, which is used to make turpentine (resin) and gum rosin [2]. Indonesia had a total pine-based production capacity of 141.150 tons in 2022 [3].

Rosin is a compound that originates from pine trees. It is obtained after the volatile component (turpentine) has been distilled. Rosin is translucent and its color ranges from brilliant yellow to brown. The substance exhibits insolubility in water but is soluble in a variety of organic solvents including alcohol, ether, acetone, benzene, chloroform, turpentine, and others [4].

Rosin comprises a complex mixture of organic acids with large molecular weights and neutral materials. Rosin acids are monocarboxylic acids derived from alkylated hydrophenanthrene nuclei that form most of the rosin. Rosin has small parts that are not acidic or neutral. These are made up of high molecular weight aldehydes, alcohols, esters, and hydrocarbons that have structures like the rosin acids. They may differ in relative proportions depending on the origin of the rosin and the degree to which it has been processed. Rosin is composed of 10% neutral and 90% acidic elements. Rosin has the chemical reactivity of a monocarboxylic acid in general. The structures of the rosin acids are shown in Figure 1 [5].

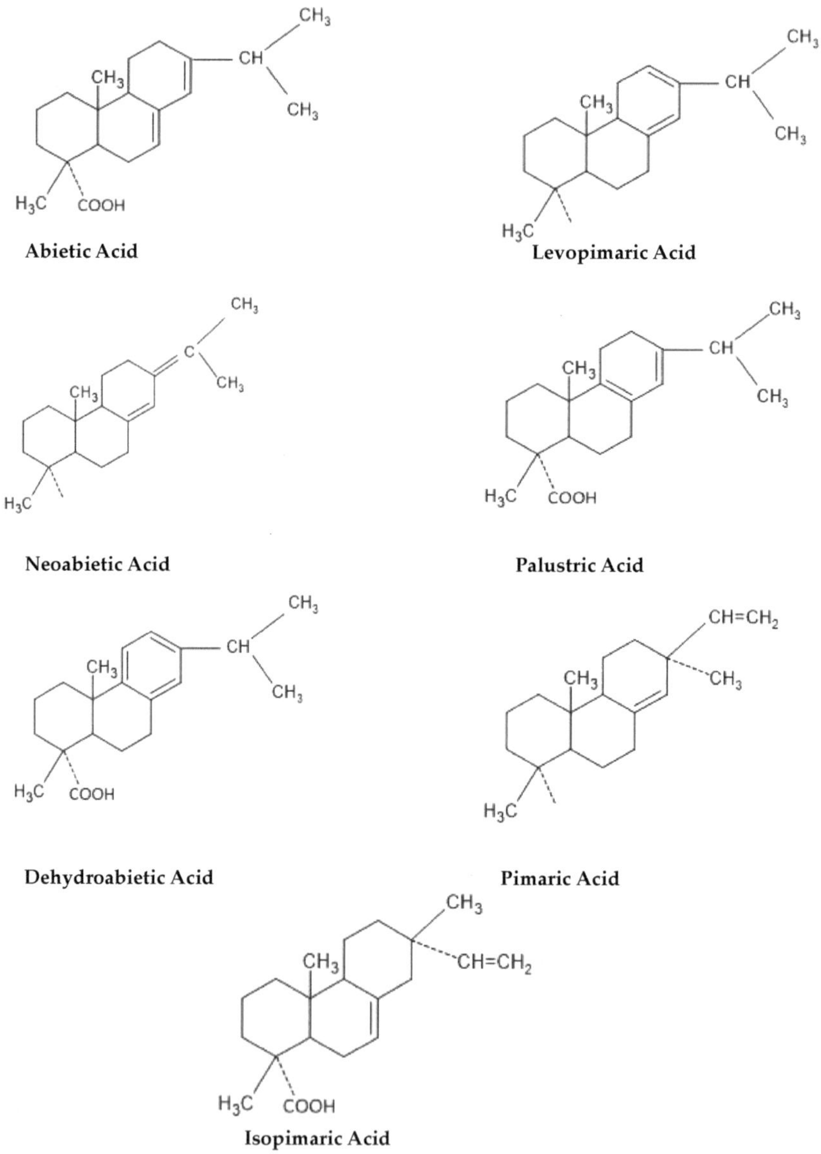

Figure 1. Structure of rosin acids [5].

Rosin is classified into numerous categories based on the portion of the tree from which it is derived, namely gum rosin, tall oil rosin, and wood rosin. Gum rosin is obtained from living pine trees, while tall oil rosin is produced as a by-product of paper pulp. Meanwhile, wood rosin is recovered from tree stumps [6]. Rosin has a variety of uses, including for antifouling [7], adhesives [8], cosmetics [9], and adsorbent [10]. Because of their biocompatibility, rosin and its by-products are being investigated in the pharmaceutical industry as an encapsulating agent for drug delivery [11]. Gum rosin is also used in solder flux [12] and road-marking paint [2,13].

Aside from the numerous advantages of rosin, the current issue is the high content of abietic acid in rosin, which causes a low softening point, has a brittle property, and leads to a dark appearance. The typical abietic acid content of gum rosin in Indonesia is 30–40%. The amount of abietic acid in gum rosin affects its quality, as it causes rosin to crystallize quickly when it is cooled from its liquid form. Crystallization causes problems in rosin production and industries that use rosin, and as such, abietic acid must be minimized to produce higher quality rosin by changing the acid to a modified form that keeps the rosin stable [14].

The structure of rosin has a carboxyl group and a conjugated double bond, and those play a big role in its reactivity. As a result, rosin is chemically modified to improve its stability and useful characteristics. Esterification-maleation [15], hydrogenation [16], dimerization [17], isomerization, and disproportionation [18] are all strategies for improving the oxidative stability of rosin [19].

Esterification is a typical modification procedure commonly used by industry and is a simple process (Figure 2). Rosin esters are formed when rosin is esterified with an alcohol or polyol. Alcohols and polyols such as glycerol, pentaerythritol, etyhlene glycol are widely used for industrial application, while PEG, methanol, allyl groups, and starch are also being continuously improved. Esterification can reduce the acid value of rosin, raise its thermal stability, and improve its resistance to both acids and alkalis. As a result, the raw material's properties allow it to be employed in more applications [20].

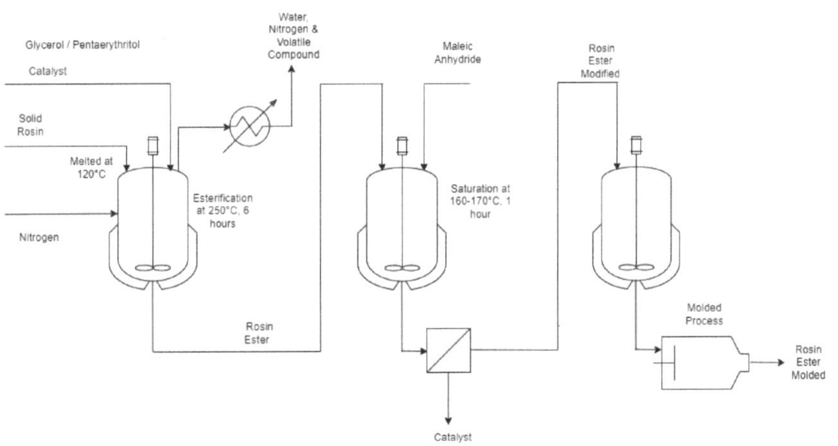

Figure 2. Flowsheet of esterification of gum rosin (processed from [2,21,22]).

The esterification process is marked by a decrease in the acid value of rosin, which enhances esterification conversion. Since rosin contains abietic acid, it will change to rosin ester when it is esterified. The rosin ester will confirm by Fourier transform infrared spectroscopy. Stretching O–H carboxylic acids is at a frequency of 3600–2500 cm^{-1} and C=O, which is an ester group, is at a frequency of 1750–1720 cm^{-1}.

The rosin ester product depends on the type of reactant used (Figure 3). For example, if methanol is used as a reactant, the rosin ester can take the form of a methyl ester, which has fuel-like properties [23]. If glycerol and pentaerythritol are used as a reactant, a solid rosin ester will be formed with a high softening point that could exceed 100 °C. This product is suitable for raw material of road marking paint [2,20]. When PEG is used in the reaction, the rosin ester will dissolve in water and this product can be used as drug delivery [24]. Esterification with an allyl group can enhance properties of biomass-based polymer material. The ester can be used as a coating or a glue adhesion agent [25]. The product using starch can be used as food packaging [26].

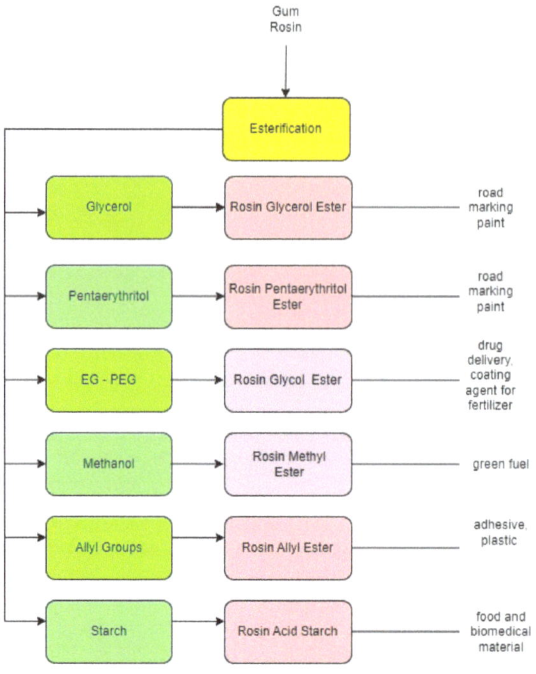

Figure 3. Agent of reactant in esterification.

This paper reviews the esterification of rosin utilizing various reactant agents such as glycerol, pentaerythritol, methanol, ethylene glycol, PEG, allyl groups and starches that are often useful in industry.

2. Rosin-Glycerol

Glycerol esterification with carboxylic acids from rosin (mostly abietic acid) is a common reaction in industry (for up to 80–90 years) [27]. Rosin glyceride is widely employed as one of the most important rosin modification products. Rosin glyceride, also known as gum ester, is formed when rosin interacts with glycerol. The higher the quality of gum esters, the lighter the color. Gum esters have a refractive index of 1.545, a relative density of 1.095, an acid value of less than 10 mg KOH g^{-1}, and a softening point over 80 °C. Gum esters are soluble in aliphatic and aromatic hydrocarbon solvents, as well as terpenes, esters, hydrocarbons, ketones, and the majority of essential oils, but are insoluble in water and alcohols of low molecular weight [21].

Glycerol, also known as 1,2,3-propanetriol, is a very viscous and dense polyalcohol with strong hygroscopic properties. It finds many uses in industries such as cosmetics, food, pharmaceuticals, and chemicals [28]. This product is obtained as a by-product of biodiesel manufacturing [29]. Biodiesel production is still in its infancy worldwide. Meanwhile, market prices for glycerol are declining rapidly and are over-available [30]. Its utilization is expected to increase as glycerol-based processes.

The following is the reaction mechanism between rosin and glycerol (Figure 4). Glycerol, which has three OH groups, allows it to react with an acid group of rosin to form three ester groups and triggers a process of dehydration/release of water molecules. The most common method used to produce esters is the reaction of a carboxylic acid with an alcohol with the release of water. Since the presence of water in the reacting mixture tends to shift the equilibrium away from rosin ester, water was continuously removed during the reaction.

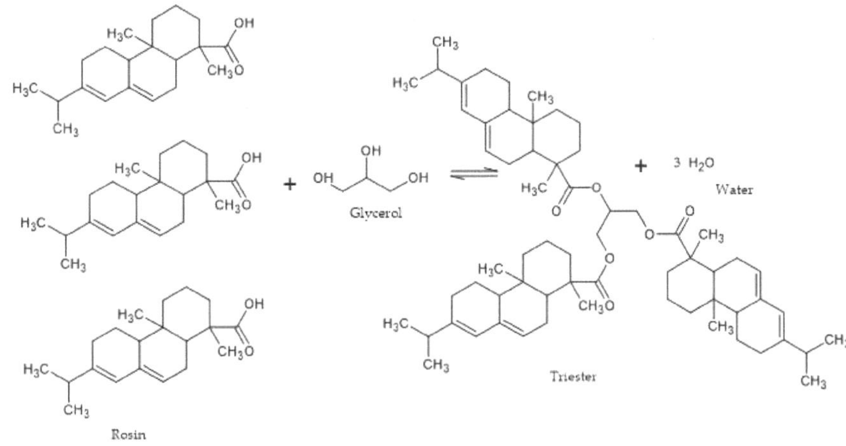

Figure 4. Mechanism of rosin reaction with glycerol (processed from [22]).

The esterification reaction can be carried out with or without a catalyst, but the conversion will be low if done without a catalyst (Table 1). The expected rosin ester with a low acid value is obtained by heating the rosin with glycerol at a temperature of 250–290 °C.

However, the use of liquid catalysts such as H_2SO_4 and H_3PO_4 in the production of rosin glycerides causes corrosion, is difficult to separate and is always environmentally harmful [22,31]. Due to the widespread use of rosin ester products in manufacturing, such as in food, medicine, printing inks and pressure-sensitive adhesives, the esterification of rosin with alcohol on heterogeneous catalysts or solid catalysts is an important reaction for industry.

Previously, hydrothermal methods were employed for the synthesis of ZSM-5. Zeolite catalytic performance was examined using rosin and glycerol as reagents for esterification. The ZSM-5 zeolites that were created had a larger specific surface area and mesoporous volume than commercial ZSM-5 zeolites, indicating a faster esterification, lower product acid values, and increased stability. The obtained esterification percentage was 93.73% [21].

Zeolite is utilized as a catalyst in several processes because of its high activity and selectivity. This is attributed to its excellent ion exchange performance, consistent pore structure, acidity, and great temperature persistence. Zeolite ZSM-5 is a catalyst that may be used for rosin and glycerol esterification processes since it its various advantages include not being corrosive to equipment, high temperature tolerance, simplicity of manufacture, high activity, and the capacity to be reused. Furthermore, solid granules of zeolite may be easily isolated from the reaction mixture [21].

Several researchers modified natural zeolite with nickel metal to boost catalytic activity. The highest conversion of 82.86% was achieved after 3 h at 240 °C and 11% ratio [14]. La metal also added to the ZSM-5 zeolite [21]. Several catalysts affect rosin esterification severity, as seen in Table 1.

La-ZSM-5 has a greater acid value and esterification rate than ZSM-5, as demonstrated in Table 1. Catalyzing the liquid phase esterification reaction requires acidification of the catalyst surface. Both the core of Bronsted acid and the center of Lewis acid can undergo the esterification reaction. The esterification reaction, on the other hand, is dependent on the Lewis acid center. Meanwhile, reaction byproducts such as ether and olefins are primarily generated in the Bronsted acid center. Based on existing catalyst characterization, La-ZSM-5 exhibits a higher concentration of Lewis acid sites compared to ZSM-5. As a result, La-ZSM-5 has higher esterification catalysis activities compared to ZSM-5 [21].

Table 1. Impact of various catalysts on the esterification process of rosin-glycerol.

Product	Reaction Time (h)	Temperature (°C)	Molar Ratio (Rosin/Glycerol)	Catalyst	Acid Value (mg KOH g^{-1})	Conversion Rate (%)	Ref.
Rosin glyceride	3.5	269	1.32	No catalyst	66.54	58.58	[32]
	3	240	11% (wt)	Ni/Zeolite	33.94	82.86%	[14]
	1 2 4 6 8 10	260	1.5 (mass ratio)	ZSM-5	69.72 38.58 20.90 13.05 11.08 10.66	58.99 77.31 87.71 92.32 93,48 93.73	[21]
	1 2 4 6 8 10	260	1.5 (mass ratio)	LaZSM-5	72.67 50.63 22.62 19.12 16.83 15.15	65.25 82.22 90.69 93.75 95.10 98.09	[21]
	2.5	240	2	Fe_3O_4/MOF-5		92.6	[22]
	3.5	269	1.32	ZnO	10.23	93.63	[32]
	3.5	269	1.32	CO_2 pressure of 3.95 MPa	8.45	94.74	[32]

The esterification reaction mechanism of rosin by glycerol using annealed Fe_3O_4/MOF-5 at temperatures ranging from ambient temperature to 900 °C was also examined [22]. Table 1 summarizes the impact of the main response parameters. In general, room temperature annealing provided the best conditions for rosin esterification utilizing Fe_3O_4/MOF-5. The development of super/subcritical CO_2 and the presence of high temperature liquid water are investigated as environmentally friendly acid catalysts in the esterification process between rosin and glycerol [32]. Experiments with response surface methods revealed that CO_2 pressure affects the yield, which is higher than that produced with a ZnO catalyst. However, the difference is barely 1%. Zinc oxide is also a common catalyst in the rosin esterification reaction. The abundance of oxygen vacancies on the surface of ZnO can boost the catalyst's Lewis acidity and catalytic activity [33].

Based on the above experiments that have been discussed and shown in Table 1, heterogeneous catalysts ZSM-5 modified by La metal achieve the highest conversions (up to 98%) but still require a higher reaction time.

3. Rosin-Pentaerythritol

Pentaerythritol is another polyhydric alcohol that frequently reacts with rosin acids. This polyalcohol has four hydroxyls. The material is a colourless crystalline compound that lacks odour and exhibits limited solubility in cold water but is readily soluble in hot water. At 30 mmHg, it has a melting point of 260 °C and a boiling temperature of 276 °C. Rosin pentaerythritol ester (RPE) is produced by esterifying rosin with pentaerythritol. RPE leads rosin glycerol ester in terms of softening point [28]. RPE is also resistant to water, alkalis, and oxidation. Therefore it can be used for paints, varnishes, as an oil lubricant additive [34], and a pressure-sensitive and hot-melt adhesive tackifier [35].

The kinetics of esterification between two polyhydric alcohols and rosin acids, glycerol and pentaerythritol, are also investigated. The data obtained by analyzing trends in the initial reaction rate and estimating kinetic constants using statistical regression with integral approach, along with microscopic observations, were utilized to construct a second order kinetic model for pentaerythritol. The rosin-glycerol system exhibits a two-stage kinetic model, wherein the identical approach yields a pseudo-first-order reaction at starting time and higher glycerol concentration and a second-order reaction for longer reaction times and lower glycerol concentration [28].

Pentaerythritol, which has four OH groups, reacts with an acid group of rosin to form tetraester. A kinetic model of RPE synthesis that included a sequence of events leading to the formation of mono-, di-, tri-, and tetra-esters (Figure 5) in addition to decarboxylation of rosin as a side reaction was provided. The reaction rates for esterification were calculated using a second-order kinetic model. The process of decarboxylation was hypothesized to follow a first-order kinetic order. The kinetic experiments were conducted within a temperature range of 260–290 °C. Pentaerythritol to rosin molar ratio was 0.8–1.2 in moles of OH/COOH. The model also correctly predicts the relationship between tetraester production and softening point, with a higher tetraester indicating a higher softening point [20].

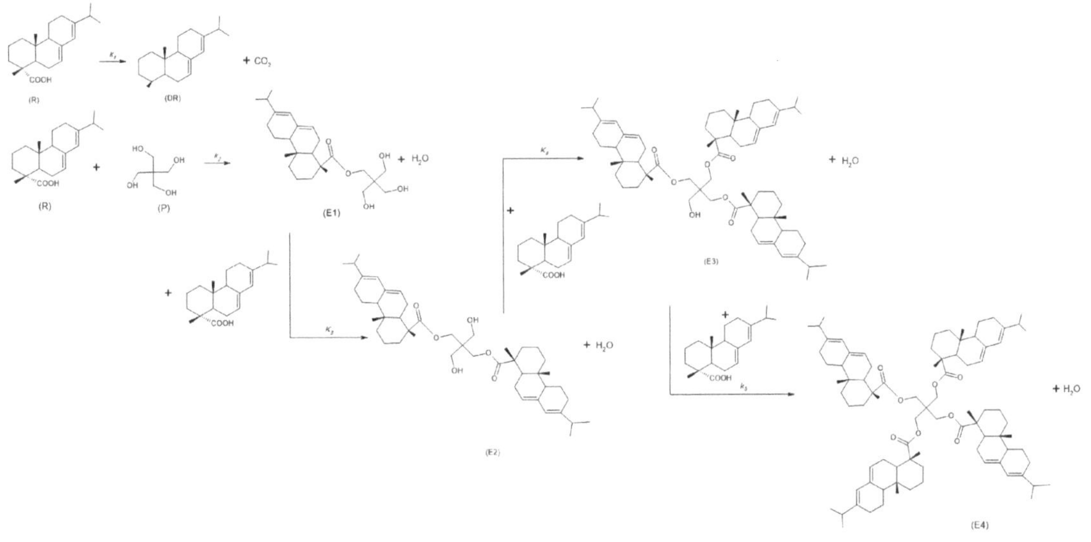

Figure 5. The steps used to change rosin and pentaerythritol into esters [20].

The use of catalyst Fe_3O_4-supported ZnO facilitated the production of pentaerythritol esters from rosin, resulting in enhanced stability, water, alkali, and oxidation resistance [36]. This catalyst was created to overcome the difficulties of catalyst recovery and the high costs associated with it in the production of rosin pentaerythritol esters. According to the results of the esterification process, Fe_3O_4@SiO_2–ZnO has a good catalytic performance and stability. The catalyst exhibited the potential for many reuses, with a maximum of five iterations observed. The optimal process parameters for producing RPE were determined as follows: zinc nitrate was used as the zinc source, with a theoretical loading of 0.45 g. The reaction was carried out at a temperature of 250 °C for a duration of 5 h, with a catalyst loading of 0.2 g. The starting materials consisted of 15 g of rosin and 0.2 g of pentaerythritol. The acid value was 25.0 mg KOH g^{-1}, with an 87.9% conversion rate [36].

The properties of RPE are summarized in Table 2. Based on the research above, the use of a catalyst in the pentaerythritol reactant as a modification agent can increase the esterification rate compared to not using a catalyst. Apart from that, the use of catalysts can reduce the mole ratio of reactant agents.

Table 2. Properties of rosin pentaerythritol ester.

Product	Reaction Time (h)	Temp. (°C)	Ratio	Catalyst	Acid Value (mg KOH g^{-1})	Soft. Point (°C)	Conversion Rate (%)	Ref.
Rosin (raw material)					190	82		[20]
	4	260–290	0.8–1.2 (Mol OH/COOH)			100–109	76–86.59	[20]
Rosin Pentaerythritol	5	250	15 g rosin, P = 1.7 g, catalyst 0.2 g		128.6		37.7	[36]
				$Fe_3O_4@SiO_2$–ZnO (Zn (Oac)$_2$)	120.0		42.0	
				$Fe_3O_4@SiO_2$–ZnO ($ZnCl_2$)	128.7		37.7	
				$Fe_3O_4@SiO_2$–ZnO ($ZnSO_4$)	71.1		65.6	
				$Fe_3O_4@SiO_2$–ZnO (Zn (NO_3)$_2$)	25		87.9	

4. Rosin-Methanol

The usage of rosin as a fuel may become increasingly popular in the future. Rosin is a desirable source for biofuel [37,38]. Oleoresin consists mostly of rosin, which makes up around 80% of its composition. Rosin is a solid substance composed of isomeric abietic acid. Due to its composition, oleoresin can't be used as a fuel in engines. Nevertheless, when rosin is changed into methyl esters, it exhibits reduced boiling and melting points and displays exceptional solubility in hydrocarbons [23].The process esterification is OH groups in methyl alcohol react with an acid group of rosin to form water and methyl ester (Figure 6).

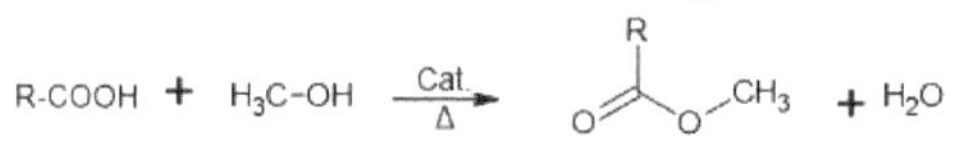

where the acids R-COOH are:

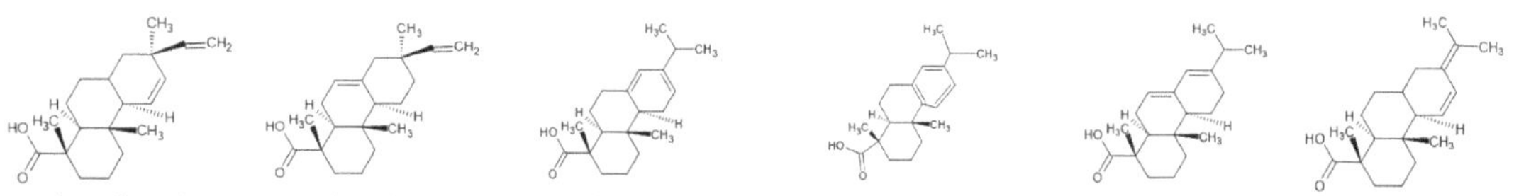

Pimaric Acid **Isopimaric Acid** **Levopimaric Acid** **Dehydroabietic Acid** **Abietic Acid** **Neoabietic Acid**

Figure 6. Scheme of rosin methyl ester reaction [23].

An investigation was conducted on the esterification of rosin with methyl alcohol utilizing acid and basic solid catalysts in the presence of different solvents [23]. The use of a calcium catalyst material and the absence of a solvent resulted in an increased conversion rate. The calcium-based substance was mostly calcium carbonate and calcium hydroxide, suggesting that basic site strength can affect catalytic performance. In five reaction cycles, a calcium-based material was utilized, and the activity was significantly reduced due to of the leaching of many catalyst phases. The activity of Amberlyst 15 for the reaction is absent, implying that being acidic does not stimulate the production of esters according to heterogeneous circumstances [23].

A set of mesoporous molecular sieves was synthesized utilizing Zr-MCM-41, Mo-MCM-41, Ti-MCM-41, and Sn-MCM-41 [39]. A group with acidic properties, such as (SO_4^{2-}, PTSA, etc.) was added to the mesoporous molecular sieves to create modified catalysts. As a result, the PTSA/ZrO_2/Mo-MCM-41 material has a mesoporous molecular sieve structure and contains many acidic sites, as confirmed by XRD, N_2 absorption-desorption isotherms, FT-IR, and Py-IR analyses. The esterification rate was 88.2% when the PTSA/ZrO2/Mo-MCM-41 catalyst was used to make rosin methyl ester, showing better catalytic performance. It is stable and can be remade and used six times without losing any of its effectiveness.

A researcher also reported on the use of spent fluid cracking catalyst (SFCCR), which is plentiful, averaging 160,000 tons per year [40]. They evaluated the process of combining rosin with methyl alcohol using subcritical CO_2 atmosphere to form an ester. Non-regenerated spent catalysts are often buried and contaminate the environment. Spent hydroprocessing catalysts has been used in construction, ceramic and refractory applications [41]. In reality, because SFCCR is predominantly made up of silica and alumina, it possesses a steady molecular sieve arrangement that has a lot of surface area and might be used as a catalytic support [42]. The findings demonstrated that CO_2 functioned as an additional catalyst, enhancing the transfer of mass and regulating the pH within the range of 3.54 to 3.91. A conversion rate of 97% was achieved during a reaction period of 5 h at a temperature of 220 °C and in the absence of a solvent.

Table 3 shows the properties of rosin methyl ester using different catalysts. Based on the data in Table 3, the ZnO metallic catalyst in SCFRR with CO_2 as an additional catalyst can achieve highest conversion of reaction.

Table 3. Properties of rosin methyl ester.

Product	React. Time (h)	Temp. (°C)	Molar Ratio (Rosin/MA)	Catalyst	Acid Value (mg KOH g^{-1})	Conversion Rate (%)	Ref.
Rosin Methyl Ester	3	64	1:748	Calcium Material (40%)		46	[23]
				MgO (20%)		23	
				TiO_2 (20%)		27	
				Kaolin (40%)		15	
				Al_2O_3 (40%)		23	
				Amberlyst 15 (40%)		~4	
	3.5	64	1:187	Calcium Material (40%)		55	
	3	180		PTSA/ZrO_2/Mo-MCM-41	28	82.7	[39]
	4	220		No Catalyst		55.42	[40]
				CO_2		66.38	
				Al_2O_3		65.52	
				ZnO		72.42	
				ZnO/SFCCR		83.28	
				ZnO/SFCCR + CO_2		94.08	
	5		1:1.5	ZnO/SFCCR + CO_2		97.01	

5. Rosin-Ethylene Glycol and Polyethylene Glycol

Rosin ethylene glycol ester is a polymeric resin that is frequently used as a tackifier raw material in the manufacture of adhesives, as well as a solid soldering flux and plasticizer. Fe_3O_4@ZnO and Fe_3O_4@SiO_2–ZnO catalyzed the reactions successfully [33]. The optimal process parameters for rosin glycol esters were established to be: a theoretical load of 0.45 g of zinc chloride as the zinc supply, a reaction temperature of 270 °C, a reaction duration of 5 h, and the use of 15 g of rosin, 3.2 g of ethylene glycol, and 0.2 g of catalyst. The result of rosin ethylene glycol ester exhibited an esterification rate of 87.2% and an acid value of 26.5 mg KOH g^{-1}. After being reused five times, the synthesized catalyst showed no significant decline in catalytic activity, indicating its superior stability and availability [33]. Different catalysts that affect rosin glycol esters synthesis (Table 4) were studied.

Table 4. Properties of rosin ethylene and polyethylene glycol ester.

Product	React. Time (h)	Temp. (°C)	Molar Ratio (Rosin/Glycol)	Catalyst	Acid Value (mg KOH g^{-1})	Soft. Point (°C)	Conversion Rate (%)	Ref.
Rosin Polyethylene Glycol 400	until the acid value is constant	220		Zinc dust (0.5%)	116.87 (brown)	52–54	30.85	[24]
Rosin Ethylene Glycol Ester	5	270		-		115.6	44	[33]
				Fe_3O_4@SiO_2–ZnO ($ZnAC_2$)		51	75.3	
				Fe_3O_4@SiO_2–ZnO ($ZnCl_2$)		26.4	87.2	
				(Zn $(NO_3)_2$)		154	25	
				Fe_3O_4@SiO_2–ZnO ($ZnSO_4$)		95.9	54	
				Fe3O4@SiO_2–ZnO ($ZnCl_2$)		33.8	83.6	
Rosin Polyethylene Glycol 1500	5	250	1:1	ZnO (0.5%)			~52	[43]
				Zn dust (0.5%)			~48	
				$SnCl_2$ (0.5%)			~40	
	9	250	2:1	ZnO (2%)			95	

Table 4. *Cont.*

Product	React. Time (h)	Temp. (°C)	Molar Ratio (Rosin/Glycol)	Catalyst	Acid Value (mg KOH g^{-1})	Soft. Point (°C)	Conversion Rate (%)	Ref.
Rosin Polyethylene Glycol 200	until the acid value is constant	240–260		Zinc dust (0.5%)	120.93 (yellow solid)	60–65	28.67	[44]
					88.19 (yellowish red solid)	50–55	47.98	
					57.71 (reddish brown solid)	45–50	65.96	

Rosin is insoluble in water, but it dissolves in ether, benzene, chloroform, and alcohol. Polyethylene glycol (PEG) is a petroleum-based polymer that represents only one group –OH that is soluble in water. Hydrophilic polymers are used in ointments and supplements as a base, as a plasticizer in film coating, as supplemental emulsifiers, and in electrical applications [43].

The properties of derivatives of rosin-polyethylene glycol from rosin and PEG was studied [43]. The target of this reaction was the carboxylic group of rosin that reacted with the hydroxyl group of PEGs to form rosin-PEG as water soluble rosin derivatives via esterification reaction. In this research, Zn-dust, ZnO and Tin chloride were selected as catalysts. Rate of reaction was significantly increased by all catalysts. Due to Lewis's acid performance, zinc oxide proved to be the most effective catalyst as it led to the largest conversion of ester after 5 h. This finding demonstrates the effectiveness of the catalyst. The optimum reaction with 95 % conversion was achieved by a temperature of 250 °C, a molar ratio of rosin to PEG of 2:1, and reaction time at 9. These derivatives were completely soluble in water, and therefore called water soluble rosin (WSR). Effects of variable reaction to conversion of esterification reaction using ethylene and polyethylene glycol are represented in Table 4.

Bioactive agent distribution via polymeric materials has received a lot of interest. PEGylation, the process of attaching PEG to any medication, peptide, polymer, or chemical component, has been studied for its potential as a pharmacological aid [44]. The ester was made by reacting to a primary alcohol, PEG 200, and rosin. The ester-adduct derivative of rosin was formed by reacting maleic anhydride with this product. The concentration of PEG 200 in the derivative preparation was varied, while the maleic anhydride content remained constant. The effect of these components on the physicochemical parameters of derivatives was compared to rosin (Figure 7). These compounds were also investigated for their capacity to release medicine in tablets and pellets. Additional studies have demonstrated that rosin can be successfully combined with PEG 400 and maleic anhydride to create a derivative of rosin that has been modified with PEG. The derivatives of rosin may be utilized to create both aqueous and organic films, which can be employed for drug delivery purposes [24].

Water soluble rosin flux (WSRF) was also synthesized from polyethylene glycol (PEG) and rosin [12]. The flux was efficiently produced by combining rosin and PEG with a 2% ZnO catalyst, using a molar ratio rosin to PEG of 2:1, and subjecting the mixture to a temperature of 250 °C for a duration of 9 h. The resulting WSRs were entirely soluble in water. The PEG3000-rosin was used for the manufacture of the WSRFs because it had the highest melting point (55.2 °C). The PEG-rosin derivatives possess the property of being soluble in water, and when combined with a low-halide-content activator, they produce solid soldering fluxes that exhibit satisfactory wetting ability. The fluxes were used in the production of solder pastes, employing the Sn-3.0Ag-0.5Cu alloy, and they exhibited the expected performance for solder paste [12].

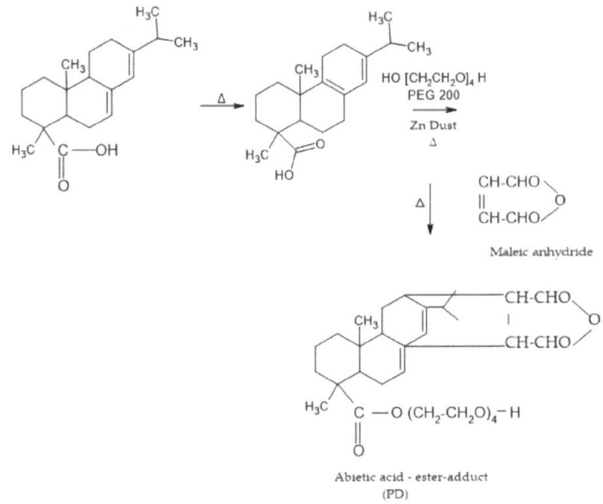

Figure 7. Reaction scheme of PEGylation [44].

The usual urea is squandered because of its significant water solubility and the ability of microorganisms to convert it, resulting in low agricultural productivity and environmental damage. The scientists created a rosin PEG adduct-coated controlled release urea fertilizer (RA-CRUF) to address these issues [45]. The synthesis process involved enhancing the properties of the rosin adduct with maleic anhydride by manipulating the reaction temperature, reaction time, and coating repetitions. The process of creating this product involves the application of rosin adduct onto urea by a spraying method. The optimum urea release rate for RA-CRUF was 510 mg L^{-1} when dissolved in water under 6.79 coating repetitions, a reaction period of 4.71 h, and a reaction temperature of 137.39 °C. The total amount of urea released was 41.76% and 47.23% in soils with a clay loam texture, respectively, compared to 100% and 82.6% for conventional urea. Regarding RA-CRUF's inquiry on the growing performance of maize plants, growth was classified as healthy with a notable rise in chlorophyll levels. Specifically, levels reached 51.11 mg g^{-1} in clay loam texture soil and 87.55 mg g^{-1} in loam texture soil.

The use of EG and the polymer, which can be dissolved in water, can indicate that it will produce rosin esters which can also be dissolved in water or released in water. Zn-based catalysts can achieve the highest conversion of up to 95% but still require a higher reaction time. To obtains high conversion and softening points; it is necessary to also pay attention to the PEG or EG molecular weight.

6. Rosin–Allyl Groups

Many other types of fine chemicals can be made from rosin. It is frequently utilized as a basic material to create functional polymers. There are two methods for incorporating rosin compounds into polymers: either by directly linking them with the polymer or by modifying suitable rosin-type monomers through the addition of active vinyl groups [46].

The first technique was introduced for synthesizing poly(vinyl benzyl abietate) from from poly(vinylbenzyl chloride)-sodium abietate and polymethacrylate from poly(glycidyl methacrylate)-abietic acid [47,48]. The addition of rosin into polymer significantly enhances mechanical and chemical qualities, including hardness and resistance to chemicals. Nevertheless, under these conditions, the rosin molecules were only attached to the surface of the polymer instead of being incorporated into the polymer's molecular structure.

The second route involves rosin monomers being added to vinyl groups [49]. Due to their large molecular weight, rosin acids frequently exhibit limited reactivity. To address this issue, allyl chloride was employed as a reactant due to its low boiling point, facilitating its removal after the reaction via vacuum distillation at low temperatures. This approach helps prevent the polymerization of the product (Figure 8). The synthesis of rosin allyl ester involved the reaction between sodium rosinate and allyl chloride as reactants. The ester yield reached a remarkable 94.7% under ideal reaction conditions [46].

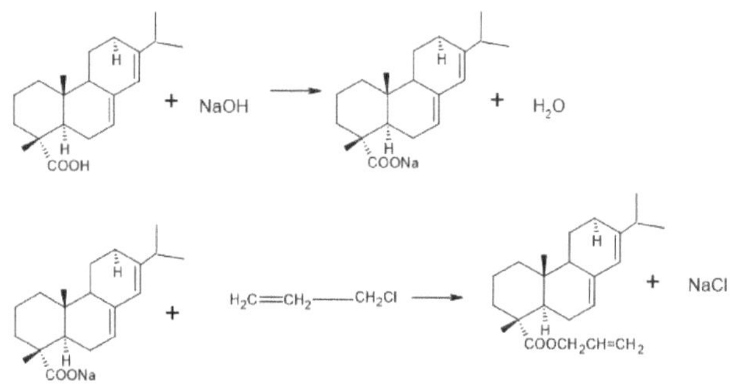

Figure 8. Rosin allyl ester synthesis [46].

A new reactive monomer allyl acryl pimarate was also made from natural rosin [25]. This material exhibits significant promise for utilization as both a biomass-derived polymer substance and a valuable fine chemical, serving as a coating or adhesive agent. Recently, a novel rosin monomer containing two allyl groups was produced using phase-transfer catalysis and the Diels-Alder process. It was thought that adding an extra active group (allyl) would increase the reaction's conversion of the monomer and encourage cross-linking between the monomers to create a mechanically stable network structure. Microwave radiation has been employed to accelerate the process. As a photoinitiator, Michlers ketone was utilized, and as a result, 90% of the monomer was converted [25].

A range of novel rosin additives can be generated by implementing a sequence of adjustments. Renewable polymers may have better adhesion, impact strength, and solvent resistance [25].

7. Enzymatic Rosin-Starch

The development of bio-based polymers to replace standard petrochemical polymers has gotten a lot of interest in recent decades [50]. Starch is the most intriguing biopolymer because of its low cost, availability, biocompatibility, and biodegradability. Native starch's use in industry has been limited due to its inherent disadvantages, such as poor surface qualities, high hydrophilicity, and poor mechanical and thermal properties. A range of physical or chemical changes to native starch have been made to improve starch characteristics [51]. Esterification is a frequent and successful approach for customizing the properties of this renewable resource [52].

The production of long-chain fatty acid starch esters involves the process of esterification, where stearic acid or lauric acid is combined with maize starch in urea/NaOH aqueous conditions [53]. An investigation of the degree of substitution (DS) of starch stearat was 0.065 and starch laurate was 0.074. The findings also indicate that starch esters exhibit superior performance compared to native starch in terms of thermostability, emulsifiability, transparency, freeze-thaw stability, and moisture absorption. Additionally, starch esters have a higher glass transition temperature (Tg) within the limited degree of substitution (DS).

Following that, new methods based on the conventional approach, such as the enzyme catalytic method, are created. Rosin acid and starch was synthesized enzymatically using lipase by researchers (Figure 9) [26]. For pretreatment, the cassava starch was dissolved in a NaOH/urea solution. The reactor was filled with prepared cassava starch, followed by the addition of DMSO, rosin, and the immobilized catalyst for 1, 2, and 4 h. After reaction, the DS ranging from 0.031–0.092 to natural starch was confirmed by FTIR. Although esterified starch showed low thermal stability and transparency compared to native starch, there was a considerable improvement in hydrophobicity, viscosity, and emulsifying capabilities [26].

Furthermore, rosin acid is quite biodegradable and biocompatible. The thermal stability and hydrophobicity of rosin acid starch have clearly increased, and it has prospective uses in waterproof coatings and plastic materials [1]. Food and biomedical products may also benefit from this use of rosin acid starch [26].

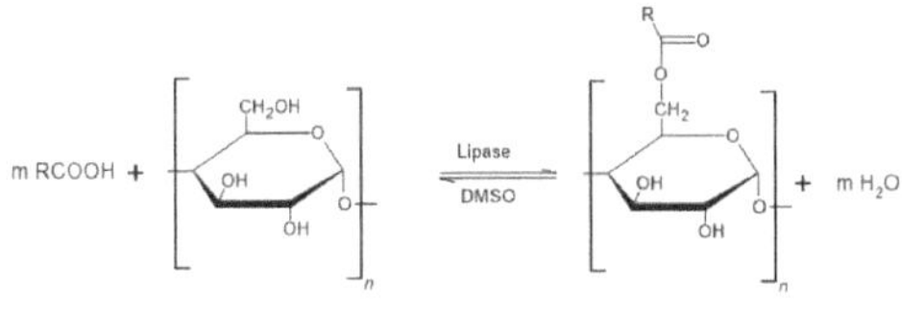

where RCOOH are

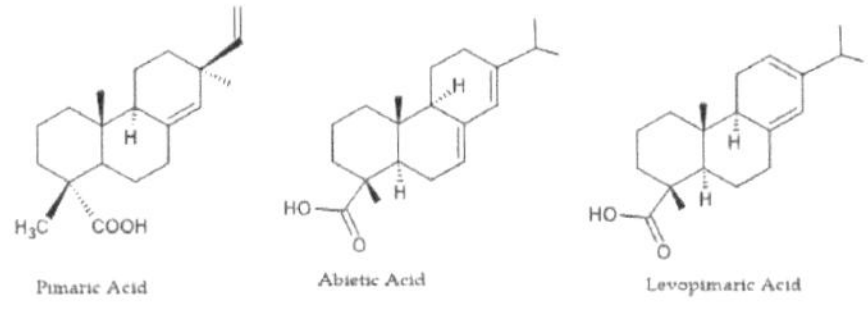

Figure 9. Esterification rosin acid and starch [26].

8. Conclusions

Rosin with its carboxyl group and conjugate double bonds has an important role in the esterification process. By using reactants or ester agents such as glycerol, pentaerythritol, methanol, glycol, allyl group, and starch, specific rosin ester will be produced.

The challenge in esterification is how to reduce the acid value or increase the rosin ester conversion, such as using heterogeneous catalysts. One of the advantages is that it can be separated. However, if the catalyst is dispersed during the reaction, it is not easy to separate. Meanwhile the use of heterogenous catalysts is a prospective way to increase the conversion of the reaction and reduce the acid value of the rosin.

Rosin is a source of raw materials that are renewable and may show potential in the future.

Author Contributions: Writing—original draft preparation M.M. and A.I.; data curation M.M. and W.W.; writing—review and editing W.W., A.I. and T.W.S.; Supervision T.W.S., W.W. and A.I., Conceptualization and Funding Acquisition A.A. and Y.A.S. All authors have read and agreed to the published version of the manuscript.

Funding: Financial support from Indonesia Oil Palm Plantation Fund Management Agency (BPDPKS): GRS-22 is gratefully acknowledged.

Data Availability Statement: Data are contained within the article.

Conflicts of Interest: The authors declare no conflict of interest.

References

1. Zhang, X.; Ma, H.; Qin, W.; Guo, B.; Li, P. Antimicrobial and improved performance of biodegradable thermoplastic starch by using natural rosin to replace part of glycerol. *Ind. Crops Prod.* **2022**, *178*, 114613. [CrossRef]
2. Aqsha, A.; Winoto, H.P.; Adhi, T.P.; Adisasmito, S.; Ramli, Y.; Siddiq, L.; Pratama, F.B.; Ramdani, M.R.; Indarto, A. Sequential Esterification—Diels-Alder Reactions for Improving Pine Rosin Durability within Road Marking Paint. *Molecules* **2023**, *28*, 5236. [CrossRef] [PubMed]
3. Prakoso, T.; Hanley, J.; Soebianta, M.N.; Soerawidjaja, T.H.; Indarto, A. Synthesis of Terpineol from α-Pinene Using Low-Price Acid Catalyst. *Catal. Lett.* **2018**, *148*, 725. [CrossRef]
4. Cabaret, T.; Boulicaud, B.; Chatet, E.; Charrier, B. Study of rosin softening point through thermal treatment for a better understanding of maritime pine exudation. *Eur. J. Wood Wood Prod.* **2018**, *76*, 1453–1459. [CrossRef]
5. Maiti, S.; Ray, S.S.; Kundu, A.K. Rosin: A renewable resource for polymers and polymer chemicals. *Prog. Polym. Sci.* **1989**, *14*, 297–338. [CrossRef]

6. Silvestre, A.J.; Gandini, A. Rosin: Major sources, properties and applications. In *Monomers, Polymers and Composites from Renewable Resources*; Elsevier: Amsterdam, The Netherlands, 2008; pp. 67–88.
7. Saat, A.M.; Yaakup, S.; Alaauldin, S.; Zawani, F.; Azaim, Z.; Isa, M.D.M.; Kamil, M.S.; Samsudin, S.; Iqbal, M. Performance of Rosin Modified Antifouling Coated on Mild Steel Surface at Various Immersion Orientation. *Int. J. Innov. Technol. Explor. Eng.* **2019**, *8*, 5562–5565. [CrossRef]
8. Lee, S.; Lee, K.; Kim, Y.-W.; Shin, J. Preparation and characterization of a renewable pressure-sensitive adhesive system derived from ε-decalactone, l-lactide, epoxidized soybean oil, and rosin ester. *ACS Sustain. Chem. Eng.* **2015**, *3*, 2309–2320. [CrossRef]
9. Kim, J.H.; Min, H.J.; Park, K.; Kim, J. Preparation and evaluation of a cosmetic adhesive containing guar gum. *Korean J. Chem. Eng.* **2017**, *34*, 2236–2240. [CrossRef]
10. Li, W.; Yu-Yu, E.; Cheng, L.-Y.; Ding, M.; Li, W.-Y.; Diao, K.-S.; Liu, S.-G.; Li, K.; Lu, H.-Q.; Lei, F.-H. Rosin-based polymer@ silica core–shell adsorbent: Preparation, characterization, and application to melanoidin adsorption. *LWT* **2020**, *132*, 109937. [CrossRef]
11. Kumar, S.; Gupta, S.K. Rosin: A naturally derived excipient in drug delivery systems. *Polim. Med.* **2013**, *43*, 45–48.
12. Phaphon, K.; Wacharasindhu, S.; Petsom, A. Preparation of PEG-rosin derivative for water soluble rosin flux. *Solder. Surf. Mt. Technol.* **2016**, *28*, 188–200. [CrossRef]
13. Mirabedini, S.; Zareanshahraki, F.; Mannari, V. Enhancing thermoplastic road-marking paints performance using sustainable rosin ester. *Prog. Org. Coat.* **2020**, *139*, 105454. [CrossRef]
14. Dewajani, H.; Chumaidi, A.; Iswara, M.A.I.; Khasanah, R.; Agustina, T.D. Synthesis ester gum through esterification reaction of rosin and gliserol using zeolite modified by nickel as catalyst. *AIP Conf. Proc.* **2019**, *2097*, 030037.
15. Prakoso, T.; Kumalasari, I.; Jiwandaru, B.; Soerawidjaja, T.H.; Azis, M.M.; Indarto, A. Synthesis of Maleic-Modified Rosin Ester from Pine Rosin. In *IOP Conference Series: Materials Science and Engineering, Volume 1143, International Seminar on Chemical Engineering Soehadi Reksowardojo (STKSR 2020), Bandung, Indonesia, 16–17 November 2020*; IOP Publishing Ltd.: Bristol, UK, 2021; p. 012071.
16. Li, Q.; Gong, S.; Yan, J.; Hu, H.; Shu, X.; Tong, H.; Cai, Z. Synthesis and kinetics of hydrogenated rosin dodecyl ester as an environmentally friendly plasticizer. *J. Renew. Mater.* **2020**, *8*, 289. [CrossRef]
17. Liu, S.; Xie, C.; Yu, S.; Liu, F. Dimerization of rosin using Brønsted–Lewis acidic ionic liquid as catalyst. *Catal. Commun.* **2008**, *9*, 2030–2034. [CrossRef]
18. Wang, L.; Chen, X.; Liang, J.; Chen, Y.; Pu, X.; Tong, Z. Kinetics of the catalytic isomerization and disproportionation of rosin over carbon-supported palladium. *Chem. Eng. J.* **2009**, *152*, 242–250. [CrossRef]
19. Lu, Y.J.; Xu, R.S.; Zhao, Z.D.; Zhang, P.H.; Wang, M.X. Recent progress on derivation and chemical modification of rosin acids. *Adv. Mater. Res.* **2013**, *785*, 1111–1116. [CrossRef]
20. Hardhianti, M.P.W.; Rochmadi; Azis, M.M. Kinetic studies of esterification of rosin and pentaerythritol. *Processes* **2021**, *10*, 39. [CrossRef]
21. Wang, X.; Guo, F.; Yu, Y.; Liu, Z.; Wang, Y.; Sun, H.; Liu, X.; Xue, Y.; Wei, X.; Guo, S. Study on the Synthesized Rosin Glyceride over LaZSM-5 Zeolite Catalyst Synthesized by the in Situ Method. *ACS Omega* **2020**, *5*, 31543–31550. [CrossRef]
22. Zhou, D.; Wang, L.; Chen, X.; Wei, X.; Liang, J.; Tang, R.; Xu, Y. Reaction mechanism investigation on the esterification of rosin with glycerol over annealed Fe_3O_4/MOF-5 via kinetics and TGA-FTIR analysis. *Chem. Eng. J.* **2020**, *401*, 126024. [CrossRef]
23. García, D.F.; Bustamante, F.; Villa, A.L.; Alarcón, E.A. Esterification of rosin with methyl alcohol for fuel applications. *Rev. Fac. Ing. Univ. Antioq.* **2021**, *100*, 10–20.
24. Morkhade, D.M.; Nande, V.S.; Barabde, U.V.; Kamble, M.U.; Patil, A.T.; Joshi, S.B. A comparative study of aqueous and organic-based films and coatings of PEGylated rosin derivative. *Drug Dev. Ind. Pharm.* **2008**, *34*, 24–32. [CrossRef] [PubMed]
25. Lu, Y.; Zhao, Z.; Chen, Y.; Wang, J.; Xu, S.; Gu, Y. Synthesis of allyl acrylpimarate by microwave irradiation and phase-transfer catalytic reaction and its UV-curing reactions as a new monomer. *Prog. Org. Coat.* **2017**, *109*, 9–21. [CrossRef]
26. Lin, R.; Li, H.; Long, H.; Su, J.; Huang, W. Structure and characteristics of lipase-catalyzed rosin acid starch. *Food Hydrocoll.* **2015**, *43*, 352–359. [CrossRef]
27. Ladero, M.; de Gracia, M.; Tamayo, J.J.; de Ahumada, I.L.; Trujillo, F.; Garcia-Ochoa, F. Kinetic modelling of the esterification of rosin and glycerol: Application to industrial operation. *Chem. Eng. J.* **2011**, *169*, 319–328. [CrossRef]
28. Ladero, M.; de Gracia, M.; Trujillo, F.; Garcia-Ochoa, F. Phenomenological kinetic modelling of the esterification of rosin and polyols. *Chem. Eng. J.* **2012**, *197*, 387–397. [CrossRef]
29. Quispe, C.A.; Coronado, C.J.; Carvalho, J.A., Jr. Glycerol: Production, consumption, prices, characterization and new trends in combustion. *Renew. Sustain. Energy Rev.* **2013**, *27*, 475–493. [CrossRef]
30. Anuar, M.R.; Abdullah, A.Z. Challenges in biodiesel industry with regards to feedstock, environmental, social and sustainability issues: A critical review. *Renew. Sustain. Energy Rev.* **2016**, *58*, 208–223. [CrossRef]
31. Sun, S.; Cheng, X.; Ma, M.; Liu, Y.; Wang, G.; Yu, H.; Liu, S.; Yu, S. High-efficient esterification of rosin and glycerol catalyzed by novel rare earth Lewis acidic ionic liquid: Reaction development and mechanistic study. *J. Taiwan Inst. Chem. Eng.* **2021**, *127*, 1–6. [CrossRef]
32. Zhou, D.; Wang, L.; Chen, X.; Wei, X.; Liang, J.; Zhang, D.; Ding, G. A novel acid catalyst based on super/subcritical CO_2-enriched water for the efficient esterification of rosin. *R. Soc. Open Sci.* **2018**, *5*, 171031. [CrossRef]
33. Li, Y.; Yu, S.; Zhang, H.; Zhang, J. Preparation of Fe_3O_4@SiO_2–ZnO catalyst and its catalytic synthesis of rosin glycol ester. *Open Chem.* **2021**, *19*, 938–944. [CrossRef]

34. Xu, Z.; Lou, W.; Zhao, G.; Zhang, M.; Hao, J.; Wang, X. Pentaerythritol rosin ester as an environmentally friendly multifunctional additive in vegetable oil-based lubricant. *Tribol. Int.* **2019**, *135*, 213–218. [CrossRef]
35. Comyn, J. Surface characterization of pentaerythritol rosin ester. *Int. J. Adhes. Adhes.* **1995**, *15*, 9–14. [CrossRef]
36. Yu, S.; Zhang, H. Preparation of Rosin Pentaerythritol Ester Over an Fe_3O_4-Supported ZnO Catalyst. *Catal. Lett.* **2020**, *150*, 3359–3367. [CrossRef]
37. García, D.; Bustamante, F.; Alarcón, E.; Donate, J.M.; Canoira, L.; Lapuerta, M. Improvements of thermal and thermochemical properties of rosin by chemical transformation for its use as biofuel. *Waste Biomass Valorization* **2020**, *11*, 6383–6394. [CrossRef]
38. Wang, L.; Huang, C.; Chen, J.; Wei, X.; Chen, X.; Liang, J. Catalyst-free biodiesel production from industrial rosin residue (dark-grade rosin) using supercritical methanol. *Waste Biomass Valorization* **2018**, *9*, 1191–1198. [CrossRef]
39. Liu, Y.; Lv, M.; Li, L.; Yu, H.; Wu, Q.; Pang, J.; Liu, Y.; Xie, C.; Yu, S.; Liu, S. Synthesis of Rosin Methyl Ester Using PTSA/ZrO_2/Mo-MCM-41 Mesoporous Molecular Sieves. *Catal. Lett.* **2019**, *149*, 1911–1918. [CrossRef]
40. Wang, X.; Wang, L.; Chen, X.; Zhou, D.; Xiao, H.; Wei, X.; Liang, J. Catalytic methyl esterification of colophony over ZnO/SFCCR with subcritical CO_2: Catalytic performance, reaction pathway and kinetics. *R. Soc. Open Sci.* **2018**, *5*, 172124. [CrossRef]
41. Marafi, M.; Stanislaus, A. Spent catalyst waste management: A review: Part I—Developments in hydroprocessing catalyst waste reduction and use. *Resour. Conserv. Recycl.* **2008**, *52*, 859–873. [CrossRef]
42. Xu, J.; Zhang, T. Fabrication of spent FCC catalyst composites by loaded V_2O_5 and TiO_2 and their comparative photocatalytic activities. *Sci. Rep.* **2019**, *9*, 11099. [CrossRef]
43. Kanlaya, P.; Sumrit, W.; Amorn, P. Synthesis and characterization of water soluble rosin-polyethylene glycol 1500 derivative. *Int. J. Chem. Eng. Appl.* **2016**, *7*, 277–281. [CrossRef]
44. Nande, V.; Barabde, U.; Morkhade, D.; Patil, A.; Joshi, S. Synthesis and characterization of PEGylated derivatives of rosin for sustained drug delivery. *React. Funct. Polym.* **2006**, *66*, 1373–1383. [CrossRef]
45. Mumtaz, I.; Majeed, Z.; Ajab, Z.; Ahmad, B.; Khurshid, K.; Mubashir, M. Optimized tuning of rosin adduct with maleic anhydride for smart applications in controlled and targeted delivery of urea for higher plant's uptake and growth efficiency. *Ind. Crops Prod.* **2019**, *133*, 395–408. [CrossRef]
46. Lu, Y.; Zhao, Z.; Gu, Y.; Chen, Y.; Bi, L. Synthesis of rosin allyl ester and its UV-curing characteristics. *Polym. J.* **2011**, *43*, 869–873. [CrossRef]
47. Kim, W.-S.; Byun, K.-R.; Lee, D.-H.; Min, K.-E.; Park, L.-S.; Seo, K.-H.; Kang, I.-K.; Park, S.-Y. Synthesis of photocrosslinkable polymers using abietic acid and their characterization. *Polym. J.* **2003**, *35*, 450–454. [CrossRef]
48. Kim, T.H.; Chang, J.Y.; Choi, J.-U.; Kim, W.-S. Synthesis and characterization of a polymethacrylate containing photoreactive abietic acid moiety. *Macromol. Res.* **2005**, *13*, 545–548. [CrossRef]
49. Zheng, Y.; Yao, K.; Lee, J.; Chandler, D.; Wang, J.; Wang, C.; Chu, F.; Tang, C. Well-defined renewable polymers derived from gum rosin. *Macromolecules* **2010**, *43*, 5922. [CrossRef]
50. Garrison, T.F.; Murawski, A.; Quirino, R.L. Bio-based polymers with potential for biodegradability. *Polymers* **2016**, *8*, 262. [CrossRef]
51. Tian, S.; Chen, Y.; Chen, Z.; Yang, Y.; Wang, Y. Preparation and characteristics of starch esters and its effects on dough physicochemical properties. *J. Food Qual.* **2018**, *2018*, 1–7. [CrossRef]
52. Lin, R.; Li, H.; Long, H.; Su, J.; Huang, W.; Wang, S. Optimization of lipase-catalyzed rosin acid starch synthesis by response surface methodology. *J. Mol. Catal. B Enzym.* **2014**, *105*, 104–110. [CrossRef]
53. Zhang, K.; Cheng, F.; Zhang, K.; Hu, J.; Xu, C.; Lin, Y.; Zhou, M.; Zhu, P. Synthesis of long-chain fatty acid starch esters in aqueous medium and its characterization. *Eur. Polym. J.* **2019**, *119*, 136–147. [CrossRef]

Disclaimer/Publisher's Note: The statements, opinions and data contained in all publications are solely those of the individual author(s) and contributor(s) and not of MDPI and/or the editor(s). MDPI and/or the editor(s) disclaim responsibility for any injury to people or property resulting from any ideas, methods, instructions or products referred to in the content.

AgriEngineering

MDPI

Article

Optimization of Black Tea Drying Temperature in an Endless Chain Pressure (ECP) Dryer

Gunaratnam Abhiram [1,2,*], Rasaiyah Diraj [1] and Rasu Eeswaran [3,*]

1 Department of Export Agriculture, Faculty of Animal Science and Export Agriculture, Uva Wellassa University, Badulla 90000, Sri Lanka
2 Environmental Sciences, School of Agriculture & Environment, Massey University, Palmerston North 4442, New Zealand
3 Department of Agronomy, Faculty of Agriculture, University of Jaffna, Ariviyal Nagar, Kilinochchi 44000, Sri Lanka
* Correspondence: abhiram@uwu.ac.lk (G.A.); rasueesw@msu.edu (R.E.)

Abstract: Drying is a crucial and energy-consuming process in black tea manufacturing that is aimed at reducing moisture content and terminating enzymatic reactions in tea leaves. In Sri Lanka, an endless chain pressure (ECP) dryer is commonly used for drying, but it consumes a significant amount of energy, necessitating the optimization of drying conditions. The current drying temperature at the Houpe tea factory in Ratnapura, Sri Lanka is 121 °C (250 °F), and it has not been optimized for a considerable period. As a result, energy consumption and wastage are high, leading to an inferior quality of black tea. To optimize factory conditions, tea leaves were dried under different temperatures: 115 (T1), 118 (T2), 121 (T3), 124 (T4), and 127 (T5) °C. Energy consumption, energy wastage, and specific energy consumption (*SEC*) for tea drying were calculated. Additionally, chemical and sensory analyses of samples of made tea were performed. *SEC* and energy wastage were significantly ($p < 0.05$) lower for treatments T1 and T2 than for other treatments. The theaflavin and thearubigin contents were significantly ($p < 0.05$) higher while total phenolic content was moderate for treatment T2. The sensory parameters of T2 outperformed other treatments. Based on these results, the optimum drying temperature for the ECP dryer was determined to be 118 °C and this temperature has been recommended for this factory.

Keywords: ECP dryer; energy consumption; organoleptic properties; theaflavins; thearubigins

Citation: Abhiram, G.; Diraj, R.; Eeswaran, R. Optimization of Black Tea Drying Temperature in an Endless Chain Pressure (ECP) Dryer. *AgriEngineering* **2023**, *5*, 1989–1999. https://doi.org/10.3390/agriengineering5040122

Academic Editors: Muhammad Sultan, Yuguang Zhou, Redmond R. Shamshiri and Muhammad Imran

Received: 1 September 2023
Revised: 2 October 2023
Accepted: 24 October 2023
Published: 1 November 2023

Copyright: © 2023 by the authors. Licensee MDPI, Basel, Switzerland. This article is an open access article distributed under the terms and conditions of the Creative Commons Attribution (CC BY) license (https://creativecommons.org/licenses/by/4.0/).

1. Introduction

The tea industry plays a vital role in Sri Lanka's economy, contributing significantly to foreign exchange earnings and domestic employment [1]. Tea represents a significant portion of Sri Lanka's agricultural revenue, contributing approximately 65% to the sector. It also plays a crucial role in the country's export revenue, accounting for around 13.5% [2]. Moreover, tea contributed approximately 2% of Sri Lanka's gross domestic product (GDP) in 2021, at over USD 1.3 billion. The tea is mainly grown in the central hills of the island and its southern foothills, with varying qualities from different elevations [3]. According to Basnayake, et al. [4], smallholders contribute to 60% of tea production, while the estate sector contributes to the rest. The country currently produces three main types of tea, namely, Ceylon black tea, Ceylon green tea, and Ceylon white tea, with higher production of orthodox black tea [5].

Drying is a unit operation in tea manufacturing that follows withering and rolling. This process aims to reduce the moisture content of the tea leaves below 7% to extend their shelf life. Furthermore, drying is an essential step to arrest enzymatic reactions in tea leaves to obtain the desired taste and color in the made tea. The method of drying and the drying temperature and drying time have a significant influence on the chemical composition and organoleptic properties of tea [6]. The under- or over-drying of tea leaves minimizes the sensory properties of the made tea. Under-dried tea increases in astringency, which creates a dry and puckering mouthfeel when consuming the tea. Furthermore, under-dried tea cannot be stored for an extended period due to its high moisture content, which promotes microbial growth, whereas over-dried tea has a brittle texture and yields an unpleasant aroma, high bitterness, and low brewing ability. Therefore, optimum drying conditions are essential to produce high-quality tea and to obtain a high auction price.

Endless chain pressure (ECP) dryers and fluidized bed dryers are commonly used to dry tea leaves at the industrial scale [7]. Meanwhile, infrared drying, microwave drying, solar drying, air–radio frequencies, etc., have been tested to produce better-quality tea in terms of volatile and flavor compounds [6,7]. However, these drying methods are still under development and are not widely used for industrial applications. In the ECP dryer, an endless chain conveyor system is used to transport the tea leaves from top to bottom while the hot air is sent through a blower from bottom to top. This design creates a temperature gradient within the tea dryer, with higher temperatures at the bottom and lower temperatures at the top. Therefore, the low-moisture tea leaves come into contact with high-temperature air, whereas high-moisture tea leaves meet low-temperature air. As a result, tea leaves with low moisture content encounter the hotter air, facilitating the removal of tightly bound moisture and preventing case hardening, where the outer layer of tea particles becomes excessively dry while the inner portion retains higher moisture levels.

Tea manufacturing is an energy-intensive process because major processes such as the withering, rolling, and drying stages consume a substantial portion of energy. The energy consumption associated with these processes corresponds to 48.9%, 21.3%, and 7.4% of the total energy consumption in tea manufacturing, respectively [8]. It has been reported that the drying process consumes around 3.5–6 kWh of thermal energy per kg of made tea [7].

In Sri Lanka, endless chain pressure (ECP) dryers and fluidized bed dryers (FBD) are predominantly used for black tea manufacturing. The high energy consumption observed in the drying process can be attributed to various factors, including the utilization of outdated dryers, inadequate maintenance practices, suboptimal drying conditions, and manual operational adjustments. These factors collectively contribute to the inefficient use of energy during the drying process. The performance of these dryers declines over time; thus, periodical optimization of the processing parameters is essential. The optimization of energy use would also help to reduce the carbon emissions of the tea sector from its current level of 2.49 kg of CO_2 per kg of tea produced in Sri Lanka, thus mitigating global warming and climate change [7].

To this end, the ECP dryer at Houpe tea factory in Ratnapura, Sri Lanka has experienced a decline in performance for about a one-month period, as indicated by several observable factors. These include the production of low-quality tea, increased energy consumption during the drying process, a higher percentage of energy wastage, and a noticeable decrease in the flavor of the dried tea. Additionally, there has been an increase in the proportion of low-grade tea being produced. Therefore, the management has requested immediate action to optimize the drying temperature, to ensure the quality of the made tea. Hence, this research was designed to optimize the drying temperature of the existing ECP dryer, with the objectives of increasing the quality of the made tea and reducing energy consumption.

2. Materials and Methods

2.1. The Experimental Setup of the ECP Dryer

An old ECP dryer with a capacity of 200 kg h^{-1} was selected for this study. This ECP dryer consists of 4 perforated trays and the average drying time at the beginning of this research was 23 min. The hot air from the furnace is fed into the dryers using a blower. The air is then released from the dryer through the top vent. There are seven temperature sensors (DS18B20, Adafruit (New York, NY, USA)) mounted at different locations to measure the inlet hot air temperature, tea drying temperature, inlet tea temperature, and ambient air temperature (Figure 1). The specifications of this temperature sensor are a temperature range of −55 °C to +130 °C, with ±0.5 °C accuracy and a 5 V excitation voltage. In order to confirm the temperature readings from sensors S2–S4, monitoring the temperature of the tea leaves, the following special care was taken: a rubber piece was glued to the bottom part of the sensor to avoid direct contact with the perforated metal drying trays and the sensors were fully immersed into the tea leaves and fixed in a set position to prevent any movement. The sensor readings were recorded on a computer with the support of an Arduino microcontroller (Adafruit) at 1-minute intervals.

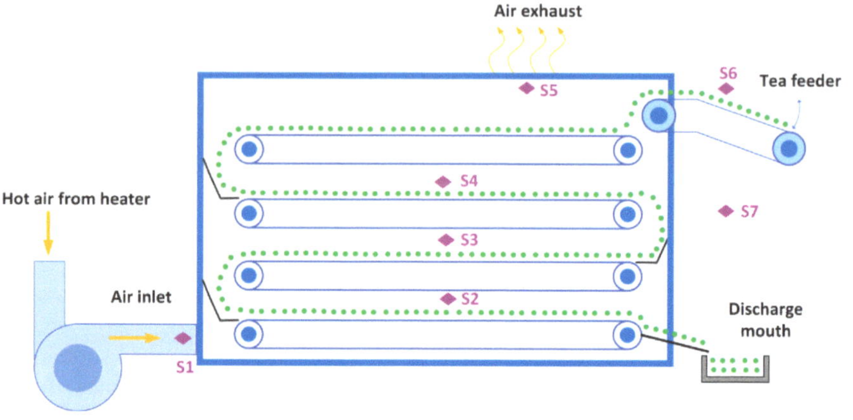

Figure 1. The schematic diagram of the experimental setup of the endless chain pressure (ECP) dryer. Temperature sensor locations and the relevant measurements are as follows. S1: inlet air temperature, S2–S4: tea leaf drying temperature, S5: exhaust air temperature, S6: the temperature of inlet tea leaves, and S7: ambient air temperature.

2.2. Drying Tea Leaves

The factory employs a drying temperature of 121 °C and, for the purposes of optimizing this temperature, temperatures were chosen that were two levels above and two levels below this temperature. In pursuit of this objective, the drying temperatures (i.e., treatments) selected were 115 °C (T1), 118 °C (T2), 121 °C (T3), 124 °C (T4), and 127 °C (T5). The corresponding drying temperatures in Fahrenheit were 240, 245, 250, 255, and 260 °F, respectively. To uphold the desired point temperature, adjustments were made to the feeding rate of firewood to the furnace and the position of the damper in the air inlet. Continuous monitoring of the inlet air temperature readings was carried out to facilitate the aforementioned operations. The ambient temperature during the experiment was 34.3 ± 1.2 °C and the relative humidity was 69.2 ± 3.4%. The airflow rate was 2.4 $m^3 s^{-1}$. For each temperature level, the experiment was repeated, and tea samples were collected. The collected samples were stored in a zip-lock bag at ambient conditions until used for analysis. The initial and final moisture contents of the tea leaves were measured using a moisture meter (MB27, OHAUS, NY, USA).

2.3. Energy Analysis

2.3.1. Energy Consumed by the Dryer

The energy of the hot air into the dryer and out of the dryer was calculated using Equations (1) and (2). To measure air velocity, anemometers were used. These were kept at 5 different points across the cross-sectional area of the air-feeding tube and the values were averaged.

$$H_{In} = \rho_{air} \times A \times V_{air} \times C_{airTin} \times (T_{in} - T_{amb}) \times DT \quad (1)$$

$$H_{Out} = \rho_{air} \times A \times V_{air} \times C_{airTout} \times (T_{out} - T_{amb}) \times DT \quad (2)$$

Here, H_{in} and H_{out} are hot air energy into and out of the system (MJ), respectively, ρ_{air} is the density of air (kg m^{-3}), A is the cross-sectional area of the air inlet port (m^2), V_{Air} is the velocity of hot air (ms^{-1}), C_{airTin} and $C_{airTout}$ are the specific heat capacity (KJ kg^{-1} K^{-1}) of hot air at T_{in} and T_{out}, respectively, T_{in} is the air inlet temperature (K), T_{out} is the air exhaust temperature (K), T_{amb} is the ambient air temperature (K), and DT is the drying time (s).

The energy consumed by the dryer (E_{Dryer}) was calculated using Equation (3).

$$E_{Dryer} = H_{In} - H_{Out} \tag{3}$$

2.3.2. Energy Consumed by Tea Leaves

The energy consumed by the tea leaves during the drying process was calculated using Equation (4). In this calculation, the average tea leaf temperature was established using the average values for temperature sensors S2–S4:

$$E_{Tea} = M_{Tea} \times C_{Tea} \times (T_{tea} - T_{amb}) \tag{4}$$

where E_{Tea} is the heat energy consumed by the tea (MJ), and M_{Tea} is the dry weight of fermented tea leaves (kg). C_{Tea} is the specific heat capacity of tea (KJ kg^{-1} K^{-1}), and T_{tea} and T_{amb} are the temperatures (K) of tea leaves and ambient air, respectively. The specific heat capacity (C_{Tea}) of black tea is 3.57 kJ kg^{-1} K^{-1} [9].

2.3.3. Energy to Heat and Evaporate the Moisture

The energy used to heat water molecules from room temperature to the average drying temperature and to evaporate the moisture from the tea leaves was calculated using Equations (5) and (6), respectively:

$$E_H = m_w \times C_w \times (T_w - T_{amb}) \tag{5}$$

where E_H is the energy (MJ) used to heat the water molecules, m_w is the weight (kg) of moisture in the tea leaves, C_w is the specific heat capacity of water (KJ kg^{-1} K^{-1}), and T_w and T_{amb} are the temperature (K) of water molecules and ambient air, respectively. The temperature of the tea leaves is nearly equal to the temperature of the water in the tea leaves; therefore, it was assumed that the tea temperature is equal to the temperature of the water:

$$E_e = m_w \times L \tag{6}$$

where E_e is the energy (MJ) used to evaporate the water molecules, m_w is the weight (kg) of moisture in the tea leaves and L is the latent heat of water evaporation (kJ kg^{-1}).

2.3.4. Specific Energy Consumption

The specific energy consumption (SEC) represents the energy needed to dry 1 kilogram of moisture from the tea leaves, and this was calculated using Equation (7).

$$SEC = \frac{E_{In}}{Amount\ of\ moisture\ removed} \tag{7}$$

2.3.5. Energy Wastage in Drying

The energy wastage of tea leaves drying was calculated using Equation (8).

$$E_{wastage} = E_{In} - (E_{Tea} + E_H + E_e) \tag{8}$$

2.4. Chemical Analysis of Dried Tea Leaves

2.4.1. Theaflavin (TF) and Thearubigin (TR) Content

The methodology outlined in the study by Abhiram et al. [10] was employed to measure the total TF and TR in tea infusions from triplicate samples. The ratio between TF and TR was then calculated.

2.4.2. Total Phenolic Content

The total phenolic content (TPC) of the dried tea leaf samples was measured using the Folin–Ciocalteu method, as described in [10]. Briefly, a 10% solution of Folin–Ciocalteu reagent was prepared, then 5 mL of this solution was mixed with 0.05 mL of tea infusion. After a few minutes, a saturated sodium carbonate solution (0.4 mL) was added and the mixture was incubated for 30 min. Similarly, a blank was prepared using water in place of the tea infusion. Then, the absorbance of the sample and blank was measured at 765 nm using a UV-Vis spectrophotometer (SHIMADZU, Kyoto, Japan). Gallic acid (0.1 g L^{-1}) was used as the standard and the TPC was expressed as the gallic acid equivalent (g GAE) per liter of infusion.

2.5. Sensory Analysis

Sensory evaluation was performed by trained tea tasters. All tea samples were evaluated for the color, flavor, and aroma of the liquor, and the appearance and color of the dried tea were rated by the tasters.

2.6. Statistical Analysis

A one-way analysis of variance (ANOVA) was performed at a 5% significance level. The mean separation was performed with the Tukey honest significant difference (HSD) post hoc test at a 5% significance level. The values were presented as the mean ± standard deviation of triplicate samples. The sensory data were subjected to analysis using the Friedman test at a significance level of 5%. To construct the radar diagram for sensory parameters, the estimated median values were utilized. All statistical analyses were performed using Minitab 18 software (Minitab, LLC., State College, PA, USA).

3. Results and Discussion

3.1. Temperature Changes throughout the Drying Time

The temperature of hot air at the inlet, hot air at the exhaust, and average tea leaf temperatures are presented in Figure 2. The average hot air inlet temperatures were 115.1 ± 0.4, 118.2 ± 0.8, 121.5 ± 0.9, 124.3 ± 0.5 and 127.7 ± 0.5 °C for treatments T1, T2, T3, T4, and T5, respectively (Figure 2a). The highest variation was observed for T3, whereas the lowest variation was observed for T1. The hot air inlet temperatures for all treatments were successfully maintained within an acceptable range of ±1 °C. However, it should be noted that maintaining a consistent temperature was challenging in this experiment, due to the absence of automatic temperature controllers for the dryer. Hence, the wood stock feeding rate and hot air damper positions were adjusted to achieve the desired temperature level. These measures were undertaken to ensure that the required temperature was maintained throughout the drying process. Average hot air temperatures at exhaust were 74.9 ± 1.1, 74.7 ± 4.3, 78.9 ± 1.4, 85.3 ± 2.2, and 82.3 ± 1.6 °C for T1, T2, T3, T4, and T5, respectively (Figure 2b). The highest variation was observed for T4 and the lowest was found for the T1 treatment. The average tea leaf temperatures were 92.5 ± 0.75, 93.0 ± 0.5, 98.7 ± 0.7, 102.2 ± 1.2, and 102.3 ± 1.2 °C for T1, T2, T3, T4, and T5, respectively (Figure 2c). Moreover, the initial moisture content of the tea leaves was 32.5 ± 2.3%. The final moisture contents were 6.5 ± 0.2, 5.6 ± 0.1, 4.8 ± 0.2, 4.3 ± 0.1, and 3.5 ± 0.3 for treatments T1, T2, T3, T4, and T5, respectively.

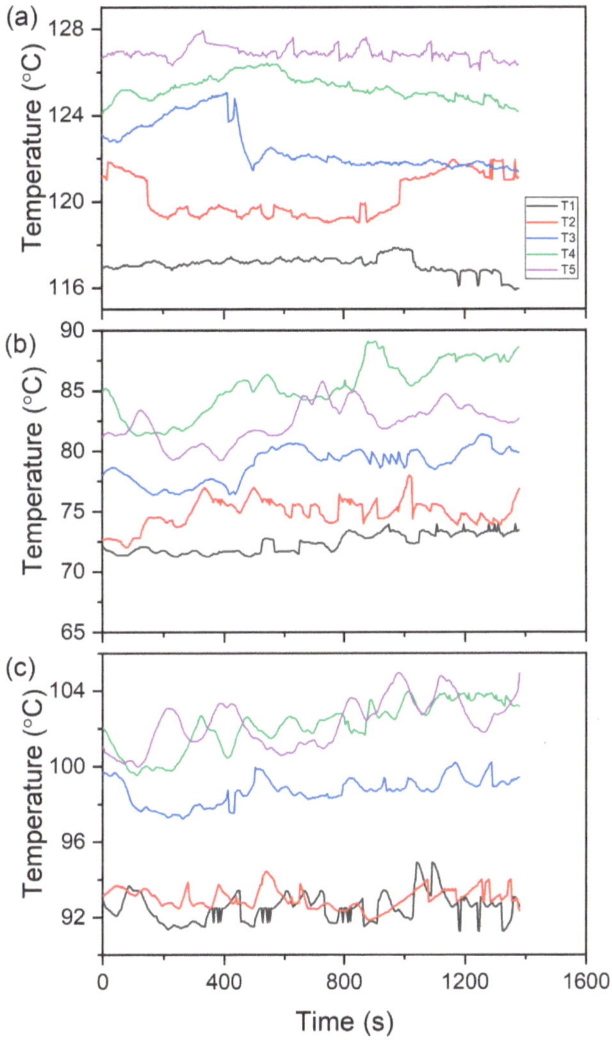

Figure 2. Temperature changes of tea drying according to time for the different treatments: (**a**) hot air inlet temperature, (**b**) hot air exhaust temperature, and (**c**) average temperature of tea leaves.

3.2. Energy Analysis

3.2.1. Energy Consumed by Dryer

The energy consumed by the dryer (*EDryer*) was calculated using the difference between the energy of the hot air moving into and that out of the dryer. It was significantly ($p < 0.05$) higher for treatment T2 (126.47 MJ) compared to other treatments (Table 1). The *EDryer* value did not significantly differ for other treatments. This showed that during tea drying under treatment T2, a high level of energy was utilized for drying. This energy demand was 9–16% higher than for other treatments. Despite the drying temperature increase from treatment T1 to T5, the E-dryer value did not increase accordingly.

Table 1. Energy analysis of the tea-drying process under different treatments.

Treatment	Energy Consumed by Dryer	Energy Consumed by Tea Leaves	Energy to Heat Moisture	Energy to Evaporate Moisture	Specific Energy Consumption (*SEC*)	Energy Wastage
	MJ	MJ	MJ	MJ	MJ/kg H_2O	MJ
T1	114.66 ± 4.19 ^{b}	4.21 ± 0.04 ^{c}	1.27 ± 0.02 ^{c}	11.73 ± 0.06 ^{e}	48.24 ± 0.22 ^{a}	233.32 ± 0.68 ^{c}
T2	126.47 ± 5.07 ^{a}	4.20 ± 0.04 ^{c}	1.31 ± 0.01 ^{c}	12.15 ± 0.06 ^{d}	47.67 ± 0.36 ^{ab}	238.82 ± 2.17 ^{bc}
T3	114.75 ± 3.70 ^{b}	4.56 ± 0.1 ^{b}	1.47 ± 0.03 ^{b}	12.53 ± 0.07 ^{c}	46.73 ± 0.73 ^{b}	240.67 ± 3.19 ^{b}
T4	106.02 ± 2.91 ^{b}	4.80 ± 0.01 ^{a}	1.58 ± 0.01 ^{a}	12.76 ± 0.05 ^{b}	47.62 ± 0.15 ^{ab}	249.76 ± 1.78 ^{a}
T5	115.62 ± 2.72 ^{b}	4.78 ± 0.03 ^{a}	1.61 ± 0.01 ^{a}	13.12 ± 0.11 ^{a}	46.87 ± 0.34 ^{b}	252.66 ± 0.72 ^{a}

Note: The different alphabetical letters within the same column are statistically different at a 5% significance level (n = 3).

3.2.2. Energy Consumed by the Tea Leaves

The energy consumed by the tea leaves was within the range of 4.2–4.8 MJ (Table 1). Only a small portion of the energy that entered the dryer was consumed by the tea leaves, ranging from 1.7 to 1.9%. As the drying temperature increased, there was a corresponding increase in the energy absorbed by the tea leaves. The T4 and T5 treatments exhibited the highest energy absorption, while the lowest absorption was observed in treatments T1 and T2. Treatment T3 demonstrated an intermediate level of energy consumption.

3.2.3. Energy to Heat and Evaporate the Moisture

Energy used to heat the moisture in tea leaves from the atmospheric temperature to the average temperature of drying was referred to as energy to heat the moisture. Treatments T4 and T5 consumed significantly ($p < 0.05$) more energy than other treatments (Table 1). The corresponding energy consumptions were 1.58 and 1.61 MJ, respectively. Treatments T1 and T2 consumed significantly ($p < 0.05$) less energy among all treatments, the values for which were 1.27 and 1.31 MJ, respectively.

The energy required to evaporate moisture (E_e) refers to the energy that water molecules consume in order to undergo the process of evaporation. The E_e significantly ($p < 0.05$) and continuously increased with drying temperature (Table 1). E_e was within the range of 11.73 to 13.12 MJ. The lowest energy was recorded for treatment T1, whereas the highest was recorded for treatment T5.

3.2.4. Specific Energy Consumption

The specific energy consumption (*SEC*) showed a decreasing trend with increasing drying temperature (Table 1). Similar findings have been reported in previous studies on drying tea [11] and herbal leaves [10]. The *SEC* of treatment T1 was 48.24 MJ/kg of moisture, which was significantly higher ($p < 0.05$) compared to T3 and T5. The *SEC* of treatments T2 and T4 were not significant ($p < 0.05$) when compared with other treatments. The observed *SEC* values for tea drying in this study fall within the range reported by other researchers [12,13].

3.2.5. Energy Wastage

Energy wastage when drying tea leaves refers to the portion of energy that remains unabsorbed by the leaves, despite being introduced into the dryer. This unabsorbed energy is considered to be a waste of energy. This energy wastage was significantly ($p < 0.05$) higher for treatment T4 and T5, while it was significantly ($p < 0.05$) lower for T1 and T2 (Table 1). The higher the drying temperature, the greater the energy wastage. The high percentage of energy wastage can be attributed to the higher specific heat capacity of a biological substance, including tea leaves.

3.3. Chemical Analysis

3.3.1. Theaflavins (TF), Thearubigins (TR), and the TF:TR Ratio

Theaflavins and thearubigins form during the fermentation process involved in the manufacturing of black tea. This process involves allowing the tea leaves to undergo oxidation, which initiates a sequence of chemical reactions, leading to the formation of these compounds [14]. Theaflavins are responsible for the distinctive flavor and aroma of black tea [15], whereas thearubigins are responsible for the color and mouthfeel [16]. Both of these chemicals possess several health benefits, such as anti-inflammatory, antioxidant, anticancer, and anti-obesity activity in humans [15]. Treatments T1 and T2 showed significantly ($p < 0.05$) higher TF and TR contents than other treatments, whereas T5 showed significantly ($p < 0.05$) lower TF and TR contents than other treatments (Figure 3a). In this study, the theaflavin and thearubigin contents significantly decreased with increasing temperature, which is in agreement with the results of other studies [17,18]. The possible hindrance of high temperature on the formation of TF and TR could explain this observation. However, it was found that the TF/TR ratio did not significantly differ among treatments T1, T2, T3, and T4 (Figure 3b). In contrast, the TF/TR ratio of treatment T5 was lower than that of the other treatments. These findings indicate that a high drying temperature can reduce the presence of chemical compounds such as TF and TR, thereby compromising the organoleptic properties of the made tea and its health benefits.

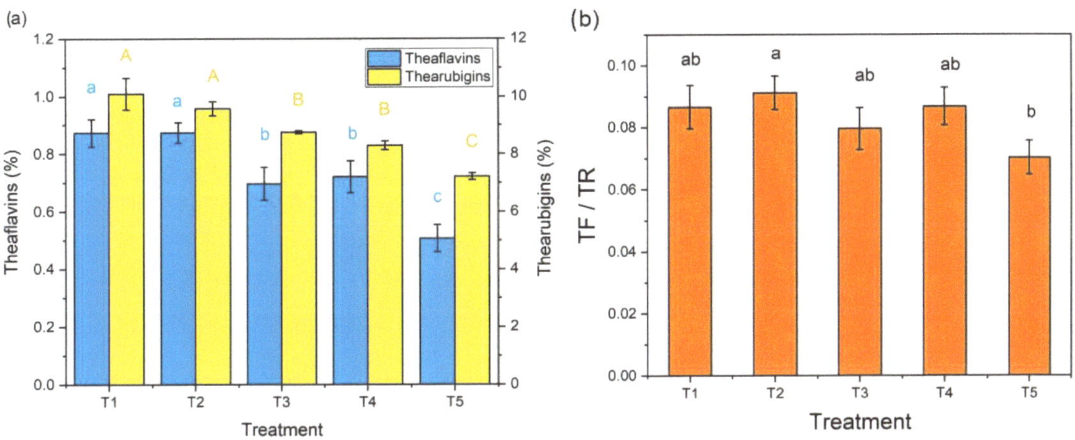

Figure 3. (**a**) Theaflavin (TF) and thearubigin (TR) contents and (**b**) the TF:TR ratio of tea leaves dried under different treatments. Note: The different alphabetical letters within the same column of the graph are statistically different at a 5% significance level (n = 3).

3.3.2. Total Phenolic Content

The total phenolic content (TPC) is responsible for made tea's antioxidant activity, anti-inflammatory, and anti-cancer properties [19]. The TPC of the dried tea leaves was between 152.1 and 200.5 mg GAE L^{-1} (Figure 4). These values are comparable with those in other studies [20]. Treatment T5 exhibited a significantly higher TPC compared to the other treatments ($p < 0.05$). Conversely, treatment T1 displayed the lowest TPC among all the treatments ($p < 0.05$). The TPC values of tea from treatments T2 and T3 were found to be comparable. The TPC values showed an increasing trend with the drying temperature, which is widely reported in the literature [10,21]. This observation could be attributed to the breakdown of extractable TPC from complex phenolic compounds [10]. However, increasing the temperature beyond a threshold level could degrade the TPC [10].

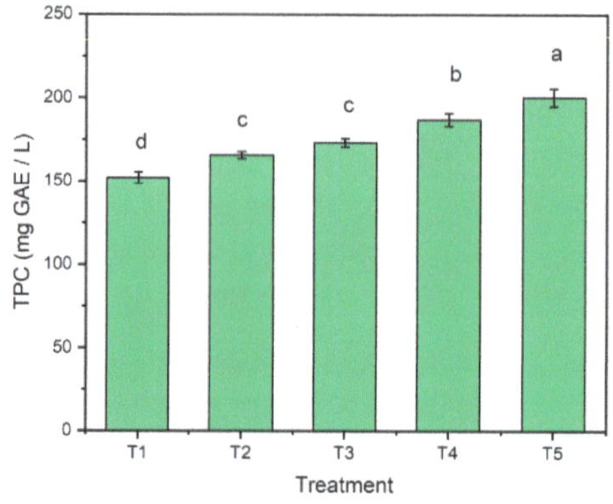

Figure 4. The total phenolic content (TPC) in tea leaves, as affected by the different treatments. Note: The different alphabetical letters in the column graph are statistically different at a 5% significance level (n = 3).

3.4. Sensory Analysis

Sensory properties determine the price of tea at auction and directly influence its appeal and suitability for consumption by customers. The tea leaves that were dried under treatment T2 showed better characteristics for liquor color, dry leaf appearance, dry leaf color, and overall acceptability than other treatments (Figure 5). The color and appearance of made tea hold significant importance from a customer's perspective when purchasing tea [20]. These visual qualities contribute to the initial impression and appeal of the product. On the other hand, when it comes to prepared tea, the liquor's properties, including its color, aroma, and flavor, become crucial factors. These sensory attributes greatly influence the overall sensory experience and quality of the tea when consumed. The results revealed that although tea leaves exposed to T2 and T3 exhibited similar liquor flavor, liquor aroma, and infused leaf color, they were considered better than other treatments. Tea leaves dried under T1 and T5 conditions showed the characteristics of under- and over-dried tea, respectively. Overall, tea leaves dried with treatments T2 and T3 exhibited more favorable sensory characteristics.

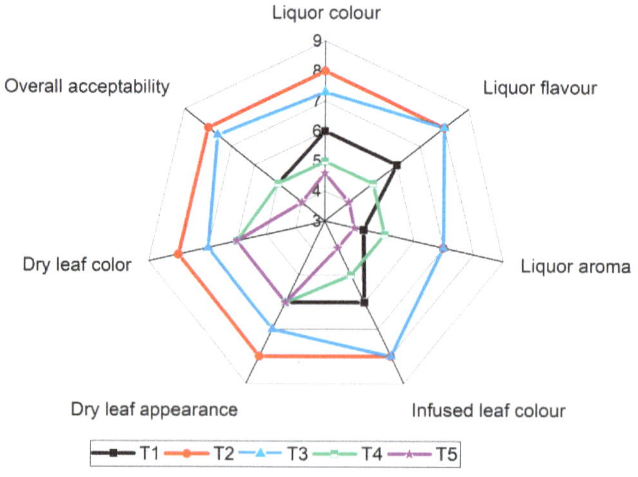

Figure 5. The sensory evaluation of tea leaves dried under different temperature treatments.

4. Conclusions

In order to optimize the drying temperature of an old endless chain pressure (ECP) dryer, tea leaves were dried under different temperatures (115, 118, 121, 124, and 127 °C). This study involved analyzing the energy consumption, conducting chemical analyses, and evaluating the sensory attributes of the dried tea leaves. It was found that drying temperature significantly ($p < 0.05$) influenced all the measured parameters. Specific energy consumption (*SEC*) increased with increasing temperature. The contents of theaflavins and thearubigins in the dried tea leaves decreased as the temperature increased, whereas the total phenolic content showed an increase with increasing temperatures during the drying process. The overall acceptability of the tea dried under T2 treatment was superior to other treatments in terms of the organoleptic properties. After considering all the analyzed parameters, T2 was determined to be the optimal drying temperature. It exhibited lower specific energy consumption (*SEC*) values and demonstrated superior chemical and sensory attributes compared to the other treatments. By adopting a drying temperature of 118 °C instead of the current temperature of 121 °C, it is estimated that a saving of 95 kJ of energy per kilogram of processed tea could be achieved. Based on the findings of this study, it is recommended to implement a drying temperature of 118 °C at the Houpe tea factory in Ratnapura, Sri Lanka. It is worth mentioning that this recommendation is currently being practiced at this factory, following this research outcome. Further research by altering both drying temperature and air-flow rate is beneficial for further optimizing the *SEC*. Given the high energy wastage in ECP dryers, reusing the waste energy by using a loop system after dehumidifying the moisture from the air can minimize energy consumption and carbon emissions. Therefore, the design and study of such systems on reusing energy will be instrumental for the tea industry.

Author Contributions: G.A.: Conceptualization, methodology, data curation, formal analysis and investigation, resources, supervision, software, and writing—original draft preparation. R.D.: Data curation, methodology, formal analysis, investigation, and writing—review and editing. R.E.: Methodology, formal analysis, and writing—review and editing. All authors have read and agreed to the published version of the manuscript.

Funding: No special funding was received for this study.

Data Availability Statement: Not applicable.

Acknowledgments: We express our sincere appreciation to Senaka Dayananda, Deputy General Manager (Engineering) of Houpe Tea Factory in Sri Lanka, as well as all the workers at the factory who provided their valuable support for this study.

Conflicts of Interest: The authors affirm that they have no known competing interests that could have influenced the work reported in this manuscript.

References

1. Abhiram, G.; Manathunga, M.; Raveendran, K.; Withanage, N. Performance Analysis of High-Efficiency Motor and Variable Speed Drive in Black Tea Processing Machinery. *J. Biosyst. Eng.* **2020**, *45*, 310–317. [CrossRef]
2. Munasinghe, M.; Deraniyagala, Y.; Dassanayake, N.; Karunarathna, H. Economic, social and environmental impacts and overall sustainability of the tea sector in Sri Lanka. *Sustain. Prod. Consum.* **2017**, *12*, 155–169. [CrossRef]
3. Wijeratne, M. Vulnerability of Sri Lanka tea production to global climate change. *Water Air Soil Pollut.* **1996**, *92*, 87–94. [CrossRef]
4. Basnayake, B.; Gunaratne, L. Estimation of technical efficiency and it's determinants in the tea small holding sector in the Mid Country Wet Zone of Sri Lanka. *Sri Lankan J. Agric. Econ.* **2002**, *4*, 137–150. [CrossRef]
5. Jayasinghe, S.L.; Kumar, L. Climate change may imperil tea production in the four major tea producers according to climate prediction models. *Agronomy* **2020**, *10*, 1536. [CrossRef]
6. Qu, F.; Zhu, X.; Ai, Z.; Ai, Y.; Qiu, F.; Ni, D. Effect of different drying methods on the sensory quality and chemical components of black tea. *LWT* **2019**, *99*, 112–118. [CrossRef]
7. Kumar, K.R.; Dashora, K.; Kumar, S.; Dharmaraja, S.; Sanyal, S.; Aditya, K.; Kumar, R. A review of drying technology in tea sector of industrial, non-conventional and renewable energy based drying systems. *Appl. Therm. Eng.* **2023**, *224*, 120118. [CrossRef]
8. De Silva, W. Status review of energy utilization by the tea industry in Sri Lanka. *Sri Lanka J. Tea Sci.* **1994**, *63*, 46–58.

9. Lian, G.; Thiru, A.; Parry, A.; Moore, S. CFD simulation of heat transfer and polyphenol oxidation during tea fermentation. *Comput. Electron. Agric.* **2002**, *34*, 145–158. [CrossRef]
10. Abhiram, G.; Briyangari, A.; Eeswaran, R. Drying of Gymnema sylvestre Using Far-Infrared Radiation: Antioxidant Activity and Optimization of Drying Conditions. *AgriEngineering* **2023**, *5*, 611–622. [CrossRef]
11. Zeng, Z.; Li, B.; Han, C.; Wu, W.; Wang, X.; Xu, J.; Zheng, Z.; Ma, B.; Hu, Z. An exergoeconomic analysis of a gas-type industrial drying system of black tea. *Entropy* **2022**, *24*, 655. [CrossRef] [PubMed]
12. Jindarat, W.; Sungsoontorn, S.; Rattanadecho, P. Analysis of energy consumption in drying process of biomaterials using a combined unsymmetrical double-feed microwave and vacuum system (CUMV)—Case study: Tea leaves. *Dry. Technol.* **2013**, *31*, 1138–1147. [CrossRef]
13. Temple, S.; van Boxtel, A. PH—Postharvest Technology: A Comparison of Dryer Types used for Tea Drying. *J. Agric. Eng. Res.* **2000**, *77*, 401–407. [CrossRef]
14. Hazarika, M.; Chakravarty, S.K.; Mahanta, P.K. Studies on thearubigin pigments in black tea manufacturing systems. *J. Sci. Food Agric.* **1984**, *35*, 1208–1218. [CrossRef]
15. Shan, Z.; Nisar, M.F.; Li, M.; Zhang, C.; Wan, C.C. Theaflavin chemistry and its health benefits. *Oxidative Med. Cell. Longev.* **2021**, *2021*, 6256618. [CrossRef]
16. Zhu, K.; Ouyang, J.; Huang, J.; Liu, Z. Research progress of black tea thearubigins: A review. *Crit. Rev. Food Sci. Nutr.* **2021**, *61*, 1556–1566. [CrossRef]
17. Temple, S.J.; Temple, C.M.; Boxtel, A.J.B.v.; Clifford, M.N. The effect of drying on black tea quality. *J. Sci. Food Agric.* **2001**, *81*, 764–772. [CrossRef]
18. Su, Y.L.; Leung, L.K.; Huang, Y.; Chen, Z.-Y. Stability of tea theaflavins and catechins. *Food Chem.* **2003**, *83*, 189–195.
19. Ding, J.; Mei, S.; Gao, L.; Wang, Q.; Ma, H.; Chen, X. Tea processing steps affect chemical compositions, enzyme activities, and antioxidant and anti-inflammatory activities of coffee leaves. *Food Front.* **2022**, 3, 505–516. [CrossRef]
20. Abhiram, G.; Jinthushayi, S.; Withanage, N. Optimization of feeding rate and pressure of Orthodox Roller in Black Tea production. *AGRIEAST J. Agric. Sci.* **2018**, *12*, 16–25. [CrossRef]
21. Selvi, K.Ç.; Kabutey, A.; Gürdil, G.A.K.; Herak, D.; Kurhan, Ş.; Klouček, P. The effect of infrared drying on color, projected area, drying time, and total phenolic content of rose (*Rose electron*) petals. *Plants* **2020**, *9*, 236. [CrossRef] [PubMed]

Disclaimer/Publisher's Note: The statements, opinions and data contained in all publications are solely those of the individual author(s) and contributor(s) and not of MDPI and/or the editor(s). MDPI and/or the editor(s) disclaim responsibility for any injury to people or property resulting from any ideas, methods, instructions or products referred to in the content.

AgriEngineering

MDPI

Article

Evaluation of the Accuracy of the Remote Determination of the Brewster Angle When Measuring Physicochemical Parameters of Soil

Gennadiy Ivanovich Linets, Anatoliy Vyacheslavovich Bazhenov, Sergey Vladimirovich Malygin, Natalia Vladimirovna Grivennaya *, Sergey Vladimirovich Melnikov and Vladislav Dmitrievich Goncharov

Department of Infocommunications, North Caucasus Federal University, Stavropol 355017, Russia; glinetc@ncfu.ru (G.I.L.); abazhenov@ncfu.ru (A.V.B.); smalygin@ncfu.ru (S.V.M.); smelnikov@ncfu.ru (S.V.M.); vldgoncharov@ncfu.ru (V.D.G.)

* Correspondence: ngrivennaia@ncfu.ru

Abstract: In precision farming technology, the moisture of the soil, its granulometric composition, specific conductivity and a number of other physical and chemical parameters are determined using remote radar sensing. The most important parameters are those measured in the area of the plant root system located well below the "air-surface" boundary. In order to create conditions for the penetration of electromagnetic waves through the "air-surface" interface with a minimum reflection coefficient, the irradiation of the Earth's surface is carried out obliquely with an angle of incidence close to the Brewster angle. The reflection coefficient, and, consequently, the Brewster angle, depend on the complex dielectric permittivity of the surface soil layer and are not known a priori. To determine the Brewster angle, the usual method is to search for the minimum amplitude of the vertically polarized signal reflected from the surface. Another approach is when the first derivative of the dependence of the modulus of the complex amplitude of a vertically polarized interference wave, taken with respect to the angle of incidence, is set equal to zero. In turn, in real dielectrics such as agricultural soils, the amplitude of the vertically polarized signal reflected from the surface is directly proportional to the reflection coefficient and does not have a pronounced minimum, which reduces the accuracy of the measurements. Based on the solution of the Helmholtz wave equation for a three-layered structure of the propagation medium (air, upper fertile soil layer, soil layer below the groundwater level), a model of the process of forming an interference wave under oblique irradiation of a planar layered dielectric with losses has been developed. Using the developed model, factors influencing the accuracy of determining the Brewster angle have been identified. For the first time, it is proposed to use the phase shift between the oscillations of the interference waves with vertical and horizontal polarization to measure the Brewster angle. A comparative assessment of the accuracy of determining the Brewster angle using known amplitude methods and the proposed phase method has been carried out. The adequacy of the method was experimentally confirmed. Recommendations have been developed for the practical application of the phase method of finding the Brewster angle for assessing the dielectric permittivity of soil and its moisture content.

Keywords: remote sensing of soil; Brewster angle; soil physicochemical parameters; soil moisture; interference waves

Citation: Linets, G.I.; Bazhenov, A.V.; Malygin, S.V.; Grivennaya, N.V.; Melnikov, S.V.; Goncharov, V.D. Evaluation of the Accuracy of the Remote Determination of the Brewster Angle When Measuring Physicochemical Parameters of Soil. *AgriEngineering* **2023**, *5*, 1893–1908. https://doi.org/10.3390/agriengineering5040116

Academic Editors: Yuguang Zhou, Redmond R. Shamshiri, Muhammad Sultan and Muhammad Imran

Received: 21 July 2023
Revised: 13 October 2023
Accepted: 17 October 2023
Published: 19 October 2023

Copyright: © 2023 by the authors. Licensee MDPI, Basel, Switzerland. This article is an open access article distributed under the terms and conditions of the Creative Commons Attribution (CC BY) license (https://creativecommons.org/licenses/by/4.0/).

1. Introduction

Modern precision agriculture is a complex process, at each stage of which it is necessary to make management decisions. The quality of the decisions taken is determined by the completeness and reliability of the available information on the condition of agricultural plants, soil composition and fertility, moisture content, and its availability to the root system, as well as a number of other indicators [1–4]. Currently, methods of remote radar measurement of the soil's physical and chemical parameters are being actively introduced using aircraft and spacecraft [5,6]. Such systems have advantages of quickness and covering large areas, but they fall significantly behind laboratory methods in terms of measurement accuracy. At the same time, laboratory studies of soil samples taken within the same soil area, as well as the use of stationary sensors, do not provide the required speed, and the results obtained do not allow for an assessment of soil conditions over vast territories [7–9].

With the emergence of unmanned aerial vehicles (UAVs), studies [10–15] are being conducted to determine the physicochemical parameters of soil using oblique irradiation of the Earth's surface in order to create conditions for the occurrence of the total (pseudo-total) refraction effect [6,16]. The advantage of this approach is that it allows for estimating the soil parameters directly in the root zone of plants. The oblique angle of incidence of an electromagnetic wave (Brewster angle), at which the effect of total (pseudo-total) refraction is observed, depends on the complex dielectric permittivity of the media that form the air–surface interface [17]. Therefore, with a known value of the Brewster angle, it is possible to determine the electrical parameters of dielectric media and estimate the moisture content, granulometric composition, and specific conductivity of soil horizons on the basis of these parameters using the known pedotransfer functions [18].

Among the methods for determining the electrical properties of dielectrics using the Brewster effect to some extent, the articles cited in [19–25] can be highlighted. The mathematical framework of these methods is based on Snell's laws and Fresnel's formulas for a flat interface between two lossless dielectrics under the condition of specular reflection of the probing signal of the radio or optical range [9,26,27]. In our opinion, the influence of various factors on the accuracy of determining the Brewster angle has not been sufficiently investigated in these methods when applied to the real, naturally occurring planar-layered dielectrics [28]. Taking into account the emergence of new technical means for oblique irradiation of the Earth's surface (e.g., UAVs) and the active integration of remote-sensing methods for measuring the physical and chemical parameters of soil into precision farming technology, the assessment of the accuracy characteristics of methods for determining the Brewster angle becomes even more relevant.

Thus, the aim of the study is to analyze the mathematical expressions that describe the process of the formation of an interference radar signal during oblique irradiation of a plane-layered dielectric, which makes it possible to evaluate the accuracy of measuring the Brewster angle. Based on the results of the modeling, it is possible to formulate recommendations for implementing approaches that can reduce measurement errors in real-world conditions.

A contradiction arises between the active use of remote-sensing methods for measuring the physico-chemical parameters of soil, including the use of the Brewster effect, and the lack of a scientific and methodological apparatus that sufficiently allows for the evaluation of the influences of different factors on the accuracy of determining the Brewster angle and other parameters based on it.

To achieve the set goal, it is necessary:

- To develop a mathematical model of electromagnetic wave interference;
- To carry out an experimental verification of the adequacy of the developed model;
- To conduct a study of the mathematical model to identify the factors influencing the accuracy of determining the Brewster angle;
- To develop recommendations for the use of the known amplitude and proposed phase methods for measuring the Brewster angle under various conditions;
- To demonstrate how, by measuring the Brewster angle, we can estimate the physical and chemical parameters of the soil using the example of determining its volumetric moisture content.

The solution to the first two problems is presented in the Research Methodology section. The Results section is dedicated to the study of the mathematical model of interference and the evaluation of its accuracy. The recommendations for using the known amplitude and newly proposed phase measurement methods for estimating the physical and chemical parameters of soil are given in the Discussion section.

2. Research Methodology

To estimate the accuracy of the Brewster angle measurement, it is proposed to use a model of electromagnetic (EM) wave interference in a three-layer propagation medium with two boundary transitions. When analyzing the physicochemical parameters of agricultural soils, these may be: the layer of air, close to vacuum in properties; the upper fertile layer of soil, represented by a dielectric with losses (due to humidity, for example); the layer of soil in which groundwater is located and which is close in its properties to conductors. In this model, the interference wave is formed by the addition of waves: a direct wave propagating only in air, a wave reflected from the air–surface interface and the sum of partial waves that have passed through the air–surface interface and are reflected from the soil layer with groundwater. To denote the corresponding electromagnetic waves, we will use the following names: direct, reflected and soil waves.

The mathematical description of complex amplitude of a direct wave follows from the solution of the Helmholtz wave equation for a wave whose phase increases with the distance from the radiation source [29] (p. 299):

$$\tilde{E}_{direct} = E(L_d)\cdot \exp(-j\gamma_1 L_d) \quad (1)$$

where $E(L_d)$ is the wave amplitude depending on the distance L_d traveled by the EM wave (in this case, the distance between the transmitting and receiving antennas) without regard to attenuation in the medium; $\gamma_1 = 2\pi/\lambda$ is the propagation coefficient of the wave; λ is the wavelength.

The complex amplitude of reflected waves is formed on the basis of the direct wave, considering the change in the amplitude and phase at the point of reflection in proportion to the reflection coefficient. Reflection of EM waves at the interface between two dielectric media differs for the vertically and horizontally polarized waves:

$$\tilde{E}v_{ref} = E\left(L_{ref}\right)\cdot\tilde{R}_v\cdot \exp\left(-j\gamma_1 L_{ref}\right), \quad (2)$$

$$\tilde{E}h_{ref} = E\left(L_{ref}\right)\cdot\tilde{R}_h\cdot \exp\left(-j\gamma_1 L_{ref}\right), \quad (3)$$

where $E\left(L_{ref}\right)$ is the wave amplitude depending on the distance L_{ref} traveled by the reflected wave without regard to attenuation; $\tilde{R}_v$, $\tilde{R}_h$ are the complex reflection coefficients for the vertically and horizontally polarized waves, which are determined by the known Fresnel formulas [29] (p. 306).

The set of soil waves is determined by the condition that each partial wave must enter the aperture area of the receiving antenna after undergoing the refraction at the "air-dielectric" interface, the reflection from the groundwater horizon (or sea water) and the reverse refraction into the air medium:

$$\tilde{E}vi_{soil} = E\left(L_{iref}\right)\cdot\tilde{T}_{v12}\cdot\tilde{R}_{v23}\cdot\tilde{T}_{v21}\cdot \exp\left(-j\gamma_1 L_{iref}\right)\cdot\exp(-j\gamma_2 2d\cdot tg\psi_i), \quad (4)$$

$$\tilde{E}hi_{soil} = E\left(L_{iref}\right)\cdot\tilde{T}_{h12}\cdot\tilde{R}_{h23}\cdot\tilde{T}_{h21}\cdot \exp\left(-j\gamma_1 L_{iref}\right)\cdot\exp(-j\gamma_2 2d\cdot tg\psi_i), \quad (5)$$

where $\tilde{T}_{v12}$, $\tilde{T}_{h12}$ are the complex refractive indices of the vertically and horizontally polarized signals for the angle of incidence φ_i and the angle of refraction ψ_i at the interface between the first and second media; $\tilde{R}_{v23}$, $\tilde{R}_{h23}$ are the complex reflection coefficients of the vertically and horizontally polarized signals from the soil layer boundary at the groundwater level; $\tilde{T}_{v21}$, $\tilde{T}_{h21}$ are the complex refractive indices of the vertically and horizontally polarized signals when passing from a denser medium (the first soil layer) into air; γ_2 is the propagation coefficient in the first soil layer, which, in the general case, is complex and consists of the phase coefficient (wave number) and attenuation coefficient; $2d \cdot tg\psi_i$ is the distance traveled by the electromagnetic wave in the second medium; d is the thickness of the first soil layer.

Thus, interference waves with the vertical and horizontal polarization at the receiving point are represented by the expressions, respectively:

$$\widetilde{E}v_{interf} = \widetilde{E}_{direct} + \widetilde{E}v_{ref} + \sum_i \widetilde{E}vi_{soil} \tag{6}$$

$$\widetilde{E}h_{interf} = \widetilde{E}_{direct} + \widetilde{E}h_{ref} + \sum_i \widetilde{E}hi_{soil} \tag{7}$$

The results of mathematical modeling in Matlab using expressions (1)–(7) are shown in Figures 1 and 2 as graphs of the moduli of the complex amplitudes included in the expressions (6) and (7).

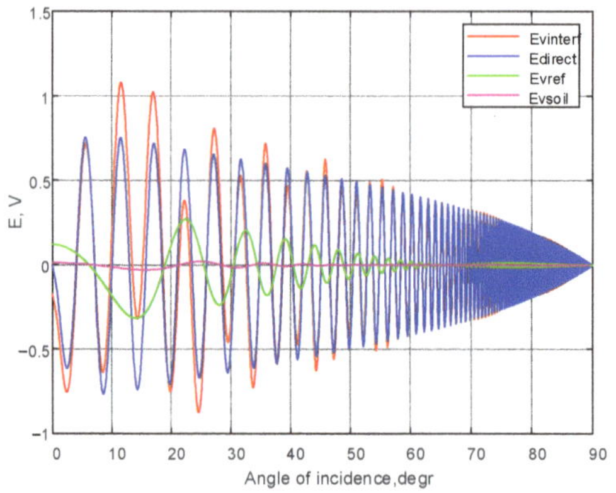

Figure 1. Vertically polarized direct, reflected and soil waves and their interference.

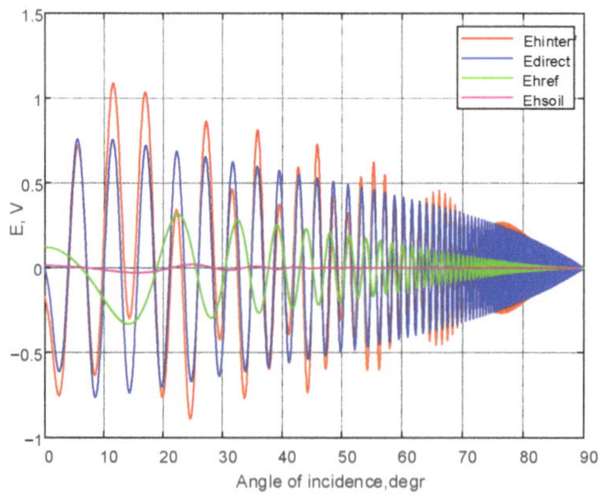

Figure 2. Horizontally polarized direct, reflected and soil waves and their interference.

The input data used for modeling correspond to the typical conditions of remote-sensing techniques used to determine the physical and chemical parameters of agricultural soils. The initial data used in the modeling include:

- Frequency of the probing signal f = 469.0 MHz;
- Thickness of the dielectric layer to be investigated d = 2 m;
- Relative dielectric permittivity of the first medium (air) $\varepsilon_{r1} = 1$;

- Specific conductivity of the first medium (air) $\sigma_1 = 0$ S/m;
- Relative permittivity of the second medium (wet loamy soil) $\varepsilon_{r2} = 3.5$;
- Specific conductivity of the second medium $\sigma_2 = 0.015$ S/m;
- Relative dielectric permittivity of the third medium (watered clay layer) $\varepsilon_{r3} = 30$;
- Specific conductivity of the third medium $\sigma_3 = 0.15$ S/m.

As follows from the obtained modeling results (see Figures 1 and 2), the interference waves have an oscillatory nature associated with the phase difference between the direct, reflected and soil waves at the observation point. The amplitude of the horizontally polarized interference wave decreases in proportion to the increase in the distance between the transmitting and receiving positions required to create different angles of incidence of the probing signal in the range from 0 to 90 degrees. The changes in the amplitude of a vertically polarized interference wave (Figure 1) are more informative; they occur under the influence of two factors. The amplitude decreases due to an increase in the distance between the transmitting and receiving positions and changes proportionally to the reflection coefficient. The second factor leads to the emergence of a minimum corresponding to the Brewster effect. The presence of the minimum of the interference wave amplitude, the proximity of the minimum to zero and its location on the abscissa axis are determined by the dielectric permittivity and the specific conductivity of the top soil layer.

Figure 3 displays graphs depicting the moduli of the complex amplitudes of the vertically and horizontally polarized interference waves.

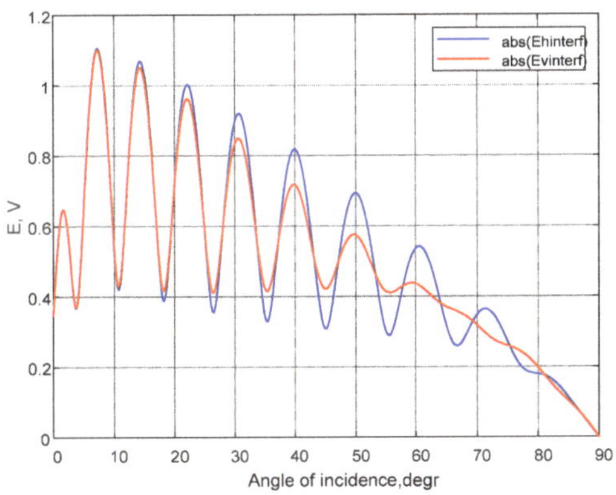

Figure 3. The resulting graphs of the moduli of interference waves with vertical and horizontal polarization.

As shown in Figure 3, when the angle of incidence is equal to the Brewster angle, a phase shift appears between the oscillations of the vertically and horizontally polarized interference waves. This, along with the minimum amplitude, is a sign of the Brewster effect.

Figure 4a shows the joint graphs of the results of mathematical modeling and experimental studies in the range of incidence angles from 45 to 75 degrees, where the Brewster effect is observed to occur in real soils. The exact value of Brewster's angle is 62.20 degrees (see Figure 4b).

The results of measuring the amplitudes of interference waves with vertical and horizontal polarization at different angles of incidence of the probing signal are presented in Table 1. The experimental conditions correspond to the modeling conditions. Additionally, the following conditions were defined:

- The transmitting antenna mounting height is 3.2 m;
- The receiving antenna mounting height is 3.0 m;
- Microstrip antennas with two types of polarization of the received (transmitted) signal.

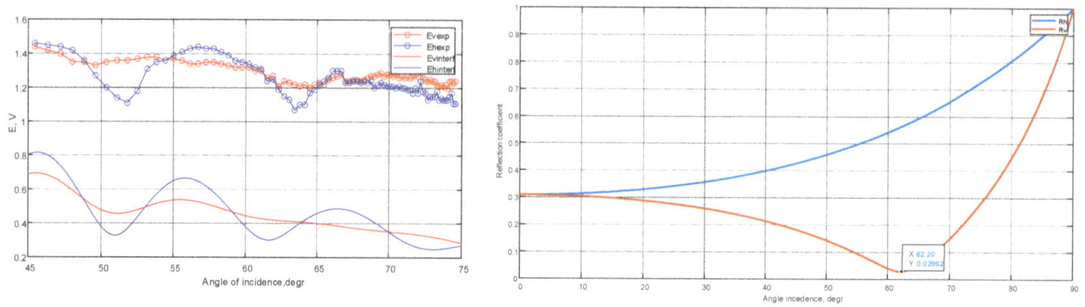

(a) (b)

Figure 4. The results of mathematical modeling and experimental studies confirming the occurrence of the Brewster effect. (**a**) The vertically and horizontally polarized waves obtained experimentally and by mathematical modeling. (**b**) The reflection coefficients of the vertically and horizontally polarized signal under the conditions of pseudo-total refraction.

Table 1. The results of measurements of the amplitudes of interference waves with vertical and horizontal polarization in the range of incidence angles from 45 to 75 degrees.

φ *, deg	Ehexp, V	Evexp, V	*φ* *, deg	Ehexp, V	Evexp, V	*φ* *, deg	Ehexp, V	Evexp, V
74.95	1.11	1.24	70.42	1.21	1.28	62.31	1.24	1.26
74.82	1.11	1.23	70.21	1.22	1.27	61.90	1.31	1.3
74.69	1.17	1.24	69.99	1.23	1.29	61.49	1.32	1.3
74.56	1.15	1.24	69.77	1.22	1.28	61.06	1.34	1.32
74.43	1.13	1.2	69.55	1.2	1.27	60.62	1.35	1.32
74.29	1.13	1.22	69.32	1.24	1.27	60.17	1.36	1.32
74.16	1.13	1.21	69.09	1.23	1.26	59.70	1.39	1.33
74.02	1.14	1.22	68.85	1.24	1.25	59.22	1.41	1.34
73.87	1.13	1.21	68.60	1.23	1.24	58.73	1.43	1.35
73.73	1.18	1.22	68.35	1.24	1.25	58.23	1.43	1.35
73.58	1.15	1.24	68.10	1.24	1.25	57.70	1.44	1.34
73.43	1.15	1.24	67.84	1.23	1.24	57.17	1.43	1.34
73.28	1.15	1.25	67.57	1.23	1.24	56.62	1.41	1.36
73.13	1.13	1.24	67.30	1.3	1.28	56.05	1.39	1.37
72.97	1.17	1.25	67.02	1.3	1.27	55.46	1.36	1.36
72.81	1.18	1.26	66.73	1.3	1.26	54.86	1.35	1.38
72.64	1.24	1.3	66.44	1.26	1.25	54.23	1.31	1.38
72.48	1.17	1.27	66.14	1.24	1.22	53.59	1.18	1.37
72.31	1.2	1.28	65.83	1.22	1.23	52.93	1.11	1.36
72.13	1.17	1.27	65.52	1.19	1.22	52.25	1.14	1.36
71.96	1.19	1.26	65.20	1.17	1.2	51.54	1.2	1.35
71.78	1.19	1.27	64.87	1.1	1.22	50.81	1.27	1.33
71.60	1.2	1.26	64.53	1.1	1.21	50.06	1.36	1.35
71.41	1.2	1.27	64.18	1.07	1.22	49.29	1.42	1.35
71.22	1.21	1.26	63.83	1.13	1.23	48.49	1.44	1.4
71.02	1.2	1.28	63.46	1.15	1.24	47.66	1.45	1.42
70.83	1.21	1.27	63.09	1.19	1.21	46.81	1.46	1.44
70.62	1.21	1.28	62.70	1.25	1.27	45.93	1.46	1.46
						45.02	1.43	1.47

* Angle of incidence.

The radio transmitter of the portable radio station TK-450S was used to form the probing signal. A portable Anritsu S362E antenna-feeder analyzer was used as a receiver.

The difference between the measured and simulated amplitudes of interference signals is determined by the presence of an electromagnetic background equal, on average, to 0.7 V at the experiment location. When comparing the experimental and modeled data, the locations of oscillation maxima and minima, as well as the intersection points of the oscillation graphs depicting the moduli of the vertically and horizontally polarized interference waves, do not differ by more than 10 percent. The experimentally measured Brewster angle is 62.60 degrees, and the magnitude obtained from the simulation results is 62.20 degrees. Thus, the good convergence between the results of modeling and experimental data allows for the use of the proposed mathematical model to study the Brewster angle measurement error.

3. Results

A comparative assessment of the accuracy characteristics was conducted for the four methods of measuring the Brewster angle: by the minimum of the modulus of the vertically polarized interference wave; by the equality to zero of the derivative of the dependence of the modulus of the vertically polarized interference wave on the angle of incidence; by the phase shift between the vertically and horizontally polarized interference waves; by the phase shift between the vertically and horizontally polarized interference waves with normalized amplitudes. The normalization is performed in order to exclude the influence of amplitude and consists of replacing all positive values of amplitude with 1 and negative values with -1.

Figures 5–7 show graphs reflecting the initial conditions for measuring the Brewster angle φ_B (relative permittivity $\varepsilon_{r2} = 3.5$, specific conductivity of the topsoil $\sigma_2 = 1.5 \times 10^{-2} S/m$) without the influence of external perturbing factors. The calculated value for the Brewster angle is 62.20 ± 0.32 degrees. The scattering of values of 0.64 degrees confirms the fact that for lossy dielectrics, the pseudo-Brewster effect occurs when the reflection coefficient does not reach zero, but only decreases to a certain minimum (when modeling $\acute{E}v_{interf} = 0.02692B$, see Figure 4b).

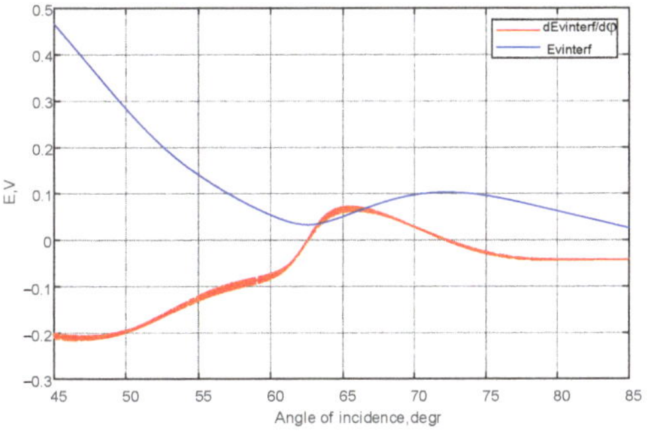

Figure 5. The results of modeling the vertically polarized interference wave and its derivative.

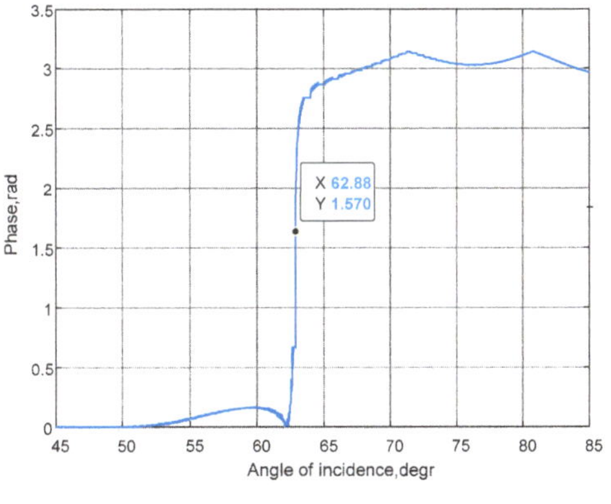

Figure 6. The results of modeling the phase difference between the vertically and horizontally polarized interference waves.

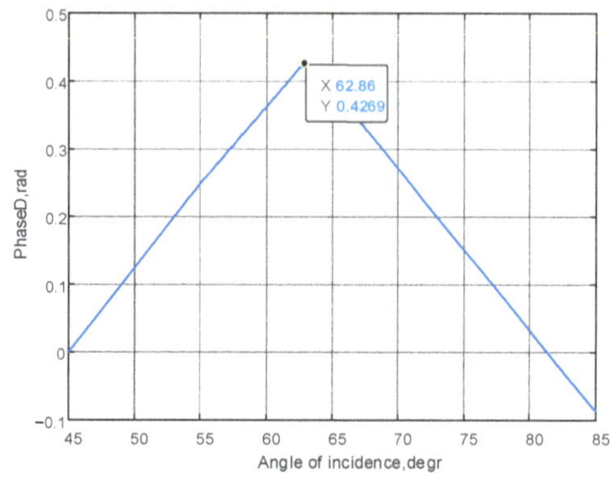

Figure 7. The results of modeling the phase difference of the vertically polarized interference waves relative to the horizontally polarized interference wave using a digital phase discriminator (discriminative characteristic).

In Figure 5, graphs of the vertically polarized interference wave and its derivative are presented. The amplitude of the vertically polarized interference wave has a plane segment of 29 identical values equal to the minimum of 0.0328 V, which corresponds to the calculated values of the model in the range of incident angles from 62.40 to 62.51 degrees, i.e., $\varphi_B = 62.46 \pm 0.05$ degrees. The absolute error was 0.26 degrees.

The derivative of the function $\overset{\sim}{Ev}_{interf}(\varphi)$ is zero at the first extremum, i.e., when the angle of incidence is equal to the Brewster angle. Through the derivative modeling, we obtained a Brewster angle of 62.54 degrees. The absolute error was 0.34 degrees, and there was no scattering of values.

The phase changes associated with the Brewster effect are estimated by the phase difference of the complex amplitudes of interference waves with vertical and horizontal polarization (see Figure 3). When the wave is reflected from the soil surface, the phase of the complex amplitude of the reflected wave changes by 180 degrees. This manifests itself in the phase of the vertically polarized interference wave. The phase difference between the vertically and horizontally polarized interference waves equal to $\pi/2$ indicates the presence of the Brewster effect.

Figure 6 shows the results of calculating the phase difference of the interference waves. The phase shift value of 1.57 rad corresponds to the angle of 62.88 degrees. Thus, the absolute measurement error is 0.68 degrees.

The normalized interference signals are processed using a digital phase discriminator. The algorithm of the digital phase discriminator works on the principle of accumulation of phase differences of the normalized amplitudes of the vertically and horizontally polarized interference waves. When the amplitudes coincide (both amplitudes are equal to 1, or both are equal to -1), the discriminator performs subtraction. In the case of a mismatch (-1 and 1 or vice versa), the discriminator performs summation. The maximum of the discriminative characteristic corresponds to the Brewster angle. Figure 7 shows the discriminative characteristic, the maximum of which is equal to 0.4269 rad and corresponds to the Brewster angle of 62.86 degrees. The absolute error is 0.66 degrees.

Thus, when measuring the Brewster angle without accounting for random noise (without perturbing factors), the absolute measurement error ranges from 0.26 to 0.68 degrees. The method based on determining the minimum of the complex amplitude modulus of the vertically polarized interference wave ensures the formation of a range of values satisfying the condition for a local minimum. The calculation of the derivative of the complex amplitude modulus of the vertically polarized interference wave and the determination of the moment when it transitions through the zero value (extremum) yield only one value with an absolute error of less than 0.50 degrees. Phase methods are found to be less accurate, but in both cases, the Brewster angle measurement error is less than 0.68 degrees.

To evaluate the precision of measuring the Brewster angle, it is proposed to analyze the perturbing factors that the authors consider as the most important ones:

- The ratio of the amplitudes and phase shifts of the direct, reflected and soil waves;
- The sizes of irregularities and their statistical characteristics at the point of incidence, as well as their influence on the type of reflection (diffuse, specular or mixed);
- The presence and magnitude of permissible random changes in the amplitude and phase (noise) separately for the direct, reflected and soil waves and the integral noise of the interference wave.

Figure 8 shows graphs of the vertically and horizontally polarized interference waves for the case when the wave amplitudes are determined by the expressions (6) and (7) (the first group of graphs) and the case when the amplitude of the direct wave is reduced tenfold (the second group of graphs).

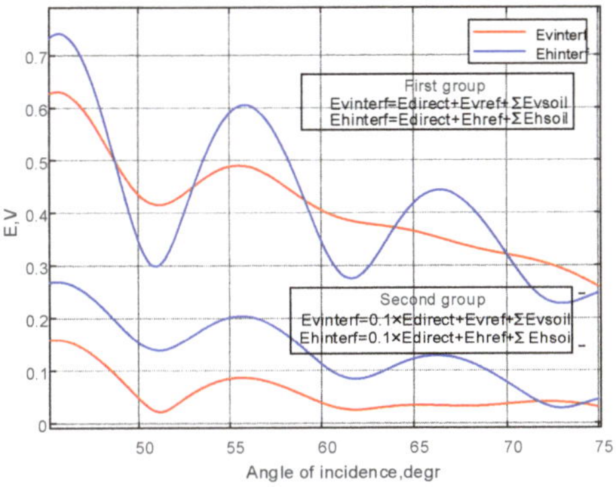

Figure 8. The results of modeling the ratio of the amplitudes of direct and reflected interference waves with vertical and horizontal polarization.

As can be seen from Figure 8, forced decreasing of the amplitude of the direct wave (e.g., by creating a specific antenna pattern) results in a decrease in the energy of interference waves and the emergence of additional minima of the amplitude modulus of the vertically polarized interference wave.

The irradiation of the Earth's surface is intended to be carried out in accordance with the Rayleigh criterion [29]:

$$h < \frac{\lambda}{8cos\varphi}.$$

The size of irregularities and their statistical characteristics at the point of incidence are taken into account by multiplying the reflection coefficient by the roughness coefficient:

$$r(h) = \exp\left(-8\pi^2\left(\frac{\Omega_h}{\lambda}\right)^2 cos^2\varphi\right), \tag{8}$$

where h is the height of random irregularities; Ω_h is the root mean square deviation of the height of random irregularities.

Figure 9a–c presents graphs that illustrate the influence of the sizes of irregularities of the investigated surface on the type of the reflected signal. According to Figure 9, when the size of the irregularity increases, the amplitude of the reflected signal decreases in the region of small angles of incidence. Outside the Brewster angle, the level of the reflected signal becomes higher. The size of the irregularities has a particularly significant impact on the amplitude of the horizontally polarized reflected signal. When $\Omega_h > 0.2\lambda$, the characteristics of the reflected signal do not allow for determining the Brewster angle.

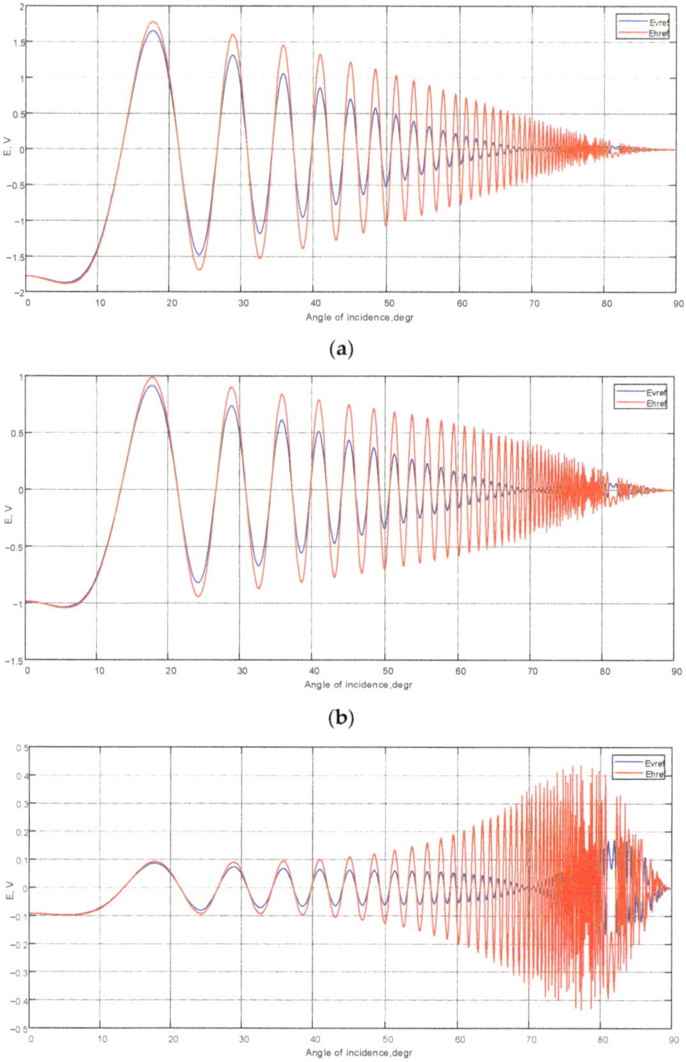

(a)

(b)

(c)

Figure 9. The influence of the root mean square deviation of the roughness height on the type of reflected signal. Amplitudes of the vertically and horizontally polarized reflected signal at the RMS deviation of the roughness height: (**a**) $\Omega_h = 0.05\lambda$; (**b**) $\Omega_h = 0.1\lambda$; (**c**) $\Omega_h = 0.2\ \lambda$.

A set of random factors causing the distortion of amplitudes and phases of the direct, reflected and ground waves is modeled by additive interference with a normal distribution and mathematical expectation equal to zero.

Figure 10a presents graphs of the vertically polarized interference wave and its derivative during the modeling of noises caused by the direct, reflected and ground waves by additive interference with a root mean square deviation of 0.0001 V. It is clear that calculating the Brewster angle using the derivative of the modulus of the complex amplitude of the vertically polarized interference wave is impossible, even in the presence of such noises. The errors of the phase methods are 0.7 degrees (Figure 10b) and 0.66 degrees (see Figure 10c), respectively.

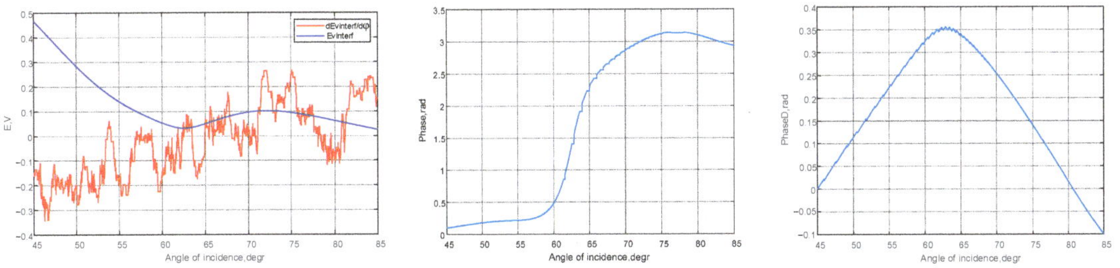

(**a**) (**b**) (**c**)

Figure 10. Measuring the Brewster angle during the modeling of noises from the direct, reflected and soil waves by additive interference with a root mean square deviation of 0.0001 V: (**a**) the dependence of the amplitude of the vertically polarized interference wave and its derivative on the angle of incidence obtained by the amplitude methods; (**b**) dependence of the phase of the interference wave on the angle of incidence obtained by the phase method; (**c**) the dependence of the phase of the interference wave on the angle of incidence obtained by the phase method using a digital phase discriminator.

Figure 11 shows the conditions for measuring the Brewster angle when modeling the noises from the direct, reflected and ground waves by additive interference with a root mean square deviation of 0.001 V.

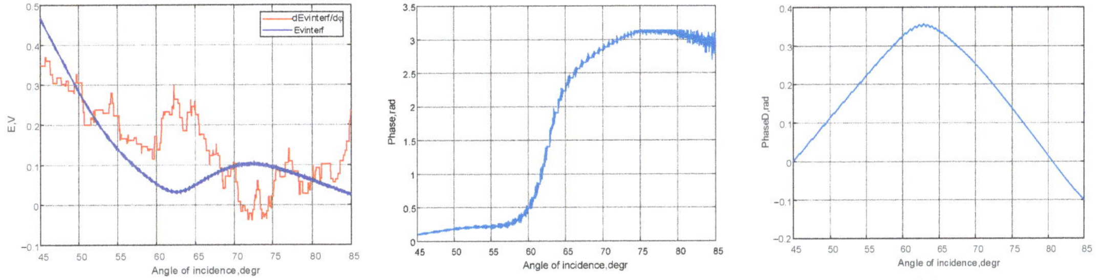

(**a**) (**b**) (**c**)

Figure 11. Measuring the Brewster angle when modeling the noises from the direct, reflected and ground waves by additive interference with a root mean square deviation of 0.001 V: (**a**) the dependence of the amplitude of the vertically polarized interference wave and its derivative on the angle of incidence obtained by the amplitude methods; (**b**) dependence of the phase of the interference wave on the angle of incidence obtained by the phase method; (**c**) the dependence of the phase of the interference wave on the angle of incidence obtained by the phase method using a digital phase discriminator.

The graph of the modulus of the complex amplitude of the vertically polarized interference wave (Figure 11a) has three local minima at the angles of incidence: 61.96; 62.62; 63.12 degrees. The absolute error for the first local minimum is 0.24 degrees, and the maximum error is 0.92. The phase shift of 1.57 rad is achieved at an angle of incidence of 62.91 (Figure 11b), which corresponds to an absolute error of 0.71 degrees. The Brewster angle measured by the digital phase discriminator is 62.86 degrees (Figure 11c), and the measurement error is 0.66 degrees.

Figure 12 shows the conditions for measuring the Brewster angle when modeling the noises from the direct, reflected and soil waves by additive interference with a root mean square deviation of 0.01 V.

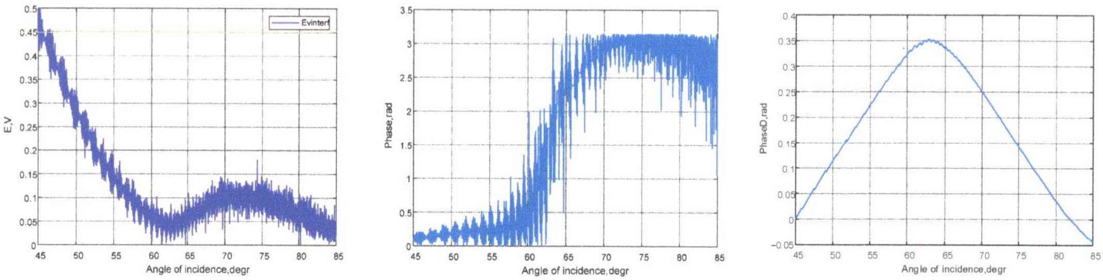

(a) (b) (c)

Figure 12. Measuring the Brewster angle when modeling the noises from the direct, reflected and ground waves by additive interference with a root-mean-square deviation of 0.01 V: (**a**) the dependence of the amplitude of the vertically polarized interference wave and its derivative on the angle of incidence obtained by the amplitude methods; (**b**) dependence of the phase of the interference wave on the angle of incidence obtained by the phase method; (**c**) the dependence of the phase of the interference wave on the angle of incidence obtained by the phase method using a digital phase discriminator.

The graph of the modulus of the complex amplitude of the vertically polarized interference wave (Figure 12a) has seven local minima at the angles of incidence: 60.77; 61.38; 62.03; 62.59; 63.16; 63.71; 64.21 degrees. The maximum error for the first minimum is 2.01 degrees. Several angles of incidence meet the condition that the phase shift is 1.57 rad (Figure 12b), and the maximum measurement error is 2.66 degrees. The Brewster angle measured by the digital phase discriminator is 62.9 degrees (Figure 12c), and the absolute error is 0.7 degrees.

Figure 13 shows the conditions for measuring the Brewster angle when modeling the noises from the direct, reflected and soil waves by additive interference with a root mean square deviation of 0.1 V.

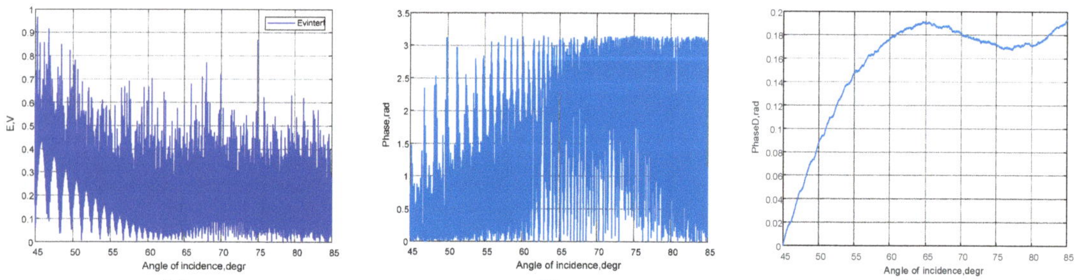

(a) (b) (c)

Figure 13. Measuring the Brewster angle when modeling the noises from the direct, reflected and ground waves by additive interference with a root mean square deviation of 0.1 V: (**a**) the dependence of the amplitude of the vertically polarized interference wave and its derivative on the angle of incidence obtained by the amplitude methods; (**b**) dependence of the phase of the interference wave on the angle of incidence obtained by the phase method; (**c**) the dependence of the phase of the interference wave on the angle of incidence obtained by the phase method using a digital phase discriminator.

As can be seen from Figure 13a,b, it is not possible to measure the Brewster angle from the minimum modulus of the complex amplitude of the vertically polarized interference wave or from the phase shift of the vertically and horizontally polarized interference waves. At the same time, the digital phase discriminator remains operational (Figure 13c). The measured value of the Brewster angle is 63.64 degrees; the absolute measurement error is 1.44 degrees.

4. Discussion

The obtained estimates of the absolute error in measuring the Brewster angle indicate that in the absence of random perturbing factors, both the amplitude and phase methods are quite accurate, with an absolute error not exceeding one degree. When studying dielectrics with losses (for example, wet, salty soils), the accuracy of measuring the Brewster angle decreases. This is due to the absence of a pronounced minimum of the vertically polarized interference signal. The increase in the size of the surface irregularities and their dispersion changes the reflection of the electromagnetic waves, making it diffuse-specular. Under certain conditions, this leads to the impossibility of determining the Brewster angle. The method in which the partial derivative of the dependence of the modulus of the complex amplitude was calculated from the angle of incidence of the electromagnetic wave turned out to be the most sensitive to the effects of additive interference.

The phase method with normalized amplitude values of the vertically and horizontally polarized interference waves proves to be the most resistant to all perturbing factors, as analyzed in the paper. When measuring the Brewster angle using this method, the error does not exceed one degree. This holds true even when dealing with diffuse-specular reflection from the upper layer of the lossy dielectric, when additive interference exceeds 10 percent of the useful signal level.

The determination of the Brewster angle φ_B makes it possible to estimate the relative permittivity ε_{r2} of the topsoil using the well-known expression [30] (p. 307):

$$\varepsilon_{r2} = (tg\varphi_B)^2. \quad (9)$$

At $\varphi_B = 62°$ and $\varepsilon_{r2} = 3.53$, the error of $1°$ ($\varphi_B = 63°$ and $\varepsilon_{r2} = 3.85$) results in a nine percent increase in relative permittivity. Thus, with the error obtained from the phase method with normalized interference waves' amplitudes of less than one degree, the error in measuring the relative permittivity does not exceed the generally accepted technical requirements [18].

The assessment of the physicochemical parameters of soil (e.g., moisture) is carried out using pedotransfer functions that are set out in the international recommendations [18].

Thus, relative dielectric permittivity with volumetric moisture content W, composition (percentage of soil particles P_{sand}, P_{clay}, P_{silt}), and soil density are related by the expression:

$$\varepsilon\prime = \left[1 + (\rho_b/\rho_s)\left(\{\varepsilon'_{sd}\}^{\alpha} - 1\right) + W^{\beta'}\left(\varepsilon'_{bw}\right)^{\alpha} - W\right]^{1/\alpha}, \quad (10)$$

where ρ_b is the bulk density of soil; ρ_s is the specific gravity of a dry mixture of soil constituents; ε'_{sd} is the component of dielectric permittivity determined by soil density, $\varepsilon'_{sd} = (1.01 + 0.44\,\rho_s)^2 - 0.062$; $\alpha = 0.65$; $\beta' = 1.2748 - 0.00519\,P_{sand} - 0.00152\,P_{clay}$; ε'_{bw} is the real part of complex relative permittivity of bound water.

Figure 14 shows the results of calculating the relative permittivity for the different values of volumetric moisture content and for several types of soil composition. On each of the graphs, two points are randomly chosen whose relative permittivity value (x-axis) does not differ by more than 10% (7.82% and 5.39%). The corresponding values of volumetric moisture (y-axis) also do not exceed ten percent (6.89% and 6.25%). As such, we can conclude that measuring the Brewster angle with an error of less than one degree allows us to determine soil moisture with an error not exceeding ten percent.

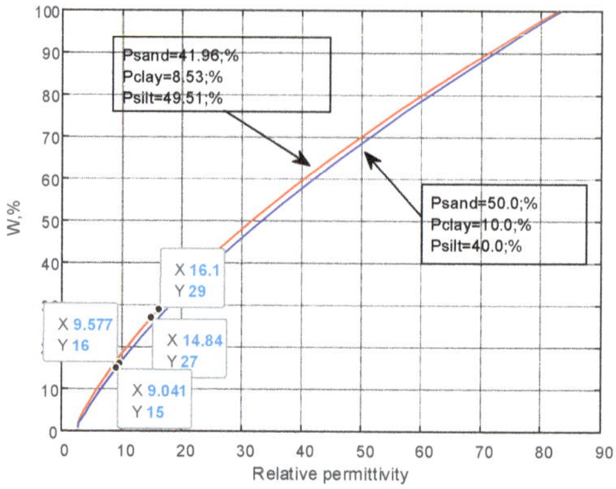

Figure 14. The results of calculating the relative permittivity for the different values of volumetric moisture for several types of soil composition.

5. Conclusions

Thus, based on the solution of the Helmholtz wave equation and Fresnel formulas, a mathematical model of electromagnetic wave interference at the boundary of lossy dielectric media was developed. Experimental confirmation has been obtained to validate the adequacy of the developed model using the most probable set of input data.

The model takes into account the root mean square deviations of the roughness heights at the air–surface interface, which affect the ratio of specularly and diffusely reflected rays, as well as the level of additive noise with the normal distribution law of individual complex amplitudes of the direct, reflected and ground waves.

The obtained results of mathematical modeling, theoretical and experimental research enable remote measurement of the soil's physical and chemical parameters, such as moisture content, with an accuracy sufficient for rational crop management using precision farming technology.

Timely remote determination of soil moisture at different stages of development of agricultural plants will allow for a reasonable choice of rational agrotechnical measures. Further research will aim to evaluate the cation exchange capacity, the particle size distribution (granulometric composition) and the humus content of soil by measuring the Brewster angle and the complex dielectric constant of soil.

One of the important results achieved through the developed model is the estimation of the error in determining Brewster's angle using various measurement methods and conditions. The findings demonstrate that information about the Brewster angle lies not only in the minimum amplitude of the vertically polarized interference signal; this information is also provided by the phase difference between the vertically and horizontally polarized interference waves. The boundary conditions determining the applicability of different methods for estimating the Brewster angle have been examined. It has been shown that the method based on the calculation of the derivative of the functional dependence of the amplitude of the vertically polarized interference signal is ineffective, even under conditions of additive noise less than 0.01% of the amplitude value. A phase method for determining the Brewster angle is proposed, based on the implementation of a digital phase discriminator algorithm for the phase difference between the vertically and horizontally polarized interference waves, normalized by amplitude. The possibility of measuring the Brewster angle with an accuracy of less than one degree using the proposed phase method has been justified. This allows for determining soil moisture with an error not exceeding ten percent.

The authors associate the prospects for further research on the accuracy of the Brewster angle measurement with the combination of amplitude and phase methods, as well as with the application of digital processing of interference signals.

Author Contributions: Conceptualization, G.I.L. and A.V.B.; methodology, S.V.M. (Sergey Vladimirovich Malygin) and A.V.B.; software, A.V.B., S.V.M. (Sergey Vladimirovich Malygin) and V.D.G.; validation, A.V.B. and S.V.M. (Sergey Vladimirovich Malygin); formal analysis, A.V.B. and S.V.M. (Sergey Vladimirovich Malygin); investigation, A.V.B., S.V.M. (Sergey Vladimirovich Malygin) and N.V.G.; data curation, A.V.B., S.V.M. (Sergey Vladimirovich Malygin), S.V.M. (Sergey Vladimirovich Melnikov) and V.D.G.; writing—original draft preparation, A.V.B. and N.V.G.; writing—review and editing, N.V.G.; visualization, A.V.B. and N.V.G.; supervision, G.I.L.; project administration, G.I.L.; funding acquisition, G.I.L. All authors have read and agreed to the published version of the manuscript.

Funding: The study was carried out as a part of the implementation of the "Priority 2030" strategic program of academic leadership with the financial support of the Ministry of Education and Science of Russia (Agreement No. 075-15-2021-1319).

Data Availability Statement: The data presented in this study are available on request from the corresponding author.

Conflicts of Interest: The authors declare no conflict of interest.

References

1. Mohamed, E.S.; Belal, A.A.; Abd-Elmabod, S.K.; El-Shirbeny, M.A.; Gad, A.; Zahran, M.B. Smart farming for improving agricultural management. *Egypt. J. Remote Sens. Space Sci.* **2021**, *24*, 971–981. [CrossRef]
2. Dubois, A.; Teytaud, F.; Verel, S. Short term soil moisture forecasts for potato crop farming: A machine learning approach. *Comput. Electron. Agric.* **2021**, *180*, 105902. [CrossRef]
3. Ibanga, O.A.; Idehen, O.F.; Omonigho, M.G. Spatiotemporal variability of soil moisture under different soil groups in Etsako West Local Government Area, Edo State, Nigeria. *J. Saudi Soc. Agric. Sci.* **2022**, *21*, 125–147. [CrossRef]
4. Idoje, G.; Dagiuklas, T.; Iqbal, M. Survey for smart farming technologies: Challenges and issues. *Comput. Electr. Eng.* **2021**, *92*, 107104. [CrossRef]
5. Srivastava, P.K.; Prasad, R.; Kumar, S.C.; Yadav, S.A.; Sharma, J.; Suman, S.; Pandey, V.; Singh, R.; Gupta, D.K. Chapter 20—Challenges in Radar remote sensing. *Radar Remote Sens.* **2022**, *484*, 377–387. [CrossRef]
6. Zajícová, K.; Chuman, T. Application of ground penetrating radar methods in soil studies: A review. *Geoderma* **2019**, *343*, 116–129. [CrossRef]
7. Francia, M.; Giovanelli, J.; Golfarelli, M. Multi-sensor profiling for precision soil-moisture monitoring. *Comput. Electron. Agric.* **2022**, *197*, 106924. [CrossRef]
8. Songara, J.C.; Patel, J.N. Calibration and comparison of various sensors for soil moisture measurement. *Measurement* **2022**, *197*, 111301. [CrossRef]
9. Shakya, A.K.; Ramola, A.; Kandwal, A.; Vidyarthi, A. Soil moisture sensor for agricultural applications inspired from state of art study of surfaces scattering models & semi-empirical soil moisture models. *J. Saudi Soc. Agric. Sci.* **2021**, *20*, 559–572. [CrossRef]
10. Pramanik, M.; Khanna, M.; Singh, M.; Singh, D.K.; Sudhishri, S.; Bhatia, A.; Ranjan, R. Automation of soil moisture sensor-based basin irrigation system. *Smart Agric. Technol.* **2022**, *2*, 100032. [CrossRef]
11. Boursianis, A.D.; Papadopoulou, M.S.; Diamantoulakis, P.; Liopa-Tsakalidi, A.; Barouchas, P.; Salahas, G.; Karagiannidis, G.; Wan, S.; Goudos, S.K. Internet of Things (IoT) and Agricultural Unmanned Aerial Vehicles (UAVs) in smart farming: A comprehensive review. *Internet Things* **2022**, *18*, 100187. [CrossRef]
12. Paul, K.; Chatterjee, S.S.; Pai, P.; Varshney, A.; Juikar, S.; Prasad, V.; Bhadra, B.; Dasgupta, S. Viable smart sensors and their application in data driven agriculture. *Comput. Electron. Agric.* **2022**, *198*, 107096. [CrossRef]
13. Peng, J.; Nieto, H.; Andersen, M.N.; Kørup, K.; Larsen, R.; Morel, J.; Parsons, D.; Zhou, Z.; Manevski, K. Accurate estimates of land surface energy fluxes and irrigation requirements from UAV-based thermal and multispectral sensors. *ISPRS J. Photogramm. Remote Sens.* **2023**, *198*, 238–254. [CrossRef]
14. Forte, E.; Pipan, M. Review of multi-offset GPR applications: Data acquisition, processing and analysis. *Signal Process.* **2017**, *132*, 210–220. [CrossRef]
15. Pan, X.; Wollschläger, U.; Gerhards, H.; Roth, K. Optimization of multi-channel ground-penetrating radar for quantifying field-scale soil water dynamics. *J. Appl. Geophys.* **2012**, *82*, 101–109. [CrossRef]
16. Chandra, M.; Tanzi, T.J. Drone-borne GPR design: Propagation issues. *C. R. Phys.* **2018**, *19*, 72–84. [CrossRef]
17. Thomson, L.I.; Osinski, G.R.; Pollard, W.H. Application of the Brewster angle to quantify the dielectric properties of ground ice formations. *J. Appl. Geophys.* **2013**, *99*, 12–17. [CrossRef]
18. Recommendation ITU-R P.527-6. Electrical Characteristics of the Surface of the Earth. P Series. Radiowave Propagation. 2019. Available online: https://www.itu.int/rec/R-REC-P.527-6-202109-I/en (accessed on 10 July 2023).
19. Zhang, S.; Liu, L.; Liu, Y. Generalized laws of Snell, Fresnel and energy balance for a charged planar interface between lossy media. *J. Quant. Spectrosc. Radiat. Transf.* **2020**, *245*, 106903. [CrossRef]

20. Nagib, N.N.; Abdallah, A.; Bahrawi, M.; Mousa, E. Novel polarimetric method for determination of the Brewster angle of a dielectric. *Results Opt.* **2020**, *1*, 100006. [CrossRef]
21. Thirion-Lefevre, L.; Guinvarc'h, R. The double Brewster angle effect. *C. R. Phys.* **2018**, *19*, 43–53. [CrossRef]
22. Azzam, R.M.A. Complex reflection coefficients of p- and s-polarized light at the pseudo-Brewster angle of a dielectric–conductor interface. *J. Opt. Soc. Am. A* **2013**, *30*, 1975–1979. [CrossRef] [PubMed]
23. Hallikainen, M.T.; Ulaby, F.T.; Dobson, M.C.; El-Rayes, M.A.; Wu, L.K. Microwave dielectric behavior of wet soil-part 1: Empirical models and experimental observations. *IEEE Trans. Geosci. Remote Sens.* **1985**, *1*, 25–34. [CrossRef]
24. Sterlyadkin, V.V. Radiometric microwave field measurements of the complex dielectric constant of the water surface. *Adv. Space Res.* **2018**, *62*, 3162–3169. [CrossRef]
25. Reppert, P.M.; Morgan, F.D.; Toksöz, M.N. Dielectric constant determination using ground-penetrating radar reflection coefficients. *J. Appl. Geophys.* **2000**, *43*, 189–197. [CrossRef]
26. Kim, K.; Kim, J.-H.; Cho, D.-S. Radar cross section analysis of marine targets using a combining method of physical optics/geometric optics and a Monte-Carlo simulation. *Ocean Eng.* **2009**, *36*, 821–830. [CrossRef]
27. Khmelinin, A.P.; Konurin, A.I.; Denisova, E.V. Simulation of electromagnetic high-frequency wave propagation processes in multilayer geo-structures. *Geohazard Mech.* 2023, *in press*. [CrossRef]
28. Kawanishi, T. The shift of Brewster's scattering angle. *Opt. Commun.* **2000**, *186*, 251–258. [CrossRef]
29. Jackson, J.D. *Classical Electrodynamics*, 3rd ed.; John Wiley & Sons: New York, NY, USA, 1999; pp. 299–307.
30. Ogilvy, J. *Theory of Wave Scattering from Random Surfaces*; Institute of Physics Publishing: Bristol, UK; Philadelphia, PA, USA, 1991; pp. 9–26, 206–213.

Disclaimer/Publisher's Note: The statements, opinions and data contained in all publications are solely those of the individual author(s) and contributor(s) and not of MDPI and/or the editor(s). MDPI and/or the editor(s) disclaim responsibility for any injury to people or property resulting from any ideas, methods, instructions or products referred to in the content.

AgriEngineering

MDPI

Article

Performance Evaluation of a Wet Medium Made of Mangosteen Peels for a Direct Evaporative Cooling System

Nattawut Chaomuang *, Thanut Nuangjamnong and Samak Rakmae

Department of Food Engineering, School of Engineering, King Mongkut's Institute of Technology Ladkrabang, 1 Soi Chalongkrung 1, Ladkrabang, Bangkok 10520, Thailand
* Correspondence: nattawut.ch@kmitl.ac.th

Abstract: The present study aimed to investigate an alternative evaporative cooling pad material made from mangosteen peel (MP) waste. Mangosteen peels were used to fill a 150 mm thick mesh container with a packing density of 180 kg/m^3. A wind tunnel was constructed and utilized to experimentally evaluate the cooling performance of this organic-waste-based pad under hot and humid conditions (31–34 °C and 55–70% RH). The performance parameters assessed included pressure drop, temperature drop, saturation effectiveness, cooling capacity, and coefficient of performance (*COP*). The influence of air velocity (0.7, 1.0, 1.4, and 1.8 m/s) on these parameters was also examined. The results revealed that the saturation effectiveness of the MP pad ranged from 53% to 77% within the considered air velocity range. The maximum temperature drop (4.6 °C), saturation effectiveness (77%), cooling capacity (0.6 kW), and *COP* (3.5) were achieved when the system operated at 1.4 m/s. A comparative study showed that, at this velocity, the MP pad provided performance nearly equivalent to that of the commercial cellulose paper pad, except for the pressure drop. This result affirms the potential of mangosteen peels as a suitable wet medium for evaporative cooling applications.

Keywords: refrigeration; cooling pad; saturation effectiveness; valorization; vegetal waste

Citation: Chaomuang, N.; Nuangjamnong, T.; Rakmae, S. Performance Evaluation of a Wet Medium Made of Mangosteen Peels for a Direct Evaporative Cooling System. *AgriEngineering* **2023**, *5*, 1865–1878. https://doi.org/10.3390/agriengineering5040114

Academic Editors: Muhammad Sultan, Yuguang Zhou, Redmond R. Shamshiri and Muhammad Imran

Received: 2 September 2023
Revised: 29 September 2023
Accepted: 10 October 2023
Published: 12 October 2023

Copyright: © 2023 by the authors. Licensee MDPI, Basel, Switzerland. This article is an open access article distributed under the terms and conditions of the Creative Commons Attribution (CC BY) license (https://creativecommons.org/licenses/by/4.0/).

1. Introduction

Among cooling technologies, vapor-compression refrigeration (VCR) is the dominant cooling system for cold storage [1]. Despite its high efficiency, this system is energy-intensive and greatly responsible for greenhouse gas emissions. Evaporative cooling systems have been deemed a viable alternative to VCR due to their simple structure, ease of maintenance, low capital and operating costs, and low environmental impacts [2]. They have been broadly implemented in various space cooling applications, such as in buildings [3,4], livestock barns [5], agricultural greenhouses [6], and horticultural product storage [7]. Numerous studies have extensively demonstrated the potential of evaporative cooling systems for the quality preservation of fruits and vegetables—for example, leafy vegetables [8], mangoes [9], peppers [10], and tomatoes [11,12]. Evaporative cooling is generally classified into two types: direct and indirect [13]. In a direct evaporative cooler, air is in direct contact with water, which can be either sprayed directly in air streams or supplied to a wet medium residing in the flow passage. The air is cooled down and humidified through water evaporation and subsequently supplied to a target space. On the other hand, indirect evaporative cooling involves two air streams. The first air stream is directly cooled and humidified through water evaporation and then used to cool the second air stream without humidification by a heat exchanger. The second air stream is then supplied to the target space. The applications of direct and indirect evaporative cooling systems rely on the humidification requirements [14]. For instance, direct evaporating cooling is preferable for cold storage of fruits and vegetables because of their high humidity requirement.

The wet medium, commonly called an "evaporative cooling pad" or simply "cooling pad", is the key determinant for the performance of both direct and indirect evaporative cooling systems. The systems incorporated with the cooling pads can apparently achieve superior cooling performance compared to those with direct water spraying [15]. Numerous studies have proposed various types of cooling pads to improve the performance of direct evaporative cooling [16]. Based on their material and configuration, the cooling pads can be categorized into fiber pads, rigid media pads, and packages or fill pads [17]. Rigid media pads made of corrugated cellulose paper are mostly used because of their large water-to-air contact areas, resulting in high saturation effectiveness and long lifespans. Franco-Salas and Peña-Fernández [18] experimentally compared the evaporative cooling performance of a brand-new rigid media pad with that of a 3-year-old used pad (same model). The results showed that the old pad achieved higher saturation effectiveness compared to the new one. The authors explained that the enhanced saturation effectiveness resulted from the salt incrustation on the pad, which caused an increase in the air–water contact time. Nada and Fouda [19] proposed a rigid media pad with a beehive structure. Its cooling performance was experimentally examined under various operating conditions (air velocity, inlet air temperature, water flow rate, water temperature, and pad thickness). It was found that the saturation effectiveness can be enhanced by increasing the pad thickness and water flow rate. However, the pad thickness should be selected with caution, because excessively thick cooling pads are often associated with high capital (i.e., price of cooling pads) and operating (i.e., energy consumption of fans) costs. Yan and He [20] experimentally investigated the influence of the cooling pad arrangements on the evaporative cooling performance. The results showed that a pair of cooling pads with a space in between them exhibited superior cooling effectiveness when compared with the case without spacing. This result suggests the possibility of using cooling pads of the same thickness to obtain the desired cooling effectiveness by leaving space between the pads.

The production of rigid media pads is complex and requires advanced machinery. Various materials have been proposed as alternatives for cooling pads, often in the form of fiber and fill pads. Table 1 presents fiber and fill pads made of different alternative materials based on natural and organic materials. Plastic-, metal-, and stone-based materials have also been investigated by numerous studies, which have been comprehensively reviewed in several papers [16,17,21].

Table 1. Experimental studies on fiber and fill pads made from natural and organic materials for direct evaporative cooling systems.

Material	Dimensions [1]	Air Velocity/ Flow Rate	Saturation Effectiveness	Cooling Capacity	Country	Ref.
Aspen fibers	$17 \times 17 \times 60$	1.4 m/s	71.6%	-	India	[22]
	$40 \times 40 \times 7.5$	5.44 m^3/min	67–75%	-	Iran	[23]
	$87 \times 61 \times 2.5$	1.4 m/s	52–67%	0.5–1.6 kW	India	[24]
	$31 \times 30 \times 5$	2.4 m/s	49.5%	-	Saudi Arabia	[25]
Bulrush	N.P.	0.1–1.2 m/s	19–35%	0.06–0.42 kW	Turkey	[21]
Coconut fibers	$30 \times 30 \times 15$	0.5–4.0 m/s	83.5–93.8	0.06–0.53 W	China	[26]
	$10 \times 13.5 \times 7$	1.9–2.8 m/s	44.1–51.5%	-	Thailand	[27]
	$17 \times 17 \times 60$	1.4 m/s	69.4%	-	India	[22]
	$92 \times 66 \times 10$	3.6–5.6 m/s	85%	-	Bangladesh	[28]
	$10 \times 10 \times 10$	0.062–0.083 kg/s	35–70%	0.1–0.5 kW	India	[29]
Eucalyptus fibers	$24^{*} \times 10$	0.03–0.08 kg/s	78%	0.81	Turkey	[30]
	N.P.	0.1–1.2 m/s	23–76%	0.12–0.68 kW	Turkey	[21]
Jute fibers	$31 \times 30 \times 5$	2.4 m/s	62.1%	-	Saudi Arabia	[25]
	$30 \times 40 \times 6$	3.0–4.5 m/s	17.4–93.8	-	Nigeria	[31]
Khus fibers	$17 \times 17 \times 60$	1.4 m/s	64.2%	-	India	[22]
Luffa fibers	$31 \times 30 \times 5$	2.4 m/s	55.1%	-	Saudi Arabia	[25]
	$96 \times 96 \times 12.5$	0.77 m/s	57.0%	-	Brazil	[32]

Table 1. *Cont.*

Material	Dimensions [1]	Air Velocity/ Flow Rate	Saturation Effectiveness	Cooling Capacity	Country	Ref.
Palash fibers	17 × 17 × 60	1.4 m/s	81.0%	-	India	[22]
Palm fibers	31 × 30 × 5 30 × 40 × 6	2.4 m/s 3.0–4.5 m/s	38.9% 49.0–98.8%	- -	Saudi Arabia Nigeria	[25] [31]
Rice husk	53 × 53 × 5.1	1–3 m/s	55.4–61.9%	-	Thailand	[33]
Sackcloth	92 × 66 × 10	3.6–5.6 m/s	69%	-	Bangladesh	[28]
Straw fiber	600 × 200 × 10	N.P.	76%	-	Sudan	[34]
Water hyacinth	20 × 50 × 5	1.5–3.6 m/s	20–40%	-	Thailand	[35]
Wood Chips	24 * × 10 10 × 10 × 10 600 × 200 × 10	0.03–0.08 kg/s 0.062–0.083 kg/s N.P.	81% 50–80% 90%	0.90 kW 0.2–0.5 kW -	Turkey India Sudan	[30] [29] [34]

[1] Width (cm) × height (cm) × thickness (cm); * diameter (mm) of cylindrical pads. N.P. = data not provided.

As shown in Table 1, materials derived from organic waste exhibit potential as cooling pad materials since they can achieve saturation effectiveness comparable to commercial cellulose pads. The present study represents another attempt to explore the cooling performance of an organic cooling pad made from mangosteen peel waste. Mangosteen (*Garcinia mangostana* L.) is renowned as one of the world's most exquisite fruits, often referred to as the "Queen of Fruits" due to its sweet–sour taste and unique appearance, with sepals arranged like a crown [36]. Mangosteen is a tropical tree species primarily cultivated in Southeast Asian countries, such as Indonesia, Malaysia, and Thailand [37]. In 2020, Thailand alone produced nearly 340,000 tons of mangosteen, with approximately 290,000 tons of fresh mangosteen intended for export, leaving about 50,000 tons for domestic consumption [38]. It was estimated that 1 kg of mangosteen fruits can yield 0.6 kg of mangosteen peels [39]. According to this estimate, at least 30,000 tons of mangosteen peels were generated and treated as waste. The valorization of mangosteen peels would be a favorable approach to waste management. For instance, Nasrullah, Saad [39] employed mangosteen peel waste as a precursor for activated carbon, demonstrating its potential through batch adsorption studies for the removal of methylene from aqueous solutions. In the present study, mangosteen peel waste was harnessed as a cooling pad material due to its high porosity and water retention capacity. The high porosity of this organic cooling pad, derived from mangosteen peels, distinguishes it from other organic cooling pads discussed in the existing literature. Its performance was experimentally characterized in terms of temperature drop, saturation effectiveness, cooling capacity, coefficient of performance (*COP*), and pressure drop. The influence of air velocity on these performance parameters was also investigated, and the optimal air velocity was determined. Finally, the performance of the mangosteen peel waste-based cooling pad was compared with that of a commercial corrugated cellulose pad.

2. Materials and Methods

2.1. Experimental Setup

In this study, mangosteen peels (MPs) were proposed and utilized as an alternative cooling pad material for a direct evaporative cooling system. Initially, the mangosteen peels were sun-dried under uniform weather conditions for several days until their water activity (a_w) reached a level below 0.6. The a_w was measured using a water activity analyzer (Aqualab 4TE, Meter group Inc., Pullman, WA, USA). Approximately 2.2 kg of the dried mangosteen peels were filled in a container (300 mm × 260 mm × 150 mm) constructed with aluminum wire mesh, resulting in a packing density of about 180 kg/m^3 (Figure 1a).

For comparative purposes, a commercial cooling pad made of corrugated cellulose papers (CPs) with identical dimensions (300 mm × 260 mm × 150 mm) was also employed and tested in the study (Figure 1b).

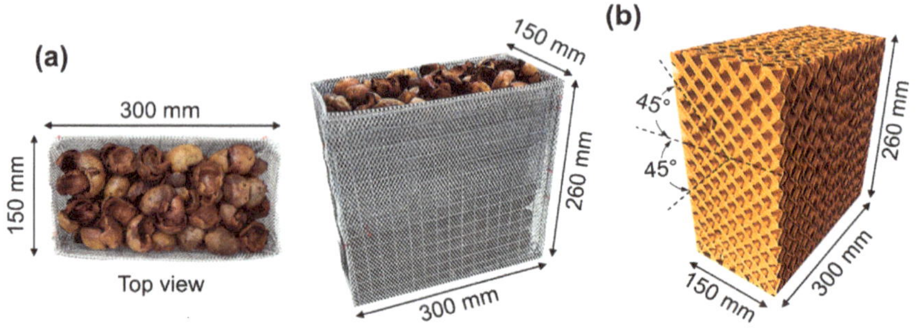

Figure 1. The cooling pads made from (**a**) mangosteen peels and (**b**) corrugated cellulose papers.

A wind tunnel with internal dimensions of 300 mm (width) × 260 mm (height) × 1500 mm (length) was constructed and used to investigate the cooling pad performance. The wind tunnel walls comprised two layers: an outer layer made of 8 mm thick acrylic plates and an inner insulation layer composed of 25 mm thick polyurethane foam boards. The insulation layer served to minimize heat losses from the system.

The test section was designed to accommodate a 150 mm thick cooling pad. This section featured a top-mounted water distribution system to evenly distribute water into the cooling pad, along with a bottom-mounted water collection system for draining excess water back into a 40 L water tank. A 40 W submersible pump (PRO-DS5000, Marine Plus Co., Ltd., Bangkok, Thailand) was employed to recirculate water from the tank to the water distribution system, maintaining a constant water flow rate of 250 L/h.

At the entrance of the wind tunnel, a 750 W centrifugal fan (SC-1304, Venz Industrial Co., Ltd., Bangkok, Thailand) was installed. The rotational speed of the fan motor was controlled using a variable frequency drive (MS300, Delta Electronics, Inc., Taiwan) to generate various air flow rates within the system. Based on a preliminary study, air velocities of 0.7, 1.0, 1.4, and 1.8 m/s were achieved by adjusting the input power frequencies to 20, 30, 40, and 50 Hz, respectively. It is important to note that these air velocities represent averages of 16 measurement points over the cross-sectional area of the wind tunnel using a hot-wire anemometer (405i, Testo SE & Co. KGaA, Titisee-Neustadt, Germany).

A depiction of the experimental arrangement can be found in Figure 2.

2.2. Experimental Procedure

Initially, the pressure drop across the cooling pad was measured for various flow rates using a digital manometer (510i, Testo SE & Co. KGaA, Germany). For each flow rate, data were recorded every 10 s for 30 min, and the average values were calculated for subsequent analysis. Subsequently, a series of experiments was conducted to evaluate the cooling pad performance. All experiments were conducted between 10:30 a.m. and 04:30 p.m. under real climate conditions in the months of June and August 2022 at our laboratory in Bangkok, Thailand (13.7269° N, 100.7722° E).

Before commencing each measurement, water circulation in the test section ran continuously for at least 30 min to ensure the saturation of the pad. In each experiment, the temperature and relative humidity of the air were measured before and after passing through the cooling pad every 30 s, using two temperature–humidity probes (probe diameter = 4 mm) connected to a data logger (176H1, Testo SE & Co. KGaA, Germany). Additionally, the temperature and relative humidity of the ambient air at two locations surrounding the wind tunnel were monitored every 1 minute using temperature–humidity data loggers (174H1, Testo SE & Co. KGaA, Germany). The specifications of the instruments used in the experiments are given in Table 2. Each experiment was replicated three times to ensure the repeatability and reliability of the results.

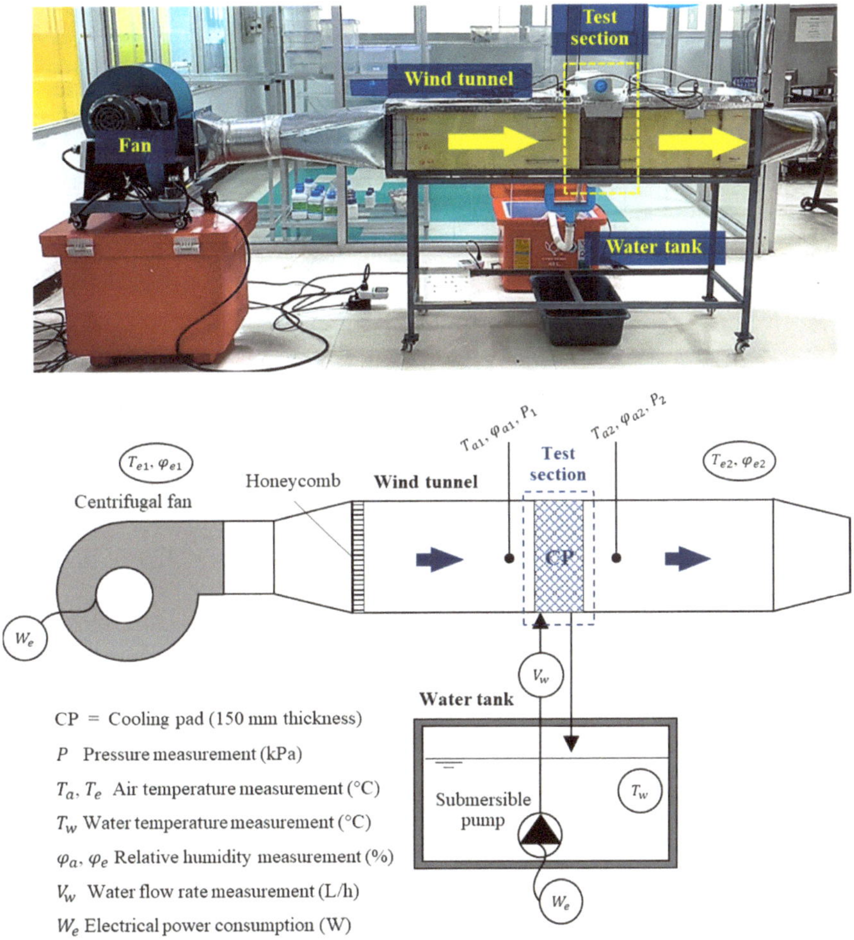

Figure 2. A schematic and photograph of experimental setup. Arrows indicate airflow direction across the test section.

Table 2. Specifications of the instruments.

Parameter	Instrument	Accuracy	Measuring Range
Air velocity	Hot-wire anemometer (Testo 405i)	±0.1 m/s ±0.3 m/s	0 to 2 m/s 2 to 15 m/s
Differential pressure	Manometer (Testo 510i)	±0.02 kPa	−15 to +15 kPa
Electrical power consumption	Wattmeter (Intertek)	±2% of measured data	0 to 3680 W
Temperature and relative humidity	Thermo-hygrometer (Testo 176H1) Thermo-hygrometer (Testo 174H)	±0.2 °C ±2.0% RH ±0.5 °C ±3.0% RH	−20 °C to +70 °C 20% to 95% −20 to +70 °C 2 to 98%

2.3. Performance Evaluation

The evaporative cooling performance of the cooling pad was evaluated based on temperature drop, saturation effectiveness, cooling capacity, *COP*, and pressure drop. The formulas used for evaluating each of these parameters are detailed in the following subsections, with the exception of the pressure drop across the pad, which was derived from the measurements as previously explained in Section 2.2.

2.3.1. Temperature Drop

Temperature drop is defined as the difference in the dry-bulb temperatures of the air entering (T_{a1}) and leaving (T_{a2}) the cooling pad, and is expressed as

$$\Delta T = T_{a1} - T_{a2} \tag{1}$$

2.3.2. Saturation Effectiveness

In the literature, various terms have been used to denote a specific cooling pad ability to lower air temperature compared to the maximum temperature reduction achievable in direct evaporative cooling systems. These terms include cooling effectiveness or efficiency [40,41], saturation effectiveness or efficiency [21,42], wet-bulb effectiveness or efficiency [30,43], and humidification efficiency [44]. In fact, these terms share the same definition. However, the term "effectiveness" is considered more precise than "efficiency" as evaporative cooling does not involve energy conversion [17]. Therefore, this study adopts the term "saturation effectiveness", which is expressed as the ratio of the difference between the dry-bulb temperatures of the inlet and outlet air to the difference between the dry-bulb and wet-bulb temperatures of the inlet air:

$$\varepsilon_s = \frac{T_{a1} - T_{a2}}{T_{a1} - T_{wb1}} = \frac{\Delta T}{T_{a1} - T_{wb1}} \tag{2}$$

An empirical equation was used to calculate the instantaneous wet-bulb temperature (T_{wb1}) associated with the dry-bulb temperature (T_a) and relative humidity (φ_a) of inlet air, expressed as follows [45]:

$$\begin{aligned} T_{wb} = T_a \tan^{-1}\left[b_1(\varphi_a + b_2)^{0.5}\right] + \tan^{-1}(T_a + \varphi_a) - \tan^{-1}(\varphi_a - b_3) \\ + b_4\varphi_a^{1.5}\tan^{-1}(b_5\varphi_a) - b_6) \end{aligned} \tag{3}$$

The values in the arctangent function are in radians and the values of the coefficients b_i (i = 1 to 6) are given in Table 3.

Table 3. The values of the coefficients used in Equation (3).

i	b_i
1	0.151977
2	8.313659
3	1.676331
4	0.00391838
5	0.023101
6	4.686035

2.3.3. Cooling Capacity

The cooling capacity of the system is calculated using Equation (4). This equation signifies the amount of sensible energy extracted from the air to lower the temperature.

$$Q_c = \dot{m}_a c_{pa}(T_{a1} - T_{a2}) = \dot{m}_a c_{pa}\Delta T \tag{4}$$

The air mass flow rate ($\dot{m}_a$) is defined as

$$\dot{m}_a = \rho_a A_c \overline{u}_a \tag{5}$$

In our case, the cross-sectional area (A_c) of the cooling pad was about 0.078 m^2, and the average air velocity ($\overline{u}_a$) was derived from measurements taken at 16 points across the cross-sectional area.

The density (ρ_a) and specific heat (c_{pa}) of moist air are calculated using the following equations [46]:

$$\rho_a = \frac{M_{da} P}{R_u (T_a + 273.15)\left(1 + \frac{M_{da}}{M_{wv}}\omega_a\right)} \tag{6}$$

$$c_{pa} = c_{pda} + \omega_a c_{pwv} \tag{7}$$

The humidity ratio (ω_a) represents the ratio of the mass of water vapor to the mass of dry air. In this study, it is determined using an empirical equation expressed as a function of the dry-bulb temperature and relative humidity [47]:

$$\omega = \frac{0.622\left[\alpha \frac{\varphi}{100} \exp\left(\frac{\beta T}{\gamma + T}\right)\right]}{10P - \left[\alpha \frac{\varphi}{100} \exp\left(\frac{\beta T}{\gamma + T}\right)\right]} \tag{8}$$

The constant parameters for Equations (6) to (8) are provided in Table 4.

Table 4. The constant parameters for Equations (6) to (8).

Parameter	Value	Unit
c_{pda}	1.006	kJ/(kg·°C)
c_{pwv}	1.84	kJ/(kg·°C)
M_{da}	28.966	kg/kmol
M_{wv}	18.015268	kg/kmol
P	101.325	kPa
R_u	8.314472	kJ/(kmol·K)
α	0.6112	kPa
β	17.62	-
γ	243.12	°C

2.3.4. Coefficient of Performance

The coefficient of performance (COP) is defined as the ratio of the achieved cooling capacity to the total electrical power supplied to operate the fan and the pump.

$$COP = \frac{Q_c}{W_{ef} + W_{ep}} \tag{9}$$

The power consumption of the pump (W_{ep}) remained constant at 6.1 W, as measured by a watt meter, because the water flow rate was not adjusted during the study. In contrast, the power consumption of the fan (W_{ef}) varied with measurements of 57 W, 95 W, 162 W, and 270 W, obtained using a watt meter when the frequency was set to 20 Hz, 30 Hz, 40 Hz, and 50 Hz, respectively.

2.4. Uncertainty Analysis

The propagation of error was used to quantify the uncertainty resulting from physical measurements of independent parameters such as temperature, relative humidity, and velocity, as described in references [48,49]. Given that y is a dependent variable expressed as a function of independent variables x_k, which are derived from measurements and/or calculations,

$$y = f(x_1, x_2 \ldots x_n) \tag{10}$$

The absolute uncertainty of the dependent variable (σ_y) is then estimated from

$$\sigma_y = \sqrt{\left(\frac{dy}{dx_1}\sigma_{x_1}\right)^2 + \left(\frac{dy}{dx_2}\sigma_{x_2}\right)^2 + \ldots + \left(\frac{dy}{dx_n}\sigma_{x_n}\right)^2} \tag{11}$$

The partial derivatives of the function with respect to the given independent variable (dy/dx_k) were numerically computed using MATLAB software (MathWorks R2021b). The uncertainty of individual independent variables (σ_{x_k}) obtained from measurements was based on the instrument accuracy (Table 2), while those obtained from calculations were determined by previously estimated uncertainties. The relative uncertainty (U_y) was subsequently estimated using Equation (12).

$$U_y = \frac{\sigma_y}{|y|} \times 100 \tag{12}$$

Table 5 provides a summary of the maximum and minimum relative uncertainties for each performance parameter. Notably, all parameters exhibited uncertainties of less than 20%. While the maximum values may seem significant when expressed as percentages, it is important to note that they correspond to relatively small absolute variations.

Table 5. The relative uncertainty (%) of the performance parameters.

Parameter	Maximum	Minimum
ΔT	11.8	5.8
ε_s	11.8	7.6
Q_c	18.2	8.7
COP	18.2	8.9

3. Results and Discussion

The cooling performance of the MP pad was evaluated by measuring the air temperature and relative humidity before and after the pad inside the wind tunnel. These measurements were conducted at four different air velocities (0.7, 1.0, 1.4, and 1.8 m/s), corresponding to mass flow rates of 0.06, 0.09, 0.12, and 0.15 kg/s, respectively. The water flow rate remained constant at 250 L/h for all experiments. The same experimental procedure was applied to the CP pad for the purpose of comparison. Each measurement was replicated three times, resulting in a total of twenty-four measurements. The collected data were used to determine the performance parameters, as explained in Section 2.3. The reported and analyzed values for each parameter represent the averages of the three replications.

3.1. *Verification of Steady State Condition*

All experiments were conducted under real climate conditions. To minimize the variations in the results, data collected during a nearly steady state period of 3 h were extracted for each experiment and used for analysis. As shown in Figure 3, the average ambient air temperature and relative humidity during this period ranged from 30 °C to 34 °C and 55% to 70%, respectively. Moreover, within a given experiment, fluctuations in ambient air temperature ($T_{a1,\max} - T_{a1,\min}$) never exceeded 2.0 °C, and relative humidity ($\varphi_{a1,\max} - \varphi_{a1,\min}$) remained within 10%.

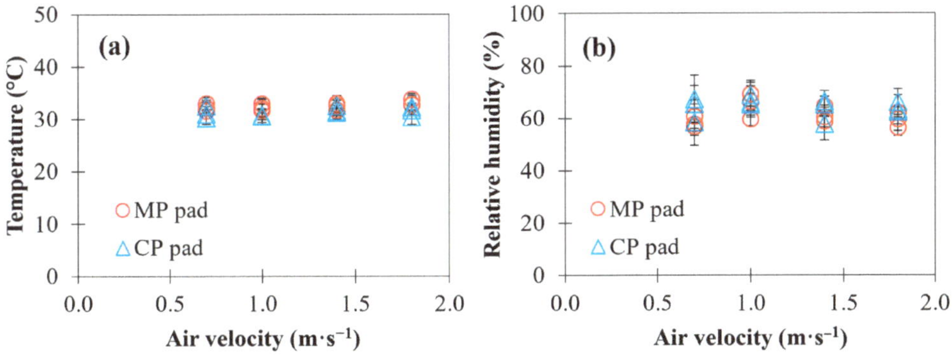

Figure 3. (**a**) Ambient air temperature and (**b**) relative humidity averaged over 3 h of the experiments.

3.2. Cooling Performance

A minimal pressure drop is considered an ideal characteristic for evaporative cooling pads [17]. As depicted in Figure 4, the pressure drop increased with higher air velocity, ranging from 14.7 Pa to 89.7 Pa. The standard deviations among replications were consistently less than 1.3 Pa. Pressure drops within this range are typical for organic-based cooling pads with similar thickness. For instance, a cooling pad made of fine fabric polyvinyl chloride (PVC) sponge mesh with a 150 mm thickness exhibited pressure drops of around 50–100 Pa at air velocities ranging from 1.5 m/s to 2.0 m/s [50]. In another study by Suranjan Salins, Reddy [29], increasing the flow rate from 0.062 kg/s to 0.083 kg/s resulted in pressure drops across the cooling pad that ranged from 47 Pa to 65 Pa for the coconut coir pad and from 50 Pa to 67 Pa for the wood shaving pad.

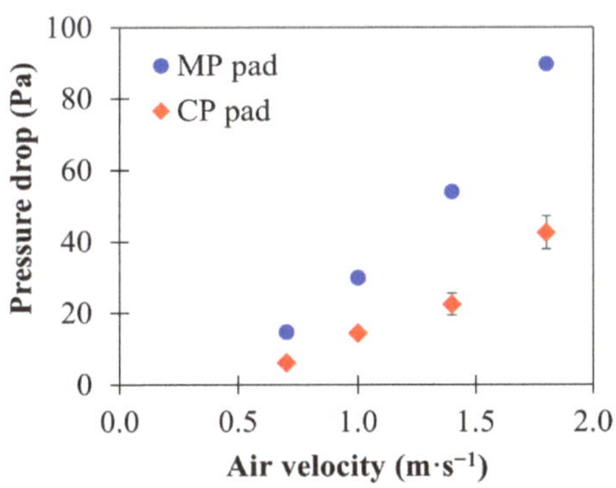

Figure 4. Pressure drops of the airflow across the cooling pads at different air velocities.

Figure 4 further illustrates that, at all considered air velocities, the MP pad produced a higher pressure drop than the CP pad (<50 Pa). This could be attributed to the relatively high packing density of the MP pad, resulting in a larger contact area and increased friction, consequently leading to greater pressure drops [29]. Ergun's correlation is commonly used to describe the pressure drop through packed beds of particles, resulting from kinetic and viscous energy losses [51]. According to Ergun's correlation, the viscous term is proportional to flow rate, while the kinetic (inertial) term is proportional to the flow rate squared. This explains the nonlinear increase in pressure drop across the MP pad with increased velocity, as shown in Figure 4.

As shown in Figure 5, the temperature drops achieved by the MP and CP pads were nearly identical, ranging between 3.4–4.6 °C for the MP pad and 3.5–4.4 °C for the CP pads.

These results align with previous findings in the literature, suggesting that air velocity has a relatively modest impact on air temperature reduction. For example, Rawangkul, Khedari [27] reported consistent temperature drops of 3.0 °C when using a coconut fiber cooling pad, even as air velocity increased from 1.9 m/s to 2.8 m/s. It is worth noting that the attainable temperature drop in evaporative cooling systems primarily hinges on the humidity level of the ambient air [52]. To achieve greater temperature drops, it is recommended to incorporate an air preconditioning unit, such as an air dehumidifier, into the system.

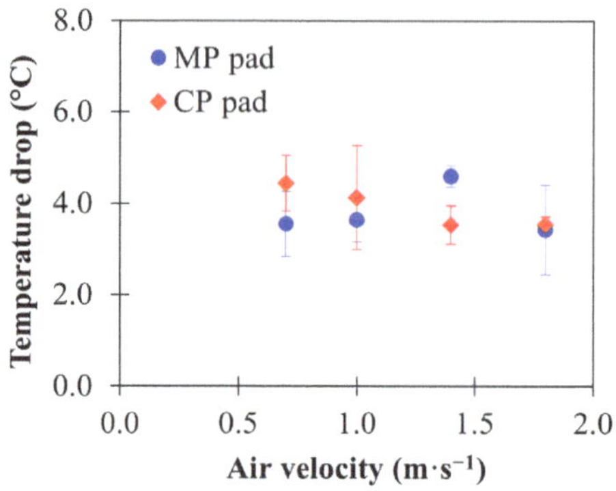

Figure 5. Variations of temperature drop across the MP and CP pads at different air velocities.

Saturation effectiveness demonstrates a direct relationship with temperature drop. As depicted in Figure 6, the saturation effectiveness of the MP pad experienced a slight increase, from 54% to 77%, as the air velocity increased from 0.7 to 1.4 m/s. However, at an air velocity of 1.8 m/s, the saturation effectiveness dropped to 53.0%. In contrast, the saturation effectiveness of the CP pad followed a declining trend with rising air velocity, decreasing from 81% at 0.7 m/s to 65% at 1.8 m/s. This decreasing trend aligns with common findings in the literature, where higher air velocity leads to shorter air-to-water interaction time, resulting in reduced saturation effectiveness [17]. It is important to note that the lower saturation effectiveness of the MP pad, compared to the CP pad at low air velocities (0.5–1.0 m/s), can be attributed to the observed pressure drop. Within this velocity range, the MP pad exhibited a higher pressure drop, resulting in shorter air-to-water interaction time compared to the CP pad.

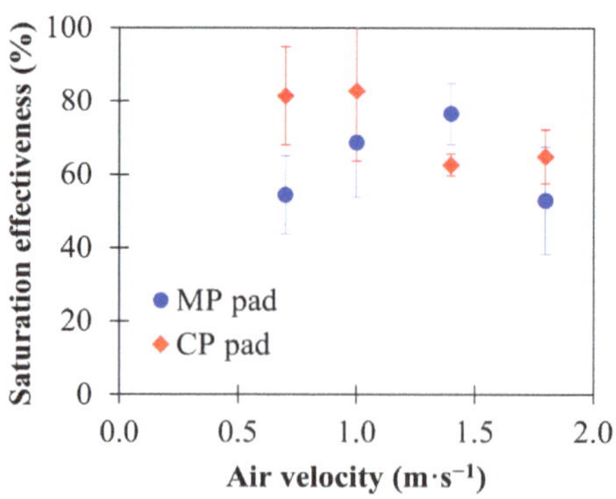

Figure 6. Saturation effectiveness of the MP and CP pads at different air velocities.

The observed contrary trend in the case of the MP pad at air velocities below 1.4 m/s is likely due to the enhanced heat and mass transfer coefficient between the air and the packed bed of particles at higher velocities. These findings are consistent with those of Gunhan, Demir [53], who reported an increasing trend in saturation effectiveness with air velocity for coarse pumice stones. Despite the need for higher air velocity, the MP pad (77% at 1.4 m/s) demonstrated saturation effectiveness nearly on par with that of the CP pad (83% at 1.0 m/s) and other fiber and fill pads made of natural and organic materials, as summarized in Table 1.

Cooling capacity is another crucial performance parameter in evaporative cooling systems. It shows a direct relationship not only with temperature drop but also with flow rate. Despite the minor differences in temperature drops within the considered velocity range, cooling capacity increased with higher air velocity, as shown in Figure 7a. Notably, there was no significant difference between the cooling capacity of the MP pad (0.2–0.6 kW) and the CP pad (0.3–0.6 kW). Under the prevailing climate conditions, the MP pad achieved a maximum cooling capacity of nearly 0.60 kW when the system operated at 1.4 m/s.

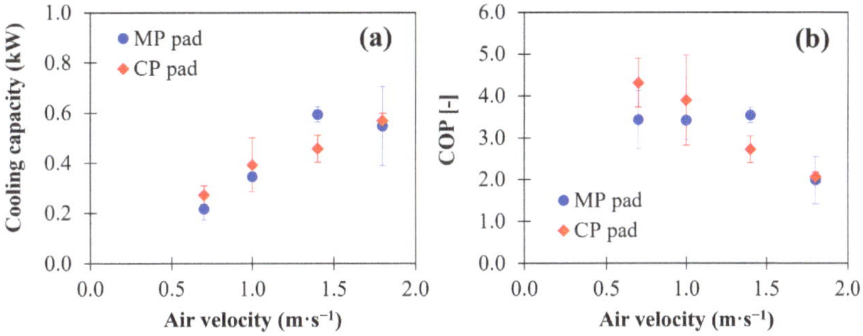

Figure 7. (**a**) Cooling capacity and (**b**) coefficient of performance (*COP*) of the system with the MP and CP pads at different air velocities.

As illustrated in Figure 7b, the *COP* of the system with the MP pad remained around 3.5 within the air velocity range of 0.4–1.4 m/s. However, due to a substantial increase in fan power consumption, the *COP* dropped sharply to 2.0 at an air velocity of 1.8 m/s. In contrast, for the CP pad, the *COP* progressively decreased with increasing air velocity. Notably, the *COP* of the MP pad closely resembled that of other organic-based materials, such as eucalyptus fibers (*COP* = 3.8 at 0.04 kg/s) [21] and wood shaving (*COP* = 2.1 at 0.0083 kg/s) [29].

Overall, the MP pad exhibited comparable performance to the commercial CP pad in terms of temperature drop, saturation effectiveness, cooling capacity, and *COP*, except for the pressure drop. These findings underscore the potential of mangosteen peels as an alternative wet medium for evaporative cooling applications. However, the organic nature of the MP pad presents certain drawbacks. Like many natural materials, it is susceptible to bacteria and fungi growth when exposed to water for extended periods, necessitating regular cleaning and sterilization [21]. To mitigate microbial growth, the use of an air filter with pore sizes of less than 0.1 μm is recommended to prevent the spread of bacteria and flora in the cooling space [54]. Moreover, prolonged use of the MP pad can lead to the production of unpleasant odors, particularly with mangosteen, which has a distinctive aroma. This drawback might limit the application of the MP pad in residential and building cooling, but it is less of an issue for storing horticultural products, such as mangosteen fruit itself.

4. Conclusions

The present study investigated an alternative wet medium made from mangosteen peels for evaporative cooling systems. Its cooling performance was experimentally investigated and compared to that of a commercial corrugated cellulose paper pad. The results revealed that the optimum air velocity to operate the evaporative cooling system with the

MP pad was 1.4 m/s. At this velocity, the MP pad achieved a maximum temperature drop of 4.6 °C, a saturation effectiveness of 77%, a cooling capacity of 0.59 kW, and a *COP* of 3.5 while maintaining a tolerable pressure drop of 54 Pa. These findings demonstrate that its performance is on par with that of the commercial cellulose paper pad, opening the door to the potential valorization of mangosteen peel waste. Given its organic nature and distinctive aroma, the MP pad is best suited for horticultural product storage or applications tolerable of odors.

Evaporative cooling systems are typically not recommended for use in areas with hot and humid climates, such as in Thailand. For future works, it is advisable to explore hybrid systems that combine evaporative cooling with a water-cooling unit or desiccant dehumidifier to enhance the performance of this organic-waste-based cooling pad.

Author Contributions: Conceptualization, N.C. and S.R.; methodology, N.C. and S.R.; software, N.C.; validation, N.C., T.N. and S.R.; formal analysis, N.C., T.N. and S.R.; investigation, N.C. and T.N.; resources, N.C.; data curation, N.C., T.N. and S.R.; writing—original draft preparation, N.C.; writing—review and editing, N.C.; visualization, N.C.; supervision, N.C. and S.R.; project administration, N.C.; funding acquisition, N.C. All authors have read and agreed to the published version of the manuscript.

Funding: This work was financially supported by King Mongkut's Institute of Technology Ladkrabang, Thailand, grant number 2564-02-01-012.

Data Availability Statement: The data used in this study can be obtained upon request from the corresponding author.

Acknowledgments: The authors appreciate and acknowledge the support from Takahashi Foundation. Special thanks are given to Denis Flick (AgroParisTech, France) for his guidance and suggestions.

Conflicts of Interest: The authors declare no conflict of interest.

References

1. Pezzutto, S.; Quaglini, G.; Riviere, P.; Kranzl, L.; Novelli, A.; Zambito, A.; Wilczynski, E. Screening of cooling technologies in Europe: Alternatives to vapour compression and possible market developments. *Sustainability* **2022**, *14*, 2971. [CrossRef]
2. Rafique, M.M.; Gandhidasan, P.; Rehman, S.; Al-Hadhrami, L.M. A review on desiccant based evaporative cooling systems. *Renew. Sustain. Energy Rev.* **2015**, *45*, 145–159. [CrossRef]
3. Fong, K.; Lee, C. Solar desiccant cooling system for hot and humid region–A new perspective and investigation. *Sol. Energy* **2020**, *195*, 677–684. [CrossRef]
4. Pandelidis, D.; Niemierka, E.; Pacak, A.; Jadwiszczak, P.; Cichoń, A.; Drąg, P.; Worek, W.; Cetin, S. Performance study of a novel dew point evaporative cooler in the climate of central Europe using building simulation tools. *Build. Environ.* **2020**, *181*, 107101. [CrossRef]
5. Kashif, M.; Niaz, H.; Sultan, M.; Miyazaki, T.; Feng, Y.; Usman, M.; Shahzad, M.W.; Niaz, Y.; Waqas, M.M.; Ali, I. Study on desiccant and evaporative cooling systems for livestock thermal comfort: Theory and experiments. *Energies* **2020**, *13*, 2675. [CrossRef]
6. Ghoulem, M.; El Moueddeb, K.; Nehdi, E.; Boukhanouf, R.; Calautit, J.K. Greenhouse design and cooling technologies for sustainable food cultivation in hot climates: Review of current practice and future status. *Biosyst. Eng.* **2019**, *183*, 121–150. [CrossRef]
7. Nkolisa, N.; Magwaza, L.S.; Workneh, T.S.; Chimphango, A. Evaluating evaporative cooling system as an energy-free and cost-effective method for postharvest storage of tomatoes (*Solanum lycopersicum* L.) for smallholder farmers. *Sci. Hortic.* **2018**, *241*, 131–143. [CrossRef]
8. Ambuko, J.; Wanjiru, F.; Chemining'wa, G.; Owino, W.; Mwachoni, E. Preservation of postharvest quality of leafy amaranth (*Amaranthus* spp.) vegetables using evaporative cooling. *J. Food Qual.* **2017**, *2017*, 5303156. [CrossRef]
9. Kipruto, K.M. Effect of near infrared reflection and evaporative cooling on quality of mangoes. *Agric. Eng. Int. CIGR J.* **2017**, *19*, 162–168.
10. Samira, A.; Woldetsadik, K.; Workneh, T.S. Postharvest quality and shelf life of some hot pepper varieties. *J. Food Sci. Technol.* **2013**, *50*, 842–855. [CrossRef]
11. Babaremu, K.; Adekanye, T.; Okokpujie, I.; Fayomi, J.; Atiba, O. The significance of active evaporative cooling system in the shelf life enhancement of vegetables (red and green tomatoes) for minimizing post-harvest losses. *Procedia Manuf.* **2019**, *35*, 1256–1261. [CrossRef]
12. Sibanda, S.; Workneh, T.S. Effects of indirect air cooling combined with direct evaporative cooling on the quality of stored tomato fruit. *CyTA-J. Food* **2019**, *17*, 603–612. [CrossRef]

13. Yang, Y.; Cui, G.; Lan, C.Q. Developments in evaporative cooling and enhanced evaporative cooling-A review. *Renew. Sustain. Energy Rev.* **2019**, *113*, 109230. [CrossRef]
14. Xuan, Y.; Xiao, F.; Niu, X.; Huang, X.; Wang, S. Research and application of evaporative cooling in China: A review (I)–Research. *Renew. Sustain. Energy Rev.* **2012**, *16*, 3535–3546. [CrossRef]
15. Naveenprabhu, V.; Suresh, M. Performance enhancement studies on evaporative cooling using volumetric heat and mass transfer coefficients. *Numer. Heat Transf. Part A Appl.* **2020**, *78*, 504–523. [CrossRef]
16. Kumar, S.; Singh, J.; Siyag, J.; Rambhatla, S. Potential alternative materials used in evaporative coolers for sustainable energy applications: A review. *Int. J. Air-Cond. Refrig.* **2020**, *28*, 2030006. [CrossRef]
17. Tejero-González, A.; Franco-Salas, A. Optimal operation of evaporative cooling pads: A review. *Renew. Sustain. Energy Rev.* **2021**, *151*, 111632. [CrossRef]
18. Franco-Salas, A.; Peña-Fernández, A.; Valera-Martínez, D.L. Refrigeration capacity and effect of ageing on the operation of cellulose evaporative cooling pads, by wind tunnel analysis. *Int. J. Environ. Res. Public Health* **2019**, *16*, 4690. [CrossRef]
19. Nada, S.; Fouda, A.; Mahmoud, M.; Elattar, H. Experimental investigation of energy and exergy performance of a direct evaporative cooler using a new pad type. *Energy Build.* **2019**, *203*, 109449. [CrossRef]
20. Yan, M.; He, S.; Li, N.; Huang, X.; Gao, M.; Xu, M.; Miao, J.; Lu, Y.; Hooman, K.; Che, J. Experimental investigation on a novel arrangement of wet medium for evaporative cooling of air. *Int. J. Refrig.* **2021**, *124*, 64–74. [CrossRef]
21. Doğramacı, P.A.; Aydın, D. Comparative experimental investigation of novel organic materials for direct evaporative cooling applications in hot-dry climate. *J. Build. Eng.* **2020**, *30*, 101240. [CrossRef]
22. Jain, J.; Hindoliya, D. Correlations for saturation efficiency of evaporative cooling pads. *J. Inst. Eng. India Ser. C* **2014**, *95*, 5–10. [CrossRef]
23. Lotfizadeh, H.; Razzaghi, H.; Layeghi, M. Experimental performance analysis of a solar evaporative cooler with three different types of pads. *J. Renew. Sustain. Energy* **2013**, *5*, 063113. [CrossRef]
24. Bishoyi, D.; Sudhakar, K. Experimental performance of a direct evaporative cooler in composite climate of India. *Energy Build.* **2017**, *153*, 190–200. [CrossRef]
25. Al-Sulaiman, F. Evaluation of the performance of local fibers in evaporative cooling. *Energy Convers. Manag.* **2002**, *43*, 2267–2273. [CrossRef]
26. Liao, C.M.; Singh, S.; Wang, T.S. Characterizing the performance of alternative evaporative cooling pad media in thermal environmental control applications. *J. Environ. Sci. Health Part A* **1998**, *33*, 1391–1417. [CrossRef]
27. Rawangkul, R.; Khedari, J.; Hirunlabh, J.; Zeghmati, B. Performance analysis of a new sustainable evaporative cooling pad made from coconut coir. *Int. J. Sustain. Eng.* **2008**, *1*, 117–131. [CrossRef]
28. Alam, M.F.; Sazidy, A.S.; Kabir, A.; Mridha, G.; Litu, N.A.; Rahman, M.A. An experimental study on the design, performance and suitability of evaporative cooling system using different indigenous materials. In Proceedings of the 7th Bsme International Conference on Thermal Engineering, Dhaka, Bangladesh, 22–24 December 2016; p. 020075.
29. Suranjan Salins, S.; Reddy, S.K.; Kumar, S. Experimental investigation on use of alternative innovative materials for sustainable cooling applications. *Int. J. Sustain. Eng.* **2021**, *14*, 1207–1217. [CrossRef]
30. Khosravi, N.; Aydin, D.; Nejhad, M.K.; Dogramaci, P.A. Comparative performance analysis of direct and desiccant assisted evaporative cooling systems using novel candidate materials. *Energy Convers. Manag.* **2020**, *221*, 113167. [CrossRef]
31. Ndukwu, M.C.; Manuwa, S.I. Techno-economic assessment for viability of some waste as cooling pads in evaporative cooling system. *Int. J. Agric. Biol. Eng.* **2015**, *8*, 151–158. [CrossRef]
32. de Melo Júnior, J.C.F.; Bamberg, J.V.M.; Machado, N.S.; Caldas, E.N.G.; Rodrigues, M.S. Evaporative cooling efficiency of pads consisting of vegetable loofah. *Comun. Sci.* **2019**, *10*, 38–44. [CrossRef]
33. Soponpongpipat, N.; Kositchaimongkol, S. Recycled high-density polyethylene and rice husk as a wetted pad in evaporative cooling system. *Am. J. Appl. Sci.* **2011**, *8*, 186–191. [CrossRef]
34. Ahmed, E.M.; Abaas, O.; Ahmed, M.; Ismail, M.R. Performance evaluation of three different types of local evaporative cooling pads in greenhouses in Sudan. *Saudi J. Biol. Sci.* **2011**, *18*, 45–51. [CrossRef] [PubMed]
35. Ubonwat, J. A Feasibility Study of Using Water Hyacinth as Wetted Media in Evaporative Cooling System. Master's Thesis, King Mongkut's Institute of Technology Thonburi, Bangkok, Thailand, 2001.
36. Choodej, S.; Koopklang, K.; Raksat, A.; Chuaypen, N.; Pudhom, K. Bioactive xanthones, benzophenones and biphenyls from mangosteen root with potential anti-migration against hepatocellular carcinoma cells. *Sci. Rep.* **2022**, *12*, 8605. [CrossRef] [PubMed]
37. Aizat, W.M.; Ahmad-Hashim, F.H.; Jaafar, S.N.S. Valorization of mangosteen,"The Queen of Fruits," and new advances in postharvest and in food and engineering applications: A review. *J. Adv. Res.* **2019**, *20*, 61–70. [CrossRef]
38. Office of Agricultural Economics. *Agricultural Statistics of Thailand 2020*; Ministry of Agriculture & Cooperatives: Bangkok, Thailand, 2020.
39. Nasrullah, A.; Saad, B.; Bhat, A.; Khan, A.S.; Danish, M.; Isa, M.H.; Naeem, A. Mangosteen peel waste as a sustainable precursor for high surface area mesoporous activated carbon: Characterization and application for methylene blue removal. *J. Clean. Prod.* **2019**, *211*, 1190–1200. [CrossRef]
40. Laknizi, A.; Ben Abdellah, A.; Faqir, M.; Essadiqi, E.; Dhimdi, S. Performance characterization of a direct evaporative cooling pad based on pottery material. *Int. J. Sustain. Eng.* **2021**, *14*, 46–56. [CrossRef]

41. Yan, M.; He, S.; Gao, M.; Xu, M.; Miao, J.; Huang, X.; Hooman, K. Comparative study on the cooling performance of evaporative cooling systems using seawater and freshwater. *Int. J. Refrig.* **2021**, *121*, 23–32. [CrossRef]
42. Velasco-Gómez, E.; Tejero-González, A.; Jorge-Rico, J.; Rey-Martínez, F.J. Experimental investigation of the potential of a new fabric-based evaporative cooling pad. *Sustainability* **2020**, *12*, 7070. [CrossRef]
43. Zhang, H.; Ma, H.; Ma, S. Investigation on indirect evaporative cooling system integrated with liquid dehumidification. *Energy Build.* **2021**, *249*, 111179. [CrossRef]
44. Kumar, S.; Salins, S.S.; Reddy, S.K.; Nair, P.S. Comparative performance analysis of a static & dynamic evaporative cooling pads for varied climatic conditions. *Energy* **2021**, *233*, 121136. [CrossRef]
45. Stull, R. Wet-bulb temperature from relative humidity and air temperature. *J. Appl. Meteorol. Climatol.* **2011**, *50*, 2267–2269. [CrossRef]
46. Gatley, D.P. *Understanding Psychrometrics*, 3rd ed.; ASHRAE: Atlanta, GA, USA, 2013; p. 259.
47. Sensirion. *Introduction to Humidity: Basic Principles on Physics of Water Vapor*; Sensirion, The Sensor Company: Chicago, IL, USA, 2009.
48. Moffat, R.J. Describing the uncertainties in experimental results. *Exp. Therm. Fluid Sci.* **1988**, *1*, 3–17. [CrossRef]
49. Taylor, J. *Introduction to Error Analysis, the Study of Uncertainties in Physical Measurements*; University Science Books: Dulles, VA, USA, 1997.
50. Liao, C.-M.; Chiu, K.-H. Wind tunnel modeling the system performance of alternative evaporative cooling pads in Taiwan region. *Build. Environ.* **2002**, *37*, 177–187. [CrossRef]
51. Coker, A.K. 4—FLUID FLOW. In *Ludwig's Applied Process Design for Chemical and Petrochemical Plants*, 4th ed.; Coker, A.K., Ed.; Gulf Professional Publishing: Burlington, VT, USA, 2007; pp. 133–302. [CrossRef]
52. Laknizi, A.; Mahdaoui, M.; Abdellah, A.B.; Anoune, K.; Bakhouya, M.; Ezbakhe, H. Performance analysis and optimal parameters of a direct evaporative pad cooling system under the climate conditions of Morocco. *Case Stud. Therm. Eng.* **2019**, *13*, 100362. [CrossRef]
53. Gunhan, T.; Demir, V.; Yagcioglu, A. Evaluation of the suitability of some local materials as cooling pads. *Biosyst. Eng.* **2007**, *96*, 369–377. [CrossRef]
54. Chen, X.; Su, Y.; Aydin, D.; Ding, Y.; Zhang, S.; Reay, D.; Riffat, S. A novel evaporative cooling system with a polymer hollow fibre spindle. *Appl. Therm. Eng.* **2018**, *132*, 665–675. [CrossRef]

Disclaimer/Publisher's Note: The statements, opinions and data contained in all publications are solely those of the individual author(s) and contributor(s) and not of MDPI and/or the editor(s). MDPI and/or the editor(s) disclaim responsibility for any injury to people or property resulting from any ideas, methods, instructions or products referred to in the content.

AgriEngineering

Article

Chicken Tracking and Individual Bird Activity Monitoring Using the BoT-SORT Algorithm

Allan Lincoln Rodrigues Siriani [1], Isabelly Beatriz de Carvalho Miranda [2], Saman Abdanan Mehdizadeh [3] and Danilo Florentino Pereira [4,*]

1 Graduate Program in Agribusiness and Development, School of Sciences and Engineering, São Paulo State University, Tupã 17602-496, SP, Brazil; allan.siriani@unesp.br
2 Undergraduate Program in Big Data, Sao Paulo College of Technology, Pompeia 17586-050, SP, Brazil
3 Department of Mechanics of Biosystems Engineering, Faculty of Agricultural Engineering and Rural Development, Agricultural Sciences and Natural Resources University of Khuzestan, Ahvaz 63417-73637, Iran; s.abdanan@asnrukh.ac.ir
4 Department of Management, Development, and Technology, School of Sciences and Engineering, São Paulo State University, Tupã 17602-496, SP, Brazil
* Correspondence: danilo.florentino@unesp.br

Abstract: The analysis of chicken movement on the farm has several applications in evaluating the well-being and health of birds. Low locomotion may be associated with locomotor problems, and undesirable bird movement patterns may be related to environmental discomfort or fear. Our objective was to test the BoT-SORT object tracking architecture embedded in Yolo v8 to monitor the movement of cage-free chickens and extract measures to classify running, exploring, and resting behaviors, the latter of which includes all other behaviors that do not involve displacement. We trained a new model with a dataset of 3623 images obtained with a camera installed on the ceiling (top images) from an experiment with layers raised cage-free in small-scale aviaries and housed in groups of 20 individuals. The model presented a mAP of 98.5%, being efficient in detecting and tracking the chickens in the video. From the tracking, it was possible to record the movements and directions of individual birds, and we later classified the movement. The results obtained for a group of 20 chickens demonstrated that approximately 84% of the time, the birds remained resting, 10% of the time exploring, and 6% of the time running. The BoT-SORT algorithm was efficient in maintaining the identification of the chickens, and our tracking algorithm was efficient in classifying the movement, allowing us to quantify the time of each movement class. Our algorithm and the measurements we extract to classify bird movements can be used to assess the welfare and health of chickens and contribute to establishing standards for comparisons between individuals and groups raised in different environmental conditions.

Keywords: animal welfare; laying hens; movement rating; precision livestock farming; YOLO

Citation: Siriani, A.L.R.; Miranda, I.B.d.C.; Mehdizadeh, S.A.; Pereira, D.F. Chicken Tracking and Individual Bird Activity Monitoring Using the BoT-SORT Algorithm. *AgriEngineering* **2023**, *5*, 1677–1693. https://doi.org/10.3390/agriengineering5040104

Academic Editors: Muhammad Sultan, Yuguang Zhou, Redmond R. Shamshiri and Muhammad Imran

Received: 14 August 2023
Revised: 25 September 2023
Accepted: 27 September 2023
Published: 29 September 2023

Copyright: © 2023 by the authors. Licensee MDPI, Basel, Switzerland. This article is an open access article distributed under the terms and conditions of the Creative Commons Attribution (CC BY) license (https://creativecommons.org/licenses/by/4.0/).

1. Introduction

The welfare of chickens in a production environment depends on several conditions on the farm and the birds' health. Van Veen et al. [1] verified with laying hen producers that the activity and movement of the hens are important indicators to assess health and well-being. In this direction, several studies have sought to better understand the behavior of animals to infer welfare [2–4].

Human monitoring of animal behavior can be exhaustive and imprecise [5–7]. In this field, computer vision has gained space for using non-invasive sensors and ensuring good accuracy in analyzes [8–12].

The ability of chickens to walk may indicate lameness [13–15], while the level of agitation of the flock may indicate stress to the rearing environment [16]. Thus, tracking chickens in a production environment has several applications for health and comfort diagnosis that can support mitigating measures and reduce losses in production [12].

Computer vision systems based on deep learning techniques make it possible to automatically discover data representations more easily, enabling the extraction of location and distance measurements [7]. The automatic detection of chickens housed on shavings bedding using computer vision cannot be considered a trivial task, as in the monitored area, there are variations in brightness, shadows, concealment by equipment, and the distortion of the camera lens itself [17]. In this sense, several studies propose different methods of detecting birds [12,16,18].

Several studies sought to track chickens in different rearing environments [6,18–22]. A common problem in tracking organisms is the loss of detection caused by traditional methods, causing ID changes during tracking. Banerjee et al. [23] tested the Strong Simple Online Real Time Tracking (Strong SORT) model on C. elegans worms and verified the effectiveness of the Id maintenance method. Young-Chul et al. [24] describe the use of a combination of association based on the proximity of Cartesian coordinates using the Hungarian algorithm and an association based on visual characteristics to maintain the trajectory of the detected object.

The BoT-SORT algorithm [25], which uses a Kalman filter and a Hungarian algorithm and is associated with the track model in Yolo, presents itself as a new possibility to track objects using a one-step detection method. We hypothesize that this tool can efficiently maintain the identification of chickens in flocks, making it possible to extract important information that may indicate states of agitation and the well-being of individuals and the flock. Our system is different from others, such as Deep-Sort [26], which is state-of-the-art in object tracking and leverages Deep Learning using a large residual network to extract features related to the appearance of an object. BoT-SORT is used to extract object movement variables such as speed, acceleration, and displacement, where these attributes were used to maintain tracking and ignore detection failures of the bird and the behavior we wanted to observe.

This study aimed to develop an intelligent system to automatically identify chickens in flocks whose housing density represents a challenge for computer vision, track the paths taken by each bird, and extract information such as distances and movement speeds of individuals.

2. Materials and Methods

2.1. Experiment and Video Recording

This study utilised a 15-minute video of 20 chickens housed in a reduced-scale shed. The camera was situated atop the aviary to capture bird images from an overhead perspective. The entire shed area was monitored by a POWER ® camera, model AP2688W, equipped with an analog Charge-Coupled Device (CCD) image sensor. Eighty layers of the Lohmann lineage were monitored from 10 June to 8 September 2020. The birds were 29 weeks old at the beginning of the experiment and were obtained from a commercial farm. The chickens were randomly divided into four groups of 20 birds each. The feed was administered daily in the amount of 110 g/bird. Access to water was free. Light management was similar to that adopted by the original farm, with a photoperiod of 17 h of light. Four models of small-scale warehouses were used, with usable dimensions of 140 cm × 300 cm and 150 cm in height, arranged in an east-west orientation. The birds were housed on a 15 cm high wood chip bed. Each henhouse had two nest boxes (40 cm × 40 cm × 40 cm) installed side by side, a pendulum feeder (Φ 30 cm), and four nipple drinkers. The complete experiment was described by Fernandes et al. [16].

2.2. Dataset and Hardware Setup

From the set of all videos recorded in the experiment, 3623 frames were randomly extracted, divided into 3123 for training, 259 for testing, and 241 for validation. The images are 352 × 240 pixels, and the capture rate was 30 fps. In each frame, all the birds appearing in the image were marked, using the Label Studio (https://labelstud.io/, accessed on 5 June 2023) tool (Figure 1). The coordinates of the birds' bounding boxes were saved in text files and, together with the images, formed the dataset for generating the Yolo model [27–31].

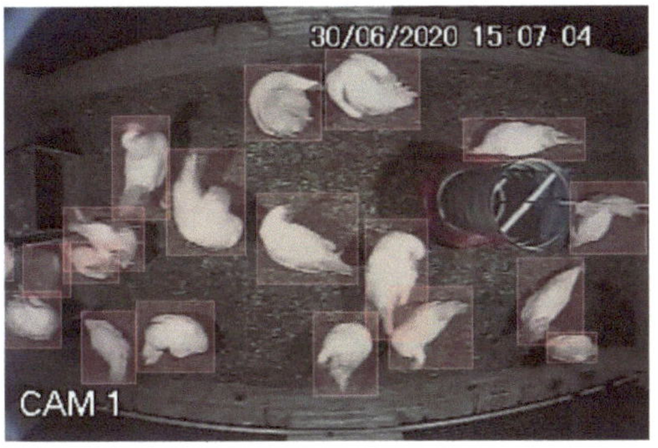

Figure 1. Image with the bounding boxes upper laying hens.

The hardware configuration of the computer utilized in this study is detailed in Table 1. The scripts for processing the videos were developed and executed using Python 3.9.16, and the Pytorch 1.13.1 framework was employed for deep learning purposes.

Table 1. Workstation hardware configuration.

CPU	CPU Intel® Core i7-10700k
CPU frequency	4.85 GHz
Cores/threads number	8/16
Memory RAM	64 GB
GPU	Nvidia RTX 2070 Super, 8 GB (2.304 CUDA cores)

2.3. Detection Model—Yolo v8n

A Yolo v8n model was generated and trained using the specified dataset for 50 epochs. Model performance was evaluated using the following metrics.

$$Precision = \frac{True\ Positive}{True\ Positive + False\ positive} \quad (1)$$

$$Recall = \frac{True\ Positive}{True\ Positive + False\ Negative} \quad (2)$$

$$mAP = \frac{1}{N}\sum_{i=1}^{N} AP_i \quad (3)$$

where AP_i is the average precision for each class and N is the number of classes. In this study, mAP equals AP.

As described in Equation (4), the loss function was utilized to evaluate how the algorithm modeled the data set and achieved the desired outcome.

$$Loss_{yolo} = l_{box} + l_{conf} + l_{class} \quad (4)$$

$$\begin{aligned} l_{box} &= \lambda_{coord}\sum_{i=0}^{S^2}\sum_{j=0}^{B} I_{i,j}^{obj}\left[(x_i - \hat{x}_i)^2 + (y_i - \hat{y}_i)^2\right] \\ &+\lambda_{coord}\sum_{i=0}^{S^2}\sum_{j=0}^{B} I_{i,j}^{obj}\left[\left(\sqrt{w_i} - \sqrt{\hat{w}_i}\right)^2 + \left(\sqrt{h_i} - \sqrt{\hat{h}_i}\right)^2\right] \end{aligned} \quad (5)$$

$$l_{conf} = \lambda_{noobj} \sum_{i=0}^{S^2} \sum_{j=0}^{B} I_{i,j}^{noobj} (C_i - \hat{C}_l)^2 \tag{6}$$

$$l_{class} = \sum_{i=0}^{S^2} \sum_{j=0}^{B} I_{i,j}^{obj} \sum_{C \in classes} (p_i(c) - \hat{p}_i(c))^2 \tag{7}$$

where l_{box}, l_{conf}, l_{class} are the losses of location, trust and class; λ_{coord} increase weight for loss in bounding box coordinates; $I_{i,j}^{obj} = 1$ if the chicken in box j and cell i match, $I_{i,j}^{obj} = 0$ where not; $I_{i,j}^{noobj}$ it's just the opposite; x_i, y_i, w_i h_i are the coordinates (x, y) of the location, width and height of the ith bouding box; λ_{noobj} reduces loss when detecting background; C is the total number of classes, which in our model is equal to 1; $\hat{p}_i(c)$ denotes the conditional class probability for the class c in the cell i.

2.4. Tracking Algorithm

The BoT-SORT multi-object tracking model of Yolo v8n was utilized in this study, which employs a state vector based on the Kalman filter and Hungarian algorithm [25].

The tracking model was elaborated through a map of stages for the execution of the system, described in Figure 2.

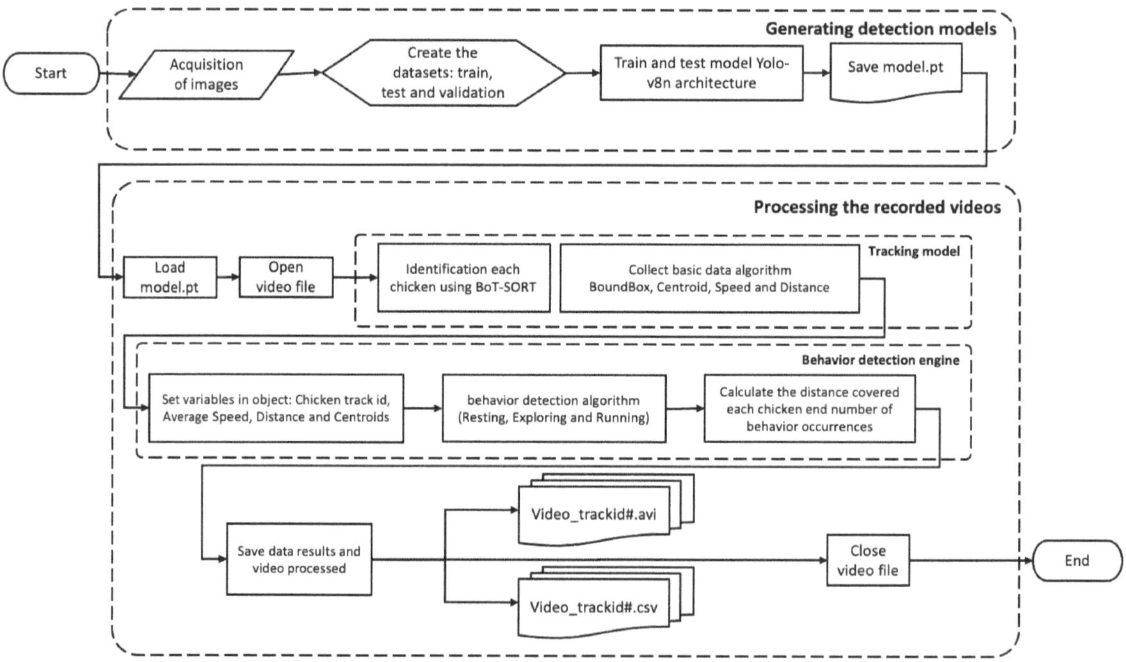

Figure 2. Stages of the tracking system development.

For the chicken tracking task, the default Kalman filter is based on a linear observation model and constant velocity movement, governed by the following linear stochastic difference equations, available at Aharon et al. [25].

$$x_k = F_k x_{k-1} + n_{k-1} \tag{8}$$

$$z_k = H_k x_k + v_k \tag{9}$$

where F_k is the transition matrix for the discrete-time of $k-1$ for k; H_k is the matrix with observed values; n_k e v_k are random variables that represent the process and measurement noise, respectively.

The process noise covariance matrix Q_k and the measurement noise covariance matrix R_k can change at each step in time. The Kalman filter consists of predicting the current time and updating over time according to the equations given below.

$$\hat{x}_{k|k-1} = F_k \hat{x}_{k-1|k-1} \tag{10}$$

$$P_{k|k-1} = F_k P_{k-1|k-1} F_k^T + Q_k$$

$$K_k = P_{k|k-1} H_k^T \left(H_k P_{k|k-1} H_k^T + R_k \right)^{-1} \tag{11}$$

$$\hat{x}_{k|k} = \hat{x}_{k|k-1} + K_k \left(z_k - H_k \hat{x}_{k|k-1} \right)$$

$$P_{k|k} = (I - K_k H_k) P_{k|k-1}$$

For each time step k, the Kalman filter estimates the state $\hat{x}_{k|k-1}$ and the covariance matrix $P_{k|k-1}$, and later updates the later estimated state $\hat{x}_{k|k}$ given the observation z_k and the covariance matrix $P_{k|k}$, calculated based on the Kalman gain K_k.

For the BoT-SORT model, the Kalman Filter state vector is described in Equation (12) and the measurement vector in Equation (13), which is implied in the corrections of the process noise covariance matrices (Q_k) and measurement noise covariance (R_k) shown in Equations (14) and (15), respectively [16].

$$x_k = \left[x_c(k),\ y_c(k),\ w(k),\ h(k),\ \dot{x}_c(k), \dot{y}_c(k), \dot{w}(k),\ \dot{h}(k) \right] \tag{12}$$

$$z_k = \left[z_{x_c}(k),\ z_{y_c}(k),\ z_w(k),\ z_h(k) \right] \tag{13}$$

$$\begin{aligned} Q_k = diag\Big(&\left(\sigma_p \hat{w}_{k-1|k-1}\right)^2, \left(\sigma_p \hat{h}_{k-1|k-1}\right)^2, \\ &\left(\sigma_p \hat{w}_{k-1|k-1}\right)^2, \left(\sigma_p \hat{h}_{k-1|k-1}\right)^2, \\ &\left(\sigma_p \hat{w}_{k-1|k-1}\right)^2, \left(\sigma_p \hat{h}_{k-1|k-1}\right)^2, \\ &\left(\sigma_p \hat{w}_{k-1|k-1}\right)^2, \left(\sigma_p \hat{h}_{k-1|k-1}\right)^2 \Big) \end{aligned} \tag{14}$$

$$R_k = diag\left(\left(\sigma_m \hat{w}_{k|k-1}\right)^2, \left(\sigma_m \hat{h}_{k|k-1}\right)^2 \right), \tag{15}$$

where x_c and y_c are the coordinates of the chicken's bounding box; w and h are the width and height of the bounding box k and $k-1$ is the current and previous state; σ_p, σ_v and σ_m are the standard deviations of the noise factors.

2.5. Measurement of Distance and Velocity of Travel

The distance covered was calculated from the coordinates of the centroids of each bird between two frames using the Euclidean method.

$$D_{(F_i - F_{i+1})} = \sqrt[2]{(x_{i+1} - x_i)^2 + (y_{i+1} - y_i)^2} \tag{16}$$

where $D_{(Fi-Fi+1)}$ is the distance between two points, the first being captured in the frame Fi and the second in the frame F_{i+1}, x_i and y_i are the coordinates of the point in the frame F_i and x_{i+1} and y_{i+1} are the coordinates of the point in the frame F_{i+1}.

Therefore, for every two frames, we calculate the distance traveled by a chicken centroid and accumulate these values during the complete shooting. However, these values are expressed in pixels and cannot be compared with other distances from other videos due to differences in camera placement and image resolution. To make the measurements comparable, we convert pixels into centimeters using a ratio rule from a known measurement of the scene.

Birds were identified as they appeared on the scene. When the bird left the scene, the path data for that identification was closed so that the same bird, when it returned to the scene, received a new identification. Thus, the distances covered were recorded for each identification.

$$D_{Id} = \sum D_{(Fi-Fi+1)} \quad (17)$$

where are D_{id} is the distance traveled by a chicken with the identification Id by the number of sequential frames that this identification was maintained.

The speed was calculated from the distances covered by the Ids, according to the equation below.

$$S_{Id} = \frac{D_{Id}}{\Delta t} \quad (18)$$

where S_{Id} is the speed of the bird identified by Id, D_{Id} is the distance traveled by the bird Id and Δt is the time between frames (in our work $\Delta t = 1/30$ s).

2.6. Distinction of Running, Exploring, and Resting Behaviors

In the process of detecting chickens in the images, different values of displacements in time were established to differentiate three movement states: resting, exploring, and running. Initially, a noise filter was applied to eliminate erroneous identifications, along with the definition of limits that demarcate the transitions from one behavior to another.

To classify the behaviors, a minimum frame count was defined to indicate the beginning of a new behavior in chickens. Specifically, "Resting" requires a minimum of 5 frames, "Exploring" requires four frames, and "Running" requires three frames. Furthermore, to qualify for detection of one of these behaviors, a bird must maintain a minimum "lifetime" of 18 frames. This implies that the identified bird, designated by its unique track_Id, must be present in the video for at least 18 consecutive frames.

Using these data, the velocity parameter was employed to distinguish the behaviors. This involved calculating the average instantaneous velocity over the last ten frames. For the "Resting" behavior, a speed of 1 cm/frame or less was used as a limit, considering that the resting chicken can move only the head or other body parts, leading to changes in the position of the centroid, even if it remains still or without displacement.

For the "Exploring" behavior, velocities were defined that vary between more than 1 cm/s and less than 100 cm/s, determined in the analysis of the results. "Running" behavior was classified when the velocity was greater than 100 cm/s. To increase accuracy and reduce false positives when identifying "Running" behavior, an additional criterion has been implemented. This criterion measured the distance between the initial and final positions within a vector containing centroids of the last ten associated frames when the reference velocity was exceeded, and we only considered when the distance measured between the extremes was greater than 20 cm. In our tests, this condition differentiated the "Running" behavior from other behaviors with fast movements and without significant movement of the hen.

3. Results

3.1. Yolo Tracking Model

Figure 3 shows the mAP (mean average precision) and Loss plots of the generated Yolo v8n model. The model achieved 98.5% accuracy. These plots depict the evolution of mAP and loss throughout training, offering a detailed view of the model's performance in chicken identification. The model demonstrated robustness, as the mAP curve remained stable above 98% despite a decrease in the loss curve starting from epoch 15. This suggests the model continues to identify chickens with high precision after an initial training phase.

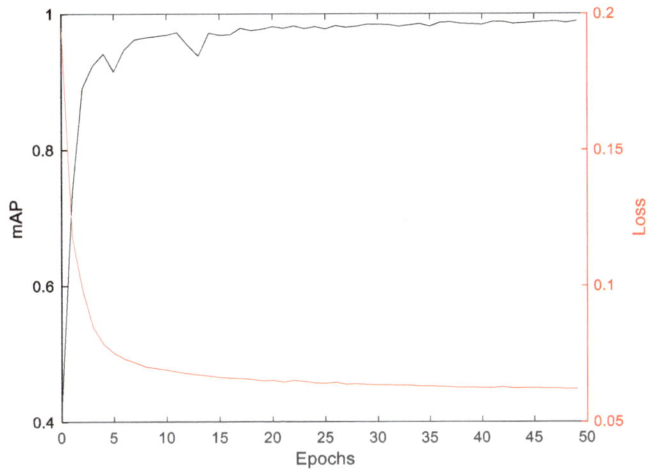

Figure 3. Graph of the evolution of the mean average precision (mAP, black line) and loss (red line) during the Yolo v8n chicken detection model training.

As illustrated in the images presented in Figure 4, our system was capable of detecting chickens within the nest, albeit with less certainty. Moreover, certain areas, such as those behind the feeder, remained concealed. Additionally, the impact of camera lens distortion on image quality at the periphery also merits consideration.

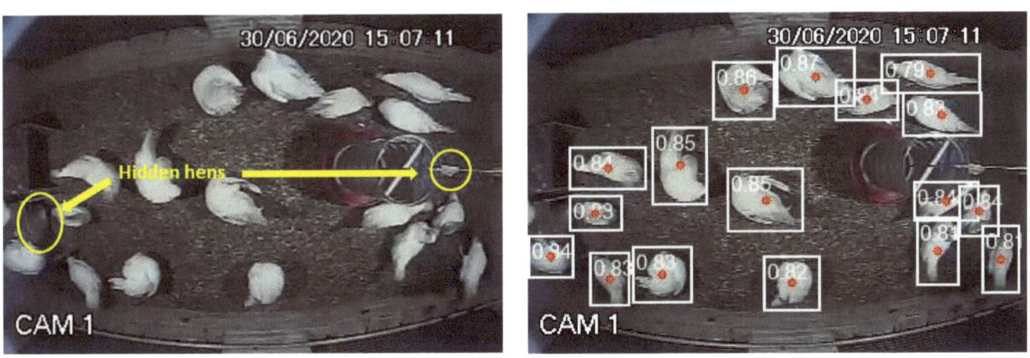

(a) (b)

Figure 4. The original frame (**a**) was processed with Yolo v8n (**b**) showing that the model's confidence value in detecting laying hens inside the nest and behind the feeder is low.

The quality of detection of chickens is fundamental for successful tracking. Chickens that can be confused with background, or vice versa, generate centroid coordinates that have the potential to interfere with the identification of birds by the Kaman filter. This is a crucial point in the development of the tracking system.

3.2. Appearing and Hiding of Chickens

We had problems hiding the chickens in the scene during our tracking tests. In Figure 4, we highlight two common places where the chicken identification is lost in the image due to obstruction of objects. In Figure 5, we show the example of the path of two chickens that appeared and disappeared in the scene near the feeder and near the nest.

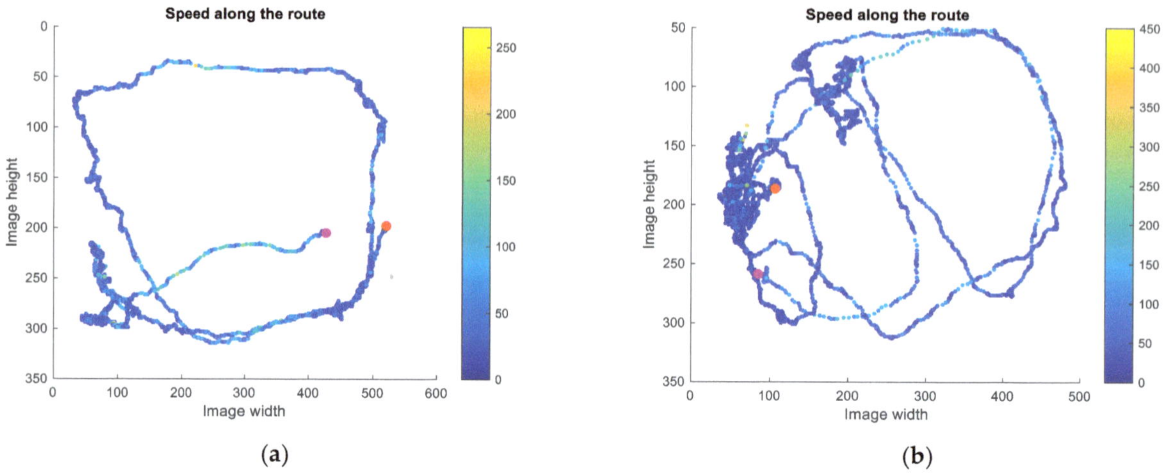

Figure 5. Path of the birds with the marks of the beginning and end of the path (magenta and red, respectively) and highlighting the moments when the bird walked faster (green and yellow dots) for the hen-line (**a**) track_Id 5 that appeared and disappeared near the feeder and (**b**) track_Id 45 that appeared and disappeared near the nest.

Markings in magenta and red correspond respectively to the first and last coordinates of chicken detection in the video, that is, the beginning and end of the mapped path. Along the way, the more yellow the marking, the faster the chicken travels. We can observe that the places of emergence and disappearance of chickens in Figure 5 are close to the nest or feeder. This was due to the already mentioned fact that we lost the identification of the bird when it left the scene, entered the nest, or passed behind the feeder. This condition is expected in the case of monitoring chickens in commercial farms, as the camera capture area will be smaller than the area of the shed, and, therefore, there will be the appearance and disappearance of chickens (same individuals or not) along the edges of the image. Thus, we decided to work with this feature, allowing our methodology to be applied in commercial breeding conditions in the future.

3.3. Direction of Displacement and Minimum Displacement

In the path graphs, we observe that there are times when the bird walks faster, and in others, there is minimal movement, suggesting that the bird is standing still, expressing some natural behavior.

As can be seen in Figure 6, our processing considered small lateral displacements, accumulating as distances covered by the birds. However, situations like the one illustrated can refer to a bird standing still, resting, or expressing some natural behavior. Therefore, our code considers distances greater than 1 cm/frame as displacement of the bird. The variation in the displacement direction also seems to contribute to the identification of the hen's state of movement, so we also started to observe this information.

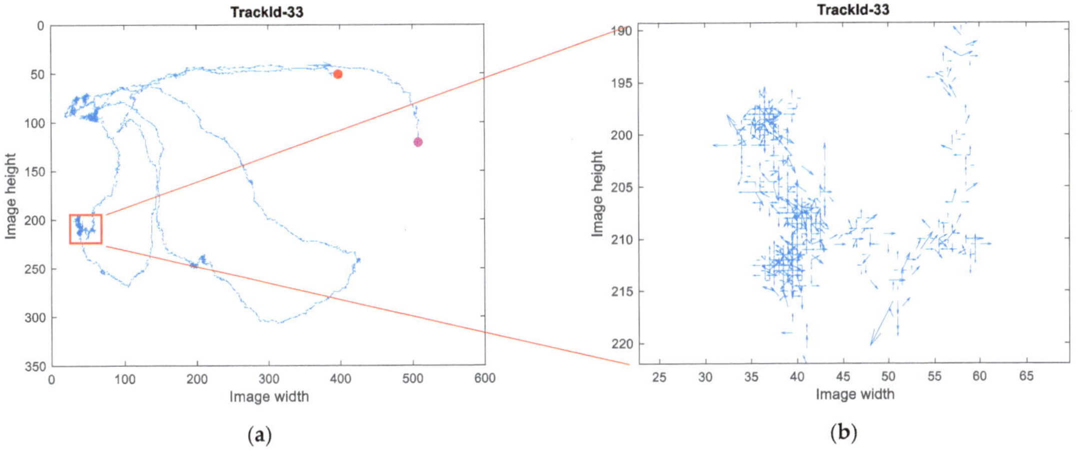

Figure 6. Arrow chart of chicken track_id 33 (**a**), highlighting a moment when the bird was still (**b**). The red and magenta dots correspond, respectively, to the beginning and end of tracking the monitored chicken.

3.4. Measures to Determine the State of Movement

Once the correction was made, we looked for metrics indicating when a bird was standing still, walking, or running. Analyzing the velocity graph in Figure 7, we observed that the velocity of this displacement varied at values below 100 cm/s when the bird was walking and peaked much above this value when the bird ran.

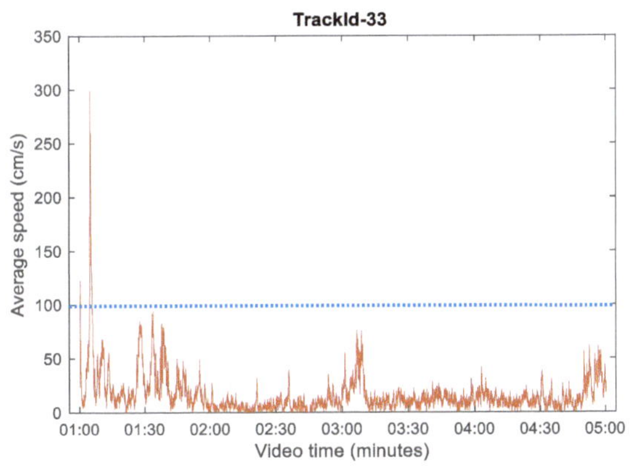

Figure 7. Graph of variation of the average speed of chicken track_Id 33 every ten frames, where the peak speed corresponds to this chicken running.

Figure 7 presents a graph depicting the variation in velocity during the time a bird was tracked by our model. This particular bird was selected for analysis as it exhibited all three behaviors during the tracking period. At the onset of tracking, a spike in speed is observed, confirmed in the original video as indicative of the running behavior. This pattern was also observed in other birds, and a velocity threshold of 100 cm/s was established for classifying running behavior in our code. Resting behavior is characterized by the bird remaining completely stationary, disregarding any minor displacements (<1 cm/frame). Consequently, any movement detected with a velocity between 0 and 100 cm/s was classified as exploring behavior. To minimize any transitional effects between behaviors, the average speed of the last ten frames (0.33 s) was considered.

3.5. Movement Analysis

Our surveillance and tracking system can identify a "new chicken" instantly. As previously reported, there are covered regions in our footage, and we found that there are chickens that remain with the same identification for a very short time. There are also those that, during the nesting behavior, even if they maintain the identification, remain still during the entire vigil. Our objective is to develop an intelligent system capable of automatically identifying chickens in flocks whose rearing density. In this sense, we focused our attention on chickens that walked more than 300 cm and were monitored by our system for more than 200 s without losing identification, and we made the graphs shown in Figure 8.

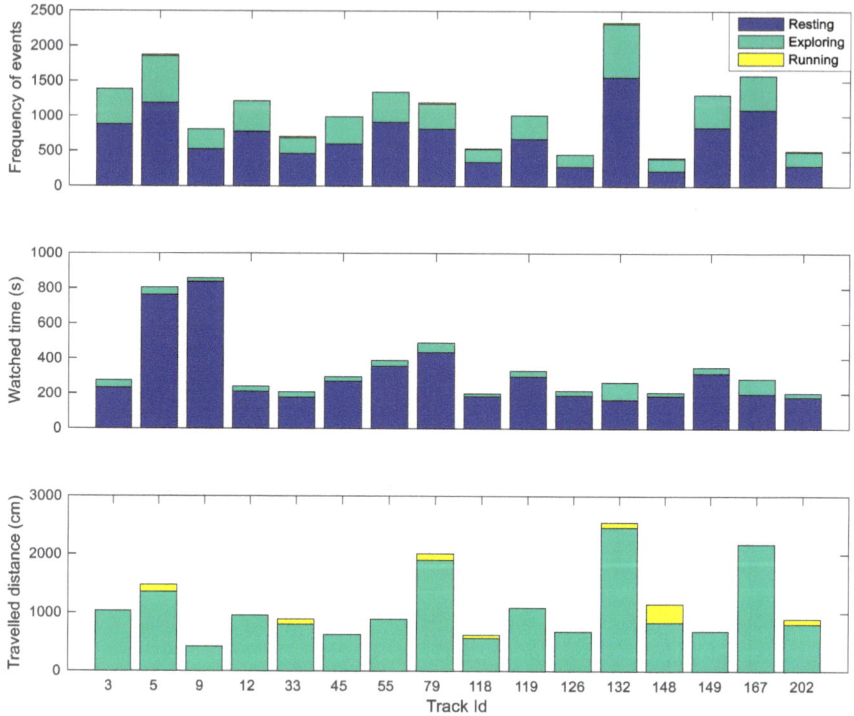

Figure 8. Graphs of frequency of occurrences, total observation time, and total distance covered for resting, exploring, and running behaviors for birds that walked more than 300 cm and were watched for more than 200 s without losing identification.

The chicken with track_Id = 148 caught our attention, which had a greater displacement per run than the other birds. Analyzing the trajectory graph, we noticed that in the region where the chicken supposedly ran, the direction vectors were misaligned. Unlike what happened with the chicken track_Id = 33, during the peak velocity, the direction vectors were aligned. We searched the video for the exact frames of each chicken for these occurrences and verified that chicken 33 ran at that moment and chicken 148 flapped its wings. Figure 9 shows the graphs and frames, highlighting the observed differences.

As observed in Figure 9, the peaks in speed are highly similar for distinct behaviors, indicating that this measure alone is insufficient for distinguishing between these behaviors. Based on these observations, the variation in the direction of movement of the birds was analyzed and compared with their speed. The tangent angle of the chicken's displacement vector for each analyzed frame indicates the intensity of the change of direction of the chicken's path. It was observed that when this route change is abrupt and associated with high speed, there is a high possibility that the bird is opening its wings. When the speed is high and there is no change in route, it is assumed that the chicken is running. When the route changes but the speed is considered normal, it is assumed that the bird has simply changed direction and continued walking.

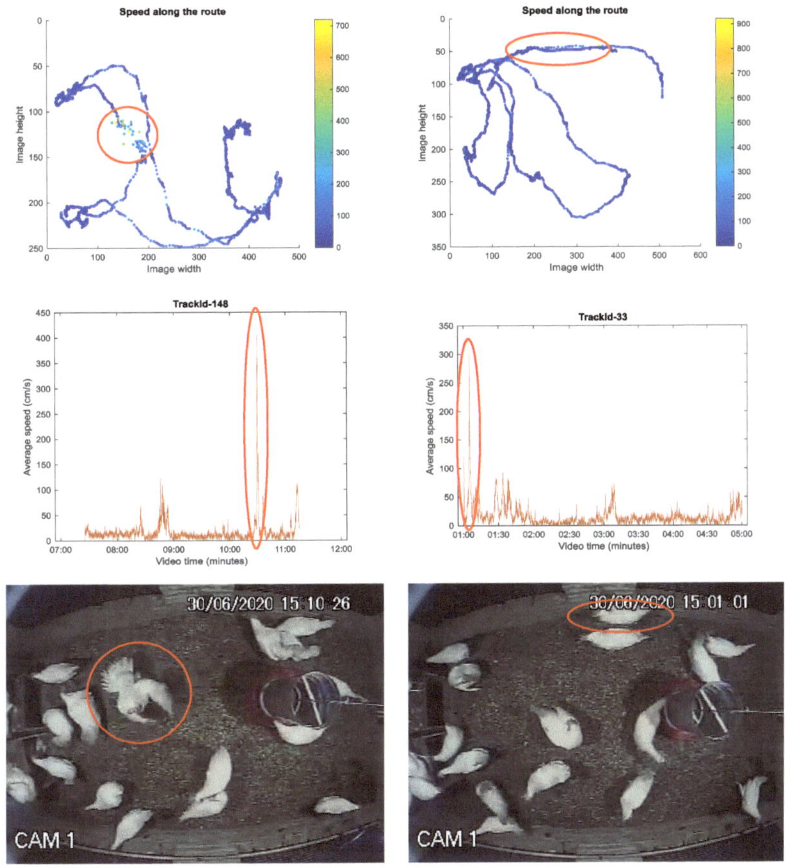

(a) **(b)**

Figure 9. Graphs of paths, variation in displacement average speed, and frame of the original video highlighting the differences in measurements observed for the chicken (**a**) track_Id = 148, which expresses the behavior of flapping wings, and (**b**) track_Id = 33, which is running. The circles mark the data in the graphs corresponding to the chickens at a given moment in the video (images) expressing different behaviors.

For the running chicken, we used the measurement of the distance between the centroids of the extreme points during the last ten frames. This approach allows the detection of the directionality of the bird's movement. Consideration of speed is of fundamental importance, as it constitutes a discriminating factor between the "Exploring" and "Running" states. In Figure 10, during flapping, the chicken identification track_Id = 79 is segregated into two identifications (track_Id = 79 and track_Id = 97). After the expression of the behavior, identification 97 ceases to exist. During flapping, there is a rapid change in the positioning of the centroid of the original identification (track_Id = 79), resulting in a speed that falls within the predefined limits for the "Running" and "Exploring" states. The generation of new identifications in different bird regions during wing flapping can also cause deviations in the centroids, mimicking the presence of another bird. The finding of a distance between the centroids of these identifications above 20 cm characterizes the

"Running" behavior. To mitigate the influence of spurious identifications, we adopted a rule requiring a minimum number of 18 frames (called "lifetime") in which the bird remains in the video. The limit of 18 frames was chosen based on observations since birds frequently performed movements that caused abrupt changes in the image centroid. This strategy aims to reduce confusion between genuine behaviors and artifactual occurrences.

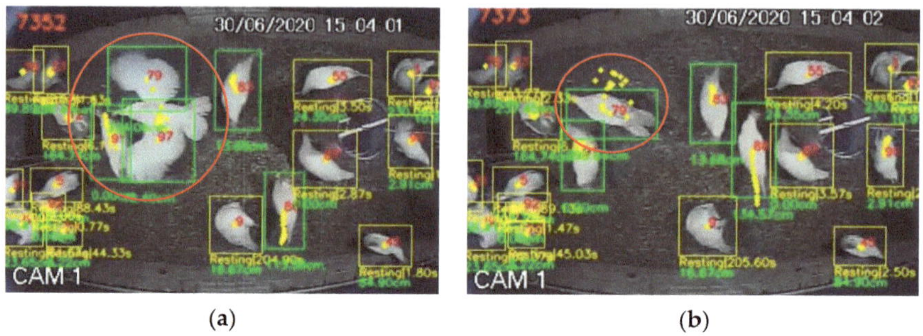

Figure 10. Capture of frames showing flapping wings. (**a**) chicken 79 was divided into two birds in identification. (**b**) chicken 97 is ceasing to exist.

We implemented these rules in our code (pseudo-code in Appendix A). The processed video that classifies bird movement with all these metric considerations is shown in the Supplementary Material.

In Table 2, we show a summary of the time spent by the birds resting, exploring, and running, which can be used to analyze the state of welfare or health of the birds in time intervals of interest.

Table 2. Summary of occurrences and times spent by birds running, exploring, and resting, disregarding Ids with a minimum lifetime.

Behavior Class	Number of Occurrences (N)	Overall Time (T) Seconds	Average Time (T/N)Seconds
Resting	808	13,537	16.8
Exploring	792	1490	1.9
Running	7	9	1.3

Table 2 presents a summary of data obtained from the entire 15-minute video. This summary can be generated for any desired time interval. It was observed that the birds spend most of their time stationary, resting or engaging in other natural behaviors that do not involve movement. The majority of the route traveled by the bird is spent exploring. A few instances of running behavior were observed, and it was verified that this behavior is expressed for very short periods.

We found no reference values for these behaviors in the literature. However, we hypothesize that by extracting precise measurements of these behaviors, it will be possible to establish reference values for each lineage, allowing the welfare of chickens submitted to different housing environments to be compared.

4. Discussion

Proposing automatic, accurate, and non-invasive approaches to monitoring animals in production is necessary to improve the efficiency and welfare of animals in housing [32]. In this sense, several studies have used convolutional neural networks (CNN) to detect and classify chicken behavior. Neethirajan [19] and Yang et al. [20] demonstrated the feasibility of using Yolo to detect chickens in several videos, suggesting applications such as counting and tracking. Yang et al. [20] classified six behaviors using Yolo-v5 with accuracies ranging from 72.3% to 95.3%. Li et al. [33] developed a computer vision system to detect chickens in feeders and drinkers. Nasiri et al. [32] estimated feeding time for broilers using Yolo-v3 to detect when heads were within the feeder area.

Individual and group behavior are important indicators of the welfare and health of birds [34–36]. Our work provides a breakthrough for tracking chickens, demonstrating that the Kalman filter and the BoT-SORT algorithm efficiently maintain the identification of birds in our videos, where the density and overlapping of bodies were a challenge for computer vision.

The BoT-SORT algorithm is highly effective in maintaining accurate identification of chickens throughout the tracking process. With our tracking algorithm, we could carry out a detailed analysis of the distances covered and the movement speeds of the chickens. This ability to identify and categorize movement behaviors makes it possible to analyze bird behavior in different environmental conditions.

Measuring the distances covered individually by the hens can indicate the activity level of the flock of individuals and can also be used to assess the gait score [37]. Campbell et al. [2] found that aviary resources were used differently by individual hens and that understanding these usage patterns by both individuals and the group will allow the optimization of cageless housing system designs. Jaihumi et al. [13] developed a Yolo v5 model to assess the mobility of broilers, and Fernández et al. [6] found an association between activity level and patterns of aviary occupancy with hock and leg injuries in broilers using computer vision.

By observing movement patterns, we could detect significant differences between individuals in the monitored group of chickens. These individual distinctions in movements are necessary to assess bird welfare and health. For example, we can identify birds exhibiting anomalous behaviors indicative of health problems or discomfort, allowing for early intervention to ensure appropriate treatment.

Collins [38] observed that some chickens walked greater distances than others. The motivation for moving laying hens may differ from that of broilers. However, in our results, we also observed that there are hens that moved more and at greater speeds than others. Taylor et al. [39] found a relationship between chickens that covered greater distances and increased welfare, and these animals had lower weight, better gait scores, and lower corticosterone responses. Considering chickens under the same rearing model, the observed differences in movement measured by our tracking system may indicate more adapted individuals.

The movement measurements obtained and the tracking algorithm developed in this study provide a solid basis for future investigations. This model can assess the welfare and health of chickens and classify and understand other natural behaviors, thus contributing to significant advances in poultry farming and animal welfare research. Such information is essential to improve poultry farming practices, ensuring a healthier and more comfortable environment for these animals.

Limitations and Suggestions for Future Studies

Our tracking method is limited in regions where the camera is obstructed, such as behind equipment, edges of images, and inside nests. Obstruction of the image by equipment was also observed in other studies [21]. In our tests, we preferred not to deal with this problem because, in commercial situations, there will be regions where the bird will leave the camera's field of view. We understand it is better to use strategies to analyze the data recorded this way, as already discussed in other works.

Our footage was low resolution and heavily distorted around the edges. The use of high-resolution cameras with less edge distortion can improve the detection accuracy, the conversion scale from pixels to centimeters, and, consequently, the measurements of the distances covered by the birds. Addressing these distortions can give greater accuracy in displacement measurements. More complex images are subject to greater chances of computational model confusion [20]. In this field, it is still necessary for work to demonstrate the stability of a flock tracking system and longer filming times.

We adopted criteria to reduce confusion in classifying the running behavior with other natural behaviors. We raised the hypothesis that these criteria can be better worked to contribute to the classification of natural behaviors. Dawkins [40] and Kashima et al. [41] proposed that behavior can be used to estimate the need and health of animals, highlighting that behavior analysis should have a more prominent role when used with new technologies. Thus, other studies that seek to classify individual or group behaviors autonomously are necessary and will constitute the future of industrial poultry for monitoring the welfare and health of chickens.

5. Conclusions

Our model obtained a mean average precision of 98.5% in detecting chickens, and the BoT-SORT algorithm was efficient in maintaining the identification of these birds during tracking. From the tracking, it was possible to analyze the distances covered and the speeds of the chickens, which made it possible to classify their movement behaviors. Movement speeds above 100 cm/s of chickens with a lifetime of more than 18 frames were sufficient to differentiate running behavior from exploring behavior. In this study, the tests showed that the chickens spent approximately 85% of their time resting, 10% of their time exploring the environment, and 6% running. Our algorithm and the measures extracted to classify chicken movement contribute to the assessment of chicken welfare and health.

Supplementary Materials: The video processed with the proposed algorithm can be downloaded at: https://www.mdpi.com/article/10.3390/agriengineering5040104/s1.

Author Contributions: The Conceptualization, Formal and Statistical Analysis, and Data Curation were made by A.L.R.S. and D.F.P.; The Software and Investigation were made by I.B.d.C.M. and D.F.P.; The Methodology, Validation, the Original Draft, Proofreading and Editing of the manuscript was carried out by A.L.R.S., S.A.M. and D.F.P.; Visualization, Supervision and Funding acquisition was made under D.F.P. All authors have read and agreed to the published version of the manuscript.

Funding: Funding was provided by the National Council for Scientific and Technological Development—CNPq (Grant # 304085/2021-9).

Institutional Review Board Statement: This study was conducted according to the guidelines of the Brazilian National Council for the Control of Animal Experimentation (CONCEA) and approved by the Sao Paulo State University's Animal Ethics Committee (protocol number 02/2020, CEUA-UNESP), of Comfort Environmental Laboratory at the School of Science and Engineering from the Sao Paulo State University (UNESP, Brazil).

Data Availability Statement: Data will be available upon request to the corresponding author.

Acknowledgments: We thank the support of the technicians of the Precision Livestock Laboratory of the School of Sciences and Engineering of UNESP, Campus at Tupã.

Conflicts of Interest: The authors declare no conflict of interest.

Appendix A

Algorithm: Chicken tracking code and movement classification

Input: Video or stream of video
Output: Video with detections and file of data about this experiment

```
# Minimum distance to consider an object has moved
min_distance = 3

# Thresholds to determine whether the object is Resting, Exploring, or Running
resting_speed_threshold = 0.8
exploring_speed_threshold = 4.17

# Minimum consecutive counts for each state
min_count_resting = 5
min_count_exploring = 4
min_count_running = 3

#Minimum distance between 10 frames in running state
distance_running_min = 25

#Lifetime from object detected in a video
lifetime = 18

# Pixel to centimeter conversion
pixel_cm = 0.8

FOR each tracked object with track_id:
    calculate average_speed using the last 10 frames
    assign average_speed to the speed property of the tracked object

    IF tracked object's lifetime is less than the lifetime:
        Discard the object's behavior
    IF the object's state is "Running" and the distance between the first and last
    frame centroids in the last 10 frames is less than distance_running_min:
        Discard the object's behavior
    IF the object's speed is less than or equal to resting_speed_threshold:
        increment count_resting
        reset counts count_exploring and count_running
        reset distance dist_resting
        IF the object's previous state was not "Resting":
            increment freq_resting
        IF count_resting is greater than or equal to min_count_resting:
            set the object's state as "Resting" and time as "Resting"
    ELSE IF the object's speed is less than exploring_speed_threshold:
        reset count_resting
        increment count_exploring
        reset count_running
        IF distance is greater than min_distance:
            increment distance dist_exploring
        IF the object's previous state was not "Exploring":
            increment freq_exploring
        IF count_exploring is greater than or equal to min_count_exploring:
            set the object's state as "Exploring" and time as "Exploring"
    ELSE IF the object's speed is greater than or equal to exploring_speed_threshold:
        reset count_resting
        reset count_exploring
        increment count_running
        IF distance is greater than min_distance:
            increment distance dist_running
        IF the object's previous state was not "Running":
            increment freq_running
        IF count_running is greater than or equal to min_count_running:
            set object's state as "Running" and time as "Running"
```

References

1. van Veen, L.A.; van den Oever, A.C.M.; Kemp, B.; van den Brand, H. Perception of laying hen farmers, poultry veterinarians, and poultry experts regarding sensor-based continuous monitoring of laying hen health and welfare. *Poult. Sci.* **2023**, *102*, 102581. [CrossRef] [PubMed]
2. Campbell, D.; Karcher, D.M.; Siegford, J.M. Location tracking of individual laying hens housed in aviaries with different litter substrates. *Appl. Anim. Behav. Sci.* **2016**, *184*, 74–79. [CrossRef]
3. Grebey, T.C.; Ali, A.B.A.; Swanson, J.C.; Widowski, T.M.; Siegford, J.M. Dust bathing in laying hens: Strain, proximity to, and number of conspecifics matter. *Poult. Sci.* **2020**, *99*, 4103–4112. [CrossRef]
4. Vasdal, G.; Muri, K.; Stubsjøen, S.M.; Moe, R.O.; Kittelsen, K. Qualitative behaviour assessment as part of a welfare assessment in flocks of laying hens. *Appl. Anim. Behav. Sci.* **2022**, *246*, 105535. [CrossRef]
5. Geffen, O.; Yitzhaky, Y.; Barchilon, N.; Druyan, S.; Halachmi, I. A machine vision system to detect and count laying hens in battery cages. *Animal* **2020**, *14*, 2628–2634. [CrossRef] [PubMed]
6. Fernandez, A.P.; Norton, T.; Tullo, E.; van Hertem, T.; Youssef, A.; Exadaktylos, V.; Vranken, E.; Guarino, M.; Berckmans, D. Real-time monitoring of broiler flock's welfare status using camera-based technology. *Biosyst. Eng.* **2018**, *173*, 103–114. [CrossRef]
7. Li, G.; Li, B.; Shi, Z.X.; Lu, G.; Chai, L.; Rasheed, K.M.; Regmi, P.; Banakar, A. Inter-Individual Distances and Orientations of Laying Hens under Eight Stocking Densities Measured by Integrative Deep Learning Techniques. *Poult. Sci.* **2023**, *102*, 103076. [CrossRef] [PubMed]
8. Yang, X.; Zhao, Y.; Gan, H.; Hawkins, S.; Eckelkamp, L.; Prado, M.; Burns, R.; Purswell, J.; Tabler, T. Modeling gait score of broiler chicken via production and behavioral data. *Animal* **2023**, *17*, 100692. [CrossRef]
9. Subedi, S.; Bist, R.; Yang, X.; Chai, L. Tracking pecking behaviors and damages of cage-free laying hens with machine vision technologies. *Comput. Electron. Agric.* **2023**, *204*, 107545. [CrossRef]
10. Fang, C.; Huang, J.; Cuan, K.; Zhuang, X. Comparative study on poultry target tracking algorithms based on a deep regression network. *Biosyst. Eng.* **2020**, *190*, 176–183. [CrossRef]
11. Guo, Y.; Aggrey, S.; Oladeinde, A.; Johnson, J.; Zock, G.; Chai, L. A Machine Vision-Based Method Optimized for Restoring Broiler Chicken Images Occluded by Feeding and Drinking Equipment. *Animals* **2021**, *11*, 123. [CrossRef] [PubMed]
12. Li, N.; Ren, Z.; Li, D.; Zeng, L. Review: Automated techniques for monitoring the behaviour and welfare of broilers and laying hens: Towards the goal of precision livestock farming. *Animal* **2020**, *14*, 617–625. [CrossRef]
13. Jaihuni, M.; Gan, H.; Tabler, T.; Prado, M.; Qi, H.; Zhao, Y. Broiler Mobility Assessment Via a Semi-Supervised Deep Learning Model and Neo-Deep Sort Algorithm. *Animals* **2023**, *13*, 2719. [CrossRef] [PubMed]
14. Fodor, I.; van der Sluis, M.; Jacobs, M.; Klerk, B.; Bouwman, A.C.; Ellen, E.D. Automated pose estimation reveals walking characteristics associated with lameness in broilers. *Poult. Sci.* **2023**, *102*, 102787. [CrossRef] [PubMed]
15. Nasiri, A.; Yoder, J.; Zhaob, Y.; Hawkins, S.; Prado, M.; Gana, H. Pose estimation-based lameness recognition in broiler using CNN-LSTM network. *Comput. Electron. Agric.* **2022**, *197*, 106931. [CrossRef]
16. Fernandes, A.M.; Sartori, D.L.; Morais, F.J.O.; Salgado, D.D.; Pereira, D.F. Analysis of Cluster and Unrest Behaviors of Laying Hens Housed under Different Thermal Conditions and Light Wave Length. *Animals* **2021**, *11*, 2017. [CrossRef]
17. Trnovszky, T.; Kamencay, P.; Orjesek, R.; Benco, M.; Sykora, P. Animal recognition system based on convolutional neural network. *Adv. Electr. Electron. Eng.* **2017**, *15*, 517–525. [CrossRef]
18. Siriani, A.L.R.; Kodaira, V.; Mehdizadeh, S.A.; Nääs, I.A.; Moura, D.J.; Pereira, D.F. Detection and tracking of chickens in low-light images using YOLO network and Kalman filter. *Neural Comput. Appl.* **2022**, *34*, 21987–21997. [CrossRef]
19. Neethirajan, S. ChickTrack—A quantitative tracking tool for measuring chicken activity. *Measurement* **2022**, *191*, 110819. [CrossRef]
20. Yang, X.; Bist, R.; Subedi, S.; Wu, Z.; Liu, T.; Chai, L. An automatic classifier for monitoring applied behaviors of cage-free laying hens with deep learning. *Eng. Appl. Artif. Intell.* **2023**, *123*, 106377. [CrossRef]
21. Yang, X.; Bist, R.; Subedi, S.; Chai, L. A deep learning method for monitoring spatial distribution of cage-free hens. *Artif. Intell. Agric.* **2023**, *8*, 20–29. [CrossRef]
22. Cui, Y.; Kong, X.; Chen, C.; Lia, Y. Research on broiler health status recognition method based on improved YOLOv5. *Smart Agric. Technol.* **2023**, *in press*. [CrossRef]
23. Banerjee, S.C.; Khan, K.A.; Sharma, R. Deep-worm-tracker: Métodos de aprendizagem profunda para detecção e rastreamento precisos para estudos comportamentais em C. elegans. *Appl. Anim. Behav. Sci.* **2023**, *266*, 106024. [CrossRef]
24. Young-Chul, Y.; Kim, D.Y.; Young-Min, S.; Yoon, K.; Jeon, M. Online Multiple Pedestrians Tracking using Deep Temporal Appearance Matching Association. *arXiv* **2020**, arXiv:1907.00831. [CrossRef]
25. Aharon, N.; Orfaig, R.; Bobrovsky, B.Z. BoT-SORT: Robust Associations Multi-Pedestrian Tracking. *arXiv* **2022**, arXiv:2206.14651v2. [CrossRef]
26. Wojke, N.; Bewley, A.; Paulus, D. Simple Online and Realtime Tracking with a Deep Association Metric. IEEE internation-al conference on image processing (ICIP). *arXiv* **2017**, arXiv:1703.07402. [CrossRef]
27. Redmon, J.; Divvala, S.; Girshick, R.; Farhadi, A. You Only Look Once: Unified, Real-Time Object Detection. In Proceedings of the IEEE Conference on Computer Vision and Pattern Recognition (CVPR), Las Vegas, NV, USA, 27–30 June 2016; pp. 779–788. [CrossRef]
28. Redmon, J.; Farhadi, A. Yolov3: An incremental improvement. *arXiv* **2018**, arXiv:1804.02767. [CrossRef]

29. Redmon, J.; Farhadi, A. Yolo9000: Better, Faster, Stronger. In Proceedings of the IEEE Conference on Computer Vision and Pattern Recognition, Honolulu, HI, USA, 21–26 July 2017; pp. 6517–6525. [CrossRef]
30. Ge, Z.; Liu, S.; Wang, F.; Li, Z.; SUN, J. Yolox: Exceeding yolo series in 2021. *arXiv* **2021**, arXiv:2107.08430. [CrossRef]
31. Yang, X.; Chai, L.; Bist, R.B.; Subedi, S.; Wu, Z. A Deep Learning Model for Detecting Cage-Free Hens on the Litter Floor. *Animals* **2022**, *12*, 1983. [CrossRef]
32. Nasiri, A.; Amirivojdan, A.; Zhao y Gan, H. Estimating the Feeding Time of Individual Broilers via Convolutional Neural Network and Image Processing. *Animals* **2023**, *13*, 2428. [CrossRef]
33. Li, G.; Zhao, Y.; Purswell, J.L.; Du, Q.; Chesser, G.D., Jr.; Lowe, J.W. Analysis of feeding and drinking behaviors of group-reared broilers via image processing. *Comput. Electron. Agric.* **2020**, *175*, 105596. [CrossRef]
34. Papageorgiou, M.; Goliomytis, M.; Tzamaloukas, O.; Miltiadou, D.; Simitzis, P. Positive Welfare Indicators and Their Association with Sustainable Management Systems in Poultry. *Sustainability* **2023**, *15*, 10890. [CrossRef]
35. Jacob, F.G.; Nääs, I.A.; Salgado, D.D.; Baracho, M.S.; Lima, N.D.S.; Pereira, D.F. Does Environmental Enrichment with Music and Strobe Light Affect Broilers' Welfare? Analyzing Their On-Farm Reaction. *Agriengineering* **2022**, *4*, 707–718. [CrossRef]
36. Pereira, D.F.; Lopes, F.A.A.; Gabriel Filho, L.R.A.; Salgado, D.D.; Mollo Neto, M. Cluster index for estimating thermal poultry stress (gallus gallus domesticus). *Comput. Electron. Agric.* **2020**, *177*, 105704. [CrossRef]
37. Li, G.; Gates, R.S.; Meyer, M.M.; Bobeck, E.A. Tracking and Characterizing Spatiotemporal and Three-Dimensional Locomotive Behaviors of Individual Broilers in the Three-Point Gait-Scoring System. *Animals* **2023**, *13*, 717. [CrossRef] [PubMed]
38. Collins, L.M. Non-intrusive tracking of commercial broiler chickens in situ at different stocking densities. *Appl. Anim. Behav. Sci.* **2008**, *112*, 94–105. [CrossRef]
39. Taylor, P.S.; Hemsworth, P.H.; Groves, P.J.; Gebhardt-Henrich, S.G.; Rault, J.L. Frequent range visits further from the shed relate positively to free-range broiler chicken welfare. *Animal* **2020**, *14*, 138–149. [CrossRef]
40. Kashima, M.A.; Green, A.R.; Sales, T.G.; Bahr, C.; Berckmans, D.; Gates, R.S. Performance of an image analysis processing system for hen tracking in an environmental preference chamber. *Poult. Sci.* **2014**, *93*, 2439–2448. [CrossRef]
41. Dawkins, M.S. Using behaviour to assess animal welfare. *Anim. Welf.* **2004**, *13*, 3–7. [CrossRef]

Disclaimer/Publisher's Note: The statements, opinions and data contained in all publications are solely those of the individual author(s) and contributor(s) and not of MDPI and/or the editor(s). MDPI and/or the editor(s) disclaim responsibility for any injury to people or property resulting from any ideas, methods, instructions or products referred to in the content.

sustainability

MDPI

Article

Analyzing Extreme Temperature Patterns in Subtropical Highlands Climates: Implications for Disaster Risk Reduction Strategies

Abdulnoor A. J. Ghanim [1], Muhammad Naveed Anjum [2,3,*], Ghulam Rasool [2], Saifullah [2], Muhammad Irfan [4], Mana Alyami [1], Saifur Rahman [4] and Usama Muhammad Niazi [5]

[1] Civil Engineering Department, College of Engineering, Najran University, Najran 61441, Saudi Arabia
[2] Department of Land and Water Conservation Engineering, Pir Mehr Ali Shah Arid Agriculture University Rawalpindi, Rawalpindi 46000, Pakistan; engrghulamrasool45@gmail.com (G.R.)
[3] State Key Laboratory of Cryospheric Science, Northwest Institute of Eco-Environment and Resources, Chinese Academy of Sciences, Lanzhou 730000, China
[4] Electrical Engineering Department, College of Engineering, Najran University, Najran 61441, Saudi Arabia
[5] Department of Mechanical Engineering Technology, National Skills University Islamabad, Islamabad 44000, Pakistan
* Correspondence: naveedwre@uaar.edu.pk

Abstract: This study utilized hot and cold indices to evaluate the changes in extreme temperature events that occurred in subtropical highland climates from 1991 to 2020. The modified Mann–Kendall (MMK) test and the Theil–Sen (TS) slope estimator were used to analyze the linear trends in the time series of the extreme temperature indices. The northern highlands of Pakistan (NHP) were considered as a case study region. The results showed that the annual maximum temperature had a slightly increasing tendency (at the rate of 0.14 °C/decade), while the annual minimum temperature had a slightly decreasing tendency (at the rate of −0.02 °C/decade). However, these trends were not significant at the 5% significance level. The decadal averages of the hot indices were the highest in the second decade (2000s), while they were the lowest in the subsequent decade (2010s). In comparison, all the cold indices except the annual minimum value of the maximum temperature (TXn) showed a persistent decline in their decadal averages throughout the 2000s and 2010s. Overall, the frequency of hot days significantly increased in the NHP during the study period. This study found that the hot days and coldest days increased over the past three decades in the NHP. However, there was a decreasing trend in the cold spell duration, cold nights, and the coldest nights over the past three decades, as demonstrated by the trends of the cold spell duration index (CSDI), the temperature of cold nights (TN10p), and the annual minimum value of the minimum temperature (TNn) indices. These changes may impact the environment, human health, and agricultural operations. The findings provide useful insights into the shifting patterns of extreme temperature events in northern Pakistan and have crucial implications for the climate-change-adaptation and resilience-building initiatives being undertaken in the region. It is suggested that the continuous monitoring of extreme temperature events is necessary to comprehend their effects on the region and devise strategies for sustainable development.

Keywords: climate change; extreme temperature; trend analysis; modified Mann–Kendall test; northern Pakistan

Citation: Ghanim, A.A.J.; Anjum, M.N.; Rasool, G.; Saifullah; Irfan, M.; Alyami, M.; Rahman, S.; Niazi, U.M. Analyzing Extreme Temperature Patterns in Subtropical Highlands Climates: Implications for Disaster Risk Reduction Strategies. *Sustainability* **2023**, *15*, 12753. https://doi.org/10.3390/su151712753

Academic Editor: Luca Salvati

Received: 22 July 2023
Revised: 11 August 2023
Accepted: 21 August 2023
Published: 23 August 2023

Copyright: © 2023 by the authors. Licensee MDPI, Basel, Switzerland. This article is an open access article distributed under the terms and conditions of the Creative Commons Attribution (CC BY) license (https://creativecommons.org/licenses/by/4.0/).

1. Introduction

Anthropogenic climate change has increased the frequency and intensity of extreme weather and climate events globally since the 1950s [1]. Moreover, the projected global warming will amplify changes in the extremes [1]. Pakistan is highly vulnerable to the impacts of extreme weather events, which are the main cause of natural disaster fatalities in the country [2]. The changing climate has reduced agricultural productivity, leading to severe food insecurity and economic losses [3]. Hence, it is essential to understand the behavior of extreme events and their sectoral implications for effective adaptation planning [4]. Standardized extreme indices can be used to characterize climate extremes and their temporal trends [4]. The Expert Team on Climate Change Detection and Indices (ETCCDI) developed a set of indices in 1999 to describe the nature of climate extremes in terms of frequency, amplitude, and persistence.

The analysis of temperature extremes is very important, particularly for subtropical climates in highland regions, because it helps with understanding the impact of climate change on the water resources, agricultural practices, and biodiversity in such climatic systems. Moreover, such investigations are necessary for highlighting climate extremes' implications for disaster risk reduction strategies. The northern highlands of Pakistan (NHP) are situated in South Asia, surrounded by Afghanistan (to the west), China (to the north), and India (to the east). The NHP hosts some of the world's highest mountains, such as K2, Nanga Parbat, and Broad Peak. This region is a vital source of freshwater for millions of people living downstream, and it influences the hydrology and climate of South Asia [5]. Climate change has caused various changes in this region, affecting both the mean and extreme climate conditions. The NHP are highly vulnerable to the impacts of climatic extremes [6]. One of the major impacts of climate change in this region is the increased frequency and intensity of extreme temperature events, which pose risks to local water resources, food security, and economic development [7]. These events can also have significant ecological and socioeconomic consequences, such as glacier melt, landslides, and heat stress on human and animal populations [8]. The changing temperature extremes in this region reflect the global pattern of enhanced extreme weather events in South Asia in recent decades [9,10].

Over the past few decades, the annual average temperature of Pakistan's mountains area has considerably increased, more than the average global warming rate [11]. This warming trend is anticipated to continue, which could have potentially catastrophic effects on the region's ecosystem, hydrology, and human populations [12]. Extreme temperature occurrences are becoming more common and severe as the direct consequence of climate change in subtropical highland climate systems. Recent studies revealed an increase in the severity, duration, and frequency of temperature extremes in the region [13,14]. Temperature extremes severely impact the region's water supply, agricultural production, and ecological systems. For instance, the melting of glaciers and snowpacks, which are essential supplies of freshwater for the region, is being accelerated by the warming trend [15]. As a result of this thawing, disasters, such as glacial lakes outburst floods (GLOFs) and landslides, are predicted to occur more often and severe [2], and water availability during the dry season is predicted to decrease. The region's ecosystem and biodiversity are under increasing threat from climate change. According to the findings of some previous studies, the negative impacts of climate change on species distribution in different altitudes could have significant ecological consequences [16]. Increasing temperatures and shifting precipitation patterns also threaten the region's forests, which are crucial for ecosystem services, including carbon sequestration and water management [17,18].

For the scientific community working on climate change impacts and variability, the use of extreme indicators allows for a better understanding of the role of extreme events and their sectoral implications. In this context, temperature extremes were extensively studied at the regional scale in various countries such as China [19], northern Europe [20], South America [21], North America [22], New Zealand [23], northwestern Africa [24], Europe [25], Nepal [26], and the Arabian Peninsula [27]. Although several authors detected a general increase in temperatures over the last decades in Pakistan [13,28–30], only a few studies focused on temperature extremes and their temporal evolution. Previously, studies on temperature extremes and their temporal evolution in Pakistan mainly focused on the trend detection of mean values using a small subset of old and generic indices. The evidence from previous studies indicated that heat waves, hot days, and warm nights are becoming more common and severe in Pakistan [9]. However, less attention was given to assessing the location-specific climate responses in the NHP. There are still knowledge gaps about the spatial and temporal variations of temperature extremes. In light of these knowledge gaps, the present study aimed to analyze the shifts in the spatial and temporal patterns of several extreme temperature indices to contribute to a better understanding of the way the climate in subtropical highland regions is changing.

This study was conducted to analyze the patterns of extreme temperature events over the past 30 years to better comprehend the changing behavior of hot and cold events in northern Pakistan. The historical shifts in frequency, intensity, and duration of extreme temperature events were assessed using data from weather stations installed in the NHP. The spatial patterns of these shifts were analyzed, with a special focus on how temperature extremes differ from one location to another. The annual and decadal changes in the extreme indices were investigated to check the temporal variations. This study provided new information about how climate change influences temperature extremes in Pakistan's northern mountainous region. The findings of this study were a valuable contribution to the existing literature on the topic of global warming. Moreover, the investigation results will also be useful for regional policymakers, disaster management authorities, farmers, and other stakeholders.

2. Materials and Methods

2.1. Study Area

The northern highlands of Pakistan contain a wide variety of topographical features, from mountainous regions to flat-topped plateaus and deep valleys. This area is distinguished by its distinctive topography, vast coverage of glaciers, and a wide range of climatic conditions, which significantly impact the local environment and regional natural resources. The spatial domain of the northern highlands of Pakistan (which covers geographic domain 33–37° N and 70–78° E) and the locations of weather stations are presented in Figure 1. Numerous glaciers and snow-capped mountains in this region are significant contributors to the local hydrology and water supply. Due to ongoing global warming, this region is most vulnerable to disasters such as flash floods and glacial lakes outburst floods (GLOFs). The climatic extremes, particularly temperature extremes, are expected to exacerbate risks of such disasters.

This part of the world is famous for its one-of-a-kind and delicate environment, which is exceptionally vulnerable to the effects of climate change. Due to its location at the confluence of the westerlies and the Indian monsoon system, this region has a complicated climate. Due to its elevation and topography, the local climate varies greatly. The amount of total annual precipitation received by this region varies from 100 mm to 1700 mm (Figure 2a), whereas the annual average temperature varies from 10 °C to 26 °C (Figure 2b). The region receives most of its precipitation during the monsoon season. The melting of snow and glaciers is the primary factor that determines how water flows through the region. The region is home to many glaciers, which serve as a significant water supply for the rivers that flow through the region. The glaciers in the NHP region are the source of the Indus River, one of Asia's most important rivers.

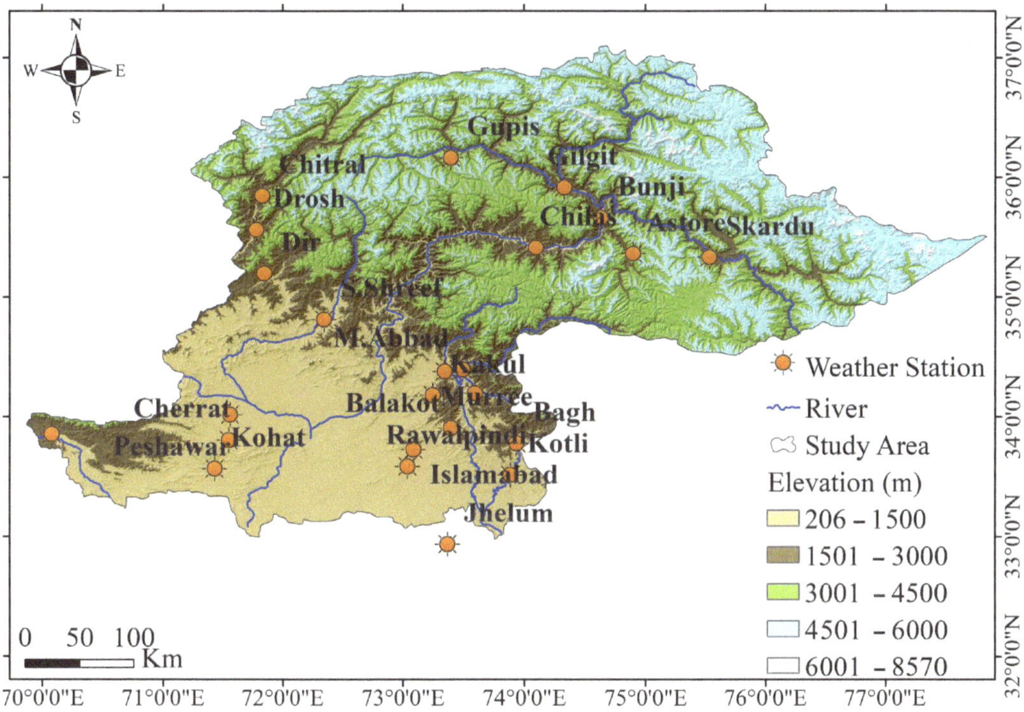

Figure 1. The spatial extent of the study area and location of the included weather stations.

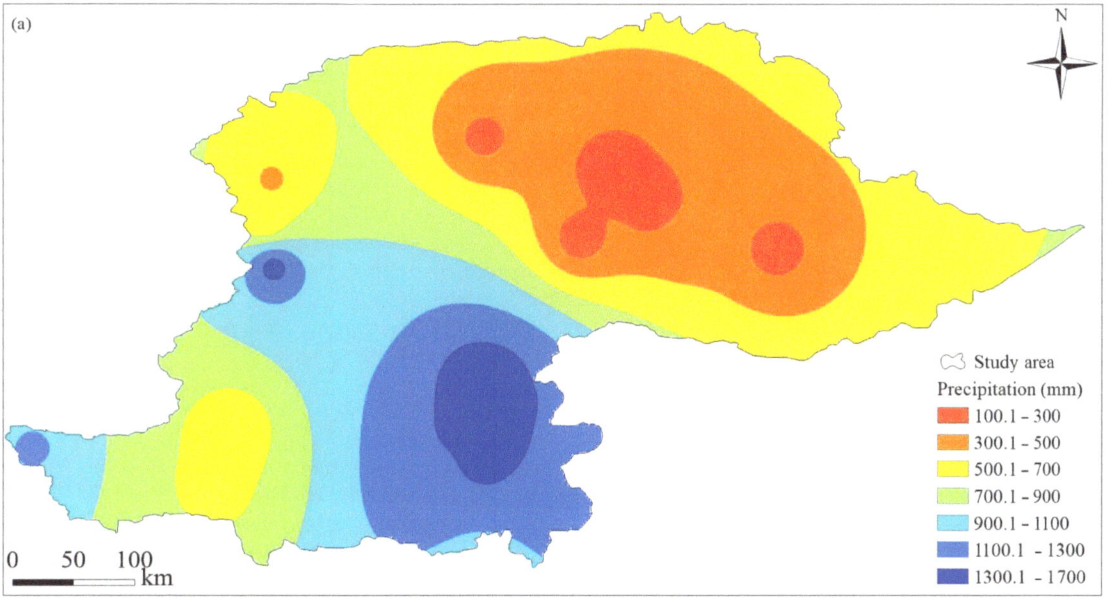

Figure 2. *Cont.*

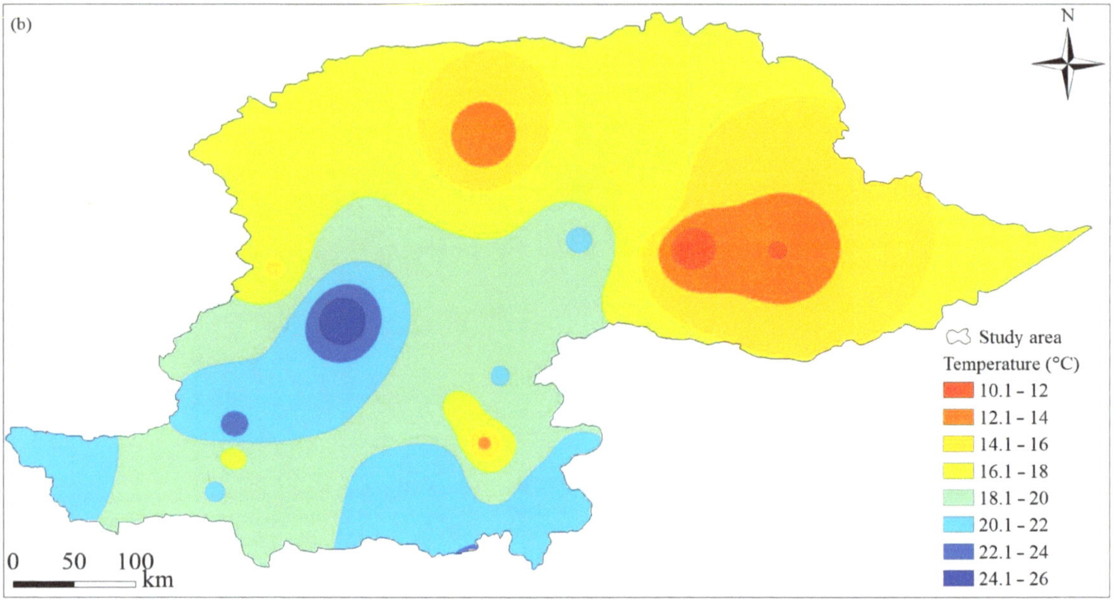

Figure 2. Spatial variability of annual average (**a**) precipitation and (**b**) temperature (during 1990–2020) in the study area.

2.2. Datasets

This research examines the spatiotemporal dynamics of extreme temperature (maximum and minimum) events over the last three decades (1991–2020). For the designed study period, daily data of maximum and minimum temperatures recorded at 24 weather stations across the northern mountainous domain of Pakistan were collected from Pakistan Meteorological Department (PMD). Annual means of maximum and minimum temperature were obtained from the daily data. Strict criteria were applied to the selection of station data for use in evaluating extreme climatic indices. The following criteria were followed in this study: (i) a monthly dataset was regarded as complete if there were no more than five missing days; (ii) a yearly dataset was considered as complete if all of its months were complete per previous criterion (i); (iii) a station's data was regarded as complete if it included no more than five missing years in its whole data. PMD previously guaranteed the consistency and accuracy of the data [31]. Moreover, RClimDex 1.0 software was used to ensure the quality of considered datasets. A threshold, three times of estimated standard deviation, was defined to identify the outliers in the time series data. The details of weather stations considered for this study are shown in Table 1.

2.3. Trends Analysis of Extreme Temperature Indices

The modified Mann–Kendall (MMK) test was used in this study to determine changes in the extremes of minimum and maximum temperatures in northern Pakistan. The MMK test is a non-parametric statistical technique widely used to identify trends in time series data. The test is an expansion of the original MK test and has the potential to be used to consider serial correlations as well as seasonal influences in the data. Hamed and Rao made modifications to the MK test in 1998 [32] by including the corrected variance of the Mann–Kendall statistics S.

Table 1. Details of considered weather stations.

Sr. No.	Weather Station	Longitude (dd)	Latitude (dd)	Altitude (m)	Average Annual Temperature (°C)	Average Annual Precipitation (mm)
1	Astore	74.90	35.37	2168	10.1	459.7
2	Bagh	74.09	33.78	2200	20.7	1165.6
3	Balakot	73.35	34.38	995.4	18.8	1475.0
4	Bunji	74.63	35.67	1470	17.7	160.7
5	Cherrat	71.55	33.81	1497.8	16.9	663.2
6	Chilas	74.10	35.42	1251	20.4	196.8
7	Chitral	71.83	35.85	1500	16.2	466.0
8	Dir	71.85	35.20	1375	15.7	1364.9
9	Drosh	71.78	35.57	1465	17.7	549.5
10	Garhi Dupatta	73.60	34.20	814	19.7	1339.0
11	Gilgit	74.33	35.92	1460	16.0	145.6
12	Gupis	73.40	36.17	2156	12.3	248.9
13	Islamabad	73.09	33.73	569	21.8	1262.5
14	Jhelum	73.37	32.94	287.2	23.9	862.1
15	Kakul	73.25	34.18	1309	16.9	1312.2
16	Kohat	71.43	33.57	513	20.1	616.3
17	Kotli	73.89	33.52	614	21.9	1183.7
18	Muzaffarabad	73.50	34.40	838	20.5	1377.7
19	Murree	73.40	33.91	2127	13.4	1694.6
20	Parachinar	70.08	33.86	1775	21.1	1114.8
21	Peshawar	71.56	34.02	360	23.1	530.3
22	Rawalpindi	73.04	33.59	540	21.7	1248.6
23	Saidu Shrif	72.35	34.82	970	26.3	1036.6
24	Skardu	75.54	35.34	2317	11.9	228.1

The significance of the result is calculated by comparing the actual value of the test statistic with its expected value under the null hypothesis of no trend. The MMK test is an effective technique for trend analysis, especially when the data are non-normal, or the trend is not linear. Since it does not need the estimation of any parameters, it is also beneficial when dealing with incomplete or missing data. In addition, the test may be used for both short-term and long-term datasets, which makes it a versatile and reliable tool for identifying trends in time series data. The following equations can be used to determine the value of S, the standardized test statistic Z, the variance V(S), and the corrected variance V*(S):

$$\mathrm{S} = \sum_{j=1}^{n-1} \sum_{k=j+1}^{n} sgn\left(x_k - x_j\right) \tag{1}$$

$$sgn\left(x_k - x_j\right) = \left\{ \begin{array}{ll} 1 & if\ x_k - x_j > 0 \\ 0 & if\ x_k - x_j = 0 \\ -1 & if\ x_k - x_j < 0 \end{array} \right\} \tag{2}$$

$$V(\mathrm{S}) = \left[n(n-1)(2n+5) - \sum_t t(t-1)(2t+5)/18\right] \tag{3}$$

$$V*(\mathrm{S}) = V(\mathrm{S})Cor \tag{4}$$

where

$$Cor = 1 + \frac{2}{n(n-1)(n-2)} \sum_{i=1}^{n-1} (n-1)(n-i-1)(n-i-2)\rho_s(i) \tag{5}$$

$$\mathrm{Z} = \left\{ \begin{array}{ll} \frac{S-1}{\sqrt{V(S)}} & if\ S > 0 \\ 0 & if\ S = 0 \\ \frac{S+1}{\sqrt{V(S)}} & if\ S < 0 \end{array} \right\} \tag{6}$$

where time series values are denoted by x_j and x_k, with k > j. n represents the total number of recordings, t represents the total length of time., and $\rho_s(i)$ is the statistically significant autocorrelation function of the ranks of recordings. Z values greater than zero indicate an upward trend, whereas values less than zero indicate a downward trend. The significance of a trend was assessed at 5% significance level. The Theil–Sen slope (TS) estimator, non-parametric method, was utilized to obtain a reliable estimate of the trend's slope. The following equation was used to determine the slope, as stated in the TS method:

$$T_i = \frac{a_j - a_k}{j - k} : for\ i = 1, 2, \ldots, N \tag{7}$$

where the sequential data values at j and k time steps (where $j > k$) are denoted by a_j and a_k, respectively. The median of N values of T_i (ordered series) gives the slope of TS slope estimator.

$$Q_i = \left\{ \begin{array}{ll} T_{\frac{N+1}{2}} & if\ N\ is\ odd \\ \frac{1}{2}\left(T_{\frac{N}{2}} + T_{\frac{(N+2)}{2}} \right) & if\ N\ is\ even \end{array} \right\} \tag{8}$$

Many different approaches have been developed to identify and classify temperature extremes. In this work, we employed a yearly examination of trends in indices of extremes in daily maximum (TMAX) and minimum (TMIN) temperatures to assess these shifts in extreme events. Spatial patterns were studied by examining them station-by-station and mapping the findings.

2.4. Extreme Temperature Indices

Table 2 explains the 13 minimum and maximum temperature indices defined by the "Expert Team on Climate Change Detection and Indices (ETCCDI)" working group. These indices can be found at http://etccdi.pacificclimate.org/list_27_indices.shtml (assessed on 2 January 2023) and were utilized in numerous studies of climatic extremes [33]. By following Alexander et al. [34], the extreme temperature indices were divided into three categories depending on the situation: (i) percentile, (ii) absolute, and (iii) duration-based indices.

Table 2. Details of the 13 minimum and maximum temperature indices used in this study.

Index	Definition	Symbol	Unit
	Duration Indices		
Cold spell duration	"Annual count of days with at least 6 consecutive days when TMIN < 10th percentile"	CSDI	Days
Warm spell duration index	"Annual count of days with at least six consecutive days when TMAX > 90th percentile"	WSDI	Days
	Percentile-Based Extreme Temperature Indices		
Cold night	"Annual count of days when TMIN < 10th percentile"	TN10p	Days
Hot night	"Annual count of days when TMIN > 90th percentile"	TN90p	Days
Cold days	"Annual count of days when TMAX < 10th percentile"	TX10p	Days
Hot days	"Annual count of days when TMAX > 90th percentile"	TX90p	Days
	Absolute Indices		
Coldest night	"Annual minimum value of TMIN"	TNn	°C
Hottest night	"Annual maximum value of TMIN"	TNx	°C
Coldest day	"Annual minimum value of TMAX"	TXn	°C
Hottest day	"Annual maximum value of TMAX"	TXx	°C
Maximum temperature means	"Annual mean daily maximum temperature"	TXmean	°C
Minimum temperature means	"Annual mean daily minimum temperature"	TNmean	°C
Diurnal temperature range	"Annual mean difference between TMAX and TMIN"	DTR	°C

The duration indices, in general, are used to characterize periods of extreme coldness (CSDI) and warmth (WSDI). The number of warmer or colder days than the highest and lowest long-term percentiles, respectively, were used to define the percentile-based indexes. These indices take into account the frequency with which cold days (TX10p), cold nights (TN10p), warm days (TX90p), and warm nights (TN90p) occurred. Maximum and minimum annual values were shown using absolute indices. The annual maximum of daily maximum temperature (TXx), annual maximum of daily minimum temperature (TNx), annual mean of maximum temperature (TX_{mean}), annual mean of minimum temperature (TN_{mean}), and daily temperature range (DTR) were all part of this dataset.

In this study, spatial interpolation was performed using the Kriging Interpolation method. Kriging is a geostatistical method that estimates values at sites that were not directly measured by taking into account the spatial correlation between data points.

3. Results

3.1. Temporal Variations in Extreme Temperature Indices

Figure 3 shows the temporal variations in the areal averaged hot extremes in the northern highlands of Pakistan during 1991–2020. The results indicate that, during the last 30 years, two hot indices (TXmean and TX90p) showed increasing tendencies at the rates of 0.14 °C/decade and 0.08 d/decade, respectively; however, the increasing tendencies were not statistically significant (at the 5% significance level). In contrast, all the other hot indices (WSDI, TN90p, TNx, and TXx) showed decreasing tendencies (at the rates of −0.09 d/decade, −1.82 d/decade, −0.35 °C/decade, and −0.48 °C/decade, respectively) over the past 30 years. Among all the hot indices, only one index (TNx) showed a significantly decreasing trend ($p < 0.05$). The decreasing trend of TNx during the past three decades indicated that the average temperature of the hottest night in the northern highlands of Pakistan was significantly decreased. Generally, the decadal average values of the hot indices were highest in the second decade (2001–2010), except for the TNx index, which showed its highest average in the 1990s and lowest in the 2010s. Overall, the decadal average values of all the hot indices consistently decreased during the past two decades. Table 3 shows the rates of the per decade changes in the hot and cold indices. The results indicated that all hot indices, except TXx, increased in the first decade (1990s). Two indices (TXmean and TX90p) showed significantly increasing trends ($p < 0.05$) at the rates of 1.70 °C/decade and 13.81 d/decade, respectively. However, no increasing trend in the hot indices was found in the second decade (2000s).

Figure 4 shows the linear trends in the cold indices over the past three decades (1991–2020). It was found that five cold indices (TNmean, CSDI, TN10p, TX10p, and TNn) decreased over the past three decades at the rates of −0.02 °C/decade, −0.08 d/decade, −1.48 d/decade, and −0.04 °C/decade. The decreasing trend of TX10p was statistically significant ($p < 0.05$), which showed that the frequency of annual average cold days was decreased in the NHP. On the other hand, two cold indices (TXn and DTR) increased at the rates of 0.50 °C/decade and 0.19 °C/decade, respectively, during the period of 1991–2020. The increasing rate of DTR was statistically significant ($p < 0.05$). The decadal average values of all cold indices, except for TXn, consistently decreased over the past two decades (2000s and 2010s). A trend analysis of the decadal times series of the cold indices indicated that the TNmean index significantly increased in the first decade (1990s), whereas it showed increasing and decreasing tendencies in the second and third decades (2000s and 2010s), respectively. However, the trends in the second and third decades were not significant at the 5% significance level. In the first decade, the CSDI index decreased at the highest rate (−13.33 d/decade), followed by TX10p and TN10p at the rates of −9.65 d/decade and −8.28 d/decade, respectively. However, these decreasing rates were not significant ($p > 0.05$). Conversely, CSDI showed the highest increasing rate in the second decade, at the rate of 16.25 d/decade. The DTR index increased during the first and second decades. Therefore, the overall increasing trend of the DTR index during the period of 1991–2020 might be due to the increasing tendency of this index during the 1990s and 2000s. The TX10p consistently decreased during the 1990s and 2000s but increased in the 2010s. The overall decreasing trend of TX10p during 1991–2020 might be due the decreasing trend of this index during the first and second decades of the study's duration.

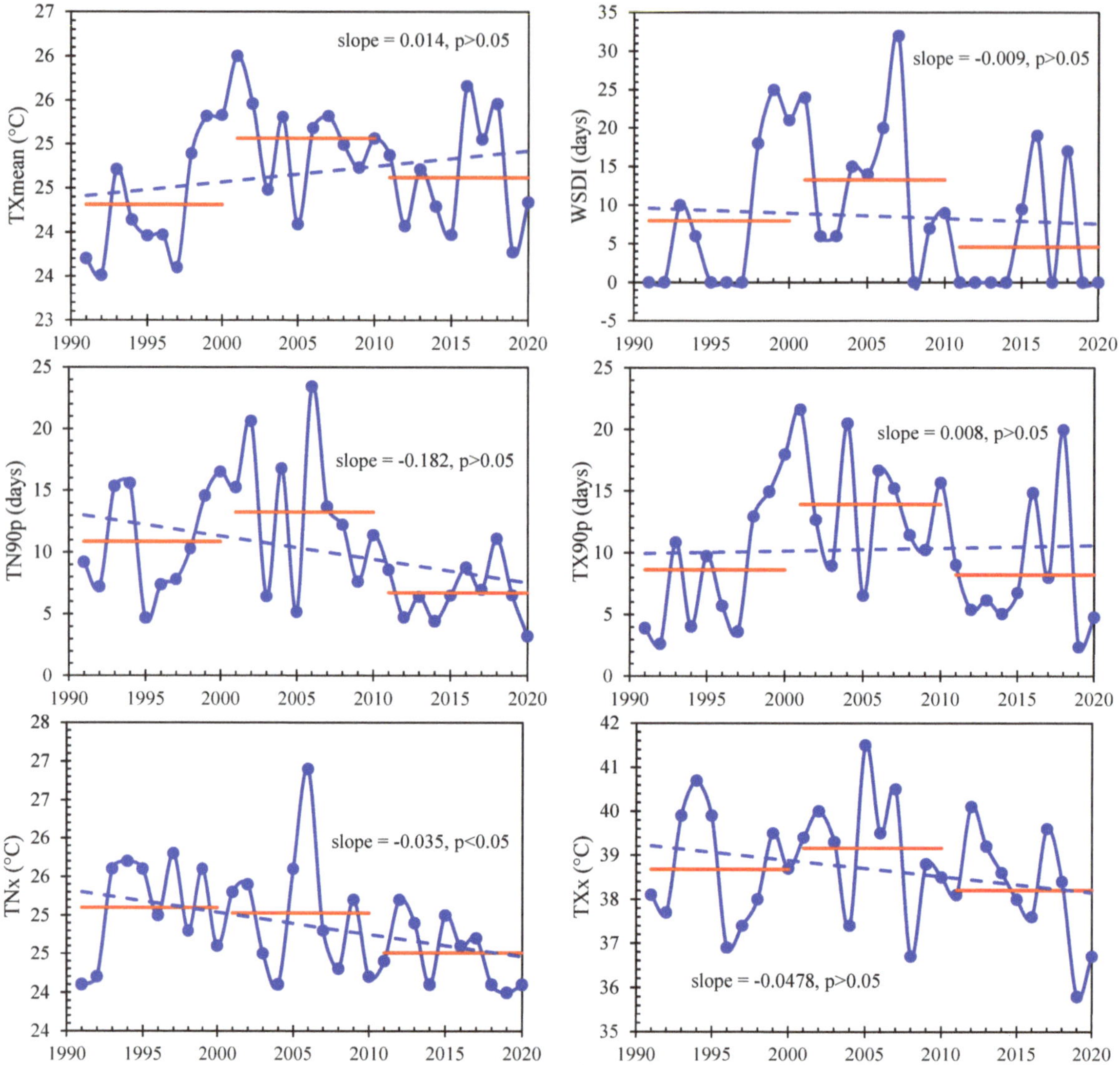

Figure 3. Linear trends in the annual average time series of hot indices. The blue dashed line represents the linear trend, while the solid red line represents the indices' decadal means.

Table 4 presents the MMK test results of six hot indices measured at 24 weather stations in northern Pakistan. The frequency and severity of extreme events and climatic shifts can be measured with these indexes.

The first index, WSDI, assesses how long warm spells last. A high value of the MMK test for the WSDI indicates an increase in the number of days with warm spells. Significantly increasing warm spells were observed at two stations, Cherat and Garhi Dupatta, as shown by their high WSDI values. The values of the MMK test of the WSDI for all other stations were non-significant at the 5% significance level.

The second index, TX90p, is the number of days for which the high temperature was at least the 90th percentile. More intense heat waves can be expected at higher TX90p levels. Five stations (Balakot, Cherat, Dir, Garhi Dupatta, and Kakul) showed significantly high values of TX90p, which indicates that they witnessed an increase in the frequency of hot days. However, two stations (Para Chinar and Skardu) exhibited a significantly decreasing trend of TX90p. Compared to the other 24 stations, the value of the MMK test for TX90p at the Cherat station was the highest (Z = 4.05), indicating that it observed the highest number of hot days. On the other hand, the value of the MMK test for the TX90p index at the Skardu (Z = −2.62) station showed that it experienced a significant decrease in the frequency of hot days.

Table 3. Rate of change in extreme temperature indices on decadal scale in the northern highlands of Pakistan.

Index	1991–2000	2001–2010	2011–2020
	Hot Indices		
TXmean (°C)	***1.70***	−0.78	−0.33
WSDI (days)	23.33	−8.00	0.00
TN90p (days)	5.15	−7.73	0.22
TX90p (days)	***13.81***	−6.61	−0.78
TNx (°C)	0.50	−0.50	−1.00
TXx (°C)	−0.14	−1.25	−3.17
	Cold Indices		
TNmean (°C)	***0.85***	−0.75	0.32
CSDI (days)	−13.33	16.25	0.00
TN10p (days)	−8.28	6.05	−4.60
TX10p (days)	−9.65	−0.33	1.50
TNn (°C)	0.40	−1.57	0.75
TXn (°C)	0.29	1.00	0.00
DTR (°C)	1.08	0.08	−0.11

The ***bold (italics)*** values show statistically significant trends.

The third index is referred to as TN90p, and it determines the percentage of days with minimum temperatures higher than the 90th percentile. High values of the MMK test for the TN90p indicate an increase in the frequency of exceptionally warm nighttime temperatures. However, when the MMK test value of TN90p is negative, it indicates a decrease in the frequency of hot nights. Most stations showed a decline in the frequency of TN90p, as indicated by the negative MMK test values of TN90p. Among the 24 stations, the frequency of hot nights significantly increased only at Bunji (Z = 2.84), while it significantly decreased at Drosh (Z = −4.30), Skardu (Z = −3.13), Balakot (Z = −2.87), Gahri Dupitta (Z = −2.55), Muzaffarabad (Z = −2.53), Kakul (Z = −2.45), Kohat (Z = −2.40), Chilas (Z = −2.14), and Dir (Z = −2.12).

TXx, the hottest day of the year, is the fourth temperature index. An increase in heatwave intensity is reflected by higher TXx values. The value of the MMK test for the TXx index was the highest at the Bunji station, which suggests that the location was subjected to more severe heat waves. The majority of stations, however, reported a declining TXx trend.

The fifth index, TNx, represents the hottest night in a year. Higher TNx levels indicate that warm nights are becoming more intense. In contrast, cooler nights are forecast when TNx readings are negative. In the NHP, most of the stations showed a decreasing tendency of the hottest nights. A significantly decreasing trend of TNx was found at Bagh, Bala Kot, Drosh, Gahri Dupitta, Kakul, Kohat, and Muzaffarabad. However, the MMK test values showed a significantly increasing trend of the hottest night at two locations (Astore and Skardu, with Z values of 2.32 and 2.00, respectively).

TXmean, a measure of the mean daily maximum temperature, is the sixth index. It was found that TXmean was significantly increased at six locations including Bala Kot (Z = 3.64), Cherat (Z = 4.66), Chilas (Z = 2.04), Gahri Dupitta (Z = 2.67), Kakul (Z = 2.41), and Kotli (Z = 2.33). However, Txmean was significantly decreased at Peshawar, as indicated by the lowest value of Z (−2.19).

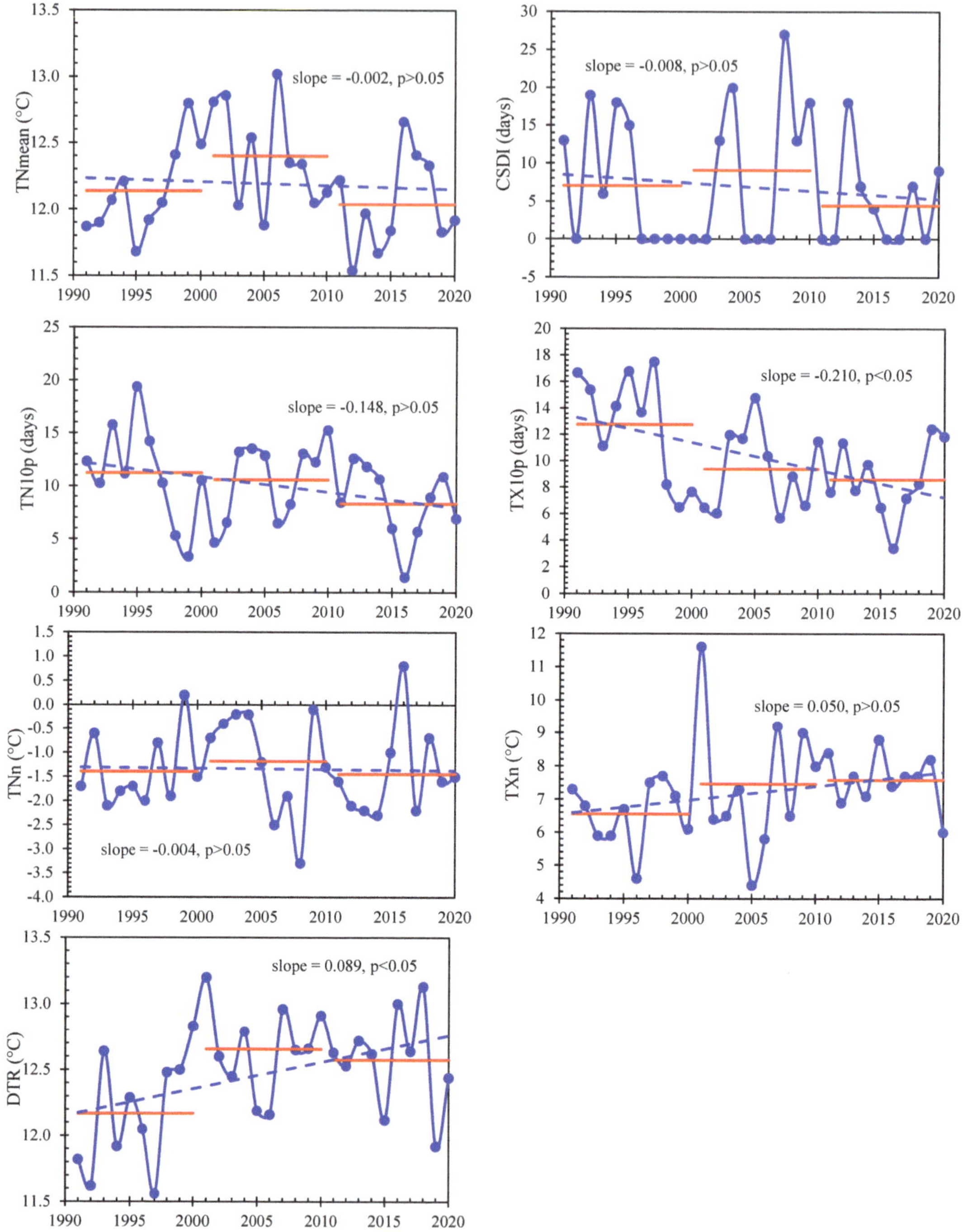

Figure 4. Linear trends in the annual average time series of maximum and minimum temperature. The linear trends are represented by dashed lines (blue color), while solid red lines represent the decadal means of indices.

Table 4. MMK test statistics values for the extreme hot indices. ***Bold values*** indicate a statistically significant trend.

Sr. No.	Station	WSDI	TX90p	TN90P	TXx	TNx	TX_{mean}
1	Astore	1.89	1.50	−0.07	1.69	***2.32***	1.84
2	Bagh	1.50	0.80	−1.94	−1.16	***−2.25***	1.55
3	Bala Kot	1.87	***3.37***	***−2.87***	−0.39	***−2.25***	***3.64***
4	Bunji	0.90	1.17	***2.84***	***2.91***	1.20	1.26
5	Cherat	***3.01***	***4.05***	−1.74	1.72	−1.58	***4.66***
6	Chilas	1.00	0.97	***−2.14***	−0.39	−0.66	***2.04***
7	Chitral	−0.06	0.54	0.56	−0.66	−0.64	0.83
8	Dir	1.36	***2.01***	***−2.12***	−0.23	−1.85	1.80
9	Drosh	0.60	−1.17	***−4.30***	***−3.06***	***−4.29***	0.54
10	Garhi Dupatta	***2.10***	***3.38***	***−2.55***	−0.05	***−3.13***	***2.67***
11	Gilgit	−0.29	−0.99	1.43	***−2.39***	−0.39	−0.51
12	Gupis	−0.44	0.31	−1.72	***−3.47***	−1.37	1.43
13	Islamabad	−0.16	−0.68	−0.68	−0.78	−0.32	0.27
14	Jhelum	0.00	1.63	0.17	−0.26	−1.16	1.07
15	Kakul	1.47	***2.92***	***−2.45***	0.87	***−2.10***	***2.41***
16	Kohat	−0.32	−0.99	***−2.40***	***−2.27***	***−2.10***	−0.43
17	Kotli	−1.01	0.78	1.77	−0.32	1.72	***2.33***
18	Muzaffarabad	−0.36	−0.48	***−2.53***	−1.81	***−2.28***	−0.46
19	Murree	0.44	1.31	−0.08	−0.27	−1.23	1.00
20	Para Chinar	0.43	***−1.97***	−1.90	0.26	0.15	1.00
21	Peshawar	−0.93	0.99	0.99	***−2.83***	−1.71	***−2.19***
22	Rawalpindi	−0.59	−0.07	−1.09	−0.82	−0.73	−0.25
23	Saidu Sharif	−0.28	0.14	0.27	0.07	0.03	0.00
24	Skardu	0.00	***−2.62***	***−3.13***	***−2.23***	***2.00***	1.39

Table 4 demonstrates how extreme weather events changed at different sites throughout northern Pakistan. There was an increase in the frequency of hot days and heat waves at some stations and a decrease in the frequency of hot nights at others. These shifts can significantly impact the environment, human health, and agricultural practices.

Table 5 displays the MMK test statistics for cold extremes. According to the results of the MMK test, the CSDI was significantly increased at three locations (Dir, Drosh, and Para Chinar), with Z values of 2.22, 2.22, and 2.59, respectively. These locations experienced an increase in the frequency of cold spells. In contrast, cold spells were significantly decreased at three other locations (Gilgit, Jhelum, and Kotli), and the values of Z at these locations were −2.34, −3.13, and −2.42, respectively. This finding indicated that these locations experienced fewer cold spells than the other locations.

The TX10p index measures the number of days per year when the location's annual maximum temperature is at or below the temperature's 10th percentile. According to the results of the MMK test, the TX10p index was decreased at most of the stations, indicating that the incidence of cold days was decreasing at these locations. However, there were a few locations where the value of the TX10p index was positive, suggesting an increase in the frequency of cold weather. Overall, TX10p and TN10p showed a decreasing trend, as indicated by the negative values at the majority of the stations.

Generally, the statistical values of the MMK test for the TXn index showed an increasing trend. This increasing trend was statistically significant at Astore, Bala Kot, Kotli, Parachinar, and Skardu. Several locations showed a decreasing trend in the TNn. Significant negative trends of TNn were found for Bagh, Chilas, Drosh, Garhi Dupatta, Islamabad, Jhelum, and Muzaffarabad. On the other hand, some stations (Para Chinar, Dir, and Murree) showed a significantly increasing trend of TNn (with values of $Z > 1.96$), indicating a significant increase in the frequency of the coldest nights. The remaining stations indicated an insignificant trend in the frequency of the coldest nights (TNn).

Table 5. MMK test statistics values for the extreme cold indices along with DTR index. ***Bold values*** indicate a statistically significant trend.

Sr. No.	Station	CSDI	TX10p	TN10P	TXn	TNn	TN_{mean}	DTR
1	Astore	−0.95	−1.38	−0.65	***2.43***	0.60	1.09	1.89
2	Bagh	−1.32	***−2.40***	−0.95	1.94	***−2.88***	−0.12	***2.67***
3	Bala Kot	0.37	***−3.20***	0.83	***2.21***	−1.38	***−2.76***	***4.49***
4	Bunji	−1.09	−0.87	***−4.08***	1.62	0.55	***3.65***	***−2.70***
5	Cherat	0.00	***−4.39***	0.29	1.87	1.52	−0.85	***3.83***
6	Chilas	1.65	***−2.63***	1.24	1.34	***−1.98***	−0.85	***4.50***
7	Chitral	0.00	−0.88	***−2.62***	−0.58	0.42	***2.07***	−0.09
8	Dir	***2.22***	***−2.21***	−0.85	1.89	***2.22***	−0.26	1.94
9	Drosh	***2.22***	−1.90	***3.50***	0.14	***−2.04***	***−3.71***	***5.32***
10	Garhi Dupatta	1.38	***−2.26***	0.08	1.02	***−3.24***	−1.14	***3.61***
11	Gilgit	***−2.34***	−0.22	***−4.08***	1.82	0.92	***3.49***	***−2.58***
12	Gupis	−1.08	***−1.97***	***−1.99***	1.30	0.32	0.24	1.24
13	Islamabad	0.94	−0.61	1.33	0.84	***−2.67***	−0.78	0.85
14	Jhelum	***−3.13***	−0.68	−1.56	−0.87	***−2.18***	0.46	1.72
15	Kakul	−0.48	***−2.21***	−1.36	1.18	0.00	−1.11	***2.69***
16	Kohat	−0.48	−0.85	−1.36	0.27	0.00	−1.09	−0.14
17	Kotli	***−2.42***	***−2.53***	***−3.43***	***2.15***	−0.26	1.95	1.05
18	Muzaffarabad	0.70	−0.99	1.63	0.31	***−3.14***	1.24	***2.50***
19	Murree	−0.84	−1.19	***−3.50***	1.46	***2.08***	***−1.99***	−1.10
20	Para Chinar	***2.59***	0.54	0.58	***2.64***	***3.23***	***−2.16***	***3.83***
21	Peshawar	−1.24	0.03	−1.22	−0.96	−0.76	1.00	−0.85
22	Rawalpindi	−0.18	−0.12	0.07	0.90	−1.00	−0.73	1.38
23	Saidu Sharif	1.19	***−2.11***	***−2.55***	0.27	0.12	1.46	***3.94***
24	Skardu	−1.02	−0.73	0.73	***2.37***	0.50	***−2.40***	1.89

Overall, TN_{mean} showed a decreasing tendency over the past 30 years. A significantly decreased trend of TN_{mean} was found at five stations (Bala Kot, Drosh, Murree, Para Chinar, and Skardu). Conversely, the trend of TN_{mean} was statistically significant at three locations (Bunji, Chitral, and Gilgit). The findings of the MMK test of DTR showed that it was significantly increasing at 10 stations. In contrast, two stations (Bunji and Gilgit) showed significantly decreasing trends in DTR. The trend of DTR was not significant at 12 locations.

To check the changes in temperature extremes for the time span between 1991 and 2020, the spatial patterns of the trend for the past 30 years' set of indicators were investigated.

3.2. Spatial Changes in Duration-Based Extreme Temperature Indices

Figure 5 displays the spatial distributions of the magnitudes of the annual average changes in the CSDI and WSDI indices in northern Pakistan. The spatial distribution of the trend of the CSDI index showed that the total number of cold days was slightly decreasing over the whole study region, and the rate of decrease was relatively higher in the northern parts compared to the southern parts. According to the spatial distribution of the CSDI index values, most of the parts of the study area, particularly the northern parts, were subjected to fewer cold spells. In contrast, the other regions, particularly the southern regions, were subjected to a decrease. The spatial variability map of the trend of the WSDI index showed that the frequency of warm days increased in most parts of the study area. Particularly, the increasing rate was more prominent in the central part of the study domain, as exhibited by the spatial variability map of the WSDI index. The spatial distribution of the WSDI index values suggested that some regions in the area are more vulnerable to heat stress than others.

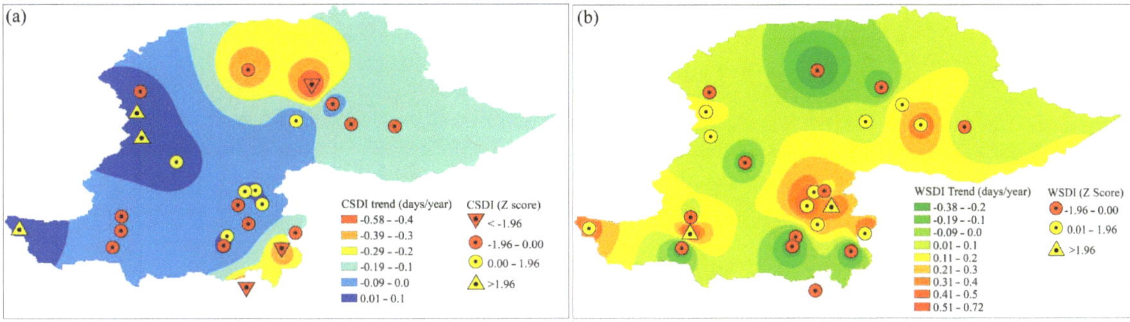

Figure 5. Spatial distribution of the magnitude of trends in duration-based extreme temperature indices ((**a**) CSDI and (**b**) WSDI) in northern Pakistan.

3.3. Spatial Changes in Percentile-Based Extreme Temperature Indices

By plotting the rate of change in TN10p values across northern Pakistan (Figure 6a), it was noticed that most stations exhibited a negative rate of change, suggesting a declining trend in the occurrence of cold nights over the last 30 years. This pattern was most pronounced in northern locations like Gilgit, Gupis, and Kotli. However, few locations indicated a positive rate of change, suggesting an increasing trend in the frequency of cold nights. The regional distribution of the rate of change in TN10p values indicated a general trend toward fewer cold nights across the study domain. However, a significant increase in the number of cold nights (indicated by a positive index value ($p < 0.05$)) was found at the Para Chinar station. There was no clear pattern in the increasing or decreasing frequency of cold nights at southern and central stations such as Islamabad, Jhelum, Kohat, Kotli, M. Abad, Murree, Peshawar, Rawalpindi, and Saidu Sharif. At these locations, a decline in the index value indicated a corresponding decline in the trend of cold nights, while an increase in the index value indicated an equivalent increase in the trend of cold nights. It was deduced that the frequency of cold nights was decreased in the northern areas, while mixed patterns were found in the central and southern areas.

The regional distribution of the TN90p index values revealed considerable variation in the occurrence of extremely hot nights at various locations. Over the reference period, the frequency of extremely hot temperatures increased at some locations. The majority of the sites with positive TN90p values, such as Bunji, Gilgit, and Kotli, were located in the northern parts of the NHP. The highest positive value of the MMK test results ($p < 0.05$) was found at the Bunji station, where the increasing rate was 0.30 days/year. The southern parts of the study area, including Islamabad, Rawalpindi, and Murree, were experiencing a decreasing trend for TN90p. The highest decreasing rate (0.64 days/year) of the TN90p index was found at the Drosh station in the northwestern part of the NHP. The spatial distribution of the TN90p index values suggested that the annual percentage of hot nights was increasing in the northeastern parts of the study domain, while in most of the other parts of the NHP.

The frequency of cold days (TX10p) was decreased at most of the stations in the entire study area, as reflected by the spatial distribution of the values of the MMK test and the TS estimator for the TX10p index (Figure 6c). Most of the stations in the study area showed positive values for the MMK test and the TS estimator for the TX90p index, suggesting a rise in the frequency of hot days in the NHP, as seen by the spatial distribution of the index. However, the results of the MMK test showed that the TX90p index was significantly decreasing at two locations, indicating that hot days were becoming less frequent in the northeastern and southwestern parts of the NHP. Some stations in the southern parts of the study area showed negative values for the index, showing a similar trend of decreasing extremely hot days.

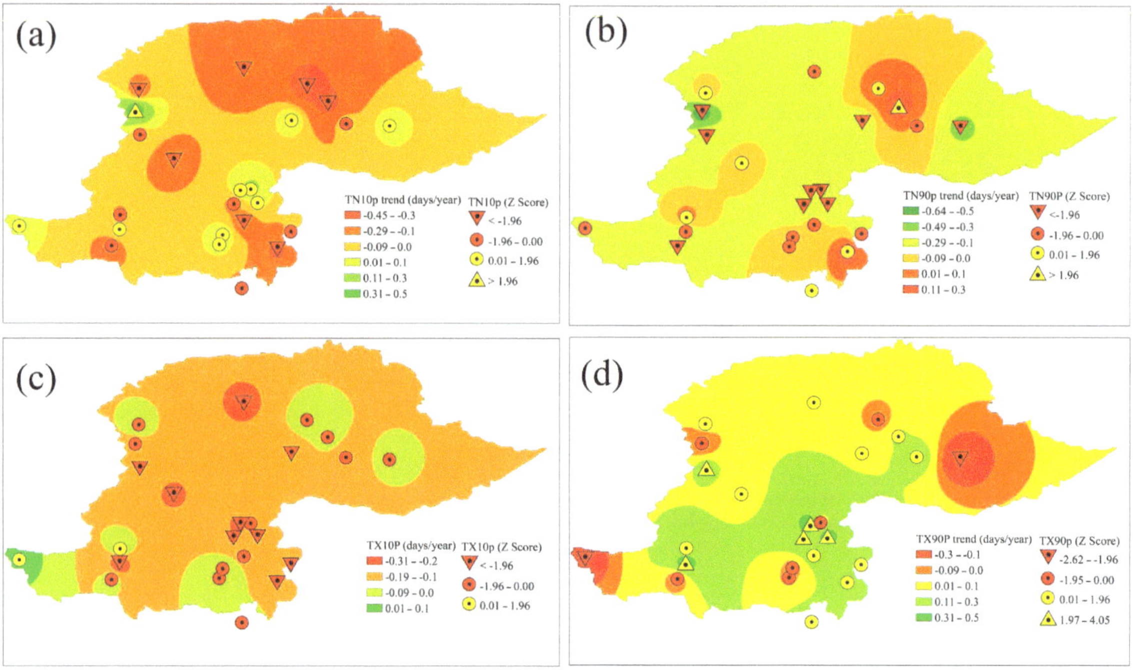

Figure 6. Spatial variability of percentile-based extreme temperature indices including (**a**) TN10p, (**b**) TN90p, (**c**) TX10p, and (**d**) TX90p.

3.4. Spatial Changes in Absolute Extreme Temperature Indices

Figure 7 explains the spatial distributions of the magnitude of changes in six absolute indices in northern Pakistan. The spatial distribution of the trend rate in the TNn index (Figure 7a) showed that most stations had negative values, suggesting a declining trend in the coldest nights in northern Pakistan. Most of the negative values (at the rate of −0.09 °C to 0.0 °C) were found in the northern sites, indicating a decreasing minimum (nighttime) temperature in that part. The opposite was true for some of the stations in the southwestern parts, where positive values indicated that the minimum temperature was rising. The spatial distribution of the rate of TNx (Figure 7b) trend revealed that the annual average temperature of the hot nights was generally decreasing. The maximum decreasing rate was −0.18 °C/year.

The spatial distribution of the annual minimum value of the maximum temperature (TXn) was increasing at most of the locations (Figure 7c). Conversely, the annual maximum value of the maximum temperature (TXx) was decreasing at most of the locations in the study area, as indicated by the spatial distribution of the TXx index (Figure 7d). The annual average of the minimum temperature index (TNmean) was decreasing at most the stations, with the maximum decreasing rate of −0.12 °C/year. In contrast, the annual average of the maximum temperature index (TXmean) was increasing at most the stations, with maximum increasing rate of 0.1 °C/year.

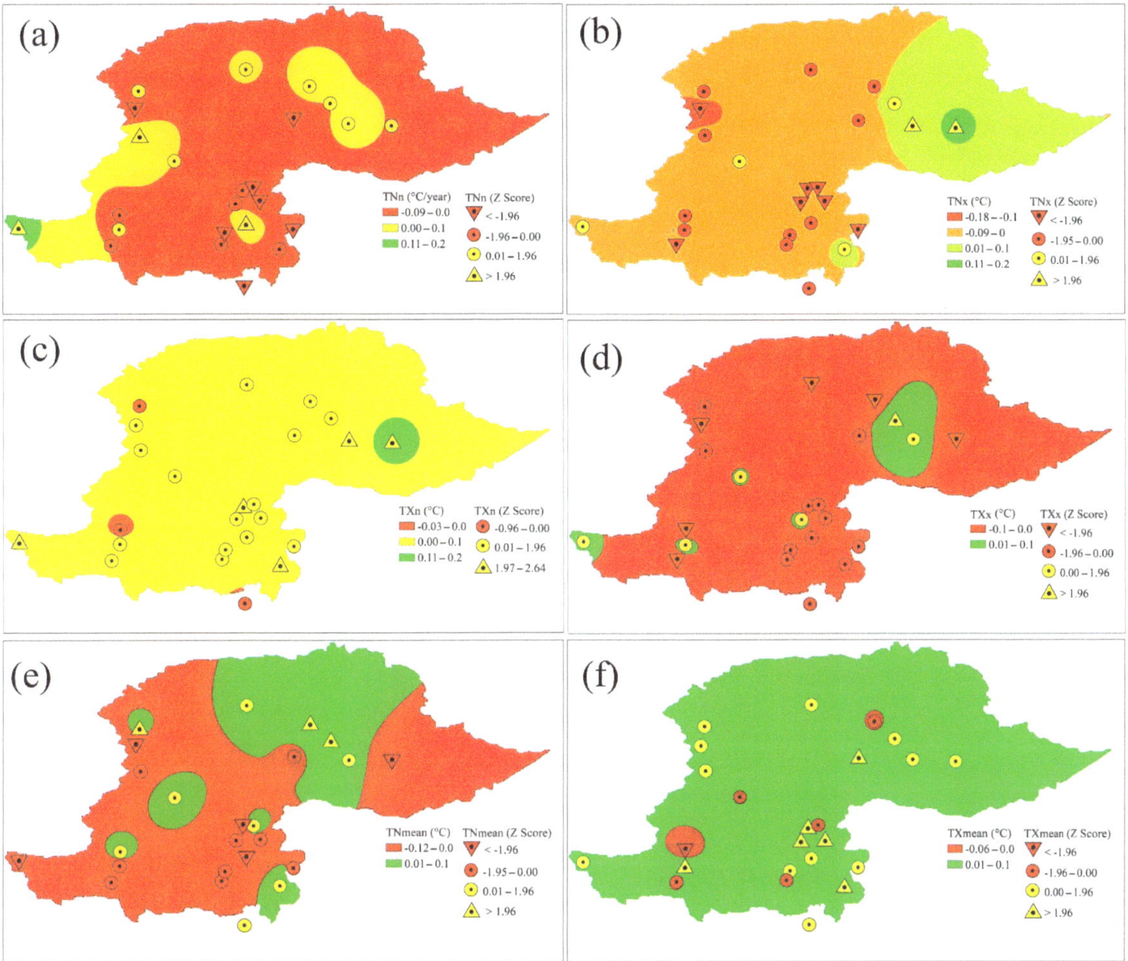

Figure 7. Spatial distributions of (**a**) TNn, (**b**)TNx, (**c**) TXn, (**d**) TXx, (**e**) TMINmean, and (**f**) TMAXmean.

4. Discussion

We analyzed the temporal and spatial evolution of extreme climate events in the northern highlands of Pakistan during 1991–2020 by selecting the indices of extreme temperature. Our study built on previous research conducted in different geographical areas around the world, which documented a rise in both the frequency and severity of the events of hot days [8,35–37]. By analyzing these indices, we found that the northern highlands of Pakistan have experienced an increase in hot days over the past 30 years. This finding was consistent with the results of Saleem et al. [27]. The results of our investigation exhibited that the daily maximum temperature increased over the NHP during the past three decades, which was in line with the findings of Hussain et al. [37] in the Karakoram region of Pakistan. Another study by Syed et al. [38] analyzed the trends in extreme temperatures in Pakistan and reported an increase in the frequency and severity of heatwaves. Our findings were consistent with their results, as we also found an increase in the frequency of extremely hot days at some stations.

Although two hot indices (hot days and the mean of the daily maximum temperature) were increasing over the past three decades, the other hot indices (hot nights, the hottest nights, the hottest days, and the warm spell duration index) were decreasing over the same study duration in the NHP. This can be explained by the distinct dynamics that mountain climates exhibit, as well as the complex interactions that take place between atmospheric and geographical aspects. The terrain and altitudes of mountainous areas are distinctive. Variations in the warming tendencies between the day and night can be caused by an elevation effect on the temperature patterns. While global warming may cause an increase in the average daily maximum temperature, the nighttime cooling impact of higher elevations may lead to a reduced number of annual average hottest days. Radiative cooling may be another potential explanation for this pattern. Radiative cooling at night increases in the mountains due to the clear sky and low humidity [39]. Therefore, it is possible for nighttime surface and air temperatures to dramatically drop, mitigating the general warming trend and resulting in a reduced number of the hottest nights.

Despite the fact that the coldest days were increasing over the past three decades in the NHP, it was found that the coldest nights, cold nights, cold days, and average of daily minimum temperature decreased over the same study period. The results of the MMK test showed that several stations, such as Garhi Dupatta, Muzaffarabad, and Islamabad, experienced a decreasing trend in the annual minimum temperature. These findings were in line with those found in an earlier study carried out by Yaseen et al. [40], which showed that the minimum temperature has decreased in northern Pakistan over the past few decades. This decreasing trend of the minimum temperature, cold nights, cold days, and cold spell duration might be due to a combination of general climate trends and mountain-specific mechanisms. The overall cooling tendencies and increased temperature variability associated with higher elevations might be the reasons for the increase in the coldest days. In addition, high-pressure systems and atmospheric blockage, both of which occur on a continental scale, can have an impact on temperature extremes in mountainous regions. The spatial variations and strength of these systems can result in changes in the air masses and the temperature characteristics that are connected with these air masses. The length of cold spells, the coldest days, and the number of cold nights may have all decreased over the previous two decades due to changes in circulation patterns. A probable additional factor in the recent downward trend of cold nights, cold days, and cold spell duration is the diminishing snow cover in the northern parts of Pakistan [11]. During the winter months, considerable snow cover is prevalent in mountainous areas. The local energy balance and temperature profiles can be affected by changes in snow cover and surface albedo (reflectivity). Reduced snow cover or altered snow qualities, like a lower albedo from dust deposition, can increase surface heating and decrease the length of cold nights, cold days, and cold spells.

This study is important because it applies the cold and hot matrices to a particular highland regional setting. As a result, it provides distinctive insights into the processes that govern the occurrence of severe temperatures. It is important to be aware of the study's limitation, which is the relatively short 30-year period record of temperature. The reported rise in the frequency of hot days and cold spell patterns is the practical ramifications. Despite noting this limitation, which includes the use of a case study approach and the limitation imposed by the length of the record, this research emphasizes its importance in highlighting the formation of strategies for climate adaptation and resilience. This study emphasizes the need for constant monitoring and well-informed decision making in the local context and similar situations.

5. Conclusions

The following findings were drawn from the investigation of temperature extremes in northern Pakistan over the period of the last three decades (1991–2020):

- The number of hot days significantly increased in the study domain. Therefore, it was deduced that the hot weather events in northern Pakistan have become more frequent and severe in the past 30 years.

- The annual maximum value of minimum temperature was significantly decreased in the study area during the past three decades.
- The difference between the maximum temperature and minimum temperature was significantly increased in the northern highlands of Pakistan during 1991–2020.
- Overall, a decreasing trend in the annual average number of cold days and cold nights was detected in the NHP.
- Only two hot indices (TXmean and TX90p) showed increasing tendencies, although their trends were not statistically significant.
- The decadal averages of all the hot indices, except for TNx, were highest in the second decade (2001–2010) and consistently lowest in the last decade (2011–2020).
- Most of the stations installed in the study domain demonstrated a decreasing trend in the level of heatwave intensity, which is reflected by the TXx index.
- The duration and frequency of cold spells increased at the Dir, Drosh, and Para Chinar locations but decreased in the central and southern parts of the study domain. These changes might be attributed to changes in large-scale atmospheric circulation patterns.
- Most of the locations in the NHP indicated an increase in the number of days with below-average temperatures. Nevertheless, there were indications of a decrease in cold weather in some areas.

Overall, this study shows significant changes in the extreme weather patterns in northern Pakistan, including longer periods of cold weather in some regions and an increase in both the severity and frequency of hot extreme weather events. The results of this investigation have important implications for regional disaster risk reduction strategies, particularly for the China Pakistan Economic Corridor (CPEC). The trends observed in the extreme temperature indices emphasize the critical importance of taking immediate action to adapt to and mitigate the effects of climate change. The combination of the slightly increasing trend in the annual maximum temperature and the downward trend in the annual minimum temperature may have far-reaching effects on a variety of sectors, including food production, human health, and the environment. In addition, an increase in the number of days with extremely high temperatures and a decrease in the number of nights with extremely high temperatures emphasize the significance of developing effective strategies to maintain resilience and sustainability in the face of shifting climatic trends. In light of these results, we urge decision makers and relevant parties in northern Pakistan to act decisively to mitigate the effects of climate change.

Author Contributions: Conceptualization, A.A.J.G., M.N.A., G.R. and S.; data curation, M.N.A., M.I., M.A. and U.M.N.; formal analysis, M.N.A. and G.R.; funding acquisition, A.A.J.G., M.I., M.A. and S.R.; investigation, M.N.A.; methodology, M.N.A., G.R. and S.; project administration, A.A.J.G., M.N.A., M.I. and M.A.; resources, S.R. and U.M.N.; software, M.N.A. and G.R.; supervision, M.N.A. and A.A.J.G.; validation, G.R. and S.; visualization, A.A.J.G., M.I., M.A. and S.R.; writing—original draft, M.N.A.; writing—review and editing, A.A.J.G. and M.N.A. All authors have read and agreed to the published version of the manuscript.

Funding: This research was funded by the Institutional Funding Committee at Najran University, the Kingdom of Saudi Arabia.

Institutional Review Board Statement: Not applicable.

Informed Consent Statement: All authors have consented to the acknowledgment.

Data Availability Statement: Not applicable.

Acknowledgments: The authors would like to acknowledge the support of this research by the Deputy for Research and Innovation Ministry of Education, the Kingdom of Saudi Arabia, through grant number (NU/IFC/2//SERC/-/2) under the Institutional Funding Committee at Najran University, the Kingdom of Saudi Arabia.

Conflicts of Interest: The authors declare no conflict of interest.

References

1. Intergovernmental Panel on Climate Change. *Climate Change 2021: The Physical Science Basis: Summary for Policymakers Working Group I Contribution to the Sixth Assessment Report of the Intergovernmental Panel on Climate Change*; IPCC: Geneva, Switzerland, 2021; ISBN 9789291691586.
2. Ghanim, A.A.J.; Anjum, M.N.; Rasool, G.; Saifullah; Irfan, M.; Rahman, S.; Mursal, S.N.F.; Niazi, U.M. Assessing Spatiotemporal Trends of Total and Extreme Precipitation in a Subtropical Highland Region: A Climate Perspective. *PLoS ONE* **2023**, *18*, e0289570. [CrossRef] [PubMed]
3. Abbas, S.; Kousar, S.; Khan, M.S. The Role of Climate Change in Food Security; Empirical Evidence over Punjab Regions, Pakistan. *Environ. Sci. Pollut. Res.* **2022**, *29*, 53718–53736. [CrossRef] [PubMed]
4. Goddard, L.; Gershunov, A. *Guidelines on Analysis of Extremes in a Changing Climate in Support of Informed Decisions for Adaptation;* World Meteorological Organization: Geneva, Switzerland, 2009; WMO-TD No. 1–55; Available online: https://library.wmo.int/index.php?lvl=notice_display&id=138 (accessed on 2 January 2023).
5. Anjum, M.N.; Ghanim, A.A.J.; Farid, H.U.; Zaman, M.; Niazi, U.M.; Rahman, S.; Alsaiari, M.A.; Irfan, M. Assessment of Climate Change and Its Impacts on the Flows of a Subtropical River Basin in the Hindu-Kush Mountain, South Asia. *Pure Appl. Geophys.* **2022**, *179*, 3841–3857. [CrossRef]
6. Tahir, A.A.; Chevallier, P.; Arnaud, Y.; Ashraf, M.; Tousif, M. Snow Cover Trend and Hydrological Characteristics of the Astore River Basin (Western Himalayas) and Its Comparison to the Hunza Basin (Karakoram Region). *Sci. Total Environ.* **2015**, *505*, 748–761. [CrossRef]
7. Muhammad Qasim, S.K. Hydro-Meteorological Characteristics of Indus River Basin at Extreme North of Pakistan. *J. Earth Sci. Clim. Chang.* **2014**, *5*, 170. [CrossRef]
8. Xu, D.; Liu, D.; Yan, Z.; Ren, S.; Xu, Q. Spatiotemporal Variation Characteristics of Precipitation in the Huaihe River Basin, China, as a Result of Climate Change. *Water* **2023**, *15*, 181. [CrossRef]
9. Abbas, S.; Yaseen, M.; Latif, Y.; Waseem, M.; Muhammad, S.; Leta, M.K.; Sher, S.; Imran, M.A.; Adnan, M.; Khan, T.H. Spatiotemporal Analysis of Climatic Extremes over the Upper Indus Basin, Pakistan. *Water* **2022**, *14*, 1718. [CrossRef]
10. Khan, F.; Ali, S.; Mayer, C.; Ullah, H.; Muhammad, S. Climate Change and Spatio-Temporal Trend Analysis of Climate Extremes in the Homogeneous Climatic Zones of Pakistan during 1962–2019. *PLoS ONE* **2022**, *17*, e0271626. [CrossRef]
11. Anjum, M.N.; Ding, Y.; Shangguan, D.; Liu, J.; Ahmad, I.; Ijaz, M.W.; Khan, M.I. Quantification of Spatial Temporal Variability of Snow Cover and Hydro-Climatic Variables Based on Multi-Source Remote Sensing Data in the Swat Watershed, Hindukush Mountains, Pakistan. *Meteorol. Atmos. Phys.* **2018**, *131*, 467–486. [CrossRef]
12. Waseem, M.; Ajmal, M.; Ahmad, I.; Khan, N.M.; Azam, M.; Sarwar, M.K. Projected Drought Pattern under Climate Change Scenario Using Multivariate Analysis. *Arab. J. Geosci.* **2021**, *14*, 544. [CrossRef]
13. Aslam, A.Q.; Ahmad, S.R.; Ahmad, I.; Hussain, Y.; Hussain, M.S. Vulnerability and Impact Assessment of Extreme Climatic Event: A Case Study of Southern Punjab, Pakistan. *Sci. Total Environ.* **2017**, *580*, 468–481. [CrossRef] [PubMed]
14. Ali, S.; Liu, Y.; Ishaq, M.; Shah, T.; Abdullah; Ilyas, A.; Din, I. Climate Change and Its Impact on the Yield of Major Food Crops: Evidence from Pakistan. *Foods* **2017**, *6*, 39. [CrossRef] [PubMed]
15. Bolch, T.; Kulkarni, A.; Kääb, A.; Huggel, C.; Paul, F.; Cogley, J.G.; Frey, H.; Kargel, J.S.; Fujita, K.; Scheel, M.; et al. The State and Fate of Himalayan Glaciers. *Science* **2012**, *336*, 310–314. [CrossRef] [PubMed]
16. Liu, Y.; Hu, X.; Wu, F.; Chen, B.; Liu, Y.; Yang, S.; Weng, Z. Quantitative Analysis of Climate Change Impact on Zhangye City's Economy Based on the Perspective of Surface Runoff. *Ecol. Indic.* **2019**, *105*, 645–654. [CrossRef]
17. Yang, T.; Li, Q.; Chen, X.; De Maeyer, P.; Yan, X.; Liu, Y.; Zhao, T.; Li, L. Spatiotemporal Variability of the Precipitation Concentration and Diversity in Central Asia. *Atmos. Res.* **2020**, *241*, 104954. [CrossRef]
18. Burić, D.; Doderović, M. Trend of Percentile Climate Indices in Montenegro in the Period 1961–2020. *Sustainability* **2022**, *14*, 12519. [CrossRef]
19. Chen, A.; He, X.; Guan, H.; Cai, Y. Trends and Periodicity of Daily Temperature and Precipitation Extremes during 1960–2013 in Hunan Province, Central South China. *Theor. Appl. Climatol.* **2018**, *132*, 71–88. [CrossRef]
20. Toll, V.; Post, P. Daily Temperature and Precipitation Extremes in the Baltic Sea Region Derived from the BaltAn65+ Reanalysis. *Theor. Appl. Climatol.* **2018**, *132*, 647–662. [CrossRef]
21. Meseguer-Ruiz, O.; Ponce-Philimon, P.I.; Quispe-Jofré, A.S.; Guijarro, J.A.; Sarricolea, P. Spatial Behaviour of Daily Observed Extreme Temperatures in Northern Chile (1966–2015): Data Quality, Warming Trends, and Its Orographic and Latitudinal Effects. *Stoch. Environ. Res. Risk Assess.* **2018**, *32*, 3503–3523. [CrossRef]
22. Insaf, T.Z.; Lin, S.; Sheridan, S.C. Climate Trends in Indices for Temperature and Precipitation across New York State, 1948–2008. *Air Qual. Atmos. Health* **2013**, *6*, 247–257. [CrossRef]
23. Caloiero, T. Trend of Monthly Temperature and Daily Extreme Temperature during 1951–2012 in New Zealand. *Theor. Appl. Climatol.* **2017**, *129*, 111–127. [CrossRef]
24. Touré Halimatou, A.; Kalifa, T.; Kyei-Baffour, N. Assessment of Changing Trends of Daily Precipitation and Temperature Extremes in Bamako and Ségou in Mali from 1961–2014. *Weather Clim. Extrem.* **2017**, *18*, 8–16. [CrossRef]
25. Croitoru, A.E.; Piticar, A. Changes in Daily Extreme Temperatures in the Extra-Carpathians Regions of Romania. *Int. J. Climatol.* **2013**, *33*, 1987–2001. [CrossRef]

26. Chapagain, D.; Dhaubanjar, S.; Bharati, L. Unpacking Future Climate Extremes and Their Sectoral Implications in Western Nepal. *Clim. Chang.* **2021**, *168*, 8. [CrossRef]
27. Salman, S.A.; Shahid, S.; Ismail, T.; Ahmed, K.; Chung, E.S.; Wang, X.J. Characteristics of Annual and Seasonal Trends of Rainfall and Temperature in Iraq. *Asia-Pac. J. Atmos. Sci.* **2019**, *55*, 429–438. [CrossRef]
28. Khan, A.J.; Koch, M.; Tahir, A.A. Impacts of Climate Change on Thewater Availability, Seasonality and Extremes in the Upper Indus Basin (UIB). *Sustainability* **2020**, *12*, 1283. [CrossRef]
29. Du, H.; Xia, J.; Yan, Y.; Lu, Y.; Li, J. Spatiotemporal Variations of Extreme Precipitation in Wuling Mountain Area (China) and Their Connection to Potential Driving Factors. *Sustainability* **2022**, *14*, 8312. [CrossRef]
30. Zhang, X.; Zwiers, F.W.; Li, G.; Wan, H.; Cannon, A.J. Complexity in Estimating Past and Future Extreme Short-Duration Rainfall. *Nat. Geosci.* **2017**, *10*, 255–259. [CrossRef]
31. Anjum, M.N.; Irfan, M.; Waseem, M.; Leta, M.K.; Niazi, U.M.; Rahman, S.U.; Ghanim, A.; Mukhtar, M.A.; Nadeem, M.U. Assessment of PERSIANN-CCS, PERSIANN-CDR, SM2RAIN-ASCAT, and CHIRPS-2.0 Rainfall Products over a Semi-Arid Subtropical Climatic Region. *Water* **2022**, *14*, 147. [CrossRef]
32. Wu, S.; Hu, Z.; Wang, Z.; Cao, S.; Yang, Y.; Qu, X.; Zhao, W. Spatiotemporal Variations in Extreme Precipitation on the Middle and Lower Reaches of the Yangtze River Basin (1970–2018). *Quat. Int.* **2021**, *592*, 80–96. [CrossRef]
33. Zhang, X.; Alexander, L.; Hegerl, G.C.; Jones, P.; Tank, A.K.; Peterson, T.C.; Trewin, B.; Zwiers, F.W. Indices for Monitoring Changes in Extremes Based on Daily Temperature and Precipitation Data. *Wiley Interdiscip. Rev. Clim. Chang.* **2011**, *2*, 851–870. [CrossRef]
34. Alexander, L.V.; Zhang, X.; Peterson, T.C.; Caesar, J.; Gleason, B.; Klein Tank, A.M.G.; Haylock, M.; Collins, D.; Trewin, B.; Rahimzadeh, F.; et al. Global Observed Changes in Daily Climate Extremes of Temperature and Precipitation. *J. Geophys. Res. Atmos.* **2006**, *111*, D05109. [CrossRef]
35. Hidalgo García, D.; Arco Díaz, J.; Martín Martín, A.; Gómez Cobos, E. Spatiotemporal Analysis of Urban Thermal Effects Caused by Heat Waves through Remote Sensing. *Sustainability* **2022**, *14*, 12262. [CrossRef]
36. Saleem, F.; Zeng, X.; Hina, S.; Omer, A. Regional Changes in Extreme Temperature Records over Pakistan and Their Relation to Pacific Variability. *Atmos. Res.* **2021**, *250*, 105407. [CrossRef]
37. Hussain, A.; Cao, J.; Hussain, I.; Begum, S.; Akhtar, M.; Wu, X.; Guan, Y.; Zhou, J. Observed Trends and Variability of Temperature and Precipitation and Their Global Teleconnections in the Upper Indus Basin, Hindukush-Karakoram-Himalaya. *Atmosphere* **2021**, *12*, 973. [CrossRef]
38. Syed, A.; Liu, X.; Moniruzzaman, M.; Rousta, I.; Syed, W.; Zhang, J.; Olafsson, H. Assessment of Climate Variability among Seasonal Trends Using in Situ Measurements: A Case Study of Punjab, Pakistan. *Atmosphere* **2021**, *12*, 939. [CrossRef]
39. Xiao, Z.; Duan, A.; Wang, Z. Atmospheric Heat Sinks over the Western Tibetan Plateau Associated with Snow Depth in Late Spring. *Int. J. Climatol.* **2019**, *39*, 5170–5180. [CrossRef]
40. Yaseen, M.; Ahmad, I.; Guo, J.; Azam, M.I.; Latif, Y. Spatiotemporal Variability in the Hydrometeorological Time-Series over Upper Indus River Basin of Pakistan. *Adv. Meteorol.* **2020**, *2020*, 5852760. [CrossRef]

Disclaimer/Publisher's Note: The statements, opinions and data contained in all publications are solely those of the individual author(s) and contributor(s) and not of MDPI and/or the editor(s). MDPI and/or the editor(s) disclaim responsibility for any injury to people or property resulting from any ideas, methods, instructions or products referred to in the content.

AgriEngineering

Article

Ventilation Operating Standard for Improving Internal Environment in Pig House Grafting Working Conditions Using CFD

Byung-Wook Oh, Hyo-Jae Seo and Il-Hwan Seo *

Department of Rural Construction Engineering, College of Agriculture & Life Sciences, Jeonbuk National University, Jeonju 54896, Republic of Korea; bwoh@jbnu.ac.kr (B.-W.O.); tjgydls@jbnu.ac.kr (H.-J.S.)
* Correspondence: ihseo@jbnu.ac.kr; Tel.: +82-103-474-5001

Abstract: Many farms utilize closed-type livestock systems to enhance productivity and facilitate effective environmental management. However, the confined nature of these closed spaces poses an increased risk of exposure to harmful gases and organic dust for both workers and livestock. Additionally, the introduction of outside air through ventilation systems can lead to temperature fluctuations within the breeding environment, resulting in potential productivity issues. This research paper employs computational fluid dynamics (CFD) to develop ventilation operation management plans that address both the working environment and the breeding environment simultaneously. The proposed plans are designed to be easily implemented in practical farm settings. The findings of this study, based on the simulation analysis, indicate that while ventilation is effective in reducing harmful gases and improving the working environment, its efficiency decreases after the initial 3 min of operation. Furthermore, uncontrolled ventilation can cause sudden temperature changes, which may adversely affect the well-being of the livestock. However, when upgraded ventilation structures are implemented, significant improvements in the working environment (an average of 27.3% improvement) can be achieved while maintaining temperature stability for the livestock. These results highlight the importance of referring to the provided ventilation operation management table before commencing work, as it enables workers to improve the working environment while minimizing the potential impact of ventilation on the breeding environment.

Keywords: ammonia; CFD; ventilation; working and breeding environments

Citation: Oh, B.-W.; Seo, H.-J.; Seo, I.-H. Ventilation Operating Standard for Improving Internal Environment in Pig House Grafting Working Conditions Using CFD. *AgriEngineering* **2023**, *5*, 1378–1394. https://doi.org/10.3390/agriengineering5030086

Academic Editors: Muhammad Sultan, Yuguang Zhou, Redmond R. Shamshiri and Muhammad Imran

Received: 12 July 2023
Revised: 3 August 2023
Accepted: 4 August 2023
Published: 14 August 2023

Copyright: © 2023 by the authors. Licensee MDPI, Basel, Switzerland. This article is an open access article distributed under the terms and conditions of the Creative Commons Attribution (CC BY) license (https://creativecommons.org/licenses/by/4.0/).

1. Introduction

The Korean livestock industry has experienced rapid growth, leading to a significant increase in both the number of livestock and meat consumption. However, this growth has also resulted in a decrease in the number of livestock farms and a shift towards large-scale and intensive operations. To manage productivity and control environmental factors such as temperature, humidity, and ventilation, the majority of farms now adopt closed-type livestock pens. While these closed spaces allow for effective management, they also present challenges such as increased livestock manure and odor, as well as higher levels of fine dust. Consequently, there is an elevated risk of workers and animals being exposed to harmful gases and organic dust within these facilities [1,2]. In light of the increasing number of occupational accidents among farmers, ensuring the working safety and welfare of workers has become a paramount concern [3]. Farmers may not possess a complete understanding of the potential hazards and safety risks associated with agricultural work, leading to a lack of adequate precautions.

Livestock facilities commonly generate harmful gases, including ammonia and hydrogen sulfide. Among these, hydrogen sulfide is highly toxic, and exposure to high concentrations can even result in fatalities. When highly concentrated hydrogen sulfide disperses from livestock excrement or slurry due to pig movement and rapid changes in internal airflow, there is a significant risk of workers being exposed to these dangerous gases [4]. Ammonia is a colorless and irritating toxic gas. Even at low concentrations, ammonia can cause irritation to the eyes and lungs, resulting in inflammation, sneezing, and excessive salivation. Inhaling harmful gases and organic dust can cause acute and chronic poisoning, as well as various serious health issues such as bronchitis, pneumonia, asthma, coughing, allergies, headaches, and dizziness [5–11].

Among the methods available for reducing harmful gases, the ventilation system offers both technical and economic advantages. It has the capability to directly remove harmful gases by discharging them outside and can be implemented using the existing infrastructure. However, the introduction of outside air through ventilation can cause rapid temperature changes in the livestock breeding environment, which can negatively impact the immune system, respiratory health, and overall productivity of the animals [12–14]. Therefore, when implementing ventilation, it is essential to analyze the resulting thermal environment changes.

There has been limited research on strategies for operating ventilation systems to enhance the working environment in the agricultural field. Moreover, studies that simultaneously consider both the breeding environment and the working environment are even scarcer. Therefore, there is a need for monitoring the internal environment and conducting simulation-based research that takes into account the working environment for workers as well as the animal-centric breeding environment. To improve the working environment, it is common practice to increase the ventilation volume to expel more harmful gases. However, increasing the ventilation volume also leads to greater temperature variations within the livestock breeding area, potentially causing issues in the breeding environment. Additionally, it is crucial to analyze the working and breeding environments concurrently since the ventilation volume and temperature inside the facility are highly correlated. Monitoring the working environment and breeding environment simultaneously poses challenges, as workers enter and exit the area periodically while the pigs remain relatively constant. Therefore, utilizing simulation techniques becomes an effective approach to gather diverse environmental data, such as temperature and location variations within the facility.

Relying solely on field experiments to gather data on the breeding and working environments presents several challenges, including limitations related to time, space, the ability to maintain stable experimental conditions, and restricted measurement points. However, by combining field measurements with computational fluid dynamics (CFD), it becomes possible to overcome these challenges and acquire a substantial amount of qualitative and quantitative data. Such data can be utilized for research purposes, such as investigating automatic environmental control within smart farms and developing management algorithms. CFD simulations have been widely used to study airflow in livestock houses. Research was conducted on the ventilation effect and changes in temperature and humidity through a 3D simulation model that included the shape of the pigs and the ventilation structure of the pig house [15–19]. Specifically, the impact of the thermal environment was analyzed, such as the effect of temperature differences on pig productivity [20–24]. While research focused on livestock workers has been less, some studies have suggested that the internal concentration of ammonia and fine dust can affect the health of the workers [17,25]. Research to monitor and improve the livestock facility environment has primarily been focused on ammonia, with various facility structures undergoing field monitoring [26–32]. Some studies used CFD to analyze the amount of ammonia released from the floor according to the shape of the floor [33,34]. Recently, uncertainty quantification (UQ) has been applied to simulations to optimize results [35–37]. However, specific research that simultaneously considers the on-site breeding environment and working environment has been extremely rare so far.

This research aims to propose a ventilation control standard that considers both the working environment and the breeding environment in a pig house simultaneously by efficiently reducing the concentration of harmful gases at the workers' height. The study involves measuring and identifying the concentration levels of harmful gases to which workers are exposed during specific tasks in the pig house. The ventilation efficiency by the combination of inlet and outlet systems is evaluated using computational fluid dynamics (CFD).

2. Materials and Methods

2.1. Experimental Pig House

The experiment was conducted in a commercial pig house with a capacity of 15,000 pigs, located in Yeonggwang-gun, Jeollanam-do, Korea. A pigpen that had a relatively high mortality rate was chosen based on interviews with the farm manager. The dimensions of the selected pigpen are 12 m in width, 15.6 m in length, and 3.2 m in height. Inside the pigpen, there is a central corridor that is 0.8 m wide, and the pigpen is divided into 12 compartments by 1 m-high walls. Each compartment can accommodate around 18 to 25 fattening pigs. The pigpen has a total capacity for 300 pigs. There are 12 windows (6 upper and 6 lower) that allow outside air to flow into the pigpen through the corridor. Additionally, two duct inlets, each with a diameter of 0.94 m, are installed on the left and right sides of the pigpen to distribute air evenly throughout (Figure 1). On the outer wall, there are two circular exhaust fans with a diameter of 1 m each, as well as three square fans with a length of 0.6 m per side. These fans are used for extracting air from the pigpen. Furthermore, two 1.5 m-diameter ducts (duct outlets) are installed on the left and right sides, extending from the outside to the center of the pigpen. These ducts serve the purpose of both removing harmful gases to the outside and allowing fresh air to enter (Figure 2). The ventilation system is operated by utilizing different combinations of inlets and outlets based on the season and internal temperature and humidity conditions.

Figure 1. Experimental pigpen with mechanical ventilation systems including five different exhaust fans and duct inlets.

2.2. Computational Fluid Dynamics

Computational fluid dynamics (CFD) encompasses three main steps: pre-processing, main computational processing, and post-processing. In the pre-processing step, the outer shape of the experimental object and the grid of the target area are designed. In the main computational processing step, the target area is analyzed using numerical analysis techniques, employing equations such as the Navier–Stokes equations. Finally, in the post-processing step, the results of the analysis are qualitatively and quantitatively evaluated. To begin the CFD analysis, the object is designed as a 3D model. Grid design and boundary conditions are then specified. The calculations are conducted using FLUENT software (version 18.1., ANSYS Inc., Rochester, NY, USA) [38], employing the principles of mass conservation, energy generation and movement, and the law of momentum conservation on each grid. Numerical analyses for fluid and energy flow are based on mass, momentum, and energy conservation laws. The conservation equations are Equations (1)–(3).

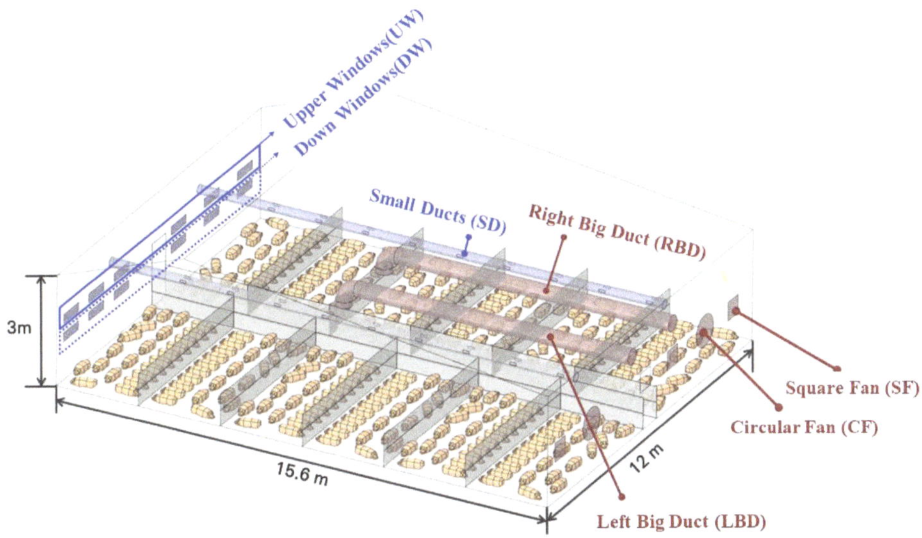

Figure 2. Ventilation system operated during spring season with combinations of various outlets (red) and inlets (blue).

Mass conservation equation:

$$\frac{\partial \rho}{\partial t}, \nabla\left(\rho \vec{v}\right) = S_m \tag{1}$$

Energy conservation equation:

$$\frac{\partial}{\partial t}(\rho E), \nabla\left(\vec{v}(\rho E, P)\right) = \nabla K_{eff} \nabla T - \sum h \vec{J}, \left(\vec{\tau}_{eff} \vec{v}\right), S_h \tag{2}$$

Momentum conservation equation:

$$\frac{\partial \rho}{\partial t}\left(\rho \vec{v}\right), \nabla\left(\rho = S_m \vec{v} \vec{v}\right) = -\nabla P, \nabla \vec{\tau}, \rho \vec{g}, \vec{F} \tag{3}$$

where ρ is the density (kg m^{-3}), $\vec{v}$ is the velocity (m s^{-1}), S_m is a mass source term based on the chemical reaction (kg m^{-2}), E is the total energy (kg m^2 s^{-2} kg^{-1}), P is the static pressure (kg m^{-1} s^{-2}), K_{eff} is the effective conductivity (kg m^{-1} s^{-3} K^{-1}), T is the temperature (K), $\vec{\tau}$ is the stress tensor (kg m^{-1} s^{-2}), $\vec{\tau}_{eff}$ is the effective stress tensor (kg m^{-1} s^{-2}), $\vec{J}$ i is the component of diffusion flux, S_h is the enthalpy rise based on the chemical reaction or radiation (kg m^{-1} s^{-3}), $\vec{g}$ is the gravitational acceleration (m s^{-2}), and $\vec{F}$ is the external force vector (kg m s^{-2}).

In this study, we used the CFD model previously validated by field experimental data [13]. The shape and boundary conditions of the CFD simulation model were designed by the experimental pig house. We used tetrahedron-shaped meshes to design complex pig shapes and structures in the pig house. The grid size is very important to simulation accuracy and computation time. We designed the grid by referring to a previous study that evaluated the grid independence test [15]. The minimum size of the grid was 0.05 cm, and a denser grid was used in the aerodynamically important part. The turbulence model was determined by a validation test in the same experimental pig house as the previous research [13]. Among the four different turbulence models, the realizable k-ε model was selected, which showed the highest correlation (R^2 value of 0.859) in terms of air velocity and temperature distribution. The specific boundary conditions used for the CFD model were outlined in Table 1.

Table 1. Boundary conditions for the CFD simulation model of the experimental pig house.

Content	Value
Number of meshes	about 4.5 million
Mesh size	Minimum size 0.05 (cm)
Mesh design	Tetrahedron
Boundary conditions	Velocity inlet and pressure outlet
Turbulence model	Realizable k-ε
Input temperature	291.8 (K)
Output temperature	289.5 (K)
Inside initial temperature	297.8 (K)
Pig heat flux	210.9 (W·m^{-1})
Ventilation rate (air exchanges)	0.55 (m^{-1})
Velocity at the inlet vent	3.71 (m·s^{-1})
NH_3 concentration at work height (1.5 m)	38.7 (ppm)

2.3. *Tracer Gas Decay Method*

The tracer gas decay (TGD) method was used to compute regional ventilation efficiency using the CFD simulation [16–19]. The conventional method used for calculating the internal ventilation rate was based on the mass flow rate calculation. This determines the air replacements per unit hour, assuming that all the air entering the pig house was used to replace the air by ventilation. While the mass flow rate calculation method offers the advantage of a straightforward calculation for determining the ventilation volume of the entire facility, it does not allow for the analysis of regional ventilation effects. Consequently, it is not suitable for evaluating the ventilation volume of individual compartments within the pig house or the dilution effect of harmful gases. It does not provide insights into the flow behavior inside the facility, which can vary based on the combination of ventilation structures.

The tracer gas decay method (TGD method) has been utilized to calculate the regional ventilation efficiency by analyzing the change in tracer gas concentration over time. The TGD method involves measuring the concentration of tracer gases such as CO_2, SF6, etc., at multiple points in field experiments to determine ventilation efficiency. By using a simulation, we can obtain more qualitative and quantitative data than in the field. In this research, the data obtained from the CFD simulation were quantitatively analyzed for regional ventilation efficiency in pig house by region using the TGD method. The air exchange rate (AER, min^{-1}) was used as a unit to denote the amount of ventilation, representing the number of times the incoming air from the outside replaces the entire indoor air. In the CFD simulation, carbon dioxide was uniformly distributed at a concentration of 2000 ppm inside the pig house, while the air entering from outside contained 400 ppm of carbon dioxide.

$$\mathrm{AER}_{MRF} = \frac{\sum v_i A_i}{V} \times 60 = \frac{\sum v_0 A_0}{V} \times 60 \tag{4}$$

$$AER_{TGD} = \frac{In \frac{C_0}{C_t}}{t - t_0} \tag{5}$$

where AER_{MRF} is the air exchange rate of mass flow (min^{-1}), AER_{TGD} is the air exchange rate of tracer gas decay(min^{-1}), V is the volume of the pig house (m^3), v_0, v_i are the velocities at the inlet and outlet (m·s^{-1}), A_0, A_i are the vent areas of the inlet and outlet (m^2), and C_0, C_t are the concentrations of the tracer gas at t_0, t time (ppm).

2.4. Experimental Procedure

A ventilation operating standard was suggested based on the field monitoring of gas concentration and CFD modeling according to various combinations of ventilation systems as shown in Figure 3. The first step involves measuring the concentration of harmful gases to which workers were exposed during various tasks performed in the pig house. The CFD model was initially initialized in the steady state, then modeled in the unsteady state for the ventilation efficiency analysis using the TGD method. Using this model, the ventilation efficiency was analyzed for different combinations of ventilation openings within the pigpen to select an upgraded ventilation operation method that effectively reduced harmful gases at the workers' height. To enhance the accuracy of the analysis, the selected ventilation system considered the ammonia emission from the floor. The analysis took into account the changes over time during ventilation, as well as the thermal environment of the pig herd. This is important because sudden temperature changes can impact the breeding environment for the pigs. By implementing the ventilation operation plan, which aimed to reduce harmful gases such as ammonia concentration, the extent of harmful gas removal by the ventilation system for improving the working environment was analyzed. Additionally, the thermal environment of the pig herd was utilized to develop the ventilation operation standard table that aimed to enhance both the working and breeding environments.

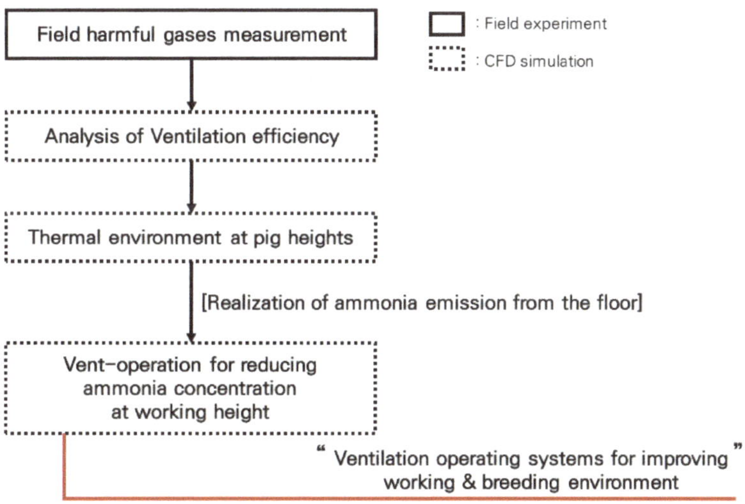

Figure 3. Research procedure for modeling ventilation efficiency considering regional thermal conditions.

2.4.1. Field Monitoring of Harmful Gas Concentrations

The concentration of harmful gases to which workers are directly exposed was measured at the height of their respiratory system during different types of work performed inside the pigpen. The measurements were conducted for both morning and afternoon daily working routines, with break time between 12 pm and 1 pm. To assess the exposure level for workers, ammonia and hydrogen sulfide sensors were attached at shoulder height, which is approximately the same height as the worker's respiratory tract (Figure 4). The working tasks were categorized as either contact work or non-contact work, depending on their interaction with the pigs. During contact work with pigs, more gases were dispersed from the floor due to animal movements. The concentrations of harmful gases were analyzed by considering the characteristics of each working task.

(a) (b)

Figure 4. Research procedure for modeling ventilation efficiency considering regional thermal conditions. (**a**) NH_3 and H_2S sensors and (**b**) installation of sensors at worker's shoulder.

2.4.2. CFD Analysis for Ventilation Efficiency and Thermal Environment

The ventilation system in the experimental pigpen was characterized by combinations of vent openings with multiple inlets and outlets that could be operated in various configurations to assess the ventilation efficiency. The outlet ventilation system was comprised of three small, square-shaped fans (SF) measuring 0.6 × 0.6 m, two circular fans (CF) with a diameter of 1.0 m, and a large duct located on the upper right side of the pigpen (RBD, right big duct). These outlet components were labeled alphabetically from A to H. The inlet ventilation system consisted of a large duct on the left-hand side (LBD, left big duct), two small, elongated ducts positioned on the upper side of the pigpen (SD, small duct), and a total of 12 windows (UW/LW, upper window and lower window) located at the end of the pigpen. These inlet components were grouped numerically from 1 to 5. The ventilation efficiency of a total of 42 ventilation systems, comprising different combinations of exhaust structures (outlets) and inflowing structures (inlets), is evaluated and presented in Table 2.

Table 2. Result of ventilation efficiency considering various combinations of ventilation structures. The meanings of symbols are as follows: SF (square fan), CF (circular fan), RBD (right big duct), LBD (left big duct), SD (small duct), UW (upper window), and LW (lower window).

Case	Outlet	Inlet	Human Height (1.5 m)			Pig Height (0.5 m)		
			Min.	Avg.	Max.	Min.	Avg.	Max.
Standard	SF, RBD	LBD, SD, UW	0.24	0.28	0.32	0.24	0.28	0.35
A-1	SF, CF, RBD	LBD, SD, UW, DW	0.25	0.35	0.45	0.25	0.35	0.47
A-2	SF, CF, RBD	SD, UW, DW	0.25	0.35	0.41	0.23	0.41	0.47
A-3	SF, CF, RBD	UW, DW	0.24	0.35	0.39	0.23	0.40	0.47
A-4	SF, CF, RBD	UW	0.26	0.31	0.34	0.26	0.28	0.34
A-5	SF, CF, RBD	DW	0.25	0.25	0.31	0.24	0.24	0.37
B-1	SF, RBD	LBD, SD, UW, DW	0.27	0.37	0.44	0.26	0.43	0.47
B-2	SF, RBD	SD, UW, DW	0.25	0.28	0.31	0.23	0.28	0.31
B-3	SF, RBD	UW, DW	0.25	0.34	0.36	0.25	0.39	0.43
B-4	SF, RBD	UW	0.23	0.29	0.32	0.23	0.31	0.32
B-5	SF, RBD	DW	0.23	0.28	0.31	0.22	0.29	0.38

Table 2. *Cont.*

Case	Outlet	Inlet	Human Height (1.5 m)			Pig Height (0.5 m)		
			Min.	Avg.	Max.	Min.	Avg.	Max.
C-1	SF	LBD, SD, UW, DW	0.28	0.36	0.43	0.26	0.41	0.46
C-2	SF	SD, UW, DW	0.26	0.35	0.42	0.26	0.40	0.47
C-3	SF	UW, DW	0.27	0.36	0.43	0.26	0.42	0.47
C-4	SF	UW	0.24	0.31	0.33	0.25	0.31	0.32
C-5	SF	DW	0.26	0.25	0.31	0.25	0.22	0.37
D-1	CF, RBD	LBD, SD, UW, DW	0.30	0.42	0.52	0.29	0.50	0.57
D-standard	CF, RBD	LBD, SD, UW	0.25	0.28	0.33	0.24	0.28	0.37
D-2	CF, RBD	SD, UW, DW	0.25	0.37	0.42	0.24	0.42	0.47
D-3	CF, RBD	UW, DW	0.26	0.37	0.44	0.25	0.42	0.48
D-4	CF, RBD	UW	0.23	0.30	0.32	0.22	0.31	0.32
D-5	CF, RBD	DW	0.26	0.27	0.31	0.25	0.25	0.38
E-1	CF	LBD, SD, UW, DW	0.26	0.39	0.45	0.46	0.44	0.25
E-2	CF	SD, UW, DW	0.29	0.33	0.47	0.28	0.35	0.49
E-3	CF	UW, DW	0.27	0.37	0.43	0.25	0.42	0.48
E-4	CF	UW	0.24	0.30	0.33	0.23	0.32	0.33
E-5	CF	DW	0.26	0.28	0.32	0.26	0.25	0.38
F-1	SF	CF, RBD, SD	0.24	0.26	0.34	0.26	0.26	0.38
F-2	SF	CF, RBD, UW, DW	0.23	0.32	0.44	0.24	0.36	0.46
F-3	SF	CF, RBD, LBD, SD, UW, DW	0.24	0.29	0.44	0.24	0.38	0.47
F-4	SF	CF, SD	0.25	0.28	0.35	0.25	0.23	0.36
F-5	SF	CF, UW, DW	0.21	0.33	0.55	0.22	0.49	0.60
F-6	SF	CF, LBD, SD, UW, DW	0.22	0.31	0.46	0.21	0.37	0.47
G-1	SF, RBD–LBD	SD, UW, DW	0.23	0.34	0.41	0.23	0.41	0.47
G-3	SF, RBD–LBD	UW, DW	0.24	0.31	0.38	0.24	0.32	0.43
G-4	SF, RBD–LBD	UW	0.22	0.29	0.33	0.21	0.29	0.31
G-5	SF, RBD–LBD	DW	0.25	0.26	0.31	0.25	0.24	0.38
H-1	SF	RBD, LBD, SD, UW, DW	0.26	0.35	0.43	0.24	0.41	0.45
H-3	SF	RBD, LBD, UW, DW	0.27	0.35	0.42	0.26	0.40	0.45
H-4	SF	RBD, LBD, UW	0.23	0.27	0.32	0.21	0.26	0.35
H-5	SF	RBD, LBD, DW	0.25	0.29	0.35	0.23	0.30	0.40

In order to improve the working environment and minimize the impact to the pig, an analysis of the thermal conditions was conducted based on the ventilation period. An optimal ventilation level was identified to provide a suitable breeding environment while ensuring the removal of harmful gases inside the pigpen. When ventilation begins, outside air enters into the pigpen, resulting in a sudden change in the thermal environment, which can lead to respiratory diseases or decreased immunity. Therefore, it is crucial to analyze both the ventilation efficiency in reducing harmful gases and the thermal conditions of the pig breeding environment. The height of the pig herd was set at 0.5 m, where the pigs breathe.

The CFD model was designed with a combination of regular tetrahedrons, triangular-shaped tetrahedrons, grids, regular hexahedrons, and square-shaped hex-dominant grids to represent the complex shapes found in the pigpen following the previous validated modeling approaches [20–22]. Dense meshes were used for aerodynamically important areas of the pigpen, such as inlets, outlets, or regions near complex structures, to enhance model accuracy [39–44]. The calculation area was subdivided into smaller sections for analyzing the ventilation efficiency and thermal conditions. The frequency analysis was performed by dividing the height of the breeding environment into a total of 630 compartments, each with a consistent size of 0.52 × 0.56 cm (Figure 5). When the temperature difference compared to the average temperature of the pigpen is less than 3 °C, it is considered a suitable temperature. Temperature difference levels were categorized into three levels: the warning level, caution level, and normal level were defined based on the proportions of compartments representing suitable temperatures, with less than 60% of the total area, less than 70%, and more than 70%, respectively.

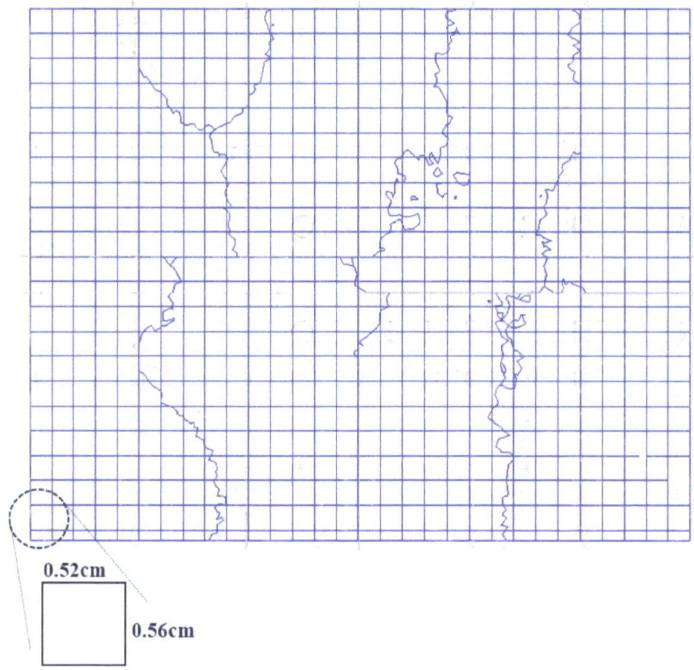

Figure 5. Divided sectors for analysis of thermal environment at 1.5 m height.

The ventilation efficiency of each region was analyzed for 42 different combinations of ventilation systems in order to identify which system could efficiently ventilate harmful gases at the worker's height. When the outlets remained the same, the ventilation efficiency was high due to the increasing inlet area by additionally opening the six lower windows located on the corridor wall. Replacing the exhaust vents from three square-shaped fans with two larger circular fans installed for the summer season led to increased ventilation efficiency at the worker's height. As high ventilation efficiency is crucial for effectively diluting harmful gases, these two ventilation schemes were selected for improving the working environment.

2.4.3. CFD Analysis for Ventilation Efficiency and Thermal Environment

To develop a ventilation operation standard that simultaneously considers the work environment and breeding environment based on CFD-computed results, it is essential to analyze the combination of ventilation system, ventilation efficiency, which can reduce harmful gases over the ventilation time, and the temperature difference affecting the pigs. Considering easy accessibility for farmers in the field, a single sheet that visualizes, simplifies, and presents the ventilation operation standard was suggested by combining the multiple analyses from the CFD model. By utilizing the double linear interpolation method through a polynomial regression analysis, the reduction rate of harmful gases in the working environment and the results of the thermal environment analysis for the breeding environment were presented together for each ventilation system.

3. Results and Discussion

3.1. Field Monitoring for Harmful Gas Concentrations

To analyze the exposure of workers to gas concentrations inside the pigpen, a comprehensive analysis was conducted using measured data from gas detectors attached at the height of the worker's shoulder and image data captured during the measurements. The work performed in the pig house was categorized into morning and afternoon work including shipment, animal manure treatment, vaccine administration, condition checking, cleaning, and marking. During the continuous monitoring of gas concentrations including ammonia and hydrogen sulfide, each datum was obtained based on the average concentrations during each task performed as shown in Figure 6. Each work was further classified as either a "contact" work if the worker directly contacted the pigs or a "non-contact" work if there is no direct contact with the pigs.

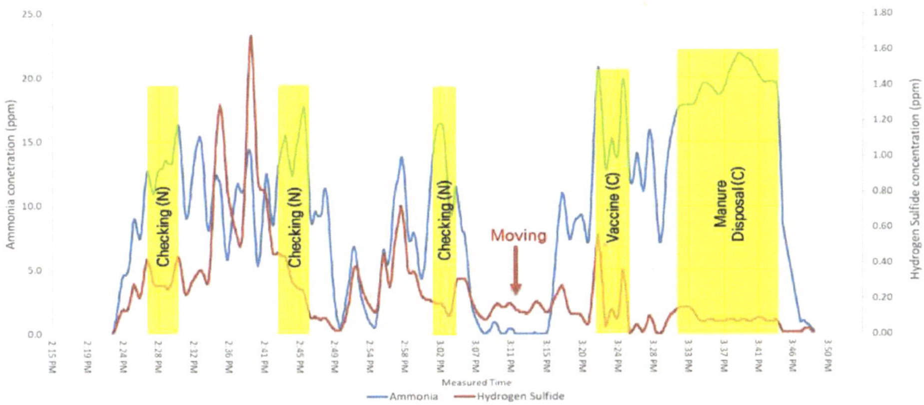

Figure 6. Ammonia and hydrogen sulfide concentrations monitored at shoulder height in pig room (yellow section at working time).

Each data point in Figures 7 and 8 represents the average ammonia and hydrogen sulfide concentrations monitored at the worker's height during the working time. During contact work such as general checking, manure treatment, and cleaning tasks, ammonia concentrations was higher compared to non-contact work. Generally, workers were exposed to ammonia concentrations ranging from late 20 ppm to 40 ppm during contact work. During the shipment task of pigs, workers were exposed to low ammonia concentrations of 5 ppm or lower due to increments of ventilation by opening whole doors in the pigpen. The average exposure to ammonia was further analyzed by the type of work. The highest ammonia concentration was observed when workers moved pigs between compartments within the pigpen, with an average exposure of 43.0 ppm. Other tasks associated with relatively high ammonia concentrations included manure treatment (31.8 ppm), condition checking (26.5 ppm), and cleaning (27.0 ppm). On the other hand, marking (8.8 ppm) and vaccine injection (6.2 ppm) resulted in relatively low ammonia exposure. The average ammonia concentration during the shipment of pigs was 3.0 ppm, indicating a low exposure level despite the direct contact with pigs. This can be attributed to the fact that the task is mainly performed in the corridor outside the pigpen. The average levels of ammonia exposure were compared between contact work and non-contact work. Contact work had an average ammonia concentration of 24.7 ppm, which was 1.58 times higher than the average concentration of 15.6 ppm observed in non-contact work.

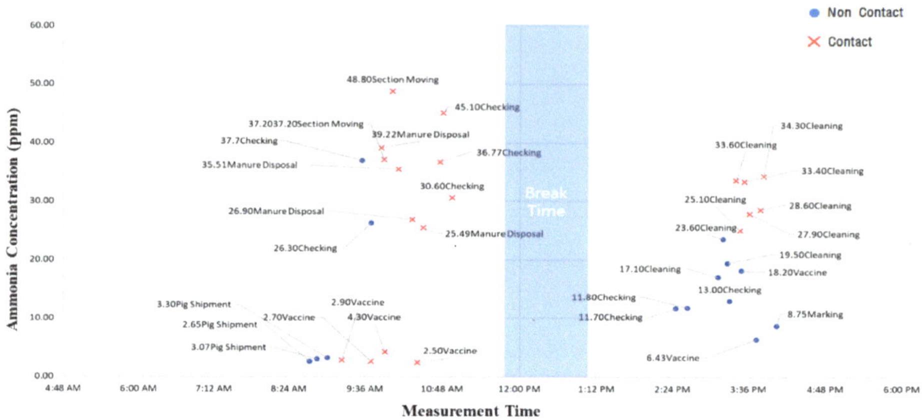

Figure 7. Average ammonia concentration during contact & non-contact works in pig house.

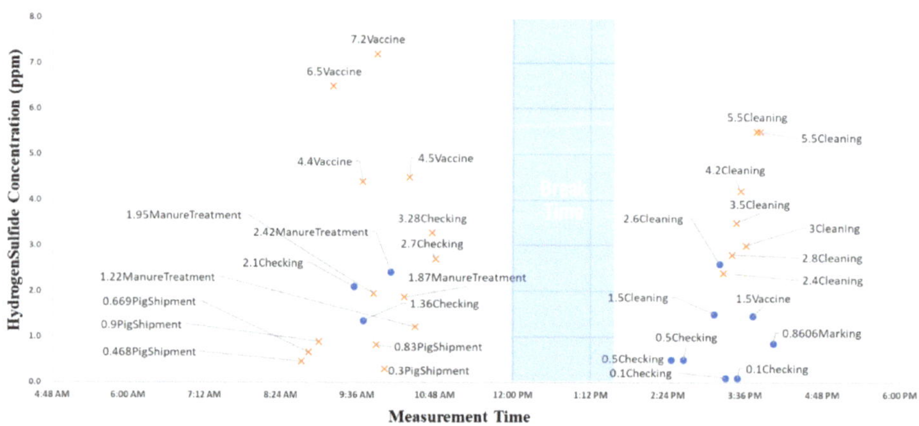

Figure 8. Average hydrogen sulfide concentration during contact & non-contact works in pig house.

The average exposure to hydrogen sulfide by working type was also analyzed (Figure 8). The highest hydrogen sulfide concentration was observed during the vaccine injection task, with an average exposure of 4.0 ppm. Internal cleaning, manure treatment, and condition checking tasks resulted in average exposures of 3.4 ppm, 1.9 ppm, and 1.5 ppm, respectively. The average levels of hydrogen sulfide exposure were compared between contact work and non-contact work. Non-contact work had an average exposure of 1.1 ppm, while direct contact work had an average exposure of 3.0 ppm, indicating an approximately 2.73-fold higher concentration in contact work. It should be noted that while hydrogen sulfide tends to be concentrated on the floor of the pigpen, the measurements in our experiment were taken at the worker's shoulder height, which corresponds to the height of the worker's respiratory system.

3.2. CFD Analysis for Ventilation Efficiency and Thermal Environment

The ventilation efficiency analysis was conducted through a CFD simulation for a total of 42 ventilation systems in the experimental pigpen at a height of 1.5 m (representing the worker's respiratory tract location) and 0.5 m (representing the pig's respiratory tract location). Table 2 provides the minimum, average, and maximum ventilation efficiencies at the corresponding heights, along with the individual analysis of ventilation efficiencies on the corridor side. Among the 50 ventilation combinations, the top 20% of structures that exhibited the highest ventilation efficiency at both the worker's height and the pig's height commonly utilized both upper and lower windows as inlets. These combinations demonstrated an average airflow exchange rate (AER·m^{-1}) of 0.34 at the worker's height and 0.4 AER·m^{-1} at the pig's height. Comparing these values with the average value of 0.28 AER·m^{-1} for the previously operated ventilation system, a 24% improvement in ventilation efficiency at the worker's height and a 43% improvement at the pig's height are observed at the same ventilation capacity with changing combinations of vents.

In cases where a single-floor window inlet structure was used, the analysis revealed higher ventilation efficiency when the upper windows were employed rather than the lower windows. Specifically, when only the upper windows were utilized, the average ventilation efficiency was 10.7% higher at the worker's height and 17.8% higher at the pig's height compared to using only the lower windows. Conversely, when both the upper and lower windows were utilized, the overall ventilation efficiency improved compared to using a single-floor window. This showed an average improvement of 25% at the worker's height and an average improvement of 45% at the pig's height. The larger improvement at the pig's height was attributed to the height of the inlet windows. Regarding the exhaust structure, the analysis results demonstrated that the ventilation efficiency improved in all ventilation structure combinations when two circular fans, which were previously only used in summer, were utilized instead of the three square-exhaust fans. On average, the efficiency improvement was 10.8% at the worker's height and 12% at the pig's height. While the three square fans were located on the walls of the left and right sides of the pigpen and in the center close to the corridors, the two circular fans were positioned in the center of the pigpen where the pigs were reared, resulting in higher ventilation efficiencies. Based on the ventilation efficiency analysis mentioned above, the airflow pattern based on the inlet and outlet combinations significantly affected the levels of ventilation efficiency for reducing harmful gases at both the worker's and the pig's heights.

Table 3 presents the thermal distribution analysis for uniformity and stability during the ventilation period. The temperature difference levels are indicated by different colors: white for normal, gray for caution, and black for warning. When the exhaust system was changed from the (B, LW) case to the (B, LW, CF) case, there was an increase in the proportion of sections representing small temperature changes of less than 3 °C. This increase was particularly significant in the case of a ventilation efficiency of 0.75 AER. The analysis indicated that increasing the inlets by opening the lower windows had an impact on the ammonia reduction rate. The change in the ventilation system using larger exhaust fans may lead to temperature drops affecting the breeding environment of the pigs.

3.3. Ventilation Operating Standard for Improving Working and Breeding Environments

When the ventilation was continuously operated for a long duration, it could lead to temperature drops affecting the thermal environment in the pigpen. The ventilation time is an important factor, not only for removing harmful gases but also for controlling internal thermal conditions. Figure 9 illustrates the ammonia reduction rate based on the ventilation capacity and time for each system. The maximum ammonia reduction rate increased significantly from 28.8% at a ventilation time of 60 s to 42.6% at a ventilation time of 180 s. However, there was no significant increase beyond 180 s, with the reduction rate remaining at 42.5% at a ventilation time of 300 s. Additionally, as the ventilation time increased, the temperature drop in the herd of pigs progressed from the caution to warning level. Therefore, the most efficient ventilation time for reducing ammonia gas can be determined to be up to 3 min, considering the balance between ammonia reduction and thermal stability.

Table 3. Temperature analysis at pig height (0.5 m) on improving ventilation system cases.

Case	Vent. R. (AER^{-1})	0.55		0.75		1.0	
	Vent. T. (°C)	x < 3	3 ≤ x < 5	x < 3	3 ≤ x < 5	x < 3	3 ≤ x < 5
Basic (Standard)	60 s	82.9	16.5	73.3	26.3	69	31
	180 s	75.9	23.7	64.6	33	59.2	38.4
	300 s	84.6	15.4	61	34.9	49.5	45.9
Upgrade-1 (B, LW)	60 s	66.3	31.6	44.1	49.8	67.8	32.2
	180 s	60.8	32.2	39.7	53	41.6	56
	300 s	66	30	33.0	56.2	39.4	58.1
Upgrade-2 (B, CF)	60 s	80.2	17.3	73.8	25.9	69.8	30.2
	180 s	81	18.9	64.9	33.7	56.2	41
	300 s	82.7	14.6	67	29.2	44.9	50.5
Upgrade-3 (B, LW, CF)	60 s	65.2	32.1	65.7	33.3	68.7	31.3
	180 s	67.1	29.4	71.4	26.7	40.5	55.7
	300 s	63.5	32.5	60.6	38.7	39.4	55.4

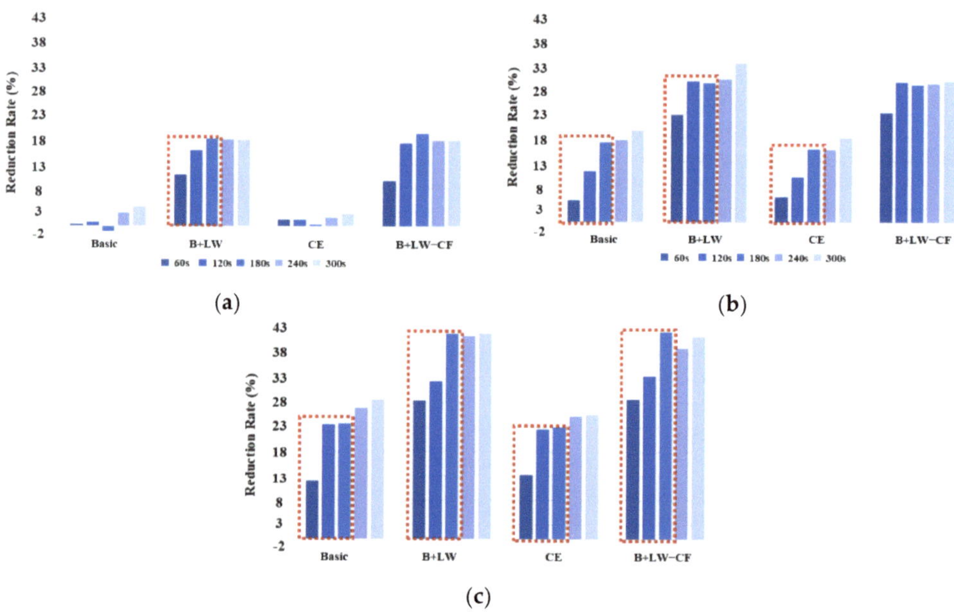

Figure 9. NH_3 reduction rate and ventilation time for each ventilation structure combination (red box is time of sharply increasing reduction rate). (**a**) 0.55AER; (**b**) 0.75AER; (**c**) 1.0AER.

Figure 10 illustrates the ventilation operating standard, which represented the appropriate ventilation volume and ventilation time in a single diagram, for the experimental pig house considering different ventilation structures for reducing harmful gases and controlling the breeding environment. It is possible to know the working and breeding environments in advance through the ventilation operating standard. For example, in Figure 10a, when ventilation was performed for 180 s with a ventilation volume of 0.85 AER/min, the removal rate of harmful gases at the worker's height was high, but the temperature at the pig's height was at the caution level. When the ventilation rate was maximum, the reduction rate of harmful gases was the highest for all cases. However, it is important to consider that the maximum ventilation rate can lead to sudden temperature drops, which can be problematic for the breeding environment. In the case of the existing ventilation system, the reduction rates of harmful gases were relatively low: 1.4% at 0.55 AER, 13.9% at 0.75 AER, and 23% at 1.0 AER. As the ventilation rate increased, the breeding environment deteriorated to the level of caution and warning in terms of thermal stability. This indicates that it is necessary to consider a balance between reducing harmful gases and maintaining a stable breeding environment when selecting the appropriate ventilation system.

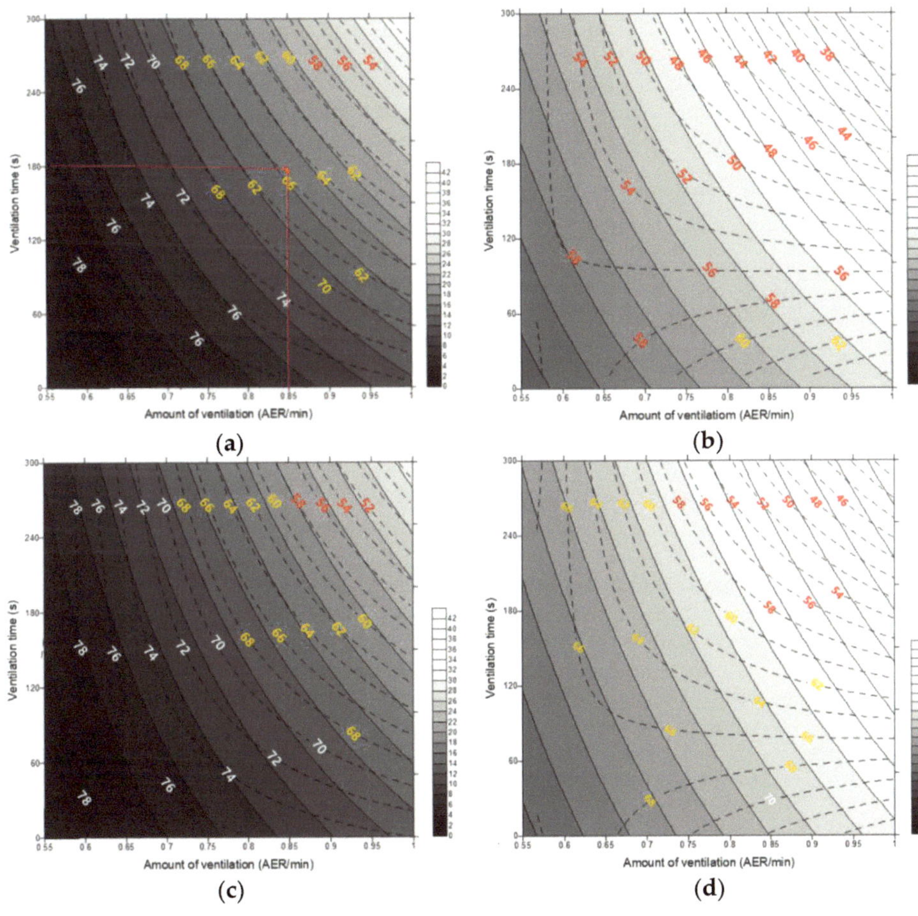

Figure 10. A ventilation diagram representing appropriate ventilation volume and time considering internal gas reduction at working height (solid line) with numeric values representing thermal suitability ratios at pig height (dot line) in colors as follows: white, yellow, and red refer to normal, caution, and warning conditions, respectively. (**a**) Basic (Standard); (**b**) Upgrade ① (B, LW); (**c**) Upgrade ② (B, CF); (**d**) Upgrade ①, ② (B, LW, CF).

Among the upgraded ventilation systems, the increased inlet area with the lower windows showed the highest ammonia gas reduction rate, averaging 27.7%, while the average reduction rates for different ventilation volumes were 16.2% at 0.55 AER, 29.3% at 0.75 AER, and 37.5% at 1.0 AER. When the ventilation efficiency was only considered to achieve the highest performance in reducing harmful gases, it also led to sudden temperature drops at the pig height. In most cases, the temperature differences reached warning levels within 1 min of ventilation time. Consequently, this ventilation system could cause problems for breeding conditions while removing harmful gases from the working environment (Figure 10b). When more exhaust fans were used with two large circular fans, the results showed a relatively low improvement in the working environment, averaging 12.1%. Specifically, this system achieved working environment improvement effects of 1.6% at 0.55 AER, 12.8% at 0.75 AER, and 22% at 1.0 AER (Figure 10c).

The ventilation system with increased inlets was effective in improving the working environment due to its successful reduction of harmful gases, while the system with increased outlets minimized the adverse effects on the breeding environment. When both upgraded measures were applied together, an overall improvement effect of 27.3% was achieved; ammonia reduction rates of 16.5% at 0.5 AER, 28.3% at 0.75 AER, and 37.2% at 1.0 AER (min^{-1}). As for temperature differences, the level did not exceed the warning threshold in most cases. By applying both upgraded ventilation structures, a high level of working environment improvement (27.3% improvement for workers) and temperature stability for the herd of pigs was achieved (Figure 10d).

4. Conclusions

This study comprehensively analyzed both the breeding environment and working environment, which should be considered simultaneously. Field measurements have shown that workers were exposed to high ammonia concentrations of 43 ppm on average when moving pigs inside the pigpen. When the worker performed a task that involved direct contact with pigs, they were exposed to a 1.58-fold higher ammonia concentration compared with that of non-contact work.

CFD (computational fluid dynamics) was used to make the optimal ventilation operation standard for improving both the working and breeding environments simultaneously in a pig house. Among 50 combinations of ventilation systems, the upgraded ventilation system was chosen considering the gas reduction rate and thermal stability. For this, it is important to appropriately adjust the ventilation volume and ventilation time, along with improvements in the ventilation structure. It is difficult to analyze this in real time on-site, and it is necessary to create standards for automatic control. Since the control standard always changes depending on the structure, it is important to set standards according to the structure. In this study, a single diagram was presented that could be evaluated on-site for the existing structure and three improved structures. The results can be used to make a livestock environment management algorithm for using IoT-based automatic precision control in the future, and for making precision agricultural systems considering upgraded working and breeding environments, simultaneously.

Author Contributions: Investigation, H.-J.S.; Writing-original draft, B.-W.O.; writing—review and editing, I.-H.S. All authors have read and agreed to the published version of the manuscript.

Funding: This research was funded by Korea Institute of Planning and Evaluation for Technology in Food, Agriculture and Forestry (IPET) through 2025 Livestock Industrialization Technology Development Program grant number [321085053SB010]. And The APC was funded by Ministry of Agriculture, Food and Rural Affairs (MAFRA).

Data Availability Statement: No new data were created or analyzed in this study. Data sharing is not applicable to this article.

Conflicts of Interest: The authors declare no conflict of interest.

References

1. Thorne, P.S.; Ansley, A.C.; Perry, S.S. Concentrations of bioaerosols, odors, and hydrogen sulfide inside and downwind from two types of swine livestock operations. *J. Occup. Environ. Hyg.* **2009**, *6*, 211–220. [CrossRef]
2. Gautam, R.; Heo, Y.; Lim, G.; Song, E.; Roque, K.; Lee, J.; Kim, Y.; Cho, A.; Shin, S.; Kim, C. Altered immune responses in broiler chicken husbandry workers and their association with endotoxin exposure. *Ind. Health* **2017**, *56*, 10–19. [CrossRef] [PubMed]

3. Park, K.-H.; Jeong, B.-Y. Characteristics and causes of musculoskeletal disorders for employees aged 50 years or older. *J. Ergon. Soc. Korea* **2009**, *28*, 139–145. [CrossRef]
4. Chapin, A.; Boulind, C.; Moore, A. Controlling odor and gaseous emission problems from industrial Swine facilities. *Yale Environ. Prot. Clin. Handb.* **1998**.
5. Bruce, J. Ventilation and Temperature Control Criteria for Pigs. *Environ. Asp. Hous. Anim. Production.* **1981**, 197–216.
6. Donham, K.J.; Scallon, L.J.; Popendorf, W.; Treuhaft, M.W.; Roberts, R.C. Characterization of dusts collected from swine confinement buildings. *Am. Ind. Hyg. Assoc. J.* **1986**, *47*, 404–410. [CrossRef] [PubMed]
7. Kim, C.-B.; Won, J.-U.; Roh, J.-H.; Lee, K.-H.; Lee, H.-J.; Lee, H.-S.; Lee, H.-J. Study on manifestation of allergic symptoms by cultivation of Cyperus exaltatus & manufacture of hwamunsok. *J. Agric. Med. Community Health* **1995**, *20*, 187–195.
8. Dutkiewicz, J. Bacteria and fungi in organic dust as potential health hazard. *Ann. Agric. Environ. Med.* **1997**, *4*, 11–16.
9. KIM, G.S.; Kim, E.A.; Lee, S.H.; Pakr, H.S. A Study on Risk Factors of Grain Dust-Induced Occupational Asthma in Grain Feedmill Workers. *Korean J. Occup. Environ. Med.* **1997**, *9*, 628–640. [CrossRef]
10. Lim, H.-S.; Bae, G.-R.; Kim, D.-H.; Jun, K.-H. Respiratory symptoms occurring among mushroom farmers. *J. Agric. Med. Community Health* **1998**, *23*, 259–268.
11. Bottcher, R.W. An environmental nuisance: Odor concentrated and transported by dust. *Chem. Senses* **2001**, *26*, 327–331. [CrossRef] [PubMed]
12. Eom, H.-J. Beware of swine respiratory diseases in winter. *Feed J.* **2006**, *4*, 106–110.
13. Oh, B.-W.; Lee, S.-W.; Kim, H.-C.; Seo, I.-H. Analysis of working environment and ventilation efficiency in pig house using computational fluid dynamics. *J. Korean Soc. Agric. Eng.* **2019**, *61*, 85–95. [CrossRef]
14. Lee, S.-W.; Kim, H.-C.; Seo, I.-H. Monitoring of Internal Harmful Factors According to Environmental Factors in Pig Farm. *J. Korean Soc. Agric. Eng.* **2020**, *62*, 105–115. [CrossRef]
15. Bjerg, B.; Svidt, K.; Zhang, G.; Morsing, S.; Johnsen, J.O. Modeling of air inlets in CFD prediction of airflow in ventilated animal houses. *Comput. Electron. Agric.* **2002**, *34*, 223–235. [CrossRef]
16. Seo, I.-H.; Lee, I.-B.; Hong, S.-W.; Hwang, H.-S.; Bitog, J.P.; Yoo, J.-I.; Kwon, K.-S.; Ha, T.-H.; Kim, H.-T. Development of a CFD model to study ventilation efficiency of mechanically ventilated pig house. *J. Korean Soc. Agric. Eng.* **2008**, *50*, 25–37. [CrossRef]
17. Seo, I.-H.; Lee, I.-B.; Moon, O.-K.; Hong, S.-W.; Hwang, H.-S.; Bitog, J.P.; Kwon, K.-S.; Ye, Z.; Lee, J.-W. Modelling of internal environmental conditions in a full-scale commercial pig house containing animals. *Biosyst. Eng.* **2012**, *111*, 91–106. [CrossRef]
18. Hong, S.-W.; Exadaktylos, V.; Lee, I.-B.; Amon, T.; Youssef, A.; Norton, T.; Berckmans, D. Validation of an open source CFD code to simulate natural ventilation for agricultural buildings. *Comput. Electron. Agric.* **2017**, *138*, 80–91. [CrossRef]
19. Li, H.; Rong, L.; Zhang, G. Reliability of turbulence models and mesh types for CFD simulations of a mechanically ventilated pig house containing animals. *Biosyst. Eng.* **2017**, *161*, 37–52. [CrossRef]
20. Wang, K.; Wang, X.; Wu, B. Assessment of hygrothermal conditions in a farrowing room with a wet-pad cooling system based on CFD simulation and field measurements. *Trans. ASABE* **2014**, *57*, 1493–1500. [CrossRef]
21. Rong, L.; Nielsen, P.V.; Bjerg, B.; Zhang, G. Summary of best guidelines and validation of CFD modeling in livestock buildings to ensure prediction quality. *Comput. Electron. Agric.* **2016**, *121*, 180–190. [CrossRef]
22. Berckmans, D. General introduction to precision livestock farming. *Anim. Front.* **2017**, *7*, 6–11. [CrossRef]
23. Kim, R.-W.; Lee, I.-B.; Ha, T.-H.; Yeo, U.-H.; Lee, S.-Y.; Lee, M.-H.; Park, G.-Y.; Kim, J.-G. Development of CFD model for predicting ventilation rate based on age of air theory using thermal distribution data in pig house. *J. Korean Soc. Agric. Eng.* **2017**, *59*, 61–71. [CrossRef]
24. Li, H.; Rong, L.; Zhang, G. Numerical study on the convective heat transfer of fattening pig in groups in a mechanical ventilated pig house. *Comput. Electron. Agric.* **2018**, *149*, 90–100. [CrossRef]
25. Kwon, K.-S.; Lee, I.-B.; Ha, T. Identification of key factors for dust generation in a nursery pig house and evaluation of dust reduction efficiency using a CFD technique. *Biosyst. Eng.* **2016**, *151*, 28–52. [CrossRef]
26. Reynolds, A. A model for predicting airborne dust concentrations within a ventilated airspace. *J. Agric. Eng. Res.* **1997**, *66*, 103–109. [CrossRef]
27. Rong, L.; Elhadidi, B.; Khalifa, H.; Nielsen, P.V. CFD modeling of airflow in a livestock building. *IAQVEC* **2010**, *2010*, 7.
28. Rong, L.; Elhadidi, B.; Khalifa, H.E.; Nielsen, P.V.; Zhang, G. Validation of CFD simulation for ammonia emissions from an aqueous solution. *Comput. Electron. Agric.* **2011**, *75*, 261–271. [CrossRef]
29. Saha, C.K.; Wu, W.; Zhang, G.; Bjerg, B. Assessing effect of wind tunnel sizes on air velocity and concentration boundary layers and on ammonia emission estimation using computational fluid dynamics (CFD). *Comput. Electron. Agric.* **2011**, *78*, 49–60. [CrossRef]
30. Wu, W.; Zhang, G.; Bjerg, B.; Nielsen, P.V. An assessment of a partial pit ventilation system to reduce emission under slatted floor—Part 2: Feasibility of CFD prediction using RANS turbulence models. *Comput. Electron. Agric.* **2012**, *83*, 134–142. [CrossRef]
31. Bjerg, B.; Cascone, G.; Lee, I.-B.; Bartzanas, T.; Norton, T.; Hong, S.-W.; Seo, I.-H.; Banhazi, T.; Liberati, P.; Marucci, A. Modelling of ammonia emissions from naturally ventilated livestock buildings. Part 3: CFD modelling. *Biosyst. Eng.* **2013**, *116*, 259–275. [CrossRef]
32. Wu, W.; Zong, C.; Zhang, G. Comparisons of two numerical approaches to simulate slatted floor of a slurry pit model–large eddy simulations. *Comput. Electron. Agric.* **2013**, *93*, 78–89. [CrossRef]

33. Zong, C.; Zhang, G. Numerical modelling of airflow and gas dispersion in the pit headspace via slatted floor: Comparison of two modelling approaches. *Comput. Electron. Agric.* **2014**, *109*, 200–211. [CrossRef]
34. Rong, L.; Nielsen, P.V.; Zhang, G. Experimental and numerical study on effects of airflow and aqueous ammonium solution temperature on ammonia mass transfer coefficient. *J. Air Waste Manag. Assoc.* **2010**, *60*, 419–428. [CrossRef] [PubMed]
35. Xia, L.; Zou, Z.-J.; Wang, Z.-H.; Zou, L.; Gao, H. Surrogate model based uncertainty quantification of CFD simulations of the viscous flow around a ship advancing in shallow water. *Ocean Eng.* **2021**, *234*, 109206. [CrossRef]
36. Cravero, C.; De Domenico, D.; Marsano, D. The use of uncertainty quantification and numerical optimization to support the design and operation management of air-staging gas recirculation strategies in glass furnaces. *Fluids* **2023**, *8*, 76. [CrossRef]
37. Cravero, C.; De Domenico, D.; Marsano, D. Uncertainty Quantification Analysis of Exhaust Gas Plume in a Crosswind. *Energies* **2023**, *16*, 3549. [CrossRef]
38. Fluent, A. R2 User's Guide. Available online: https://forum.ansys.com/uploads/846/SCJEU0NN8IHX.pdf (accessed on 23 June 2021).
39. Seo, I.-H.; Lee, I.-B.; Moon, O.-K.; Kim, H.-T.; Hwang, H.-S.; Hong, S.-W.; Bitog, J.; Yoo, J.-I.; Kwon, K.-S.; Kim, Y.-H. Improvement of the ventilation system of a naturally ventilated broiler house in the cold season using computational simulations. *Biosyst. Eng.* **2009**, *104*, 106–117. [CrossRef]
40. Kwon, K.-S.; Lee, I.-B.; Hwang, H.-S.; Bitog, J.P.; Hong, S.-W.; Seo, I.-H.; Choi, J.-S.; Song, S.-H.; Moon, O.-K. Analysis on the optimum location of an wet air cleaner in a livestock house using CFD technology. *J. Korean Soc. Agric. Eng.* **2010**, *52*, 19–29. [CrossRef]
41. Lee, S.-Y.; Lee, I.-B.; Kwon, K.-S.; Ha, T.-H.; Yeo, U.-H.; Park, S.-J.; Kim, R.-W.; Jo, Y.-S.; Lee, S.-N. Analysis of Natural Ventilation Rates of Venlo-type Greenhouse Built on Reclaimed Lands using CFD. *J. Korean Soc. Agric. Eng.* **2015**, *57*, 21–33. [CrossRef]
42. Lee, S.-Y.; Lee, I.-B.; Kim, R.-W. Evaluation of wind-driven natural ventilation of single-span greenhouses built on reclaimed coastal land. *Biosyst. Eng.* **2018**, *171*, 120–142. [CrossRef]
43. Sun, H.; Keener, H.M.; Deng, W.; Michel, F.C., Jr. Development and validation of 3-d CFD models to simulate airflow and ammonia distribution in a high-rise™ hog building during summer and winter conditions. *Agric. Eng. Int. CIGR J.* **2004**, 6.
44. Rong, L.; Aarnink, A. Development of ammonia mass transfer coefficient models for the atmosphere above two types of the slatted floors in a pig house using computational fluid dynamics. *Biosyst. Eng.* **2019**, *183*, 13–25. [CrossRef]

Disclaimer/Publisher's Note: The statements, opinions and data contained in all publications are solely those of the individual author(s) and contributor(s) and not of MDPI and/or the editor(s). MDPI and/or the editor(s) disclaim responsibility for any injury to people or property resulting from any ideas, methods, instructions or products referred to in the content.

sustainability

MDPI

Article

Identifying the Role of Biostimulants in Turnip (*Brassica rapa* L.) Production Compared with Chemical Fertilization

Qurat-Ul-Ain Raza [1,†], Muhammad Amjad Bashir [1,*,†], Abdur Rehim [1,*], Yucong Geng [2,*], Hafiz Muhammad Ali Raza [1,3], Sajjad Hussain [4], Ijaz Ahmad [5] and Muhammad Wasif [3]

1 Department of Soil Science, Faculty of Agricultural Sciences and Technology, Bahauddin Zakariya University, Multan 60800, Pakistan
2 School of History, Geography and Tourism, Chengdu Normal University, Chengdu 610000, China
3 Department of Soil Science, College of Agriculture, University of Layyah, Layyah 31200, Pakistan
4 Department of Horticulture, Faculty of Agricultural Sciences and Technology, Bahauddin Zakariya University, Multan 60800, Pakistan
5 Department of Agronomy, Engro Fertilizers Limited, Lahore 54000, Pakistan
* Correspondence: amjad.bashir941@gmail.com (M.A.B.); abdur.rehim@bzu.edu.pk (A.R.); tiger86gyc@gmail.com (Y.G.)
† These authors contributed equally to this work.

Abstract: Chemical fertilizers play an integral role in advancing food production to combat the increasing food challenges and ensure food security. Unfortunately, the overuse of these fertilizers has negatively influenced the soil and the environment. Considering this aspect, two pot experiments were performed to evaluate the efficacy of biostimulants in vegetable production systems. The first experiment compared the effects of chemical fertilizers (CF) with glycine (GL), aspartic acid (AA), lysine (LY), and vitamin B complex (VB). The plant's physiological and morphological attributes and yield were studied. The results confirmed that VB has the potential to improve the rate of transpiration (26%), total chlorophyll content (27%), root diameter (213%), and dry matter (289%) compared with CF. In the second experiment, the effects of chemical fertilizers (CF) were compared with Isabion® (I), 25% CF + GL + LY (B1), 25% CF + GL + AA (B2), and 25% CF + AA + LY (B3). Similar attributes were analyzed to identify the influence of the applied treatments on turnip production. The results demonstrated that B2 enhanced the rate of photosynthesis (963%), transpiration (254%), and stomatal conductance (76%). Moreover, B1 improved the plant's fresh weight (6%) and moisture contents (4%) compared to CF. In conclusion, biostimulants (LY, VB, and B1) are capable of improving turnip performance and production compared to CF. Future studies must focus on the efficiency of biostimulants against the long-term application effects on soils, nutrient-use efficiency, and crop production. Furthermore, the mechanism of action needs to be addressed in the future.

Keywords: amino acids; leaf extract; vitamins; lysine; foliar application

Citation: Raza, Q.-U.-A.; Bashir, M.A.; Rehim, A.; Geng, Y.; Raza, H.M.A.; Hussain, S.; Ahmad, I.; Wasif, M. Identifying the Role of Biostimulants in Turnip (*Brassica rapa* L.) Production Compared with Chemical Fertilization. *Sustainability* **2023**, *15*, 11851. https://doi.org/10.3390/su151511851

Academic Editor: Luigi Ledda

Received: 28 June 2023
Revised: 21 July 2023
Accepted: 27 July 2023
Published: 1 August 2023

Copyright: © 2023 by the authors. Licensee MDPI, Basel, Switzerland. This article is an open access article distributed under the terms and conditions of the Creative Commons Attribution (CC BY) license (https://creativecommons.org/licenses/by/4.0/).

1. Introduction

Vegetables are a good source of biologically active substances [1], calories, and nutrients; thus, they are highly recommended for humans' daily diet. Plant products are a major part of the human diet and provide dietary fiber, minerals, and phytochemicals that are beneficial for human health. Studies reported that a vegetable-rich diet helps maintain bone strength, blood pressure, heart and mental health, and hormonal production, in addition to providing numerous other health benefits. Moreover, vegetables can also be used to fight against undernutrition, hunger, and poverty and grow the economy even if grown and consumed locally [2]. With the increasing world population, it has been estimated that approximately 9.7 billion people will need to be fed by 2050 [3]. During the last two decades (1997–2017), vegetable production has doubled (from 0.55 to 1.09 billion); therefore, it is essential to provide favorable soil and environmental conditions to fulfill the demands for vegetables in the future [4,5]. The turnip (*Brassica rapa*) is a vital root vegetable and forage crop. It is indigenous to Asia, Europe, Russia, and the Near East and is now widely cultivated throughout the globe as both a vegetable and oil source [6]. In Pakistan, the turnip was cultivated on 9609 hectares of land with 167,065 tons of production in 2020–2021, making it an important vegetable locally [7]. Furthermore, it has also been reported to have various medicinal benefits, including acting as a therapeutic agent against kidney and liver diseases and other ailments. It has also been discovered that turnip has antimicrobial, anti-inflammatory, antitumor, cardio-protective, antidiabetic, analgesic, and nephron-protective properties and aids against metabolic syndrome and obesity [8]. Moreover, it improves growth in response to heavy fertilization [9], which is also a cause of non-point pollution.

Modern agriculture is highly dependent on fertilizers' application for achieving higher crop productivity and is a crucial tool to target food safety challenges. Chemical fertilizers greatly contribute to improving crop productivity and soil nutrient pools. However, the troublesome impacts of high-dose chemical fertilization cannot be ignored in environmental deterioration. The excessive use of chemical fertilizers has adversely affected the soil eco-profile; altered the soil's biological and physicochemical properties; declined soil organic matter; reduced soil fertility; disturbed biodiversity; and polluted the air, water, and soil, respectively [10]. In addition, the excessive use of nitrogenous fertilizers has increased the concentration of nitrates in many vegetables, becoming a significant source of nitrate for humans [11]. The scientific community and agricultural sectors face a massive challenge to enhance crop productivity to feed the growing population worldwide and reduce the hazardous impacts on the environment and human health [12].

Various technologies have been introduced to boost vegetables' growth and minimize the toxic agriculture inputs, i.e., fertilizers and pesticides [13–15]. One of the emerging technologies to proliferate sustainable agriculture production is the use of biostimulants [16]. According to the new European Union regulation (EU; 2019/1009), biostimulants are a product that has the potential to improve the plant's nutritional processes and achieve the following properties to improve nutrient uptake: tolerance against biotic/abiotic stresses, enhanced quality traits, and ability to contribute to nutrients' availability in the rhizosphere [17]. Biostimulants are a mixture of organic and inorganic materials that can improve a crop's growth, productivity, quality, and nutrient uptake. Biostimulants cannot replace fertilizers because they are not nutrients but effectively reduce the requirement for fertilizers and overcome nutrient deficiency [18]. The global market of biostimulants reached USD 2.5 billion in 2019 and is one of the fastest-growing agriculture-associated industries, ultimately reducing the chemical fertilizers market [19].

The beneficial responses of biostimulants regarding improved crop yield have been reported in recent years. Combining zinc–lysine chelates and zinc-solubilizing bacteria significantly improved maize (*Zea mays* L.) yield and grain biofortification [20]. Moreover, growing plants hydroponically improved the pigment content in leaves and positively influenced yellow pepper (*Capsicum annum* L.) yield with biostimulants [21]. Without chemical fertilizers, radish (*Raphanus sativus* L.) root and shoot biomass increased with the integrated use of biostimulants under greenhouse conditions [18]. It is also reported that strawberry (*Fragaria* × *ananassa*) plants can benefit from quality and yield traits with biostimulants under insufficient nutrients [22]. Tomato *Solanum lycopersicum* L. plants also responded positively to exogenously applied amino-acids-based biostimulants, which promoted plant-growth processes [23]. Vitamin B was also reported to improve plants' oxidative stress, ultimately enhancing yield [24] and increasing root growth [18]. Thus, various biostimulants have been reported to improve yield in numerous crops, including the bean (*Phaseolus vulgaris* L.), with an increased number of pods (26%) [25]; Lavandin (*Lavandula* × *intermedia*), with improved fresh weight (47%) and dry weight (38%); white radish, with enhanced fresh weight (478%) [3]; and red radish, with increased root biomass (65%) [18]. However, the impact on the nutritional and functional properties of edible parts might differ depending upon the species, dose, properties of biostimulants, and time of application [8].

The need for this study arises from the necessity to find sustainable solutions for turnip cultivation under stress-less conditions and find an eco-friendly approach to conventional agricultural practices, especially over-fertilization. This study aimed to identify the use of biostimulants for the performance of turnips and improve their performance compared with chemical fertilization. With this aim, two pot experiments were performed with sole and combined application of biostimulants. Our study hypothesized that applying biostimulants would help minimize the use of chemical fertilizers. Prior to the extent of the authors' knowledge, experiments evaluating the effect of exogenously applied biostimulants on turnip growth with reduced fertilization in stress-less environments have not been reported yet. Moreover, various alternatives to chemical fertilization have been reported, but the use of biostimulants and their combinations under normal soil, environmental, and climatic conditions to support turnip growth and its impacts on plant physiology in the absence or limitation of chemical fertilizers endows it with the distinction of being a novel study. Hence, the objectives are to (i) evaluate the influences of biostimulants on turnip growth, photosynthetic activity, and yield, (ii) study the effects of biostimulants in the absence/presence of chemical fertilizers, and (iii) reduce the application of chemical fertilizers to minimize its negative impacts on the environment.

2. Materials and Methods

This study was organized into two pot experiments to evaluate the response of turnips with the application of biostimulants.

2.1. Experimental Details

The first experiment was performed at the experimental station of the Department of Soil Science (30.258° E, 71.515° N), FAS&T, Bahauddin Zakariya University (BZU) Multan, Pakistan. Multan is located in a semi-arid region with hot summers and mild winters. The maximum temperature during the growing season was 36 °C, and the minimum temperature was 7 °C. Multan is the pronounced part of Bari Doab, Punjab plains, and its elevation is 114 to 135 m above sea level, geographically located in the center of Pakistan [26]. This study was conducted during the growing season of 2021 (September–December).

The second study was performed at the experimental area of the College of Agriculture, BZU, Bahadur Sub-campus Layyah (30.97° E, 70.96° N) Pakistan. Layyah City lies in a desert climate with hot days, cold nights, and little annual rainfall. The average annual temperature is 25 °C in Layyah, and during the experimental season, it ranged from 6–36 °C. The trial was conducted from October to January 2021.

2.1.1. Experiment 1

The pots were lined using plastic sheets, and 20 kg of soil was added. The physicochemical properties of the soil used in this experiment are given in Table 1. This study was focused on five treatments with three replicates and was organized in a completely randomized design (CRD). The following treatments were tested:

- Chemical fertilizer (CF; N 62 kg ha^{-1}, P 49 kg ha^{-1}, K 62 kg ha^{-1});
- Glycine (GL; 2.47 g pot^{-1});
- Lysine (LY; 4.79 g pot^{-1});
- Aspartic Acid (AA; 4.37 g pot^{-1});
- Vitamin B complex (VB; 0.46 g pot^{-1}).

The treatments' doses were kept the same according to nitrogen (N) requirements (62 kg ha^{-1}) of turnip. In the case of CF, urea, di-ammonium phosphate (DAP), and sulphate of potash (SOP) were used as a source of N, phosphorus (P), and potassium (K). Chemical fertilizers were not applied to any treatment except CF. Turnip seeds were soaked in water overnight before sowing, and five seeds (*cultivar*; Purple top white globe) per pot were sown on 24 September 2021. Later, two plants per pot were maintained after thinning. Once the seedlings were established, four foliar applications of each treatment were made at ten days intervals on 15 October, 25 October, 5 November, and 15 November 2021. The crop was harvested at maturity on 9 December 2021. Plants were irrigated according to the requirement by maintaining the field capacity level, and weeds were eradicated manually. Plant root and shoot samples were collected at maturity and used to calculate the yield, physiology, and morphological attributes.

Table 1. Physio-chemical properties of soil.

Soil Properties	Experiment 1	Experiment 2
Soil pH	7.7	7.8
Electrical conductivity ($dS{\cdot}m^{-1}$)	0.5	0.1
Organic matter (%)	0.4	0.7
Saturation (%)	34	28
Texture (USDA System)	Loam	Sandy Loam
Soil total N ($mg\ kg^{-1}$)	221	450
Soil extractable P ($mg\ kg^{-1}$)	8.2	7.1
Soil extractable K ($mg\ kg^{-1}$)	30.1	62.34

2.1.2. Experiment 2

Soil (8 kg) was filled in plastic sheet-lined pots. The soil physicochemical characteristics in this experiment are given in Table 1. Five treatments with three replicates included in the experiment are as follows:

- Recommended chemical fertilizer (62 kg ha^{-1} N, 49 kg ha^{-1} P, K 62 kg ha^{-1} K) (CF);
- Isabion® (contains 100 g L^{-1} amino acids; 1 L ha^{-1}) (I);
- Treatment with 25% CF + Glycine (5 g L^{-1}) + Lysine (1 g L^{-1}) (B1);
- Treatment with 25% CF + Glycine (5 g L^{-1}) + Aspartic acid (2 g L^{-1}) (B2);
- Treatment with 25% CF + Lysine (1 g L^{-1}) + Aspartic acid (2 g L^{-1}) (B3).

Treatment application methods and practices were identical to the first experiment. Five seeds were sown on 21 October 2021 in each pot. Later two healthy plants per pot were later maintained, and the smaller ones were removed. Foliar applications of treatments were carried out on 16 November, 26 November, 6 December, and 16 December 2021. Irrigation was applied according to the crop need, while weed eradication was conducted manually. Turnip roots and shoots were harvested at maturity on 13 January 2022, and all the samples were collected, washed, and used for the required analyses. GL, LY, and AA were purchased from Sigma Aldrich (St. Louis, MO, USA), while VB was purchased from Martin Dow Market Ltd. (Karachi, Pakistan).

2.2. Laboratory Analysis

Before harvesting, two healthy leaves from each replication were selected, and plant photosynthetic parameters, i.e., photosynthetic and transpiration rates, internal CO_2 concentrations, and stomatal conductance were measured in the morning (between 08:30 and 9:30 a.m.) using an infrared gas analyzer (IRGA; Analytical Development Company, Hoddesdon, UK). Photosynthetic pigments, including chlorophyll, were estimated using the colorimetric method. One healthy leaf from each replication was collected, cut into small pieces, dipped in ethanol, and stored in the dark for 24 h for chlorophyll extraction. The spectrophotometer (UV 1600 Shimadzu, Kyoto, Japan) was used to determine chlorophyll a and b contents at 665 and 649 nm wavelengths. Analysis was performed in two replicates to minimize the error [27]. Plant fresh weight was determined immediately after harvesting, while the plants were oven dried at 65 °C until a constant weight was achieved. Plant fresh and dry weights were recorded using a weighing balance. Moreover, turnip root diameter was determined using digital Vernier calipers.

2.3. Statistical Analysis

The one-way analysis of variance (ANOVA) was carried out to analyze the significance of data using Statistix 9. The Least Significance Difference (LSD) at $p \leq 0.05$ was used for pairwise comparison. R studio (2022.12.0) and Microsoft Excel (2016) were the critical software for data processing and graphs preparation.

3. Results

3.1. Experiment 1

The results demonstrated that biostimulants are potentially important in sustainable agriculture (Figure 1). A significant improvement in photosynthetic rate with the use of AA (165%), followed by LY (164%), VB (158%), and GL (128%), compared to CF, was observed (Figure 1A). Transpiration was also increased significantly with VB (26%) and AA (12%) (Figure 1B), whereas stomatal conductance and internal CO_2 were reduced with biostimulants compared to CF (Figure 1C,D). In addition, the photosynthetic pigments (chlorophyll a + b) were significantly improved with the application of VB (26.95%), followed by AA (18%), LY (14%), and GL (5%) compared to CF (Figure 2A). CF significantly enhanced moisture content, followed by GL, LY, AA, and VB (Figure 2B).

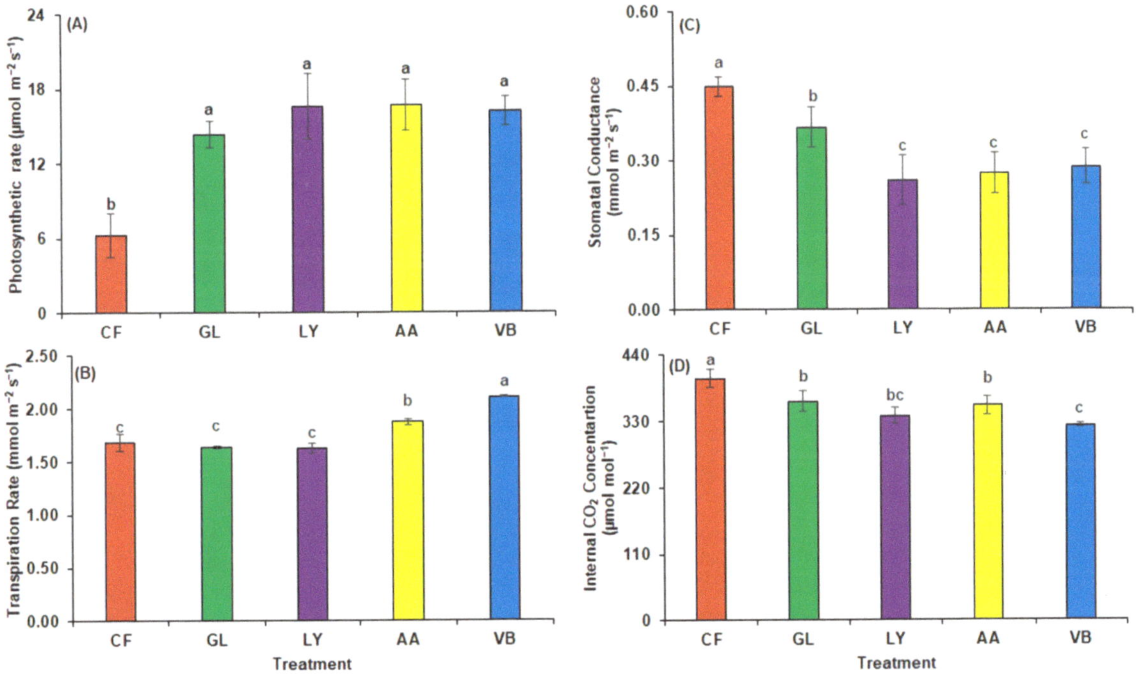

Figure 1. Effect of biostimulants on (**A**) photosynthetic rate, (**B**) transpiration rate, (**C**) stomatal conductance, and (**D**) internal CO_2 concentration, where CF is chemical fertilizer, GL is glycine, LY is lysine, AA is aspartic acid, and VB is vitamin B complex. The data sets are presented as means ± SD (n = 3). Columns that share the same letter(s) are not statistically significant at $p \leq 0.05$.

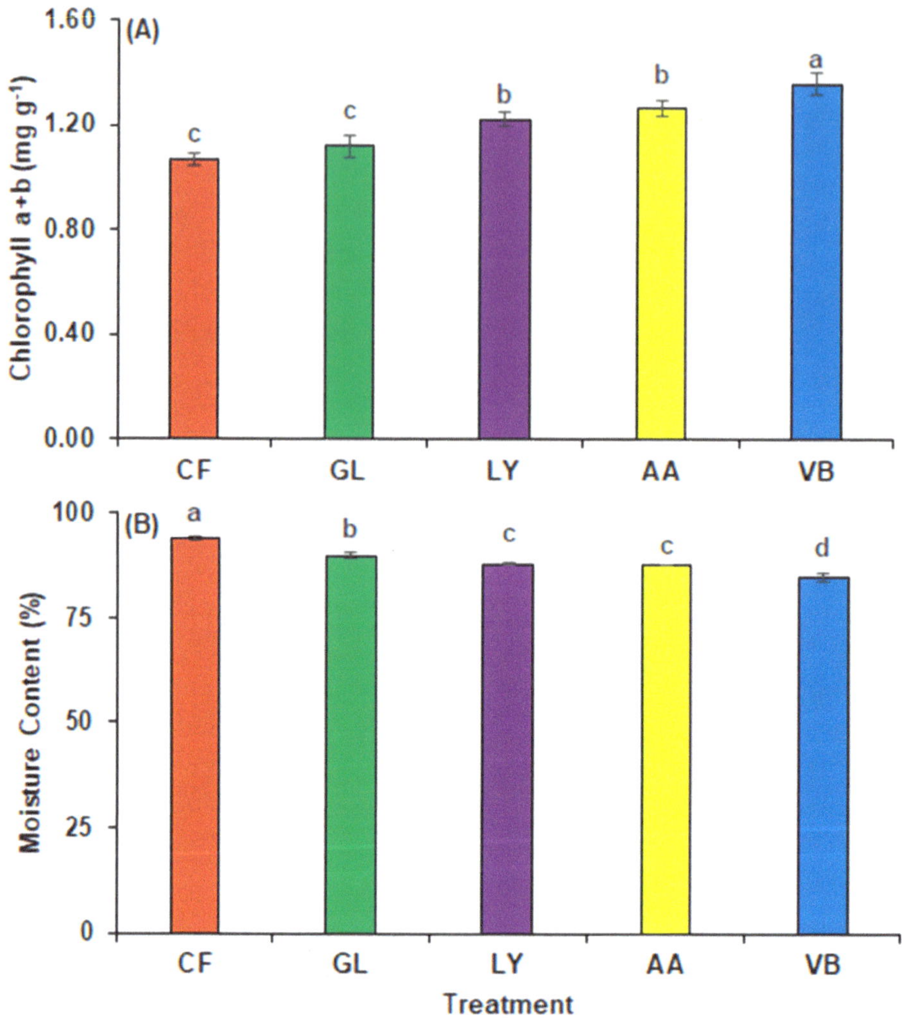

Figure 2. Pigment and moisture content analysis: (**A**) photosynthetic pigments and (**B**) moisture content. CF is chemical fertilizer, GL is glycine, LY is lysine, AA is aspartic acid, and VB is vitamin B complex. The data sets are presented as means± SD (n = 3). Columns that share the same letter(s) are not statistically significant at $p \leq 0.05$.

The plant fresh weight (shoot + root) metric was significantly improved using biostimulants (Figure 3A). LY improved the fresh weight (82%), followed by AA (72%), VB (54%), and GL (51%) in comparison to CF. An increase in plant dry weight was observed with the use of VB (289%), LY (271%), AA (257%), and GL (160%), respectively (Figure 3B). The root diameter metric was increased in VB (213%), AA (185%), GL (164%), and LY (60%) compared with CF (Figure 3C).

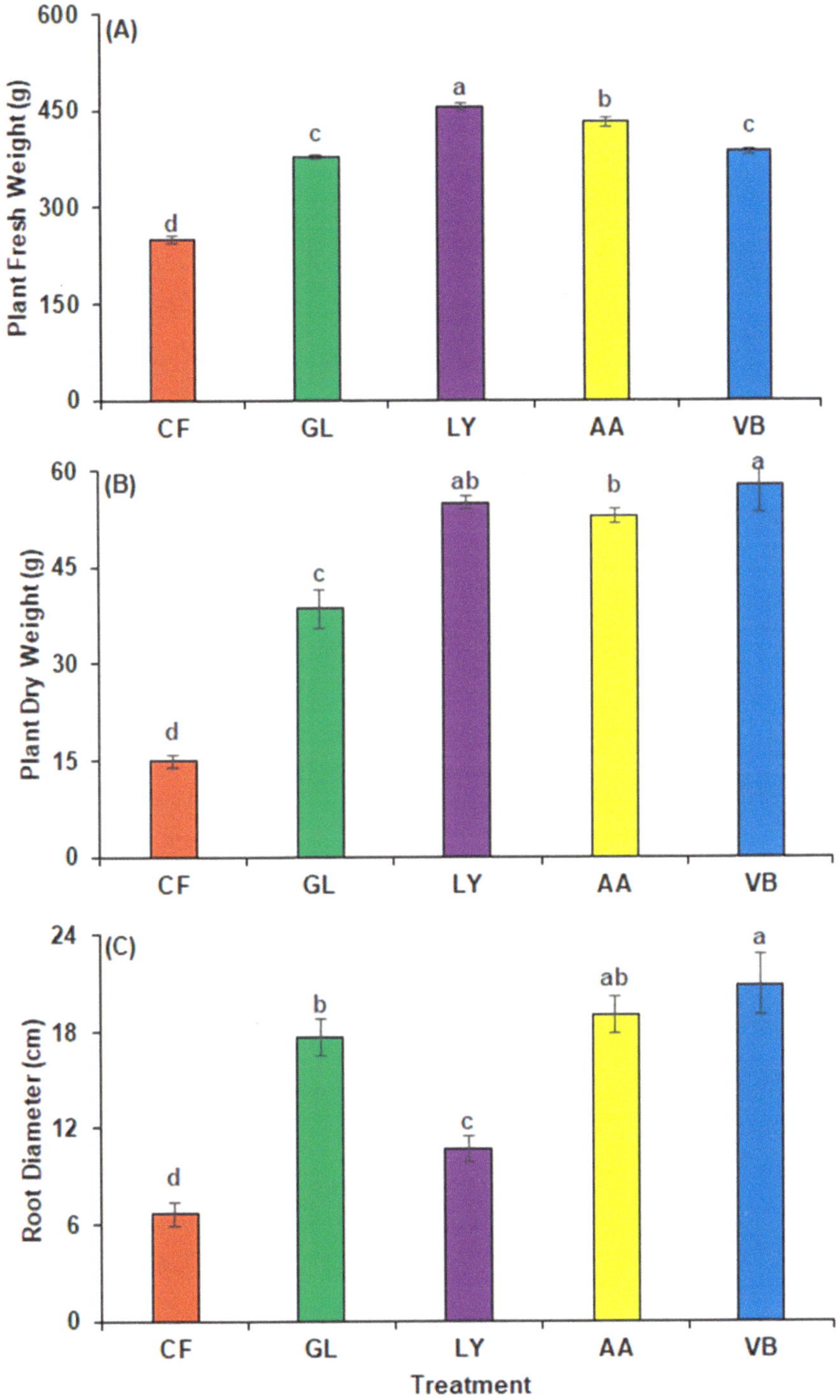

Figure 3. Plant biomass and morphology: (**A**) plant fresh weight, (**B**) plant dry weight, (**C**) root diameter. CF is chemical fertilizer, GL is glycine, LY is lysine, AA is aspartic acid, and VB is vitamin B complex. The data sets are presented as means± SD (n = 3). Columns that share the same letter(s) are not statistically significant at $p \leq 0.05$.

3.2. Experiment 2

The integrated application of biostimulants and chemical fertilizers significantly improved the photosynthetic rates in turnip plants (Figure 4A). Compared to CF, the photosynthetic rate was highest in B2 (963%), followed by B3 (491%), B1 (100%), and I (21%), respectively (Figure 5A). Transpiration was also increased by 254%, 234%, 188%, and 169% in B2, B3, B1, and I, respectively (Figure 4B). Stomatal conductance also experienced significant benefits with the use of B2 (76%), followed by B3 (58%), I (40%), and B1 (29%), respectively (Figure 4C). Internal CO_2 concentration declined as CF (341 µmol mol^{-1}) and I (346 µmol mol^{-1}) showed the highest values (Figure 4D).

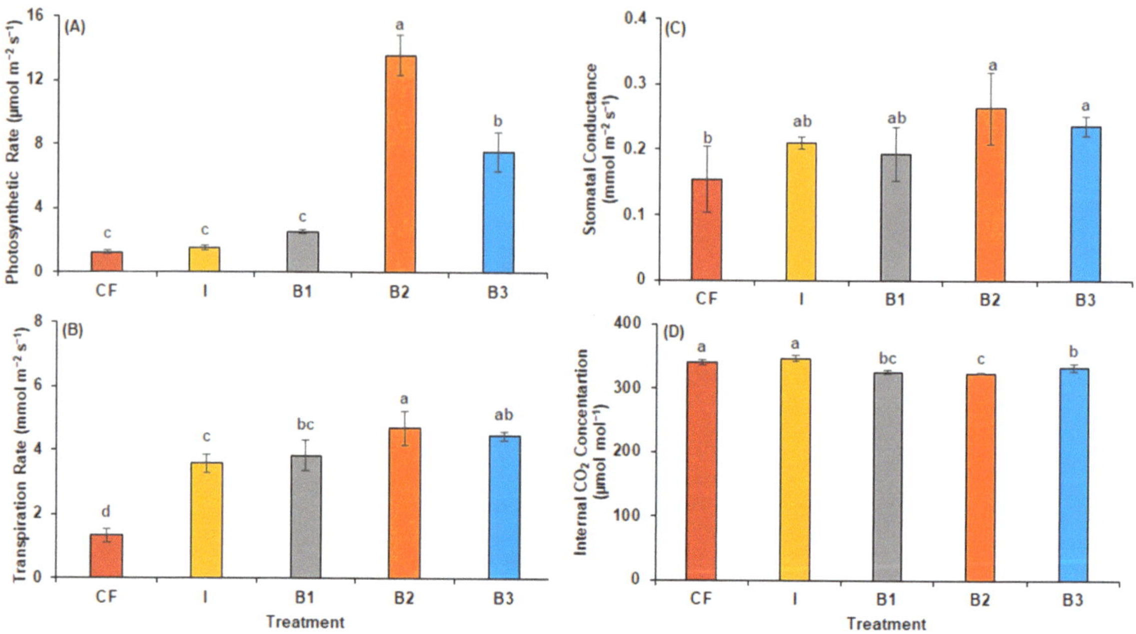

Figure 4. Effect of biostimulants on (**A**) photosynthetic rate, (**B**) transpiration rate, (**C**) stomatal conductance, and (**D**) internal CO_2 concentration, where CF is chemical fertilizer, I is Isabion, B1 is 25% CF + glycine + lysine, B2 is 25% CF + glycine + aspartic acid, and B3 is 25% CF + aspartic acid + lysine. The data sets are presented as means± SD (n = 3). Columns that share the same letter(s) are not statistically significant at $p \leq 0.05$.

Photosynthetic pigments (chlorophyll a + b) were improved with the application of I (98%), B2 (83%), B3 (75%), and B1 (73%), respectively, compared to CF (Figure 5A). Moisture contents were increased with the application of B1, I, and B3 by 4%, 3%, and 2% compared to CF (Figure 5B). Plant fresh weight (shoot + root) was increased in B1 (6%), while the other treatments, I (67.00 g), B2 (50.00 g), and B3 (47.67 g), showed negative results as compared to CF (102.67 g), respectively (Figure 6A). Moreover, dry weight was highest in CF, followed by B1, B2, B3, and I (Figure 6B). In addition, CF and B1 improved the root diameter, while the other treatments, I (2.88 cm), B2 (3.31 cm), and B3 (3.05 cm), showed negative results compared to CF (3.79 cm), respectively (Figure 6C).

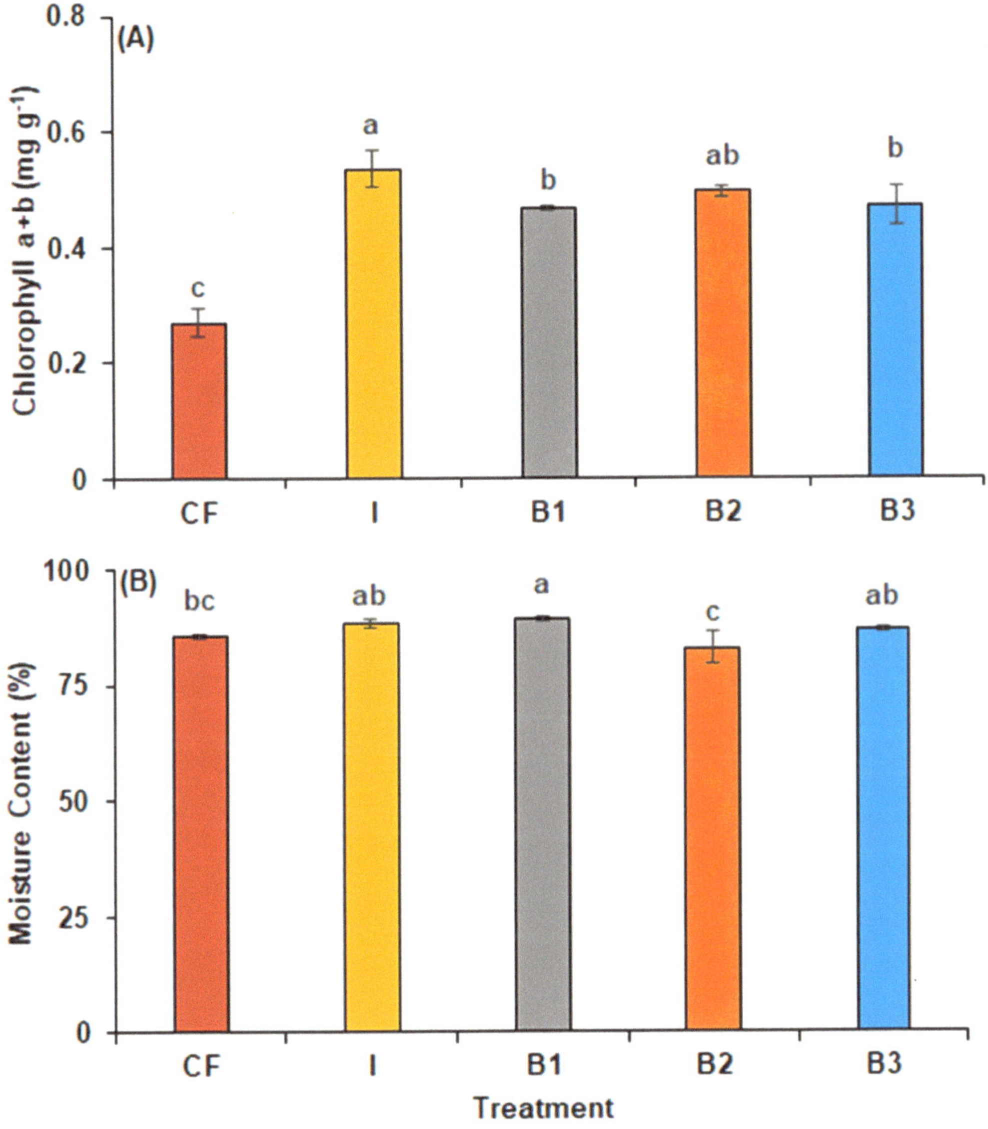

Figure 5. Pigment and moisture content analysis: (**A**) photosynthetic pigments and (**B**) moisture content. CF is chemical fertilizer, I is Isabion, B1 is 25% CF + glycine + lysine, B2 is 25% CF + glycine + aspartic acid, and B3 is 25% CF + aspartic acid + lysine. The data sets are presented as means± SD (n = 3). Columns that share the same letter(s) are not statistically significant at $p \leq 0.05$.

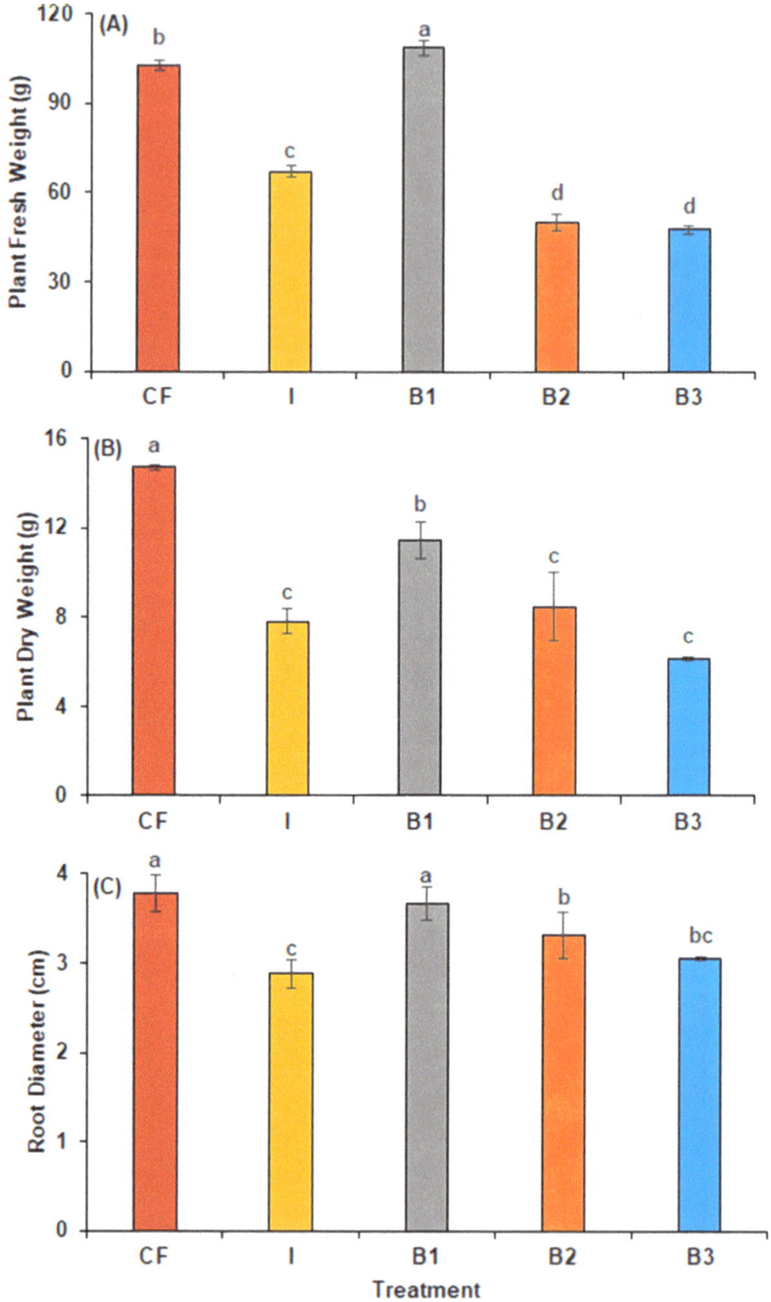

Figure 6. Plant biomass and morphology: (**A**) plant fresh weight, (**B**) plant dry weight, and (**C**) root diameter. CF is chemical fertilizer, I is Isabion, B1 is 25% CF + glycine + lysine, B2 is 25% CF + glycine + aspartic acid, and B3 is 25% CF + aspartic acid + lysine. The data sets are presented as means± SD (n = 3). Columns that share the same letter(s) are not statistically significant at $p \leq 0.05$.

3.3. Pearson Correlation

The results demonstrated a moderately positive correlation of photosynthetic rate with stomatal conductance, fresh and dry weight, chlorophyll content, and root diameter. At the same time, it has a weak negative association with internal CO_2 assimilation, transpiration, and moisture content. The transpiration rate strongly correlates negatively with fresh weight, dry weight, and root diameter. It negatively relates to chlorophyll content, internal CO_2 assimilation, and weak correlation with stomatal conductance and moisture content. Stomatal conductance was positively correlated with internal CO_2 assimilation, plant fresh and dry weight, root diameter, and chlorophyll content, while it showed a positive weak relation with moisture content. Internal CO_2 assimilation is positively associated with fresh and dry weight, root diameter, chlorophyll content, and moisture content. A strong positive influence of chlorophyll content on plant biomass and morphology was observed, while moderate positive relation with moisture content was observed. A strong positive association was observed between fresh and dry weight and root diameter, while a slightly positive correlation with moisture content. In addition, moisture content also presented a weak positive correlation with root diameter (Figure 7).

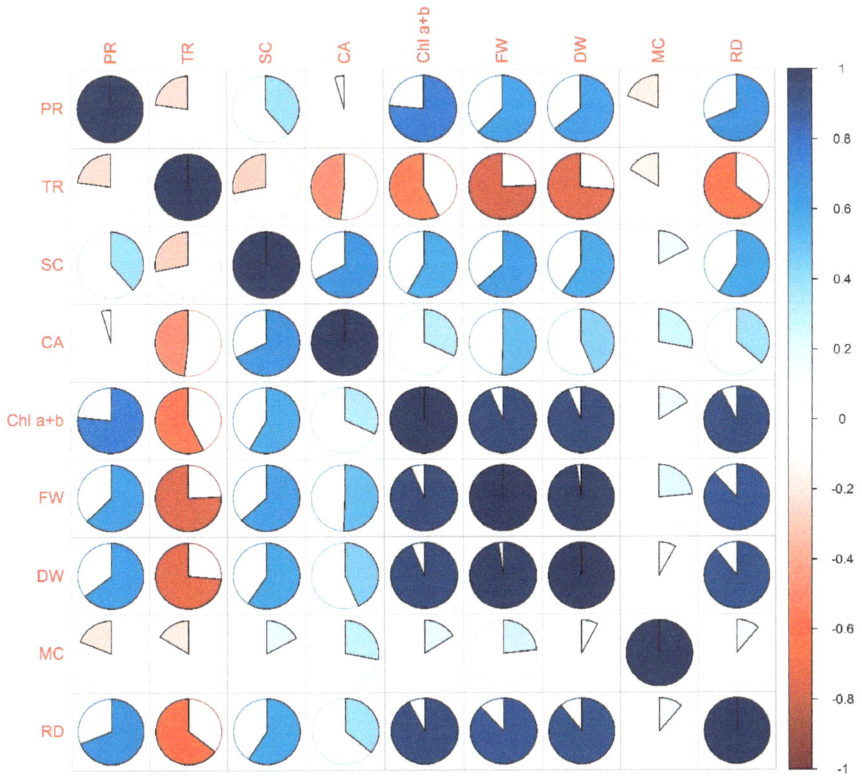

Figure 7. Pearson correlation: PR is photosynthetic rate, TR is transpiration rate, SC is stomatal conductance, CA is internal CO_2 concentration, Chl a + b is chlorophyll content, FW is plant fresh weight, DW is plant dry weight, MC is moisture content, and RD is root diameter.

4. Discussion

Biostimulants can improve crop productivity and decrease the use of chemical fertilizers. Our studies revealed that the foliar application of biostimulants significantly improved crop nutrition, growth, and productivity. In experiment 1, biostimulants improved plants' photosynthesis (Gl, Ly, AA, and VB) and transpiration rates (AA and VB), possibly via the composition of biostimulants and their role in plant mechanisms. Vitamins are significant in cellular metabolism and participate in more than 140 biochemical reactions. It also contributes to the biosynthesis of amino acids, fatty acids, plant hormones, organelle-specific compounds, and neurotransmitters. Furthermore, it is involved in the breakdown of storage compounds, reactive oxygen species scavenging, and chlorophyll synthesis and ultimately benefit photosynthesis [28]. In addition, experiment 2 showed that B1 improved photosynthesis in plants. This shows that the combination of GL and AA helps improve plant physiology. It might be due to amino acids that play a role in primary and secondary plant metabolic processes. It not only impacts plant physiological processes, plant vegetative growth, fruit maturity, seed germination, defense against a/biotic stresses, and reactive oxygen species but also acts as a reverse source of N [29].

Plants are capable of producing amino acids and vitamins, but energy is required for the processes; therefore, the foliar application of such biostimulants aids plants in improving their growth and development [30]. The results revealed that biostimulants improved turnip growth and physiological activities. In both experiments, the turnip growth and yield were significantly enhanced with the foliar feeding of biostimulants. It could be related to improved photosynthetic efficiency, cell enlargement, chlorophyll formation, and participation in the formation of growth hormones, including auxin, gibberellin, and/or cytokine [18]. In experiment 1, LY, AA, and VB improved the chlorophyll contents, whereas all combinations were helpful in enhancing photosynthetic pigments in the second experiment. The substances in biostimulants might be effective against chlorophyll deterioration and benefit the biosynthesis of the thylakoid membrane, thus resulting in improved photosynthesis [12]. In addition, chlorophyll a is directly involved in photosynthesis, whereas chlorophyll b allows plants to absorb light at a broader range of wavelengths, enhances cell metabolism, and involves in the synthesis of plant secondary metabolites [31].

A similar case was observed in both experiments. The combined use of amino acids-based biostimulants significantly improved the plant biomass and root diameter. It might be associated with improved fertilizer assimilation, water and nutrient uptake, and plant photosynthesis [32]. Amino acids also act as chelating agents [30] and contribute to improving nutrient uptake and nutrient use efficiency for macro- and micro-nutrients [33]. Additionally, the plants promptly absorb amino acids, become a stable source of molecule precursors, and contribute to plant metabolic activities [34]. Vitamins also play a significant role in root growth and development, act as an antioxidant against osmotic stress, and involve protein biosynthesis [24].

Amino acids also contribute to improving nutrient uptake and nutrient use efficiency for macro- and micro-nutrients [33]. Various published material reported that the use of amino acids and vitamin-based biostimulants improved radish root and shoot biomass [18], sweet pepper yield [35], physiological and nutritional responses in tomato [29], energy value in pea seeds [36], physiology and leaf properties of the sweet cherry tree [37], and cucumber growth and yield [38]. Moreover, commercial amino acid-based biostimulants have shown improved yield even with the half dose of N fertilizer [33]. It can be associated with stimulating hormone activity, nitrogen and carbon metabolism enzymes, plant biochemical (sugar synthesis, antioxidant enzyme production), and physiological attributes [39].

Lysine plays numerous roles in plant growth and development; it has been reported to induce the jasmonate signaling pathway, energy metabolism, serotonin accumulation, systemic immunity, and multiple roles in plant metabolism [40]. Glycine also helps plants to improve cell membrane stability and protection against peroxidation [41]. It has also been reported that glycine can stabilize the oxygen-evolving activity of the photosystem II complex, maintain cell membrane integrity and its components, and synthesize ATP [42]. Amino acids play a key role in N metabolism by regulating nitrate and ammonium uptake, nitrate reduction, ammonium incorporation, nitrogen remobilization, and protein synthesis and help plant grow even in nutrient-less conditions [41].

The excessive use of chemical fertilizers has adversely affected the soil environment, and this ongoing practice will reduce the soil's ability to serve humanity in the long run with quality food products [43]. Using biostimulants to boost nutrient uptake and plant growth can be helpful in reducing the rates, ultimately minimizing the detrimental impacts of chemical fertilizers on soil and the environment. Our studies reveal that biostimulants can improve crop yield, physiology, and productivity. Moreover, the combined use of biostimulants (B1 with a 75% reduction in fertilizer) can achieve outcomes similar to chemical fertilizers. The impacts of biostimulants on crop biochemistry and quality need to be addressed. Moreover, other application methods, such as root or soil application, could be another aspect to study. The scientific community also needs to testify to the long-term impacts and the response of various crops in field conditions to serve the farming community better.

5. Conclusions

The Green Revolution was a practical approach to food and nutrition security. Moreover, it alleviated extreme poverty and malnourishment. But the excessive use of chemical fertilizers deteriorated the soil and environment. In this scenario, biostimulants can be an alternative method for sustainable production and minimize fertilizer applications to overcome agricultural non-point pollution. The foliar application of biostimulants is a feasible strategy to implement. However, its beneficial impacts depend on multiple aspects, including the dose and time of application and the crop species. Our studies conclude that biostimulants can improve the plant biomass, gaseous exchange attributes, root morphology, and chlorophyll contents significantly in turnip. Moreover, its impacts on other food crops, its doses, and its time of application might differ from our reported results. Our study confirms that the foliar application of biostimulants can improve turnip crop physiology and yield and confirms our hypothesis that biostimulants can reduce the use of chemical fertilizers. The potential benefits of our research state that small-scale farmers and kitchen gardeners can implement biostimulants and grow turnips with less use of fertilization. Furthermore, the use of biostimulants in field conditions and their response on a larger scale needs to be addressed. Its mechanism of action is still a research gap that scientists can identify with the latest analytical techniques and bioinformatics in the future. In addition, biostimulants are economical because they improve the yield, whereas our study lacks sufficient investigation and requires further examination.

Author Contributions: Conceptualization, M.A.B., A.R. and Y.G.; Data curation, M.A.B., Y.G., S.H. and I.A.; Formal analysis, Q.-U.-A.R. and M.W.; Funding acquisition, Y.G.; Investigation, Q.-U.-A.R. and M.W.; Methodology, Q.-U.-A.R., I.A. and H.M.A.R.; Resources, A.R., S.H. and H.M.A.R.; Software, M.A.B.; Supervision, A.R.; Writing—original draft, Q.-U.-A.R.; Writing—review and editing, M.A.B., A.R., S.H., I.A., M.W., H.M.A.R. and Y.G. All authors have read and agreed to the published version of the manuscript.

Funding: The support was received from the Chengdu Normal University youth doctor training program fund.

Data Availability Statement: All the required data have been added to the manuscript in the form of figures and table. Figure data can be taken from the corresponding author upon request.

Acknowledgments: This study was supported by Higher Education Commission (HEC) Pakistan under the indigenous Ph.D. 5000 fellowship program. Qurat-Ul-Ain Raza (520(PH-II) 2AV6-075/HEC/IS/2020) acknowledges HEC for providing financial support to fulfill her degree requirements at BZU.

Conflicts of Interest: The authors declare that they have no competing interest that could have appeared to influence the work reported in this article.

References

1. Bursać Kovačević, D.; Brdar, D.; Fabečić, P.; Barba, F.J.; Lorenzo, J.M.; Putnik, P. Strategies to Achieve a Healthy and Balanced Diet: Fruits and Vegetables as a Natural Source of Bioactive Compounds. In *Agri-Food Industry Strategies for Healthy Diets and Sustainability: New Challenges in Nutrition and Public Health*; Academic Press: Cambridge, MA, USA, 2020; pp. 51–88. ISBN 9780128172261.

2. Buturi, C.V.; Mauro, R.P.; Fogliano, V.; Leonardi, C.; Giuffrida, F. Mineral Biofortification of Vegetables as a Tool to Improve Human Diet. *Foods* **2021**, *10*, 223. [CrossRef]
3. Raza, Q.-U.-A.; Bashir, M.A.; Rehim, A.; Ejaz, R.; Raza, H.M.A.; Shahzad, U.; Ahmed, F.; Geng, Y. Biostimulants Induce Positive Changes in the Radish Morpho-Physiology and Yield. *Front. Plant Sci.* **2022**, *13*, 2475. [CrossRef] [PubMed]
4. Dong, J.; Gruda, N.; Li, X.; Tang, Y.; Zhang, P.; Duan, Z. Sustainable Vegetable Production under Changing Climate: The Impact of Elevated CO2 on Yield of Vegetables and the Interactions with Environments—A Review. *J. Clean. Prod.* **2020**, *253*, 119920. [CrossRef]
5. Magama, P.; Chiyanzu, I.; Mulopo, J. A Systematic Review of Sustainable Fruit and Vegetable Waste Recycling Alternatives and Possibilities for Anaerobic Biorefinery. *Bioresour. Technol. Rep.* **2022**, *18*, 101031. [CrossRef]
6. Sun, R. Economic/Academic Importance of *Brassica rapa*. In *The Brassica rapa Genome*; Springer: Berlin/Heidelberg, Germany, 2015; pp. 1–15. ISBN 9783662479018.
7. Government of Pakistan, Ministry of National Food Security & Research Economic Wing Islamabad. 2021. Available online: http://www.mnfsr.gov.pk/Detail/NzJlNWE2N2MtNzY5Ni00ZWUwLWFkNTItZjk4MmM4MWY1MWE1 (accessed on 12 May 2023).
8. Paul, S.; Geng, C.A.; Yang, T.H.; Yang, Y.P.; Chen, J.J. Phytochemical and Health-Beneficial Progress of Turnip (*Brassica rapa*). *J. Food Sci.* **2019**, *84*, 19–30. [CrossRef]
9. Nawaz, M.Q.; Ahmed, K.; Qadir, G.; Rizwan, M.; Nawaz, M.F.; Sarfraz, M. Growth and Yield of Turnip (*Brassica rapa* L.) in Response to Different Sowing Methods and Nitrogen Levels in Salt-Affected Soils. *Pakistan J. Agric. Res.* **2020**, *33*, 126–134. [CrossRef]
10. Pahalvi, H.N.; Rafiya, L.; Rashid, S.; Nisar, B.; Kamili, A.N. Chemical Fertilizers and Their Impact on Soil Health. In *Microbiota and Biofertilizers*; Springer: Cham, Switzerland, 2021; Volume 2, pp. 1–20.
11. Salehzadeh, H.; Maleki, A.; Rezaee, R.; Shahmoradi, B.; Ponnet, K. The Nitrate Content of Fresh and Cooked Vegetables and Their Health-Related Risks. *PLoS ONE* **2020**, *15*, e0227551. [CrossRef]
12. Rouphael, Y.; Colla, G. Editorial: Biostimulants in Agriculture. *Front. Plant Sci.* **2020**, *11*, 40. [CrossRef] [PubMed]
13. Jamal, A.; Saeed, M.F.; Mihoub, A.; Hopkins, B.G.; Ahmad, I.; Naeem, A. Integrated Use of Phosphorus Fertilizer and Farmyard Manure Improves Wheat Productivity by Improving Soil Quality and P Availability in Calcareous Soil under Subhumid Conditions. *Front. Plant Sci.* **2023**, *14*, 1034421. [CrossRef]
14. Khan, I.; Amanullah; Jamal, A.; Mihoub, A.; Farooq, O.; Farhan Saeed, M.; Roberto, M.; Radicetti, E.; Zia, A.; Azam, M. Partial Substitution of Chemical Fertilizers with Organic Supplements Increased Wheat Productivity and Profitability under Limited and Assured Irrigation Regimes. *Agriculture* **2022**, *12*, 1754. [CrossRef]
15. Geng, Y.; Bashir, M.A.; Zhao, Y.; Luo, J.; Liu, X.; Li, F.; Wang, H.; Raza, Q.U.A.; Rehim, A.; Zhang, X.; et al. Long-Term Fertilizer Reduction in Greenhouse Tomato-Cucumber Rotation System to Assess N Utilization, Leaching, and Cost Efficiency. *Sustainability* **2022**, *14*, 4647. [CrossRef]
16. Hamid, B.; Zaman, M.; Farooq, S.; Fatima, S.; Sayyed, R.Z.; Baba, Z.A.; Sheikh, T.A.; Reddy, M.S.; Enshasy, H.E.; Gafur, A.; et al. Bacterial Plant Biostimulants: A Sustainable Way towards Improving Growth, Productivity, and Health of Crops. *Sustainability* **2021**, *13*, 2856. [CrossRef]
17. Bashir, M.A.; Rehim, A.; Raza, Q.-U.-A.; Muhammad Ali Raza, H.; Zhai, L.; Liu, H.; Wang, H. Biostimulants as Plant Growth Stimulators in Modernized Agriculture and Environmental Sustainability. In *Technology in Agriculture [Working Title]*; IntechOpen: London, UK, 2021; ISBN 978-1-83881-922-4.
18. Rehim, A.; Amjad Bashir, M.; Raza, Q.-U.-A.; Gallagher, K.; Berlyn, G.P. Yield Enhancement of Biostimulants, Vitamin B12, and CoQ10 Compared to Inorganic Fertilizer in Radish. *Agronomy* **2021**, *11*, 697. [CrossRef]
19. Rašovský, M.; Pačuta, V.; Ducsay, L.; Lenická, D. Quantity and Quality Changes in Sugar Beet (*Beta vulgaris* Provar. Altissima Doel) Induced by Different Sources of Biostimulants. *Plants* **2022**, *11*, 2222. [CrossRef]
20. Bashir, S.; Basit, A.; Abbas, R.N.; Naeem, S.; Bashir, S.; Ahmed, N.; Ahmed, M.S.; Ilyas, M.Z.; Aslam, Z.; Alotaibi, S.S.; et al. Combined Application of Zinc-Lysine Chelate and Zinc-Solubilizing Bacteria Improves Yield and Grain Biofortification of Maize (*Zea mays* L.). *PLoS ONE* **2021**, *16*, e0254647. [CrossRef]
21. Paradiković, N.; Vinković, T.; Vinković Vrček, I.; Žuntar, I.; Bojić, M.; Medić-Šarić, M. Effect of Natural Biostimulants on Yield and Nutritional Quality: An Example of Sweet Yellow Pepper (*Capsicum annuum* L.) Plants. *J. Sci. Food Agric.* **2011**, *91*, 2146–2152. [CrossRef] [PubMed]
22. Soppelsa, S.; Kelderer, M.; Casera, C.; Bassi, M.; Robatscher, P.; Matteazzi, A.; Andreotti, C. Foliar Applications of Biostimulants Promote Growth, Yield and Fruit Quality of Strawberry Plants Grown under Nutrient Limitation. *Agronomy* **2019**, *9*, 483. [CrossRef]
23. Alfosea-Simón, M.; Simón-Grao, S.; Zavala-Gonzalez, E.A.; Cámara-Zapata, J.M.; Simón, I.; Martínez-Nicolás, J.J.; Lidón, V.; Rodríguez-Ortega, W.M.; García-Sánchez, F. Application of Biostimulants Containing Amino Acids to Tomatoes Could Favor Sustainable Cultivation: Implications for Tyrosine, Lysine, and Methionine. *Sustainability* **2020**, *12*, 9729. [CrossRef]
24. Asensi-Fabado, M.A.; Munné-Bosch, S. Vitamins in Plants: Occurrence, Biosynthesis and Antioxidant Function. *Trends Plant Sci.* **2010**, *15*, 582–592. [CrossRef]
25. Kocira, A.; Lamorska, J.; Kornas, R.; Nowosad, N.; Tomaszewska, M.; Leszczyńska, D.; Kozłowicz, K.; Tabor, S. Changes in Biochemistry and Yield in Response to Biostimulants Applied in Bean (*Phaseolus vulgaris* L.). *Agronomy* **2020**, *10*, 189. [CrossRef]

26. Abbas, M.; Atangana Njock, P.G.; Wang, Y. Influence of Climate Change and Land-Use Alteration on Water Resources in Multan, Pakistan. *Appl. Sci.* **2022**, *12*, 5210. [CrossRef]
27. Yuan, Z.; Cao, Q.; Zhang, K.; Ata-Ul-Karim, S.T.; Tan, Y.; Zhu, Y.; Cao, W.; Liu, X. Optimal Leaf Positions for SPAD Meter Measurement in Rice. *Front. Plant Sci.* **2016**, *7*, 719. [CrossRef]
28. Parra, M.; Stahl, S.; Hellmann, H. Vitamin B6 and Its Role in Cell Metabolism and Physiology. *Cells* **2018**, *7*, 84. [CrossRef]
29. Alfosea-Simón, M.; Simón-Grao, S.; Zavala-Gonzalez, E.A.; Cámara-Zapata, J.M.; Simón, I.; Martínez-Nicolás, J.J.; Lidón, V.; García-Sánchez, F. Physiological, Nutritional and Metabolomic Responses of Tomato Plants After the Foliar Application of Amino Acids Aspartic Acid, Glutamic Acid and Alanine. *Front. Plant Sci.* **2021**, *11*, 2138. [CrossRef] [PubMed]
30. Popko, M.; Michalak, I.; Wilk, R.; Gramza, M.; Chojnacka, K.; Górecki, H. Effect of the New Plant Growth Biostimulants Based on Amino Acids on Yield and Grain Quality of Winter Wheat. *Molecules* **2018**, *23*, 470. [CrossRef]
31. Toscano, S.; Romano, D.; Patanè, C. Effect of Application of Biostimulants on the Biomass, Nitrate, Pigments, and Antioxidants Content in Radish and Turnip Microgreens. *Agronomy* **2023**, *13*, 145. [CrossRef]
32. Haghighi, M.; Saadat, S.; Abbey, L. Effect of Exogenous Amino Acids Application on Growth and Nutritional Value of Cabbage under Drought Stress. *Sci. Hortic.* **2020**, *272*, 109561. [CrossRef]
33. Halpern, M.; Bar-Tal, A.; Ofek, M.; Minz, D.; Muller, T.; Yermiyahu, U. The Use of Biostimulants for Enhancing Nutrient Uptake. *Adv. Agron.* **2015**, *130*, 141–174. [CrossRef]
34. Rouphael, Y.; Colla, G. Toward a Sustainable Agriculture through Plant Biostimulants: From Experimental Data to Practical Applications. *Agronomy* **2020**, *10*, 1461. [CrossRef]
35. Rodrigues, M.; Baptistella, J.L.C.; Horz, D.C.; Bortolato, L.M.; Mazzafera, P. Organic Plant Biostimulants and Fruit Quality—A Review. *Agronomy* **2020**, *10*, 988. [CrossRef]
36. Biel, W.; Podsiadło, C.; Witkowicz, R.; Kępińska-Pacelik, J.; Stankowski, S. Effect of Irrigation, Nitrogen Fertilization and Amino Acid Biostimulant on Proximate Composition and Energy Value of *Pisum sativum* L. Seeds. *Agriculture* **2023**, *13*, 376. [CrossRef]
37. Serapicos, M.; Afonso, S.; Gonçalves, B.; Silva, A.P. Exogenous Application of Glycine Betaine on Sweet Cherry Tree (*Prunus avium* L.): Effects on Tree Physiology and Leaf Properties. *Plants* **2022**, *11*, 3470. [CrossRef] [PubMed]
38. Zargar Shooshtari, F.; Souri, M.K.; Hasandokht, M.R.; Jari, S.K. Glycine Mitigates Fertilizer Requirements of Agricultural Crops: Case Study with Cucumber as a High Fertilizer Demanding Crop. *Chem. Biol. Technol. Agric.* **2020**, *7*, 19. [CrossRef]
39. Al-Karaki, G.N.; Othman, Y. Effect of Foliar Application of Amino Acids Biostimulants on Growth, Macronutrient, Total Phenols Contents and Antioxidant Activity of Soilless Grown Lettuce Cultivars. *S. Afr. J. Bot.* **2023**, *154*, 225–231. [CrossRef]
40. Yang, Q.; Zhao, D.; Zhang, C.; Sreenivasulu, N.; Sun, S.S.M.; Liu, Q. Lysine Biofortification of Crops to Promote Sustained Human Health in the 21st Century. *J. Exp. Bot.* **2022**, *73*, 1258–1267. [CrossRef]
41. Mohammadipour, N.; Souri, M.K. Beneficial Effects of Glycine on Growth and Leaf Nutrient Concentrations of Coriander (*Coriandrum sativum*) Plants. *J. Plant Nutr.* **2019**, *42*, 1637–1644. [CrossRef]
42. Díaz-Zorita, M.; Fernández-Canigia, M.V.; Grosso, G.A. Applications of Foliar Fertilizers Containing Glycinebetaine Improve Wheat Yields. *J. Agron. Crop Sci.* **2001**, *186*, 209–215. [CrossRef]
43. Kopittke, P.M.; Menzies, N.W.; Wang, P.; McKenna, B.A.; Lombi, E. Soil and the Intensification of Agriculture for Global Food Security. *Environ. Int.* **2019**, *132*, 105078. [CrossRef]

Disclaimer/Publisher's Note: The statements, opinions and data contained in all publications are solely those of the individual author(s) and contributor(s) and not of MDPI and/or the editor(s). MDPI and/or the editor(s) disclaim responsibility for any injury to people or property resulting from any ideas, methods, instructions or products referred to in the content.

AgriEngineering

Technical Note

Milking Machine Settings and Liner Design Are Important to Improve Milking Efficiency and Lactating Animal Welfare—Technical Note

Shehadeh Kaskous [1,*] and Michael W. Pfaffl [2]

[1] Department of Research and Development, Siliconform, Schelmengriesstrasse 1, 86842 Türkheim, Germany
[2] Department of Animal Physiology and Immunology, TUM School of Life Sciences, Technical University of Munich, Weihenstephaner Berg 3, 85354 Freising, Germany; michael.pfaffl@tum.de
* Correspondence: skaskous@siliconform.com

Abstract: The purpose of milking machines is to harvest milk at optimal quality and speed, while maintaining animal comfort and teat defense mechanisms against invading mastitis pathogens. Therefore, the milking machine is a very important piece of equipment on dairy farms to maintain a long healthy lactation by following the physiological conditions of the udder. The mechanical forces during long-term machine milking processes lead to changes in the teat tissue. This effect is related to the degree of adaptation of the milking machines to the physiological requirements of the individual udder anatomy and the physiological conditions of the lactating animals. If both, milking machine settings and liner design are not suitable for all teats and animals on the farm, some animals will not be fully milked, the teat condition will deteriorate over time and in the end, they may suffer from mastitis. Therefore, maintaining healthy udders and teats during milking is a central key component of an effective milking machine to produce good milk yield with higher quality by preventing mastitis and maintaining animal health and welfare. On large and thick teats, conventional liners often fit too tight, causing a massive mechanical stress load on the tissue. On small teats, however, they often do not adhere sufficiently close to the teat which can cause a considerable air admission and hence liner slips. The new liners, "Stimulor® StressLess" (Siliconform, Türkheim, Germany), have a wave-like lip construction and adapt well to the different teat sizes in a herd, thus ensuring consistent milking of lactating animals. A proper milking machine accommodates all teat sizes and forms, has a low vacuum to effectively open the teat and to stimulate physiological milk release and letdown. In addition, the right pulsation rate will maintain a stable vacuum on the teat area during milking. In conclusion, an ideal milking machine adapts to the morphological, anatomical, and physiological characteristics of the udder and teats of the lactating animals and it should achieve a physiologically ideal milking process that meets high animal welfare standards and increases milk production with a high quality standard.

Keywords: lactating animals; liner; milking machine; Stimulor® StressLess; teat cup

Citation: Kaskous, S.; Pfaffl, M.W. Milking Machine Settings and Liner Design Are Important to Improve Milking Efficiency and Lactating Animal Welfare—Technical Note. *AgriEngineering* **2023**, *5*, 1314–1326. https://doi.org/10.3390/agriengineering5030083

Academic Editors: Muhammad Sultan, Yuguang Zhou, Redmond R. Shamshiri and Muhammad Imran

Received: 14 June 2023
Revised: 21 July 2023
Accepted: 26 July 2023
Published: 28 July 2023

Copyright: © 2023 by the authors. Licensee MDPI, Basel, Switzerland. This article is an open access article distributed under the terms and conditions of the Creative Commons Attribution (CC BY) license (https://creativecommons.org/licenses/by/4.0/).

1. Introduction

An increase in milk yield through increased milk synthesis, milk secretion and a long and stable persistence (a slower drop in yield over the course of the lactation) is possible if the healthy udder is emptied frequently and completely. Bruckmaier and Blum [1] have emphasized that optimal machine milking removes the stored milk from the udder cistern quickly and completely while maintaining udder health with the aim of harvesting high milk yield, maintaining animal health, and achieving low milking costs. Several researchers have also shown that milking machine settings such as pulsation frequency and ratio, vacuum level and automatic teat cup removal have a major impact on milk yield, milking time, and overall udder health [2–7]. However, the milking machine is known to affect the milk produced and its quantity and quality in lactating animals [5,8–10].

An unsuitable milking machine is cited as the cause of milking that is too slow or too rough, not physiological, and therefore incomplete. Crooked teat cups, no proper pulsator, higher vacuum and wrongly dimensioned teat liners are factors resulting in an unsuitable milking process for the lactating animals. The reason for this may be low milk yield, longer milking times, and impaired udder health. Maintaining udder health during milking is very important to achieve high yield with good milk quality. It is well known that the mechanical forces of an improperly used machine milking results in poor teat tissue changes [11] and allows pathogenic bacteria to penetrate the udder [12], particularly via the skin of the teat end, which will increase the risk of new mastitis infections and thereafter increase somatic cell count (SCC) in the milk [13–15].

Moreover, it should be noted that insufficient pre-stimulation can lead to poor milking performance, especially in cows, camels, and buffalo, which have small udder cisterns and large amount of alveolar milk [1,16–19]. A new study has shown that efficient machine milking requires an optimal interaction between the alveolar milk ejection in the udder induced by hormonal posterior hypophyseal oxytocin release, smooth muscle contraction of the alveolar myoepithelial cells and thereafter the milk removal by the milking machine [19]. The timely onset of this milk ejection is particularly important for animals and species with low levels of cistern milk. Most milk in sheep and goats can be obtained before ejection, because the proportion of cistern milk is significantly high. The cistern milk can account for up to 70% of the stored milk (depending on the breed) [20]. However, having an insufficient pre-stimulation in cattle results in delayed oxytocin release and milk ejection, and a separate removal of the cisternal milk before milk ejection is required. This causes the well-known milk flow pattern in bovines [17]. Therefore, mechanical udder preparation must be ensured in order to trigger the physiological milk ejection at the right moment [21]. In addition, it was observed that the average milk flow was higher when milk was removed within the first two minutes of milking, and the duration of incline and time until peak milk flow were shorter with pre-stimulation than without pre-stimulation [19]. Sandrucci et al. [22], reported that a proper udder preparation, including teat fore-stripping and pre-dipping resulted in better milking performances compared to a poor preparation. This results in a higher milk yield per milking, shorter milking time, lower bimodality, lower SCC in the milk, and better animal welfare.

Many milking machines offer modern technology at a high level. However, the milking system chosen is not always suitable for all animals in the herd, since the physiological control mechanisms and oxytocin release in animals are individual and cannot be changed by humans. The milking routine and technology must be adapted to the animal anatomy and physiology in order to achieve optimal milk removal [23]. Gasparik et al., [5] reported that optimal values for milking machine settings to better adapt to animal physiology were still discussed among scientists. For this reason, many farms suffer with their own milking machines because the demands of the lactating animals are not yet met. This means that the milking machine must act in a natural and gentle way similar to suckling the calf, and must not be uncomfortable or painful for the animals. Observations on suckling calves showed that the calf can drink all the milk while the udder remained healthy [23]. Interestingly, it was reported that Siliconform's milking philosophy is to increase the milk yield, improve the quality, protect animal welfare, and reduce production costs. Therefore, the milking machines "MultiLactor" for cows and "StimuLactor" for camels were developed [24]. Field studies have shown that cows on the farm remained healthy with a MultiLactor quarter-individual milking system throughout the study period. The reason for this lies in the use of the right milking machine, milking program, and optimally fitting liners, e.g., suitable teat cups with an air inlet valve (also known as Bio-Milker; Siliconform), suitable silicone liner, low vacuum level, correct pulsation rate and pulsation ratio [10,25].

Furthermore, when the lactating animals are milked, the teat cup liner is the immediate interface between the technical aggregate of the milking machine and the highly sensitive organ, the teat skin. Therefore, the liner and milking machine settings have a significant influence on the duration of milk removal, the degree of udder emptying, the maintenance of udder health and finally, on the overall quality of the raw milk [14,15,24,26–28]. In the following, the importance of the milking machine settings and liner design for optimal milking performance and welfare of the lactating animals are summarized and discussed.

2. Influence of Milking Machine Settings on Milk Removal

The ideal milking system meets the physiological needs of lactating animals to increase milk yield, achieve better milk quality and maintain healthy udders. Therefore, the settings of the milking machine are very critical to lactating animals [5,29]. Individual quarter pulsation milking systems may prevent overmilking and improve teat-end condition [10,25,30]. Besides the liner, there are three operating parameters that regulate mechanical milking: vacuum level, pulsation rate and pulsation ratios. The milking system needs to provide stable vacuum, adequate pulsation, and gentle milking action. Inside the teat liner there is a continuous negative pressure (vacuum) between 36 and 50 kPa, which causes the teat cup to adhere to the teat, overcoming the teat canal resistance and transporting the milk that has escaped during the suction phase. Therefore, the vacuum directly "under the teat" is subject to dynamic changes due to cyclic liner movement and stop-and-flow of the milk [27]. However, the pressure in the pulsation chamber of the teat cup alternates (controlled by the pulsator) between operating vacuum (suction phase) and atmospheric pressure (relief phase) [31]. Reinemann and Mein, ref. [32] reported that the right choice of a liner, vacuum level and pulsation settings are essential for balanced milking that needs to be fast, smooth, and complete. The mechanical stress on the teat is always present and is essentially caused by the teat liner shaft [21]. This mechanical stress becomes harmful during the relief phase, if a high differential pressure quickly builds up between the teat end vacuum and inside the pulsation chamber and the liner shaft compresses the teats at high speed [5]. This is almost always the case during blind milking and at low milk flow rates at the end of the milking period. The resulting differential pressure is greater, the higher the milking vacuum was set. For example, if you milk with a vacuum of more than 40 kPa, the teats of the cows are unnecessarily stressed. In nature, the teats must only withstand the relatively gentle massage pressure of the suckling calf's tongue. On the other hand, teat liners are often significantly harder, less flexible and put more stress on the teat tips. However, this is exactly where the sphincter muscle and the teat canal are anatomically located, which are both supposed to take over the essential infection barrier function.

2.1. *Influence of the Vacuum Level in the Milking Machine on Milk Removal*

Obviously, the vacuum from the milking machine opens the teat canal and milk flows out during suction phase, but blood and lymph from the teat wall are also drawn to the end of the teat. The collapsing liner exerts a mechanical force on the end of the teat during the relief phase, closing the teat canal and transporting blood and lymph back to the udder. However, the mechanical stress on the teats during mechanical milking, both through teat stretching due to the high vacuum in phases when the liner is open and through compression due to the closed liner, leads to a temporary occlusion of the blood vessels in the teat wall [11]. Further, vacuum and pulsation ratio are important operating parameters that affect milking performance by milking machine [33]. The results also clearly demonstrated the importance of increased vacuum in the mouthpiece chamber for teat end congestion and the consequent reduction in cross-sectional area [34].

It has been observed that teat end hyperkeratosis increases with increasing liner pressure of the milking machine, and this furthermore depends on the length of the teat cup and the type of milking situation [35]. When the liner closes in the relief phase, a vacuum overpressure from the milking machine is exerted negatively on the teat. Vacuum levels and milking machine settings are known to affect milking behavior and teat tissue condition [36]. It has been found that high vacuum (50 kPa) in the milking machine compared to lower vacuum (42 kPa) can lead to teat tissue damage, milk duct damage and hyperkeratosis [37]. New research emphasized that the frequency of bimodalities with both pre-stimulation and without pre-stimulation was higher at high vacuum level than at low vacuum level [19]. The use of high milking vacuum for camels could lead to udder health problems reflected by a high somatic cell count in the produced milk and a negative impact on the health status of the teats [18]. A positive relationship between increasing working vacuum and somatic cell counts in the milk has been found in buffalo [38] and other dairy species [14,39–41]. Normally, the vacuum pump creates a vacuum for the whole milking system (vacuum line, milk line and all connected tubes and equipment) and determines the vacuum in the claw and pulsator chambers. The milking vacuum depends on many factors, e.g., the vacuum required to pump the milk (high line), the length and diameter of the milk tube and the number and type of devices integrated in the system [42,43]. Furthermore, the vacuum in the milking machine is the main force required to keep the liner on the teats during milking. An interesting aspect is that the lack of milk flow during milking at the beginning and end of the milking process causes increased vacuum forces on the teats, which can lead to greater teat damage [43]. Conversely, when the milk flow is high during the plateau phase, the loss of vacuum is particularly high compared to the beginning or end of milking when the milk flow is low [36,41]. It is noteworthy that a drop in claw vacuum during milking can reduce milking efficiency by reducing milk removal efficiency. Several lines of evidence indicate that modern milking machines with Bio-Milker liners (Siliconform) clearly show, during the suction phase with low milk line, that the vacuum in the inner chamber remains stable or slightly higher than in the pulsation chamber [7].

Figure 1 shows the vacuum conditions in the Bio-Milker (Siliconform) teat cup. With this teat cup, the pressure on the teats is not too high during the relief phase. As can be seen from the diagram, the vacuum in the inner chamber during the relief phase is low compared to a normal teat cup. As a result, the blood flows easily and no congestion is seen. Bio-Milker (Siliconform) teat cups are known to have a stable vacuum in the teat area and consequently minimized vacuum loss, had no engorgement and drastically reduced the risk of infection [23]. Normal milking machines without the Bio-Milker teat cup show different results, namely: in the absence of milk flow, the vacuum in the inner chamber is similar to the full system vacuum level, whereas it drops as soon as milk is available and is transported in the milk tubes [36]. However, it should be noted that in high-line milking systems, the vacuum level in all milking machines decreases during the period of high milk flow rate [7,44]. Therefore, a vacuum drop in the inner chamber during machine milking can affect milking performance [41] and increase the probability of liner slips [43].

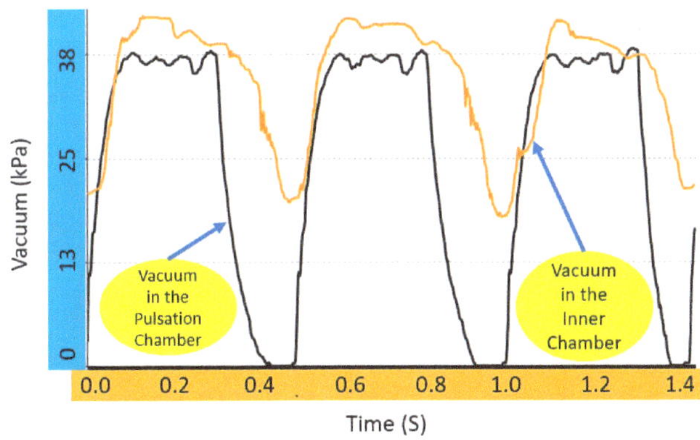

Figure 1. Cyclical vacuum fluctuations in the pulsation and inner chambers of the teat cup (Bio-Milker teat cup liners, Siliconform) during milking in the sheep milking machine with the following settings: vacuum level 38 kPa, pulsation rate 120 cycles/min and pulsation ratio 60:40 [7].

2.2. Influence of Pulsation Rate and Pulsation Ratios in the Milking Machine on Milk Removal

A pulsator is a device for generating cyclic pressure changes in the pulse system of a milking machine. As a switch, it alternately connects the impulse chamber of the teat cup with the vacuum supply of the milking system or with atmospheric air. The periodically changing pressure conditions in the two-chamber teat cup control the movement of the teat liner and thus the change between milk flow (suction phase, b-phase) and circulatory relief of the teat by pressing the teat liner (relief phase, d-phase) [5,42,45] (Figure 2).

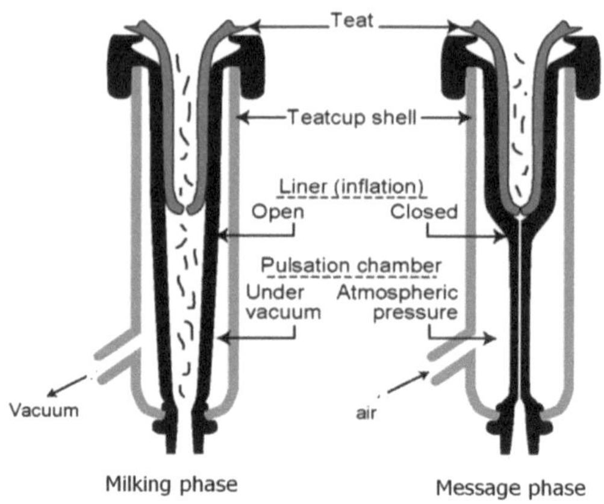

Figure 2. Suction phase (milking phase) and relief phase (massage phase) in mechanical milk removal according to Krömker [31] with some changes.

The main purpose of the pulsation is to limit the formation of congestion and oedema in the teat tissue during machine milking [46]. However, the function of the pulsation in the teat cup is to squeeze the end of the liner, which supports the return flow of blood and lymph through the veins and lymphatics during milking [47]. Thus, vacuum pulsation and the pulsation ratio are important operating parameters that affect milking performance [5,34]. An ideal pulsation setting on the milking machine helps to maintain a high milk flow, limit animal discomfort, reduce the occurrence of new intramammary infections and stimulate the milk ejection reflex [29]. Interestingly, individual quarter-pulsation milking systems can prevent overmilking and reduce hyperkeratosis in Holstein cows [48]. Field experiments and research have shown that a relatively narrow range of pulsation rates and ratios is required to ensure teat-end health, udder health and to optimize milking speed [49]. However, it is well known that the pulsation ratio has a great influence on the milk flow rate, milking time, teat state and SCC of milk [4,5]. It was observed that the mean and maximum milk flow rate increased with increasing broader ratio (from 60:40 to 65:35 and to 70:30) [33]. It has also been observed that a wide pulsator ratio results in faster milking than a narrow pulsator ratio [50,51]. The praxis results showed that when using a quarter-individual milking system "MultiLactor" (Siliconform), milk yield per milking, average milk flow rate and machine-on time were significantly ($p < 0.001$) different after changing the pulsation ratio from 60:40 to 65:35 (Table 1) [6].

Table 1. Least square means for milk yield per milking, average milk flow rate and machine-on time after changing the pulsation ratio from 60:40 to 65:35 while using a "MultiLactor" (Siliconform) quarter individual milking system [6].

Pulsation Ratio	Total Milk Yield/Milking [kg]	Average Flow Rate [kg/min]	Machine-On Time [min]
60:40	13.67 ^{a} ± 0.33	1.75 ^{a} ± 0.05	8.36 ^{a} ± 0.25
65:35	14.00 ^{b} ± 0.33	1.86 ^{b} ± 0.05	8.03 ^{b} ± 0.25

^{a,b} Means with different superscripts on the same column differ ($p < 0.001$).

Most modern milking equipment manufacturers set their pulsators within a relatively small range around the 60:40 ratio [43], as the milking speed has been optimized by keeping the liner open for about 0.5 to 0.6 s at each cycle [52]. However, the peak milk flow rate reaches its maximum at a pulsation ratio in the range of 60:40 to 70:30, depending on the properties of the liners used [46]. In addition, it was found that large ratios such as 70:30 resulted in shorter machine on-time and higher milk yields compared to ratios of 60:40 and 50:50 [50]. Likewise, increasing the pulsator ratio from 60:40 to 77:33 had no negative effects on teat tissue and had a positive impact on milk yield and shorter milking time [53]. Furthermore, studies have shown that peak milk flow rate consistently decreases after 70:30 to 80:20. The reason could be insufficient duration of pressure loading to overcome teat congestion [46]. It is worth noting that the correct setting of the pulsation ratio allows the pulsation chamber to recover to full atmospheric pressure for at least 150 milliseconds (d-phase) (15%) on each cycle. This helps to avoid teat congestion when using vacuum milking machines [47]. However, post-milking teat thickness tends to increase with the pulsation ratio, whereas it decreases significantly when the pulsation rate is between 20 and 80 cycles/min [29,35,47]. It has also been found that pulsation rates below 55 cycles/min affect udder health by increasing the risk of new intramammary infections [54].

3. Influence of the Teat Cup Liners on Teat Condition and Udder Health

Milking machine malfunctions, e.g., not the right liner and/or pulsators, the vacuum impeding blood and lymph circulation, lead to an unnatural strain on the teat tissue and thus to pain and immune reactions in the lactating animals. These artificial physiological and pathological changes in the circulatory system damage the teat ends, reduce the effectiveness of the udder's defense systems [53] and cause mastitis [34,55]. Thus, changes in teat condition can occur [29,34], which can be divided into short-, medium-, and long-term changes [56]. The most important short-term changes caused by circulatory disorders are discolored teats, ringing, swelling and openness of the teat canal [27,34]. These changes are reversible from one milking to another or even within a few hours. Swelling as a short-term effect of the milking machine on the teats can be measured using ultrasound [37]. However, the histological investigations of Stauffer et al., [11] have shown that enlargement of thick-walled veins in the teat tissue is the main reason for teat wall thickening in response to machine milking. It is noteworthy that each individual milking stimulates keratin growth in the teat canal to replace the keratin lost during milking [27]. If an unsuitable milking machine or misfitting teat liners are used for a long period, the milking machine attacks the teat end tissue, forms a ring of callus around the teat opening causing hyperkeratosis [24,30,35,37]. In addition, it has been observed that increased strength and thickness of the teat wall occurs as a result of congestion and oedema [37]. The blockage of the blood vessels leads to an accumulation of intravascular fluid [57] and could be a prerequisite for the development of an oedema called extravascular fluid build-up [58].

Gleeson et al. [53] have clearly shown that liner design had a greater impact on teat tissue changes and milking characteristics than pulsation settings. The magazine DairyNZ [59] confirmed in milked cows that liners have a major impact on milking performance, udder, and teat health. Furthermore, a teat cup liner must be specially designed for the animal's teat to prevent clogging of the teats and to allow proper blood flow. Therefore, it can be stated that a good liner achieves the following:

1. Seals both ends of the shell airtight.
2. Provides a properly sized mouthpiece and shaft that fits a range of teat shapes and sizes and minimizes liner slippage, clusters dropping and damage that can lead to mastitis.
3. Enables milking to be as quick and complete as possible to minimize congested teats, discomfort, and injury.
4. Is easy to clean and lasts a long time without changes.

New results from Holst et al., [60] showed that milk flow was higher with round liners than with triangular ones and the main milking time was shorter. As a result, it is better for the health and condition of the teats. Furthermore, previous studies have reported that calf suckling influences teat swelling less than a machine milking [61] (Table 2). However,

SCC in lactating cows was slightly increased when claves suckled the cow in addition to machine milking compared to cows milked only by machine [62].

Table 2. Relative change in teat parameters, measured on ultrasonographic views, after machine milking and calf suckling Referring to Neijenhuis, Ref. [61].

Parameters	Calf Suckling	Machine Milking
Wall thickness (%)	6	26–50
Duct length (%)	7	19–28
Cistern diameter (%)	−9	−27–−65

It is noteworthy that the mechanical stress on the teat is significantly higher in machine milking than in the suckling calf [27], although the vacuum level and the pressure of the calf's mouth on the teat is significantly higher during suckling than in machine milking [63]. The explanation for these contradictions is that the cyclic compression of the liner during milking and the continuous vacuum are responsible for the stresses in machine milking, while when the calf suckles the pressure between the mouth and teat tissue drops to zero after each sip of milk [11].

4. Influence of Teat Cup Liners on the Milking Process

The teat cup liner is constantly in motion during the milking process. Therefore, it is one of the most active parts of the milking machine. Throughout their lifetime, they withstand the effects of cleaning agents and disinfectants as well as high temperatures [28]. However, Silicone material is largely temperature-neutral between −50 to +50 °C (no change in this temperature range in terms of flexibility and milking behavior) [23]. In addition, the more complex the milking system, the more the liner must perform throughout the milking process and over its lifetime. Optimum milking performance can only be achieved when all variables are considered, and material and design are perfectly matched [25]. Thus, for optimal milk removal, the continuous availability of milk from the cluster attachment to the point of extraction is crucial. The liner plays a big part in this as its movement provides continuous stimulation during milking, resulting in oxytocin release and continuous milk ejection until cluster detachment [1,64]. In all cases it has been shown that liners, milk yield and blind milking must be considered as risk factors for teat damage as they affect milking speed, overall vacuum levels, peaking and over-milking [29]. A balance should be struck between optimal milking speed and the risk of teat damage.

Many lactating animal farms have different teat sizes and teat shapes in their herds. With large teats, the liners often close tightly and strain the tissue in the teat, while with small teats, air gets into the teat cup and the cluster falls [24]. However, problems with teat cup attachment or poor adhesion to the teat during milking can result in milk particles splashing back onto the teat at high speed and becoming a source of infection if unsuitable liners are used. Such milk particles are still in the milking cluster of the previously milked lactated animal. It was observed that after milking a cow infected with streptococcus, these germs can be transmitted on to the next six to eight cows [65]. Therefore, several reports have shown that the teat cup liner transfers the shortcomings and mistakes in milking technique directly to the teat. In addition, using the wrong liner increases the risk of mastitis due to longer milking times, reduced milk flow and a higher stripping milk amount. Hillerton [66] observed that teat liners showed a significant impact on milking performance, cow behavior during milking, udder health and teat reaction. Thus, successful milking really requires perfectly fitting liners for milking machines and animal welfare of the entire herd. The new results showed that the use of the new liner "Stimulor® StressLess" (Siliconform, Germany) on a cow farm in Austria led to a significant increase in milk yield in all stages of lactation [24] (Figure 3).

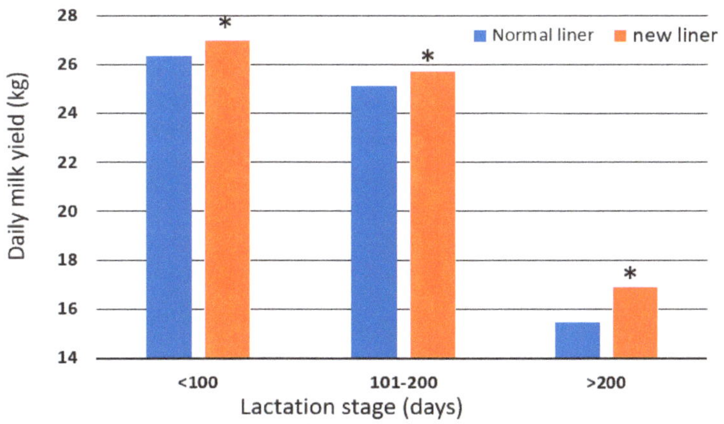

Figure 3. Daily milk yield before and after using the new liners in all examined cows (n = 42) in different stages of lactation on a dairy farm in Austria. * Mean significantly different ($p < 0.05$) in the same stage of lactation.

These new liners have unique properties that are excellent for maintaining teat health during machine milking for all lactating animals. One of their distinguishing features is the presence of the wavy construction design of its mouthpiece. Therefore, it adapts well to different cows and individual teat sizes in a herd. Furthermore, the liner supports positive welfare of lactating animals and the head vacuum is kept in the physiological range as it adheres particularly well to the teat.

5. How Do you Choose the Right Liner for Your Lactating Animals?

Choosing a liner is a complex and difficult task and an extremely important one. Therefore, when choosing the right liner, the milking machine, its vacuum settings, and the personal anatomy of the lactating animals must be considered. If the proposed liners fit the milking equipment, the milking performance of your lactating animals will in principle be comparable to other liners aimed at fast, efficient milking without negative effects. A stronger liner can also justify a gear change in this case. However, the design and management of the entire milking parlor affects the performance of the liners. If the liner is small, it narrows the milk drainage pathways and prevents high milk flow. On the other hand, if a small teat is milked with a liner that is too large, the tissue is overstretched and results in subsequent accumulation of connective tissue. The results are that the teats become hard and the milking ability of the cow deteriorates. Milking machines should be evaluated in their entirety to optimize the milking process. Teat cup liners must be carefully selected to fit the herd's teats and accommodate the machines. As a rule, teat liners are classified according to shaft length and shape, material, and construction. The head of the liner is designed to ensure a good fit to the base of the teat without restricting the connection between the gland- and teat cistern. Therefore, the correct liner selection must include the following objectives:

1. Maintains animal health: Ensures the teat is not damaged which can lead to problems of mastitis.
2. Increases milking efficiency: The right teat cup liners help keep the clusters in place and milk the animals faster.
3. Reduces animal stress: Ensure the liner does not cause pain leading to animal discomfort and animal handling issues.

It is noteworthy that the effects of teat liners that are not adapted to the herds udder shape are often underestimated. It is known that the teat cup liner takes on a massage function in machine milking. By collapsing (below the tip of the teat) the blood and liquid collected in the end of the teat is to be massaged back towards the bottom of the udder. However, teat liner shafts that are much too soft or too stiff cannot transmit sufficient force to the teat tip [67].

There are many high-quality milking machine liners on the market today, that can be used in conventional milking parlors or in automatic milking systems. Unfortunately, there are also cheap, poor-quality liners on the market, which many farmers complain about. It has long been known that silicone liners are better than synthetic liner for the following reasons [65]:

- Compared to synthetic rubber, silicone liners have good chemical and mechanical properties. This keeps them stable under different conditions.
- The extremely smooth surface does not give bacteria a chance to penetrate the silicone material during machine milking.
- Silicone teat cups are absolutely boil-proof and largely heat-stable. This means that cleaning solutions on these liners have no effect compared to synthetic liners.
- Silicone teats last three times as long as rubber materials and therefore remain economical despite being twice the price.

An interesting aspect is the liner shape and head structure. A round liner is the most common shape worldwide. Recent studies have shown that milking with triangular mouthpieces' chamber-ventilated liner leads to increased milk foam, which could potentially have a negative impact on milk quality [60]. A previous study showed that the type of liner used had a significant impact on keratinization and the frequency of cracked teat ends [68]. This means that herds milked with round teats had a 0.43 higher total keratinization score than herds milked with square teats. In addition, cows milked with round liners have on average 20% more cracked teat ends than cows milked with square liners.

Observations in practice have shown that teat shape and teat size vary greatly from animal to animal, including within the herd on the same farm, especially in camels [69] (Figure 4).

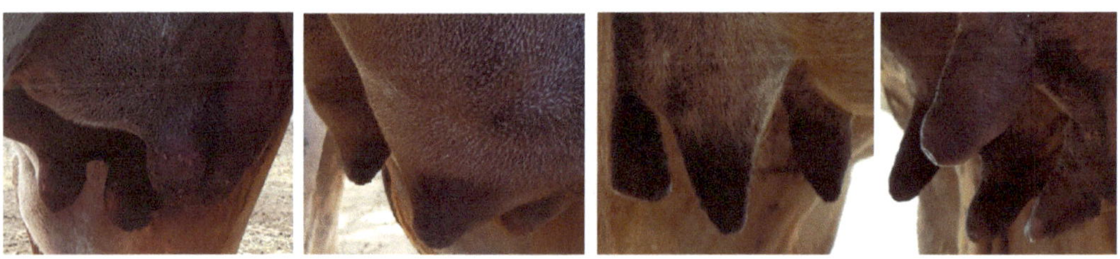

Figure 4. Different teat sizes and dimensions in lactating camels in the same herd [24].

The question of which liners best suits my milking machine, and my herd, can only be answered if the liners fit the morphological and anatomical conditions of the lactating udders. In practice, there is often no choice of liners according to the teat dimensions, but standard teat cup liners are often used. Most manufacturers of milking equipment have always strived to produce well-fitting teat liners. However, we cannot use an additional teat cup liner for each lactating animal on the farm. Hence, there were always some animals on the farms suffering from clinical or subclinical mastitis.

There is no teat liner that fits all lactating animals in a herd. Under these conditions, the company Siliconform, Germany, set itself the task of developing liners for all lactating animals, as the company has been in milking technology for 50 years and has a great amount of experience with all lactating animals such as cows [10,25], sheep, goats [7] and camels [69]. The company has development an exceptional teat liner for all lactating species. The name of the new liner is "Stimulor® StressLess" (Siliconform) (Figure 5). These new liners have unique properties that are excellent for maintaining teat health during milking in all lactating animals. One of its distinguishing features is the presence of the wave-shaped design of its head, which allows the "StressLess Stimulor" (Siliconform) to adapt well to the different teat sizes in a herd, thus ensuring consistent milking of lactating animals. In this way, the aim of the milking technology was achieved, which is to imitate the rhythmic process of calf suckling during machine milking as far as possible.

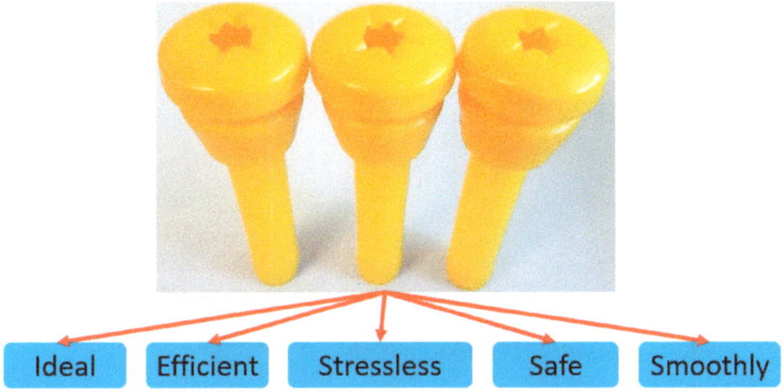

Figure 5. New teat cup liners (Stimulor® StressLess, Siliconform) for lactating animals.

Different teat sizes can only be optimally milked if the milking machine uses the integrated adaptive lip of the liner, which reacts to a pressure difference and allows air to flow in at the right moment. As a result, the vacuum of the liner head is kept in the physiological range. The new special silicone liner (Stimulor® StressLess, Siliconfrom) is ideal for all teat sizes and measurements of lactating animals and its design is of great importance for the acceptance of female lactating animals during the milking process. In practice, this means protection of the sensitive teat tissue and more comfort during milking. Furthermore, the new liners adhere particularly well and do not cause any slurping noises during milking. Finally, the new liner prevents injuries and thus makes a decisive contribution to better animal welfare and improved animal health.

6. Conclusions

The milking system should not adversely affect milk release and letdown which can reduce yield and alter composition.

Incorrect pulsator setting and/or high vacuum level and/or wrong pulsation ratios can result in disruptive machine milking and/or damage to the teat ends, which can lead to mastitis and reduced milk yield.

Liners are the only part of the milking machine that meet the teat of lactating animals, and their design is critical in udder health and for efficient milking.

The Stimulor® StressLess liner has a wavy construction. With the integrated adaptive lip, different teat sizes can be milked with the same liner.

Author Contributions: S.K. conceptualization and writing the original manuscript; M.W.P. revision of content and language improvement of the original manuscript. All authors have read and agreed to the published version of the manuscript.

Funding: This research received no external funding.

Data Availability Statement: Not applicable.

Conflicts of Interest: The authors declare no conflict of interest.

References

1. Bruckmaier, R.M.; Blum, J.W. Oxytocin release and milk removal in ruminants. *J. Dairy Sci.* **1998**, *81*, 939–949. [CrossRef] [PubMed]
2. Parilova, M.; Stadnik, L.; Jezkova, A.; Stolc, L. Effect of milking vacuum level and overmilking on cow's teat characteristics. *Acta Univ. Agric. Et Silvic. Aemendelianae Brun.* **2011**, *59*, 193–202. [CrossRef]

3. Edwards, J.P.; O'Brien, B.; Lopez-Villalobos, N.; Jago, J.G. Overmilking causes deterioration in teat-end condition of dairy cows in late lactation. *J. Dairy Res.* **2013**, *80*, 344–348. [CrossRef]
4. Ferneborg, S.; Svennersten-Sjaunja, K. The effect of pulsation ratio on teat condition, milk somatic cell count and productivity in dairy cows in automatic milking. *J. Dairy Res.* **2015**, *82*, 453–459. [CrossRef]
5. Gasparik, M.; Duchacek, J.; Stadnik, L.; Novakova, V. Impact of milking settings optimization on milk quality, milking time, and milk yield in Holstein cows. *IOP Conf. Ser. Mater. Sci. Eng.* **2018**, *420*, 012073. [CrossRef]
6. Kaskous, S. Optimization of the pulsation ratio during the course of milk removal after using a quarter individual milking system "MultiLactor". *Int. J. Agric. Innov. Res.* **2018**, *6*, 284–289.
7. Kaskous, S. Laboratory Tests to Optimize the Milking Machine Settings with Air Inlet Teat Cups for Sheep and Goats. *Dairy* **2022**, *3*, 29–46. [CrossRef]
8. Caroprese, M.; Albenzio, M.; Marino, R.; Muscio, A.; Zezza, T.; Sevi, A. Behaviour, milk yield, and milk composition of machine-and Hand-milked Murgese mares. *J. Dairy Sci.* **2007**, *90*, 2773–2777. [CrossRef]
9. Baines, J. The role of machine milking in quality milk production and ensuring dairy cow welfare. *Cattle Pract.* **2010**, *18*, 48–52.
10. Kaskous, S. The effect of using quarter individual milking system "MultiLactor" on improvement of milk performance and milk quality of different dairy cows breeds in different farms. *Emir. J. Food Agric.* **2018**, *30*, 57–64. [CrossRef]
11. Stauffer, C.; Van der Vekens, E.; Stoffel, M.; Schweizer, D.; Bruckmaier, R.M. Increased teat wall thickness in response to machine milking. *J. Dairy Sci.* **2021**, *104*, 9082–9092. [CrossRef] [PubMed]
12. Neijnhuis, F. Teat Condition in Dairy Cows. Ph.D. Thesis, Faculty of Veterinary Medicine, Utrecht University, Utrecht, The Netherlands, 2004; pp. 1–189.
13. Hamann, J.; Osteras, O. Teat Tissue Reactions to machine milking and new infection risk. *Bull. Int. Dairy Fed.* **1994**, *297*, 34–41.
14. Mein, G.A.; Williams, D.M.D.; Reinemann, D.J. Effects of milking on teat end hyperkeratosis: 1. Mechanical forces applied by the teat cup liner and responses of the teat. In Proceedings of the 42nd Annual Meeting of the National Mastitis Council, Fort Worth, TX, USA, 26–29 January 2003; pp. 26–29.
15. Bobic, T.; Andrasic, D.; Mijic, P.; Gregic, M.; Baban, M.; Gantner, V. The effect of teat cup liner type on the teat condition of Holstein primiparous cows. *J. Cent. Eur. Agric.* **2018**, *19*, 766–771. [CrossRef]
16. Thomas, C.S.; Svennersten-Sjaunja, K.; Bhosrekar, M.R.; Bruckmaier, R.M. Mammary cisternal size, cisternal milk and milk ejection in Murrah buffaloes. *J. Dairy Res.* **2004**, *71*, 162–168. [CrossRef] [PubMed]
17. Kaskous, S.; Bruckmaier, R.M. Best combination of pre-stimulation and latency period duration before cluster attachment for efficient oxytocin release and milk ejection in cows with low to high udder-filling levels. *J. Dairy Res.* **2011**, *78*, 97–104. [CrossRef]
18. Kaskous, S. Physiology of lactation and machine milking in dromedary she-camel. *Emir. J. Food Agric.* **2018**, *30*, 295–303. [CrossRef]
19. Tuor, M.; Wellnitz, O.; Bruckmaier, R.M. The interplay of continuous milk ejection and milking system with and without prestimulation at different vacuum settings. *J. Dairy Sci.* **2023**, *106*, 3615–3624. [CrossRef]
20. Bruckmaier, R.M. Laktationsphysiologie. In *Kurzes Lehrbuch Milchkunde und Milchhygiene*; Krömker, V., Ed.; Parey in MVS Medizinverlage Stuttgart: Stuttgart, Germany, 2007; pp. 6–22.
21. Borghese, A.; Rasmussen, M.; Thomas, C.S. Milking management of dairy buffalo. *Ital. J. Anim. Sci.* **2007**, *6* (Suppl. S2), 39–50. [CrossRef]
22. Sandrucci, A.; Tamburini, A.; Bava, L.; Zucali, M. Factors affecting milk flow traits in dairy cows: Results of a field study. *J. Dairy Sci.* **2007**, *90*, 1159–1167. [CrossRef] [PubMed]
23. Kaskous, S. MultiLactor Melksystem, eine Revolution in der Melktechnik. In *Forschungsbericht, Abteilung der Forschung und Entwicklung*; Siliconform: Türkheim, Germany, 2016; pp. 1–40.
24. Kaskous, S. Importance of liner design and milking machine settings for optimal milking performance and welfare in camels. In Proceedings of the 6th Conference of the International society of Camelid Research and Development (ISOCARD), "The role of camel in food security and economic development" King Faisal University, Al Ahsa, Saudi Arabia, 12–16 March 2023; p. 37.
25. Kaskous, S. Optimization of milk performance and quality in dairy farms by using a quarter individual milking system "MultiLactor". *Int. J. Environ. Agric. Biotechnol.* **2020**, *5*, 943–952. [CrossRef]
26. Sagkob, S.; Rudovsky, H.-J.; Pache, S.; Herrmann, H.-J.; Bernhardt, H. Effects of different cluster types on teat end condition and milk release. *Landtechnik* **2010**, *65*, 27–30.
27. Odorcic, M.; Rasmussen, M.D.; Paulrud, C.O.; Bruckmaier, R.M. Review: Milking machine settings, teat condition and milking efficiency in dairy cows. *Animal* **2019**, *13*, 94–99.
28. Miekley, B. (K)ein Mythos, Bedeutung und Limitierung von Zitzengummis im Melkprozess. In Proceedings of the 22. Jahrestagung, Wissenschaftliche Gesellschaft der Milcherzeugerberater (WGM) e.V., an der Bayerischen Landesanstalt für Landwirtschaft in Grub, Bayern, Germany, 4–6 October 2022; pp. 11–13.
29. Vermaak, P.; Petzer, I.M.; Karzis, J. Effects of milking machine settings and teat liners on bovine udder health. *South Afr. J. Anim. Sci.* **2022**, *52*, 421–432. [CrossRef]
30. Neijenhuis, F.; Barkema, H.W.; Hogeveen, H.; Noordhuizen, J.P.T.M. Classification and longitudinal examination of callused teat ends in dairy cows. *J. Dairy Sci.* **2000**, *83*, 2795–2804. [CrossRef]
31. Krömker, V. Melken. In *Kurzes Lehrbuch Milchkunde und Milchhygiene*; Krömker, V., Ed.; Parey in MVS Medizinverlage Stuttgart: Stuttgart, Germany, 2007; pp. 23–46.

32. Reinemann, D.J.; Mein, G.A. Unravelling the mysteries of liner compression. In Proceedings of the Countdown Symposium, Tackling Strep Uberis, Melbourne, VIC, Australia, 8 June 2011; p. 10.
33. Spencer, S.B.; Shin, J.-W.; Rogers, G.W.; Cooper, J.B. Short communication: Effect of vacuum and ratio on the Performance of a Monoblock silicone milking liner. *J. Dairy Sci.* **2007**, *90*, 1725–1728. [CrossRef] [PubMed]
34. Penry, J.F.; Upton, J.; Mein, G.A.; Rasmussen, M.D.; Ohnstad, I.; Thompson, P.D.; Reinemann, D.J. Estimating teat canal cross-sectional area to determine the effects of teat-end and mouthpiece chamber vacuum on teat congestion. *J. Dairy Sci.* **2017**, *100*, 821–827. [CrossRef]
35. Hamann, J.; Burvenich, C.; Mayntz, M.; Osteras, O.; Haider, W. Machine induced changes in the status of the bovine teat tissue with respect to new infection risk. *Bul. IDF* **1994**, *2*, 13–22.
36. Ambord, S.; Bruckmaier, R.M. Milk flow-dependent vacuum loss in high-line milking systems: Effect on milking characteristics and teat tissue condition. *J. Dairy Sci.* **2010**, *93*, 3588–3594. [CrossRef]
37. Neijnhuis, F.; de Koning, K.; Barkema, H.; Hogeveen, H. The effect of machine milking on teat condition. *ICAR Tech. Ser.* **2001**, *7*, 33–40.
38. Pazzona, A.; Murgia, L. Effect of milking vacuum on leukocyte count in Buffalo milk. In Proceedings of the 24th International Conference on Agricultural Mechanization, Zaragoza, Spain, 1–4 April 1992; pp. 691–694.
39. Hamann, J. Effect of machine milking on teat end condition with special emphasis on infection risk. *World Rev. Anim. Prod.* **1990**, *25*, 9–12.
40. Sinapis, E.; Vlachos, I. Influence du niveau de vide de la machine a traire et des facteurs zootechniques sur le comptages de cellules somatiques chez les chevres locale grecque. In Proceedings of the 6th International Symposium on the Machine Milking of Small Ruminants, Athens, Greece, 26 September–1 October 1998; European Association for Animal Production (EAAP) Wageningen Academic Publishers: Wageningen, The Netherlands, 1999. No. 95. pp. 513–518.
41. Rasmussen, M.D.; Madsen, N.P. Effects of milk line vacuum, pulsator airline vacuum, and cluster weight on milk yield, teat condition, and udder health. *J. Dairy Sci.* **2000**, *83*, 77–84. [CrossRef] [PubMed]
42. Spencer, S.B. Milking machine, Principles, and design. In *Encyclopaedias of Dairy Sciences*, 2nd ed.; Fuquay, J.W., Ed.; Academic Press: San Diego, CA, USA, 2011; pp. 941–951.
43. Besier, J.; Bruckmaier, R.M. Vacuum levels and milk-flow-dependent vacuum drops affect machine milking performance and teat condition in dairy cows. *J. Dairy Sci.* **2016**, *99*, 3096–3102. [CrossRef] [PubMed]
44. Besier, J.; Lind, O.; Bruckmaier, R.M. Dynamics of teat-end vacuum during machine milking: Types, causes, and impacts on teat condition and udder health-A literature review. *J. Appl. Anim. Res.* **2016**, *44*, 263–272.
45. Dolezal, O.; Hlasny, J.; Jilek, F.; Hanus, O.; Vegricht, J.; Pytloun, J.; Matous, E.; Kvapilik, J. *Milk, Milking, Milking Parlours*; Agrospoj: Prague, Czechia, 2000; p. 241.
46. Mein, G.; Reinemann, D.; O'Callaghan, E.; Ohnstad, I. Where the rubber meets the teat and what happens to milking characteristics. In 100 years with liners and pulsators in machine milking. *Bull. Int. Dairy Fed.* **2004**, *388*, 431–446.
47. Hamann, J.; Mein, G.A. Teat thickness changes may provide biological test for effective pulsation. *J. Dairy Res.* **1996**, *63*, 179–189. [CrossRef]
48. Sterrett, A.E.; Wood, C.L.; McQuerry, K.J.; Bewley, J.M. Changes in teat-end hyperkeratosis after installation of an individual quarter pulsation milking system. *J. Dairy Sci.* **2013**, *96*, 4041–4046. [CrossRef]
49. Upton, J.; Penry, J.F.; Rasmussen, M.D.; Thompson, P.D.; Reinemann, D.J. Effect of pulsation rest phase duration on teat end congestion. *J. Dairy Sci.* **2016**, *99*, 3958–3965. [CrossRef]
50. Thomas, C.V.; Force, D.K.; Bremel, D.H.; Stressen, S. Effect of pulsation ratio, pulsation rate and teat cup liner design on milking rate and milk production. *J. Dairy Sci.* **1991**, *74*, 1243–1249. [CrossRef]
51. O'Callaghan, E.J. Effects of pulsation characteristics on machine yield, milking time, and cluster stability. *Ir. J. Agric. Food Res.* **1998**, *37*, 201–207.
52. Williams, D.M.; Mein, G.A.; Brown, M.R. Biological responses of the bovine teat to milking: Information from measurement of milk flow rate within individual pulsation cycles. *J. Dairy Res.* **1981**, *48*, 7–21. [CrossRef]
53. Gleeson, D.E.; O'Callaghan, E.J.; Rath, M.V. Effect of liner design, pulsator setting, and vacuum level on bovine teat tissue changes and milking characteristics as measured by ultrasonography. *Ir. Vet. J.* **2004**, *57*, 289–296. [CrossRef] [PubMed]
54. Osteras, O.; Ronningen, O.; Sandvik, L.; Waage, S. Field Studies shows associations between pulsator characteristics and udder health. *J. Dairy Res.* **1995**, *62*, 1–13. [CrossRef]
55. Bray, D.R.; Fowler, P.A.; Fialho, F.B.; Bucklin, R.A.; Yeralan, S.; Tran, T.; Braun, R.K. An automated system for monitoring milking system parameters. In Proceedings of the 37th National Mastitis Council Annual Meeting, St. Louis, MO, USA, 25–28 January 1998; pp. 127–136.
56. Mein, G.A.; Neijnhuis, F.; Morgan, W.F.; Reinemann, D.J.; Hillerton, J.E.; Baines, J.R.; Ohnstad, I.; Rasmussen, M.D.; Timmus, L.; Britt, J.S.; et al. Evaluation of bovine teat condition in commercial dairy herds: 1. Non-infectious factors. In Proceedings of the 2nd International Symposium on Mastitis and Milk Quality, Vancouver, BC, Canada, 13–15 September 2001; pp. 362–366.
57. Penry, J.F.; Crump, P.M.; Ruegg, P.L.; Reinemann, D.J. Short communication: Cow-and quarter-level milking indicators and their associations with clinical mastitis in an automatic milking system. *J. Dairy Sci.* **2017**, *100*, 9267–9272. [CrossRef]
58. Reinemann, D.J. The smart position on teat condition. In Proceedings of the New Zealand Milk Quality Conference, Hamilton, New Zealand, 18–19 June 2012; pp. 124–131.

59. DairyNZ. Correct teat cup liners will keep cows healthy and happy. *Management-Dairy News, 1 February 2018*. Available online: www.ruralnewsgroup.co.nz/dairy-management (accessed on 7 January 2018).
60. Holst, G.E.; Adrion, F.; Umstätter, C.; Bruckmaier, R.M. Type of teat cup and cluster ventilation affect vacuum conditions in the liner and milking performance in dairy cows. *J. Dairy Sci.* **2021**, *104*, 4775–4786. [CrossRef]
61. Neijnhuis, F. Spenen onder spanning. Tepels van de koe een goede indicator van melker en melkmachine. *Veeteelt* **1999**, *16*, 40–42.
62. Barth, K. Effects of suckling on milk yield and milk composition of dairy cows in cow-calf contact systems. *J. Dairy Res.* **2020**, *87*, 133–137. [CrossRef]
63. Rasmussen, M.D.; Mayntz, M. Pressure in the teat cistern and the mouth of the calf during suckling. *J. Dairy Res.* **1998**, *65*, 685–692. [CrossRef]
64. Bruckmaier, R.M.; Schams, D.; Blum, J.W. Continuously elevated concentrations of oxytocin during milking are necessary for complete milk removal in dairy cows. *J. Dairy Res.* **1994**, *61*, 323–334. [CrossRef] [PubMed]
65. Model, I.; Rudovsky, H.-J. Welcher Zitzengummi ist der richtige? *Milchpraxis* **1999**, *37*, 209–212.
66. Hillerton, J.E. Do liners differ? In Proceedings of the 44th Annual Meeting of the National Mastitis Council, Madison, WL, USA, 16–19 January 2005; pp. 133–138.
67. Reinemann, D.; Mein, G.A. Schonend(er) melken. *Elite Magazin für Milcherzeuger* **2013**, *6*, 64–66. Available online: www.elite-magazin.de (accessed on 7 January 2018).
68. Schukken, Y.H.; Petersson, L.G.; Rauch, B.J. Liners and teat end health. *NMC Annu. Meet. Proc.* **2006**, *45*, 183–196.
69. Kaskous, S. A new milking technology: "STIMULACTOR" for lactating camels. *J. Camel Pract. Res.* **2021**, *28*, 1–9. [CrossRef]

Disclaimer/Publisher's Note: The statements, opinions and data contained in all publications are solely those of the individual author(s) and contributor(s) and not of MDPI and/or the editor(s). MDPI and/or the editor(s) disclaim responsibility for any injury to people or property resulting from any ideas, methods, instructions or products referred to in the content.

agriculture

MDPI

Review

Analysis of Hotspots in Subsurface Drip Irrigation Research Using CiteSpace

Yatao Xiao [1,2,†], Chaoxiang Sun [2,†], Dezhe Wang [2], Huiqin Li [2,*] and Wei Guo [1,*]

1 Institute of Farmland Irrigation of CAAS, Key Laboratory of High-Efficient and Safe Utilization of Agriculture Water Resources, Chinese Academy of Agricultural Sciences, Xinxiang 453003, China
2 College of Mechanical and Electrical Engineering, Henan Agricultural University, Zhengzhou 450002, China
* Correspondence: lihuiqin@henau.edu.cn (H.L.); guowei02@caas.cn (W.G.); Tel./Fax: +86-371-56558040 (H.L.); +86-373-3393277 (W.G.)
† These authors contributed equally to this work.

Abstract: To investigate the research hotspots and development trends of subsurface drip irrigation (SDI) over the past 20 years, this study analyzed relevant literature from the Web of Science Core Collection spanning from 2002 to 2022. The data were visualized using CiteSpace, showcasing the publication volume trends, countries, keywords, cited references, authors, and affiliated institutions. Based on 1079 articles, the annual publication volume showed an overall upward trend. The United States had the most extensive research coverage and highest publication volume, whereas China had the fastest growing publication rate in recent years. However, relatively little cooperation occurred among research teams and institutions. Over time, research topics became increasingly diverse, with water conservation and yield increases being the primary research objectives. In addition to improving irrigation and fertilizer use efficiency, SDI has also been applied in research on the safe utilization of unconventional water resources (wastewater and salt water) and the optimization of soil conditions. Among these, aerated irrigation technology—aimed at improving root growth in the rhizosphere—may become a new branch of SDI research. Currently, the main research focus in the field of SDI is the diffusion and distribution of water in the crop root zone, for which Hydrus model simulation is a particularly important method.

Keywords: subsurface drip irrigation; CiteSpace software; hotspot analysis; frontier analysis

Citation: Xiao, Y.; Sun, C.; Wang, D.; Li, H.; Guo, W. Analysis of Hotspots in Subsurface Drip Irrigation Research Using CiteSpace. *Agriculture* **2023**, *13*, 1463. https://doi.org/10.3390/agriculture13071463

Academic Editors: Redmond R. Shamshiri, Muhammad Imran, Muhammad Sultan and Yuguang Zhou

Received: 13 June 2023
Revised: 20 July 2023
Accepted: 21 July 2023
Published: 24 July 2023

Copyright: © 2023 by the authors. Licensee MDPI, Basel, Switzerland. This article is an open access article distributed under the terms and conditions of the Creative Commons Attribution (CC BY) license (https://creativecommons.org/licenses/by/4.0/).

1. Introduction

The rapid rate of urbanization and population growth continually increases the global imbalance between water supply and demand for agriculture, industry, and the environment. The scarcity of irrigation water has become a major challenge for agricultural development worldwide [1]. According to projections by the United Nations, the global population is expected to reach 9.55 billion by 2050, corresponding to a 70% increase in food demand and 19% increase in water use in agricultural irrigation [2–4]. Traditional irrigation methods, such as flood irrigation, exhibit low water resource utilization efficiency and exacerbate water wastage [3]. Therefore, the development of water-saving irrigation technologies has become a priority. Simultaneously, improving grain yield while reducing agricultural water consumption has emerged as a prominent topic in contemporary agricultural development [5].

In 1920, Lee O. Charles from California developed and patented a basic subsurface drip irrigation (SDI) system that utilized tile pipes [6]. However, owing to technological limitations at the time, the SDI system had no practical applications. In recent years, owing to escalating environmental pollution, water resource crises, and technological and material advances, SDI technology has been implemented for various crops with considerable economic benefits. Generally, an SDI system comprises a water source, head pivot, main pipe, capillary channels, and emitters. A head pivot typically includes a pumping station, fertilization device, testing device, and several filtering devices to remove impurities from the effluent and provide pressure for transporting water over long distances. This ensures uniform water flow to the field and facilitates the supply of water and fertilizer to the root zone of crops through capillary channels and emitters [7].

SDI technology is renowned for its water- and fertilizer-saving abilities and high degree of automation. Water and fertilizer are delivered directly to crop roots through pipes buried in the tilled layer. As shown in Figure 1, compared with other irrigation methods, SDI can significantly reduce water loss through surface evaporation, enhance irrigation efficiency, improve crop net photosynthesis, and reduce crop transpiration [8]. In addition, SDI can effectively control weed growth, inhibit the proliferation of soil diseases and insect pests, reduce pesticide use, and significantly improve crop yield and quality [9,10]. In this study, we analyzed the literature on SDI in the core collection of the Web of Science to provide insights into research trends and emerging issues in the field. Our study elucidates the current issues of concern and provides informed directions for development in the field of SDI.

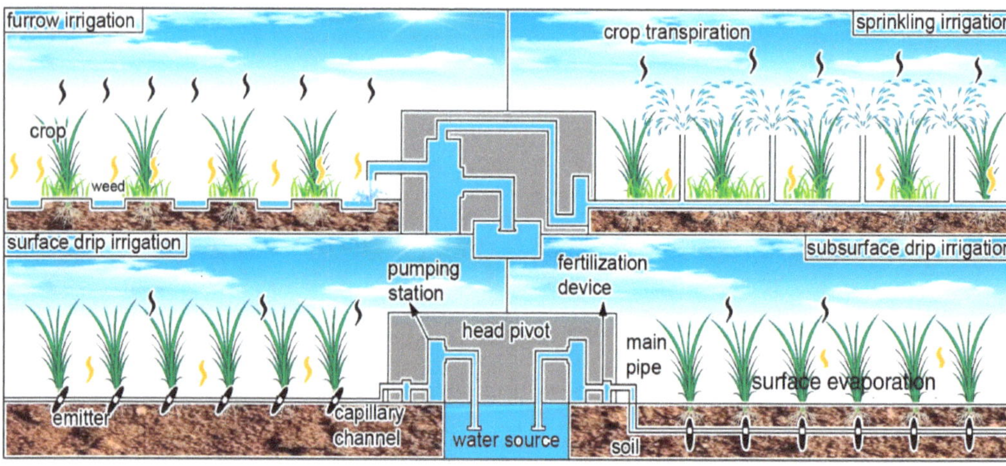

Figure 1. Schematic diagram of four different irrigation methods.

2. Materials and Methods

2.1. Data Sources

To ensure the accuracy and authority of the data, we used the core collection of the Web of Science to retrieve data based on the topic (with the retrieval term "subsurface drip irrigation") and limited the retrieval period to between January 2002 and December 2022. The article type was set to "article" and "review article". A total of 1080 articles were obtained and the article data, including full records and cited references, were imported into CiteSpace (v.6.1.R6, 64-bit) in text (.txt) format.

2.2. Research Methods

CiteSpace is a data visualization software developed by Prof. Chen Chaomei using the Java platform. CiteSpace can analyze input data from papers and generate a visual map of the focus and relationships between papers in a given field of research.

We screened the imported literature in CiteSpace, with the time slice set to one year and threshold (top N per slice) set to 50; the Pathfinder clipping method was used to simplify the atlas. During this step, an article that was officially published in 2023 but had been pre-published online was excluded. In the following results, we elucidated the development trend and widespread application of research interest in subsurface drip irrigation through annual publication volume and countries of origin. We examined the collaboration among various research teams in the field of SDI through the analysis of institutions and authors. The topic hotspots and research directions were expressed through the examination of keywords, while the research scope and orientations were explored through reference co-citation analysis. Apart from specific mentions, all the analyses were conducted based on the data from 41 reviews and 1038 articles.

In the knowledge maps, the greater the node range, the more frequently an object appears or is referenced. The thickness of the lines connecting the nodes indicates the degree of co-occurrence or co-citation. To assess the importance of the nodes in the network, betweenness centrality (BC) was estimated, as shown in purple on the knowledge map, with a value between 0 and 1. The median is generally considered to have a BC > 0.1. Clustering is another way to analyze research areas and distinguish unique topics. In clustering analysis, a modularity (Q) > 0.3 indicates that the clustering results are significant, while a silhouette (S) > 0.5 indicates good clustering and (S) > 0.7 indicates great clustering [11]. Finally, the burstiness of keywords can be used to determine the research hotspots (citation bursts) and analyze the direction of SDI development.

3. Results

3.1. Analysis of Annual Publication Volume and Countries of Origin

Annual publication volume is a critical indicator of a particular research field's developmental trends, based on the level of research activity and interest in a specific area [12]. Figure 2 illustrates the annual publication volumes for studies relating to SDI and the associated countries. Overall, the number of publications steadily increased over time. However, a slight decrease in the number of publications was observed in 2005, 2012, and 2017. The highest number of publications occurred in 2021, suggesting that the issue of agricultural water shortages is gaining increasing attention, and research on SDI is expanding.

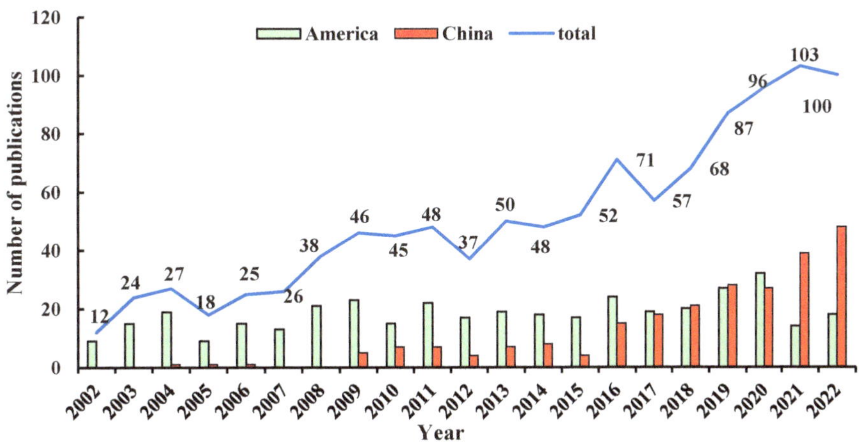

Figure 2. Annual circulation of publications.

Table 1 presents the number of publications according to location over the past 20 years. The United States led research on SDI, with the highest centrality and largest circulation of publications. Chinese publications ranked second in circulation, starting in 2004 and slowly growing until 2016 when publication rates surged. By 2022, the number of Chinese publications reached 48, accounting for ~50% of the total number of published articles that year. As the most populous country in the world, China produces 25% of the world's food with only 10% of the world's arable land and thus faces considerable challenges in agricultural water usage; thus, food security and water shortage may be the reasons behind China's substantial research investment in water-saving agriculture [13–15].

Table 1. Publication volume per country.

Rank	Number of Papers	Percentage %	Centrality	Burst	Country
1	386	35.78	0.57	28.42	USA
2	241	22.34	0.17		China
3	73	6.77	0.25		Spain
4	72	6.67	0.12	5.67	Brazil
5	56	5.19	0.07	4.02	Egypt
6	48	4.45	0.12	3.68	Israel
7	47	4.35	0.08		India
8	46	4.25	0.14		Iran
9	45	4.15	0.04	8.08	Australia
10	40	3.70	0.02		Turkey
11	40	3.70	0.14		Saudi Arabia
12	28	2.59	0.11		Italy
13	23	2.13	0.00	3.68	Canada
14	19	1.76	0.10		Germany
15	17	1.57	0.06	4.05	Greece
16	15	1.39	0.01		Pakistan
17	14	1.29	0.09		England
18	14	1.29	0.10		South Africa
19	13	1.20	0.14		France
20	13	1.20	0.13		Portugal

3.2. *Analysis of Publishing Institution and Authors*

The analysis of publishing institutions and authors is essential for understanding the structural developments and changes in the research field [16]. As shown in Table 2, 12 institutions published more than 20 publications each. Regarding publication volume and centrality, the US Department of Agriculture Agricultural Research Service was the most productive institution and published 9.54% of all publications. The institution with the second-highest publication volume was Northwest A&F University, which also had the highest publication volume and citation burst in the past five years. Several research institutes in the United States have published in the field of SDI, such as the University of California, Davis; the University of Arizona; and the University of California, Riverside.

Table 2. Top 12 institutions with the largest number of publications in SDI research.

Frequency	Centrality	Burst	Institution
103	0.25	7.27	US Department of Agriculture Agricultural Research Service
49	0.01	9.74	Northwest A&F University
34	0.04		University of California, Davis
30	0.06		China Agricultural University
28	0.07		Chinese Academy of Science
27	0.04	4.78	University of Arizona
26	0.08		University of California, Riverside
24	0.02	5.04	Texas A&M University
22	0.04		Ben Gurion University of the Negev
22	0.03		King Saud University
21	0.02		China Institute of Water Resources and Hydropower Resources
20	0.02		Kansas State University

Most of the institutions with the highest number of publications were from the United States or China, emphasizing their significant roles in SDI research.

Figure 3 illustrates the author contributions, consisting of 645 nodes and 581 lines with a density of 0.0028. Notably, although numerous research teams appeared in the author visualization map, each team operated independently. The fragmented nature of SDI research calls for strengthened communication and cooperation among institutions to form an interactive support network that utilizes the unique strengths of each research group and ultimately enhances research efficiency and depth.

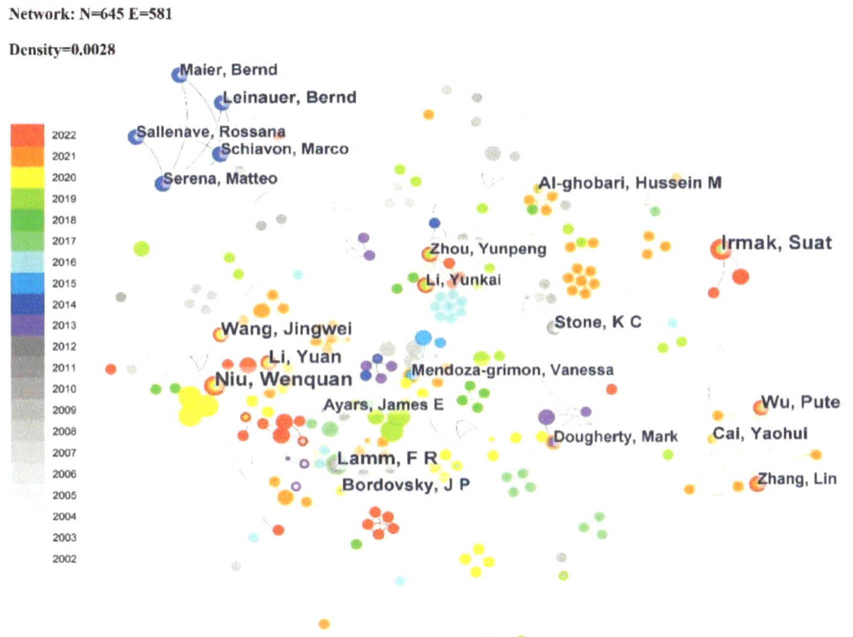

Figure 3. Author co-occurrence map for authors producing literature relating to SDI.

3.3. *Keyword Analysis*

3.3.1. Exploration of Hot Topics in Review Articles

To comprehensively explore the research hotspots and directions in the field of SDI, we manually analyzed the keyword information from 41 reviews based on the literature data and presented the results in Table 3 (the numbers following "*" represent the frequency of the keyword occurrence). Review articles are based on primary research data and information, representing the perspectives of other scholars on the advancements of subsurface drip irrigation in the respective field. The frequency of keywords in review papers can reflect the research focuses of other scholars in the SDI domain. A higher occurrence of keywords of the same type indicates a higher level of attention towards that particular area. Through the analysis of review paper keywords, we observed that, besides the category "irrigation methods", which is closely related to the search term, the distribution of soil moisture, environmental pollution, and crop cultivation have received the most attention. This suggests that SDI has been widely adopted in numerous crop cultivations, while also potentially posing challenges in controlling agricultural pollution. Investigating the water distribution and hydraulic characteristics of SDI under different irrigation and soil conditions is a popular research topic. In contrast, the topic of economic benefits discussion appears to be relatively limited, indicating that economic factors receive less attention in the research and development of SDI. Moreover, advancements in new technologies have provided opportunities for SDI, generating extensive discussions on enhancing water use efficiency, utilizing unconventional water resources, and increasing crop yields.

Table 3. Statistical analysis of keyword information in review articles.

Tags	Keywords	Count
Irrigation method	subsurface drip irrigation × 9; irrigation × 6; micro irrigation × 5; irrigation methods × 4; trickle irrigation × 3; deficit irrigation × 3; drip irrigation × 3; irrigation systems × 2; SDI × 2; sprinkler irrigation × 2; surface irrigation × 2; limiting flow; furrow irrigation; spray irrigation; subsurface irrigation	45
Soil moisture distribution and changes	water-table; soil moisture distribution pattern; wetting pattern; transport; drying-rewetting frequency; soil water; water movement; hydraulic conductivity; integrated water; variably saturated flow; one dimensional infiltration; surface point source; dependent linearized infiltration; unsaturated hydraulic conductivity; steady state flows; groundwater; soil water potential; soil water potential threshold; soil hydraulic properties; root water uptake; soil management; root distribution; root water; root zone water; fine root; sorption; evapotranspiration;	27
Environmental issues and pollution	microbial contamination; environmental health; microbial activity; oxide emissions; denitrification; nitrogen-fixation; similarity solutions; soil properties; heavy metals; boron toxicity; steady state; excess boron; chemigation; soil contamination; pathogens; health risks; soil contamination; ammonia volatilization; N_2O; CO_2; CH_4; C_2H_2	22
Crop	lettuce; cotton; wheat cropping system; winter-wheat; rice; grassland soils; pomegranate; tomato root distribution; cropping systems; cotton; sugarcane; plant; vitis vinifera; wine; plant growth; plant conditioners; corn; interspecific interaction; niche differentiation; plant breeding; plant roots; winter wheat	22
New technologies	remote sensing × 3; discharge rate; technical parameters; wireless sensor networks; soil-moisture; precision agriculture; screenhouse; automation; innovation and technology; low energy precision application; tensiometer; soil water potential sensor; walled carbon nanotubes; ground penetrating radar; electrical conductivity	17
Water use efficiency	water use efficiency × 7; water saving × 2; water requirements; water use productivity; crop water productivity; water use; crop productivity; water productivity; water resources	16
Nonconventional water resource utilization	treated wastewater × 3; water quality; wastewater reuse; low quality water; virtual water; unconventional water resources; site waste water; wastewater; salinity; soil salinity; nonuniform transient salinity; coal salt tolerance; n-mineralization	15
Yield	Yield × 3; crop production × 2; aril colour; fruit quality; plant growth; physical properties; waste of water and feed; crop quality; yield response; grain-yield	13
Irrigation strategy and management	water management × 3; management; management strategies; irrigation management strategies; design; operation systems; irrigation schemes; irrigation management × 2; irrigation scheduling	12

Table 3. *Cont.*

Tags	Keywords	Count
Agricultural development	Sustainability × 3; conservation agriculture × 2; protected cultivation; residue management; zero tillage; cereal system; sustainable irrigation; agriculture	11
Climate change and water resources	climate change × 3; groundwater depletion; drought; environmental impact; micro-climate regulation; heat stress; temperature	9
Region and terrain	coastal-plain; India; basin tillage; ceramic pots; sub-Sahara; ogalla aquifer; indo-gangetic plains of India	7
Fertilizers and nutrition	Fertigation × 2; controlled release fertilizers; macronutrients; nutrients; nutrient source	6
Economic benefits	payoff period; runoff; water pricing and rationing	3
Others	LEPA; uniformity; prisma-p; semi-permeable membrane; nitrogen isotope; semifield; conservative numerical-solution; finite-element method; conservative numerical-solution; finite-element method, meta-analysis; bibliometric analysis; Green-ampt analysis	13

3.3.2. Analysis of Keyword Co-Occurrence

Analyzing the keywords of a certain field of research can be useful in elucidating research hotspots and core content within that field [17]. Figure 4 shows a keyword co-occurrence map containing 596 nodes, 2509 connections between the nodes, and a density of 0.0142. Table 4 presents high-frequency keywords, where the top 10 keywords have a frequency > 100. By combining these keywords, we observed that water use efficiency and yield were prominent research directions related to SDI.

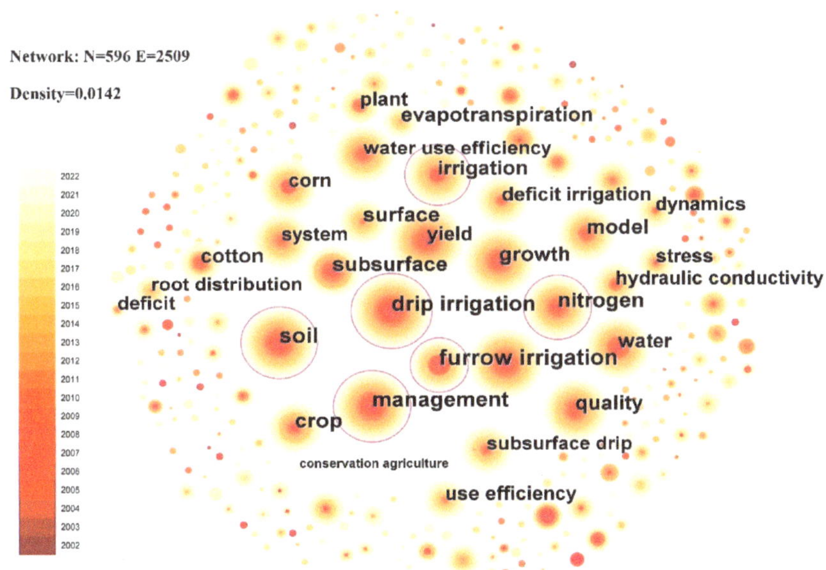

Figure 4. Keyword co-occurrence map displaying the most frequently used words associated with SDI research.

Table 4. Top 30 most frequently cited keywords used in literature pertaining to SDI.

Rank	Count	Centrality	Burst	Keyword
1	380	0.02		subsurface drip irrigation
2	244	0.06		yield
3	174	0.05		water use efficiency
4	170	0.11		drip irrigation
5	161	0.06		growth
6	155	0.10		management
7	151	0.17		soil
8	122	0.06		water
9	117	0.06		deficit irrigation
10	113	0.09		quality
11	93	0.08		use efficiency
12	88	0.05		system
13	71	0.10		irrigation
14	67	0.05		model
15	66	0.10		nitrogen
16	61	0.08		crop
17	59	0.07		subsurface drip
18	55	0.05		corn
19	53	0.01		simulation
20	51	0.07		surface
21	48	0.05		cotton
22	44	0.04		fruit quality
23	43	0.04		dynamics
24	43	0.03		field
25	43	0.07		subsurface
26	41	0.07		evapotranspiration
27	41	0.04		hydraulic conductivity
28	39	0.05		stress
29	38	0.03	5.36	flow
30	35	0.13	5.48	furrow irrigation

Keywords with a high BC represent the most important research areas in SDI. Figure 4 shows six keywords with high centrality: soil (0.17), furrow irrigation (0.13), drip irrigation (0.11), irrigation (0.10), management (0.10), and nitrogen (0.10). Soil is a fundamental component of crop production. The physical and chemical properties of soil have a considerable influence on the effectiveness of SDI. Based on the degree of permeability, sandy soil is more suitable for SDI compared to the more compact clayey soil, which is also vulnerable to uneven water and fertilizer distribution [18]. Soil temperature, pH, and salt content affect the ability of plants to absorb and utilize water and nutrients [19,20]. Furrow, drip, and irrigation were closely associated with the search terms used in this study. Compared with other irrigation methods, SDI yields higher water use efficiency, less soil erosion, and greater water conservation and is more conducive to improving crop yield and quality. Furthermore, it also reduces labor intensity [21,22]. SDI management encompasses irrigation scheduling, fertilization, system maintenance, and soil crop management. Effective management strategies are essential for enhancing crop yield and quality, maintaining soil health, improving water and fertilizer use efficiency, and achieving sustainable crop production [23]. Nitrogen is an important nutrient in crop growth and directly affects yield. Variations in irrigation methods can significantly affect nitrogen use efficiency. Studies have demonstrated that SDI can improve nitrogen utilization efficiency while reducing nitrogen loss [24,25]. Additionally, SDI can be combined with other measures, such as soil testing, mulch planting, and crop rotation, to optimize nitrogen use efficiency [24–26].

3.3.3. Keyword Clustering Relating to SDI Literature According to Time of Citation

A keyword clustering map is a valuable tool for characterizing nodes with similar themes [17]. As shown in Figure 5, the Q of keyword clustering was 0.4584, and the S was 0.7546. The selection of keywords and the log-likelihood ratio utilized by the clustering algorithm generated 14 labels, in which 588 out of 596 nodes were classified; each node's label was unique. The thickness of the lines between nodes within a cluster indicates the correlation between research directions (the denser the lines, the stronger the correlation) [27].

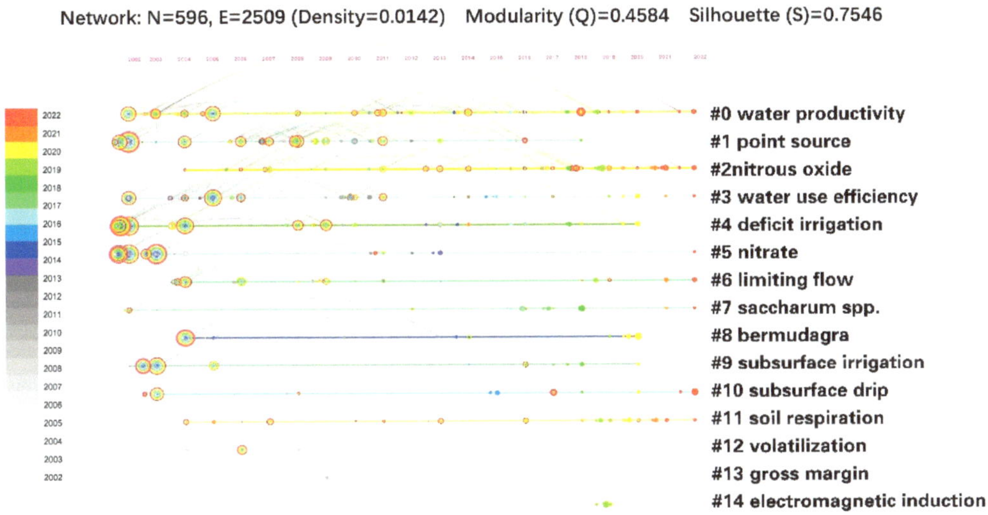

Figure 5. Keywords timeline clustering map for keywords relating to SDI research.

To further refine the research content and identify research hotspots, tags with the same theme were summarized into nine parts (tags #9 and #10 were excluded because of their close association with the search term). Part 1 (#0 water productivity and #3 water use efficiency) includes 126 nodes, and the main years of #0 and #3 were 2012 and 2011, respectively. Part 2 (#2 nitrous oxide and #5 nitrate) contained 109 nodes, and the main years for #2 and #5 were 2016 and 2007, respectively. Part 3 (#4 deficit irrigation and #6 limiting flow) had 90 nodes, with the main year being 2010. Part 4 (#7 *Saccharum* spp. and #8 Bermudagra (Bermuda grass)) had a total of 73 nodes, and the main years of #7 and #8 were 2013 and 2012, respectively. Part 5 (#1 point source) appeared in 2009 and contained 65 nodes. Part 6 (#11 soil respiration) had 30 nodes and appeared in 2015. Part 7 (#12 volatilization) had 19 nodes, and the main year was 2005. Part 8 (#13 gross margin) had six nodes, and its main year was 2009. Finally, Part 9 (#14 electromagnetic induction) comprised six nodes, and its primary year was 2019.

Among the nine parts, water use efficiency in Part 1 contained the highest number of nodes, indicating that crop water absorption and utilization efficiency has been a major focus of research in SDI. Part 2, which considers nitrogen, is divided into two periods. The earlier period focused on the effect of nitrogen fertilizer on crop yield, while the latter dealt with the avoidance of the adverse effects of nitrate on the environment. The precise control of nitrogen fertilizer dosage through SDI can promote crop growth while avoiding pollution and uncontrolled emissions of greenhouse gases (CO_2 and N_2O, among others) [26,28]. Part 3, deficit irrigation, is an important water-saving strategy in arid and semiarid regions [29] that aims to limit input while promoting yield and quality, thereby achieving an efficient balance between irrigation volume (input) and crop yield (output) for maximum profit [30–32]. Part 4, crop, indicates the beneficial impact of SDI on *Saccharum* spp. and Bermudagra (Bermuda grass) crop yield and quality while reducing water and fertilizer usage [33–36]. Part 5, the study of point sources, relates to the design of SDI systems and crop planting methods for optimized efficiency. The distribution of point sources not only has a significant impact on soil water depletion and moisture peak, but also limits the amount of tilling and crop spacing [37–40]. However, the timeline node distribution indicated that point source research was initially a hot topic but has gradually become less popular in recent research. Part 6, soil respiration, was evenly distributed throughout the research period, and its popularity increased slightly with the development of oxygation (oxygenated irrigation water). In general, irrigation saturates the spaces between soil particles and reduces air content, resulting in reduced respiration in the crop root zone. The oxygenation of irrigation water can effectively improve respiration in the root zone, thus increasing yield and quality [41–44]. Part 7, node distribution and connection, indicated that the study of evapotranspiration is outdated, albeit closely related to other parts. Evaporation is an important parameter for determining crop water demand. Based on crop water demand and local water supply conditions, efficient planning for precipitation and irrigation can achieve the best economic, social, and ecological outcomes [45–47]. In Part 8, profit studies were observed to be rare and do not overlap with other categories. However, from a commercial perspective, the complicated operating mode of SDI requires higher investment costs, which may inhibit its application and marketing [21]. Part 9, the application of sensor technology in agriculture, developed rapidly. Sensor technology provides a means to locate and monitor the distribution of water and nutrients in the soil and automate the functioning of the SDI system. The combined use of electromagnetic induction technology can further optimize irrigation and fertilization strategies to improve water and nutrient use efficiency and crop yield [48,49].

3.3.4. Analysis of Keyword Bursts

A keyword burst refers to a significant change in the frequency of a keyword over time. Keywords with strong and long-lasting burst intensities may represent a new research direction [50].

Figure 6 shows that the burst intensity for words before 2007 were strongly correlated with the search terms used in this study. "Trickle irrigation" had the highest burst intensity, whereas "subsurface" had the longest burst duration. Between 2002 and 2022, research focused on the characteristics of SDI. In 2010, three keywords showed a high burst intensity, among which "irrigation" appeared in 2004 and was associated with several burst words of the previous publication period. The studies of drip distribution in irrigation systems and water infiltration distribution also gained attention during this period. Finally, the close association in the burst timing of "water productivity", "grain yield", and "performance" represents the latest hot topics in SDI research. Therefore, we expect that SDI system research and development will continue to focus on improving yield and water conservation.

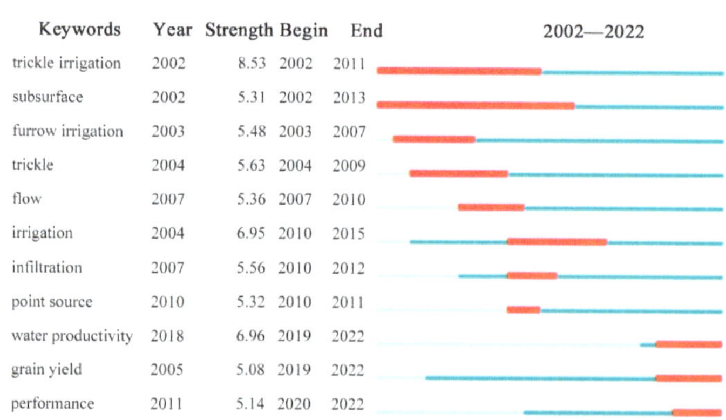

Figure 6. Top 11 keywords relating to SDI with the strongest citation bursts over the period between 2002 and 2022.

3.4. Analysis of Reference Co-Citation

We assessed the topic trends and temporal development of SDI research through a reference co-citation analysis. Reference co-citation refers to the overall number of citations for a particular paper [16]. Because some references were published before the papers screened in our study, the analysis of reference co-citation includes some studies before 2002. Table 5 highlights citations with high frequency or strong centrality, and Figure 7 shows a cluster analysis of citations over time. Each node in the figure represents a cited article and its size reflects the proportion of citations [51].

Table 5. Publications on SDI with high co-citation frequencies.

Frequency	Centrality	Burst	Author	Source	Year
29	0.05	10.43	Ayars, J.E.	*Agr. Water Manag.*	2015
25	0.00	10.27	Lamm, F.R.	*Trans. ASABE*	2016
19	0.01	6.66	Ben-Noah, I.	*Agr. Water Manag.*	2016
18	0.08	6.36	Simunek, J.	*Vadose Zone J.*	2016
17	0.02	5.29	Du, Y.D.	*Agr. Water Manag.*	2018
16	0.00	7.49	Li, Y.	*Soil Sci. Soc. Am. J.*	2016
15	0.25	7.40	Kandelous, M.M.	*Agr. Water Manag.*	2010
15	0.17	5.24	Irmak, S.	*Irrig. Sci.*	2016
13	0.15	4.04	Cai, Y.H.	*Agr. Water Manag.*	2017
13	0.24	7.93	Lamm, F.R.	*Irrig. Sci.*	2003
11	0.22	5.65	Simunek, J.	*Vadose Zone J.*	2008
10	0.20	4.40	Dabach, S.	*Agr. Water Manag.*	2015
10	0.11	5.79	Gardenas, A.I.	*Agr. Water Manag.*	2005
8	0.22	4.79	Kandelous, M.M.	*Soil Sci. Soc. Am. J.*	2011
6	0.21	3.13	Payero, J.O.	*Agr. Water Manag.*	2006
6	0.13	3.59	Kandelous, M.M.	*Agr. Water Manag.*	2012
4	0.15		Thompson, T.L.	*Soil Sci. Soc. Am. J.*	2002
4	0.20		Ajdary, K.	*Agr. Water Manag.*	2007
4	0.13		Evett, S.R.	*Agr. Water Manag.*	2002
3	0.22		Abou Lila, T.S.	*Irrig. Sci.*	2013
3	0.12		Bekele, S.	*Agr. Water Manag.*	2007
3	0.11		Selim, T.	*Soil Sci. Soc. Am. J.*	2013
2	0.12		Thompson, T.L.	*Soil Sci. Soc. Am. J.*	2002

We divided the timeline map into four parts. The first part covered the period before 2002, during which the research directions included #7 clay deposit, #13 potato crop, #16 precision flow control, and #19 *Prunus dulcis*. The second part spans 2002–2008, comprising the research directions #5 stem yield, #9 soil moisture sensors, #3 Wyoming, #10 gross margin, #12 antioxidant activity, #15 quality assurance standard (QAS), #14 economic benefits, and #18 energy balance. The third part spans 2008–2014, comprising the research directions #11 soil water balance, #4 irrigation water productivity, #6 crop coefficients, #1 aerated irrigation, #2 alfalfa, #20 *Saccharum officinarum* L., and #22 organic fertilizer application. Finally, the research directions of the fourth part, spanning 2014–2022, comprised #0 partial root-zone drying, #1 aerated irrigation, #2 alfalfa, #8 Hydrus, #17 redistribution, and #21 subsurface drainage.

Analysis of the timeline map revealed that the scope of SDI research has broadened over time. The earliest node of import occurred in Category #7, which pertains to the proper management of clay deposits. This is a key factor in the effective use of SDI because clay deposits may clog emitters and reduce the flow of water and nutrients to crop roots [52]. Al-Eter et al. [16] demonstrated that the use of clay deposit amendments in sandy soils improved soil water storage and promoted root growth and yield. Three categories produced key nodes in the second period. The earliest category was #5, which addressed stem yield and showed that SDI contributed considerably to the yield of root and tuber crops (sweet potatoes, onions, and potatoes). Although #10 gross margin produced three nodes of importance, the keyword analysis in Section 3.3.2 suggested that profit is not the most important factor in the development of subsurface drip irrigation. This is likely related to the high input cost, high technical threshold of operation and maintenance, and long income cycle of SDI. Furthermore, the optimization of and improvement in SDI equipment mainly focuses on enterprise transformation, with relevant information stored as part of the core enterprise data. Considering the distribution of nodes over time, #3 Wyoming emerged because of the impact of coalbed sodic water on agriculture in Wyoming, United States. Because of large-scale coal mining in the area, salty water has been injected into the coal seam and has forced farmers to use slightly saltier water for irrigation [53–55]. SDI, with its completely closed pipeline and precisely controlled flow rate, offers significant advantages in the use of nonconventional water resources (wastewater and saltwater) for irrigation. This setup can prevent the transfer of pollutants while maintaining the levels of nutrients, such as nitrogen, phosphorus, potassium, and organic matter, required for crop growth [56,57]. Brackish water irrigation is mainly applied to sandy loam with good drainage and leaching conditions in arid areas. In salinized soils, a decrease in soil water content promotes soil salinity, but brackish water irrigation can increase soil water content and dilute soil salinity [15,58–61]. The third period encompassed the largest number of research categories. The two key nodes were #6 (crop coefficients) and #4 (irrigation water productivity). The research focus during this period was on developing appropriate irrigation strategies according to the water demand cycles of different crops. Aerated irrigation, a new branch of research, emerged during this period. The fourth phase focused on the infiltration distribution of irrigation water in the root zone of crops, which is commonly associated with the Hydrus model for water movement simulation in SDI.

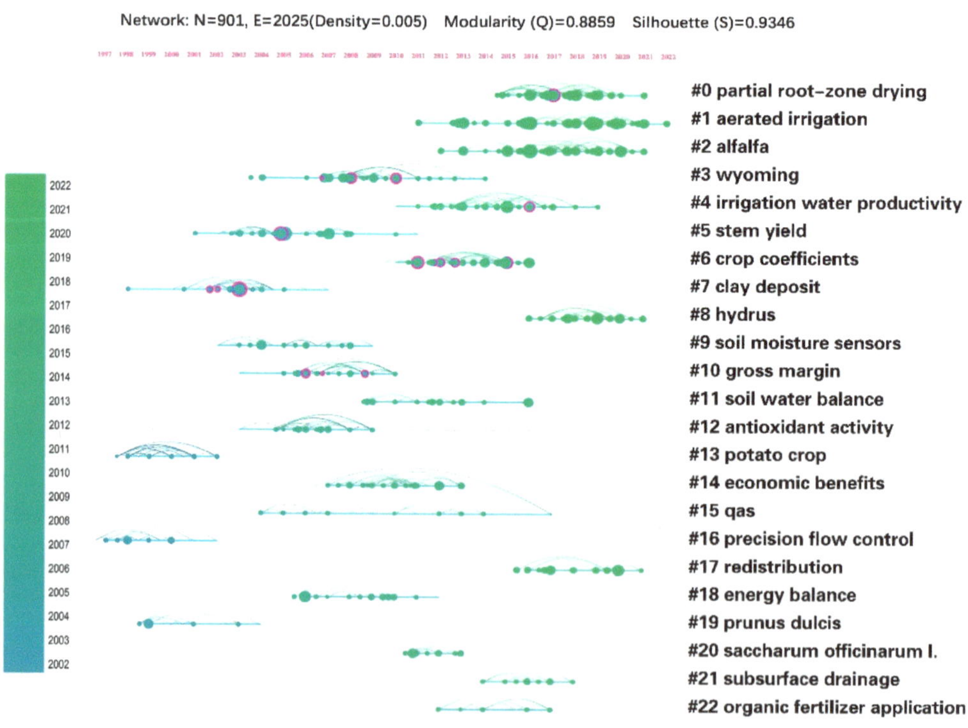

Figure 7. Reference clustering according to time.

Early research on SDI primarily focused on traditional techniques, such as precise control, water conservation, yield increase, and soil and water conservation. In recent years, with the emergence and development of new technologies, some new research methods and branches have entered the research scope of SDI. For example, Hydras is a simulation software for groundwater and soil water movement, mainly used to study and simulate soil moisture, infiltration and transportation, hydrological processes, and soil hydrochemical reactions [62–64]. SDI research now considers different soil types and water management practices to predict and optimize soil moisture and ultimately enhance agricultural productivity [65]. Aerated irrigation is an innovative development that increases the feeding capacity of crop roots by infusing irrigated water with oxygen or air, allowing plants to absorb more water and nutrients, and reducing the need for frequent watering. This results in water conservation and improved plant health [4,44,52,66,67]. Among application crops, alfalfa has become a hot topic in SDI research. Alfalfa thrives in cool and humid climates, and SDI can reduce the occurrence of soil surface crusting and soil erosion, which are unfavorable for the growth of alfalfa. Therefore, SDI has broad prospects for application in alfalfa planting and the forage industry [68,69].

4. Discussion

In recent years, the application of subsurface drip irrigation in water-saving agriculture has increased considerably. In the United States, the area of SDI application increased by over 167% between 2003 and 2018, mainly for crops such as corn, onions, cotton, and tomatoes [21]. The integration of water and fertilizer is one of the main advantages of SDI, which enables consistent and evenly distributed fertilizer and pesticide application throughout the entire growth cycle [24]. SDI is characterized by low water flow rates, where water is released from emitters and diffuses through capillary action and gravity to the root zone of crops. This closed irrigation method minimizes the occurrence of water spreading on the soil surface, effectively reducing the leaching and runoff of fertilizers and pesticides. Consequently, SDI mitigates their movement within the soil, thereby mitigating pollution and environmental contamination [9]. This can improve fertilizer and water use efficiency while limiting nutrient loss, which ultimately promotes yield and limits the impact on the environment. SDI has demonstrated excellent performance in saline water irrigation. Under favorable drainage conditions, continuous SDI induces leaching of the soil beneath the emitters, preventing salt accumulation near the root zone and mitigating the negative impact of salinity on crops [70]. A large number of research results in the above research fields have been published in journals related to agronomy, water resources, and environmental sciences, such as *Agricultural Water Management*, *Agronomy*, *Agriculture*, *Water*, and other excellent journals.

The development of internet technology has led to the intelligent and precise transformation of SDI. Against the backdrop of current agricultural production shifting towards intensification and automation, there has been rapid progress in the development of extensive, non-destructive, and fast detection techniques for assessing soil moisture, nutrient levels, and crop losses in agricultural fields [71]. The utilization of big data and satellite remote sensing technology enables the refinement and intelligentization of irrigation strategies. Through real-time monitoring and data analysis, timely information on meteorological changes, soil temperature and moisture, and surface evaporation can be obtained. By evaluating the water status of agricultural fields in conjunction with crop growth and water requirements, targeted irrigation recommendations can be provided [49,72]. Soil sensors and the Internet of Things have significantly improved the monitoring, management, and maintenance of SDI systems [17,73]. Because the working parts of an SDI system, such as pipes and irrigators, are buried underground, blockages and pipeline leakages can be challenging to detect. Delayed maintenance can lead to small-scale droughts or flooding, which reduces irrigation efficiency and results in crop loss. Enhanced monitoring devices can accurately sense and control the operational status of SDI systems and assess soil health, thereby providing effective assistance in system management and irrigation strategy formulation.

Enhancing the operational lifespan poses a significant challenge to the development of SDI systems. Compared to other irrigation methods, SDI has higher maintenance and operating costs, and the economic viability of SDI systems relies on achieving a lifespan of 15 to 20 years [22]. Blockages and pipeline damage are frequently encountered issues in SDI systems. Clogging primarily occurs in the emitters of subsurface drip irrigation systems and is caused by particle blockage, deposition from chemical reactions between fertilizers, and the aggregation of microorganisms in confined spaces [74]. Regular flushing or periodic flushing can effectively prevent clogging in pipelines and emitters [75]. The anti-clogging performance of emitters depends on factors such as channel shape, dimensions, materials, and other factors. Increasing the fluid velocity and turbulence inside the emitters can contribute to improved anti-clogging performance. Computational Fluid Dynamics (CFD)-based numerical simulation techniques have been widely applied in the design and development of emitters, playing a significant role in optimizing channel design and enhancing anti-clogging capabilities [76]. Pipeline damage can result from multiple factors: aging of the pipeline material, breakage caused by expansion and contraction under varying temperatures, damage inflicted by plowing equipment, and damage from rodents [20]. Implementing appropriate pipeline layout and land cultivation strategies can help prevent damage to SDI systems during plowing operations. However, traditional SDI pipe materials are plastic and rubber, which are susceptible to UV- and temperature-mediated aging. Composite and ceramic materials have recently emerged as alternatives for manufacturing SDI pipes and emitters [77]. These materials can reduce the long-term losses and maintenance costs of SDI, while improving environmental friendliness and energy efficiency.

The economic benefits of SDI systems should be evaluated based on specific circumstances and considerations. SDI is recognized for its water-saving effects; however, it may not deliver superior economic benefits when compared to other conventional irrigation methods [78]. As revealed by keyword mapping, relatively few studies have assessed the economic benefits of SDI. From the perspective of economic competitiveness, SDI may not be suitable for all crops. For example, root crops (such as sweet potatoes, onions, and potatoes), as well as certain vegetable crops (such as tomatoes), have demonstrated significant increases in yield using SDI; however, cereals have not exhibited sufficient yield increases to justify the advantages of SDI over other irrigation methods [79–81]. Notably, there has been an increase in research on the use of SDI in alfalfa cultivation. This is due to the precise water and nutrient control offered by SDI. Excessive or inadequate soil moisture levels and fertilization can reduce alfalfa yields. The application of SDI systems allows for effective control of production costs and enhances forage production [82]. The application of SDI systems involves considerable financial investment and requires the design, construction, and management of specialized equipment by technically trained personnel. Moreover, SDI systems rely on electricity to drive water pumps and valves, among other devices, which increases running costs. Since subsurface drip irrigation systems are susceptible to damage from plowing equipment, they are more suitable for low-tillage or no-tillage cropping models [78]. Crop rotation may not be suitable for SDI because the fixed burial depth and spacing of drip heads may be inconsistent with crop row spacing and planting depth, thus reducing irrigation efficiency. In practice, considering crop type, land topography, and water resources. Further research is needed to evaluate the economic benefits of SDI systems under different scenarios [83].

5. Conclusions

In this study, CiteSpace was used to examine the trends in the research field of subsurface drip irrigation. Research on SDI has garnered increasing attention in recent decades, its applications have continued to expand, and it has demonstrated considerable potential in improving the yields of a variety of crops. Based on statistical analysis of publication numbers, the United States and China have been the most influential countries in SDI research. Research on SDI is relatively dispersed among various research institutions and teams. It is anticipated that by enhancing collaboration, research efficiency and depth can be improved. Current SDI research focuses on improving water use efficiency and increasing crop yield, with the distribution of water in the crop root zone being a research hotspot. The Hydrus model plays a significant role in studying the movement of soil moisture peaks and nutrient substances under SDI conditions. To optimize the productivity of the root zone, aerated irrigation may emerge as a new branch of SDI. The significant potential demonstrated by SDI in water-saving irrigation and the nonconventional water resources (wastewater and saltwater) positions it as a promising solution for increased application in arid regions. The development of new technologies and materials presents opportunities for the widespread adoption of SDI, facilitating the formulation of precision irrigation strategies and extending system longevity. In order to control costs and increase yields, SDI is likely to be increasingly applied in the cultivation of forage crops, such as alfalfa, in the livestock industry. Despite certain limitations, such as high investment costs and complex management and maintenance, SDI shows great potential in the development of water-saving and production-increasing technologies to meet future food demands, while limiting costs related to both long-term operation and environmental impact.

Author Contributions: Y.X. conceived the study and closely supervised the writing of the manuscript. C.S. wrote and revised the manuscript. D.W. was responsible for the document retrieval. H.L. and W.G. contributed to the editing of the manuscript. All authors have read and agreed to the published version of the manuscript.

Funding: This research was funded by the "Open Fund of the Key Laboratory for the Efficient and Safe Utilization of Agricultural Water Resources (Farmland Irrigation Research Institute, CAAS)", grant number FIRI 202003-0101 and FIRI 2021030201, the "China Postdoctoral Science Foundation", grant number 2021M701103, the "Key Technology Research Program of the Higher Education Institutes of Henan Province, China", grant number 22A416008, and the "Agricultural Science and Technology Innovation Program", grant number CAAS-ASTIP.

Institutional Review Board Statement: Not applicable.

Data Availability Statement: The literature data used in this study are based solely on the results retrieved from the Web of Science Core Collection. Due to limitations in the search timeframe and the access permissions for other databases, we acknowledge that the search results may not be comprehensive. Furthermore, it is important to acknowledge the valuable research results published in other databases or in outstanding journals in different languages. We extend our gratitude for their contributions to the field.

Acknowledgments: The authors would like to acknowledge Chao Mei Chen for developing CiteSpace and allowing public access to it for free. The authors also would like to acknowledge the contribution of anonymous reviewers for their comments and suggestion.

Conflicts of Interest: The authors declare that they have no conflict of interest.

References

1. Kandelous, M.M.; Šimůnek, J. Comparison of numerical, analytical, and empirical models to estimate wetting patterns for surface and subsurface drip irrigation. *Irrig Sci.* **2010**, *28*, 435–444. [CrossRef]
2. Vollset, S.E.; Goren, E.; Yuan, C.W.; Cao, J.; Smith, A.E.; Hsiao, T.; Bisignano, C.; Azhar, G.S.; Castro, E.; Chalek, J.; et al. Fertility, mortality, migration, and population scenarios for 195 countries and territories from 2017 to 2100: A forecasting analysis for the Global Burden of Disease Study. *Lancet* **2020**, *396*, 1285–1306. [CrossRef]
3. Hanjra, M.A.; Qureshi, M.E. Global water crisis and future food security in an era of climate change. *Food Policy* **2010**, *35*, 365–377. [CrossRef]
4. Du, Y.-D.; Niu, W.-Q.; Gu, X.-B.; Zhang, Q.; Cui, B.-J.; Zhao, Y. Crop yield and water use efficiency under aerated irrigation: A meta-analysis. *Agric. Water Manag.* **2018**, *210*, 158–164. [CrossRef]
5. Dirwai, T.L.; Mabhaudhi, T.; Kanda, E.K.; Senzanje, A. Moistube irrigation technology development, adoption and future prospects: A systematic scoping review. *Heliyon* **2021**, *7*, e06213. [CrossRef] [PubMed]
6. Lee, C.O. Irrigation Tile. U.S. Patent No. 1,350,229, 17 August 1920.
7. Lamm, F.R.; Bordovsky, J.P.; Schwankl, L.J.; Grabow, G.L.; Enciso-Medina, J.; Peters, R.T.; Colaizzi, P.D.; Trooien, T.P.; Porter, D.O. Subsurface drip irrigation: Status of the technology in 2010. *Trans. ASABE* **2012**, *55*, 483–491. [CrossRef]
8. Umair, M.; Hussain, T.; Jiang, H.; Ahmad, A.; Yao, J.; Qi, Y.; Zhang, Y.; Min, L.; Shen, Y. Water-saving potential of subsurface drip irrigation for winter wheat. *Sustainability* **2019**, *11*, 2978. [CrossRef]

9. McHugh, A.D.; Bhattarai, S.; Lotz, G.; Midmore, D.J. Effects of subsurface drip irrigation rates and furrow irrigation for cotton grown on a vertisol on off-site movement of sediments, nutrients and pesticides. *Agron. Sustain. Dev.* **2008**, *28*, 507–519. [CrossRef]
10. Sutton, K.F.L.; Lanini, W.T.; Mitchell, J.P.; Miyao, E.M.; Shrestha, A. Weed control, yield, and quality of processing tomato production under different irrigation, tillage, and herbicide systems. *Weed Technol.* **2006**, *20*, 831–838. [CrossRef]
11. Sabe, M.; Pillinger, T.; Kaiser, S.; Chen, C.; Taipale, H.; Tanskanen, A.; Tiihonen, J.; Leucht, S.; Correll, C.U.; Solmi, M. Half a century of research on antipsychotics and schizophrenia: A scientometric study of hotspots, nodes, bursts, and trends. *Neurosci. Biobehav. Rev.* **2022**, *136*, 104608. [CrossRef]
12. Qin, Y.; Zhang, Q.; Liu, Y. Analysis of knowledge bases and research focuses of cerebral ischemia-reperfusion from the perspective of mapping knowledge domain. *Brain Res. Bull.* **2020**, *156*, 15–24. [CrossRef]
13. Li, T.; Baležentis, T.; Cao, L.; Zhu, J.; Kriščiukaitienė, I.; Melnikienė, R. Are the changes in China's grain production sustainable: Extensive and intensive development by the LMDI approach. *Sustainability* **2016**, *8*, 1198. [CrossRef]
14. Yang, H.; Chen, W.; Chen, Y.; Zhang, F.; Yang, X. Assessing the impact of shallow subsurface pipe drainage on soil salinity and crop yield in arid zone. *PeerJ* **2021**, *9*, e12622. [CrossRef]
15. Wu, W.Y.; Hu, Y.Q.; Guan, X.Y.; Xu, L.J. Advances in research of reclaimed water irrigation in China *. *Irrig. Drain.* **2020**, *69*, 119–126. [CrossRef]
16. Small, H. Co-citation in scientific literature-new measure of relationship between 2 documents. *J. Am. Soc. Inf. Sci.* **1973**, *24*, 265–269. [CrossRef]
17. Zhu, Y.; Kim, M.C.; Chen, C. An Investigation of the Intellectual Structure of Opinion Mining Research. *Inf. Res. Int. Electron. J.* **2017**, *22*, 739.
18. Sheta, A.S.; Al-Omran, A.M.; Falatah, A.M.; Al-Harbi, A.R. Effect of clay deposits on physicochemical and intermittent evaporation characteristics of torripsamment. *Arid. Land Res. Manag.* **2006**, *20*, 295–307. [CrossRef]
19. Jia, Y.; Gao, W.; Sun, X.; Feng, Y. Simulation of soil water and salt balance in three water-saving irrigation technologies with HYDRUS-2D. *Agronomy* **2023**, *13*, 164. [CrossRef]
20. Nolz, R.; Loiskandl, W.; Kammerer, G.; Himmelbauer, M.L. Survey of soil water distribution in a vineyard and implications for subsurface drip irrigation control. *Soil Water Res.* **2016**, *11*, 250–258. [CrossRef]
21. Lamm, F.R.; Colaizzi, P.D.; Sorensen, R.B.; Bordovsky, J.P.; Dougherty, M.; Balkcom, K.; Zaccaria, D.; Bali, K.M.; Rudnick, D.R.; Peters, R.T. A 2020 Vision of subsurface drip irrigation in the U.S. *Trans. ASABE* **2021**, *64*, 1319–1343. [CrossRef]
22. Lamm, F.R.; Rogers, D.H. Longevity and performance of a subsurface drip irrigation system. *Trans. ASABE* **2017**, *60*, 931–939. [CrossRef]
23. Volschenk, T. Water use and irrigation management of pomegranate trees-A review. *Agric. Water Manag.* **2020**, *241*, 106375. [CrossRef]
24. El-Beltagi, H.S.; Hashem, F.A.; Maze, M.; Shalaby, T.A.; Shehata, W.F.; Taha, N.M. Control of gas emissions (N_2O and CO_2) associated with applied different rates of nitrogen and their influences on growth, productivity, and physio-biochemical attributes of green bean plants grown under different irrigation methods. *Agronomy* **2022**, *12*, 249. [CrossRef]
25. Irmak, S.; Mohammed, A.T.; Kukal, M.S. Maize response to coupled irrigation and nitrogen fertilization under center pivot, subsurface drip and surface (furrow) irrigation: Growth, development and productivity. *Agric. Water Manag.* **2022**, *263*, 107457. [CrossRef]
26. Hamad, A.A.A.; Wei, Q.; Wan, L.; Xu, J.; Hamoud, Y.A.; Li, Y.; Shaghaleh, H. Subsurface drip irrigation with emitters placed at suitable depth can mitigate N_2O emissions and enhance Chinese cabbage yield under greenhouse cultivation. *Agronomy* **2022**, *12*, 745. [CrossRef]
27. Chen, C. *CiteSpace: A Practical Guide for Mapping Scientific Literature*; Nova Science Publishers: Hauppauge, NY, USA, 2016.
28. Li, S.; Tan, D.; Wu, X.; Degré, A.; Long, H.; Zhang, S.; Lu, J.; Gao, L.; Zheng, F.; Liu, X.; et al. Negative pressure irrigation increases vegetable water productivity and nitrogen use efficiency by improving soil water and NO_3^--N distributions. *Agric. Water Manag.* **2021**, *251*, 106853. [CrossRef]
29. Bordovsky, J.P. Preplant and early-season cotton irrigation timing with deficit amounts using subsurface drip (SDI) systems in the Texas High Plains. *Irrig. Sci.* **2020**, *38*, 485–499. [CrossRef]
30. Lipan, L.; Carbonell-Pedro, A.A.; Cárceles Rodríguez, B.; Durán-Zuazo, V.H.; Franco Tarifa, D.; García-Tejero, I.F.; Gálvez Ruiz, B.; Cuadros Tavira, S.; Muelas, R.; Sendra, E.; et al. Can Sustained Deficit Irrigation Save Water and Meet the Quality Characteristics of Mango? *Agriculture* **2021**, *11*, 448. [CrossRef]
31. Li, X.; Ma, J.; Zheng, L.; Chen, J.; Sun, X.; Guo, X. Optimization of the Regulated Deficit Irrigation Strategy for Greenhouse Tomato Based on the Fuzzy Borda Model. *Agriculture* **2022**, *12*, 324. [CrossRef]
32. Schiavon, M.; Leinauer, B.; Serena, M.; Maier, B.; Sallenave, R. Plant growth regulator and soil surfactants' effects on saline and deficit irrigated warm-season grasses: I. Turf quality and soil moisture. *Crop. Sci.* **2014**, *54*, 2815–2826. [CrossRef]
33. Resende, R.S.; Nascimento, T.; Carvalho, T.B.; Amorim, J.R.A.; Rodrigues, L. Reducing sugarcane irrigation demand through planting date adjustment in Alagoas State, Brazil. *Rev. Bras. Eng. Agric. Ambient.* **2021**, *25*, 75–81. [CrossRef]
34. Silva, A.L.B.D.O.; Pires, R.C.M.; Ribeiro, R.V.; Machado, E.C.; Blain, G.C.; Ohashi, A.Y.P. Development, yield and quality attributes of sugarcane cultivars fertigated by subsurface drip irrigation. *Rev. Bras. Eng. Agric. Ambient.* **2016**, *20*, 525–532. [CrossRef]
35. Elliott, M.L.; McInroy, J.A.; Xiong, K.; Kim, J.H.; Skipper, H.D.; Guertal, E.A. Taxonomic diversity of rhizosphere bacteria in golf course putting greens at representative sites in the Southeastern United States. *HortScience* **2008**, *43*, 514–518. [CrossRef]

36. Souza, C.F.; Bizari, D.R. Soil solution distribution in subsurface drip irrigation in sugarcane. *Eng. Agric.* **2018**, *38*, 217–224. [CrossRef]
37. Vishwakarma, D.K.; Kumar, R.; Kumar, A.; Kushwaha, N.L.; Kushwaha, K.S.; Elbeltagi, A. Evaluation and development of empirical models for wetted soil fronts under drip irrigation in high-density apple crop from a point source. *Irrig. Sci.* **2022**, 1–24. [CrossRef]
38. Subbaiah, R. A review of models for predicting soil water dynamics during trickle irrigation. *Irrig. Sci.* **2013**, *31*, 225–258. [CrossRef]
39. Elmaloglou, S.; Diamantopoulos, E. The effect of intermittent water application by surface point sources on the soil moisture dynamics and on deep percolation under the root zone. *Comput. Electron. Agric.* **2008**, *62*, 266–275. [CrossRef]
40. Bin Zainal Abidin, M.S.; Shibusawa, S.; Ohaba, M.; Li, Q.; Bin Khalid, M. Capillary flow responses in a soil–plant system for modified subsurface precision irrigation. *Precis. Agric.* **2014**, *15*, 17–30. [CrossRef]
41. Pendergast, L.; Bhattarai, S.P.; Midmore, D.J. Benefits of oxygation of subsurface drip-irrigation water for cotton in a Vertosol. *Crop. Pasture Sci.* **2013**, *64*, 1171–1181. [CrossRef]
42. Zhu, Y.; Cai, H.; Song, L.; Wang, X.; Shang, Z.; Sun, Y. Aerated irrigation of different irrigation levels and subsurface dripper depths affects fruit yield, quality and water use efficiency of greenhouse tomato. *Sustainability* **2020**, *12*, 2703. [CrossRef]
43. Chen, X.; Dhungel, J.; Bhattarai, S.P.; Torabi, M.; Pendergast, L.; Midmore, D.J. Impact of oxygation on soil respiration, yield and water use efficiency of three crop species. *J. Plant Ecol.* **2011**, *4*, 236–248. [CrossRef]
44. Zhu, Y.; Dyck, M.; Cai, H.-J.; Song, L.-B.; Chen, H. The effects of aerated irrigation on soil respiration, oxygen, and porosity. *J. Integr. Agric.* **2019**, *18*, 2854–2868. [CrossRef]
45. Yang, M.D.; Leghari, S.J.; Guan, X.K.; Ma, S.C.; Ding, C.M.; Mei, F.J.; Wei, L.; Wang, T.C. Deficit subsurface drip irrigation improves water use efficiency and stabilizes yield by enhancing subsoil water extraction in winter wheat. *Front. Plant Sci.* **2020**, *11*, 508. [CrossRef]
46. Çetin, Ö.; Üzen, N.; Temiz, M.G.; Altunten, H. Improving cotton yield, water use and net income in different drip irrigation systems using real-time crop evapotranspiration. *Pol. J. Environ. Stud.* **2021**, *30*, 4463–4474. [CrossRef]
47. Mohammed, A.T.; Irmak, S. Maize response to coupled irrigation and nitrogen fertilization under center pivot, subsurface drip and surface (furrow) irrigation: Soil-water dynamics and crop evapotranspiration. *Agric. Water Manag.* **2022**, *267*, 107634. [CrossRef]
48. Ganjegunte, G.; Leinauer, B.; Schiavon, M.; Serena, M. Using electro-magnetic induction to determine soil salinity and sodicity in turf root zones. *Agron. J.* **2013**, *105*, 836–844. [CrossRef]
49. De Oliveira, L.A.; Woodbury, B.L.; de Miranda, J.H.; Stromer, B.S. Using electromagnetic induction technology to identify atrazine leaching potential at field scale. *Geoderma* **2020**, *375*, 114525. [CrossRef]
50. Chen, C. A glimpse of the first eight months of the COVID-19 literature on Microsoft academic graph: Themes, citation contexts, and uncertainties. *Front. Res. Metr. Anal.* **2020**, *5*, 607286. [CrossRef] [PubMed]
51. Chen, C.; Song, M. Visualizing a field of research: A methodology of systematic scientometric reviews. *PLoS ONE* **2019**, *14*, e0223994. [CrossRef]
52. Niu, W.; Guo, Q.; Zhou, X.; Helmers, M.J. Effect of aeration and soil water redistribution on the air permeability under subsurface drip irrigation. *Soil Sci. Soc. Am. J.* **2012**, *76*, 815–820. [CrossRef]
53. Al-Eter, A.L.I.; Nadeem, M.; Wahb-Allah, M.A.; Al-Harbi, A.R.; Al-Omran, A.M. Impact of irrigation water quality, irrigation systems, irrigation rates and soil amendments on tomato production in sandy calcareous soil. *Turk. J. Agric. For.* **2010**, *34*, 59–73.
54. Bern, C.R.; Breit, G.N.; Healy, R.W.; Zupancic, J.W. Deep subsurface drip irrigation using coal-bed sodic water: Part II. Geochemistry. *Agric. Water Manag.* **2013**, *118*, 135–149. [CrossRef]
55. Palacios-Diaz, M.D.P.; Fernández-Vera, J.R.; Hernández-Moreno, J.M.; Amorós, R.; Mendoza-Grimón, V. Effect of Irrigation Management and Water Quality on Soil and Sorghum bicolor Payenne Yield in Cape Verde. *Agriculture* **2023**, *13*, 192. [CrossRef]
56. Allende, A.; Monaghan, J. Irrigation water quality for leafy crops: A perspective of risks and potential solutions. *Int. J. Environ. Res. Public Health* **2015**, *12*, 7457–7477. [CrossRef]
57. Palacios-Díaz, M.P.; Mendoza-Grimón, V.; Fernández-Vera, J.R.; Rodríguez-Rodríguez, F.; Tejedor-Junco, M.T.; Hernández-Moreno, J.M. Subsurface drip irrigation and reclaimed water quality effects on phosphorus and salinity distribution and forage production. *Agric. Water Manag.* **2009**, *96*, 1659–1666. [CrossRef]
58. He Xinlin, L.H.; Jianwei, Y.; Guang, Y.; Mingsi, L.; Ping, G.; Aimaiti, A. Comparative Investigation on Soil Salinity Leaching under Subsurface Drainage and Ditch Drainage in Xinjiang Arid Region. *Int. J. Agric. Biol. Eng.* **2013**, *9*, 109–118.
59. Hanson, B.R.; May, D.E.; Simūnek, J.; Hopmans, J.W.; Hutmacher, R.B. Drip irrigation provides the salinity control needed for profitable irrigation of tomatoes in the San Joaquin valley. *Calif. Agric.* **2009**, *63*, 131–136. [CrossRef]
60. Chase, C.A.; Stall, W.M.; Simonne, E.H.; Hochmuth, R.C.; Dukes, M.D.; Weiss, A.W. Nutsedge control with drip-applied 1,3-dichloropropene plus chloropicrin in a sandy soil. *Horttechnology* **2006**, *16*, 641–648. [CrossRef]
61. Sevostianova, E.; Leinauer, B.; Sallenave, R.; Karcher, D.; Maier, B. Soil salinity and quality of sprinkler and drip irrigated cool-season turfgrasses. *Agron. J.* **2011**, *103*, 1503–1513. [CrossRef]
62. Zamani, S.; Fatahi, R.; Provenzano, G. A comprehensive model for hydraulic analysis and wetting patterns simulation under subsurface drip laterals. *Water* **2022**, *14*, 1965. [CrossRef]

63. Yang, T.; Šimůnek, J.; Mo, M.; McCullough-Sanden, B.; Shahrokhnia, H.; Cherchian, S.; Wu, L. Assessing salinity leaching efficiency in three soils by the HYDRUS-1D and -2D simulations. *Soil Tillage Res.* **2019**, *194*, 104342. [CrossRef]
64. El-Nesr, M.N.; Alazba, A.A.; Šimůnek, J. HYDRUS simulations of the effects of dual-drip subsurface irrigation and a physical barrier on water movement and solute transport in soils. *Irrig. Sci.* **2014**, *32*, 111–125. [CrossRef]
65. Kandelous, M.M.; Šimůnek, J.; van Genuchten, M.T.; Malek, K. Soil water content distributions between two emitters of a subsurface drip irrigation system. *Soil Sci. Soc. Am. J.* **2011**, *75*, 488–497. [CrossRef]
66. Niu, W.-Q.; Fan, W.-T.; Persaud, N.; Zhou, X.-B. Effect of post-irrigation aeration on growth and quality of greenhouse cucumber. *Pedosphere* **2013**, *23*, 790–798. [CrossRef]
67. Chen, H.; Shang, Z.; Cai, H.; Zhu, Y. Irrigation combined with aeration promoted soil respiration through increasing soil microbes, enzymes, and crop growth in tomato fields. *Catalysts* **2019**, *9*, 945. [CrossRef]
68. Alam, M.; Trooien, T.P.; Dumler, T.J.; Rogers, D.H. Using subsurface drip irrigation for alfalfa. *J. Am. Water Resour. Assoc.* **2002**, *38*, 1715–1721. [CrossRef]
69. Wang, Y.D.; Kou, D.; Muneer, M.A.; Fang, G.J.; Su, D.R. The effects of irrigation regimes on soil moisture dynamics, yield and quality of Lucerne under subsurface drip irrigation. *Appl. Ecol. Environ. Res.* **2020**, *18*, 4179–4194. [CrossRef]
70. Zhang, Y.J.; Chang, T.T.; Guan, Y.L.; Shao, X.H.; Zhang, J.; Li, M.H. Soil Moisture Content, Soil Salinity And Water Use Efficiency Under Surface Drip and Flood Irrigation In Continuous Cropping Plastic Greenhouse of Eastern China. *Fresenius Environ. Bull.* **2018**, *27*, 6668–6676.
71. Qu, D.Y.; Wang, X.B.; Kang, C.P.; Liu, Y. Promoting agricultural and rural modernization through application of information and communication technologies in China. *Int. J. Agric. Biol. Eng.* **2019**, *11*, 1–4. [CrossRef]
72. Nolz, R.; Loiskandl, W. Evaluating soil water content data monitored at different locations in a vineyard with regard to irrigation control. *Soil Water Res.* **2017**, *12*, 152–160. [CrossRef]
73. Jadoon, K.Z.; Moghadas, D.; Jadoon, A.; Missimer, T.M.; Al-Mashharawi, S.K.; McCabe, M.F. Estimation of soil salinity in a drip irrigation system by using joint inversion of multicoil electromagnetic induction measurements. *Water Resour. Res.* **2015**, *51*, 3490–3504. [CrossRef]
74. Zhangzhong, L.L.; Yang, P.L.; Zheng, W.G.; Liu, Y.; Guo, M.J.; Yang, F.R. Effects of Drip Irrigation Frequency on Emitter Clogging using Saline Water for Processing Tomato Production. *Irrig. Drain.* **2019**, *68*, 464–475. [CrossRef]
75. Du, P.S.; Li, Z.Q.; Wang, C.C.; Ma, J.J. Analysis of the Influence of the Channel Layout and Size on the Hydraulic Performance of Emitters. *Agriculture* **2022**, *12*, 541. [CrossRef]
76. Wang, C.C.; Li, Z.Q.; Ma, J.J. Influence of Emitter Structure on Its Hydraulic Performance Based on the Vortex. *Agriculture* **2021**, *11*, 508. [CrossRef]
77. Cai, Y.; Wu, P.; Zhu, D.; Zhang, L.; Zhao, X.; Gao, X.; Ge, M.; Song, X.; Wu, Y.; Dai, Z. Subsurface irrigation with ceramic emitters: An effective method to improve apple yield and irrigation water use efficiency in the semiarid Loess Plateau. *Agric. Ecosyst. Environ.* **2021**, *313*, 107404. [CrossRef]
78. Yadav, A.; Sharma, N.; Upreti, H.; Singhal, G.D. Techno-economic analysis of irrigation systems for efficient water use in the backdrop of climate change. *Curr. Sci.* **2022**, *122*, 664–673. [CrossRef]
79. Ma, X.; Sanguinet, K.A.; Jacoby, P.W. Direct root-zone irrigation outperforms surface drip irrigation for grape yield and crop water use efficiency while restricting root growth. *Agric. Water Manag.* **2020**, *231*, 105993. [CrossRef]
80. Lamm, F.R. Cotton, Tomato, corn, and onion production with subsurface drip irrigation: A review. *Trans. ASABE* **2016**, *59*, 263–278. [CrossRef]
81. Badr, M.A.; Abou Hussein, S.D.; El-Tohamy, W.A.; Gruda, N. Efficiency of subsurface drip irrigation for potato production under different dry stress conditions. *Gesunde Pflanz.* **2010**, *62*, 63–70. [CrossRef]
82. Trejo, J.A.M.; Aguiluz, H.W.A.; Ramirez, J.O.; Lopez, A.R.; Gonzalez, M.R.; Rangel, P.P.; Trejo, I.D.M.; Castruita, M.A.S.; Vidal, J.A.O.; Coronado, P.Y. Water use in alfalfa (*Medicago sativa*) with subsurface drip irrigation. *Rev. Mex. de Cienc. Pecu.* **2010**, *1*, 145–156.
83. Diotto, A.V.; Irmak, S. Embodied energy and energy return on investment analyses in maize production for grain and ethanol under center pivot, subsurface drip, and surface (furrow) irrigation with disk tillage and no-till practices. *Trans. ASABE* **2016**, *59*, 873–884. [CrossRef]

Disclaimer/Publisher's Note: The statements, opinions and data contained in all publications are solely those of the individual author(s) and contributor(s) and not of MDPI and/or the editor(s). MDPI and/or the editor(s) disclaim responsibility for any injury to people or property resulting from any ideas, methods, instructions or products referred to in the content.

AgriEngineering

Article

Opportunities of Digital Transformation in Post-Harvest Activities: A Single Case Study of an Engineering Solutions Provider

Daniel Schmidt [1], Maria Angela Butturi [2] and Miguel Afonso Sellitto [1,*]

1 Production and Systems Engineering Graduate Program, Universidade do Vale do Rio dos Sinos (UNISINOS), Av. Unisinos, 950—Cristo Rei, São Leopoldo 93022-000, Brazil; daniel.panambi@gmail.com
2 Department of Sciences and Methods for Engineering, University of Modena and Reggio Emilia (UNIMORE), 42122 Reggio Emilia, Italy; mariaangela.butturi@unimore.it
* Correspondence: sellitto@unisinos.br

Abstract: The purpose of this article is to identify opportunities that digital transformation in post-harvest activities offers to an engineering solution provider. The research method is a simple case study. The object is a company based in southern Brazil that provides engineering-integrated digital solutions to grain producers, including products and services. The specific objectives are to describe the company's digital products and services, identify opportunities and players, and discuss how players can take advantage of opportunities owing to business process digitalization. The main results include separating products into three technological layers and identifying five types of opportunities (financing, commercialization, operation, logistics, traceability, and insurance), eight types of players, and the main opportunities for each player. The most significant opportunities are risk reduction in insurance contracts, improvement in grain quality, increments in food safety, and accurate information on grain movements. The main implication of the study is that grain producers and other players can explore opportunities, and solution providers can evolve toward complete digitalization by integrating service into the current offerings of post-harvest engineering solutions.

Keywords: digital transformation; agribusiness; post-harvest; food safety; food quality; engineering solutions

Citation: Schmidt, D.; Butturi, M.A.; Sellitto, M.A. Opportunities of Digital Transformation in Post-Harvest Activities: A Single Case Study of an Engineering Solutions Provider. *AgriEngineering* **2023**, *5*, 1226–1242. https://doi.org/10.3390/agriengineering5030078

Academic Editors: Muhammad Sultan, Yuguang Zhou, Redmond R. Shamshiri and Muhammad Imran

Received: 28 April 2023
Revised: 3 July 2023
Accepted: 6 July 2023
Published: 12 July 2023

Copyright: © 2023 by the authors. Licensee MDPI, Basel, Switzerland. This article is an open access article distributed under the terms and conditions of the Creative Commons Attribution (CC BY) license (https://creativecommons.org/licenses/by/4.0/).

1. Introduction

The world population should rise from 6.9 billion in 2010 to more than 9.4 billion in 2050 [1]. At the same time the population grows, the global availability of arable land per capita decreases. While in 1950 there was approximately 0.52 hectares per capita, by 2050, this figure will decrease to 0.17 hectares [2]. Furthermore, aggravated by climate change, the per capita demand for water has also risen, completing a picture and highlighting the need for more productivity in agricultural production systems [3]. The use of digital technologies can contribute to increasing such productivity [4]. The literature offers many cases in developed and developing countries that support the success of digital transformation in agri-food system management [5].

Agriculture, specifically the production of grains and cereals, plays an essential role in producing food for human and animal consumption. The literature estimates that 1 kg of beef, pork, and poultry production requires 7, 4, and 2 kg of grains, respectively [6]. The interest of the authors of this article includes the so-called grain post-harvest activities.

Post-harvest activities include companies from the metal-mechanical industry that provide engineering solutions for the sector. Such companies deliver solutions to maintain the quality of grains or cereals after harvesting until used by the agro-industry, retailers, and final consumers [7]. For vendors of post-harvest supporting equipment, digital transformation is an opportunity for a rapid evolution in business models. The income of such companies has depended almost exclusively on the sale of physical goods, such as grain silos, conveyors, storage facilities, or dryers. Currently, those companies, based on digital technologies, also offer engineering solutions that incorporate services into the current portfolio of products [8]. Nonetheless, such companies should still understand how digital transformation influences their business models. It is necessary to identify what capabilities to develop as well as the main opportunities digital transformation may convey to provide new sources of revenue [9]. Identifying such opportunities is the research gap this article aims to bridge.

The purpose of this article is to identify opportunities that digital transformation in post-harvest activities offers to an engineering solution provider. The research method is a simple case study. The object is a company based in southern Brazil that provides engineering-integrated digital solutions to grain producers, integrating products and services. The specific objectives are to describe the company's digital products and services, identify the main types of opportunities and the interested players, and discuss how the players can benefit from opportunities. The main expected implication is to offer post-harvest solutions vendors a guideline to explore new opportunities and change their business models by integrating digital services and physical goods. The structure of the rest of this article is as follows. Sections 2–5 contain the literature review, methodology and results, types of opportunities, and final remarks.

2. Literature Review

2.1. Digital Technologies

Cutting-edge digital technologies merge information, computing, communication, and connectivity to disrupt business models [10] and reshape relationships within business networks [11]. The key characteristic of digital technologies lies in their ability to integrate digital capabilities into objects that were previously purely physical, like equipment, appliances, or vehicles [12]. Furthermore, digital technologies introduce advanced functionalities, such as autonomy and tracking capabilities, which drastically enhance performance [13]. In essence, digital innovation initiatives blend digital and physical elements within a layered modular architecture, yielding novel products accompanied by unprecedented services that cater to the end user's needs [14]. Such initiatives often lead to profound alterations in products [15], organizational structures, and process management [16]. Ultimately, a sociotechnical digital transformation unfolds in social and institutional contexts, rendering digital technologies integral to the infrastructure [17].

Digital innovation possesses distinctive traits, including re-programmability, data homogenization, and a self-referential nature. Layered modular architectures form the organizing logic behind digitized products, allowing them to function simultaneously as both products and platforms. An exemplary instance appears in devices reliant on cloud storage. The layered architecture empowers companies to compete in specific layers, such as equipment, while fostering collaboration in others, such as services [18], aligning with the established concept of coopetition [19]. The convergence of digital technologies is apparent through interconnected yet distinct elements like artifacts, platforms, software, and databases that leverage the same digital infrastructure [14].

By embedding digital artifacts into physical devices, information can be stored, enabling programmability, addressability, communication, traceability, and association [20]. The separation of form and function enables artifacts to swiftly acquire new features at relatively insignificant costs. Digital infrastructures, such as social media, data analytics, cloud computing, and 3D printing, offer tools for rapid scalability [21] and international expansion [22]. In recent years, trailblazing pioneers like Google, Amazon, and Meta have risen to prominence, spearheading a new era of platform-based competition [23,24].

Prominent digital companies and their platforms have expedited the process of digitalization in the business realm. A seminal article from 1991 [25] introduced the concept of a pervasive computing environment, envisioning a future where revolutionary technologies seamlessly blend into everyday life, becoming indistinguishable from commonplace activities, like reading a book. A decade later, a pivotal study [26] predicted the proliferation of ubiquitous computing as mobile computing merged with pervasive computing, which integrates natural movements and interactions within physical and social environments. Subsequently, digitalization emerged as the foundation for immersive experiential computing, recognized as a sociotechnical process [17]. This process leverages advancements in digital infrastructures to analyze, interpret, and shape transformations within social and institutional contexts. In essence, digitalization is increasingly regarded as an entrepreneurial journey, wherein new business models undermine existing advantages, giving rise to more valuable or rapidly growing companies [14,15,18,20,21].

2.2. *Digital Transformation in Companies*

Digital transformation serves as a catalyst for companies to engage in experimentation and develop new business models [27]. Its impact can be far-reaching, transforming entire industries (e.g., passenger transport and accommodation), unifying products and services (books and document copying), spawning new businesses (cargo tracking), or presenting novel value propositions (e-commerce offering speed, affordability, and personalized delivery) [28]. In certain traditional sectors, digital transformation becomes imperative to safeguard established advantages [29–31]. It differs from other forms of strategic evolution primarily due to the rapid pace of change [23,32]. As digital transformation unfolds, it introduces heightened volatility, complexity, and uncertainty, necessitating adjustments in business models, organizational structures, and processes [33–35]. Digitalization opens opportunities for customer interaction, often leading to unforeseen innovations in business models [36–39].

Organizations equipped with transformative capabilities typically foster agile and entrepreneurial mindsets, emphasizing external networking [40]. Transformative capabilities support strategic renewal processes that involve adapting assets and structures to ensure responsiveness in swiftly changing digital environments [41,42]. Digital transformation presents challenging trade-offs, such as building innovation capabilities while preserving existing products, innovating not only products but also processes; balancing conflicts involving customers, employees, and suppliers; and establishing governance structures that ensure flexibility and control simultaneously [32]. For instance, in e-commerce, buyers and suppliers engage in online commercial transactions [43], creating a wealth of options and new expectations that prompt companies to reassess or augment their transactional value propositions [44–48].

Digital technologies disrupt the traditional logic of business models by elevating customer expectations for complementary products or services [47,49–51]. In response to customer demands, many internet-based businesses prioritize value creation through exceptional customer service over immediate profit capture, occasionally leading to flawed business models [24]. Consequently, established companies often encounter significant barriers to business model innovation that can impede their journey toward digitalization [29,36,50,52–57].

A business model encompasses a company's mechanisms for creating, delivering, and capturing value [47], encompassing strategic priorities, such as cost, revenue, and profit [58]. Innovation-based business models may incorporate elements, such as learning [52], shifts in management approaches [59], evolution [60], replication [38], reconfiguration [61], modularization, scalability [62], and digital transformation [36]. Certain business contexts, such as agri-food supply chains, require considerations for sustainability, including eco-design, where the business model targets not only financial performance but also environmental impact [63].

Difficulties in implementing innovation-based models are not uncommon. One reason is the belief that standard targets, such as profit margins or revenues, remain stable and can be pursued until achieved. When targets are not met, the common perception is that more effort needs to be invested [64,65]. However, in innovation-driven markets, new alternatives can swiftly disrupt performance parameters, for better or worse, thereby posing risks to even well-established business plans [44,51]. Another challenge arises from the trade-offs that emerge over time between static and dynamic models, necessitating more agile methods and monitoring of additional control variables [56,66]. An illustrative example of a model based on trade-offs is the servitization model, where a product company offers an associated service. Often, to ensure long-term service sales, the product must be sold under less favorable conditions, introducing a trade-off between the product and service [64,67]. This trade-off is commonly observed among providers of post-harvesting engineering solutions. Path dependency represents another barrier to innovation-based business models, as successful models from the past tend to be perpetuated [68,69]. However, exogenous shocks to performance can help balance the endogenous dependence on models that have previously been effective but are currently approaching exhaustion [70]. Disruptive business models may be deemed unlikely for certain companies [71], as they prefer incremental digitalization over disruptive modifications to existing activities [53,72,73].

The digital transformation of the business model necessitates a convergence of corporate and business unit models, requiring interdependent decision making [36,56]. To manage strategic complexity, companies often rely on previous experiences rather than entirely new approaches [70]. This decision-making process gives rise to conflicting demands, necessitating a delicate balance between agility and stability [69], certainty and uncertainty [74], or short- and long-term benefits [52]. In some cases, companies adopt a rational approach by addressing challenges in successive stages. They fully overcome one challenge before confronting a contradictory one, as exemplified by product servitization. Once the product-based business model is firmly established (low uncertainty), the company introduces the service component (high uncertainty). Subsequently, the focus may shift back to refining the product and so on, as the process continues.

2.3. *Enabling Technologies for Digital Transformation in Post-Harvesting Activities*

Digital transformation processes must rely on integrating single digital technologies, encompassing the automation of processes and intelligent interconnections of machines. The integration results in cyber–physical systems. Such production systems are simultaneously physical, providing a flow of physical material, and logic, performing supervisory and control tasks [75]. Many recent authors have listed and analyzed the most relevant technologies supporting digital transformation processes, the so-called enabling technologies [76]. Among recent authors, relying on updated references, [77] cite automation (AUT), cyber–physical systems (CPS), big data analysis (BGA), radio frequency identification (RFID), cloud computing (CC), internet of things (IoT), additive manufacturing (AM), virtual reality and augmented reality (VRAR), and simulation (SIM) as enabling technologies for digital transformation in engineering solutions for primary activities.

Enabling technologies can contribute in different ways to post-harvest engineering solutions. AUT and CSF mainly help to reduce variability in processes [78]. The BDA mainly helps to develop models and find new behavior patterns for key variables in decision-making processes. CC mainly helps monitor the physical equipment's life-cycle behavior [79]. RFID mainly improves reliability in logistical operations, especially transportation [80]. The IoT helps to improve flexibility in key post-harvest processes, such as transportation and warehousing [81]. AM mainly assists in reducing material losses in machinery design and promotes reverse logistics activities, mainly managing shavings and leftovers produced by equipment manufacturers [82]. VRAR helps develop an eco-design and servitization principle: improving the performance of a product by incorporating environmental concerns and associated services [63]. Finally, SIM can help reduce the likelihood of making wrong choices in decision-making processes, including sales and purchases processes [83].

3. Results

3.1. Methodology

The research strategy was a single case study. The object of study was a Brazilian company that provides engineering solutions for post-harvest activities. The primary research technique was non-participant direct observation, in which the researchers directly monitored the main activities in real time without interfering with their results.

The study included the following steps:

(i) The researchers collected and studied related documents issued by the engineering solution provider company;

(ii) The researchers took guided tours of the company's facilities and two nearby customer installations, accompanied by company practitioners;

(iii) The researchers interviewed three portfolio managers of the company;

(iv) In a final meeting, one of the researchers presented the notes to the managers, who eventually amended, adjusted, and finally confirmed the findings, ensuring reliability.

This study does not encompass losses quantification; instead, it solely focuses on describing potential opportunities that may be viable for future innovation initiatives. Owing to the use of a single case study strategy, the findings do not aim at providing external validity, meaning that the findings are only expected to be valid within the research scope. For broader external validity, future research should encompass the entire industry rather than focusing on a single company. The final meeting with the participants, during which they had the opportunity to review and amend certain aspects of the findings, enhances the internal validity and reliability of the study. This research is exploratory and qualitative, representing the initial approach to the problem. Consequently, the results involve a descriptive analysis of phenomena without the use of mathematical models to explain them. As is typical of exploratory studies, these findings serve to stimulate further research that can delve deeper into the subject matter.

3.2. Post-Harvest Main Activities

Post-harvesting activities play a key role in managing the quality and safety of agricultural grains after the harvesting process. They are responsible for ensuring the integrity of the grains throughout the entire production chain [84]. Grains, being living products, require careful handling and storage, with specific considerations for cleanliness, temperature, humidity, and continuous monitoring [85]. Agricultural grains are prone to rapid deterioration and can be affected by the growth of fungi, yeasts, bacteria, and harmful mycotoxins. Failure to prevent these deterioration processes poses a risk to human consumption, including the consumption of animal products derived from grain-based feeds [86]. One of the key objectives and benefits of post-harvest activities is the loss prevention in the food production system. Various stages of the production cycle can contribute to significant qualitative (in terms of grain quality) and quantitative (in terms of volume) losses during the grain's journey within a grain storage unit (GSU). Figure 1 illustrates the primary issues related to the safety and quality of stored grains.

After the harvest, the grains are transported to a grain storage unit (GSU) using various logistical modes, such as road, waterway, or rail transport. In a GSU, there are nine critical control points (PCC) that are crucial for preventing qualitative and quantitative losses in grains. Figure 2 depicts the points. Additionally, Figure 3 highlights a representative GSU equipped by the company under study, highlighting the specific control points.

This article focuses on the post-harvest process depicted in Figures 2 and 3. The process begins with the reception of grains (PCC1), followed by the discharge into the hoppers (PCC2). The grains then undergo precleaning (PCC3) to remove coarse impurities before being transferred to buffer silos (PCC4) for intermediate warehousing. If needed, the grains route to dryers (PCC5) for a high-temperature treatment to eliminate moisture and ensure safe humidity levels for long-term storage. Next, the grains pass through cleaning machines (PCC6) to remove fine impurities that may affect their quality. Subsequently, the grains are stored in permanent storage silos (PCC7) until they are ready for commercialization. Once the sale is completed, the shipping process (PCC8) begins. Throughout the entire storage process, various crossing points (PCC9) contribute to the losses. The implementation of digital technologies can effectively address the PCCs, aiming to minimize losses in the overall process.

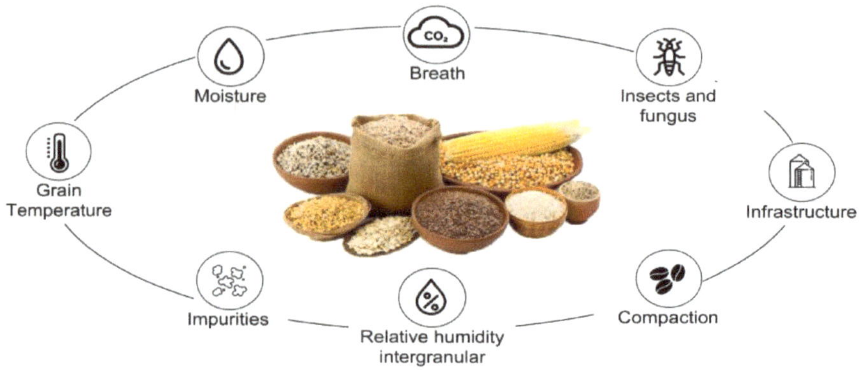

Figure 1. Factors that influence the safety and quality of stored grains.

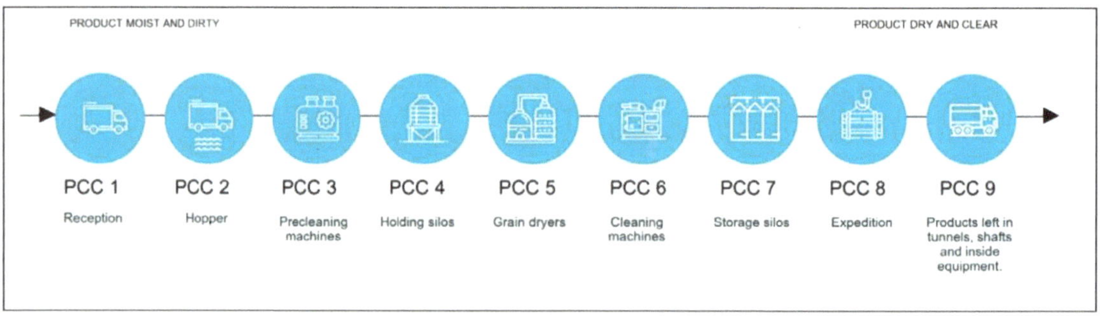

Figure 2. Qualitative and quantitative losses in the post-harvesting process.

Figure 3. Grain Storage Unit (GSU) including reception (1), hopper (2), precleaning (3), holding silos (4), dryers (5), cleaning (6), storage silos (7), expedition (8), and internal movement (9) processes.

A thorough study delved into over 300 cases of loss reduction initiatives in post-harvesting [87]. The findings revealed a potential to decrease production losses from 2% to 28%, with an average reduction of approximately 7% and a standard deviation of around 11%. The study also uncovered that more than 80% of initiatives aimed at curbing losses focus on implementing storage technology interventions for farmers and 6% for traders.

Consequently, it seems reasonable to anticipate that digital technologies could have an average impact of around 7% on the overall volume of produce. This estimate is comparable to reports that indicate loss reductions between 5 and 8% in production volume due to the introduction of technological management elements [88].

In a grain storage unit (GSU), the drying and storage processes are pivotal. Grain dryers are equipped with various sensors and systems to ensure efficient operation, such as pressure sensors to balance mixtures in the drying air, level sensors to prevent the equipment from operating when empty, an exhaust air temperature sensor to detect early signs of fire, a frequency inverter to adjust drying speed and time, and an automatic fuel supply system to stabilize the drying temperature by regulating burning. To monitor grain quality, the system provides real-time measurements of grain moisture, drying air temperature, and grain mass. Real-time alert systems enhance safety and equipment efficiency. For storage, the recommendation is at least one temperature sensor per 150 m^3 of grain. The system transmits in real time through remote platform intragranular relative humidity sensors and meteorological data. The system tracks the storage evolution and grain mass temperatures, and it enables the creation of rules that link the aeration system to air renewal based on climatic conditions. Safety and quality objectives may vary based on drying, cooling, conservation methods, local climate, and the customer's strategy [84].

Dry matter loss or technical breakdown occurs during grain storage, leading to weight loss caused by chemical oxidation reactions that consume energy stored in organic compounds, such as sugars and starches. The acceptable level of dry matter loss varies, but authors suggest values between 0.1% and 0.5% [89,90]. In Brazil, field evidence establishes an official technical breakage rate of 0.3% per month of storage [91].

3.3. Digital Technology Products and Services for Post-Harvesting

The company offers a range of products and services that utilize digital technologies integrated with fixed equipment throughout the entire set of processes in the GSUs. The offers include receiving, handling, precleaning, drying, post-cleaning storage and preservation, and grain shipping. Various digital field technologies assist farmers in real-time monitoring of grain temperature and humidity [92]. Additionally, virtual reality is utilized for training and inspection activities [93], while AI and CC are leveraged to predict the behavior of key variables in storage operations [94]. The literature provides a recent overview of the enabling technologies used in Agriculture 4.0 [95]. To enhance understanding, the company categorizes its offerings into three layers of products, with the first layer being closer to the physical space and the third layer closer to cyberspace.

The first layer encompasses sensors that collect data related to physiological conditions, such as temperature and humidity. It also involves gathering data on machine productivity and maintenance, utilizing sensors for flow, temperature, movement, bushing, and mechanical component alignment. This layer includes motors, standardized sequencing, and protection instrumentation, such as inductive and capacitive sensors, as well as protection relays. Finally, the layer incorporates specialized instrumentation, such as thermometry systems for monitoring temperature and humidity sensors for drying processes.

The second layer comprises control panels and supervisory systems that employ various automation architectures to manage the performance and efficiency of the entire process through preset management and process parameter control. In fully automated units, the panels can interact with each other (Machine to Machine—M2M) without human intervention, leveraging IoT technology. This capability is crucial as it supports the concept of decentralizing the automation process, enabling individual and independent control of each piece of equipment. It also allows a single control room to oversee multiple GSUs. Depending on the arrangement, the control room may or may not be located near the field. It is not uncommon for customers to control multiple GSUs from a single control room. The company exclusively sells physical equipment and add-ons from the first two layers, whether they are included in the overall engineering solution or integrated into the machinery. The equipment warranty is contingent upon the safe operating conditions ensured by the automation, control, and instrumentation systems of the first two layers.

The third layer encompasses remote digital services, typically hosted in the cloud. The company has incorporated such a service into its engineering solution since 2019. The cloud platform communicates with the second layer through gateways and internet access infrastructure. The platform features a highly adaptable interface that efficiently retrieves real-time data and information generated by the equipment's sensors. It records operational history, serving as a valuable resource for facilitating timely decision-making processes for both customers and equipment suppliers.

In terms of enabling technologies for the digital transformation of industrial processes, the company primarily relies on automation (AUT) and radio frequency identification (RFID) in the first layer. The second layer is predominantly supported by cyber–physical systems (CPS), the internet of things (IoT), and virtual reality and augmented reality (VRAR). The third layer relies mainly on big data analysis (BGA), cloud computing (CC), and simulation (SIM). Additive manufacturing (AM) plays a role in expediting the supply of spare parts and prototyping while also facilitating the development of new technologies. It is conceivable that AM will be utilized in future machinery development, particularly due to the market's requirements for modularity and scalability. Figure 4 provides an illustration of the content of the three layers.

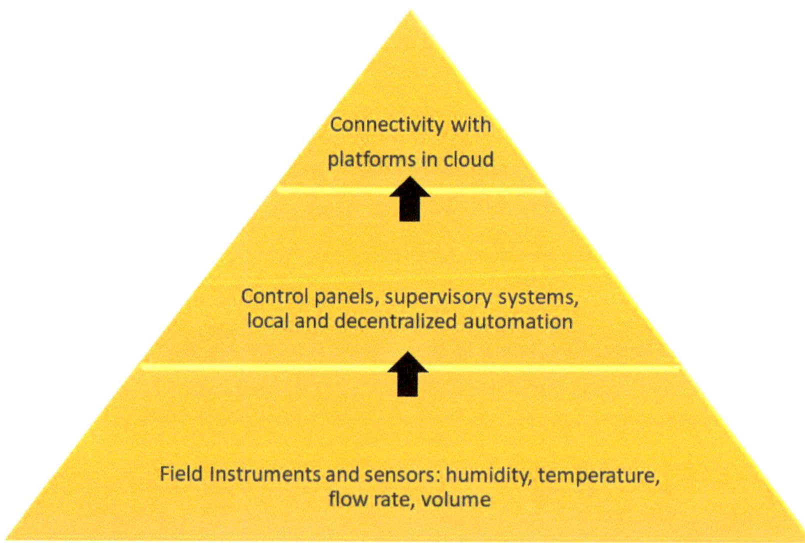

Figure 4. The three layers of digital products for post-harvesting that the company provides.

4. Types of Opportunities

The competitive priorities for providers of post-harvest solutions undergo changes as the market evolves in implementing and adopting GSU technology products. The company under study identifies five categories of opportunities for customers and stakeholders: financing, commercialization, operation and logistics, insurance, and traceability. The opportunities attract the interest of eight key players, each representing a potential customer for new products and services. The primary player is the agricultural producer, encompassing all types of opportunities. The other players include trading companies, agro-industries involved in grain purchases, financial institutions, input suppliers, insurance companies, food retailers, and end consumers. Each player establishes unique competitive priorities for every viable opportunity, aligning them with their respective value perspectives. Figure 5 provides an illustration of the different opportunity types.

Figure 6 represents a technology platform view that integrates the five types of opportunities and a sixth element, the solution provider.

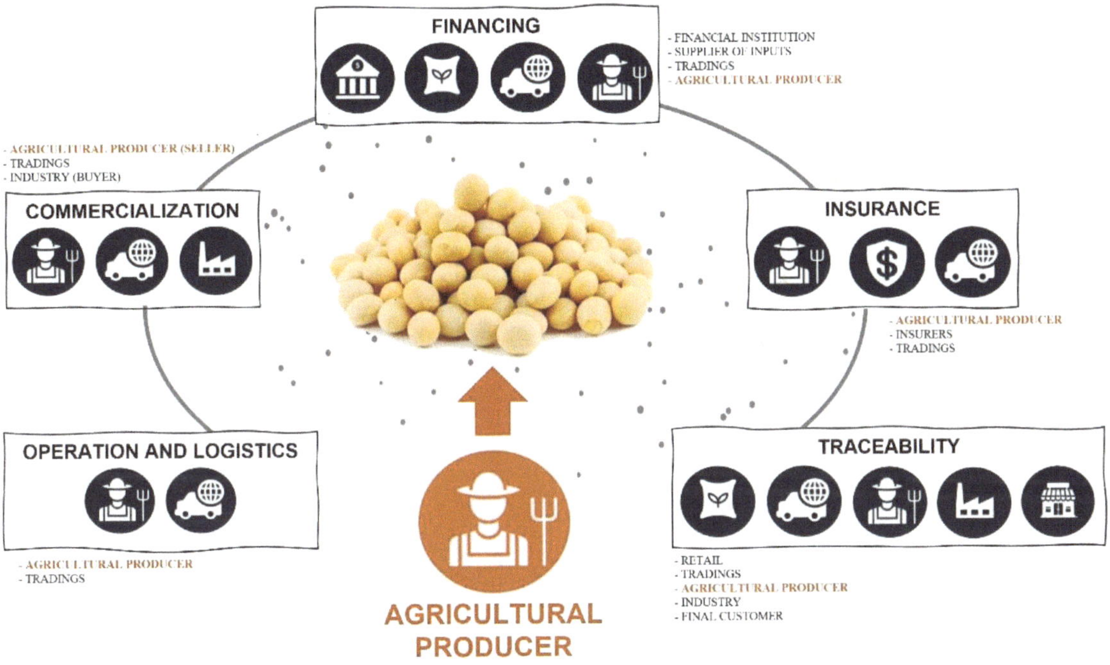

Figure 5. Opportunities for post-harvest solution providers.

The key component of the system is the big data machine, which harnesses data from IoT solutions integrated into the equipment. It serves as a vital link that connects to the machine-learning device, enabling the generation of essential information for the development of new products and services linked to the platform. With the assistance of a digital twin driven by artificial intelligence and other advanced technologies, the system seamlessly integrates the automation solutions within each client's equipment. The integration not only enhances the concept of autonomous equipment but also strengthens the manufacturing industry's drive toward performance enhancement. By utilizing digital technologies to extract valuable insights and knowledge from vast amounts of data, solution providers can establish continuous connectivity with customers, even beyond the delivery, installation, and commissioning of systems. Offering digital services positions solution providers as consultants, providing support and guidance to customers in their quest for enhanced efficiency and preservation of grain quality. Digital services open new and sustainable revenue streams, justifying investments in the development of innovative products. The studied company addresses the specific requirements of each opportunity type, ensuring tailored solutions for their customers.

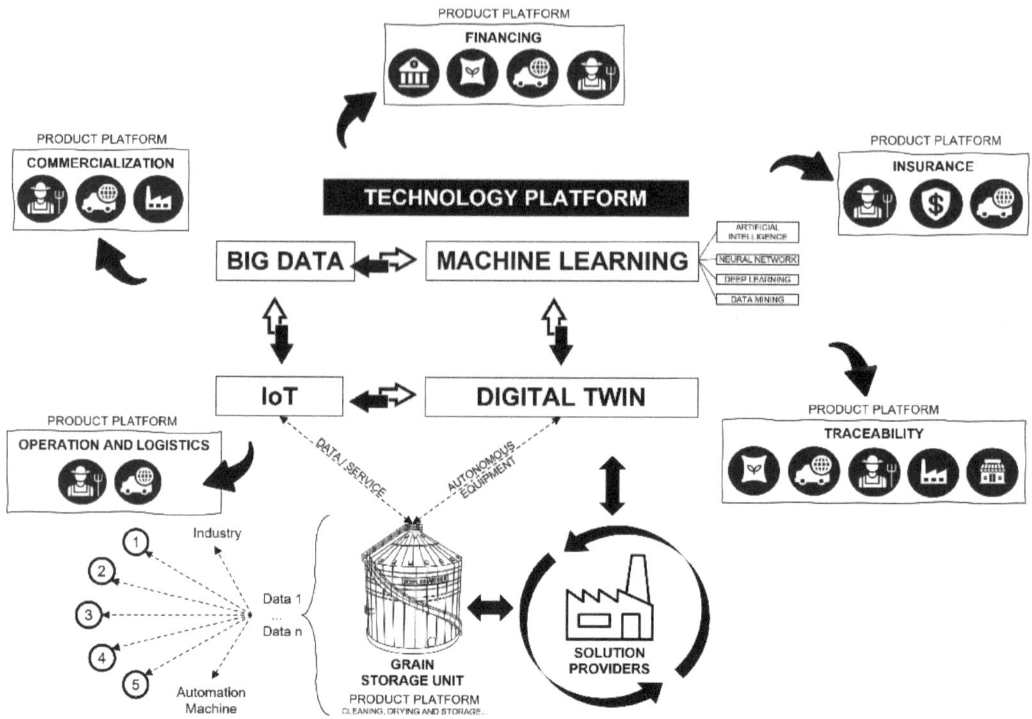

Figure 6. Technologic platform.

Exploring the Opportunities

As for financing, the main opportunity lies in digitizing the credit application process for agricultural producers, with the grain harvest serving as collateral for the loans. Financial institutions, input suppliers, and trading companies focus on minimizing default risks, while agricultural producers seek fast, affordable, and secure credit options. The platform effectively addresses the uncertainties associated with credit information and risk assessment, as well as the challenge of tracking grains covered by warranties to prevent fraud. In essence, the primary challenges faced by these stakeholders revolve around credit approval uncertainties, difficulties in monitoring warranted grain, and the risk of fraud. Additionally, agricultural producers face the additional challenge of convincing financial institutions to have faith in the effectiveness of their production. Historical data that showcase the producer's performance and punctuality become crucial in securing lower interest rates, gaining preferential access to financial resources, and becoming eligible for more stringent yet government-subsidized financing options.

As for commercialization, trading companies, agro-industries, and producers share a common need for improved reliability in purchase and sale operations, along with agility and security in transactions. Trading companies and buyers prioritize reducing transaction costs and mitigating uncertainties associated with receiving goods, which involves accessing information about input and grain quality, market price quotations, minimizing storage expenses, expanding storage capacity, and establishing connections with new grain producers. Platforms effectively address these limitations by providing the necessary tools and information. On the other hand, producers aim to secure the most favorable commercial conditions in terms of price, receipt timeframe, and contract closure speed. They require access to information that reveals the correlation between grain price and quality, streamlined transaction processes, and connections to new grain buyers. Platforms effectively address these limitations by providing the necessary resources and connections to facilitate efficient and favorable trade for producers. In summary, platforms play a vital role in meeting the needs of trading companies, agro-industries, and producers by enhancing reliability, reducing costs, mitigating uncertainties, and facilitating efficient and secure transactions.

As for operations and logistics, producers and trading companies have specific requirements. Producers are looking to expand their warehousing capacity and maintain their current equipment. They also seek solutions for maintaining grain quality through advanced quality control methods and monitoring services that improve efficiency and reduce waste along the entire grain supply chain. By integrating e-commerce solutions with IoT data from customer-installed equipment, proactive measures can anticipate preventive and predictive maintenance needs, optimizing services and parts. Additionally, the technology platform-backed e-commerce system can identify new business opportunities, such as expanding capacities in GSUs and offering equipment and technology acquisitions to both existing and potential customers. Producers face challenges, including the high costs associated with acquiring and maintaining silos, a lack of tools for grain quality control, heavy reliance on manual labor, time-consuming and costly transshipment, and transportation stages that often result in contract non-compliance, fines, and deterioration of grain quality during prolonged storage. On the other hand, trading companies encounter difficulties due to the absence of quality control tools for stored and in-transit grain, excessive reliance on manual labor in transportation, and the lack of grain monitoring during transshipment and transportation, leading to waste and losses. Such challenges ultimately have an impact on the agro-industry, food retailers, and consumers. To overcome limitations, the use of automated, integrated strategic platforms can prove beneficial. Such platforms offer comprehensive solutions that address the specific needs of producers and trading companies, streamlining operations, enhancing efficiency, and ensuring the maintenance of grain quality throughout the entire supply chain.

As for insurance, one of the main challenges is to digitize the insurance contracting process and leverage technology for intelligent risk analysis and remote monitoring of risks associated with harvest, storage, and transportation. The key stakeholders are insurance companies, trading companies, and producers. For insurance companies, it is crucial to mitigate risks by conducting more efficient and cost-effective inspections of agricultural properties to reduce fraud and cargo theft. Insurers often face difficulties due to the high-risk nature of the agricultural activity, which leads to high inspection costs, low inspection efficiency, and a prevalence of fraud and cargo theft incidents. Trading companies, on the other hand, face the challenge of enhancing security against fraud, theft, and returns. They strive to protect their interests and minimize potential financial losses resulting from these risks. Producers consider insurance as essential for mitigating uncertainties and ensuring a financial return on their crop investments. However, high insurance policy prices pose a significant difficulty for them. An integrated platform provides producers with the ability to compare prices and benefits offered by different insurers. Additionally, by presenting reliable information on historical productivity, producers can diminish the insurer's risk perception, resulting in lower-priced insurance policies for them. Overall, digitalization and the integration of platforms in the insurance sector address challenges by streamlining processes, enhancing risk analysis capabilities, reducing fraud incidents, and providing greater transparency and cost efficiency for all parties involved. By embracing technology and digital solutions, insurance companies, trading companies, and producers can benefit from improved efficiency and effectiveness in managing insurance contracts and mitigating risks associated with agricultural activities.

As for traceability, the primary challenge lies in establishing a digital platform capable of integrating information across the entire supply chain, which includes cleaning, drying, and storage processes. By incorporating comprehensive and integrated information, it becomes possible to establish standardized commercialization norms and certify the product's origin and quality for the market. The key stakeholders are producers, trading companies, buyers, and retailers. For producers, trading companies, and buyers, the certification of grain origin and journey is of utmost importance to ensure product quality. This includes aspects such as verifying the appropriate use of pesticides in harvest management. Producers face challenges related to adopting product quality standards and pressures from various stakeholders to identify and reduce pesticide usage while embracing environmentally sustainable practices. Trading companies and buyers struggle with limited monitoring capabilities throughout the transportation, storage, and processing of grains due to a multitude of suppliers and product mixes along the supply chain. Additionally, they face increasing pressure to adopt sustainable practices and ensure environmental safety. For retailers and consumers, the supply of nutritious, healthy, and environmentally safe food is paramount. It is equally essential to have monitoring data and assurance of food safety for the end consumers. Overcoming these challenges relies on reliable tracking systems and accurate information regarding the origin, journey, and quality of the food products. Addressing these challenges requires the development and implementation of robust traceability systems that track and record essential data points throughout the supply chain. By leveraging digital platforms and integrating information, stakeholders can ensure transparency, reliability, and compliance with quality and safety standards. This enables consumers to make informed choices about the food while promoting sustainable practices and fostering trust throughout the supply chain.

Table 1 depicts the main implications of the study and the opportunities that digital transformation conveys for players interested in the solutions provided by the company.

Table 1. Synthesis of the implication of the study.

	Opportunities				
Players	**Financing**	**Commercialization**	**Operation and Logistics**	**Traceability**	**Assurance**
Producers	Credit cheap and fast	Increased reliability	Increased grain quality	Origin certified	Reduced cost
Trading companies	Reduced default risk	Increased reliability	Increased grain quality	Origin certified	Reduced risk
Agro-industry		Increased reliability	Increased grain quality	Origin certified	
Financial institutions	Reduced default risk				
Suppliers	Reduced default risk				
Insurance companies				Information	Reduced risks
Food retailers			Increased food quality	Safer food	
Consumers			Increased food quality	Safer food	

5. Final Remarks

Digital transformation is the process an organization applies to integrate digital technologies into its business, fundamentally changing how it delivers value to customers. Digital transformation can, at the same time, increase productivity, improve the customer experience, and reduce operating costs. In post-harvest activities, particularly in GSUs, digital technologies have given equipment autonomy to communicate with other machinery and collect data required by automatic platforms of strategic management.

The purpose of this article was to identify opportunities that digital transformation in post-harvest activities offers to an engineering solution provider. The research method was a simple case study. The object of study was a company based in southern Brazil that provides integrated engineering digital solutions, including products and services, to grain producers. The primary findings included the differentiation of the company's products into three technological layers, the identification of five important opportunity kinds and eight players, and the potential contributions of each technology to the top players in each opportunity type. The types of opportunities are financing, commercialization, operations and logistics, traceability, and insurance. The players are grain producers, trading companies, the agro-industry that purchases grains, financial institutions, input suppliers, insurance companies, food retailers, and final consumers. The digital transformation presents a host of opportunities, many of which are already implemented, across various aspects of the production chain for post-harvest services. The implications include the following:

(i) Digitalization of credit taking: The adoption of digital processes and platforms for credit applications and approvals, streamlining and expediting the financing process for producers;
(ii) Digitalization of purchase operations: Implementing digital solutions to facilitate purchase transactions, improving efficiency, and reducing costs for trading companies and agro-industries;
(iii) Digitalization of sale operations: Utilizing digital platforms to enhance sales processes, enabling faster and more secure transactions for both producers and trading companies;
(iv) Digitalization of grain quality control: Leveraging digital technologies to monitor and control grain quality throughout the supply chain, ensuring higher standards and reducing quality-related risks for all stakeholders;
(v) Digitalization of food safety control: Implementing digital systems and technologies to enhance food safety protocols, enabling better traceability and ensuring the delivery of safe and high-quality food products to consumers;
(vi) Digitalization of grain movement information: Utilizing digital platforms to track and monitor grain movement, providing real-time information on storage, transportation, and logistics, resulting in increased transparency and efficiency in the supply chain;
(vii) Digitalization of insurance contracting processes: Adopting digital solutions for insurance procedures, simplifying and streamlining the contracting process, reducing costs, and improving risk assessment and management for insurers, producers, and trading companies.

Integrating field equipment with management platforms is a crucial aspect of this digital transformation, enabling data collection, analysis, and decision making. By embracing digital opportunities, the actors in the post-harvest services production chain can unlock significant value, improving efficiency, reducing risks, and enhancing overall performance. According to the Brazilian Company of Food Supply (CONAB), Brazil's grain production for the 2022/23 harvest may surpass 313 million tons [96]. The company highlights that each percentage point of error, for instance, in humidity control, may lead to a loss exceeding 3 million tons, which emphasizes the economic feasibility of employing technology to preserve the physiological and quantitative integrity of grains. By utilizing monitoring and automation systems capable of autonomous decision making with minimal human intervention, losses diminish while augmenting the global grain supply.

This study makes room for additional research. One possibility is constructing a framework or roadmap that guides a company's digital transformation and provides post-harvest activity solutions. Another option is to survey the company's customers (there are more than a thousand rural producers around the globe already served by the studied company) to understand the digitalization stage the industry is in and, mainly, what competitive priorities rural producers aim to meet with the digitization. Finally, multiple case studies should describe the peculiarities and difficulties of customers with digitalized products and processes and the digital transformation of the implemented solutions.

Author Contributions: Conceptualization, D.S.; methodology, D.S.; validation, M.A.S. and M.A.B.; formal analysis, M.A.S.; investigation, D.S.; writing—original draft preparation, D.S.; writing—review and editing, M.A.S.; visualization, M.A.B.; supervision, M.A.B.; project administration, D.S.; funding acquisition, M.A.S. All authors have read and agreed to the published version of the manuscript.

Funding: This research was partially funded by CNPq, the Brazilian research agency, under the grant number 303496/2022-3. The APC was funded by AgriEngineering.

Data Availability Statement: Not applicable.

Conflicts of Interest: The authors declare no conflict of interest.

References

1. Alexandratos, N.; Bruinsma, J. World Agriculture towards 2030/2050: The 2012 Revision. 2012. Available online: https://www.fao.org/3/ap106e/ap106e.pdf (accessed on 27 April 2023).
2. Chandra, R.; Collis, S. Digital agriculture for small-scale producers: Challenges and opportunities. *Commun. ACM* **2021**, *64*, 75–84. [CrossRef]
3. Hrustek, L. Sustainability is driven by agriculture through digital transformation. *Sustainability* **2020**, *12*, 8596. [CrossRef]
4. Di Vaio, A.; Boccia, F.; Landriani, L.; Palladino, R. Artificial intelligence in the agri-food system: Rethinking sustainable business models in the COVID-19 scenario. *Sustainability* **2020**, *12*, 4851. [CrossRef]
5. Qin, T.; Wang, L.; Zhou, Y.; Guo, L.; Jiang, G.; Zhang, L. Digital Technology-and-Services-Driven Sustainable Transformation of Agriculture: Cases of China and the EU. *Agriculture* **2022**, *12*, 297. [CrossRef]
6. Godfray, H.; Beddington, J.; Crute, I.; Haddad, L.; Lawrence, D.; Muir, J.; Pretty, J.; Robinson, S.; Thomas, S.; Toulmin, C. Food security: The challenge of feeding 9 billion people. *Science* **2010**, *5967*, 812–818. [CrossRef]
7. Hossen, M.; Talukder, M.; Mamun, M.; Rahaman, H.; Paul, S.; Rahman, M.; Miarudin, M.; Ali, A.; Islam, M. Mechanization status, promotional activities, and government strategies of Thailand and Vietnam in comparison to Bangladesh. *AgriEngineering* **2020**, *2*, 489–510. [CrossRef]
8. Dyck, G.; Hawley, E.; Hildebrand, K.; Paliwal, J. Digital Twins: A novel traceability concept for post-harvest handling. *Smart Agric. Technol.* **2022**, *3*, 100079. [CrossRef]
9. Kayikci, Y.; Subramanian, N.; Dora, M.; Bhatia, M. Food supply chain in the era of Industry 4.0: Blockchain technology implementation opportunities and impediments from the perspective of people, process, performance, and technology. *Prod. Plan. Control* **2022**, *33*, 301–321. [CrossRef]
10. Sibanda, W.; Ndiweni, E.; Boulkeroua, M.; Echchabi, A.; Ndlovu, T. Digital technology disruption on bank business models. *Int. J. Bus. Perform. Manag.* **2020**, *21*, 184–213. [CrossRef]
11. Bharadwaj, A.; El Sawy, O.; Pavlou, P.; Venkatraman, N. Digital business strategy: Toward a next generation of insights. *MIS Q.* **2013**, *37*, 471–482. [CrossRef]
12. Warner, K.; Wäger, M. Building dynamic capabilities for digital transformation: An ongoing process of strategic renewal. *Long. Range Plann.* **2019**, *52*, 326–349. [CrossRef]
13. Yoo, Y.; Boland, R., Jr.; Lyytinen, K.; Majchrzak, A. Organizing for innovation in the digitized world. *Organ. Sci.* **2012**, *23*, 1398–1408. [CrossRef]
14. Nambisan, S. Digital entrepreneurship: Toward a digital technology perspective of entrepreneurship. *Enterpren. Theor. Pract.* **2017**, *41*, 1029–1055. [CrossRef]
15. Autio, E.; Nambisan, S.; Thomas, L.; Wright, M. Digital affordances, spatial affordances, and the genesis of entrepreneurial ecosystems. *Strat. Entrepreneur. J.* **2018**, *12*, 72–95. [CrossRef]
16. Weber, E.; Büttgen, M.; Bartsch, S. How to take employees on the digital transformation journey: An experimental study on complementary leadership behaviors in managing organizational change. *J. Bus. Res.* **2022**, *143*, 225–238. [CrossRef]
17. Tilson, D.; Lyytinen, K.; Sørensen, C. Research commentary-digital infrastructures: The missing IS research agenda. *Inform. Syst. Res.* **2010**, *21*, 748–759. [CrossRef]
18. Henfridsson, O.; Yoo, Y. The liminality of trajectory shifts in institutional entrepreneurship. *Organ. Sci.* **2013**, *25*, 932–950. [CrossRef]
19. Sellitto, M.; Luchese, J. Systemic cooperative actions among competitors: The case of a furniture cluster in Brazil. *J. Ind. Compet. Trade* **2018**, *18*, 513–528. [CrossRef]
20. Ancillai, C.; Sabatini, A.; Gatti, M.; Perna, A. Digital technology and business model innovation: A systematic literature review and future research agenda. *Technol. Forecast. Soc.* **2023**, *188*, 122307. [CrossRef]
21. Agnusdei, L.; Krstić, M.; Palmi, P.; Miglietta, P. Digitalization as a driver to achieve circularity in the agro-industry: A SWOT-ANP-ADAM approach. *Sci. Total Environ.* **2023**, *882*, 163441. [CrossRef]
22. Monaghan, S.; Tippmann, E. Becoming a multinational enterprise: Using industry recipes to achieve rapid multinationalization. *J. Int. Bus. Stud.* **2018**, *49*, 473–495. [CrossRef]
23. Sebastian, I.; Ross, J.; Beath, C.; Mocker, M.; Moloney, K.; Fonstad, N. How big old companies navigate digital transformation. *MIS Q. Exec.* **2017**, *16*, 197–213.
24. Teece, D.; Linden, G. Business models, value capture, and the digital enterprise. *J. Org. Des.* **2017**, *6*, 1–14. [CrossRef]
25. Weiser, M. The computer for the 21st century. *Sci. Am.* **1991**, *265*, 94–105. [CrossRef]
26. Lyytinen, K.; Yoo, Y. Research commentary: The next wave of nomadic computing. *Inf. Syst. Res.* **2002**, *13*, 377–388. [CrossRef]
27. Westerman, G.; Bonnet, D. Revamping your business through digital transformation. *MIT Sloan Manag. Rev.* **2015**, *56*, 10.
28. Thompson, N.; Bonnet, D.; Ye, Y. Why innovation's future isn't (just) open. *MIT Sloan Manag. Rev.* **2020**, *61*, 55–60.
29. Berman, S. Digital transformation: Opportunities to create new business models. *Strat. Leader.* **2012**, *40*, 16–24. [CrossRef]
30. Fitzgerald, M. How digital acceleration teams are influencing Nestlé´s 2000 brands. *MIT Sloan Manag. Rev.* **2014**, *55*, 1–5.
31. Gray, P.; El Sawy, O.; Asper, G.; Thordarson, M. Realizing strategic value through center-edge digital transformation in consumer-centric industries. *MIS Q. Exec.* **2013**, *12*, 1–17.

32. Svahn, F.; Mathiassen, L.; Lindgren, R. Embracing digital innovation in incumbent firms: How Volvo cars managed competing concerns. *MIS Q. Exec.* **2017**, *41*, 239–254. [CrossRef]
33. Dattée, B.; Alexy, O.; Autio, E. Maneuvering in poor visibility: How firms play the ecosystem game when uncertainty is high. *Acad. Manag. J.* **2018**, *91*, 466–498. [CrossRef]
34. Loonam, J.; Eaves, S.; Kumar, V.; Parry, G. Towards digital transformation: Lessons learned from traditional organizations. *Strat. Change* **2018**, *27*, 101–109. [CrossRef]
35. Matt, C.; Hess, T.; Benlian, A. Digital transformation strategies. *Bus. Inf. Syst. Eng.* **2017**, *57*, 339–343. [CrossRef]
36. Aspara, J.; Lamberg, J.; Laukia, A.; Tikkanen, H. Corporate business model transformation and inter-organizational cognition: The case of Nokia. *Long Range Plan.* **2013**, *46*, 459–474. [CrossRef]
37. Chesbrough, H. Business model innovation: Opportunities and barriers. *Long Range Plan.* **2010**, *43*, 354–363. [CrossRef]
38. Khanagha, S.; Volberda, H.; Oshri, I. Business model renewal and ambidexterity: Structural alteration and strategy formation process during transition to a Cloud business model. *RD Manag.* **2014**, *44*, 322–340. [CrossRef]
39. Wirtz, B.; Schilke, O.; Ullrich, S. Strategic development of business models: Implications of the Web 2.0 for creating value on the Internet. *Long Range Plan.* **2010**, *43*, 272–290. [CrossRef]
40. Day, G.; Schoemaker, P. Adapting to fast-changing markets and technologies. *Calif. Manag. Rev.* **2016**, *58*, 59–77. [CrossRef]
41. Agarwal, R.; Helfat, C. Strategic renewal of organizations. *Organ. Sci.* **2009**, *20*, 281–293. [CrossRef]
42. Teece, D. The foundations of enterprise performance: Dynamic and ordinary capabilities in an (economic) theory of firms. *Acad. Manag. Perspect.* **2014**, *28*, 328–352. [CrossRef]
43. Zott, C.; Amit, R.; Massa, L. The business model: Recent developments and future research. *J. Manag.* **2011**, *37*, 1019–1042.
44. Aversa, P.; Haefliger, S.; Reza, D. Building a winning business model portfolio. *MIT Sloan Manag. Rev.* **2017**, *58*, 49–54.
45. Baden-Fuller, C.; Haefliger, S. Business models and technological innovation. *Long Range Plan.* **2013**, *46*, 419–426. [CrossRef]
46. Casadesus-Masanell, R.; Ricart, J. From strategy to business models and tactics. *Long Range Plan.* **2010**, *43*, 195–215. [CrossRef]
47. Teece, D. Business models, business strategy, and innovation. *Long Range Plan.* **2010**, *43*, 172–194. [CrossRef]
48. Weill, P.; Woerner, S.L. Optimizing your digital business model. *MIT Sloan Manag. Rev.* **2013**, *54*, 71–78. [CrossRef]
49. El Sawy, O.; Kræmmergaard, P.; Amsinck, H.; Vinther, A. How LEGO built the foundations and enterprise capabilities for digital leadership. *MIS Q. Exec.* **2016**, *15*, 174–201.
50. Iansiti, M.; Lakhani, K. The truth about blockchain. *Harv. Bus. Rev.* **2017**, *95*, 118–127.
51. McGrath, R. Business models: A discovery-driven approach. *Long Range Plan.* **2010**, *43*, 247–261. [CrossRef]
52. Demil, B.; Lecocq, X. Business model evolution: In search of dynamic consistency. *Long Range Plan.* **2010**, *43*, 227–246. [CrossRef]
53. Kim, S.; Min, S. Business model innovation performance: When does adding a new business model benefit an incumbent? *Strat. Entrepren. J.* **2015**, *9*, 34–57. [CrossRef]
54. Markides, C. Disruptive innovation: In need of better theory. *J. Prod. Innovate. Manag.* **2006**, *23*, 19–25. [CrossRef]
55. Nylén, D.; Holmström, J. Digital innovation strategy: A framework for diagnosing and improving digital product and service innovation. *Bus. Horiz.* **2015**, *58*, 57–67. [CrossRef]
56. Velu, C.; Stiles, P. Managing decision-making and cannibalization for parallel business models. *Long Range Plan.* **2013**, *46*, 443–458. [CrossRef]
57. Weill, P.; Woerner, S. Thriving in an increasingly digital ecosystem. *MIT Sloan Manag. Rev.* **2015**, *56*, 27–35.
58. Teece, D. Business models and dynamic capabilities. *Long Range Plan.* **2018**, *51*, 40–49. [CrossRef]
59. Velu, C. A systems perspective on business model evolution: The case of an agricultural information service provider in India. *Long Range Plan.* **2017**, *50*, 603–620. [CrossRef]
60. Dunford, R.; Palmer, I.; Benveniste, J. Business model replication for early and rapid internationalization: The ING direct experience. *Long Range Plan.* **2010**, *43*, 655–674. [CrossRef]
61. Calia, R.; Guerrini, F.; Moura, G. Innovation networks: From technological development to business model reconfiguration. *Technovation* **2007**, *27*, 426–432. [CrossRef]
62. Gauss, L.; Lacerda, D.; Sellitto, M. Module-based machinery design: A method to support the design of modular machine families for reconfigurable manufacturing systems. *Int. J. Adv. Manuf. Tech.* **2019**, *102*, 3911–3936. [CrossRef]
63. Borchardt, M.; Poltosi, L.; Sellitto, M.; Pereira, G. Adopting ecodesign practices: A case study of a midsized automotive supplier. *Environ. Qual. Manag.* **2009**, *19*, 7–22. [CrossRef]
64. Ritter, T.; Lettl, C. The wider implications of business-model research. *Long Range Plan.* **2018**, *51*, 1–8. [CrossRef]
65. Sosna, M.; Trevinyo-Rodríguez, R.; Velamuri, S. Business model innovation through trial-and-error learning: The Naturhouse case. *Long Range Plan.* **2010**, *43*, 383–407. [CrossRef]
66. Smith, W.; Binns, A.; Tushman, M. Complex business models: Managing strategic paradoxes simultaneously. *Long Range Plan.* **2010**, *43*, 448–461. [CrossRef]
67. Martinez, V.; Neely, A.; Velu, C.; Leinster-Evans, S.; Bisessar, D. Exploring the journey to services. *Int. J. Prod. Econ.* **2017**, *192*, 66–80. [CrossRef]
68. Cavalcante, S.; Kesting, P.; Ulhøi, J. Business model dynamics and innovation: (Re)establishing the missing linkages. *Manag. Decis.* **2011**, *49*, 1327–1342. [CrossRef]

69. Doz, Y.; Kosonen, M. Embedding strategic agility. *Long Range Plan.* **2010**, *43*, 370–382. [CrossRef]
70. Laudien, S.; Daxböck, B. The influence of the industrial Internet of things on business model design: A qualitative-empirical analysis. *Int. J. Innovat. Manag.* **2016**, *20*, 1640014. [CrossRef]
71. Karimi, J.; Walter, Z. The role of dynamic capabilities in responding to digital disruption: A factor-based study of the newspaper industry. *J. Manag. Inf. Syst.* **2015**, *32*, 39–81. [CrossRef]
72. Foss, N.; Saebi, T. Fifteen years of research on business model innovation: How far have we come, and where should we go? *J. Manag.* **2017**, *43*, 200–227. [CrossRef]
73. Foss, N.; Saebi, T. Business models and business model innovation: Between wicked and paradigmatic problems. *Long Range Plan.* **2018**, *51*, 9–21. [CrossRef]
74. Gavetti, G.; Levinthal, D. Looking forward and looking backward: Cognitive and experiential search. *Adm. Sci. Q.* **2000**, *45*, 113–137. [CrossRef]
75. Ghobakhloo, M.; Iranmanesh, M.; Grybauskas, A.; Vilkas, M.; Petraitė, M. Industry 4.0, innovation, and sustainable development: A systematic review and a roadmap to sustainable innovation. *Bus. Strateg. Environ.* **2021**, *30*, 4237–4257. [CrossRef]
76. Klingenberg, C.; Borges, M.; Antunes, J., Jr. Industry 4.0 as a data-driven paradigm: A systematic literature review on technologies. *J. Manuf. Technol. Manag.* **2019**, *32*, 570–592. [CrossRef]
77. Kadel, N.; Silva, J.; Ghobakhloo, M.; Ritter, A.; Silva, D.; Sellitto, M. Industry 4.0 technologies and sustainability challenges: A case study in the coal mining industry. *Int. J. Prod. Lifecycle Manag.* **2022**, *14*, 186–205.
78. Abdulnour, S.; Baril, C.; Abdulnour, G.; Gamache, S. Implementation of Industry 4.0 Principles and Tools: Simulation and Case Study in a Manufacturing SME. *Sustainability* **2022**, *14*, 6336. [CrossRef]
79. Ng, T.; Lau, S.; Ghobakhloo, M.; Fathi, M.; Liang, M. The application of industry 4.0 technological constituents for sustainable manufacturing: A content-centric review. *Sustainability* **2022**, *14*, 4327. [CrossRef]
80. Agnusdei, G.; Coluccia, B.; Elia, V.; Miglietta, P. IoT technologies for wine supply chain traceability: Potential application in the Southern Apulia Region (Italy). *Procedia Comput. Sci.* **2022**, *200*, 1125–1134. [CrossRef]
81. Javaid, M.; Haleem, A.; Singh, R.; Suman, R. Enabling flexible manufacturing system (FMS) through the applications of industry 4.0 technologies. *Internet Things Cyber-Phys. Syst.* **2022**, *2*, 49–62. [CrossRef]
82. Agnusdei, L.; Del Prete, A. Additive manufacturing for sustainability: A systematic literature review. *Sustain. Futures* **2022**, 100098. [CrossRef]
83. Baierle, I.; Sellitto, M.; Frozza, R.; Schaefer, J.; Habekost, A. An artificial intelligence and knowledge-based system to support the decision-making process in sales. *S. Afr. J. Ind. Eng.* **2019**, *30*, 17–25. [CrossRef]
84. Müller, A.; Nunes, M.; Maldaner, V.; Coradi, P.; Moraes, R.; Martens, S.; Leal, A.; Pereira, V.; Marin, C. Rice drying, storage, and processing: Effects of post-harvest operations on grain quality. *Rice Sci.* **2022**, *29*, 16–30. [CrossRef]
85. Jagannathan, B.; Vijayakumar, P. The need for prevention-based food safety programs for fresh produce. *Food Prot. Trends* **2019**, *39*, 572–579.
86. Faibil, D.; Agyemang, M.; Amponsah, O.; Gupta, H.; Kusi-Sarpong, S. Assessing drivers of post-harvest losses: Tangible and intangible resources' perspective. *Environ. Dev. Sustain.* **2021**, *23*, 15785–15829. [CrossRef]
87. Stathers, T.; Holcroft, D.; Kitinoja, L.; Mvumi, B.M.; English, A.; Omotilewa, O.; Kocher, M.; Auli, J.; Torero, M. A scoping review of interventions for crop postharvest loss reduction in sub-Saharan Africa and South Asia. *Nat. Sustain.* **2020**, *3*, 821–835. [CrossRef]
88. Benyam, A.A.; Soma, T.; Fraser, E. Digital agricultural technologies for food loss and waste prevention and reduction: Global trends, adoption opportunities and barriers. *J. Clean. Prod.* **2021**, *323*, 129099. [CrossRef]
89. Atungulu, G.G.; Thote, S.; Wilson, S. Dry matter loss for hybrid rough rice stored under reduced-oxygen conditions. *Cereal Chem.* **2017**, *94*, 497–501. [CrossRef]
90. Suleiman, R.; Bern, C.; Brumm, T.; Rosentrater, K. Impact of moisture content and maize weevils on maize quality during hermetic and non-hermetic storage. *J. Stored Prod. Res.* **2018**, *78*, 1–10. [CrossRef]
91. Meneghetti, V.; Biduski, B.; Tibola, C.; Junior, A.; Miranda, M.; Lima, M.; Guarienti, E.; Gutkoski, L. Evaluation of losses and quality maintenance of wheat during storage in a commercial unit in Brazil. *J. Sci. Food Agric.* **2022**, *102*, 1569–1575. [CrossRef]
92. Lutz, É.; Coradi, P.; Jaques, L.; Carneiro, L.; Teodoro, L.; Teodoro, P.; Meneghetti, V.; Souza, G. Real-time equilibrium moisture content monitoring to predict grain quality of corn stored in silo and raffia bags. *J. Food Process Eng.* **2022**, *45*, e14076. [CrossRef]
93. Anjos, F.; Rocha, L.; Silva, D.; Pacheco, R. Impacts of the Application of Virtual and Augmented Reality on Teaching-Learning Processes in Engineering Courses: A Systematic Literature Review About Learning and Satisfaction on Students. *Int. J. Virtual Pers. Learn. Environ.* **2022**, *12*, 1–19. [CrossRef]
94. Sonwani, E.; Bansal, U.; Alroobaea, R.; Baqasah, A.; Hedabou, M. An Artificial Intelligence Approach Toward Food Spoilage Detection and Analysis. *Front. Public Health* **2022**, *9*, 2254. [CrossRef]

95. Silva, F.; Baierle, I.; Correa, R.; Sellitto, M.; Peres, F.; Kipper, L. Open Innovation in Agribusiness: Barriers and Challenges in the Transition to Agriculture 4.0. *Sustainability* **2023**, *15*, 8562. [CrossRef]
96. CONAB. Brazilian Grain Harvest. 2023. Available online: https://www.conab.gov.br/info-agro/safras/graos (accessed on 7 January 2023). (In Portuguese)

Disclaimer/Publisher's Note: The statements, opinions and data contained in all publications are solely those of the individual author(s) and contributor(s) and not of MDPI and/or the editor(s). MDPI and/or the editor(s) disclaim responsibility for any injury to people or property resulting from any ideas, methods, instructions or products referred to in the content.

agriculture

MDPI

Article

Spatiotemporal Estimation of Reference Evapotranspiration for Agricultural Applications in Punjab, Pakistan

Hadeed Ashraf [1], Saliha Qamar [2], Nadia Riaz [1], Redmond R. Shamshiri [3,*], Muhammad Sultan [1,*], Bareerah Khalid [4], Sobhy M. Ibrahim [5], Muhammad Imran [6] and Muhammad Usman Khan [7]

1 Department of Agricultural Engineering, Bahauddin Zakariya University, Multan 60800, Pakistan; hadeedashraf15@gmail.com (H.A.); nadiariaz018@gmail.com (N.R.)
2 Department of Energy Systems Engineering, Faculty of Agricultural Engineering and Technology, PMAS Arid Agricultural University, Rawalpindi 46000, Pakistan; salihaqamar31@gmail.com
3 Department of Engineering for Crop Production, Leibniz Institute for Agricultural Engineering and Bioeconomy, 14469 Potsdam, Germany
4 Agricultural Systems and Engineering, Department of Food, Agriculture and Bioresources, School of Environment, Resources and Development, Asian Institute of Technology, Klong Luang, Pathum Thani 12120, Thailand; bareerahkhalid@gmail.com
5 Department of Biochemistry, College of Science, King Saud University, P.O. Box 2455, Riyadh 11451, Saudi Arabia; syakout@ksu.edu.sa
6 Department of Mechanical, Biomedical and Design Engineering, College of Engineering and Physical Sciences, Aston University, Birmingham B4 7ET, UK; m.imran12@aston.ac.uk
7 Department of Energy Systems Engineering, Faculty of Agricultural Engineering and Technology, University of Agriculture, Faisalabad 38040, Pakistan; usman.khan@uaf.edu.pk
* Correspondence: rshamshiri@atb-potsdam.de (R.R.S.); muhammadsultan@bzu.edu.pk (M.S.)

Abstract: Estimation of reference evapotranspiration (ET_o) is a key element in water resources management and crop water requirement which, in turn, affects irrigation scheduling. ET_o is subject to the influence of various climatic parameters including minimum temperature (T_{min}), maximum temperature (T_{max}), relative humidity (RH), windspeed (WS), and sunshine hours (SH). Usually, the influence of the climatic parameters and a dominating climatic factor influencing ET_o is estimated on yearly basis. However, in diverse climatic regions, ET_o varies with the varying climate. Therefore, this study aims to estimate the spatiotemporal variation in the influence of the climatic parameters on ET_o in Punjab, Pakistan, for the period 1950–2021, specifically focusing on decennial, annual, and monthly patterns. The study area was divided into five agroclimatic zones. The Penman–Monteith model was used to estimate ET_o. The influence was assessed using geographic weighted regression (GWR) and multiscale geographic weighted regression (MGWR) as the primary methods. As per results from MGWR, ET_o in Punjab was highly influenced by the T_{min}, T_{max}, and WS. Additionally, annual ET_o exhibited a higher value in southern Punjab in comparison to northern Punjab, with a range of 2975 mm/year in the cotton–wheat zone to 1596 mm/year in the rain-fed zone. Over the course of the past seventy years, Punjab experienced an average increasing slope of 5.18 mm/year in ET_o. T_{min} was the highest monthly dominant factor throughout the year, whereas WS and SH were recorded to be the dominant factor in the winters, specifically. All in all, accurate estimation of ET_o, which serves as an essential component for crop water requirement, could potentially help improve the irrigation scheduling of crops in the agroclimatic zones.

Keywords: reference evapotranspiration; geographic weighted regression; multiscale geographic weighted regression; Pakistan

Citation: Ashraf, H.; Qamar, S.; Riaz, N.; Shamshiri, R.R.; Sultan, M.; Khalid, B.; Ibrahim, S.M.; Imran, M.; Khan, M.U. Spatiotemporal Estimation of Reference Evapotranspiration for Agricultural Applications in Punjab, Pakistan. *Agriculture* **2023**, *13*, 1388. https://doi.org/10.3390/agriculture13071388

Academic Editor: Margarita García-Vila

Received: 23 May 2023
Revised: 10 July 2023
Accepted: 11 July 2023
Published: 12 July 2023

Copyright: © 2023 by the authors. Licensee MDPI, Basel, Switzerland. This article is an open access article distributed under the terms and conditions of the Creative Commons Attribution (CC BY) license (https://creativecommons.org/licenses/by/4.0/).

1. Introduction

The determination of reference evapotranspiration (ET_o) holds significant importance in the realm of remote sensing applications and the regulation of agricultural water usage. In crop water management, it is customary to employ a theoretical grass reference crop possessing specific attributes as a benchmark surface for the purpose of approximating ET_0 [1]. The estimation of ET_0 holds significant value in the management of hydrological and agricultural water systems. Nonetheless, it is imperative to conduct a thorough evaluation of the quantity of water necessary for the growth of any given crop [1]. Accurately determining ET_0 across different land use regions is crucial for efficient management of water resources [2]. Methods for transferring water vapor, such as soil water balance [3,4] and remote sensing techniques [5], have been employed to quantify ET_0. In addition, several hybrid techniques have been used for forecasting ET_0 [6,7].The accurate computation of ET_0 is a complicated undertaking that entails the consideration of multiple meteorological variables. Several regions around the world have reported the spatiotemporal variation in the climatic parameters, mainly including temperature [8], which ultimately impacts ET_0. Consequently, the development of a precise empirical model that can effectively account for the complexities of the process is a challenging task, as noted in reference [9]. Additionally, climate change plays a significant role in disturbing the surrounding environment in different ways [10], including depleting water levels [11,12], a disturbed hydrological cycle which includes ET_0 [13,14] as an important component, and ultimately economic loss [15].

A variety of interpolation techniques are employed in GIS software, each distinguished by the models utilized to forecast unknown values on the basis of sample points. The inverse distance weighted (IDW) and Kriging techniques are among the most frequently employed methods. The inverse distance weighting (IDW) method is a deterministic interpolation technique that relies on nearby reference point measurements or established mathematical formulae to evaluate the texture of the resulting interpolation area [16]. Kriging is a geostatistical method that incorporates the spatial configuration of sample points by assigning weights that differ according to their respective locations. Kriging methodology not only generates an interpolated or predicted surface but also provides a level of precision or confidence in the estimates, which is in contrast to the IDW interpolation technique. The spatiotemporal estimation of ET_0 across 41 stations in Pakistan was investigated by Shah et al. [17] during the period of 2006 to 2015. The research discovered that ET_0 demonstrated variability and manifested both positive and negative trends across various seasons in multiple regions of Pakistan. Geographically weighted regression (GWR) could potentially serve as a more appropriate technique for addressing this issue with greater precision. The authors Fotheringham et al. (2017) [18] presented a method called multiscale geographically weighted regression (MGWR) to tackle the issues arising from nonstationary connections and heterogeneity. This particular model has the ability to precisely replicate the effects of different independent variables on a dependent variable with regards to spatial variability.

The MGWR method has garnered considerable interest from diverse fields owing to its efficacy in modeling the spatial associations among multiple dependent and independent variables [18]. The effectiveness of ET_0 estimation models is inherently connected to the specific meteorological conditions of a given location, as stated in reference [19]. Therefore, it is crucial to ascertain appropriate models that are contingent on the accessibility of meteorological data. The significance of ET_0 on hydrological processes and water resources is particularly noteworthy in predominantly arid regions of Pakistan, surpassing its impact in other climatic zones. The exacerbation of droughts and aridity presents a formidable challenge to the sustainable development of agriculture and water resources in Pakistan, a nation heavily reliant on agriculture. Accurate estimation of ET_0 is essential for efficient monitoring and management of droughts and aridity. The inadequacy of meteorological information has posed a difficulty in accurately determining ET_0 across various regions of Pakistan. Therefore, it is crucial to investigate suitable ET_0 models that can be suggested for ET_0 estimation in diverse areas. The weather in Pakistan varies between different seasons on a monthly level, which makes it difficult to understand the influence of climatic parameters and the dominant factor influencing ET_0 in each month. One of the major difficulties in water resources management is the accurate estimation and forecasting of ET_0. However, the most recent estimation of ET_0 and the dominant factor influencing it on a monthly level are missing, which negatively impacts water resources management ultimately impacting the end-user.

According to the studied literature, it can be safely assumed that up-to-date estimations of the reference evapotranspiration and, especially, of the monthly influence of a dominant factor (out of minimum temperature, maximum temperature, relative humidity, windspeed, and sunshine hours) on ET_0 for Punjab, Pakistan, are missing. Therefore, this study investigates the spatiotemporal variation of reference evapotranspiration across Punjab, Pakistan. Furthermore, spatial variation in the monthly impact of influencing parameters on ET_0 and the dominant factor have been estimated.

2. Materials and Methods

2.1. Study Area

The selection of Punjab, Pakistanm for the analysis of the impact of influencing parameters on ET_0 is justified by the fact that agriculture constitutes a significant proportion of the country's overall exports [20], with Punjab contributing to 60% of this portion [20], as depicted in Figure 1. Moreover, Punjab has attempted to tackle the problems of food security for many years; therefore, the assessment of the influence of climatic conditions on crop production in such an agriculture-rich region is imperative. Furthermore, accurate ET_0 estimation in this region can aid in decision making about irrigation scheduling, water allocation for different crops, and the scheduling of other agricultural activities [21]. Moreover, this study assesses the behavior of ET_0 in five different agroclimatic zones of Punjab, with 36 selected study points, i.e., districts of the province. Different approaches can be used for the classification of Punjab districts into regions, such as dividing Punjab into three regions, north, central, and south, or disintegrating rural Punjab into north and south. In ref. [22], Amjad et al. used the methodology of Pinckney [23] to classify rural areas of Punjab province into five agroclimatic or crop zones. The agroclimatic zones of the Punjab were classified according to the published report by the International Food Policy Research Institute.

The categorization of various zones is primarily determined by the primary kharif season crop, as wheat dominates the rabi season in most regions. The kharif season in Pakistan encompasses the time frame from April to September, during which monsoon rains are prevalent. In contrast, the rabi season occurs from October to March, featuring cooler temperatures and reduced precipitation. Agricultural activities during the rabi season heavily rely on irrigation methods for crop cultivation. In regions where irrigation is practiced, rice and cotton are the predominant kharif crops of utmost importance. Rice cultivation exhibits a propensity for prevalence in regions characterized by elevated water tables, dense soils, and copious precipitation, whereas cotton production is commonly observed in arid locales. Therefore, a significant differentiation can be observed between regions that are suitable for rice cultivation and those that are suitable for cotton cultivation. Based on these two crops, i.e., cotton and rice, alternating with the dominant crop, wheat, two zones, namely, the cotton–wheat zone and rice–wheat zone, were created. The cotton–wheat zone, located in the southern and southwestern regions of Punjab, exhibits a climatic pattern characterized by hot summers and cool winters. The precipitation in this region is comparatively lower than that in the rice–wheat zone, rendering it conducive for the cultivation of cotton during the kharif season and wheat during the rabi season. The rice–wheat zone, predominantly situated in the central and eastern regions of Punjab, exhibits a climate distinguished by warm summers and mild winters. The monsoon season is characterized by a substantial precipitation pattern that occurs during the kharif season, creating favorable conditions for the cultivation of rice. In contrast, the rabi season is characterized by lower levels of precipitation, making it conducive for the cultivation of wheat.

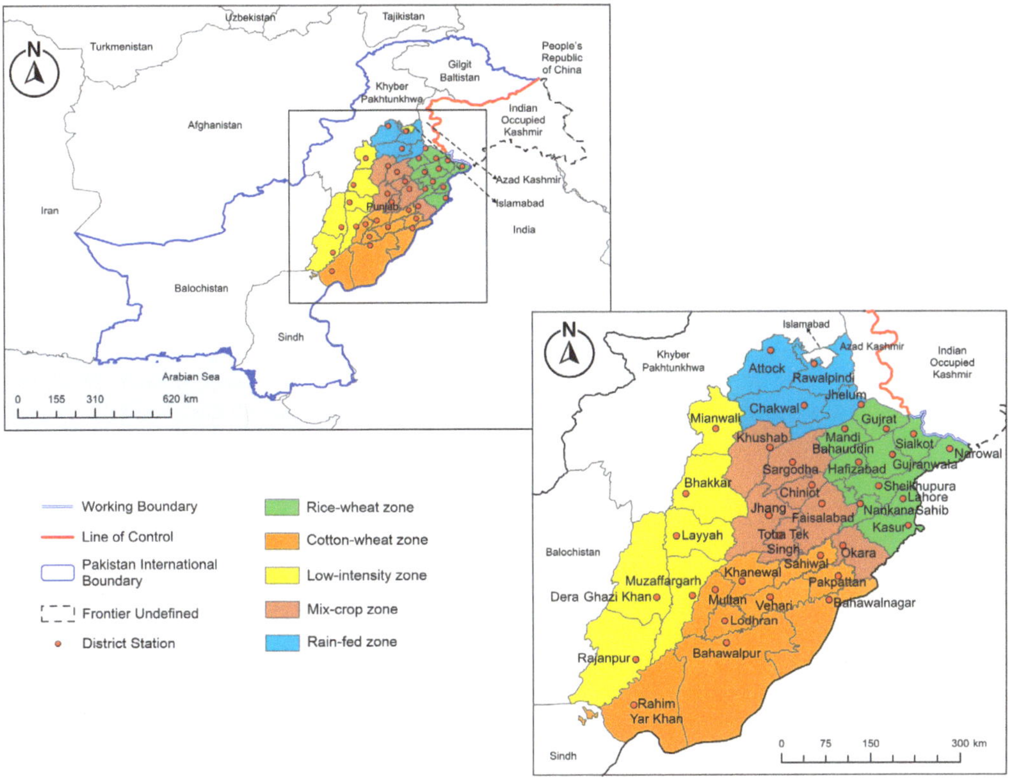

Figure 1. Illustration of the study area and location of the 36 district stations and delineation of the five agroclimatic zones in Punjab, Pakistan.

A distinct geographical region, with Faisalabad at its center, exhibits favorable conditions for cultivating multiple kharif crops, without any single crop exerting dominance. This region has been categorized as the mixed-crop zone. The region exhibits a confluence of climatic conditions observed in different zones.

The rain-fed zone or the barani region is deemed distinctive owing to its dependence on precipitation for agricultural purposes. The rain-fed zone situated in the northern and northwestern regions of Punjab exhibits a climate characterized by higher aridity. The precipitation levels in this particular region are constrained.

Furthermore, the region adjacent to the left bank of the Indus River in Punjab, characterized by limited irrigation infrastructure and lower levels of agricultural activity, is recognized as a distinct area referred to as the low-intensity zone. The climate in this zone is characterized by aridity, resulting in minimal precipitation. It is imperative to acknowledge that the aforementioned descriptions offer a comprehensive outline of the climatic conditions prevalent in the agroclimatic zones of Punjab. However, it is crucial to recognize that there may be deviations within each zone due to localized factors, including but not limited to elevation, proximity to water bodies, and microclimatic circumstances. The delineation of the five agroclimatic zones, namely, the low-intensity zone, the rice–wheat zone, the mixed-crop zone, the cotton–wheat zone, and the rain-fed zone, is attributed to the variations in growing climatic conditions and agricultural practices observed from north to south. Table 1 presents the average values of the five climatic parameters influencing ET_o for the period 1950–2021 for the five agroclimatic zones.

Table 1. Summary of the climatic parameters (1950–2021) for the five agroclimatic zones in Punjab, Pakistan.

Agroclimatic Zone	Maximum Temperature T_{max} (°C)	Minimum Temperature T_{min} (°C)	Relative Humidity RH (%)	Sunshine Hours SH (h)	Windspeed WS (m/s)	Reference Evapotranspiration ET_o (mm/day)
Cotton–wheat zone	32.97	18.44	40.4	11.99	2.85	6.49
Rice–wheat zone	30.47	17.35	67.49	11.99	3.47	5.47
Low-intensity zone	32.35	17.86	44.99	11.99	3.29	6.52
Mixed-crop zone	31.8	18.02	57.37	11.99	3.62	6.15
Rain-fed zone	28.89	15.45	74.17	12	4.04	5.03

The low-intensity zone includes the districts Bhakkar, Dera Ghazi Khan, Layyah, Mianwali, Muzaffargarh, and Rajanpur, with 32.3 °C, 17.8 °C, 44.9%, 11.9 h, 3.2 m/s, and 6.5 mm/day of T_{max}, T_{min}, RH, SH, WS, and ET_o, respectively, with an average for the period of 1950–2021. The rice–wheat zone includes districts Gujranwala, Gujrat, Hafizabad, Kasur, Lahore, Mandi Bahauddin, Nankana Sahib, Narowal, Sheikhupura, and Sialkot, with 30.4 °C, 17.3 °C, 67.4%, 11.9 h, 3.4 m/s, and 5.4 mm/day of T_{max}, T_{min}, RH, SH, WS, and ET_o, respectively. The mixed-crop zone includes the districts Chiniot, Faisalabad, Jhang, Khushab, Okara, Sargodha, and Toba Tek Singh, with 31.8 °C, 18.0 °C, 57.3%, 11.9 h, 3.6 m/s, and 6.1 mm/day of T_{max}, T_{min}, RH, SH, WS, and ET_o, respectively. The cotton–wheat zone includes the districts Bahawalnagar, Bahawalpur, Khanewal, Lodhran, Multan, Pakpattan, Rahim Yar Khan, Sahiwal, and Vehari, with 32.9 °C, 18.4 °C, 40.4%, 11.9 h, 2.8 m/s, and 6.4 mm/day of T_{max}, T_{min}, RH, SH, WS, and ET_o, respectively. The rain-fed zone includes the districts Attock, Chakwal, Jhelum, and Rawalpindi, with 28.8 °C, 15.4 °C, 74.1%, 12.0 h, 4.0 m/s, and 5.0 mm/day of T_{max}, T_{min}, RH, SH, WS, and ET_o, respectively.

2.2. Data Collection

For the estimation of ET_o, satellite data were used. The data of 5 different climatic parameters were obtained from 2 online sources (summarized details of the data sources are presented in Table 2). For minimum temperature (T_{min}) and maximum temperature (T_{max}), the satellite data were obtained from the Climatic Research Unit gridded Time Series (CRU–TS). However, for relative humidity (RH) and windspeed (WS), the satellite data were obtained from National Centers for Environmental Prediction (NCEP) and the National Center for Atmospheric Research (NCAR) (NCEP–NCAR) Reanalysis 1 for Punjab. For further analyses, data points were extracted for the 36 districts (categorized into 5 agroclimatic zones as per [22,23]) in the Punjab from 1950 to 2021. Figure 2 illustrates the workflow of the presented study. The present study assessed the decadal patterns of ET_o spanning from 1950 to 2021. In order to analyze the trend of reference evapotranspiration, the available data were divided into seven distinct decades spanning 1950 to 2021. Additionally, the data were further categorized into five distinct zones, namely, the cotton–wheat zone, the low-intensity zone, the rice–wheat zone, the mixed-crop zone, and the rain-fed zone, all of which are located within the Punjab province of Pakistan.

Table 2. Summary of the data sources used in this study.

Dataset Name	Variables	Temporal Extent	Spatial Resolution	Ref.
CRU–TS	T_{min}, T_{max}	1901–2021	All land areas (excluding Antarctica) at 0.5° resolution	[24]
NCEP–NCAR Reanalysis 1 (Data provided by the NOAA PSL, Boulder, Colorado, USA, from their website at https://psl.noaa.gov (accessed on 28 January 2023))	RH and WS	1948–2021	All land areas (excluding Antarctica) at 2.5° resolution	[25]
Sunshine hours (calculated)	SH	–	–	–

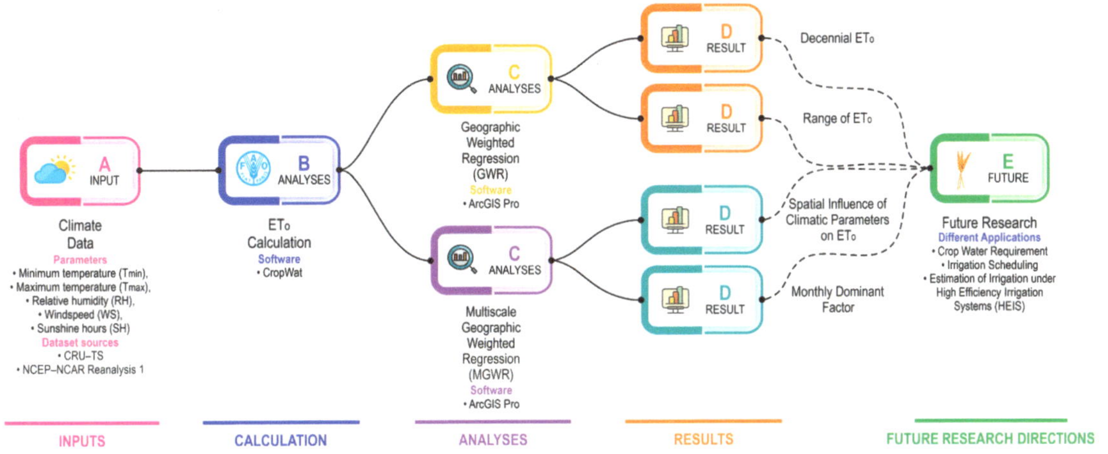

Figure 2. Workflow of the presented study for the estimation of ET_o.

2.2.1. ET_o Estimation

In this study, the standard Penman–Monteith (PM) model was utilized to calculate the reference evapotranspiration in mm/day. It was developed in the mid-1960s by Allen and Tideman, and it combines both energy and mass balance principles [26]. The PM ET_o model is given as Equation (1). Details of the parameters are available in the literature [26].

$$ET_o = \frac{0.408\Delta(R_n - G) + \gamma\frac{900}{T+273}u_2(e_s - e_a)}{\Delta + \gamma(1 + 0.34u_2)} \quad (1)$$

where ET_o is the reference evapotranspiration in mm day^{-1}; Δ is the slope of the vapor pressure curve in kPa $°C^{-1}$; R_n is the net radiation at the crop surface in MJ m^2 day^{-1}; G is the soil heat flux density in $MJm^2.day^{-1}$, which is considered as 0 [1]; γ is the psychrometric constant in kPa $°C^{-1}$; T is the mean air temperature, which is determined from minimum and maximum temperatures in °C; u_2 is the wind speed at a height of 2 m height in ms^{-1}; e_s is the saturation vapor pressure in kPa; and e_a is the actual vapor pressure in kPa.

2.2.2. Trend Analysis

The present investigation employed a prevalent technique for trend analysis, namely, linear regression, due to its ability to detect and measure patterns in data across temporal dimensions. Through the application of a linear model to the dataset, researchers are able to approximate the pace and orientation of variation in a given variable that is of significance. The aforementioned approach proves to be especially advantageous in the examination of extended patterns in climatic parameters, such as ET_o. In this study, the Mann–Kendall (MK) test was employed to identify fluctuations in the timeseries data [27]. The Sen's slope technique has been utilized in combination with the MK test to examine patterns in ET_o data. The technique, formulated by Sen in 1968, offers a computation of the pace of alteration in a given variable across a period [28]. This study employs linear trend analysis to examine the Punjab region, utilizing the linear regression method for the period spanning 1950 to 2021.

2.2.3. Data Quality and Integrity Assessment

In the current scenario, data were analyzed through basic statistical methods such as minimum and maximum values, range, mean, standard deviation, skewness, and kurtosis. Table 3 presents an overview of the descriptive statistics of the climatic parameters for the estimation of ET_o for Punjab, Pakistan, for the period 1950–2021. The analysis conducted and presented in Table 3 demonstrates that the studied variables exhibited varying skewness values, encompassing both positive and negative values. This suggests the presence of asymmetry in the distributions of these variables. Moreover, the kurtosis values indicate that the climatic variables demonstrate either a distribution with lower peak and heavier tails (platykurtic) or a distribution with higher peak and lighter tails (leptokurtic).

Table 3. Descriptive statistics of the climatic parameters for the estimation of ET_o for Punjab, Pakistan, for the period of 1950–2021.

Variable	Mean	St. Dev.	Minimum	Maximum	Range	Skewness	Kurtosis
T_{min} (°C)	17.631	0.516	16.426	18.774	2.348	−0.14	−0.47
T_{max} (°C)	31.496	0.499	30.168	32.412	2.243	−0.25	−0.46
RH (%)	55.747	2.529	49.407	61.944	12.537	−0.33	0.02
WS (m/s)	3.3819	0.5137	2.3519	4.3133	1.9615	−0.32	−1.19
ET_o (mm/year)	2339.7	149.1	2094.1	2691.3	597.1	0.63	−0.34

2.2.4. Spatial Distribution of ET_o

The utilization of spatial visualization techniques for ET_o distribution analysis facilitates the identification of regions with elevated water demand and potential depletion or overuse of water resources [1]. Furthermore, it has the potential to identify regions with favorable temperature [29] and humidity conditions for cultivating various crops, thereby facilitating crop selection and land-use management. The utilization of Inverse Distance Weighting (IDW) for the purpose of interpolating ET_o data and creating spatial representations of ET_o across various regions of Pakistan has been extensively documented in numerous studies [30]. As a result, the ET_o data underwent spatial interpolation using the IDW interpolation technique in the ArcGIS software, as opposed to more intricate methods such as Kriging. This process yielded evapotranspiration maps for various climatic periods spanning from 1950 to 2021, which are presented in Figure 3 for the purpose of change analysis.

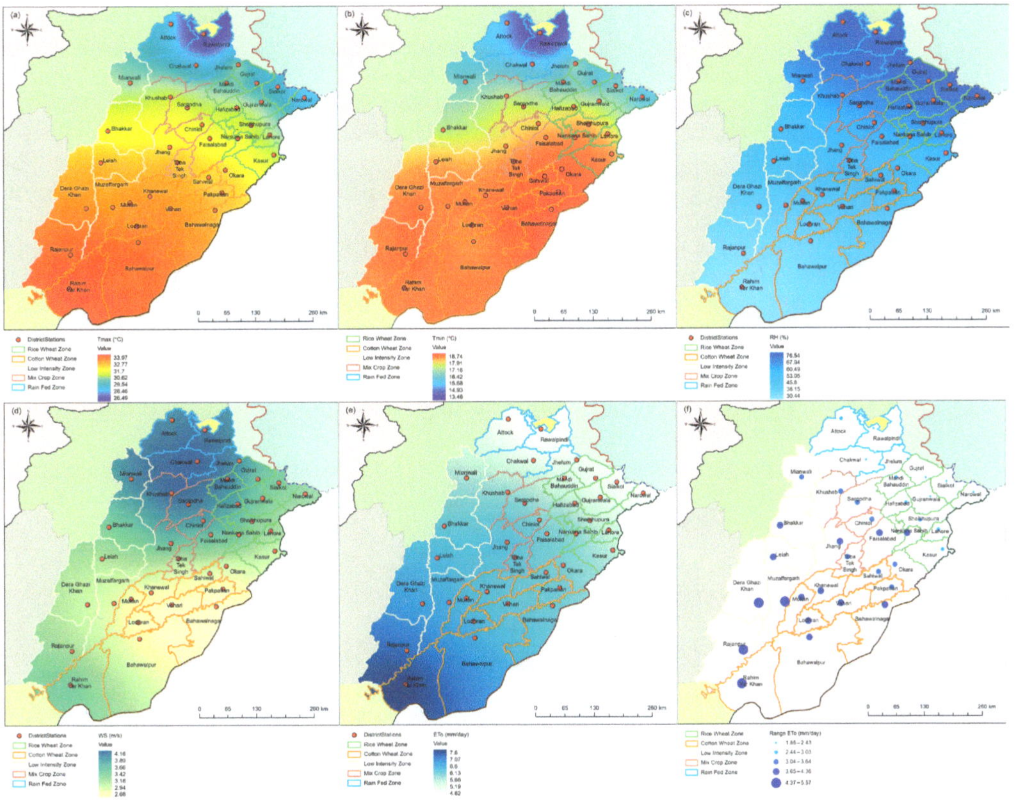

Figure 3. Spatial distribution of average annual (1950–2021) (**a**) T_{max} (°C), (**b**) T_{min} (°C), (**c**) RH (%), (**d**) WS (m/s), (**e**) ET_o (mm/day), and (**f**) range of ET_o (mm/day). Note: range represents the difference between the highest and lowest ET_o (mm/day) at a specific station throughout the study period (1950–2021).

2.2.5. Spatial Relationship between Climatic Variables and ET_o

The present study employed Pearson's correlation coefficient to establish the correlation between the influencing parameters and the reference evapotranspiration. This coefficient was preferred over other methods of evaluating linear relationships due to its distinct advantages. The utilization of Pearson's correlation coefficient (R) is widespread in the assessment of models in studies related to ET_o, as evidenced by several studies [31]. The requisite quantity of data for establishing a correlation is comparatively lower than in other techniques of data assessment, typically ranging from three to seven variable pairs [32]. Furthermore, it has the capability to conduct an analysis on a considerable number of variables without causing substantial overstatement of outcomes [32]. A sensitivity analysis was carried out to assess the impact of ambient weather variables on the computed ET_o. Bois et al. [33] have discussed several techniques for performing sensitivity analyses, which have been documented in the literature. In this study, a monthly analysis was performed to establish a correlation between the influencing parameters and ET_o. The mathematical expression denoting the Pearson correlation is represented by Equation (2):

$$r = \frac{\sum_{i=1}^{N}\left(x - \bar{x}\right)\left(y - \bar{y}\right)}{\sqrt{\sum_{i=1}^{N}\left(x - \bar{x}\right)^2 \sum_{i=1}^{N}\left(y - \bar{y}\right)^2}} \quad (2)$$

In this study, the observed climatic variable was denoted as x and the estimated ET_o was represented as y; $\bar{x}$ and $\bar{y}$ are their subsequent mean values. Similarly, the coefficients were used to determine the most influential climatic variables on ET_o in the study areas.

3. Results

3.1. Trend of ET_o

Figure 4 depicts the results of the Mann–Kendall trend analysis applied to annual ET_o in the five agroclimatic zones of Punjab for analyzing the heterogeneity of the timeseries ET_o. In Figure 4a, a heterogeneity in the year 2000 was observed in ET_o for Punjab during the period of 1950 to 2021. The sudden fluctuation in the trend of ET_o could be attributed to the changing climate, which ultimately impacts the climatic parameters which influence ET_o. Figure 4b shows cumulative trend of ET_o for the period of 1950–2021, which is increasing through the time period. A lowest ET_o of 2094.12 mm/year was observed in the year 1963 for the Punjab, whereas the highest ET_o, i.e., 2691.26 mm/year, was observed in 2021, which clearly illustrates the impact of rising temperatures. Similarly, trends of the ET_o for the five agroclimatic zones are presented in Figure 4; all of the zones show rapidly increasing trends of ET_o except in the low-intensity zone, which could be attributed to lower precipitation in the area. The agroclimatic zones comprising the respective districts are also presented in Figure 4.

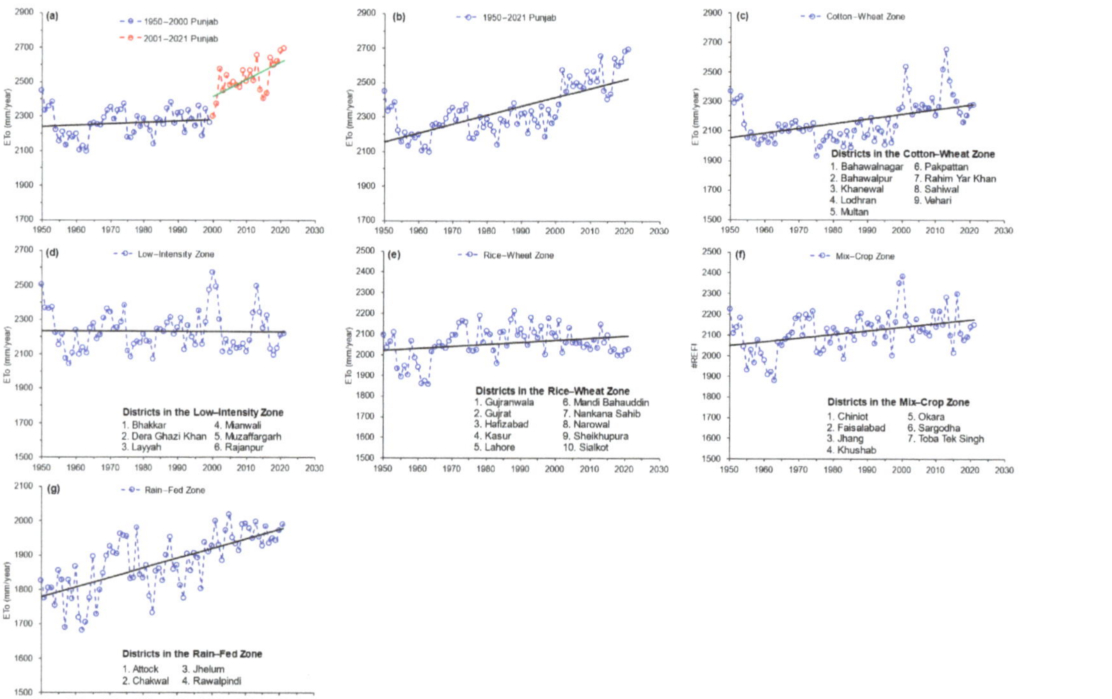

Figure 4. Illustration of (**a**) homogeneity test of ET_o in Punjab, (**b**) overall trend of ET_o in Punjab, (**c**) ET_o trend in the cotton–wheat zone, (**d**) ET_o trend in the low-intensity zone, (**e**) ET_o trend in the rice–wheat zone, (**f**) ET_o trend in the mixed-crop zone, and (**g**) ET_o trend in the rain-fed zone using the Mann–Kendall trend test for the period 1950–2021. Note: names of the districts included in respective agroclimatic zones are presented in each graph.

3.2. Decennial ET_0

The data presented in Figure 5 indicate a decadal pattern that demonstrates a rise in ET_0 throughout the duration of the study. The cartographic representation of the 1950s (depicted in Figure 5a) revealed a diminished ET_0 in the rain-fed and rice–wheat zones (i.e., 1328.03 mm/year), whereas the southwestern region exhibited relatively elevated values (i.e., 2370.56 mm/year). Figure 5b illustrates the fluctuation of difference in ET_0 during the second decade (1960–1969) throughout Punjab with respect to the ET_0 base year 1950s. The study noted that the highest difference in the ET_0 values across the northern and southern regions were found in the cotton–wheat zone, with Rahim Yar Khan particularly exhibiting a 10.19 mm/year elevated ET_0 compared to the base year. The mixed crop zone, also known as the central region, typically exhibited a moderate-to-low level of ET_0. In majority of the cotton–wheat zone, the difference ET_0 was negative (up to −91.31 mm/year) compared to the base year (illustrated in Figure 5b). The results obtained during the time frame of 1970–1979 indicate a rise in the difference of ET_0 in the rice–wheat zone and rain-fed zone (i.e., up to 116.26 mm/year) when compared to the base year. This is illustrated in Figure 5c. During the fourth decade spanning 1980 to 1989, there was an observed increase in the difference of ET_0 in comparison to the base year 1950s, as illustrated in Figure 5d. The highest positive difference in observed ET_0 was 84.77 mm/year in the rice–wheat zone, rain-fed zone, and somewhat in the mixed-crop zone compared to the base year. As depicted in Figure 5e, the rate of difference in ET_0 exhibited an increasing trend from the northern region to the southern region (i.e., from 115.74 mm/year to −79.35 mm/year, respectively), or from the rain-fed zone to the cotton–wheat and low-intensity zone during the period 1990–1999. Drastically elevated trends in the difference of ET_0 were observed in the year 2000–2009 compared to the base year 1950s, as shown in Figure 5f. Typically, rain-fed zone was the most affected zone (i.e., 234.73 mm/year), followed by the rice–wheat zone (i.e., 200.8 mm/year), compared to the base year 1950s. According to Figure 5g, the highest difference in ET_0 was observed in the northern region (i.e., 734.57 mm/year) compared to the southern region (320.29 mm/year) in the year 2010–2021 relative to the base year 1950s. Finally, Figure 5h illustrates the ET_0 levels in the period of 2010–2021. i.e., 2690.85 mm/year in the southern region of the area, whereas 2042.86 mm/year was recorded in the northern regions of the area. Over the course of the past decade, the cotton–wheat and low-intensity zones have exhibited the highest ET_0 values, whereas the mixed-crop zone has demonstrated a moderate rate (i.e., 2400 mm/year to 2500 mm/year) with slight anomaly in the Jhang district. Conversely, the rice–wheat zone in the northeastern region and the rain-fed zone in the northwestern region have displayed relatively low ET_0 rates; however, they were the most affected zones when compared to the base year 1950s due to climate change. The rate of reference evapotranspiration is directly proportional to temperature, resulting in a significant fluctuation in ET_0 rate during the month of September as supported by the findings of previous research [34]. As per the data presented in Figure 5, the highest evapotranspiration rate recorded was 2690.8 mm/year in the year 2010–2021. Prior research has identified comparable patterns that align with the results of the current investigation. Table 4 presents the decennial pattern of ET_0 in relation to the five agroclimatic zones in Punjab, Pakistan, spanning from 1950 to 2021.

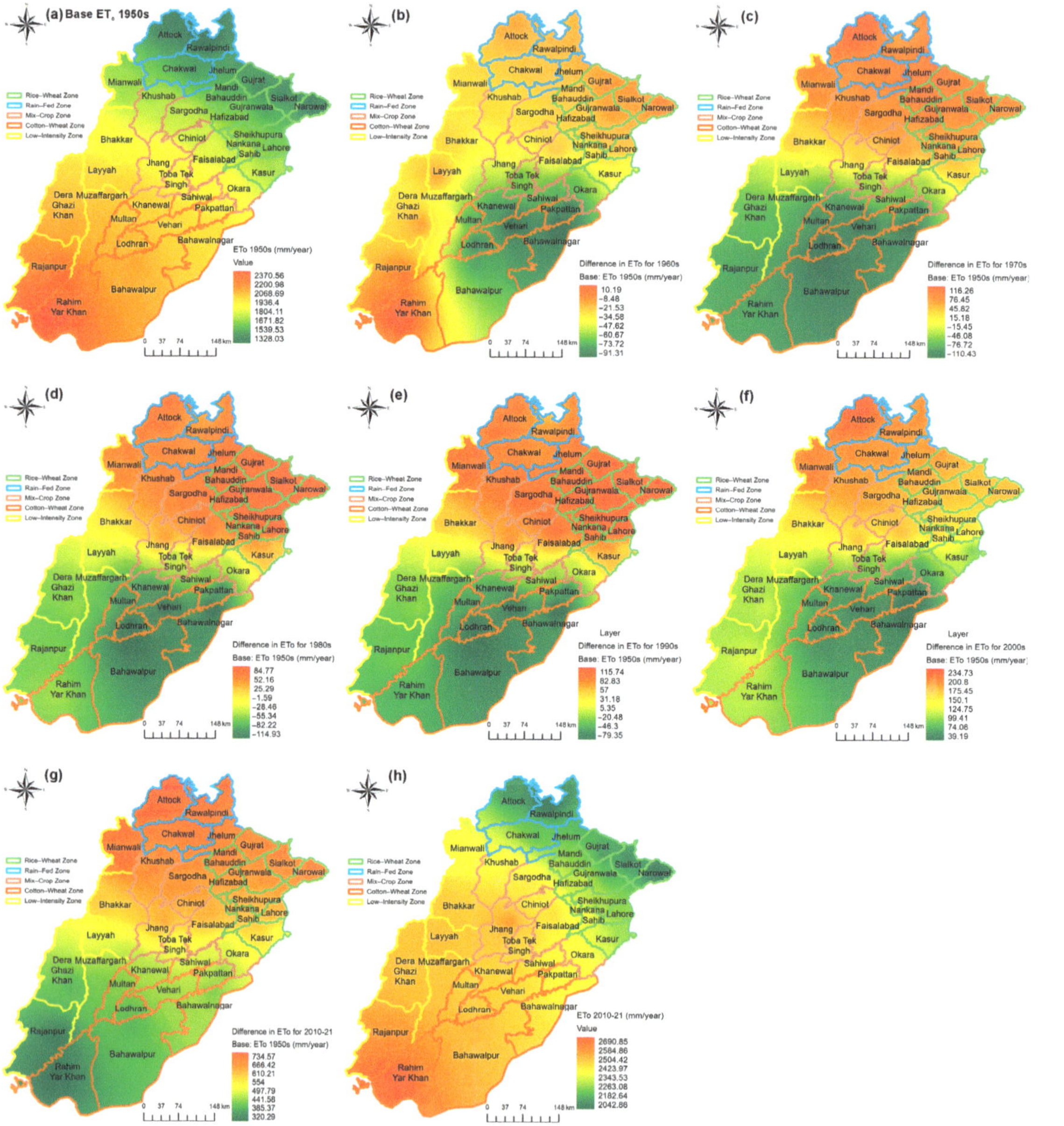

Figure 5. Decadal patterns of ET_o spanning from 1950 to 2021 for Punjab, Pakistan: (**a**) base ET_o for 1950s; (**b**) difference in ET_o for 1960s; (**c**) difference in ET_o for 1970s; (**d**) difference in ET_o for 1980s, (**e**) difference in ET_o for 1990s; (**f**) difference in ET_o for 2000s; (**g**) difference in ET_o for 2010–21; and (**h**) ET_o for 2010–21. Note: the difference in ET_o is relative to the ET_o of base year 1950s.

Table 4. Decennial trend of ET_o against each of the five agroclimatic zones from 1950 to 2021 in Punjab, Pakistan.

Agroclimatic Zone	ET_o (mm/year) per Decade							ET_o Trend (mm/year)
	1950–1950	1960–1969	1970–1979	1980–1989	1990–1999	2000–2009	2010–2021	1950–2021
Low-intensity zone	2243	2230	2221	2220	2287	2227	2237	
Cotton–wheat zone	2157	2096	2068	2072	2123	2293	2323	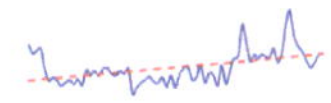
Rice–wheat zone	1999	1990	2100	2097	2105	2070	2046	
Rain-fed zone	1801	1805	1904	1850	1873	1956	1965	
Mixed-crop zone	2064	2044	2117	2114	2173	2164	2149	

3.3. *Sensitivity Analysis*

3.3.1. Geographic Weighted Regression (GWR)

In order to examine the relationship between the distribution of ET_o and its influencing factors, a correlation analysis was conducted using Geographically Weighted Regression (GWR) at 36 district points. Subsequently, maps were generated using Inverse Distance Weighting (IDW) gridding. Figure 6 presents the resultant maps that illustrate the correlation between ET_o and various climatic parameters for the time period 1950–2021. A correlation ranging from 0.29 to 0.89 with standard deviation of 0.1 was observed between ET_o and T_{min} reported in Figure 6a. According to Figure 6a, low correlation (i.e., 0.29) was observed in the southern region of the area covering most of the cotton–wheat zone and the low-intensity zone. This could be attributed to the gradual change in elevation from north to south, which is known to impact ET_o. In Figure 6b, a good correlation ranging 0.78 to 0.91 with standard deviation of 0.03 was observed between ET_o and T_{max}. According to Figure 6b, the correlation trend follows a similar direction as that of ET_o and T_{min}; however, the minimum value of the correlation is raised to 0.78, covering most of the cotton–wheat zone, the low-intensity zone, and the mixed-crop zone. This could be attributed to the changing climate and rising temperatures across the country. A correlation ranging from 0.67 to 0.89 with standard deviation of 0.05 was observed between ET_o and RH reported in Figure 6c. A minimum correlation of 0.67 was observed between ET_o and RH in the central region of the mixed-crop zone, whereas a correlation of 0.7 to 0.8 was observed between ET_o and RH in most of the cotton–wheat zone, the low-intensity zone, and the mixed-crop zone. In the rain-fed and rice–wheat zones, a correlation of 0.8 to 0.89 was observed between ET_o and RH. This trend could be attributed to the frequent rainfall events in the northern region of the area leading to increased RH compared to the infrequent rainfall events in the southern region leading to drier conditions and lesser RH. According to Figure 6d, a correlation ranging from 0.29 to 0.63, with standard deviation 0.06, was observed between ET_o and WS. Minimum correlation, i.e., 0.29, was observed in the southern region of the cotton–wheat zone and the low-intensity zone due to lesser windspeed in these southern regions of the area. A moderate correlation of 0.6 was observed between ET_o and WS in the central Punjab region, comprising some districts of the cotton–wheat, low-intensity, and rice–wheat zones. This could be attributed to typically higher windspeeds in these regions due to their geographical location and the location of the Himalayan ranges. In the northern region of the area, namely, the rain-fed zone and the rice–wheat zone, a correlation of 0.3 to 0.4 was observed between ET_o and WS.

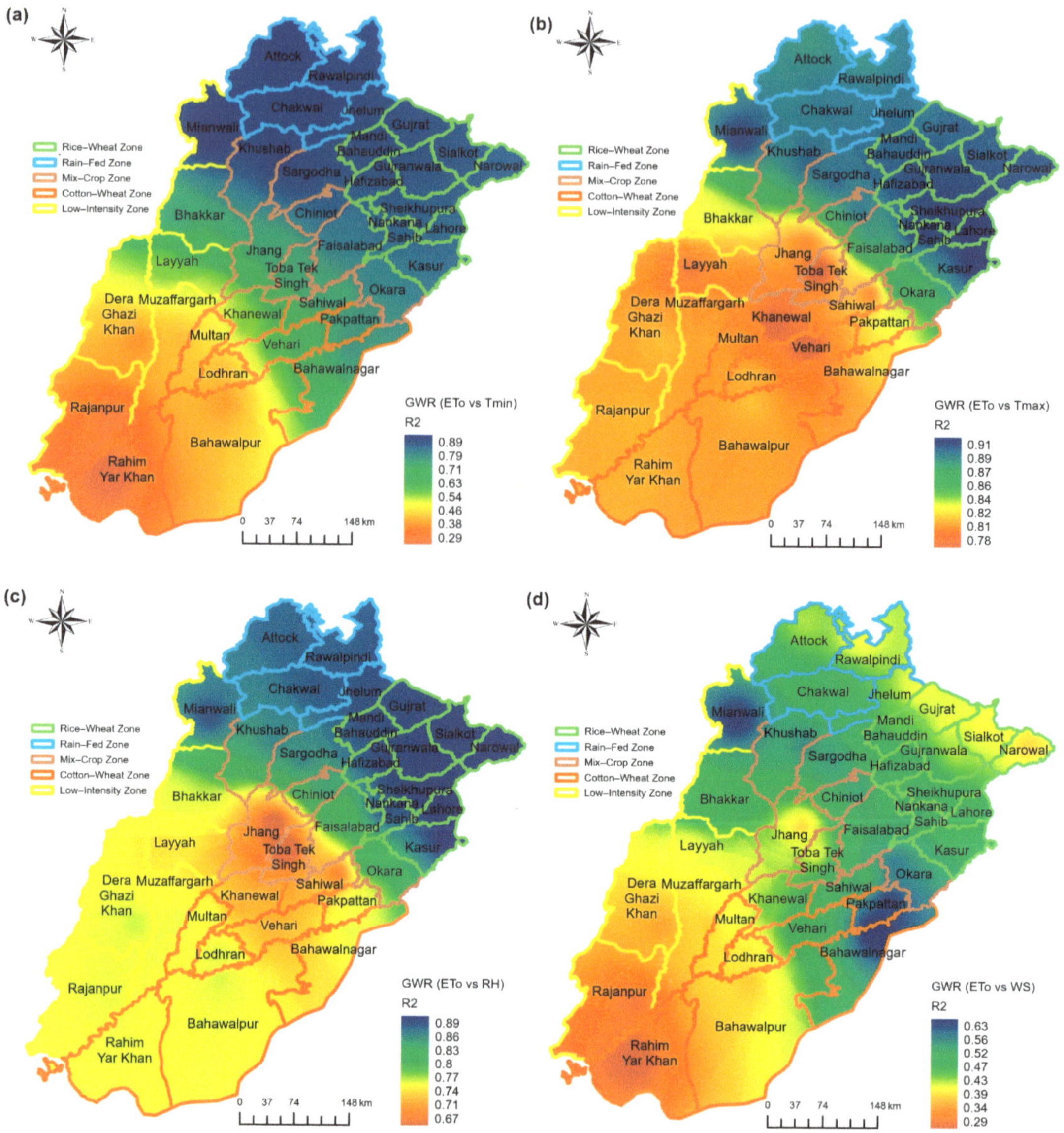

Figure 6. Spatial variation of Pearson's correlation coefficient between ET_o and (**a**) T_{max}, (**b**) T_{min}, (**c**) RH, and (**d**) WS, using geographic weighted regression for 1950–2021 in Punjab, Pakistan.

3.3.2. Multiscale Geographic Weighted Regression (MGWR)

The MGWR model was utilized to estimate ET_o in Punjab from 1950–2021, and Figure 7 displays the monthly distribution of the relative contribution of minimum temperature (T_{min}) in this estimation. The findings presented in Figure 7a demonstrate the varying levels of contribution of T_{min} during the month of January across different regions, ranging from 0.17% to 0.75%. The study revealed that ET_o was significantly impacted by the rain-fed and low-intensity zones, whereas the rice–wheat and mixed-crop zones exhibited a relatively lower level of influence. Figure 7b demonstrated a comparable pattern where the minimum temperature (T_{min}) had an impact on ET_o during February. The range of this impact was between 0.18% and 0.76%, indicating a significant level of contribution in both the rain-fed and cotton–wheat zones. During the month of March, the contribution of T_{min} varied between 0.15 to 0.74%. The rain-fed zone was observed to be the most affected, as depicted in Figure 7c. The impact of T_{min} during the month of April is depicted in Figure 7d, revealing a significant influence on the rain-fed, low-intensity, and cotton–wheat zones, with a range of 0.17% to 0.71%. The rice–wheat zone, rain-fed zone, low-intensity zone, and mixed-crop zone experienced significant impact in May, with an overall range of 0.22% to 0.76%, as illustrated in Figure 7e. During the month of June, the T_{min} had an impact on the value, which varied between 0.31% and 0.85%. The rice–wheat zone, low-intensity zone, and rain-fed zone were the areas that were most affected, as shown in Figure 7f. The data presented in Figure 7g indicate that the values observed for July ranged between 0.25% and 0.80%. The rice–wheat zone and mixed-crop zones exhibited comparatively lower levels of influence, whereas the remaining zones were subject to significant impact. According to Figure 7h, the contribution of T_{min} during the month of August varied across different regions, with values ranging from 0.32% to 0.85%. The study revealed that the rain-fed zone had a significant impact on ET_o, whereas the remaining zones exhibited a comparatively lower level of contribution. Figure 7i illustrates the impact of T_{min} on ET_o for the month of September. The observed values exhibited a significant level of contribution within the rain-fed zone, ranging from 0.27% to 0.86%. The rain-fed zone in October exhibited a similar pattern, with the highest impact ranging from 0.22% to 0.72%, as shown in Figure 7j. The influence of T_{min} on ET_o was observed in November, with values ranging from 0.16% to 0.72%, as depicted in Figure 7k. The rice–wheat zone, cotton–wheat zone, low-intensity zone, and mixed-crop zone were found to have experienced the least amount of impact. Finally, it should be noted that the impact of T_{min} on ET_o in the rain-fed region remained consistently significant throughout the month of December, exhibiting the greatest level of influence. The observed values exhibited a range between 0.16% and 0.75%, with the mixed-crop zone displaying the least amount of impact, as depicted in Figure 7l.

The findings derived from Figure 7 demonstrate that the impact of climatic variables on ET_o exhibits variability on a monthly basis. Figure 8 depicts the dominant factor that exerted the highest influence on ET_o on a monthly basis from 1950 to 2021 in the region of Punjab, Pakistan.

According to Figure 8, the dominant factor varied across space and time. In terms of monthly variation, WS was the dominant factor across most of the Punjab in January, as shown in Figure 8a. However, in February, SH was observed to be the dominant factor across majority of the 36 districts in Punjab, as shown in Figure 8b. Similar to January, WS dominated the influence on ET_o across Punjab in March, as shown in Figure 8c. According to Figure 8d, the WS and SH were the most prominent dominant factors out of the five climatic parameters influencing ET_o in April across Punjab. In May, T_{min}, SH, and WS were the most dominant factors spread across Punjab (shown in Figure 8e). This could be linked to longer solar days, higher wind speed across the plains, and temperature variation due to start of summer months. In June, T_{min} was the most dominant factor influencing ET_o across Punjab, shown in Figure 8f. Similarly, in July, T_{min} and T_{max} were the driving factors influencing reference evapotranspiration across Punjab as shown in Figure 8g. In August, T_{min} and WS were the most dominant factors shown in Figure 8h. According to Figure 8i, in September, T_{min}, T_{max}, and WS were the most dominant factors influencing different agroclimatic zones of the Punjab. According to Figure 8j, in October, in the low intensity zone, SH was the driving factor, whereas in the upper Punjab, RH was the driving factor influencing ET_o. In other zones, T_{min} and SH were observed to be influencing ET_o. In November, a mix–up of T_{min}, T_{max}, WS, and SH influenced ET_o at different stations across Punjab, as shown in Figure 8k; however, WS and SH were the most common dominant factors at majority of the stations. According to Figure 8l, WS and SH were the dominant factors impacting ET_o across Punjab in December.

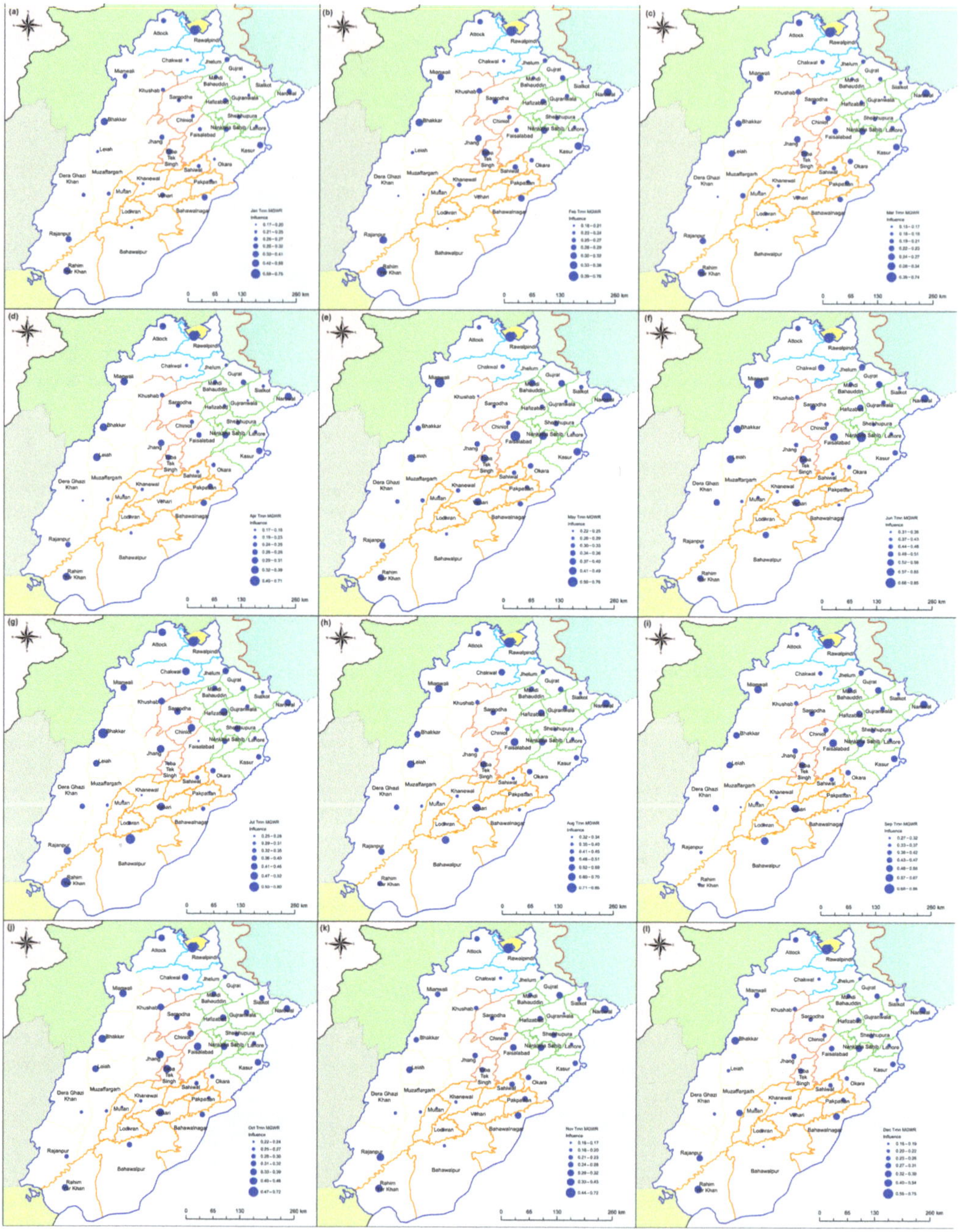

Figure 7. Spatial and monthly influence of T_{min} on ET_o using multiscale geographic weighted regression for the month of (**a**) January, (**b**) February, (**c**) March, (**d**) April, (**e**) May, (**f**) June, (**g**) July, (**h**) August, (**i**) September, (**j**) October, (**k**) November, (**l**) December for 1950–2021 in Punjab, Pakistan.

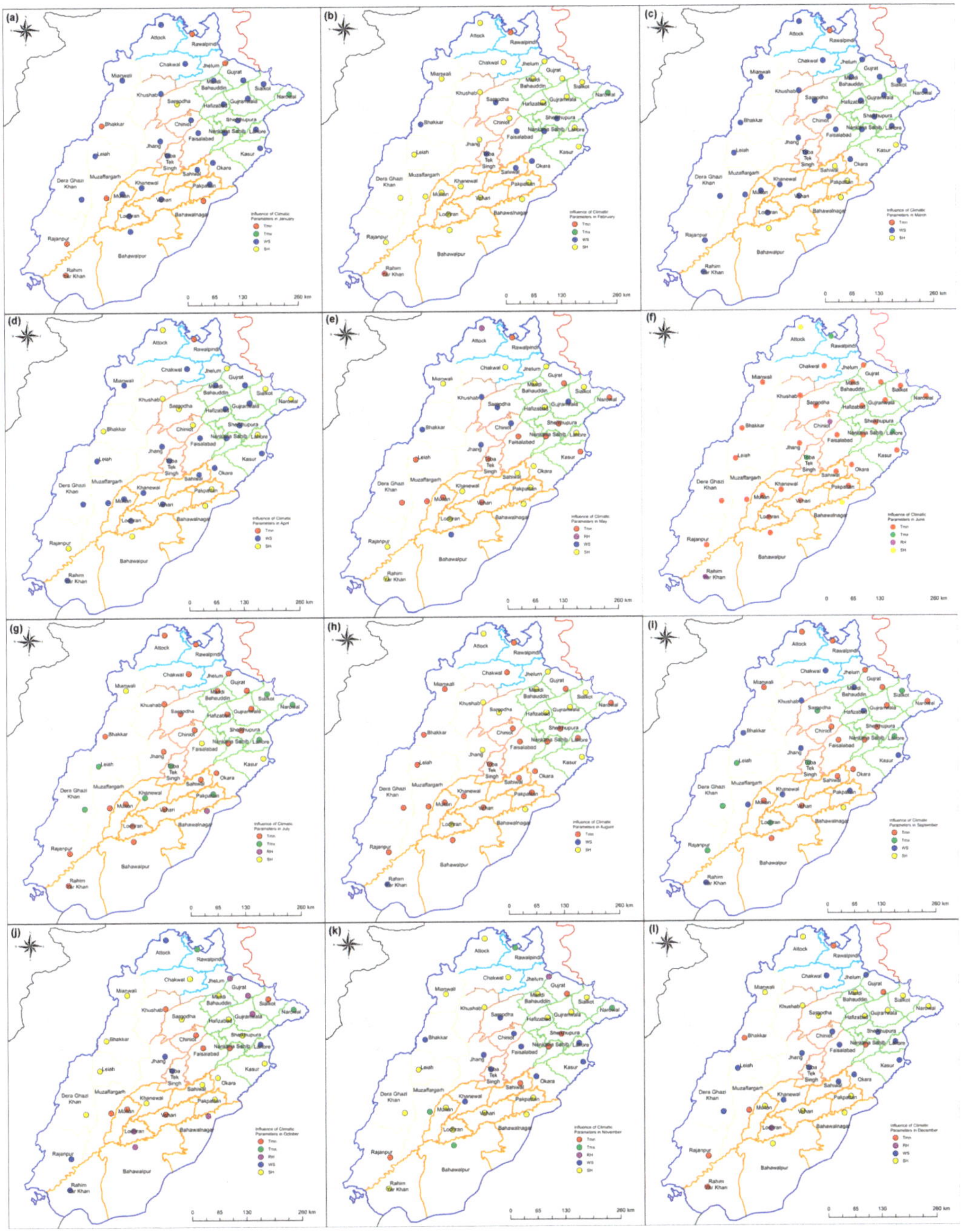

Figure 8. Illustration of the spatial and monthly variation in the dominant factor out of the five climatic parameters on ET_o using multiscale geographic weighted regression (MGWR) for the month of (**a**) January, (**b**) February, (**c**) March, (**d**) April, (**e**) May, (**f**) June, (**g**) July, (**h**) August, (**i**) September, (**j**) October, (**k**) November, (**l**) December for the period 1950–2021.

4. Discussion

T_{min} series showed more significant trends, which is in accordance with the global trend over the previous decades. The overall pattern assessment revealed that annual T_{min} had a considerable influence on ET_o in the rain-fed zone. ET_o is impacted in the heavy rainfall [35,36] seasons due to variations in relative humidity. To address the varying influence, the GWR technique was used, which can be effective in allowing model parameters to vary across space [18]. This approach can provide valuable insights into the complex spatial patterns of ET_o and its influencing factors, which can aid in developing more accurate and localized water management strategies. The correlation of climatic factors to ET_o findings demonstrates that all parameters have a substantial influence on ET_o [37]. According to the findings by Hupet et al. [38], it has been observed that radiation has a significant influence on ET_o. Since minimum temperature remains unchanged as much during the wet season as it does during the period of dryness, it has a greater influence on ET_o during the dry season. Therefore, the impact of T_{min} was observed on ET_o via MGWR and its subsequent influence from the month of January to December for the study period 1950–2021. According to Zhang et al. [39], ET_o was highly influenced by T_{min} and T_{max} across China on an annual basis. The study provided a detailed dominant factor analysis throughout the study period (1970–2014) in China. However, the study failed to correlate the impact of monthly variation of weather conditions/seasons on the influencing ability of the climatic parameters. The present study assesses this impact on a monthly basis. Furthermore, the estimated reference evapotranspiration could be correlated with measured ET_o from any of the stations for reference. Additionally, in future research directions, the estimated spatiotemporal ET_o (which serves as an essential component in the crop water requirement estimation) could help in the estimation of zone- and crop-specific irrigation scheduling across Punjab. Ultimately, based on the study, the estimation of the amount of water required for irrigation purpose would help the relevant policymakers to better allocate the water resources in the country.

5. Conclusions

Maximum T_{max}, T_{min}, RH, WS, and ET_o (i.e., 33.9 °C, 18.7 °C, 76.5%, 4.2 m/s, and 7.6 mm/day, respectively) were observed across the study area. The highest ET_o was observed in the southern regions of the area. The range of ET_o was also higher in the southern region (i.e., up to 5.57 mm/day). As per GWR, T_{max} had the highest correlation (i.e., 0.91) with ET_o in the mixed-crop zone. As per the MGWR, T_{min}, T_{max}, and WS had the highest influence on ET_o throughout different regions in Punjab. Throughout the study period, ET_o showed an increasing trend. In addition, a heterogeneity was observed in the ET_o trend in year 2001. Out of the five agroclimatic zones, the cotton–wheat zone observed the highest ET_o due to the higher temperatures in the region. In terms of a monthly dominant factor, the highest overall dominant factor was T_{min}, whereas WS and SH were the highest observed dominant factors in the winters, specifically. It was concluded that T_{min} and T_{max} had the highest overall influence on ET_o in Punjab. In addition to these parameters, WS was also observed to be the highest dominant factor influencing ET_o on a monthly basis. In terms of future research directions, the ultimate goal of the proposed study was to investigate the zone- and crop-specific irrigation scheduling which incorporates ET_o as an essential component.

Author Contributions: Conceptualization, H.A. and S.Q.; Data curation, H.A., S.Q. and N.R.; Formal analysis, H.A., S.Q. and B.K.; Funding acquisition, R.R.S., M.S., S.M.I. and M.I.; Investigation, H.A.; Methodology, H.A., M.S. and M.U.K.; Project administration, M.S., S.M.I. and M.I.; Resources, M.S.; Software, H.A., S.Q., N.R. and M.U.K.; Supervision, R.R.S., M.S., M.I. and M.U.K.; Validation, R.R.S. and M.S.; Visualization, H.A., S.Q., R.R.S., M.S., S.M.I., M.I. and M.U.K.; Writing—original draft, H.A., S.Q. and N.R.; Writing—review and editing, S.Q., N.R., R.R.S., M.S., B.K., S.M.I., M.I. and M.U.K. All authors have read and agreed to the published version of the manuscript.

Funding: This work was supported by Researchers Supporting Project number (RSP2023R100), King Saud University, Riyadh, Saudi Arabia.

Institutional Review Board Statement: Not applicable.

Informed Consent Statement: Not applicable.

Data Availability Statement: The data are available within the article.

Acknowledgments: This work was supported by Researchers Supporting Project number (RSP2023R100), King Saud University, Riyadh, Saudi Arabia. The authors would like to express their gratitude to Innovate UK's Energy Catalyst Programme (Ayrton Funding provided by the Foreign, Commonwealth Development Office through their Transforming Energy Access Programme) and UK aid for the support and funding provided for the research project titled 'Hybrid Energy Powered Smart Irrigation System for Smallholder Farmers' (Project number: 10039507). This financial support from Innovate UK's Energy Catalyst Programme has played a pivotal role in facilitating the successful execution of our research. The authors acknowledge the financial support by the Open Access Publication Fund of the Leibniz Association, Germany.

Conflicts of Interest: The authors declare no conflict of interest.

References

1. Allen, R.G.; Pereira, L.S.; Raes, D.; Smith, M. *Crop Evapotranspiration—Guidelines for Computing Crop Water Requirements—FAO Irrigation and Drainage Paper 56*; FAO: Rome, Italy, 1998; Volume 300, p. D05109.
2. Wu, B.; Quan, Q.; Yang, S.; Dong, Y. A social-ecological coupling model for evaluating the human-water relationship in basins within the Budyko framework. *J. Hydrol.* **2023**, *619*, 129361. [CrossRef]
3. Zakir-Hassan, G.; Punthakey, J.F.; Shabir, G.; Yasmeen, F.; Sultan, M.; Ashraf, H.; Sohoo, I.; Majeed, F. Physicochemical Investigation of Rainfall for Managed Aquifer Recharge in Punjab (Pakistan). *Water* **2022**, *14*, 2155. [CrossRef]
4. Zakir, G.; Kahlown, M.A.; Punthakey, J.F.; Shabir, G.; Aziz, M.; Sultan, M.; Ashraf, H.; Majeed, F. Evaluation of hydraulic efficiency of lined irrigation channels—A case study from Punjab, Pakistan. *Hydrol. Res.* **2023**, *54*, 523–546. [CrossRef]
5. Elnmer, A.; Khadr, M.; Kanae, S.; Tawfik, A. Mapping daily and seasonally evapotranspiration using remote sensing techniques over the Nile delta. *Agric. Water Manag.* **2019**, *213*, 682–692. [CrossRef]
6. Karbasi, M.; Jamei, M.; Ali, M.; Malik, A.; Yaseen, Z.M. Forecasting weekly reference evapotranspiration using Auto Encoder Decoder Bidirectional LSTM model hybridized with a Boruta-CatBoost input optimizer. *Comput. Electron. Agric.* **2022**, *198*, 107121. [CrossRef]
7. Malik, A.; Jamei, M.; Ali, M.; Prasad, R.; Karbasi, M.; Yaseen, Z.M. Multi-step daily forecasting of reference evapotranspiration for different climates of India: A modern multivariate complementary technique reinforced with ridge regression feature selection. *Agric. Water Manag.* **2022**, *272*, 107812. [CrossRef]
8. Chen, Z.; Liu, Z.; Yin, L.; Zheng, W. Statistical analysis of regional air temperature characteristics before and after dam construction. *Urban Clim.* **2022**, *41*, 101085. [CrossRef]
9. Muhammad Adnan, R.; Chen, Z.; Yuan, X.; Kisi, O.; El-Shafie, A.; Kuriqi, A.; Ikram, M. Reference Evapotranspiration Modeling Using New Heuristic Methods. *Entropy* **2020**, *22*, 547. [CrossRef] [PubMed]
10. Pei, Y.; Qiu, H.; Zhu, Y.; Wang, J.; Yang, D.; Tang, B.; Wang, F.; Cao, M. Elevation dependence of landslide activity induced by climate change in the eastern Pamirs. *Landslides* **2023**, *20*, 1115–1133. [CrossRef]
11. Huang, S.; Lyu, Y.; Sha, H.; Xiu, L. Seismic performance assessment of unsaturated soil slope in different groundwater levels. *Landslides* **2021**, *18*, 2813–2833. [CrossRef]
12. Li, Q.; Song, D.; Yuan, C.; Nie, W. An image recognition method for the deformation area of open-pit rock slopes under variable rainfall. *Measurement* **2022**, *188*, 110544. [CrossRef]
13. Li, J.; Wang, Z.; Wu, X.; Zscheischler, J.; Guo, S.; Chen, X. A standardized index for assessing sub-monthly compound dry and hot conditions with application in China. *Hydrol. Earth Syst. Sci.* **2021**, *25*, 1587–1601. [CrossRef]
14. Zhou, J.; Wang, L.; Zhong, X.; Yao, T.; Qi, J.; Wang, Y.; Xue, Y. Quantifying the major drivers for the expanding lakes in the interior Tibetan Plateau. *Sci. Bull.* **2022**, *67*, 474–478. [CrossRef]
15. Xie, X.; Huang, L.; Marson, S.M.; Wei, G. Emergency response process for sudden rainstorm and flooding: Scenario deduction and Bayesian network analysis using evidence theory and knowledge meta-theory. *Nat. Hazards* **2023**, *117*, 3307–3329. [CrossRef]
16. Samanta, S.; Pal, D.K.; Lohar, D.; Pal, B. Interpolation of climate variables and temperature modeling. *Theor. Appl. Climatol.* **2012**, *107*, 35–45. [CrossRef]
17. Saeed Shah, S.M.; El-Morshedy, M.; Mansoor, W. Spatial-Temporal Interpolation of Reference Evapotranspiration for Pakistan. *Math. Probl. Eng.* **2022**, *2022*, 5488725. [CrossRef]
18. Fotheringham, A.S.; Brunsdon, C.; Charlton, M. *Geographically Weighted Regression: The Analysis of Spatially Varying Relationships*; John Wiley & Sons: Hoboken, NJ, USA, 2002.

19. Muhammad, M.K.I.; Nashwan, M.S.; Shahid, S.; Ismail, T.b.; Song, Y.H.; Chung, E.-S. Evaluation of Empirical Reference Evapotranspiration Models Using Compromise Programming: A Case Study of Peninsular Malaysia. *Sustainability* **2019**, *11*, 4267. [CrossRef]
20. Finance Division, Government of Pakistan. *Pakistan Economic Survey 2021-22*; Finance Division, Government of Pakistan: Islamabad, Pakistan, 2022; p. 548.
21. Nagappan, M.; Gopalakrishnan, V.; Alagappan, M. Prediction of reference evapotranspiration for irrigation scheduling using machine learning. *Hydrol. Sci. J.* **2020**, *65*, 2669–2677. [CrossRef]
22. Amjad, R.; Arif, G.M.; Mustafa, U. Does the labor market structure explain differences in poverty in rural Punjab? *Lahore J. Econ.* **2008**, *SE*, 139–162. [CrossRef]
23. Pinckney, T.C. *The Demand for Public Storage of Wheat in Pakistan*; International Food Policy Reseasrch Institute: Washington, DC, USA, 1989; Volume 77.
24. Zhang, Q.; Kong, D.; Singh, V.P.; Shi, P. Response of vegetation to different time-scales drought across China: Spatiotemporal patterns, causes and implications. *Glob. Planet. Chang.* **2017**, *152*, 1–11. [CrossRef]
25. Kalnay, E.; Kanamitsu, M.; Kistler, R.; Collins, W.; Deaven, D.; Gandin, L.; Iredell, M.; Saha, S.; White, G.; Woollen, J.; et al. The NCEP/NCAR 40-Year Reanalysis Project. *Bull. Am. Meteorol. Soc.* **1996**, *77*, 437–472. [CrossRef]
26. Allen, R. Penman–Monteith Equation. In *Encyclopedia of Soils in the Environment*; Hillel, D., Ed.; Elsevier: Oxford, UK, 2005; pp. 180–188.
27. Mann, H.B. Nonparametric Tests Against Trend. *Econometrica* **1945**, *13*, 245–259. [CrossRef]
28. Sen, P.K. Estimates of the Regression Coefficient Based on Kendall's Tau. *J. Am. Stat. Assoc.* **1968**, *63*, 1379–1389. [CrossRef]
29. Gao, C.; Wang, Z.; Ji, X.; Wang, W.; Wang, Q.; Qing, D. Coupled improvements on hydrodynamics and water quality by flowing water in towns with lakes. *Environ. Sci. Pollut. Res.* **2023**, *30*, 46813–46825. [CrossRef] [PubMed]
30. Ashraf, A.; Ahmed, A.; Mukhtar, A.; Iqbal, M. Changing Pattern of Agroclimate in Pakistan and Measures for Agriculture Sustainability. *Pak. J. Meteorol.* **2023**, *2023*, 15.
31. Zhang, L.; Cui, Y.; Xiang, Z.; Zheng, S.; Traore, S.; Luo, Y. Short-term forecasting of daily crop evapotranspiration using the 'Kc-ETo' approach and public weather forecasts. *Arch. Agron. Soil Sci.* **2018**, *64*, 903–915. [CrossRef]
32. Harrell, F.E. *Regression Modeling Strategies*; Springer Nature Switzerland AG: Cham, Switzerland, 2015; Volume 330, p. 14.
33. Bois, B.; Pieri, P.; Van Leeuwen, C.; Wald, L.; Huard, F.; Gaudillere, J.P.; Saur, E. Using remotely sensed solar radiation data for reference evapotranspiration estimation at a daily time step. *Agric. For. Meteorol.* **2008**, *148*, 619–630. [CrossRef]
34. Zhang, F.; Geng, M.; Wu, Q.; Liang, Y. Study on the spatial-temporal variation in evapotranspiration in China from 1948 to 2018. *Sci. Rep.* **2020**, *10*, 17139. [CrossRef]
35. Wu, X.; Guo, S.; Qian, S.; Wang, Z.; Lai, C.; Li, J.; Liu, P. Long-range precipitation forecast based on multipole and preceding fluctuations of sea surface temperature. *Int. J. Climatol.* **2022**, *42*, 8024–8039. [CrossRef]
36. Zhu, G.; Liu, Y.; Wang, L.; Sang, L.; Zhao, K.; Zhang, Z.; Lin, X.; Qiu, D. The isotopes of precipitation have climate change signal in arid Central Asia. *Glob. Planet. Chang.* **2023**, *225*, 104103. [CrossRef]
37. Huang, Y.; Bárdossy, A.; Zhang, K. Sensitivity of hydrological models to temporal and spatial resolutions of rainfall data. *Hydrol. Earth Syst. Sci.* **2019**, *23*, 2647–2663. [CrossRef]
38. Hupet, F.; Vanclooster, M. Effect of the sampling frequency of meteorological variables on the estimation of the reference evapotranspiration. *J. Hydrol.* **2001**, *243*, 192–204. [CrossRef]
39. Zhang, L.; Traore, S.; Cui, Y.; Luo, Y.; Zhu, G.; Liu, B.; Fipps, G.; Karthikeyan, R.; Singh, V. Assessment of spatiotemporal variability of reference evapotranspiration and controlling climate factors over decades in China using geospatial techniques. *Agric. Water Manag.* **2019**, *213*, 499–511. [CrossRef]

Disclaimer/Publisher's Note: The statements, opinions and data contained in all publications are solely those of the individual author(s) and contributor(s) and not of MDPI and/or the editor(s). MDPI and/or the editor(s) disclaim responsibility for any injury to people or property resulting from any ideas, methods, instructions or products referred to in the content.

AgriEngineering

MDPI

Article

Evaluation of Body Surface Temperature in Pigs Using Geostatistics

Maria de Fátima Araújo Alves *, Héliton Pandorfi, Abelardo Antônio de Assunção Montenegro, Rodes Angelo Batista da Silva, Nicoly Farias Gomes, Taize Calvacante Santana, Gledson Luiz Pontes de Almeida, Gabriel Thales Barboza Marinho, Marcos Vinícius da Silva * and Weslley Amaro da Silva

Department of Agricultural Engineering, Federal Rural University of Pernambuco, Recife 52171-900, Brazil; heliton.pandorfi@ufrpe.br (H.P.); abelardo.montenegro@ufrpe.br (A.A.d.A.M.); rodesangel@gmail.com (R.A.B.d.S.); nicoly_farias@hotmail.com (N.F.G.); taize.calvacante@ufrpe.br (T.C.S.); gledson.almeida@ufrpe.br (G.L.P.d.A.); gabriel.bmarinho@ufrpe.br (G.T.B.M.); weslleyamaro.s@gmail.com (W.A.d.S.)

* Correspondence: fatima.alves@ufrpe.br (M.d.F.A.A.); marcos.viniciussilva@ufrpe.br (M.V.d.S.)

Abstract: This paper explores the potential of infrared thermography and geostatistics in animal production and presents the results of the application of the combination of these techniques, contributing significantly to efforts to obtain animals' responses to the environments in which they are located and thereby ensuring improvements in productivity and animal welfare. The objective was to verify the variability in surface temperature in pigs submitted to different climate control systems using geostatistics. Three growing animals per stall were selected. Dry bulb temperature (Tbd, °C), relative humidity (RH, %) and thermal images were recorded at 08:00 and 12:00 h. To analyze the data, semivariograms were made, the theoretical model was validated and kriging maps were constructed. The mean temperature of the pigs in the pen with adiabatic evaporative cooling (AEC) ranged from 32.40 to 36.25 °C; for the pigs in the forced ventilation (FV) pen, the range of variation was from 32.51 to 36.81 °C. In the control group (Con), with natural ventilation, the average temperature was 37.51 to 38.45 °C. The geostatistical analysis provided a mathematical model capable of illustrating the variation in temperature in the caudal–dorsal regions of the pigs according to the environments to which the animals were subjected.

Keywords: climatization systems; thermal comfort; piggery

Citation: Alves, M.d.F.A.; Pandorfi, H.; Montenegro, A.A.d.A.; Silva, R.A.B.d.; Gomes, N.F.; Santana, T.C.; Almeida, G.L.P.d.; Marinho, G.T.B.; Silva, M.V.d.; Silva, W.A.d. Evaluation of Body Surface Temperature in Pigs Using Geostatistics. *AgriEngineering* **2023**, *5*, 1090–1103. https://doi.org/10.3390/agriengineering5020069

Academic Editors: Muhammad Sultan, Yuguang Zhou, Redmond R. Shamshiri and Muhammad Imran

Received: 31 March 2023
Revised: 7 June 2023
Accepted: 13 June 2023
Published: 19 June 2023

Copyright: © 2023 by the authors. Licensee MDPI, Basel, Switzerland. This article is an open access article distributed under the terms and conditions of the Creative Commons Attribution (CC BY) license (https://creativecommons.org/licenses/by/4.0/).

1. Introduction

The consumption of animal protein has increased considerably, according to a survey conducted by the Brazilian Association of Animal Protein [1], and, worldwide, Brazil is the fourth-largest consumer [2]. Among the main demands of consumers is a guarantee of the welfare of confined animals, thus requiring the application of ethical actions in the systems that are assigned to animal production [3]. As for the meaning of animal welfare, the Terrestrial Animal Health Code [4] defined it as the physical and mental state of the animal in relation to the conditions in which it lives and dies. Ref. [4] reported five conditions that animals need to experience under human management: the absence of hunger, malnutrition and thirst; freedom from fear and anguish; the absence of heat stress or physical discomfort; freedom from pain, injury and disease; and freedom to express normal patterns of behavior.

Thus, due to the increase in pig production, the comfort and well-being of pigs have been undergoing changes, either due to reductions in space, the impairment of movement or changes in social interactions, all of which contribute to the occurrence of problems related to thermal comfort and consequently to productivity, thus requiring that breeders show greater dedication to ensuring favorable animal production systems [5].

Ref. [6] points out that one of the ways to quantify animal welfare, from a thermal point of view, is by the evaluation of climatic variables. The climatic variables that most affect pig production are air temperature, relative humidity, radiation and wind speed [7,8]. High temperatures cause heat stress in animals and result in negative impacts that affect their physiological reactions as well as their behavior and the amount of food that they ingest [9].

According to [10], air quality control in pig facilities is also important because it improves productivity levels and decreases damage to the health of pigs. Ref. [11] illustrates the negative impacts on animal production by citing those pigs that have difficulty dissipating heat in hot environments and in conditions of high humidity; these factors limit evaporative losses through animal respiration, which decreases the appetite of the animals.

Several studies have been developed in order to describe the ideal production environment in terms of thermal comfort for animals, highlighting the study of the temperature and humidity index (THI) researched by [12], the black globe temperature and humidity index (BGTHI) developed by [13], the radiant heat load (RHL) recommended by [14] and the enthalpy (h) defined by [8,15].

The thermal condition of housing is an important determining factor for good pig productivity. A low capacity for heat regulation in pig housing ends up causing heat loss in a more critical way, as when there is a drop in temperature the pigs strive to increase their heat production. This happens through the thermogenesis of muscle tremors, which is associated with social and individual thermoregulatory attitudes; thus, it can be verified through their posture whether animals are comfortable or uncomfortable in their current environment [16].

Ref. [17] states that the best way to evaluate confined animal production environments is through non-invasive, innovative methods. According to [18], an innovative, non-invasive technique that was disclosed in a Brazil–Italy workshop is the use of infrared thermography. This technique has great potential for analyzing animal thermal comfort in production environments and works by mapping an animal or a part of its body and distinguishing areas of different temperatures through the artificial visualization of light within the infrared spectrum [18]. Ref. [19] reported that this is a non-invasive technique by means of which you can measure the animal's body surface temperature without needing to have contact with the animal. Ref. [20] used infrared thermography to assess the stress condition of pigs. According to [21], infrared thermography has become one of the most popular technologies in the area of animal production and has been used in several studies because it is a non-invasive technique. It involves extracting temperature information through the energy emitted by the animal's body surface and transforms it into a thermal image, and it has been evaluated as a highly effective and safe technique. Ref. [22] reports that infrared thermography has been used in both human and veterinary medicine with precision and practicality, since it can be used for the prevention, diagnosis and prognosis of diseases.

The results of infrared thermography can be interpreted by geostatistics, which, according to [23], is a statistical tool that assumes that the probability of variation of the sampled attributes, separated by the same distance, is equal, assuming stationarity within the range of spatial continuity. Ref. [24] stated that geostatistics is one of the procedures for analyzing space–time data that takes into account the existing interactions between spatial and temporal components, allowing interpolations in time and space. Among the advantages of using geostatistics, according to [25], is the fact that it is a tool that can use as much information as is available with respect to the variable of interest—which is difficult to collect—to estimate its values in non-sampled locations through the generation of a continuous surface of estimated points.

Ref. [26] used geostatistical analysis to evaluate the temporal dependence of the variables analyzed in the work and to determine the parameters of the theoretical models that best fitted the experimental semivariograms. Ref. [27] points out that such a tool is able to express the spatial variability within animal housing facilities by showing the distributions of environmental variables.

Such tools are of great importance in pig farming, since they allow the monitoring of animals without the need for direct contact with them, avoiding stress and contributing to thermal comfort, for the evaluation of their health and well-being, as well as to assess the quality of the environment in which they are being raised. Through thermal imaging, it is possible to visualize the body temperature distribution of animals, and valuable information can be obtained regarding thermal comfort and disease prevention. Geostatistics, on the other hand, can indicate the spatial variability of the surface temperature of animals and can also be used to evaluate the effectiveness of management practices and sanitary control measures, allowing the identification of areas that require further attention and monitoring. The application of thermal images and geostatistics in animal production has become increasingly relevant and promising, especially with regard to the evaluation of animal welfare and the identification of areas with higher productive potential. Thus, this study is important because it explores the potential of these techniques and presents the results of the application of the combination of these techniques, contributing significantly to efforts to obtain responses from animals in the environments in which they are located, thereby ensuring improvements in productivity and animal welfare.

Much research has been developed using thermal imaging of various parts of animals' bodies; for example, the authors of [28] recorded the surface temperatures of lactating sows with thermal imaging. The authors of [29] collected the temperatures of the backs, eyes, vulvas and ears of 18 pigs using infrared thermography to predict rectal temperature (°C). In [30], the authors used infrared thermography to record the temperatures of the ear and back regions of pigs. However, research on the variation in the dorsal–caudal region in pigs exposed to environments with different grading systems using infrared thermography is scarce.

This work aims to verify the temperature variation in the dorsal–caudal region of pigs submitted to different acclimatization systems using thermal images and geostatistics.

2. Materials and Methods

2.1. Experiment Location

The research data were collected at the Bioterium of the Experimental pigs Facility on the campus of the Federal Rural University of Pernambuco in the municipality of Serra Talhada—the domain of the members of the university's Ambience Research Group. The municipality of Serra Talhada is located in a region of arid climatic conditions which has an average altitude of 492 m and coordinates of latitude and longitude of 07°59′31″ S and 38°17′54″ W. According to the Köppen classification, the climate is BSwh (hot and dry) [31] and presents an average annual temperature of 24.8 °C and relative humidity near 62.5% [32].

2.2. Animal Facilities

Data collection was performed from 1 August to 30 December 2018 and lasted 92 days. The experiment was approved by the CEUA/UFRPE (Ethics Committee on Animal Use) under protocol no. 23082.021090/2016-81.

The animal housing used in this study was built along the longitudinal axis in the east–west direction; the roof was covered with ceramic tiles, the walls were built of masonry, and the floor was made of concrete and had an area of 6 m^2. All stalls were equipped with feeders and drinking fountains installed 30 cm from the floor. The animals received water and balanced feed appropriate to their life stages. Figure 1 illustrates the arrangement of the stalls.

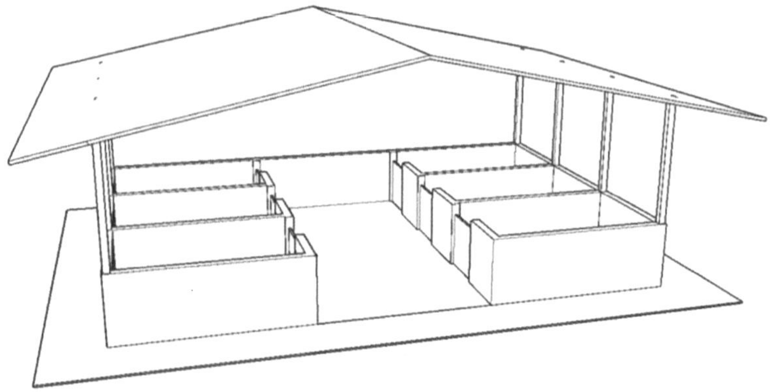

Figure 1. Structure of the animal facilities used in this study. Source: The authors.

Three stalls were used, one for each treatment. Different acclimatization systems were installed in the three experimental pens: three stalls were fitted with an acclimatizer to study the effect of adiabatic evaporative cooling (AEC) on the animals; three stalls were fitted with an axial fan to study the effects of forced ventilation (FV) on the animals; and the last three stalls had no acclimatization system. The animals were analyzed under natural ventilation (Con). The air conditioners used in the stalls had an average flow rate of 3 $L.h^{-1}$, with independent motors and a propeller rotation of 1750 RPM, with a central disk of 3450 RPM. The flow rate of the fans was 1200 $m^3 h^{-1}$, at 1780 RPM, and the propeller diameter was 11″. The climate control systems were randomly distributed in the pens to reduce the influence of external factors.

We used 9 animals that were available in the university's experimental vivarium for research, which were distributed in threes in each bay. The animals were 3/4 Duroc, 1/4 Pietrain commercial strains in the growth phase. The growth phase comprises the physiological phase of the animal after weaning (about 21 to 28 days), and the animal remains in this phase from 63 to 110 days of life. The average weight of the pigs was approximately 35 kg, and the ages ranged from 28 to 110 days. In terms of feeding, the pigs received a balanced and controlled diet. They were fed a combination of commercially available pig feed consisting of grains, protein sources, vitamins and minerals. The feed was carefully formulated to meet the nutritional requirements of the pigs at their respective ages and weights. Water was available throughout the experiment to ensure adequate hydration.

2.3. Meteorological Variables of the Environment

The meteorological variables collected were dry bulb temperature (Tbd; °C), relative humidity (RH, %), black globe temperature (Tgb; °C) and wind speed (Ws, $m.s^{-1}$). Tbd, RH and Tgb were recorded using a HOBO® datalogger model U12-012 (Onset Computer Corporation Bourne, USA, MA,) at 08:00 and 12:00 h. The instrument was installed according to the recommendations of [33], in which it is recommended that the equipment should preferably be installed in the central region of the shed. The instrument was installed according to the recommendations of [33], in which it is recommended that the equipment should preferably be installed in the central region of the shed. The Ws ($m\ s^{-1}$) was collected using a digital thermo-anemometer, model TAFR-180, with a scale of 0.1 to 20.0 $m\ s^{-1}$ and a resolution of 0.1 $m\ s^{-1}$, at 1.50 m from the ground. The measurement times were selected based on [34], the authors of which took into consideration a time with a lower temperature and a time with a higher temperature. These variables were collected inside the stalls by means of a datalogger and outside by means of a meteorological shelter, both at 1.50 m height, and were used to verify the variations between treatments. The thermal characterization of the stalls was determined through a reclassification of the BGTHI for pigs, considering the minimum optimal temperature value (15 °C), the upper critical temperature (27 °C), the average values of the thermoneutral zone for the finishing phase (15 to 18 °C) mentioned by the authors of [35] in their reference study and the relative air humidity value range of 50 to 70% cited in [36] as a basis for the reclassification of the BGTHI. These temperature and humidity values were used to estimate the dew-point temperature (DPT, °C) using the Magnus–Tetens [37] equation (Equation (1)). Subsequently, the BGTHI was estimated using the equation given in [38], which was developed for cattle.

$$\mathrm{DPT} = (c \times \gamma \times (\mathrm{Tbd}, \mathrm{RH}))/(b - \gamma(\mathrm{Tbd}, \mathrm{RH})) \quad (1)$$

where DPT is the dew-point temperature, γ(Tbd, RH) is the intermediate result, Tbd is the temperature, RH is the relative humidity, b is a constant with a value of 238.88 and c is a constant with a value of 17.368.

2.4. Physiological Variables of the Animals

To evaluate the effects of the different acclimatization systems on the animals' thermal comfort, the following physiological variables were collected from the animals: rectal temperature (RT), respiratory rate (RR) and surface skin temperature (SST), at 08:00 and 12:00 h, once a week, throughout the experimental period. The RR was measured by counting the movement of the flank in 1 (one) minute, using a stopwatch, and then calculating the number of movements per minute. The animals' body temperature was checked by means of thermal images obtained with a Flir i60 thermal imager. For thermal image registration, the camera angle was standardized, and the distance between the equipment and the animal was maintained at 1 m, as in [39,40]. The camera emissivity was set to 0.98, which is suitable for biological tissues [41], for image correction using Flir software, version 1.9.

The region selected in the animals was the dorsal–caudal region. This region in pigs was also studied by the authors of [42], who evaluated the temperatures of various body regions. Ref. [5] evaluated the welfare and surface temperatures of piglets submitted to different heating systems. Ref. [20] analyzed heat-stress indicators by means of eye temperature. Ref. [43] analyzed the influence of the environment on sow reproduction. Ref. [34] studied the body surface temperature of heifers using thermal imaging.

The extraction step for the coordinates of the selected region was performed with a software system developed in Python language with an interface using the Tkinter library. This step is of great importance for geostatistical analysis because when working with pixel-by-pixel temperatures it is necessary to know the location of each temperature value in the image.

When using the Flir software to perform image information extraction, one notices that it only extracts the temperature values as a whole or from a certain area within the image. This prompted the development of the software.

The routine used for the development consisted of importing and loading the image on the screen and opening a square, where it was moved to the desired region, considering the movement of the mouse over the image, selecting the ROI (lateral array) and displaying the coordinates of the selected window. The window was then closed and the program finished (Figure 2).

The software for extraction of the region of interest in the images was developed in Python language, using the Tkinter library to develop the interface and Opencv for analysis. Initially, by means of the "Open" button, the image is loaded from the user's computer, which allows for its visualization—a feature developed by the function cv2.imread from the Opencv library.

After loading the image, multiple regions of interest can be selected using the "Select ROI" button, which uses the select ROIs function. Using this function in OpenCV, you can manually select precisely the areas of interest, which allows you to perform many tasks for that specific area of the image. It also makes it possible to use the specific area as input for another task, such as drawing a tracing figure (rectangle) using the coordinates or cutting the image with precision and freedom.

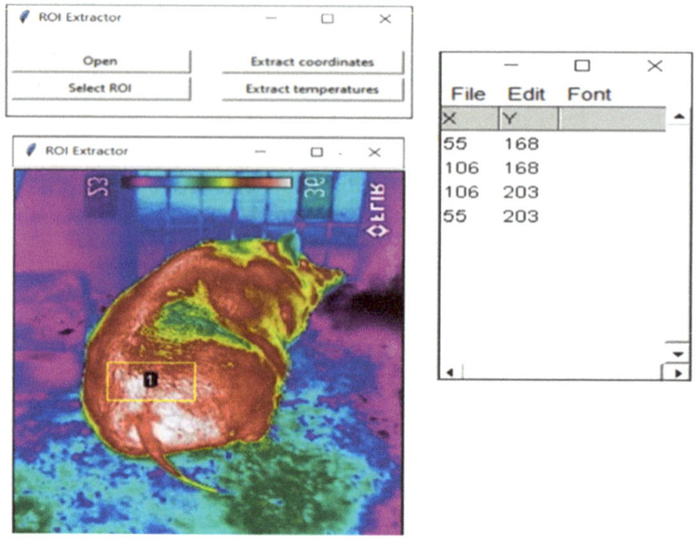

Figure 2. Interface developed for extracting the coordinates from the thermal image. Note: The number 1 represents a point sampled in the caudal dorsal region.

2.5. Statistics

The animal surface temperature data were submitted to exploratory and geostatistical analysis. The exploratory analysis included descriptive statistics and normality analysis, both performed in Excel 2016 software. The parameters determined were the mean, median, mode, standard deviation (SD), variance, coefficient of variation (CV), kurtosis, asymmetry, minimum value, maximum value and the point count. The coefficient of variation was classified according to the proposal of [44], ranging from low (CV < 12%) to medium (12% > CV < 24%) to high (CV > 24%). To study normality, the Kolmogorov–Smirnov test was used, with $\alpha = 0.05$.

Geostatistics was applied according to the methodology proposed in [45] to characterize the temperature variation in the dorsal–caudal regions of the 3 animals in each of the 3 treatments at 8:00 a.m. and 12:00 p.m. We used GS+ software, version 7.0 [46] to generate the graphs.

Once the semivariograms were elaborated, the models that best fitted the data were verified. The models tested were the spherical, exponential and Gaussian models, which, according to [47], were the most relevant. To select the model that best fitted the data, the R^2 value and the degree of spatial dependence (DSD) were determined through the relationship $C/(C0 + C1)$, following the criteria recommended by [48], which classified DSD as strong spatial dependence (DSD < 25%), moderate dependence ($25 \leq$ DSD < 75%) and weak dependence (DSD $\geq$ 75%). The fitted semivariograms were selected and validated using the cross-validation criterion proposed by [49], in which it was observed whether the means and standard deviations of the errors between the observed and estimated data were close to 0 and 1, respectively—a step considered to be of great importance [50].

With the parameters of the semivariogram models selected, kriging maps were made using Surfer 13 software [51].

3. Results and Discussion

The characterization of the pens regarding thermal comfort at 08:00 and 12:00 h was evaluated through the classification of BGTHI for pigs developed in this study. The thermal comfort characterization of the stalls at 8 and 12 h was evaluated through the classification of the BGTHI for pigs developed in this study. The classification is presented in Table 1.

Table 1. BGTHI reclassification for pigs.

Class	Interval
Cold stress	BGTHI ≤ 67
Comfort	68 < BGTHI ≤ 74
Alert	74 < BGTHI ≤ 76
Emergency	BGTHI > 76

The reclassification of BGTHI for pigs took into account critical air temperature and humidity values for pigs. The threshold for the cold-stress class was the minimum optimal temperature, the threshold for the comfort class was the average of the thermoneutral zone for pigs, and the thresholds for the alert and emergency phases were based on the fragmentation of the upper critical temperature. The humidity range varied from 50 to 70%.

The BGTHI values were presented as comfortable in the pens with AEC at 8 and 12 h, ranging from 71.24 to 71.02. However, in the pens with FV, the BGTHI was evaluated as comfortable at 08:00 h, with a value of 73.69, but with stress the BGTHI reached 80.24 at 12:00 h. In the pens with Tbd, the condition was also evaluated as emergency at both time points, with a range of 73.63 to 81.43. These results were similar to those found according to the classification of [52] regarding the black globe temperature and humidity index (BGTHI) for the same facilities. The following Figure 3 presents the classification results.

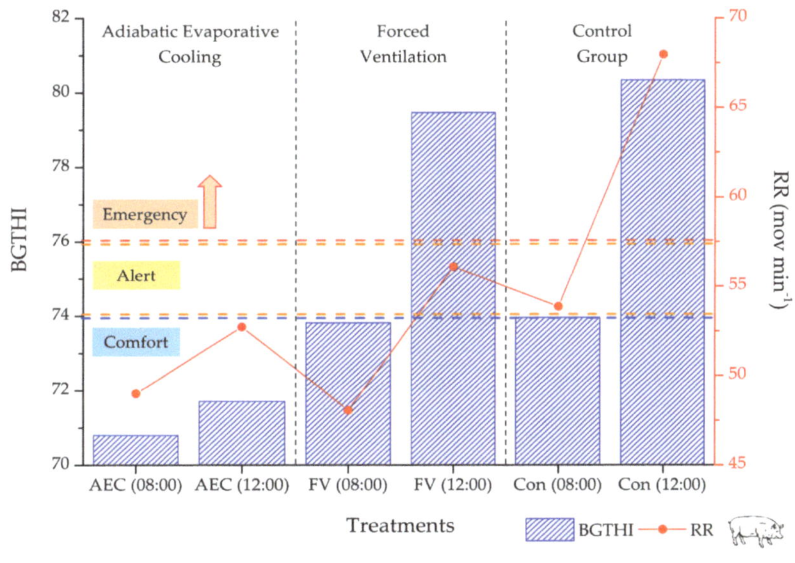

Figure 3. Classification of BGTHI in animal premises.

As observed, the effect of high temperatures in the facilities influenced the respiratory rate of the pigs, indicating possible heat stress. In the FV and Con treatments, as the BGTHI increased, the respiratory rate of the animals also increased.

The results for Tbd and RH as functions of the environments of the treatments, of the external environment and of the times are shown in Table 2.

Table 2. Temperatures and humidities of the air inside the pens and in the external environment.

Time	AEC		FV		Tbd		External Environment	
	Tbd °C	RH%	Tbd °C	RH%	Tbd °C	RH%	Tbd °C	RH%
08:00 a.m.	22.96	70.7	24.87	67.93	24.63	72.5	24.44	65.25
12:00 p.m.	29.46	90.5	32.79	42.06	33.05	47.29	33.31	39.07

The ambient air temperatures and humidities in the stalls with AEC, FV and Tbd treatments at 08:00 a.m. were within the limits considered optimal for pigs in the growing phase, according to the recommendations of [52], which considered a temperature range of 20 to 27.3 °C to be comfortable for pigs in the growth phase and a temperature above 27 °C to be critical, from which it can be inferred that the animals were subjected to periods of heat stress at 12:00 p.m. However, the microclimate in the AEC treatment was already under the influence of the climate control equipment in the first hour. The microclimate of the stall with AEC had a lower value of 1.9 °C at 08:00 a.m., while in the other stalls there were no significant differences at this time. This fact shows the attenuation caused by the air conditioner, while in the stall with the fan it was not possible to detect a greater variation. However, the temperatures in this room were low due to the low solar radiation at this time.

Observing the environments, after 4 h of operation of the climatization systems, it was possible to notice, in relation to the control experiment, that the air conditioner reduced the temperature by up to 3.59 °C, while the fan operation decreased it by only 0.26 °C—a fact found by the authors of [53], who studied thermal comfort in the same facilities and found that the fan had low efficiency with respect to reducing the Tbd.

It can be seen that, in relation to the external environment, the acclimatization provided a high reduction in the Tbd in the hottest hours of the day compared with the non-acclimatized and non-shaded environment, which is favorable when dealing with pigs, which are homeothermic animals. This is explained by the characteristics of the climatizer that uses water evaporation to cool the air that circulates around the animals and reduces the temperature of the environment, providing greater thermal comfort for the pigs. According to [34], animals submitted to environments with temperatures outside the thermal comfort zone demand more energy for the maintenance of body temperature, which can affect the performance and, in more severe cases, cause the death of the animals.

Statistical Analysis

The descriptive statistics for the animals' skin surface temperatures extracted from the thermal images and the KS test analysis are presented in Table 3 as a function of the treatments and their respective times.

Table 3. Descriptive statistics for dorsal–caudal surface temperatures of the pig.

Treatment	Time	Mean	Median	SD	CV	Kustosis	Asymmetry	KS
AEC	08:00 a.m.	32.40	32.6	2.21	6.81	−1.34	−0.20	0.36
AEC	12:00 p.m.	36.25	36.44	0.73	2.00	5.7	−2.16	0.30
FV	08:00 a.m.	32.51	32.89	1.03	3.17	−1.34	−0.41	0.28
FV	12:00 p.m.	36.81	36.80	0.17	0.48	0.80	0.32	0.40
Con	08:00 a.m.	37.53	37.57	0.25	0.68	−0.79	−0.08	0.24
Con	12:00 p.m.	38.45	38.53	0.28	0.73	0.42	−1.03	0.29

The means and medians for each of the animals analyzed showed close values, which shows that the data were close to a normal distribution. Similar results were found in [6]. The coefficient of variation (CV) in statistical analyses serves as a first indicator of whether or not data heterogeneity exists. Considering the research of [44], the dorsal–caudal surface temperatures of the pigs in the three treatments showed low variability (CV < 12%), and the mean was representative. Most animals presented kurtosis and asymmetry values close to 0, except for the temperature collected at 12:00 p.m. in the AEC treatment, with values of 5.7 and −2.16, respectively. However, the Kolmogorov–Smirnov test results were not significant for all the animals analyzed.

It was observed that the animals in the stall with AEC had the lowest mean temperatures in the dorsal–caudal region (32.40 and 36.25 °C) when compared with the animals submitted to the FV treatment, which had values of 32.51, 32.59 and 36.81, and those under the Con treatment, which had values of 37.15, 37.53 and 38.45 °C, reflecting the importance of the acclimatization with AEC during the hours of 08 a.m. and 12 p.m. These results were close to those obtained by the authors of [42], who analyzed the temperatures of various body regions of pigs and found an average temperature for the caudal–dorsal region of 34.6 °C. For [54], in housing conditions within the thermoneutral zone, the skin temperature (SST) of pigs should be close to 34 ± 1 °C. Thus, the use of descriptive statistics was able to distinguish the animals submitted to the different acclimatization systems. Ref. [5] found the SSTs of pigs in concrete facilities to be similar to those recorded in this research, ranging from 39 to 41 °C. In the findings of [55], it was mentioned that body surface temperatures in the lumbar and dorsal regions have a high correlation with RT, which is considered the most suitable for evaluating temperatures in animals. Ref. [20] cited dorsal temperature as an indicator for describing the environment the animal is inhabiting.

Table 4 illustrates a proposed model fitted to the experimental data; the nugget effect, the sill, the range, the degree of spatial dependence (DSD) and the R^2 value for each of the animals were analyzed.

Table 4. Semivariogram parameters.

Treatments	Time	Models	Nugett	Sill	Range	DSD	R^2
AEC	08:00 a.m.	Gaussian	0.510	4.600	05.100	87.800	0.810
AEC	12:00 p.m.	Gaussian	0.120	0.500	22.600	75.500	0.930
FV	08:00 a.m.	Gaussian	0.200	1.660	27.450	98.800	0.980
FV	12:00 p.m.	Gaussian	0.030	0.100	43.200	99.230	0.990
Con	08:00 a.m.	Gaussian	0.010	0.030	28.800	71.600	0.970
Con	12:00 p.m.	Gaussian	0.003	0.140	54.200	97.740	0.990

The Gaussian model was the theoretical model with the best fit to the temperature data for the dorsal–caudal regions of the animals in all climatization systems. Such a model was validated by presenting the means between the actual values and those estimated by the model closer to zero and the standard deviation of one. According to [56], the validation of semivariograms is performed using the residuals, which is the way to check the assumptions of the model used in kriging, making the comparison between the actual values and the kriged ones, where such residuals show the performance of the model. The Gaussian model was also observed in the results of [34]—a study on the variability of body SST in cows—and in [57]—a study on the variability of udder temperature in cows.

For all the treatments analyzed, no pure nugget effect was detected, showing that there was a correlation between the different types of climatization and the SSTs of the animals; however, the results showed the low precision of the sampled grid [48]. In the AEC treatment, the nugget effect ranged from 0.51 to 0.12; in FV, the range was 0.20 to 0.03; and in Con, it was 0.01 to 0.001.

The sill indicates the point from which the samples become independent because of the distance separating them. In the facilities with AEC, the sill varied from 4.60 to 0.50. In the facilities with FV, it ranged from 0.66 to 0.10, and in the facilities with Tbd treatment, the sill ranged from 0.03 to 0.14. Similar plateau values were found in [56], which revealed that the lower the sill value, the more similar the neighboring values are.

The range indicates the distance (the region on the surface of the pig) over which the sample points (temperatures) are more similar to each other and could be used for estimating values with closer spacings than those separated by larger distances. In the AEC, the largest value was 22.60 mm at 12:00 p.m., and the smallest value was 05.10 mm at 08:00 a.m. The average value was 13.8 mm. In the FV treatment, the range was 27.20 to 43.20 mm at 08:00 a.m. and 12:00 p.m., and the average value was 35.2 mm. The Con treatment was similar to the FV treatment, with a greater range of 28.80 (08:00 a.m.) to 54.20 mm (12:00 p.m.) and an average value of 41.50 mm. The highest ranges obtained were for the Con treatment, followed by the FV treatment, and the lowest range was found for the AEC treatment.

According to the classification proposed in [48], the dorsal–caudal surface temperatures fell into the moderate-to-strong class of spatial dependence in all proposed models. This result was also found by the authors of [34], who, when studying temperature variability in cows, detected that the model with the best theoretical fit was the Gaussian model with a strong degree of spatial dependence.

The spatial variations of the temperatures analyzed in the dorsal–caudal region (Figure 4) of the animals are illustrated in Figures 5–7.

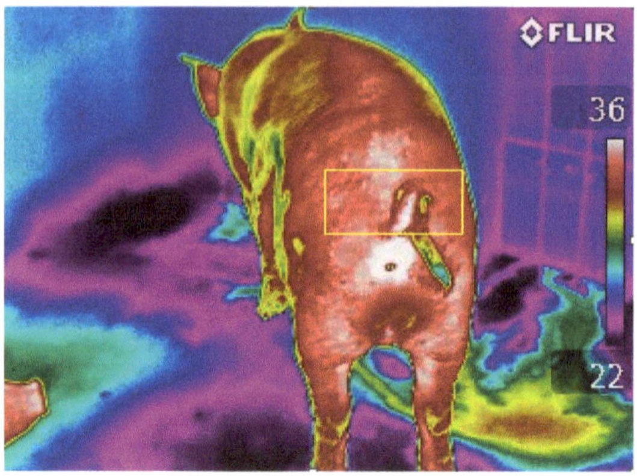

Figure 4. Dorsal–caudal region studied.

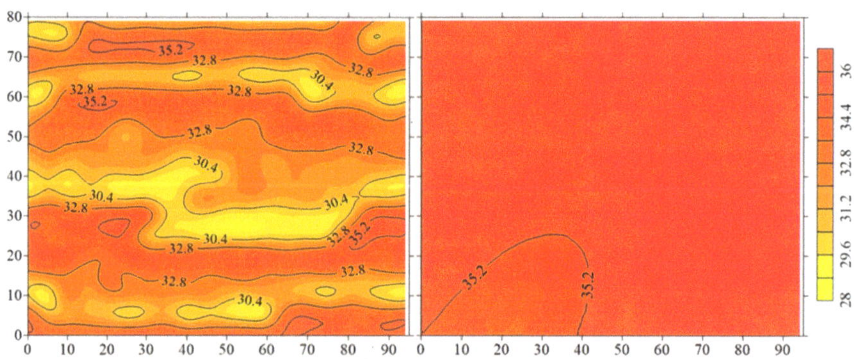

Figure 5. Kriging maps of the dorsal–caudal region for the AEC treatment at 8:00 a.m. and 12:00 p.m.

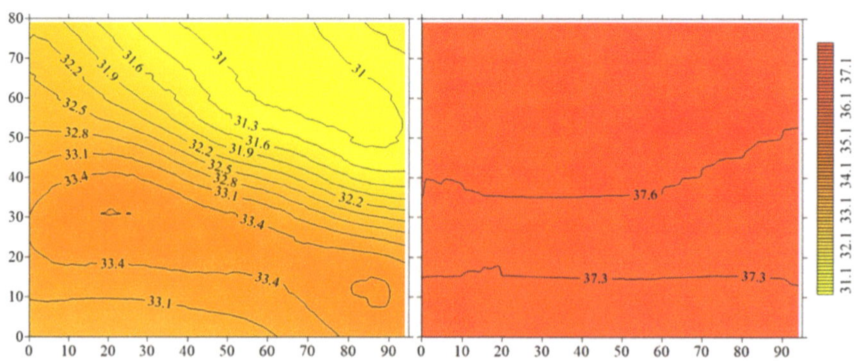

Figure 6. Kriging maps of the dorsal–caudal region for the FV treatment at 08:00 a.m. and 12:00 p.m.

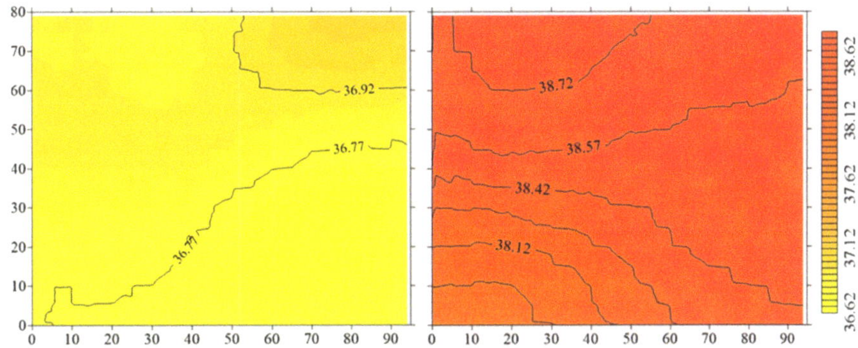

Figure 7. Kriging maps of the dorsal–caudal region for the T treatment at 08:00 a.m. and 12:00 p.m.

These figures show the kriging maps, which are widely used in geostatistics to estimate unknown values at unsampled locations based on collected samples. In the context of the surface temperature of animals, kriging maps can be applied to estimate the spatial distribution of surface temperatures over the skin. The values estimated by kriging were similar to those found through the measurements of the authors of [39], obtained using infrared thermography in pigs, who found values ranging from 32 to 34 °C, similar to the values recorded for the animals in the AEC and FV treatments at 08:00 h in the morning. The authors of [39] also found temperature values in the caudal–dorsal region ranging from 35.8 to 37.7 °C, which were similar to the values found in this study for the animals at 12:00 h, validating the results of the kriging.

The elaborated kriging maps show the temperature distributions in the dorsal–caudal regions of the animals. Along the X- and Y-axes are the coordinates of the dorsal–caudal regions studied in millimeters. The isolines represent the estimated temperature values. These values varied from 28 to 36°C. The elaborated kriging maps showed the temperature distributions in the dorsal–caudal regions of the animals. In [24], the kriging maps also presented the distribution of air temperature values in the state of Minas Gerais. In this way, it was possible to visualize the highest temperature values in the hottest hours of the analyzed period (redder colors) in all treatments, showing the relationship between the animal's skin temperature values and the environment.

Therefore, it can be seen that the highest temperatures found were for the animals in the stall with the Con treatment (natural acclimatization), followed by the animals in the stall with the FV treatment and, lastly, the animals in the stall with the AEC treatment. The animal housing for the Con treatment, by absorbing solar radiation in its structure, heats the microclimate, and as this, in turn, is not being influenced by air conditioning equipment capable of attenuating the temperature, it ends up heating the animal's skin, thus increasing the temperature in the caudal–dorsal region. What happened in the FV treatment is explained by the inefficiency of the ventilator in controlling the temperature, while in the stall with AEC, the animals remained at temperatures considered optimal throughout the analyzed period, justifying the importance of acclimatization for the production environment.

In general, it was possible to visualize the information on the spatial dependence of the semivariograms expressed in the kriging maps. Such information reflects the variability in animal surface skin temperatures related to climate systems and is useful for identifying whether housings require more or less care.

Regarding the rectal temperatures of the animals subjected to AEC, the average data showed values of 38.3 °C and 38.8 °C for the animals at 08:00 and 12:00 h, respectively. The animals in the FV treatment had values of 38.5 °C and 38.7 °C at 08:00 and 12:00 h, while the animals in the control group had average values of 38.7 °C and 39.7 °C. As for the average respiratory rate, the animals in the AEC treatment had values of 49 and 52.73 breaths per minute, those in the FV treatment had values of 48.7 and 56.10 breaths per minute, and those in the Con treatment had values of 53.87 and 67.97 breaths per minute. These results indicate that the animals under the effect of the air conditioner showed lower physiological values compared to the other animals, with lower respiratory rates.

4. Conclusions

The results show that thermal imaging and geostatistical analysis are able to provide valuable information about the thermal behavior of pigs and the characteristics of their environments. Furthermore, a combination of these tools contributes to the implementation of appropriate management measures and the improvement of animal welfare. The geostatistical analysis provided a mathematical model capable of illustrating the variation in temperature in the caudal–dorsal regions of the pigs according to the environments to which the animals were subjected, showing the potential of these tools for monitoring the body temperature of pigs in different environments and combining this data with information on spatial variability transmitted by geostatistical analysis.

The work highlights the importance of using thermal imaging and geostatistics in animal production, highlighting their potential to provide accurate and auxiliary information on animal performance and welfare, as well as to assist in decision making regarding production management. The use of these tools can contribute significantly to the promotion of more sustainable and efficient practices in animal production, benefiting both producers and consumers.

Author Contributions: Conceptualization, M.d.F.A.A. and H.P.; methodology, A.A.d.A.M.; software, R.A.B.d.S.; validation, N.F.G., T.C.S. and G.L.P.d.A.; formal analysis, G.T.B.M.; investigation, M.V.d.S.; resources, W.A.d.S.; data curation, M.V.d.S.; writing—original draft preparation, T.C.S.; writing—review and editing, T.C.S.; visualization, H.P.; supervision, M.V.d.S.; project administration, M.d.F.A.A.; funding acquisition, T.C.S. All authors have read and agreed to the published version of the manuscript.

Funding: This research received no external funding.

Institutional Review Board Statement: The research was approved by the CEUA/UFRPE (Animal Ethics Committee) under protocol number 23082.021090/2016-81, date of license issuance: 5 December 2016.

Informed Consent Statement: Not applicable.

Data Availability Statement: Not applicable.

Acknowledgments: The authors wish to acknowledge the *Programa de Pós-Graduação em Engenharia Agrícola* (PGEA) and the *Grupo de Pesquisa em Ambiência* (GPESA) of the *Universidade Federal Rural de Pernambuco* (UFRPE) for the necessary support for the development of this study. The authors would also like to acknowledge the *Coordenação para o Aperfeiçoamento do Pessoal do Ensino Superior* (CAPES—Financial Code 001) and the *Fundação de Apoio à Investigação do Estado de Pernambuco* (FACEPE) for the provision of research scholarships.

Conflicts of Interest: The authors declare no conflict of interest.

References

1. ABPA. Relatório Anual 2021. Available online: https://abpa-br.org/quem-somos/abpa-relatorio-anual/ (accessed on 24 March 2021).
2. Embrapa Embrapa Suínos e Aves. Available online: https://www.embrapa.br/suinos-e-aves/cias/estatisticas (accessed on 1 March 2022).
3. Galvão, A.T.; da Silva, A.d.S.L.; Pires, A.P.P.; de Morais, Á.F.F.; Mendonça Neto, J.S.N.; de Azevedo, H.H.F. Bem-Estar Animal na Suinocultura: Revisão. *Pubvet* **2019**, *13*, 148. [CrossRef]
4. OIE. Terrestrial Animal Health Code. Available online: https://www.oie.int/en/standard-setting/terrestrial-code/access-online (accessed on 6 March 2021).
5. Damasceno, F.A.; Magela Soares, C.; Eduardo, C.; Oliveira, A.; Freire, L.; Damasceno, B.; Ferreira, P.; Ferraz, P. Evaluation of Behavior in Piglets Using Digital Image Processing. *Energ. Agric.* **2019**, *34*, 217–229. [CrossRef]

6. Medeiros, B.B.; Moura, D.J.D.; Massari, J.M.; Curi, T.M.D.C.; Maia, A.P.D.A. Uso da Geoestatística na Avaliação de Variáveis Ambientais em Galpão de Suínos Criados Em Sistema" Wean to Finish" Na Fase de Terminação. *Eng. Agríc.* **2014**, *34*, 800–811. [CrossRef]
7. Soares, E.A.; Leal, A.F.; Malheiros Filho, J.R.; Camerini, N.L.; Nascimento, J.W.B.; Dermeval, A.F. Zoneamento Bioclimático Para Produção de Suínos na Maternidade no Município de Areia—PB. Available online: https://www.confea.org.br/sites/default/files/uploads-imce/Contecc2019/Agronomia/DIAGNOSTICO%20BIOCLIMATICO%20PARA%20PRODU%C3%87%C3%83O%20DE%20SUINOS%20NO%20MUNICIPIO%20DE%20AREIA-PB.pdf (accessed on 29 March 2023).
8. Borges, P.H.M.; de Mendoza, Z.M.S.H.; Morais, P.H.M.; dos Santos, R.L. Artificial Neural Networks for Predicting Animal Thermal Comfort. *Eng. Agríc.* **2018**, *38*, 844–856. [CrossRef]
9. dos Santos, T.C.; Carvalho, C.D.C.S.; da Silva, G.C.; Diniz, T.A.; Soares, T.E.; Moreira, S.d.J.M.; Cecon, P.R. Influência do Ambiente Térmico no Comportamento e Desempenho Zootécnico de Suínos. *Rev. Ciênc. Agroveter.* **2018**, *17*, 241–253. [CrossRef]
10. Pandorfi, H.; Almeida, G.L.P.; Guiselini, C. Zootecnia de Precisão: Princípios Básicos e Atualidades na Suinocultura. *Rev. Bras. Saúde Produção Anim.* **2012**, *13*, 558–568. [CrossRef]
11. Pandorfi, H.; Da Silva, I.J.O.; Piedade, S.M.S. Conforto Térmico para Matrizes Suínas em Fase de Gestação, Alojadas em Baias Individuais e Coletivas. *Rev. Bras. Eng. Agríc. Ambient.* **2008**, *12*, 326–332. [CrossRef]
12. Thom, E.C. The Discomfort Index. *Weatherwise* **1959**, *12*, 57–61. [CrossRef]
13. Buffington, D.E.; Collazo-Arocho, A.; Canton, G.H.; Pitt, D.; Thatcher, W.W.; Collier, R.J. Black Globe-Humidity Index (BGHI) as Comfort Equation for Dairy Cows. *Trans. ASAE* **1981**, *24*, 711–714. [CrossRef]
14. Esmay, M.L. *Principles of Animal Environment*, 2nd ed.; American Society of Agricultural Engineers: St. Joseph, MI, USA, 1982.
15. Albright, L.D. *Environment Control for Animals and Plants*; American Society of Agricultural Engineers: St. Joseph, MI, USA, 1990; Volume 4.
16. Nasirahmadi, A.; Hensel, O.; Edwards, S.A.; Sturm, B. A New Approach for Categorizing Pig Lying Behaviour Based on a Delaunay Triangulation Method. *Animal* **2017**, *11*, 131–139. [CrossRef]
17. Castro, J.D.O.; Campos, A.T.; Ferreira, R.A.; Júnior, T.Y.; Tadeu, H.C. Uso de Ardósia na Construção de Celas de Maternidade: I—Efeito Sobre o Ambiente e Comportamento de Suínos. *Eng. Agríc.* **2011**, *31*, 458–467. [CrossRef]
18. Villa, F.; Nivaldo, A.; Junior, K.; De Oliveira, C.C. *Aplicações Da Termografia Por Infravermelho (TIV) Na Bovinocultura de Corte*; Repositorio de Información Tecnológica de Embrapa: Brasília, Brazil, 2020.
19. Warriss, P.D.; Pope, S.J.; Brown, S.N.; Wilkins, L.J.; Knowles, T.G. Estimating the Body Temperature of Groups of Pigs by Thermal Imaging. *Vet. Rec.* **2006**, *158*, 331–334. [CrossRef] [PubMed]
20. Pulido-Rodríguez, L.F.; Titto, E.A.L.; Henrique, F.L.; Longo, L.S.; Hooper, H.B.; Pereira, T.L.; Pereira, A.M.F.; Titto, C.G. Infrared Thermography of the Ocular Surface as Stress Indicator for Piglets Postweaning. *Pesqui. Vet. Bras.* **2017**, *37*, 453–458. [CrossRef]
21. Vieira, F.M.; Gouveia, A.B.V.S.; de Paulo, L.M.; Sampaio, S.A.; Borges, K.F.; da Silva, N.F.; dos Santos, F.R.; Minafra, C.S. Termografia infravermelha na avicultura. *Veter. Zootec.* **2022**, *29*, 1–21. [CrossRef]
22. da Silva, T.C.S.; Mariz, T.M.d.A.; Escodro, P.B. Use of Thermography in Clinical and Sports Evaluations of Equine Animals: A Review. *Res. Soc. Dev.* **2022**, *11*, e13911530532. [CrossRef]
23. Cardoso, L.A.S.; Farias, P.R.d.S.; Soares, J.A.C. Geoestatística Aplicada à Fitossanidade: Estado da Arte e Perspectivas Futuras para o Estado do Pará, Brasil. Available online: https://www.journalijdr.com/geoestat%C3%ADstica-aplicada-%C3%A0-fitossanidade-estado-da-arte-e-perspectivas-futuras-para-o-estado-do-par%C3%A1 (accessed on 29 March 2023).
24. Viana, R.S.M.; dos Santos, G.R.; Moreira, D.S.; Louzada, J.M.; Rosa, L.M.F. O Uso da Geoestatística Espaço-Temporal na Predição da Temperatura Máxima do Ar. *Rev. Bras. Geogr. Fís.* **2019**, *12*, 96–111. [CrossRef]
25. Marques, S.d.F.; Pitombo, C.S. Intersecting Geostatistics with Transport Demand Modeling: A Bibliographic Survey. *Rev. Bras. Cartogr.* **2020**, *72*, 1004–1027. [CrossRef]
26. Da Silva, H.M.; Sarnighausen, V.C.R. Análise Geoestatística de Séries Temporais de Temperatura do Ar, Evapotranspiração de Referência e Precipitação Pluvial. *Rev. Ibero-Am. Ciênc. Ambient.* **2021**, *12*, 710–717. [CrossRef]
27. Silva Neto, S.P.; Silva, R.G.; Santos, A.C.; Gama, F.R.; Guerra, M.S.S.; Brito, M.J.D. Padrões Espaciais de Deposição de Fezes por Bovinos de Corte Em Áreas de Pastagem. *Rev. Bras. Saúde Produção Anim.* **2011**, *12*, 538–550.
28. Nääs, I.D.A.; Garcia, R.G.; Graciano, D.E.; De Santana, M.R.; Caldara, F.R.; Dourados, G. Temperatura Superficial de Porcas em Lactação Submetidas ao Resfriamento Adiabático. *Encicl. Biosf.* **2013**, *9*, 2006–2013.
29. Jia, G.; Li, W.; Meng, J.; Tan, H.; Feng, Y. Non-Contact Evaluation of Pigs' Body Temperature Incorporating Environmental Factors. *Sensors* **2020**, *20*, 4282. [CrossRef] [PubMed]
30. Basak, J.K.; Okyere, F.G.; Arulmozhi, E.; Park, J.; Khan, F.; Kim, H.T. Artificial Neural Networks and Multiple Linear Regression as Potential Methods for Modelling Body Surface Temperature of Pig. *J. Appl. Anim. Res.* **2020**, *48*, 207–219. [CrossRef]
31. Bezerra, A.C.; da Silva, J.L.B.; Silva, D.A.d.O.; Batista, P.H.D.; Pinheiro, L.d.C.; Lopes, P.M.O.; Albuquerque, G.B. Monitoramento Espaço-Temporal da Detecção de Mudanças em Vegetação de Caatinga por Sensoriamento Remoto no Semiárido Brasileiro. *Rev. Bras. Geogr. Fís.* **2020**, *13*, 286–301. [CrossRef]
32. Jardim, A.M.d.R.F.; de Queiroz, M.G.; Júnior, G.D.N.A.; da Silva, M.J.; da Silva, T.G.F. Estudos Climáticos Do Número de Dias de Precipitação Pluvial para o Município de Serra Talhada-PE. *Rev. Eng. Agric.—REVENG* **2019**, *27*, 330–337. [CrossRef]
33. Barbosa Filho, J.A.D.; Silva, I.J.O.; Silva, M.A.N.; Silva, C.J.M. Avaliação Dos Comportamentos de Aves Poedeiras Utilizando Seqüência de Imagens. *Eng. Agríc.* **2007**, *27*, 93–99. [CrossRef]

34. Batista, P.H.D.; de Almeida, G.L.P.; Pandorfi, H.; da Silva, M.V.; da Silva, R.A.B.; da Silva, J.L.B.; Santana, T.C.; Rodrigues, J.A.d.M. Thermal Images to Predict the Thermal Comfort Index for Girolando Heifers in the Brazilian Semiarid Region. *Livest. Sci.* **2021**, *251*, 104667. [CrossRef]
35. Baêta, F.C.; Souza, S.C. *Ambiência Em Edificações Rurais—Conforto Animal*; UFV: Abbotsford, BC, Canada, 1997; Volume 246.
36. Campos, J.A.; Tinôco, I.d.F.F.; Baêta, F.d.C.; Cecon, P.R.; Mauri, A.L. Air Quality, Thermal Environment and Performance of Pigs Raised in Nurseries with Different Dimensions. *Agric. Eng.* **2009**, *29*, 339–347. [CrossRef]
37. Lawrence, M.G. The Relationship between Relative Humidity and the Dewpoint Temperature in Moist Air: A Simple Conversion and Applications. *Bull. Am. Meteorol. Soc.* **2005**, *86*, 225–233. [CrossRef]
38. Teixeira, D.L.; Boyle, L.A.; Enríquez-Hidalgo, D. Skin Temperature of Slaughter Pigs With Tail Lesions. *Front. Vet. Sci.* **2020**, *7*, 198. [CrossRef]
39. Moreira, V.E.; Veroneze, R.; Teixeira, A.D.R.; Campos, L.D.; Lino, L.F.L.; Santos, G.A.; Silva, B.A.N.; Campos, P.H.R.F. Effects of Ambient Temperature on the Performance and Thermoregulatory Responses of Commercial and Crossbred (Brazilian Piau Purebred Sires × Commercial Dams) Growing-Finishing Pigs. *Animals* **2021**, *11*, 3303. [CrossRef]
40. Mostaço, G.M.; Miranda, K.O.D.S.; da S Condotta, I.C.F.; Salgado, D.D. alessandro Determination of Piglets' Rectal Temperature and Respiratory Rate through Skin Surface Temperature under Climatic Chamber Conditions. *Eng. Agric.* **2015**, *35*, 979–989. [CrossRef]
41. Rauber, S.M. *Termografia Infravermelho na Avaliação de Estratégias de Alimentação de Precisão Para Suínos*; LUME: Porto Alegre, Brazil, 2016.
42. Nääs, I.d.A. A Iinfluência do Meio Ambiente na Reprodução. In Proceedings of the Simpósio Internacional de Suinocultura, São Paulo, Brazil, 9–12 July 2000.
43. Warrick, A.W.; Nielsen, D.R. Spatial Variability of Soil Physic Properties in the Field. *N. Y. Acad.* **1980**, 655–675. [CrossRef]
44. Vieira, S.R. Geoestatística em Estudos de Variabilidade Espacial do Solo. *Tópicos Ciência Solo. Viçosa Soc. Bras. Ciência Solo* **2000**, *1*, 1–54.
45. *Gamma Design Software GS+—Geostatistics for the Environmental Sciences*; Gamma Design Software: Plainwell, MI, USA, 2012. Available online: http://www.gammadesign.com/ (accessed on 16 October 2022).
46. Obroślak, R.; Dorozhynskyy, O. Selection of a Semivariogram Model in the Study of Spatial Distribution of Soil Moisture. *J. Water Land Dev.* **2017**, *35*, 161–166. [CrossRef]
47. Cambardella, C.A.; Moorman, T.B.; Novak, J.M.; Parkin, T.B.; Karlen, D.L.; Turco, R.F.; Konopka, A.E. Field-Scale Variability of Soil Properties in Central Iowa Soils. *Soil Sci. Soc. Am. J.* **1994**, *58*, 1501–1511. [CrossRef]
48. Vauclin, M.; Vieira, S.R.; Vachaud, G.; Nielsen, D.R. The Use of Cokriging with Limited Field Soil Observations. *Soil Sci. Soc. Am. J.* **1983**, *47*, 175–184. [CrossRef]
49. Santarosa, L.V.; Manzione, R.L. Soil Variables as Auxiliary Information in Spatial Prediction of Shallow Water Table Levels for Estimating Recovered Water Volume. *RBRH* **2018**, *23*, e24. [CrossRef]
50. Golden Software. *Golden Software Surfer for Windows: Realese 7.0, Contouring and 3D Surface Mapping for Scientist's Engineers User's Guide*; Golden Software: Golden, CO, USA, 2016.
51. Rinaldo, D.; Le Dividich, J.; Noblet, J. Adverse Effects of Tropical Climate on Voluntary Feed Intake and Performance of Growing Pigs. *Livest. Prod. Sci.* **2000**, *66*, 223–234. [CrossRef]
52. Gomes, N.F.; Barnabe, J.M.C.; Silva, W.A.; Pandorfi, H.; Guiselini, C.; Almeida, G.L.P. Acondicionamento Térmico para Suínos na Fase de Crescimento no Semiárido Pernambucano. In Proceedings of the Simpósio Internacional de Ambiência e Engenharia na Produção Animal; Simpósio Internacional de Ambiência e Engenharia na Produção Animal, Lavras, Brazil, 5–7 June 2019.
53. Malmkvist, J.; Pedersen, L.J.; Kammersgaard, T.S.; Jørgensen, E. Influence of Thermal Environment on Sows around Farrowing and during the Lactation Period. *J. Anim. Sci.* **2012**, *90*, 3186–3199. [CrossRef]
54. Magano, D.A.; Machado, M.R.R.; Jerónimo, J.A.; Guedes, J.V.C.; Doberstein, A.P.S. Modelagem Geostatistica Aplicada a Distribuição Espacial de Lagartas Presentes na Cultura da Soja. *Braz. J. Anim. Environ. Res.* **2023**, *6*, 20–29. [CrossRef]
55. da Silva, R.Â.B.; Pandorfi, H.; de Almeida, G.L.P.; de Assunção Montenegro, A.A.; da Silva, M.V. Spatial Dependence of Udder Surface Temperature Variation in Dairy Cows with Healthy Status and Mastitis. *Rev. Bras. Saude Prod. Anim.* **2019**, *20*. [CrossRef]
56. de Carvalho, J.R.P.; Assad, E.D.; Evangelista, S.R.M.; Pinto, H.D.S. Estimation of Dry Spells in Three Brazilian Regions—Analysis of Extremes. *Atmos. Res.* **2013**, *132–133*, 12–21. [CrossRef]
57. Ricci, G.D.; da Silva-Miranda, K.O.; Titto, C.G. Infrared Thermography as a Non-Invasive Method for the Evaluation of Heat Stress in Pigs Kept in Pens Free of Cages in the Maternity. *Comput. Electron. Agric.* **2019**, *157*, 403–409. [CrossRef]

Disclaimer/Publisher's Note: The statements, opinions and data contained in all publications are solely those of the individual author(s) and contributor(s) and not of MDPI and/or the editor(s). MDPI and/or the editor(s) disclaim responsibility for any injury to people or property resulting from any ideas, methods, instructions or products referred to in the content.

sustainability

Article

Wheat Straw Biochar Produced at a Low Temperature Enhanced Maize Growth and Yield by Influencing Soil Properties of *Typic calciargid*

Muhammad Aon [1,2,*], Zeshan Aslam [2], Shahid Hussain [1], Muhammad Amjad Bashir [3,*], Muhammad Shaaban [1], Sajid Masood [1], Sidra Iqbal [4], Muhammad Khalid [2], Abdur Rehim [1], Walid F. A. Mosa [5], Lidia Sas-Paszt [6], Samy A. Marey [7] and Ashraf Atef Hatamleh [8]

1 Department of Soil Science, Faculty of Agricultural Sciences and Technology, Bahauddin Zakariya University, Multan 60800, Pakistan
2 Institute of Soil and Environmental Sciences, University of Agriculture, Faisalabad 38000, Pakistan
3 Department of Agronomy, Engro Fertilizers Ltd., Lahore 54000, Pakistan
4 Department of Plant Breeding and Genetics, University of Agriculture Faisalabad, Sub-Campus Depalpur, Okara 38040, Pakistan
5 Plant Production Department (Horticulture-Pomology), Faculty of Agriculture, Saba Basha, Alexandria University, Alexandria 21531, Egypt
6 The National Institute of Horticultural Research, Konstytucji 3 Maja 1/3, 96-100 Skierniewice, Poland
7 King Saud University, P.O. Box 2455, Riyadh 11451, Saudi Arabia
8 Department of Botany and Microbiology, College of Science, King Saud University, P.O. Box 2455, Riyadh 11451, Saudi Arabia
* Correspondence: dr.maon@bzu.edu.pk (M.A.); amjad.bashir941@gmail.com (M.A.B.)

Abstract: Arid and semi-arid soils are low in organic matter and have poor fertility, making them a serious threat to crop production. Most organic amendments, such as crop residues and farmyard manure, are short lived because of rapid decomposition. Incubation and pot studies were conducted to assess the impact of wheat straw biochar (produced at 350 °C) on temporal changes in soil microbial biomass and fertility status and to evaluate the efficacy of biochar for maize production in the top layer of *Typic calciargid*. The incubation study compared four levels of biochar (control, 0.5, 1.0 and 2.0% on a w/w basis of soil) and two fertilizer rates, i.e., unfertilized (no NPK fertilizer) and fertilized (nitrogen, P_2O_5 and K_2O with rates of 125, 80 and 52.5 mg kg^{-1} soil, respectively). After incubation, the 2.0% biochar significantly improved the soil cation exchange capacity, organic carbon and microbial biomass carbon by up to 35, 59 and 26%, respectively, while decreasing the soil pH by up to 1.5% compared to that of the control treatment. When fertilized, the 2.0% biochar improved the soil's available phosphorous, extractable potassium and total nitrogen by up to 59, 39 and 28%, respectively, compared to those of the control. The results from the pot experiment showed that using the 1% biochar with fertilizer significantly increased the maize dry biomass and grain yield by up to 57 and 72%, respectively, compared to those of the control. Additionally, the nitrogen and phosphorus recoveries from the mineral fertilizers improved significantly (up to 26 and 38%, respectively) when using the 1.0% biochar compared to those of the control. Conclusively, the addition of 1.0% biochar significantly improved maize growth and yield by enhancing nutrient recovery from mineral fertilizer and improving soil properties.

Keywords: wheat straw biochar; alkaline calcareous soil; microbial biomass carbon; nutrient recovery; soil organic carbon

Citation: Aon, M.; Aslam, Z.; Hussain, S.; Bashir, M.A.; Shaaban, M.; Masood, S.; Iqbal, S.; Khalid, M.; Rehim, A.; Mosa, W.F.A.; et al. Wheat Straw Biochar Produced at a Low Temperature Enhanced Maize Growth and Yield by Influencing Soil Properties of *Typic calciargid*. *Sustainability* **2023**, *15*, 9488. https://doi.org/10.3390/su15129488

Academic Editor: Jan Hopmans

Received: 2 April 2023
Revised: 31 May 2023
Accepted: 6 June 2023
Published: 13 June 2023

Copyright: © 2023 by the authors. Licensee MDPI, Basel, Switzerland. This article is an open access article distributed under the terms and conditions of the Creative Commons Attribution (CC BY) license (https://creativecommons.org/licenses/by/4.0/).

1. Introduction

Biochar may persist in soils and sediments for hundreds of years [1–3] and has the potential to be used in agronomic crop production [4]. Its high surface area and porosity enable it to retain water and essential nutrients for a longer period and to provide the best habitat for soil microorganisms [5,6]. Soil applications of various organic amendments may have some limitations (i.e., early decomposition). However, a biochar application may be an effective strategy to overcome such limitations.

Recent research has reported the usefulness of soil amendments to improve agronomic and fertilizer use efficiencies. Manure, mulches and cover crops have been proven to be effective in supporting the rapid cycling of plant nutrients in the soil through the microbial population and supplying these essential plant nutrients to different crops [7–9]. However, the commonly used organic amendments are usually short lived and mineralized to CO_2. Thus, to sustain soil productivity, repeated applications of organic amendments are recommended [10]. Biochar, on the other hand, stays in soils for longer periods. It could help fulfill the growing need to turn marginal and unproductive lands into fertile and productive lands.

In soil environments, biochar interacts with several plant nutrients. The binding ability of plant essential cations is significantly improved by the addition of soil biochar [11,12]. Thus, improved cationic retention (i.e., NH_4^+ retention) and slower ion release in biochar-amended soil may be because of its high cation exchange capacity (CEC). Biochar has the potential to reduce the losses of mineral nitrogen (N) added as ammoniacal N. The addition of biochar is also a source of labile carbon (C) and as a result, the N immobilization of microbial biomass is another mechanism that contributes to the retention of N in top soil [13,14]. The availability of potassium (K) to plants is mainly affected by the physical properties of the soil, including its hydraulic conductivity, infiltration rate, bulk density and soil aggregation [15]. Biochar applications improve the physical properties of soil by increasing the soil organic matter status, which acts as a soil conditioner [16]. Therefore, biochar can play a significant role in K availability to plants. Biochar may also have a high anion exchange capacity, so it may also enhance phosphorus (P) availability by controlling the activity of the cations that interact with soil P [17]. Biochar applications to alkaline soils may be an effective strategy to reduce soil pH [18]. Phosphorus redistribution, mineralization and immobilization in soils depend on the physicochemical properties of soils [19,20]. Although both the phosphate ions and biochar surface may have negative charges, biochar can still absorb the phosphate ions on it through an interaction that can overcome electrostatic repulsion [21].

Various factors, such as the temperature and type of feedstock used in the pyrolysis process, influence the characteristics of biochar. These factors contribute to a wide range of values for properties such as pH, specific surface area, nutrient content, CEC, ash content and C content [22]. The temperature at which pyrolysis occurs has a significant impact on both the physical and chemical properties of biochar. This is because the structure and chemical bonds undergo stagewise decomposition during the pyrolysis process [23]. The pH of biochar generally falls within a range from acidic to alkaline, with lower-pH biochars often being overlooked. However, as the pyrolysis temperature increases, the pH of biochar tends to rise due to the depletion of acidic functional groups at higher temperatures [24]. Nonetheless, it should be noted that biochars produced at lower temperatures can still exhibit neutral to acidic properties [25,26]. Higher pyrolysis temperatures tend to yield biochar with elevated pH levels but lower CEC values and N content [22,27]. A few works from the literature have suggested that the lower pH resulting from lower temperature pyrolysis can be neutral to acidic, initially increasing the availability of nutrients for plants in arid calcareous soils [24]. Similarly, neutral biochars may exhibit different behaviors compared to commonly available alkaline biochars when introduced into soil during environmental processes.

Alkaline calcareous (*Typic calciargid*) soils generally have a high pH with a low content of organic matter. This type of soil lacks macronutrients (i.e., N and P) and micronutrients (i.e., zinc, iron and boron), and their deficiency is a common phenomenon in this type of soil. The sustainability of crop yields is a prerequisite to a boom in the economy of a country. In alkaline calcareous soils, the application of biochar may play a crucial role in sustaining soil health, its microbial population and nutrient levels [28]. Studies have revealed the outcomes of biochar application on soils and plant productivity [29,30]. However, very limited information is available about the effect of wheat straw biochar produced at a low temperature on nutrient availability, microbial biomass and other properties of alkaline calcareous soil within the great *Typic calciargid* group. Therefore, studies are needed to better understand the effects of biochar produced at a low temperature on the chemical, microbial and nutritional properties of *Typic calciargid* and crop growth. Therefore, a research project was planned: (i) to assess the response of wheat straw biochar produced at a low temperature on the soil properties of *Typic calciargid;* and (ii) to evaluate wheat straw biochar produced at a low temperature for maize production.

2. Materials and Methods

2.1. Production and Characterization of Wheat Straw Biochar

For the production of wheat straw biochar, wheat straw feedstock was collected from a local farmer, sun-dried in a dust-free environment and then oven-dried at 65 °C in an air-derived oven (Tokyo Rikakikai, Eyela WFO-600ND, Tokyo, Japan) until a constant weight was obtained. The dried material was crushed into small pieces (with a size of 5–10 mm) followed by pyrolysis in a muffle furnace (Gallonhop, London, UK) at 350 °C [31]. The gradual increase in furnace temperature from the starting room temperature was set at 8 to 9 °C per minute. Twenty minutes of residence time was set after achieving 350 °C. After allowing it to cool at room temperature, biochar was collected from the furnace, and it was then finely ground until it had a particle size of $\leq$2 mm.

Sub-samples of the produced biochar were analyzed in the laboratory for key parameters (Table 1). Moisture contents in the biochar were determined gravimetrically. The pH and electrical conductivity (EC) were measured in a supernatant of a 1:20 mixture of biochar in distilled water. Ammonium acetate compulsory displacement method was followed to determine the CEC of biochar [32]. For the determination of ash contents, biochar samples were heated in a muffle furnace at 750 °C for 5 h [33]. Hydrogen (H), C, N and sulfur (S) contents were measured through Vario-Micro Elemental (CHNS-O) Analyzer (Elementar Analysensysteme-GmbH, Langenselbold, Germany). Biochar was digested in sulfuric acid and hydrogen peroxide [34] followed by the determination of other plant nutrients. The digests were run on a flame photometer (PFP7, Jenway, Essex, UK) for K and on an atomic absorption spectrophotometer (AAnalyst-100, Perkin-Elmer, Norwalk, CT, USA) for Mg, Ca, Fe, Mn and Zn analyses. Vanadate-molybdate method [35] was followed to measure P concentration using a UV–visible spectrophotometer (UV-1201, Shimadzu, Tokyo, Japan). Oxygen contents in the biochar were calculated by using the following equation:

$$O\ (\%) = 100\ - \%(H + N + C + Ash)$$

Different elemental (C:P, C:S and C:N) and molar (O:C, H:C and (O+N):C) ratios were also calculated to categorize the produced biochar (Table 1).

Table 1. Characterization of biochar produced at 350 °C in a muffle furnace.

Properties	Unit	Value
	Physical/Chemical Characteristics	
Ash Content	%	11.53
Water content	%	03.12
Conversion efficiency	%	57.28
CEC	$cmol_c\ kg^{-1}$	56.34
Electrical Conductivity (1:20)	$dS\ m^{-1}$	01.27
pH (1:20)	–	07.56

Table 1. *Cont.*

Properties	Unit	Value
	Nutritional/Elemental characteristic	
Carbon (C)	%	57.54
Hydrogen (H)	%	04.37
Oxygen (O)	%	25.27
Nitrogen (N)	%	01.29
Sulfur (S)	g kg^{-1}	04.05
Phosphorus (P)	g kg^{-1}	01.98
Potassium (K)	g kg^{-1}	10.17
Magnesium (Mg)	g kg^{-1}	04.74
Calcium (Ca)	g kg^{-1}	03.53
Iron (Fe)	mg kg^{-1}	89.60
Zinc (Zn)	mg kg^{-1}	43.76
Manganese (Mn)	mg kg^{-1}	61.50
	Elemental ratio characteristic	
C:N	–	44.6
C:P	–	291
C:S	–	142
	Molar ratios characteristic	
H:C	–	0.91
O:C	–	0.33
(O+N):C	–	0.35

Values are means of three replications (n = 3).

2.2. Analysis of Soil

A bulk soil sample (from 0 to 15 cm soil depth) was collected from the Research Area of the Institute of Soil and Environmental Sciences, University of Agriculture, Faisalabad, Pakistan (31°26′ N and 73°06′ E latitude, 184.4 m altitude) and classified as *Typic calciargid*. After air-drying, it was passed through a 2 mm sieve. A sub-sample of sieved soil was used to determine its physicochemical characteristics. For the determination of soil texture, the hydrometer method proposed by [36] was followed. The soil texture class was sandy clay loam with 58.1, 19.4 and 22.5% of sand, silt and clay, respectively. Calomel glass electrode assembly was used for the determination of the pH of saturated soil paste, which was 7.93. An extract of saturated soil paste had an EC of 1.30 dS m^{-1}. Walkley–Black method [37] was used for the determination of soil organic carbon (SOC) content which was 3.98 g kg^{-1} soil. The CEC of the soil was determined as proposed by [38], which was 16.3 $cmol_c$ kg^{-1}. The method in [39] was followed to determine extractable soil K, which was 110 mg kg^{-1}. Available soil P was determined in $NaHCO_3$ extract [40], which was 7.34 mg kg^{-1}.

2.3. Incubation Experiment

Four biochar rates (0.0, 0.5, 1.0 and 2.0% on w/w basis of soil) in combination with control or recommended doses of NPK were thoroughly mixed into soil in plastic pots. Each pot had 400 g of the sieved soil, and the treated pots were arranged according to a 2-factorial completely randomized design (CRD) with twelve replicates. In the respective treatments, basal doses of 250 kg ha^{-1} N (125 mg kg^{-1} soil), 160 kg ha^{-1} P_2O_5 (80 mg kg^{-1} soil) and 105 kg ha^{-1} K_2O (52.5 mg kg^{-1} soil) were added, respectively, as urea, single super phosphate and potassium sulfate.

After the application of treatments, pots were covered with polythene sheets with small holes to allow gases to exchange between soil and atmosphere and to limit moisture loss. The pots were incubated at 25 ± 3 °C, and moisture contents in the soil were maintained at 80% field capacity.

After 25, 50, 75 and 100 days of the application of the treatment, from each treatment, 3 pots were randomly selected for sampling. The collected soil samples were sealed in plastic bags and immediately preserved in an ultra-low-temperature freezer (Robus Technologies, Dublin, Ireland) at −40 °C to avoid chemical, biological and nutritional changes. Soil chemical and nutritional analyses were determined by the following standard procedure explained below.

The chloroform fumigation method [41] was used for the determination of microbial biomass C in soil samples, and then the resulting flush of oxidizable C (analyzed by spectrophotometer) was adjusted by the factor, which was 0.45 [42]. The Ninhydrin method [43] was followed for the determination of microbial biomass N and adjusted by a factor of 6.47 [44]. Computations were performed to determine the C:N ratio using soil organic matter [45].

2.4. Pot Experiment

A pot study was conducted at a glasshouse at the institute. For this, 24 polyethylene-lined plastic pots (with dimensions of 51 × 33 cm) were each filled with 17 kg soil. Treatments comprised possible combinations of four biochar (control, 0.5, 1.0 and 2.0%) and two fertilizer (control and fertilized) rates. In a factorial arrangement, the treatments were arranged according to CRD in triplicates. Control pots received no fertilization, but the remaining pots received N (125 mg kg^{-1} soil), P_2O_5 (80 mg kg^{-1} soil) and K_2O (52.5 mg kg^{-1} soil), respectively, as urea, single super phosphate and potassium sulfate. Nitrogen was added in three equal splits, i.e., before sowing, 15 days after germination and 30 days after germination. After the application of fertilizers and biochar, the soil of each pot was thoroughly mixed followed by equilibration for 7 d. After that, five pre-soaked (for 24 h) healthy seeds of maize (Syngenta-8441, Lahore, Pakistan) were sown per pot. Ten days after the germination, the maize seedlings were thinned to a single plant pot^{-1}. Throughout the experimental period, to maintain moisture content at 90% of field capacity, tap water was used for irrigation in all the pots. For the peak vegetative growth, plant physiological traits (water use efficiency, photosynthetic rate, stomatal conductance and vapour pressure deficit) were recorded using a portable photosynthesis system (CIRAS-3, PP Systems-Hitchin, Hertfordshire, UK) between 09:30 am and 10:30 am. For this purpose, the fully expanded upper leaf (2nd leaf from the top) of each plant was selected.

Maize plants were harvested at maturity (after 115 days of sowing). Root and shoot samples were collected and washed with tap water. After washing, samples were air-dried and then oven-dried at 65 °C in an air-driven oven, till a constant weight was obtained. After recording shoot, cob, grain and root dry matter yields, the plant samples were finely ground in a metal-free grinder. The known weights of the samples were digested in sulfuric acid and hydrogen peroxide [34]. Kjeldhel method was used for the determination of N from plant samples [46]. For P determination, after developing yellow color by the vanadate-molybdate method [35], samples were run on a UV–visible spectrophotometer. For K determination, a flame photometer was used.

Computations of nutrient uptake were based on the data of dry matter yield and nutrient contents in plant tissues. Nutrient recovery was calculated using the following equation [47]:

$$\text{Nutrient Recovery } (\%) = \frac{(\text{Nutrient Uptake in Biochar Treated Pots} - \text{Nutrient Uptake in Control Pots})}{\text{Nutrient Added through Inorganic Fertilizer and Biochar}} \times 100$$

Post-harvest soil samples were collected after three days of harvesting. Collected soil samples were sun-dried and crushed to pass through a 2 mm sieve. The soil samples were analyzed for various chemical and nutritional characteristics of the soil. The details of the procedures followed are described below in the soil analysis.

2.5. Statistical Analysis

General data computations were performed in Microsoft Excel 365® (Microsoft Corporation, Redmond, WA, USA). Statistical software Statistix 8.1® was used to check the significance of different treatments by analysis of variance test followed by Tukey's multiple comparison test ($p \leq 0.05$).

Research resources: The funds for the research's resources were provided by the Higher Education Commission, Islamabad (Pakistan) through the Project No. 20-2431.

3. Results

3.1. Biochar Characterization

The biochar had an ash content of 11.53%, a moisture content of 3.12% and a conversion efficiency of 57.3%. The CEC, EC and pH of the biochar were 56.34 cmolc kg^{-1}, 01.27 dS m^{-1} and 07.56, respectively. In terms of its nutritional and elemental characteristics, the biochar contained 57.54% C, 4.37% hydrogen (H), 25.27% oxygen (O) and 1.29% N. It also contained 4.05 g kg^{-1} of sulfur (S), 1.98 g kg^{-1} of P, 10.17 g kg^{-1} of potassium (K), 4.74 g kg^{-1} of magnesium, 3.53 g kg^{-1} of calcium, 89.60 mg kg^{-1} of iron, 43.76 mg kg^{-1} of zinc and 61.50 mg kg^{-1} of manganese. The elemental ratios revealed that the C:N ratio was 44.6, the C:P ratio was 291 and the C:S ratio was 142. Additionally, for the molar ratios, the H:C ratio was 0.91, the O:C was 0.33 and (O+N):C was 0.35.

3.2. Incubation Experiment

3.2.1. Chemical and Nutritional Properties of Soil

A significant decrease in soil pH was observed with incremental biochar rates (Figure 1a). Overall, the highest pH value (7.92) was observed in the control treatment without fertilizer. The lowest value (7.82) for the soil pH was observed after 100 days of incubation, when a 2% biochar rate was used. Soil chemical properties (CEC and TOC) were also significantly improved with the addition of biochar in the soil (Figure 1b,c). The maximum soil CEC (of 24.5 $cmol_c$ kg^{-1}) and TOC (of 9.1 g kg^{-1}) were observed after 100 days of incubation (Figure 1b).

The biochar addition at 2.0% improved the total soil N and available P concentration, significantly increasing over time (Figure 2a). As compared to the first sampling, the highest concentration of available P (36.93 mg kg^{-1}) was also observed after 100 days of incubation in the treatment for which the 2.0% biochar was added.

3.2.2. Biological Properties of Soil

Microbial biomass C was significantly increased with an increase in the incubation duration after the addition of the biochar to the soil (Figure 3a). The maximum concentration of microbial biomass C (451.65 µg g^{-1}) was in the treatment for which the 2.0% biochar was added to the soil and after 50 days of the incubation period. Therefore, an increase of 26% was observed in microbial biomass C as compared with that of the respective control treatment.

After 50 days of incubation, a significant decrease in microbial biomass N was observed; however, after 75 and 100 days, no further statistical changes in microbial biomass N were observed (Figure 3b). Overall, the highest amount of microbial biomass N (30.0 µg g^{-1}) was observed after 25 days of incubation in the amended 2.0% biochar treatment with fertilizer. This was 43% greater than the microbial biomass N measured after 100 days of incubation in the amended 2.0% biochar treatment.

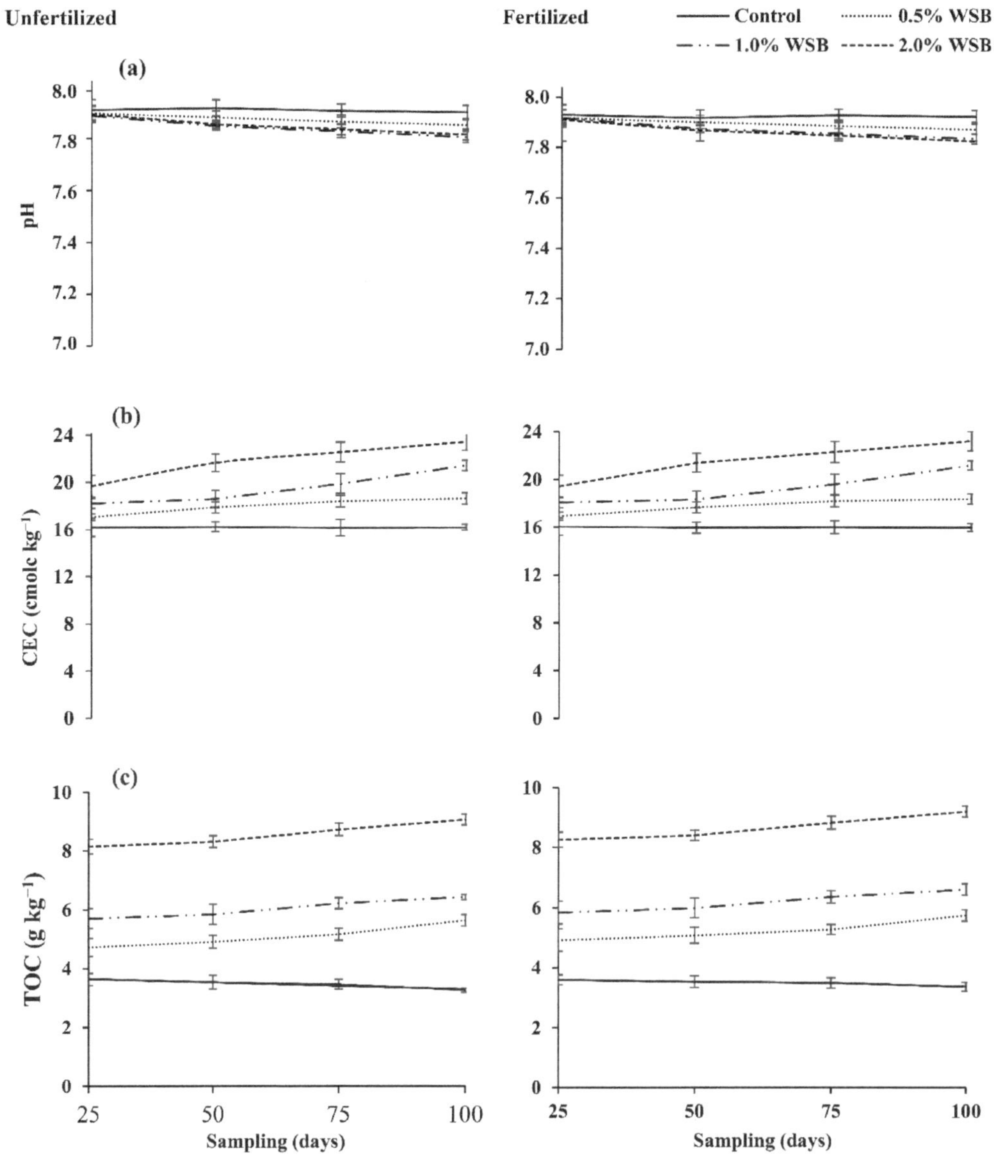

Figure 1. Temporal variation in soil pH, CEC and total organic C at different biochar rates under fertilized and unfertilized conditions. Wheat straw biochar (WSB) was added at 0.0 (control), 0.5, 1 and 2% of dry weight of soil on *w*/*w* basis. Standard deviations are shown as error bars (n = 3). Whereas, the figures indicates (**a**); variations in soil pH (**b**); variations in CEC, and (**c**) total organic carbon.

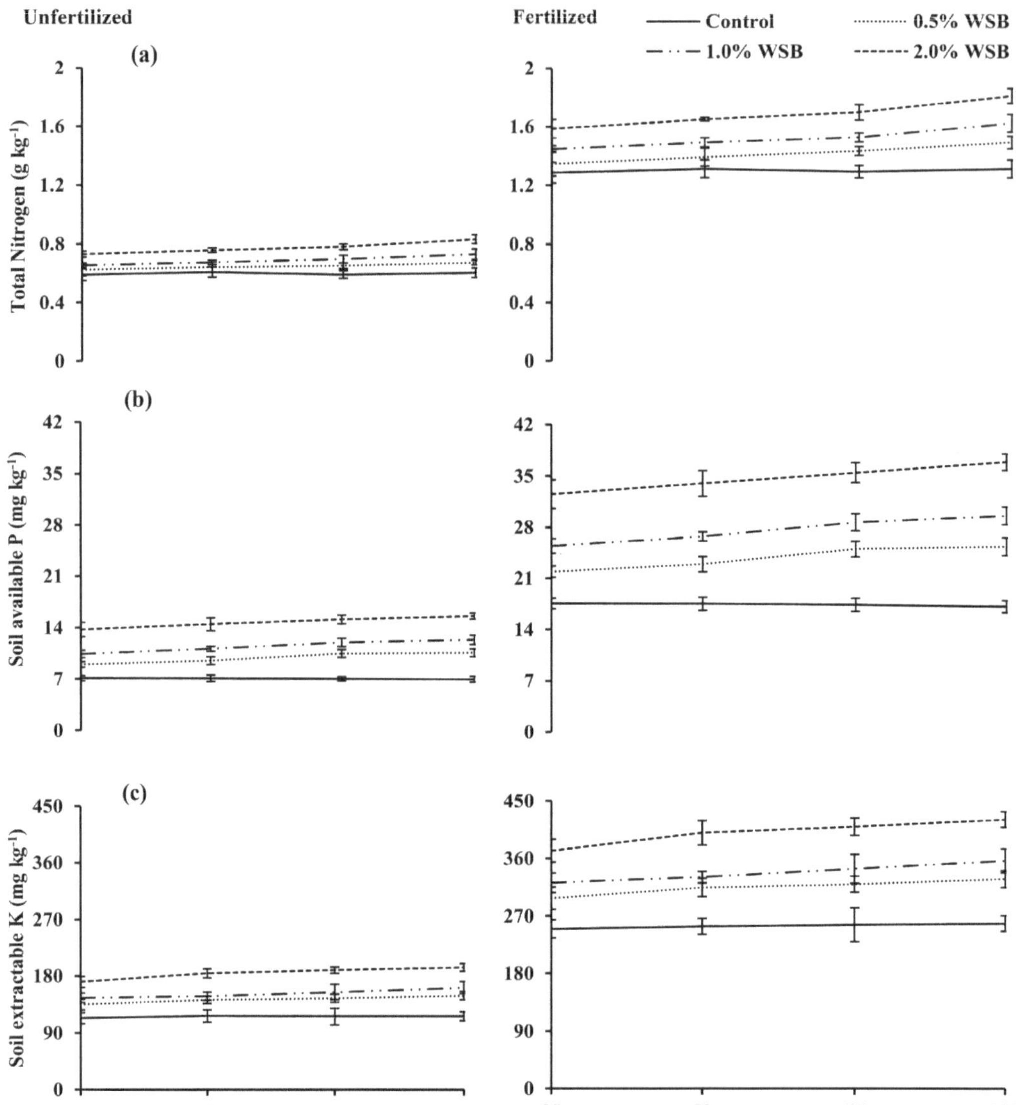

Figure 2. Temporal variation in nutritional characteristics of soil at different biochar rates under fertilized and unfertilized conditions. Wheat straw biochar (WSB) was added at 0.0 (control), 0.5, 1 and 2% of dry weight of soil on w/w basis. Standard deviations are shown as error bars (n = 3). Whereas, the figures shows (**a**) total nitrogen; (**b**) soil available P; and (**c**) soil extractable K.

Under unfertilized and fertilized conditions, at various sampling durations, a significant increase in the C:N ratio was observed with the application of different rates of biochar, compared to those of the respective control treatments (Figure 3c). Under the unfertilized condition, the highest C:N ratio (44.8) was observed after 100 days of the treatment's application. Under the fertilized condition, significant increases in the C:N ratio were observed in the treatments with respective biochar rates compared to the treatments with different biochar rates after 50, 75 and 100 days of the treatment's application.

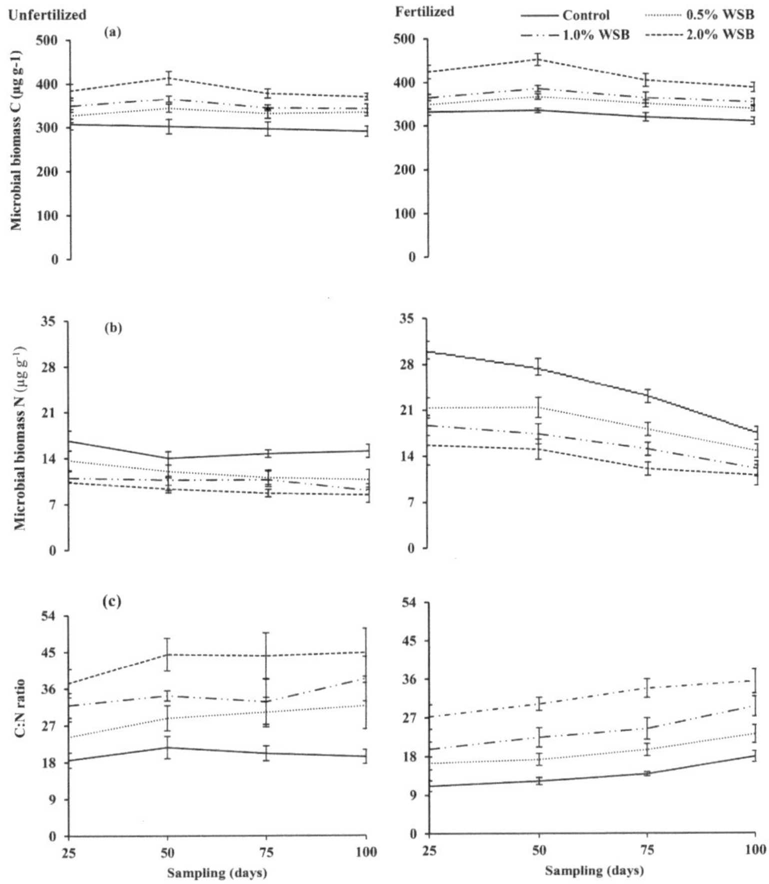

Figure 3. Temporal variation in biological characteristics of soil at different biochar rates under fertilized and unfertilized conditions. Wheat straw biochar (WSB) was added at 0.0 (control), 0.5, 1 and 2% of dry weight of soil on *w*/*w* basis. Standard deviations are shown as error bars (n = 3). Whereas the figures indicates the (**a**) microbial biomass C; (**b**) microbial biomass N; and (**c**) C:N ratio.

3.3. Pot Experiment

3.3.1. Physiological and Agronomic Parameters

The main effects of the biochar and fertilizer were significant for the maize's physiological traits (Table 2). The stomatal conductance (gs), water use efficiency and photosynthetic rate (A) of the maize plants were improved significantly by increasing the biochar rates and by the addition of NPK, and their maximum values (248.00 mmol m^{-2} s^{-1}, 6.30 mmol CO_2 mol^{-1} H_2O and 27.00 µmol m^{-2} s^{-1}) were measured at the 1.0% biochar rate added in the soil along with the recommended NPK. The vapour pressure deficit in the maize plants was decreased significantly by the addition of biochar to the soil and by the addition of NPK. A minimum vapour pressure deficit of 2.42 KPa was also found in the treatment in which the 1.0% biochar was added along with NPK, and it was up to 47% less than that of the control treatment.

The maize's agronomics parameters (dry biomass, root dry weight, grain and straw yield) were significantly improved by the addition of biochar and fertilizer (Table 3). The wheat straw with the addition of 1.0% biochar in combination with the recommended NPK produced the highest yields, and the maximum values for the dry biomass, grain yield, stover yield and root dry weight were 236, 79, 157 and 64 g pot^{-1}, respectively.

Table 2. Effect of wheat straw biochar and NPK fertilizer on physiological characteristics of maize at maximum vegetative growth stage.

Biochar Rates	**Fertilizer Dose**			
	0% RFD *	**100% RFD**	**0% RFD**	**100% RFD**
	Photosynthetic Rate (μmol m^{-2} s^{-1})		**Water Use Efficiency (mmol CO_2 mol^{-1} H_2O)**	
Control	11.67 ± 1.45 f	19.33 ± 0.88 cd	3.32 ± 0.13 d	5.63 ± 0.10 b
0.5%	13.67 ± 0.88 ef	22.67 ± 1.45 bc	3.70 ± 0.13 c	5.70 ± 0.09 b
1.0%	14.67 ± 0.88 ef	27.00 ± 1.15 a	3.95 ± 0.12 c	6.30 ± 0.16 a
2.0%	15.66 ± 0.88 de	25.67 ± 1.45 ab	3.99 ± 0.12 c	5.90 ± 0.13 b
	Stomatal conductance (mmol m^{-2} s^{-1})		Vapour pressure deficit (KPa)	
Control	133.67 ± 5.82 e	207.33 ± 7.36 c	4.59 ± 0.25 a	3.06 ± 0.12 c
0.5%	143.67 ± 5.21 e	228.67 ± 4.92 b	4.03 ± 0.11 b	2.88 ± 0.06 cd
1.0%	163.67 ± 5.05 d	248.00 ± 4.59 a	3.76 ± 0.19 b	2.42 ± 0.11 e
2.0%	172.33 ± 6.77 d	240.00 ± 3.79 ab	3.74 ± 0.12 b	2.62 ± 0.07 de

* Recommended fertilizer rate, i.e., 225, 160 and 110 kg ha^{-1} of N, P and K, respectively. All values are means of the three replications ± standard deviation (n = 3). According to the Tukey–HSD test, the significance between treatments is represented through different letters in each column at $p \leq 0.05$.

Table 3. Effect of wheat straw biochar and NPK fertilizer on agronomical traits of maize.

Biochar Rates	**Fertilizer Dose**			
	0% RFD *	**100% RFD**	**0% RFD**	**100% RFD**
	Grain Yield in g pot^{-1}		**Stover Yield in g pot^{-1}**	
Control	22.3 ± 5.67 d	078.9 ± 7.13 c	078.9 ± 7.13 c	152.7 ± 9.33 a
0.5%	22.3 ± 3.53 d	097.8 ± 5.83 bc	097.8 ± 5.83 bc	157.4 ± 6.56 a
1.0%	31.3 ± 4.33 cd	107.4 ± 5.97 b	107.4 ± 5.97 b	156.6 ± 7.02 a
2.0%	35.0 ± 2.31 c	114.4 ± 2.95 a	114.4 ± 2.95 b	159.9 ± 4.58 a
	Dry biomass yield in g pot^{-1}		Root dry weight in g pot^{-1}	
Control	101.3 ± 9.47 e	205.7 ± 6.88 b	15.4 ± 2.63 d	50.5 ± 1.69 b
0.5%	120.2 ± 3.31 d	219.4 ± 6.39 ab	20.7 ± 2.82 d	53.2 ± 2.36 b
1.0%	138.8 ± 6.21 c	235.9 ± 4.12 a	24.3 ± 3.49 cd	63.5 ± 4.12 a
2.0%	149.4 ± 5.16 c	231.9 ± 4.40 a	31.2 ± 1.37 c	63.0 ± 6.38 a

* Recommended fertilizer rate, i.e., 225, 160 and 110 kg ha^{-1} of N, P and K, respectively. All values are means of the three replications ± standard deviation (n = 3). According to the Tukey–HSD test, the significance between treatments is represented through different letters in each column at $p \leq 0.05$.

3.3.2. Nutrient Concentration and Uptake in Maize Plant

The nitrogen and K contents in the stover and grains of the maize plant were significantly enhanced when biochar was added, and the same trend was observed with NPK (Table 4). A significant interactive effect of the biochar and fertilizer was observed for the P concentration in different parts of the maize plant. The addition of the 1.0% biochar along with the recommended NPK resulted in the maximum concentrations of N, P and K in the stover and grains of the maize plants. The maximum concentrations were 0.43% N, 1.83 g P kg^{-1} and 6.51% K in the stover, and 1.47% N, 4.83 g P kg^{-1} and 5.79% K in the grains.

A significant interactive effect of biochar and fertilizer was observed on the nutrient uptake in the maize grains (Table 5). The highest values of 1.17, 0.38 and 0.46 g pot^{-1} for, respectively, the N, P and K uptake were measured with the additions of the NPK fertilizer and 1.0% biochar. Significant effects of the biochar and fertilizer were also observed on the N, P and K uptake in the maize stover with their values being 0.67, 0.29 and 1.02 g pot^{-1}, respectively, for the 1.0% biochar in combination with the recommended dose of NPK fertilizer.

Table 4. Effect of wheat straw biochar and NPK fertilizer on nutrient concentration of maize.

Biochar Rates	**Fertilizer Dose**			
	0% RFD *	**100% RFD**	**0% RFD**	**100% RFD**
	Nitrogen Concentration in Grain (%)		**Nitrogen Concentration in Stover (%)**	
Control	0.55 ± 0.02 d	1.25 ± 0.06 b	0.13 ± 0.02 d	0.36 ± 0.03 a
0.5%	0.62 ± 0.06 d	1.24 ± 0.02 b	0.17 ± 0.02 cd	0.38 ± 0.03 a
1.0%	0.77 ± 0.03 c	1.47 ± 0.02 a	0.24 ± 0.03 bc	0.43 ± 0.02 a
2.0%	0.77 ± 0.02 c	1.37 ± 0.04 a	0.27 ± 0.02 b	0.42 ± 0.03 a
	Phosphorus concentration in grain (g kg^{-1})		Phosphorus concentration in stover (g kg^{-1})	
Control	1.63 ± 0.06 c	3.46 ± 0.05 b	0.63 ± 0.06 f	1.16 ± 0.05 c
0.5%	1.77 ± 0.05 c	3.61 ± 0.04 b	0.72 ± 0.05 e	1.21 ± 0.05 b
1.0%	1.85 ± 0.05 c	4.83 ± 0.04 a	0.78 ± 0.05 e	1.43 ± 0.09 a
2.0%	1.65 ± 0.27 c	4.65 ± 0.04 a	0.98 ± 0.07 d	1.25 ± 0.11 b
	Potassium concentration in grain (%)		Potassium concentration in stover (%)	
Control	3.60 ± 0.06 d	4.63 ± 0.05 bc	4.28 ± 0.06 e	5.46 ± 0.05 c
0.5%	3.69 ± 0.05 d	5.30 ± 0.04 ab	4.83 ± 0.05 d	6.05 ± 0.04 b
1.0%	4.29 ± 0.05 cd	5.79 ± 0.04 a	5.17 ± 0.05 cd	6.51 ± 0.04 a
2.0%	4.33 ± 0.07 cd	5.29 ± 0.04 ab	4.99 ± 0.07 d	6.26 ± 0.04 ab

* Recommended fertilizer rate, i.e., 225, 160 and 110 kg ha^{-1} for N, P and K, respectively. All values are means of the three replications ± standard deviation (n = 3). According to the Tukey–HSD test, the significance between treatments is represented through different letters in each column at $p \leq 0.05$.

Table 5. Effect of wheat straw biochar and NPK fertilizer on nutrient uptake in maize grain and stover.

Biochar Rates	**Fertilizer Dose**			
	0% RFD *	**100% RFD**	**0% RFD**	**100% RFD**
	Nitrogen Uptake in Grain (g pot^{-1})		**Nitrogen Uptake in Stover (g pot^{-1})**	
Control	0.13 ± 0.035 e	0.66 ± 0.045 c	0.11 ± 0.025 c	0.560 ± 0.073 a
0.5%	0.14 ± 0.029 e	0.76 ± 0.034 c	0.17 ± 0.026 bc	0.599 ± 0.061 a
1.0%	0.24 ± 0.042 de	1.17 ± 0.047 a	0.26 ± 0.019 bc	0.673 ± 0.048 a
2.0%	0.27 ± 0.013 d	0.99 ± 0.063 b	0.30 ± 0.010 b	0.674 ± 0.070 a
	Phosphorus uptake in grain (mg pot^{-1})		Phosphorus uptake in stover (mg pot^{-1})	
Control	036.52 ± 09.55 f	183.83 ± 11.87 d	048.92 ± 00.74 f	222.52 ± 06.33 c
0.5%	039.27 ± 05.07 f	223.92 ± 09.31 c	076.25 ± 09.27 e	253.79 ± 15.27 b
1.0%	058.15 ± 08.62 e	383.30 ± 19.47 a	090.77 ± 01.15 de	286.61 ± 14.13 a
2.0%	056.63 ± 06.62 e	334.85 ± 12.82 b	112.82 ± 11.23 d	264.23 ± 08.54 ab
	Potassium uptake in grain (g pot^{-1})		Potassium uptake in stover (g pot^{-1})	
Control	0.082 ± 0.025 e	0.245 ± 0.013 c	0.341 ± 0.05 d	0.84 ± 0.084 b
0.5%	0.083 ± 0.015 e	0.328 ± 0.009 b	0.471 ± 0.17 cd	0.95 ± 0.053 ab
1.0%	0.136 ± 0.026 de	0.460 ± 0.031 a	0.553 ± 0.21 c	0.10 ± 0.070 a
2.0%	0.153 ± 0.021 d	0.380 ± 0.011 b	0.570 ± 0.20 c	1.00 ± 0.032 a

* Recommended fertilizer rate, i.e., 225, 160 and 110 kg ha^{-1} for N, P and K, respectively. All values are means of the three replications ± standard deviation (n = 3). According to the Tukey–HSD test, the significance between treatments is represented through different letters in each column at $p \leq 0.05$.

3.3.3. Nutrient Recovery from Mineral Fertilizers

There was a significant effect of the biochar application on the N and P recovery from the urea and single super phosphate fertilizers, but a non-significant effect was observed for the K recovery from the potassium sulfate (Figure 4). Overall, for all of the biochar rates, a maximum percentage of K recovery (99%) was observed for the 1% biochar application. The maximum N and P recoveries, up to 74 from the urea and 42% from the single superphosphate, were achieved with the 1.0% biochar application. Hence, there was an improved N recovery (26%) and P recovery (38%) from the urea and single superphosphate compared to those of the control treatment without biochar.

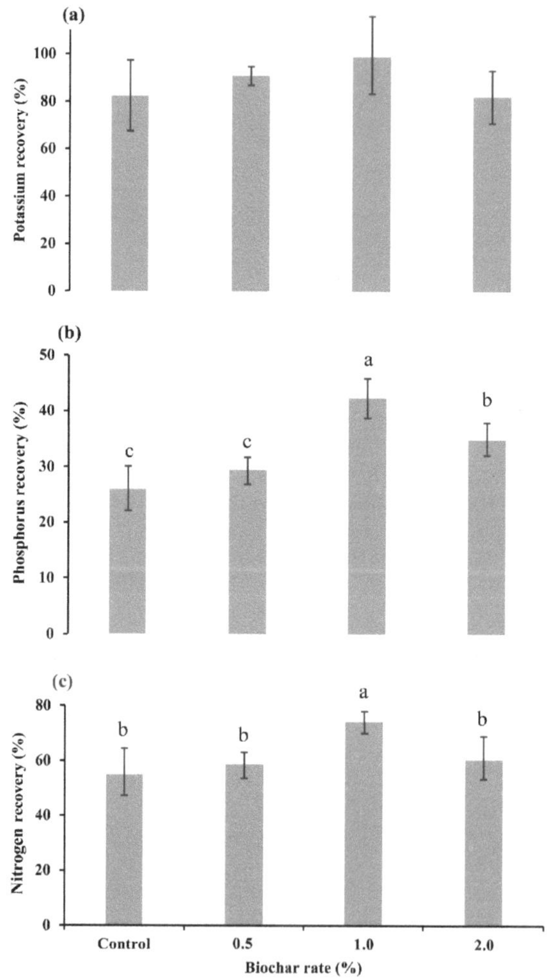

Figure 4. Response of wheat straw biochar (biochar) on the recovery of nutrients (K, P and N) from mineral fertilizers. According to the Tukey–HSD test, the significance between treatments is represented through different letters. In case of K recovery, treatment differences were found to be non-significant. The figure indicate (**a**) potassium recovery; (**b**)phosphorus recovery; and (**c**) nitrogen recovery.

3.3.4. Chemical and Nutritional Properties of Post-Harvest Soil

The soil's chemical properties were significantly affected in response to the addition of biochar (Table 6). The highest statistical CEC, total organic C and EC were 28 $cmol_c$ kg^{-1},

8.9 g kg^{-1} and 2.05 dS m^{-1}, respectively, while the minimum soil pH (7.81) was measured with the 2.0% biochar.

Table 6. Effect of wheat straw biochar and fertilizer rates on soil chemical properties after maize harvesting.

Biochar Rates	Fertilizer Dose			
	0% RFD *	100% RFD	0% RFD	100% RFD
	Soil EC (dS m^{-1})		Soil pH	
Control	1.24 ± 0.07 f	1.38 ± 0.04 e	7.92 ± 0.03 a	7.91 ± 0.02 ab
0.5%	1.55 ± 0.06 d	1.66 ± 0.07 cd	7.85 ± 0.01 cd	7.87 ± 0.01 bc
1.0%	1.73 ± 0.04 c	1.92 ± 0.03 ab	7.81 ± 0.01 d	7.81 ± 0.01 cd
2.0%	1.88 ± 0.04 b	2.05 ± 0.04 a	7.80 ± 0.01 d	7.81 ± 0.01 cd
	Soil CEC ($cmol_c$ kg^{-1})		Total soil organic C (g kg^{-1})	
Control	16.37 ± 0.62 e	15.30 ± 1.03e	3.97 ± 0.14 d	3.94 ± 0.15 d
0.5%	20.57 ± 0.54 d	21.60 ± 1.05 cd	5.26 ± 0.32 c	5.27 ± 0.26 c
1.0%	23.40 ± 0.51 c	25.50 ± 0.44 b	7.17 ± 0.30 b	7.23 ± 0.23 b
2.0%	26.97 ± 0.75 ab	28.40 ± 0.53 a	8.80 ± 0.24 a	8.91 ± 0.12 a

* Recommended fertilizer rate, i.e., 225, 160 and 110 kg ha^{-1} of N, P and K, respectively. All values are means of the three replications ± standard deviation (n = 3). According to the Tukey–HSD test, the significance between treatments is represented through different letters in each column at $p \leq 0.05$.

A significant interactive effect was observed for the biochar × fertilizers on the soil's total N concentration, while only the main effects of the biochar and fertilizers were found to be significant for the soil's available P concentration (Table 7). The maximum total N and available P (i.e., 0.074 mg N kg^{-1} and 9.27 mg P kg^{-1}) were measured with the 2.0% biochar along with the NPK fertilizers. However, only a non-significant effect of the biochar and fertilizers was observed for the soil extractable K concentration.

Table 7. Effect of different fertilizer rates and wheat straw biochar on soil nutritional properties after maize harvesting.

Biochar Rates	0% RFD *	100% RFD
	Soil N (mg pot^{-1})	
Control	0.060 ± 0.0009 d	0.060 ± 0.0006 d
0.5%	0.061 ± 0.0009 d	0.065 ± 0.0006 c
1.0%	0.064 ± 0.0006 c	0.069 ± 0.0010 b
2.0%	0.066 ± 0.0006 c	0.074 ± 0.0009 a
	Available soil P (mg kg^{-1})	
Control	7.55 ± 0.3 f	8.15 ± 0.11 e
0.5%	8.25 ± 0.1 de	8.57 ± 0.10 cd
1.0%	8.72 ± 0.1 bc	9.03 ± 0.11 ab
2.0%	9.02 ± 0.1 ab	9.27 ± 0.10 a
	Extractable soil K (mg kg^{-1})	
Control	111.60 ± 05.06 e	109.71 ± 04.03 e
0.5%	143.56 ± 04.82 d	163.89 ± 06.37 c
1.0%	165.62 ± 04.38 c	188.60 ± 05.18 b
2.0%	174.15 ± 05.97 bc	206.97 ± 05.40 a

* Recommended fertilizer rate, i.e., 225, 160 and 110 kg ha^{-1} of N, P and K, respectively. All values are means of the three replications ± standard deviation (n = 3). According to the Tukey–HSD test, the significance between treatments is represented through different letters in each column at $p \leq 0.05$.

4. Discussion

Biochar produced at a low pyrolysis temperature yields a greater recovery of nutrients, which are usually lost at higher temperatures [48]. Due to biochar's addition, a significant improvement in the agronomic and physiological parameters (Tables 2 and 3) of maize growth may be linked to the high nutrient contents in biochar produced at a low temperature (Table 1). The alkalinity of biochar can be increased by higher pyrolysis temperatures, which lead to the elimination of acidic functional groups, such as hydroxyl, formyl or carboxyl and the enrichment of basic cations [2,44]. *Typic calciargid* soils are already alkaline in pH [49]. In this type of soil, biochars produced at high pyrolysis temperatures may not be a suitable option, so, to avoid this scenario, in the current study, biochar was produced at a low pyrolysis temperature with aim of obtaining a product with a comparatively neutral pH (Table 1).

The biochar produced at a low temperature may have the advantage of a high CEC [22]. Biochars produced at low temperatures retain acidic functional groups, such as phenol, lactonic and carboxylic acid. These functional groups not only contribute to achieving a pH near to neutral or even acidic for the biochar, but the pore surface negative charge of these functional groups also contributes to a better CEC that helps to retain cations, i.e., ammonium [50]. In the current incubation and pot studies, a significant improvement in the soil CEC (Figure 1b, Table 6) was observed, especially when 1 or 2% biochar was added. In the incubation study, a temporal increase in the soil CEC might be the result of the oxidation of the functional groups present on the biochar surfaces [51]. Over a short time period, a minor temporal increase in soil CEC has been reported under acidic soils (oxisols) [52]; however, this phenomenon was more prominent in our case, in which a significant temporal increase in the CEC of *Typic calciargid* soil was observed over the period of 100 days (Figure 1b). The CEC of biochar plays a crucial role in enhancing soil nutrient retention and minimizing fertilizer losses [53]. Soil with a greater CEC has a greater available nutrient concentration. In our study, the better availability of essential nutrients from biochar-amended soil resulted in an improved efficiency of nutrient use as compared with that of the control (Figure 4; Table 5). Hence, better nutrient retention in the topsoil because of the addition of biochar resulted in improved nutrient recoveries from the mineral fertilizers. This increase in nutrient availability from the applied mineral fertilizers may also be linked to improved soil chemical properties, i.e., soil CEC and TOC. The addition of biochar increased maize growth and the nutritional parameters of maize, which may also be linked to the increase in the availability of nutrients from the mineral fertilizers (Figures 1 and 2; Table 4). Biochar applications improve various other soil attributes (including its physical properties and fertility status) which may ultimately contribute to the improved productivity of crop plants [43].

The chemical properties of soil, including CEC, SOC, EC and pH, are very important for nutrient bioavailability in plants. The utilization of high pyrolysis temperatures (600–700 °C) in biochar production leads to a reduction in the abundance of acidic functional groups, particularly carboxylic functional groups, while introducing more basic functional groups [54,55]. Biochar produced at elevated temperatures displays a highly aromatic composition with well-structured carbon layers. Nevertheless, in the present investigation, biochar was produced at low pyrolysis temperatures, resulting in a biochar pH (7.56, Table 1) lower than that of the *Typic calciargid* soil (7.93) used in the current study [22]. The optimum availability of nutrients to plants ranges from a soil pH of 6.5 to 7.5. In current study, a minor decrease in soil pH was observed as a result of the biochar application (Table 6), which may oxidize the other elements present on the soil surface through microbial as well as chemical activity [56]. The oxidization thus produces carboxylic functional groups [11] that can reduce the alkalinity and decrease the soil pH as shown in the incubation experiment (Figure 1). However, this decrease was very low (up to 0.2), but at least the biochar application produced at a low temperature did not show any further increase in the pH of the already alkaline soil (the alkaline behavior of biochar has been observed in the majority of studies conducted all over the world); thus, we can conclude that the biochar produced at a low pyrolysis temperature can be used as to amend the alkaline nature of soils.

The better root growth and its proliferation in the soil might also be because of the greater ethylene concentration produced in the biochar-amended soil as compared with that in the control [57]. As compared to the control, a minor improvement in yield with the addition of charcoal was observed; however, the yield increase was greater with the combined application of charcoal and mineral fertilizers. From the meta-analysis, it was concluded that the application of biochar with inorganic fertilizers may increase plant yield by 25.3% [58]. The addition of soil biochar may improve soil microbial biomass and microbial C (Figure 3), which may ultimately improve the organic C status of soils [8]. A significant amount of organic matter is added with biochar, which increases the concentration of organic matter in the soil [59]. A significant improvement in the microbial biomass C in soil was observed with increasing biochar rates up to 50 days during the incubation study; after that, it decreased up to 75 days and then became constant for up to 100 days (Figure 3). The increase in the microbial biomass C was probably due to the labile C fraction present in biochar [60–62]. With time, non-significant changes have been reported in microbial biomass C for up to 4 years, which were improved initially with biochar applications. The reason behind this is the recalcitrant nature of biochar, i.e., the resistance of biochar constituents to oxidation [63,64]. The higher organic C status of soils amended with biochar has also been well documented [41,65,66]. This provides a suitable and safe habitat for the growth of soil microorganisms for longer periods [67,68]. In current study, the N and P recovery was improved as a result of the addition of 1% biochar; however, the K recovery was statistically non-significant with the addition of biochar (Figure 4). Biochar increases the availability of essential soil nutrients (such as K, P and dissolved organic matter), adsorbs harmful compounds, improves soil moisture and pH, and ultimately influences soil microbial activity [59,66,69].

After 50 days of the treatments, a significant decrease in the microbial biomass N in the soil was observed with the increasing rates of biochar (Figure 3). Zhang et al. [8] also observed a significant reduction in microbial biomass N and the microbial biomass C:N ratio because of NH_4^+ adsorption on biochar surfaces. This shows that biochar may initially limit microbial activity in soil at a certain level by slowing down the mineralization of soil organic N [8,70]. Over time, the adsorbed NH_4^+ is released from the biochar surface and becomes available to plants and microorganisms. This was evident in our incubation study with an increase in soil microbial biomass N after 50 days of incubation (Figure 3). Similar findings have also been reported and link this phenomenon with the nutrient cycling in soil, which results in better crop growth and improved soil fertility status [71].

5. Conclusions

In both the greenhouse experiment and incubation study, the addition of 1.0 and 2.0% biochars produced at a low temperature significantly improved the maize growth and yield by enhancing the nutrient recovery (P and K) from mineral fertilizers in *Typic calcargid* soil. The biochar increased the soil total N, available P, extractable K, soil organic C and CEC, but a minor decrease was observed in the soil pH. Further, the incubation study confirmed the temporal enhancement in the soil microbial biomass C in biochar-amended *Typic calciargid* soil. The positive effects of biochar were more prominent with the addition of 1% biochar and were associated with improvements in the soil's chemical, biological and nutritional properties.

Author Contributions: Conceptualization, M.A., Z.A., M.A.B., S.I., M.K., A.R., L.S.-P. and S.A.M.; Data curation, S.H., M.A.B. and A.A.H.; Formal analysis, M.A., Z.A. and S.M.; Funding acquisition, W.F.A.M. and S.A.M.; Investigation, M.A. and S.M.; Methodology, S.H. and M.S.; Project administration, L.S.-P.; Resources, A.R., W.F.A.M. and A.A.H.; Software, M.A.B. and M.S.; Supervision, M.K.; Validation, S.I.; Visualization, S.H.; Writing—original draft, M.A.; Writing—review and editing, Z.A., S.H., M.A.B., M.S., S.M., S.I., M.K., A.R., W.F.A.M., L.S.-P., S.A.M. and A.A.H. All authors have read and agreed to the published version of the manuscript.

Funding: This work was funded by the Researchers Supporting Project number (RSPD2023R752), King Saud University, Riyadh, Saudi Arabia.

Data Availability Statement: Data will be made available on request to corresponding author.

Acknowledgments: The authors extend their appreciation to the Researchers Supporting Project (RSPD2023R752) King Saud University, Riyadh, Saudi Arabia.

Conflicts of Interest: The authors declare no conflict of interest.

References

1. Hardy, B.; Sleutel, S.; Dufey, J.E.; Cornelis, J.T. The Long-Term Effect of Biochar on Soil Microbial Abundance, Activity and Community Structure Is Overwritten by Land Management. *Front. Environ. Sci.* **2019**, *7*, 110. [CrossRef]
2. Blanco-Canqui, H. Does Biochar Improve All Soil Ecosystem Services? *GCB Bioenergy* **2021**, *13*, 291–304. [CrossRef]
3. Kalu, S.; Simojoki, A.; Karhu, K.; Tammeorg, P. Long-Term Effects of Softwood Biochar on Soil Physical Properties, Greenhouse Gas Emissions and Crop Nutrient Uptake in Two Contrasting Boreal Soils. *Agric. Ecosyst. Environ.* **2021**, *316*, 107454. [CrossRef]
4. Lévesque, V.; Oelbermann, M.; Ziadi, N. Biochar in Temperate Soils: Opportunities and Challenges. *Can. J. Soil Sci.* **2020**, *102*, 1–26. [CrossRef]
5. Liu, L.; Deng, G.; Shi, X. Adsorption Characteristics and Mechanism of P-Nitrophenol by Pine Sawdust Biochar Samples Produced at Different Pyrolysis Temperatures. *Sci. Rep.* **2020**, *10*, 5149. [CrossRef]
6. Ajeng, A.A.; Abdullah, R.; Ling, T.C.; Ismail, S.; Lau, B.F.; Ong, H.C.; Chew, K.W.; Show, P.L.; Chang, J.S. Bioformulation of Biochar as a Potential Inoculant Carrier for Sustainable Agriculture. *Environ. Technol. Innov.* **2020**, *20*, 101168. [CrossRef]
7. Oloo, K.P.; Asbon, O.P. Depletion of Phosphate Rock Reserves and World Food Crisis: Reality or Hoax? *Afr. J. Agric. Res.* **2020**, *16*, 1223–1227. [CrossRef]
8. Zhang, Y.; Wu, L.; Zhang, X.; Deng, A.; Abdulkareem, R.; Wang, D.; Zheng, C.; Zhang, W. Effect of Long-Term Organic Amendment Application on the Vertical Distribution of Nutrients in a Vertisol. *Agronomy* **2022**, *12*, 1162. [CrossRef]
9. Saleem, M.; Pervaiz, Z.H.; Contreras, J.; Lindenberger, J.H.; Hupp, B.M.; Chen, D.; Zhang, Q.; Wang, C.; Iqbal, J.; Twigg, P. Cover Crop Diversity Improves Multiple Soil Properties via Altering Root Architectural Traits. *Rhizosphere* **2020**, *16*, 100248. [CrossRef]
10. Bonanomi, G.; De Filippis, F.; Zotti, M.; Idbella, M.; Cesarano, G.; Al-Rowaily, S.; Abd-ElGawad, A. Repeated Applications of Organic Amendments Promote Beneficial Microbiota, Improve Soil Fertility and Increase Crop Yield. *Appl. Soil Ecol.* **2020**, *156*, 103714. [CrossRef]
11. Huang, H.; Niu, Z.; Shi, R.; Tang, J.; Lv, L.; Wang, J.; Fan, Y. Thermal Oxidation Activation of Hydrochar for Tetracycline Adsorption: The Role of Oxygen Concentration and Temperature. *Bioresour. Technol.* **2020**, *306*, 123096. [CrossRef] [PubMed]
12. Ahmed, A.A.; Fawaz, P.Q.; Eupkouk, A.W. Biochar Amended Soils and Crop Productivity Improvement. *Environ. Sci. Technol.* **2016**, *42*, 5137–5143.
13. Sohi, S.P.; Krull, E.; Lopez-Capel, E.; Bol, R. A Review of Biochar and Its Use and Function in Soil. *Adv. Agron.* **2010**, *105*, 47–82. [CrossRef]
14. Kizito, S.; Luo, H.; Lu, J.; Bah, H.; Dong, R.; Wu, S. Role of Nutrient-Enriched Biochar as a Soil Amendment during Maize Growth: Exploring Practical Alternatives to Recycle Agricultural Residuals and to Reduce Chemical Fertilizer Demand. *Sustainability* **2019**, *11*, 3211. [CrossRef]
15. Fanai, L.; David, A.A.; Thomas, T.; Swaroop, N.; Hassan, A.; David, A. Assessment of Potassium and Sulphur on the Soil Properties, Growth and Yield of Onion (*Allium cepa* L.). *Pharma Innov. J.* **2021**, *10*, 2508–2512.
16. Liang, J.; Li, Y.; Si, B.; Wang, Y.; Chen, X.; Wang, X.; Chen, H.; Wang, H.; Zhang, F.; Bai, Y.; et al. Optimizing Biochar Application to Improve Soil Physical and Hydraulic Properties in Saline-Alkali Soils. *Sci. Total Environ.* **2021**, *771*, 144802. [CrossRef]
17. DeLuca, T.H.; Gundale, M.J.; MacKenzie, M.D.; Jones, D.L. Biochar Effects on Soil Nutrient Transformations. In *Biochar for Environmental Management*; Routledge: Oxfordshire, UK, 2015; pp. 453–486.
18. Hailegnaw, N.S.; Mercl, F.; Pračke, K.; Száková, J.; Tlustoš, P. Mutual Relationships of Biochar and Soil PH, CEC, and Exchangeable Base Cations in a Model Laboratory Experiment. *J. Soils Sediments* **2019**, *19*, 2405–2416. [CrossRef]
19. Clausing, S.; Polle, A. Mycorrhizal Phosphorus Efficiencies and Microbial Competition Drive Root P Uptake. *Front. For. Glob. Chang.* **2020**, *3*, 54. [CrossRef]
20. Rawal, N.; Pande, K.R.; Shrestha, R.; Vista, S.P. Phosphorus and Potassium Mineralization as Affected by Phosphorus Levels and Soil Types under Laboratory Condition. *Agrosyst. Geosci. Environ.* **2022**, *5*, e20229. [CrossRef]
21. Hollister, C.C.; Bisogni, J.J.; Lehmann, J. Ammonium, Nitrate, and Phosphate Sorption to and Solute Leaching from Biochars Prepared from Corn Stover (*Zea mays* L.) and Oak Wood (*Quercus* spp.). *J. Environ. Qual.* **2013**, *42*, 137–144. [CrossRef]
22. Tomczyk, A.; Sokołowska, Z.; Boguta, P. Biochar Physicochemical Properties: Pyrolysis Temperature and Feedstock Kind Effects. *Rev. Environ. Sci. Biotechnol.* **2020**, *19*, 191–215. [CrossRef]
23. Huang, H.; Reddy, N.G.; Huang, X.; Chen, P.; Wang, P.; Zhang, Y.; Huang, Y.; Lin, P.; Garg, A. Effects of Pyrolysis Temperature, Feedstock Type and Compaction on Water Retention of Biochar Amended Soil. *Sci. Rep.* **2021**, *11*, 7419. [CrossRef]
24. Ippolito, J.A.; Laird, D.A.; Busscher, W.J. Environmental Benefits of Biochar. *J. Environ. Qual.* **2012**, *41*, 967–972. [CrossRef] [PubMed]

25. Hagner, M.; Kemppainen, R.; Jauhiainen, L.; Tiilikkala, K.; Setälä, H. The Effects of Birch (*Betula* spp.) Biochar and Pyrolysis Temperature on Soil Properties and Plant Growth. *Soil Tillage Res.* **2016**, *163*, 224–234. [CrossRef]
26. Zhang, X.; Gao, B.; Creamer, A.E.; Cao, C.; Li, Y. Adsorption of VOCs onto Engineered Carbon Materials: A Review. *J. Hazard. Mater.* **2017**, *338*, 102–123. [CrossRef] [PubMed]
27. Zhang, H.; Chen, C.; Gray, E.M.; Boyd, S.E. Effect of Feedstock and Pyrolysis Temperature on Properties of Biochar Governing End Use Efficacy. *Biomass Bioenergy* **2017**, *105*, 136–146. [CrossRef]
28. Song, D.; Tang, J.; Xi, X.; Zhang, S.; Liang, G.; Zhou, W.; Wang, X. Responses of Soil Nutrients and Microbial Activities to Additions of Maize Straw Biochar and Chemical Fertilization in a Calcareous Soil. *Eur. J. Soil Biol.* **2018**, *84*, 1–10. [CrossRef]
29. Kapoor, A.; Sharma, R.; Kumar, A.; Sepehya, S. Biochar as a Means to Improve Soil Fertility and Crop Productivity: A Review. *J. Plant Nutr.* **2022**, *45*, 2380–2388. [CrossRef]
30. Khan, Z.; Zhang, K.; Khan, M.N.; Bi, J.; Zhu, K.; Luo, L.; Hu, L. How Biochar Affects Nitrogen Assimilation and Dynamics by Interacting Soil and Plant Enzymatic Activities: Quantitative Assessment of 2 Years Potted Study in a Rapeseed-Soil System. *Front. Plant Sci.* **2022**, *13*, 853449. [CrossRef]
31. Sánchez, M.E.; Lindao, E.; Margaleff, D.; Martínez, O.; Morán, A. Pyrolysis of Agricultural Residues from Rape and Sunflowers: Production and Characterization of Bio-Fuels and Biochar Soil Management. *J. Anal. Appl. Pyrolysis* **2009**, *85*, 142–144. [CrossRef]
32. Gaskin, J.W.; Steiner, C.; Harris, K.; Das, K.C.; Bibens, B. Effect of Low-Temperature Pyrolysis Conditions on Biochar for Agricultural Use. *Trans. ASABE* **2008**, *51*, 2061–2069. [CrossRef]
33. Slattery, W.J.; Ridley, A.M.; Windsor, S.M. Ash Alkalinity of Animal and Plant Products. *Aust. J. Exp. Agric.* **1991**, *31*, 321–324. [CrossRef]
34. Wolf, B. A Comprehensive System of Leaf Analyses and Its Use for Diagnosing Crop Nutrient Status. *Commun. Soil Sci. Plant Anal.* **1982**, *13*, 1035–1059. [CrossRef]
35. Chapman, H.D.; Pratt, P.F. *Methods of Analysis for Soils, Plants and Water*; Univ. California: Berkeley, CA, USA, 1961.
36. Gee, G.; Bauder, J. Particle-Size Analysis. In *Methods of Soil Analysis, Part 1: Physical and Mineralogical Methods*, 2nd ed.; Agronomy Monogram; American Society of Agronomy Madison: Madison, WI, USA, 1986; pp. 383–411.
37. Nelson, D.; Sommers, L.E. Total Carbon, Organic Carbon, and Organic Matter. In *Methods of Soil Analysis, Part 2: Chemical and Microbiological Properties*; American Society of Agronomy: Madison, WI, USA, 1982; pp. 570–571.
38. Sumner, M.E.; Miller, W.P. Cation Exchange Capacity and Exchange Coefficients. In *Methods of Soil Analysis: Part 3 Chemical Methods*; American Society of Agronomy: Madison, WI, USA, 2018; pp. 1201–1229. [CrossRef]
39. Richards, L.A. Diagnosis and Improving of Saline and Alkaline Soils. In *Agriculture Handbook. 60 US, Salinity Laboratory Staff*; US Department of Agriculture: Washington, DC, USA, 1954.
40. Olsen, S.R.; Sommers, L.E. Phosphorus. In *Methods of Soil Analysis, Argon No 9, Part 2: Chemical and Microbiological Properties*, 2nd ed.; American Society of Agronomy: Madison, WI, USA, 1982; pp. 403–430.
41. Ladd, J.N.; Amato, M. Relationship between Microbial Biomass Carbon in Soils and Absorbance (260 nm) of Extracts of Fumigated Soils. *Soil Biol. Biochem.* **1989**, *21*, 457–459. [CrossRef]
42. Wu, J.; Joergensen, R.G.; Pommerening, B.; Chaussod, R.; Brookes, P.C. Measurement of Soil Microbial Biomass C by Fumigation-Extraction-an Automated Procedure. *Soil Biol. Biochem.* **1990**, *22*, 1167–1169. [CrossRef]
43. Joergensen, R.G.; Brookes, P.C. Ninhydrin-Reactive Nitrogen Measurements of Microbial Biomass in 0.5 m K_2SO_4 Soil Extracts. *Soil Biol. Biochem.* **1990**, *22*, 1023–1027. [CrossRef]
44. Sparlig, G.P.; Gupta, V.V.S.R.; Zhu, C. Release of Ninhydrin-Reactive Compounds during Fumigation of Soil to Estimate Microbial C and N. *Soil Biol. Biochem.* **1993**, *25*, 1803–1805. [CrossRef]
45. Chatterjee, A.; Lal, R.; Wielopolski, L.; Martin, M.Z.; Ebinger, M.H. Evaluation of different soil carbon determination methods. *Crit. Rev. Plant Sci.* **2009**, *28*, 164–178. [CrossRef]
46. Jackson, M.L. Soil Chemical Analysis. *J. Agric. Food Chem.* **1960**, *7*, 138. [CrossRef]
47. Mengel, K.; Kirkby, E.A. *Principles of Plant Nutrition*, 5th ed.; Kluwer: Dordrecht, The Netherlands, 2001.
48. Keiluweit, M.; Nico, P.S.; Johnson, M.; Kleber, M. Dynamic Molecular Structure of Plant Biomass-Derived Black Carbon (Biochar). *Environ. Sci. Technol.* **2010**, *44*, 1247–1253. [CrossRef]
49. Naeem, M.A.; Khalid, M.; Ahmad, Z.; Naveed, M. Low Pyrolysis Temperature Biochar Improves Growth and Nutrient Availability of Maize on Typic Calciargid. *Commun. Soil Sci. Plant Anal.* **2016**, *47*, 41–51. [CrossRef]
50. Shaaban, A.; Se, S.M.; Mitan, N.M.M.; Dimin, M.F. Characterization of Biochar Derived from Rubber Wood Sawdust through Slow Pyrolysis on Surface Porosities and Functional Groups. *Proc. Procedia Eng.* **2013**, *68*, 365–371. [CrossRef]
51. Liu, X.H.; Han, F.P.; Zhang, X.C. Effect of Biochar on Soil Aggregates in the Loess Plateau: Results from Incubation Experiments. *Int. J. Agric. Biol.* **2012**, *14*, 975–979.
52. Domingues, R.R.; Sánchez-Monedero, M.A.; Spokas, K.A.; Melo, L.C.A.; Trugilho, P.F.; Valenciano, M.N.; Silva, C.A. Enhancing Cation Exchange Capacity Ofweathered Soils Using Biochar: Feedstock, Pyrolysis Conditions and Addition Rate. *Agronomy* **2020**, *10*, 824. [CrossRef]
53. Kharel, G.; Sacko, O.; Feng, X.; Morris, J.R.; Phillips, C.L.; Trippe, K.; Kumar, S.; Lee, J.W. Biochar Surface Oxygenation by Ozonization for Super High Cation Exchange Capacity. *ACS Sustain. Chem. Eng.* **2019**, *7*, 16410–16418. [CrossRef]

54. Sharma, P.; Abrol, V.; Sharma, V.; Chaddha, S.; Srinivasa Rao, C.; Ganie, A.Q.; Ingo Hefft, D.; El-Sheikh, M.A.; Mansoor, S. Effectiveness of Biochar and Compost on Improving Soil Hydro-Physical Properties, Crop Yield and Monetary Returns in Inceptisol Subtropics. *Saudi J. Biol. Sci.* **2021**, *28*, 7539–7549. [CrossRef] [PubMed]
55. Mwadalu, R.; Mochoge, B.; Danga, B. Assessing the Potential of Biochar for Improving Soil Physical Properties and Tree Growth. *Int. J. Agron.* **2021**, *2021*, 6000184. [CrossRef]
56. Liu, X.-H.; Zhang, X.-C. Effect of Biochar on PH of Alkaline Soils in the Loess Plateau: Results from Incubation Experiments. *Int. J. Agric. Biol.* **2012**, *14*, 65–70.
57. Spokas, K.A. Review of the Stability of Biochar in Soils: Predictability of O:C Molar Ratios. *Carbon Manag.* **2010**, *1*, 289–303. [CrossRef]
58. Bai, S.H.; Omidvar, N.; Gallart, M.; Kämper, W.; Tahmasbian, I.; Farrar, M.B.; Singh, K.; Zhou, G.; Muqadass, B.; Xu, C.Y.; et al. Combined Effects of Biochar and Fertilizer Applications on Yield: A Review and Meta-Analysis. *Sci. Total Environ.* **2022**, *808*, 152073. [CrossRef]
59. Datta, R.; Sahoo, K.; Rahman, M.M.; Hafeez, A.; Pan, T.; Tian, J.; Cai, K. Modified Biochars and Their Effects on Soil Quality: A Review. *Environments* **2022**, *9*, 60. [CrossRef]
60. Ameloot, N.; Graber, E.R.; Verheijen, F.G.A.; De Neve, S. Interactions between Biochar Stability and Soil Organisms: Review and Research Needs. *Eur. J. Soil Sci.* **2013**, *64*, 379–390. [CrossRef]
61. Singh, B.P.; Cowie, A.L. Long-Term Influence of Biochar on Native Organic Carbon Mineralisation in a Low-Carbon Clayey Soil. *Sci. Rep.* **2014**, *4*, 3687. [CrossRef] [PubMed]
62. Gasim, M.F.; Choong, Z.Y.; Koo, P.L.; Low, S.C.; Abdurahman, M.H.; Ho, Y.C.; Mohamad, M.; Suryawan, I.W.K.; Lim, J.W.; Oh, W. Da Application of Biochar as Functional Material for Remediation of Organic Pollutants in Water: An Overview. *Catalysts* **2022**, *12*, 210. [CrossRef]
63. Aon, M.; Khalid, M.; Naeem, M.A.; Zafar-ul-Hye, M.; Hussain, S.; Hussain, M.; Aslam, Z. Peanut-Waste Biochar and Buffalo Manure Decreased Nitrogen and Phosphorus Requirement of Maize Grown in an Alkaline Calcareous Soil. *Int. J. Agric. Biol.* **2018**, *20*, 2661–2668.
64. Zafar-Ul-hye, M.; Wasim, M.M.; Munir, T.M.; Aon, M.; Shaaban, M.; Abbas, M.; Hussain, M.; Ahmad, M. Co-Application of Sugarcane Bagasse Biochar, Farmyard Manure and Mineral Nitrogen Improved Growth Indices of Corn Grown in Alkaline Calcareous Soil. *J. Plant Nutr.* **2020**, *43*, 1293–1305. [CrossRef]
65. Wang, Y.; Ren, Q.; Li, T.; Zhan, W.; Zheng, K.; Liu, Y.; Chen, R. Influences of Modified Biochar on Metal Bioavailability, Metal Uptake by Wheat Seedlings (*Triticum aestivum* L.) and the Soil Bacterial Community. *Ecotoxicol. Environ. Saf.* **2021**, *220*, 112370. [CrossRef]
66. Lehmann, J.; Rillig, M.C.; Thies, J.; Masiello, C.A.; Hockaday, W.C.; Crowley, D. Biochar Effects on Soil Biota—A Review. *Soil Biol. Biochem.* **2011**, *43*, 1812–1836. [CrossRef]
67. Wang, Z.; Guo, H.; Shen, F.; Yang, G.; Zhang, Y.; Zeng, Y.; Wang, L.; Xiao, H.; Deng, S. Biochar Produced from Oak Sawdust by Lanthanum (La)-Involved Pyrolysis for Adsorption of Ammonium (NH_4^+), Nitrate (NO_3^-), and Phosphate (PO_4^{3-}). *Chemosphere* **2015**, *119*, 646–653. [CrossRef]
68. Kocsis, T.; Ringer, M.; Biró, B. Characteristics and Applications of Biochar in Soil–Plant Systems: A Short Review of Benefits and Potential Drawbacks. *Appl. Sci.* **2022**, *12*, 4051. [CrossRef]
69. Adebajo, S.O.; Oluwatobi, F.; Akintokun, P.O.; Ojo, A.E.; Akintokun, A.K.; Gbodope, I.S. Impacts of Rice-Husk Biochar on Soil Microbial Biomass and Agronomic Performances of Tomato (*Solanum lycopersicum* L.). *Sci. Rep.* **2022**, *12*, 1787. [CrossRef] [PubMed]
70. Dempster, D.N.; Gleeson, D.B.; Solaiman, Z.M.; Jones, D.L.; Murphy, D.V. Decreased Soil Microbial Biomass and Nitrogen Mineralisation with Eucalyptus Biochar Addition to a Coarse Textured Soil. *Plant Soil* **2012**, *354*, 311–324. [CrossRef]
71. Zhang, Q.Z.; Dijkstra, F.A.; Liu, X.R.; Wang, Y.D.; Huang, J.; Lu, N. Effects of Biochar on Soil Microbial Biomass after Four Years of Consecutive Application in the North China Plain. *PLoS ONE* **2014**, *9*, e102062. [CrossRef] [PubMed]

Disclaimer/Publisher's Note: The statements, opinions and data contained in all publications are solely those of the individual author(s) and contributor(s) and not of MDPI and/or the editor(s). MDPI and/or the editor(s) disclaim responsibility for any injury to people or property resulting from any ideas, methods, instructions or products referred to in the content.

agriculture

MDPI

Review

Mitigating the Impacts of the COVID-19 Pandemic on Crop Farming: A Nanotechnological Approach

Fadekemi O. Akinhanmi [1,*], Opeyemi I. Ayanda [1], Eze F. Ahuekwe [1] and Gabriel A. Dedeke [2]

1 Department of Biological Sciences, Covenant University, Ota 1023, Nigeria
2 Department of Pure and Applied Zoology, Federal University of Agriculture, Abeokuta 111101, Nigeria
* Correspondence: hazelfadekemi@gmail.com or fadekemi.akinhanmipgs@stu.cu.edu.ng

Abstract: COVID-19 is a highly infectious respiratory disease that resulted in a global pandemic that has affected every stage and sector of life. Although it is mainly seen as a health issue, its impacts and ripple effects also resonated in the education, technology, agriculture, and research fields, creating socio-economic disruptions across the globe. In a bid to curb the wide spread of the disease, diverse sudden restriction measures were adopted, which had implications on food security and food availability via supply shortages and agricultural disruptions. Scientific studies such as those regarding nanotechnological developments, which had been underway for improving food quality and crop improvement, were also slowed down due to the complexities of the pandemic and global restrictions. Nanotechnology is a developing and promising field for further development of crop productivity by enhancing the proficiency of agricultural resources, thereby increasing food yield and food security. The application of nanotechnology crop farming involves the use of nano-scale materials that can be formulated into nano-emulsion, nano-capsule, nano-fertilizer, nano-pesticide, and nano-biosensor applications for improved agricultural productivity. In as much as the challenges of nanotoxicity could raise health and environmental concerns, advances in the biosynthesis of nanomaterials potentially allay such fears and concerns. Furthermore, these ideas will help in bridging the gap created by the pandemic on food availability, food security, and agriculture. This review focuses on the implications of the COVID-19 pandemic on nanotechnological applications for improved crop productivity and nanotechnological mitigation strategies on the impacts of the COVID-19 pandemic, risk assessment, and regulatory issues surrounding nano-crop farming, and this study provides an insight into future research directions for nanotechnological improvements in crop farming and the sustainable development of nano-enabled agriculture.

Keywords: COVID-19; crop improvement; food shortage; nanotechnology

Citation: Akinhanmi, F.O.; Ayanda, O.I.; Ahuekwe, E.F.; Dedeke, G.A. Mitigating the Impacts of the COVID-19 Pandemic on Crop Farming: A Nanotechnological Approach. *Agriculture* **2023**, *13*, 1144. https://doi.org/10.3390/agriculture13061144

Academic Editors: Muhammad Sultan, Yuguang Zhou, Redmond R. Shamshiri and Muhammad Imran

Received: 1 December 2022
Revised: 7 February 2023
Accepted: 7 February 2023
Published: 29 May 2023

Copyright: © 2023 by the authors. Licensee MDPI, Basel, Switzerland. This article is an open access article distributed under the terms and conditions of the Creative Commons Attribution (CC BY) license (https://creativecommons.org/licenses/by/4.0/).

1. Introduction

The recent coronavirus disease 2019 (COVID-19) put the world into sudden chaos with the need for survival as the threat was combated on all fronts. Although several coronavirus diseases existed before this time, only six have reportedly caused mild to severe respiratory tract infections in humans [1]. The coronavirus is a highly infectious virus that causes respiratory illnesses with a high probability of morbidity and mortality [2]. COVID-19 is a highly contagious infection caused by a new strain of coronavirus disease (SARS-CoV-2), which results in severe acute respiratory tract symptoms [3]. First discovered and recorded in Wuhan, China, in December 2019, this virus is highly infectious and was spread worldwide within a short time [4]. In light of this, the World Health Organization (WHO) declared it a global pandemic on 11 March 2020 [5]. About a month later, the WHO had reported over 2.1 million cases of COVID-19 in 213 countries, with the most affected countries including the United States of America, Italy, China, Spain, Germany, France, the United Kingdom, Canada, Turkey, Brazil, Belgium, and Russia [6]. Since then, the number of cases has continued to rise globally and has become a critical threat to public health.

As much as COVID-19 is a health issue, it has also affected every other stage and sector of life. For instance, the education sector suffered with the distance education process and researchers becoming active technological educators, which instructors had previously avoided. Laboratories that ought to have been coordinating experimental studies, reporting more scientific research, and conducting field trials of different technical applications were suspended due to the movement restrictions and lockdowns. However, this period of 'pause' is expected to provide new research approaches to new and developing research areas [7]. This global pandemic has also resulted in recognizable socio-economic disturbances and has defied racial issues and geographic discrimination [8]. Governments worldwide attempted to curb the spread of COVID-19 by implementing various curtailing measures such as quarantines and lockdowns, travel restrictions to the complete shutdown of many international airports, border shutdowns, business closures, and regulations of several social and economic activities. Unfortunately, these sudden measures, via multiple dynamics, have had profound implications on food security and availability, leading to supply shortages amplified by agricultural disruption [9]. Before the pandemic outbreak, factors such as climate change, increased global population, rapid urbanization, and economic growth in most developed countries led to limited natural resources responsible for agricultural activities and productivity. About two billion people face food insecurity and malnutrition at the moderate or severe level, with the number increasing since 2014, rising by 60 million over five years [10]. The measures engineered to curtail and curb the spread of the virus also led to agricultural disruption via labor shortages and delays in the food supply. Scientific studies, which had been underway for enhancing practicable and applicable technologies to improve crop productivity, food quality, and quantity, were also slowed down due to the complexities of the global restrictions [10]. However, mitigating the challenges of food insecurity and the setbacks caused by the pandemic requires continuous, practical, productive, and sustainable agricultural development. In addition, the rising awareness of consumers of food and agrarian items and their health advantages is convincing researchers, industrialists, and farmers to discover methods that potentially enhance food quantity and quality while maintaining its vital nutritional benefits [11]. The developing field of nanotechnology has shown latent capability for further crop farming development by enhancing agricultural resources' proficiency and proffering solutions to agrarian challenges for improving food yield and security [12].

Nanotechnology is a multidisciplinary field of science that can be described as the fabrication of carefully organized particles formed from atoms that can change their material properties at a particular size of 1 nm–100 nm (nanoscale) [13]. Several industries and sectors, such as the medical services and pharmaceutics, textile, food and agricultural, information technology, and renewable energy fields, have had significant contributions from nanotechnology because of its promising capability to make most products more proficient. Moreover, nanotechnology approaches are still being studied to further combat the spread and management of COVID-19 through personalized medicine, artificial intelligence, and nano-therapies [14]. Recently, the recognition of nanotechnology in crop farming, a rapidly developing field, has gained significant ground because of its capacity to offer prospective answers essential for viable and sustainable agriculture [15]. Although progress in nanotechnological advancements in crop farming has been steadily increasing, the issue of delay in field applications and, more so, pandemics could serve as a 'clog in the wheel' of progress. Nanotechnology manages nanomaterials that have a minimum of one dimension on the nanoscale. Tested nanomaterials in crop farming include inorganic (metal and metal oxide nanoparticles), organic (naturally occurring nanoparticles), and combined (i.e., nanohybrids) materials [16]. The use of nanotechnology in crop farming accentuates the goal of the development of clean, safe, precise, and environmentally friendly nanomaterials, utilizing biologically compatible and nontoxic solvents, biodegradable and natural grids, and energy-productive and viable procedures [17,18]. Nano-carriers are equipped to enhance active ingredients' solubility while protecting them from volatilization and degradation. Improving proficiency can produce better outcomes, utilizing lower dosages and several other applications, thus reducing environmental pollution and risks to human wellbeing [19,20]. Various nanomaterial formulations are utilized in crop farming, including nano-herbicides, nano-fungicides, nano-pesticides, nano-fertilizers, growth-promoting substances, and nanobiosensors, among other applications [21]. Engineering nanomaterials is the state-of-the-art research track that upholds the advancement of innovative farming fields by offering a more extensive surface area, which is vital for the viable improvement in agri-business [22]. Accordingly, nanotechnology can decrease the incertitude and organize the management techniques of agricultural production as a replacement for traditional methods while attenuating the lag caused by the COVID-19 pandemic. This present review discusses the applications of nanotechnology in alleviating the impacts of the COVID-19 pandemic on crop farming and crop improvement in battling food shortages and food security.

2. Implications of the COVID-19 Pandemic on Nanotechnological Advancement in Crop Farming

Despite ongoing research on nanotechnological applications in improving crop production and combatting food shortages, the COVID-19 pandemic undermined these efforts due to multiplex dynamics caused by the lockdowns and other actions intended to curb the spread of the disease. Although technology progressed at the speed of the scientific data obtained, the sudden outbreak of the COVID-19 pandemic hindered progress in scientific studies and related projects worldwide [23]. Instead, research efforts focused on ending the pandemic and economic giants were also geared towards investing in necessary experiments to permanently end the pandemic [24]. To this end, research is ongoing to harness the potential of nanotechnology to proffer solutions to the raging pandemic. The primary focus of the scientific community has been continuous investigation and development to improve the characteristics of already-in-use preventive, diagnostic, and treatment measures for the virus through the application of nanomaterials [25].

The impact of the pandemic also created conditions for significant disruption to food systems, giving rise to a dramatic increase in food insecurity and hunger [26]. The measures implemented to curb the spread of the disease severely impacted crop production, jeopardizing global food security [27]. Furthermore, the pandemic's impacts were felt in the form of rising food prices, unemployment, and income losses [28]. The effects of COVID-19 on crop production can be described as follows: (1) disruption of the input supply chain; (2) shortage of labor and mobility; and (3) a shortage of food occasioned by travel restrictions and the ban on movement of persons and food distribution from the production site to the consumers [7]. These inconsistencies exacerbated food security challenges, price hikes, and increased food losses. Although the restrictions did not directly affect agricultural activities, the agrarian sector was indirectly affected via its connections to the rest of the economy. For instance, businesses such as hotels, restaurants, and catering services, which are majorly fed by agricultural production, were significantly affected by the COVID-19 policy responses [9]. Moreover, farming activities require high demand for labor, but the restriction of movement and lockdown limited access to farmlands by laborers and farmers [29]. In addition, the lack of sufficient viable seeds and the delay in input procurement due to the disruption of the agricultural input supply chain, such as seeds, agrochemicals, and fertilizers, further reduced crop yields and the quality of agro-products [7].

Food quality and safety were also compromised by the increasing rate of COVID-19 cases, which reduced available human resources, necessitating inexperienced farm workers' engagement, poor farm management, poor crop yields, and decreased quality of the crops produced [28]. Crop productivity could also be affected in the future, especially if the virus is not fully contained and the restriction measures continue. With the vast array of applications of nanotechnology, it could also serve as a bridge to close the lapses and setbacks that resulted from the pandemic. Although nanomaterials have the potential to increase and promote sustainable crop production, the progress in agricultural nanotechnology is slow because of toxicity and safety concerns [30].

3. Mitigating COVID-19's Impacts on Food Security through Nanotechnology

The direct and indirect impacts of the COVID-19 pandemic on food security, farmers' income, food availability, and crop production have prompted measures to mitigate the further spread of the virus among crucial players in crop production and the food supply chain. In a recent publication by the WHO, recommendations were made on measures to hamper SARS-CoV-2 transmission and manage COVID-19 occupational outbreaks through worksite restrictions, physical distancing, isolation of infected persons, regular worksite disinfection, regular hand hygiene, environmental monitoring, and appropriate use of personal protective equipment (PPE) [31]. While these measures are pertinent, the complimentary application of technologies such as nanotechnology is important in crop production scenarios considering its large workforce and labor mobility issues. Major nanotechnological innovations, such as precision farming and artificial intelligence, nano-biofortification of food crops, genetic cargoes, and smart delivery systems, are emerging in crop production [32]. Although still in their infancy, these technological advancements have the potential to cushion the impacts of the COVID-19 pandemic on food availability, quality, and crop production. For instance, precision farming aims at the targeted delivery of nano agrochemicals, thereby improving nutrient-use efficiency, crop production and yield, plant resilience, and soil health. Nanotechnology-based processes such as nano-nutraceuticals and nanofood are emerging to increase the nutrient quality of crop yield, provide physiological health benefits, and enhance protection against chronic diseases [33]. Nanotechnology can also be applied to food packaging to improve its shelf life and maintain freshness for extended periods. Essentially, nano-based materials have the potential to double up crop productivity and availability, thereby alleviating the food shortage caused by the pandemic.

Some educational institutions in the United States are currently developing easy-to-use portable smart nanosensors to detect and monitor food quality [34]. Scientists working with the National Institute of Food and Agriculture (NIFA) are also developing eco-friendly processes for fabricating silver NPs that can be used in antimicrobial films and food packaging. While the potential risks of nanomaterials are still being studied to facilitate public acceptance, manufacture, and use, public awareness is being raised to incite society, industry, and governments to fund, fast-track the development of, commercialize and adopt nanotechnology globally.

Nanomaterials possess enhanced physical and chemical properties such as in color, hydrophobicity, melting point, flexibility, and reactivity. The applicability and efficiency of nanostructures are the changes in their properties with size and "size effects" [35]. The surface-area-to-volume ratio increases with the reduced size of the materials, thereby enhancing their surface reactivity, otherwise known as surface effects; alterations in the bandgap enhance their chemical properties due to quantum size effects [36]. These improved properties, coupled with their size, have made nanomaterials attractive candidates for the attention of researchers in every field. The application of nanomaterials has been reported in different areas, such as medicine, nano-therapy, energy production and storage, virology, communication systems, food and beverages, and so on. In addition, there are reports that nanotechnology has the potential to address the three significant challenges arising from the COVID-19 outbreak: disease prevention, prompt and early diagnosis, and treatment [14]. The COVID-19 pandemic triggered measures that have had and still have ripple effects on crop production and food supply, leading to food insecurity, hunger, and malnutrition. Nevertheless, innovation and research progress in nanotechnological advances can alleviate these challenges by safeguarding food security, safety, and preservation. Detailed discussions on nanomaterials and their applications in crop farming are stated as follows.

3.1. Nanomaterials and Their Classification

Nanomaterials (NMs) are objects whose measurement is between 1 nm and 100 nm (nanoscale). They may exist as particles, strands, wires, composites, flakes, chips, tubes, cylinders, or films [37]. One way of classifying nanomaterials is according to their dimension, as shown in Figure 1. Additionally, due to size control and the enormous surface area of these materials at the nano-scale, they are valuable and have outstanding performance in numerous fields of utilization. Nanomaterials display varying properties from their bulk analogs, representing their unique component, including a large surface-to-volume ratio and explicit physicochemical and natural characteristics [38].

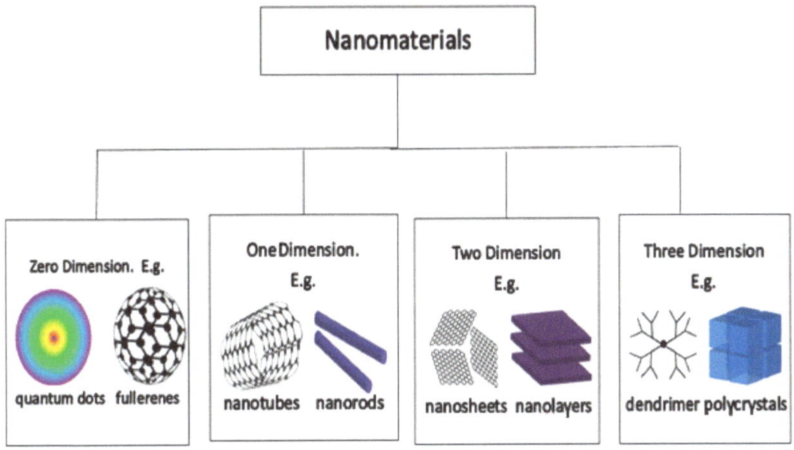

Figure 1. Classification of nanomaterials [37].

Synthesis of Nanomaterials

Some nanomaterials exist in the natural world due to processes of photochemical reactions, forest fires, volcanic eruptions, erosion, plant, animals, or even microbes. There are three basic approaches used in synthesizing nanomaterials.

1. Physical synthesis, also described as the top-down approach. This technique involves the breakdown of bulk materials into smaller and tiny particles, and this process is energy-consuming and takes more time to achieve thermal stability. Several methods applied in the physical synthesis of nanomaterials include the tube furnace, laser ablation, thermal dehydration, and thermal decomposition [38].
2. Chemical synthesis, also described as the bottom-up approach. This involves the buildup of nanomaterials from atoms and smaller particles. The chemicals and reducing agents used in this process pose a toxicity risk during their application. Nanomaterials' chemical synthesis methods include physicochemical reduction, wet chemical synthesis, and electrochemical techniques.
3. Biological synthesis, also referred to as "green" synthesis. This involves the production of nanomaterials from biogenic materials or substances. The biosynthesis of nanomaterials can be from the following: The use of microorganisms such as fungi, yeast, and bacteria; the use of plant extracts or enzymes; or the use of templates such as DNA, membranes, viruses, or diatoms. This process is cheaper, easier, and produces safer and more environmentally friendly nanomaterials [37].

3.2. Dimensions of Nanomaterials in Crop Farming

Nanotechnology is an emerging technology with potential applications in our daily existence, implementing the change in society. The rapid advancement in nanotechnology has been aiding the transformation of the conventional crop farming framework with the help of different innovative components such as nanoparticles, nano-biosensors, nanosensors, antivirals, antimicrobials, and nano-fertilizers [11]. Nanotechnology has been suggested to positively influence crop science and nutrition by improving crops' shelf-life, supporting tracking systems for contaminants, shaping better techniques for food storage, and aiding in the development of value-adding nutritional supplements or antimicrobial agents into food crops. Hence, nanomaterials may improve crop production with the help of different innovative methods and materials such as nanoparticles, nanoencapsulation, nanoemulsion, nanofiber, and carbon nanotube [39].

3.2.1. Nanoparticles

Nanoparticles (NPs) are objects or structures whose dimensions range between 1 and 100 nm (nanoscale). They may be fibers, flakes, wires, composites, films, and tubes. There are also structured nanomaterials such as quantum dots, nanoclusters, nanocrystals, nanowires, and nanotubes [38]. Because of their size and superior surface area, nanoparticles are promising tiny substances with enormous influence in the engineering, medicine, electronics, photonics, biotechnology, and agriculture sectors. They possess translucent or formless characteristics, which can act as conveyors for liquid droplets or gases and should be regarded as a unique state of matter other than solid, gaseous, liquid, and plasma states [37].

3.2.2. Nanoencapsulation

Recent progressions in nanotechnology include in the encapsulation methodology, which was initially used in biotechnology to improve quality production measures and to prompt the disintegration of generative cells alongside their by-products [40]. Eventually, encapsulation was introduced to nanotechnological advancements, which were discovered to be profoundly effective compared to regular encapsulation. Encapsulation is the innovative strategy of enclosing different active substances in matrices, ceding their payload under particular conditions at a specific rate. The techniques involved in encapsulation can secure the bioactive ingredients from environmental conditions and natural effects until the controlled delivery of the active ingredients is achieved. In other words, nanoencapsulation can be defined as the enclosing of active components inside nanometer-sized shells or capsules, also known as nanoshells or nanocapsules [41]. Nanocapsules vary in size and shape, depending on the materials and techniques used to produce the capsules or shells [42]. The design of encapsulated active components predominantly relies on the selected capsule material (pectin, zein protein, chitosan, lecithin, etc.) and the nanoencapsulation technique [42]. Furthermore, nanoencapsulation enhances the bioavailability of enclosed bioactive components and improves biological efficiencies [43]. Agrochemicals such as pesticides, herbicides, and fertilizers can be enclosed in nanocapsules for controlled release and targeted delivery of their active compounds for disease control, nutrient uptake, enhanced growth, and the development of crops [44].

3.2.3. Nanoemulsion

Nanotechnology applications have been investigated to improve global food security, such as crop conservation and protection, by providing consumable covering or seed preparation, essentially improving food quality [45]. Nanoemulsions are novel bioactive conveyance systems made up of emulsified water and oil stage (w/o) or oil and water stage (o/w) with droplet sizes within the range of 50 nm to 100 nm. The merits of a tiny droplet size in nanoemulsions include the assurance of even circulation and dissemination of the active components through the surfaces. Moreover, nanoemulsions further develop infiltration properties of bioactive components because of the low surface tension and high surface area of the emulsion [46]. The droplet size of nanoemulsions has merits over conventional emulsions because of better collection stability, better optical clarity, and improved bioactivity of emulsified ingredients [47]. The use of thymol nanoemulsion-loaded quinoa protein/edible chitosan coating was recently described to conserve the quality and safety of strawberries with reduced yeast and fungus load. Additionally, weight reduction was observed during the whole storage period [45].

3.2.4. Nanofiber

Most nanotechnology research has been limited to a few particles or molecules, but massive fabrication techniques and large-scale devices may also be developed for profit-oriented commodities. Nanofiber technology is a technique involving the synthesis and application of fibers with nanoscale dimensions. This technology assembles one- and two-dimensional nanomaterials for the large-scale production of macroscale nanostructures [48]. Nanofibers can be described as fibers with a diameter equal to or less than 100 nm. Nanofibers are essential and significant due to their specific surface area and adaptability, making them favored material structures for varying drug delivery and crop production applications. The adequacy of nanofiber application has been tested in tomato and lettuce seeds coated with nanofiber. It was observed that the rapid growth in the seeds resulted in improved seedling biomass for both plants, suggesting accurate delivery of agrochemicals, which further improves germination and seedling biomass for model seeds [49].

3.2.5. Carbon Nanotubes (CNTs)

Engineered nanomaterials have recently been implicated in nanotechnology advancement because of their extensive scope of utilization in different fields. Nanotubes belong to this category of engineered nanomaterials, with carbon nanotubes being an essential group [50]. Carbon nanotubes are cylindrically shaped rolled-up graphite sheets whose length is at the micrometer scale, with a diameter of around 100 nm. They are byproducts of carbon fibers and fullerenes with particles of 60 atoms of carbon organized in particularly enfolded cylinders. CNTs are of two kinds: single-walled carbon nanotubes and multi-walled carbon nanotubes. Where a highly detailed ratio is necessary for biological applications, CNTs are appropriate, and they are equally considered suitable for many applications [51]. Prominent applications of CNTs in crop farming include seed priming, rapid plant development, biosensor diagnostics and analysis, and pesticides [52]. It has been reported that CNTs can permeate the thick seed coat and prompt development [53]. In addition, optical sensors based on CNTs have been developed to screen the real-time detection of pathogenic microbes, organophosphate chemical warfare instruments, and pesticides [54].

3.3. Applications of Nanotechnology in Crop Farming

The recent advancements in nanotechnology can help us to accomplish sustainable agricultural development to further develop resource proficiency and efficiency, fortify resilience, and secure the agriculture field's responsibility. This technology, delivered via different techniques such as precision farming, crop improvement, and soil remediation, will be an economic force to be reckoned with to enhance current agricultural activities. Figure 2 elaborates on the application avenues in the agricultural sector. In addition, various delivery techniques of NMs include foliar spraying, seed priming and coating, irrigation, and hydroponics.

Foliar spraying describes NP entry through the leaves and sometimes through other non-reproductive parts of the plant. Upon contact with the plant, NP uptake occurs through natural plant openings (stomata, lenticels, etc.), and they are translocated via the apoplastic and symplastic pathways [32]. The apoplastic pathway promotes the radial movement of NPs through the cell membrane and is determined by capillary forces and osmotic pressure. On the other hand, the symplastic pathway enhances intracellular movement through the plasmodesmata, determined by endocytosis or ion channels. Although seed priming and coating with NPs have been investigated in *Oryza sativa*, *Glycine max*, and *Cucumis sativus* [55], the mechanism of NP action is still understudied. NPs can also be applied in hydroponics, wherein an adequate environment with accurate parameters can be maintained for enhanced uptake of NPs. For instance, ZnO NPs were administered by hydroponic treatment to improve the morphological and physiological parameters of *Nicotiana tabacum* seedlings [56]. Nanotechnology can reduce the utilization and misuse of mass agrochemicals and proffer more affordable solutions and environmentally friendly alternatives.

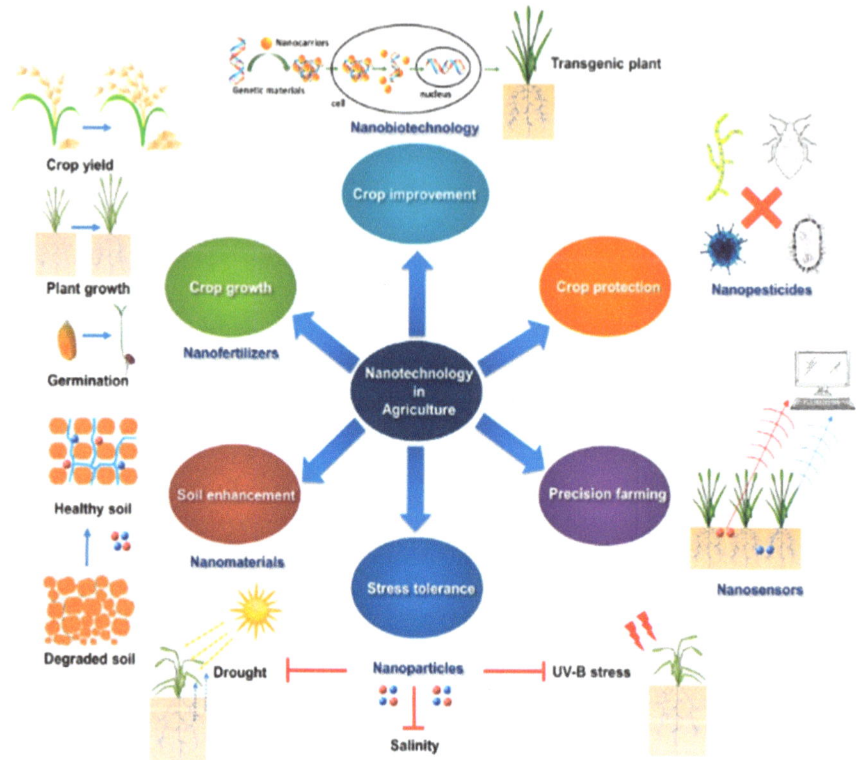

Figure 2. Nanotechnology applications in crop farming [57].

3.3.1. Precision Farming and Artificial Intelligence

Precision farming has boosted crop productivity and quality by utilizing reduced quantities of agrochemicals such as insecticides, herbicides, and fertilizers while targeting particular actions. This highly profound tool has merits over the use of supercomputers and remote sensing gadgets to detect crop problems, eventually resulting in cost-effective production and enormous advantages for farmers. Precision farming relies on thorough comprehension of modern innovations that enable harvesters to decipher the number of various components responsible for the low yield of crops while creating designs that will steer prompt decisions regarding the application and timing of inputs [57]. Moreover, precision farming is expected to minimize agricultural waste production and thus would be more eco-friendly compared to established farming practices. Nanotechnology is viewed as one of the main drivers of precision farming since the characteristics of nanomaterials will enable the improvement of more proficient strategies for crop management [58].

A recent publication suggested that artificial intelligence (AI) and machine learning modeling could drive progress in nano-enabled precision farming [59]. Although still in their infancy, these approaches are gaining regulatory acceptance for the safety assessment of nanomaterials, facilitating safe-by-design nanomaterials for a range of consumer products. The integration of AI in precision farming could accelerate development and provide the insights needed to overcome the current barriers in nano-enabled crop farming [59].

For instance, AI may predict the impact of NMs on the agricultural ecosystem and their performance in enhancing agricultural production by integrating experimental data from different soil conditions and plant species while minimizing pollution and ensuring a high safety profile for nanomaterials in soil and minimal residues in edible plants.

Precision agriculture based on nanotechnology uses devices such as the green seeker or normalized difference in vegetative index (NDVI) sensor. This device generates red and near-infrared light using nano-based light-emitting diodes and calculates NDVI values accordingly. This idea is based on the knowledge that healthy plants' chlorophylls absorb more red light during photosynthesis and reflect more near-infrared light than unhealthy plants [60]. Values for the NDVI are computed based on this difference. The agricultural sector may be positively impacted by precision farming, which enables real-time monitoring of field factors such as soil conditions, soil fertility, moisture level, temperature, crop growth, crop nutrient status, insect attacks, various plant diseases, and so on [60].

3.3.2. Improvement of Crop Productivity

Nanotechnological advancements have promising prospects for improving and enhancing crop productivity by ensuring superior plant management. This advancement involves the development of nano-fertilizers, nano-pesticides, nano-herbicides, nano-fungicides, nano-biosensors for pathogen and contaminant detection, etc. (Table 1).

Nanopesticide

Pesticides can be chemical or biogenic substances or compounds used to manage pests for enhanced quantity and quality of crop production [57]. Conventional pesticides are known for their poor specificity, high toxic potential, and environmental persistence attributes [61]. They pose severe ecological and public health hazards due to these factors. Since the main components of pesticides are prone to decomposition by varying environmental factors, abiotic factors, including salinity, temperature, and light, may alter pesticides' biological effectiveness and chemical dynamics [47]. Biotic factors, such as microorganisms, play the most important role in the decomposition and pesticide persistence in the environment [12]. The environmental problems caused by the overuse of conventional pesticides have led to the advent of nano-pesticides, potentially overcoming these issues through a possible reduction in chemical usage.

Moreover, nano-pesticides show valuable properties, such as optimized resistance, porosity, thermal strength, solubility, translucency, and biodegradability, which are important and necessary for viable agricultural production [62]. Essentially, the controlled and slow release of active components minimizes the chemical load of pesticides required to control pests. Over the years, various nanopesticide formulations have been developed and applied by scientists for more specific actions. Compared to conventional pesticides, nanopesticides have more effective control of pests while still being environmentally friendly [63]. Previous research has successfully tested the insecticidal efficacy of various nanoparticles against agricultural pests. Zheng et al. [64] reported the insecticidal efficacy of nanoparticles against red fire ants, *Solenopsis invicta*, which are insect pests of agricultural commodities. Their results showed an alteration and inhibition of digestive functions, eventually leading to death.

Nanoherbicides

The efficacy of conventional herbicides against weeds is limited to the shoot part, leaving the root area as a new source of weeds in the forthcoming season [22]. Nanotechnology has improved the effectiveness of chemical herbicides by precise delivery to receptors in the underground regions of weeds [71]. These nanoherbicides provide new entrance routes and active transport to various parts of the roots, which will eventually inhibit food glycolysis backup in the root tissues [80]. In addition, this technique will reduce the use of toxic chemical components such as imidazolinones and glyphosate. NMs, such as polymeric NPs, have been used extensively as nanocapsules or nanospheres for enclosing various conventional herbicides [70]. Research on two plant species, *Brassica* sp. and *Zea mays*, suggested potent growth inhibition of *Brassica* sp., which is the actual target organism, using nanoherbicide formulation.

Table 1. Applications of nanomaterials for crop productivity.

Nanopesticide	Agricultural Use	References
PQ-loaded CS-gated porous carbon nanoparticles	User-safe and efficient chitosan-gated porous carbon nanopesticides and nanoherbicides.	[61]
Zinc oxide nanoparticles (ZnO)	Nanoparticles enhance thiamethoxam insecticidal activity against *Spodoptera litura* larvae	[65]
Silicon oxide nanoparticles (SiO_2)	Insecticide properties against leaf worm (*Spodoptera littoralis*)	[66]
SiO_2	Maize nanofertilizer and low-dose pesticide for pests that can affect maize during storage post-harvest (*Sitophilus oryzae, Rhizopertha dominica, Tribolium castaneum, Orizaephilus surinamenisis*)	[67]
Silicon nanoparticles (SiNP)	Nanoosilicon enhances maize resistance against oriental armyworm (*Mythimna separata*) by activating the biosynthesis of chemical defenses	[68]
Si	Biological silicon nanoparticles maximize the efficiency of nematicides against biotic stress induced by *Meloidogyne incognita* in eggplant	[69]
Nanoherbicides		
PQ-loaded CS-gated porous carbon nanoparticle	User-safe and efficient chitosan-gated porous carbon nanopesticides and nanoherbicides	[61]
Poly(epsilon-caprolactone) nanoparticles	Application of poly(epsilon-caprolactone) nanoparticles containing atrazine herbicide as an alternative technique to control weeds and reduce damage to the environment	[70]
Chitosan/tripolyphosphate nanoparticles	Nanoparticles loaded with paraquat herbicide: An environmentally safer alternative for weed control.	[71]
Nanoatrazine	Nanoencapsulated atrazine inhibited the photosynthetic activity of the weed *Alternanthera tenella*	[72]
Nanofungicide		
Zinc oxide NP (ZnO)	Fungicidal activity against multiple pathogenic fungi of apple orchards (*Alternaria mali, Botryosphaeria dothidea, Diplodia seriata*).	[73]
Manganese oxide NP (MnO)	Antifungal activity against soil-borne pathogens (*P. nicotianae, T. basicola*) with a possibility to control other plant pathogens	[74]
Titanium oxide NP (TiO_2)	Antifungal activity against wheat rust.	[75]
Copper oxide NP (CuO)	Antifungal activity against the plant pathogen *Colletotrichum gloeoesporioide*	[76]
Nanofertilizer		
Zeolite/Fe_2O_2	Nanofertilizer with less toxic effect toward humans compared to other fertilizers.	[77]
Copper (Cu) nanowires	Fertilizer for improved plant physiological performance and agronomical parameters	[78]
Urea-loaded mesoporous $ZnAl_2SiO$	Nanofertilizer for slow delivery of urea and zinc	[79]
SiO_2	Maize nanofertilizer and low-dose pesticide for pests that can affect maize during storage post-harvest (*Sitophilus oryzae, Rhizopertha dominica, Tribolium castaneum, Orizaephilus surinamenisis.*	[67]

Furthermore, it has been reported that NP application could proficiently minimize the mobility of atrazine in the soil, reducing soil contamination [70]. NPs have also been used for the encapsulation of three triazine herbicides to minimize concerns regarding environmental contamination and toxicity [81]. Nanoencapsulation of these herbicides reportedly increased their efficacy and stability. Further investigations suggested the controlled and timely release of triazines from the nanocapsules, with reduced genotoxic impacts for the formulated nanoherbicides [82]. Likewise, chitosan/tripolyphosphate nanoparticles were utilized for encapsulating paraquat, and inorganic nanoparticles such as silica have equally been employed for nanoherbicide formulation [71,82]). Agricultural waste-based nanomaterials such as nano absorbents rice husk were sequestered and used as a nanocarrier for 2,4-dichloro phenoxy acetic acid herbicide. The results showed high herbicide sorption with optimal activity [83].

Nanofungicide

Fungal infestations of crops practically result in huge losses and low quality in crops yearly [12]. Nanomaterials such as cobalt and nickel ferrite nanoparticles have been developed with the potential to solve the problem of loss in crop productivity. These nanoparticles have reportedly shown antifungal efficacy against specific plant fungal pathogens, *Dematophora necatrix*, *Fusarium oxysporum*, and *Colletotrichum gloeosporiodes*, and can therefore be applied for the management of fungal plant disease [84]. Biologically synthesized copper nanoparticles were also used as a proficient antimycotic component under certain experimental conditions against *Neofusicoccum* sp., *Fusarium oxysporum*, and *Fusarium solani*. The generation of intracellular reactive oxygen species (ROS) was suggested as the cause of such antifungal activity [85]. Recently, nanocomposite utilization has broadened the scope of plant protection because of its high effectiveness and environmentally friendly nature. In one instance, a fungicide, antracol, was formulated with silver-incorporated chitosan nanocomposites. Increased antimycotic activity was observed compared to each component alone [86]. Copper nanoparticles were also tested against *Phytophthora infestans*, and the results showed their antifungal efficacy in tomatoes compared to the currently available non-nano copper formulations [87].

Nanofertilizers

Conventional fertilizers have availability and delivery problems, mainly due to a lack of synchronization between nutrient availability from the fertilizer and nutrient demand from the plant [88]. This eventually results in low nutrient-use efficiency, one of the solutions nanoformulations deliver. Nanofertilzer formulations are smartly designed delivery systems that can strategically deliver their payload of nutrients and substantially improve crop productivity by enhancing effectiveness, improving soil organic matter, and minimizing unbalanced propagation and nutrient deficiencies [89]. Nanofertilizers are predominantly produced by integrating nutrients with nanomaterials but can be applied in different formulations, depending on the peculiarity of use, as shown in Figure 3. The importance of iron in plant physiology is key to respiration and the formation of chlorophyll, and its deficiency in plants may result in chlorosis. Thus, iron delivery and availability in plants are necessary for their growth. In this regard, iron-based nanofertlizers have shown promising potential. A recent study on the treatment of *Catharantus roseus* with iron oxide nanoparticles showed significant enhancements [89]. It has also been reported that a nano iron (Fe) complex showed an optimum plant growth effect on tobacco under low iron supply [90]. Similarly, one of the vital micronutrients that plays a key role in plant nitrogenous assimilation is Molybdenum (Mo). Thus, Mo deficiency can result in increased nitric oxide accumulation in a plant, which is harmful to human consumption [91].

Mo-based NPs have been produced to enhance their proficient use. Synthesized Mo NPs have been observed to be more effective in increasing nitrate reductase (NR) activity compared with elemental Mo in *Spinacia oleracea*. As a result, nitric oxide accumulation was also minimized [92]. Osman et al. [93] also reported an increased seed yield of *Phaseolus vulgaris* after applying molybdenum oxide nanoparticles. Cabbage productivity was observed to be low in Iraq on a large scale [94], and this led to a study being carried out to examine the improvement in cabbage growth and yield by NPK nanofertilizers, boosted by commercial iron, zinc, and cerium oxide (CeO_2) nanoparticles (NPs) [95]. Metal nanoparticles enhanced the NPK nanofertilizer when used to increase the growth and yield of cabbage. The plants also observed higher chlorophyll content and enhanced NPK nanofertilizer uptake. As previously highlighted, polymeric nanoparticles have been used for coating biofertilizers, and these formulations have been observed to show resistance against leaching and desiccation [57]. This led to the development of water-in-oil emulsification technology that has been used to encapsulate and deliver nutrients to target sites. This procedure also minimizes evaporation, enhances cell viability, and improves controlled release dynamics.

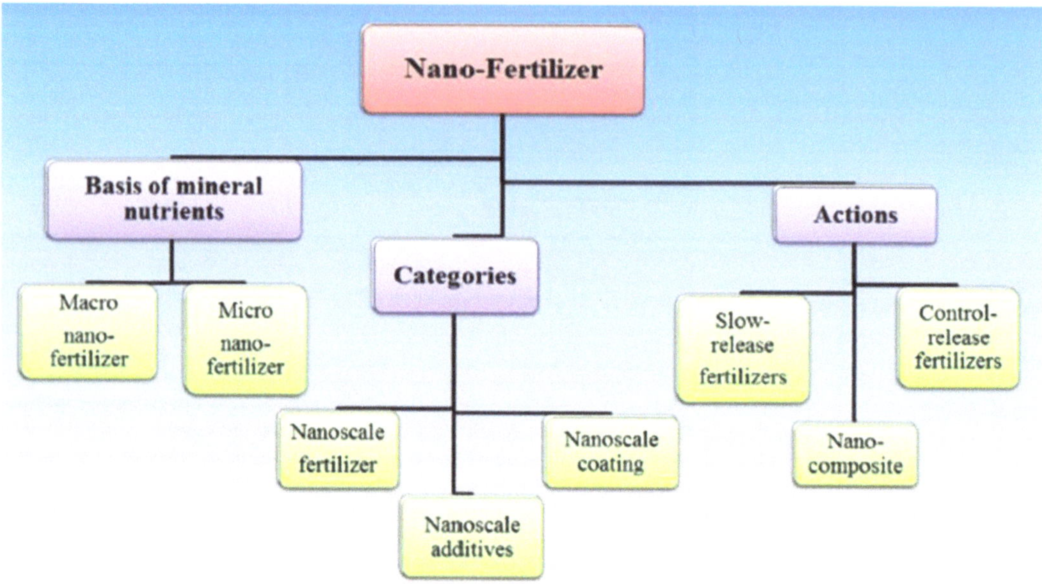

Figure 3. Types of nanofertilizers [88].

Furthermore, advances in green nanotechnology have changed global agriculture dynamics, and nanomaterials' use as nanofertilizers has aroused the potential to meet the projection of global food demand and viable agriculture. For instance, the application of carbon NP fertilizer can improve the grain yields of wheat, maize, soybean, rice vegetables, and maize [57]. Applying chitosan-NPK fertilizer increased crop yield and mobilization indices of determined wheat yield variables compared with control yield variables [96]. The application of nanotechnology in plant biology stimulates vital facets of study, as plants' primary nutrient access points are the plant root and leaf surfaces, which are more permeable at the nano dimensions. Essentially, nanofertilizer application can enhance the uptake of nutrients through these pores or by exploiting endocytosis or ion channels [97]. Moreover, it has been observed from various reports that the miniature size of nanomaterials facilitates an increased surface mass ratio of particles; thus, a large quantity of nutrients reaches the plants slowly and steadily for an extended period [98]. Hence, nanofertilizers ensure the balanced nutrition of crops throughout the growth cycle, eventually resulting in optimal agricultural production.

Nanobiosensors

Some devices can convert biological signals to electrical signals for quantitative data analysis via a microprocessor [99]. These gadgets are called biosensors. The vital component of this device utilizes biological molecules and living organisms, which serve as a communication device (bio-recognition motifs), an adequate transducer, and a signal detector. In a nanobiosensor, an analyte with a biorecognition element is exposed to the biorecpetor coated with NMs for enhanced sensitivity, and signals received are converted by a transducer to an electronic signal on the detector [57]. Some unique features of nanobiosensors include immobilization, fictionalization, and scaling-down that incorporates biological components of a transduction system into complex engineering to enhance the analytical capacity of NMs [98]. Nanobiosensors depend on an on/off mechanism, analyte detection within parts per trillion (ppt), and hinder the dissected matrix reliant on nano-formulation [100]. Using nanobiosensors permits easy detection compared to other regular biosensing methods with enhanced specificity and sensitivity, reducing rapid and exact detection time, and, more importantly, enabling real-time signal surveillance utilizing miniature compact devices. The significant applicability of nanobiosensors in the agricultural sector includes pesticide detection, food-borne pathogenic microbes, poisonous pollutants, heavy metal ions, etc. In addition, nanotechnology has been expanded to track and survey soil conditions, plant stress, plant growth, available nutrients, and food quality [101]. Fluorescent nanosensors were reported for organochlorine pesticide detection [102]. A gold nanoparticle-based colorimetric nanosensor for organophosphorus pesticide detection has also been reported [103].

Nanotechnology has similarly been utilized in electronic noses (e-noses), viewed as artificial intelligence frameworks and next-generation sensing devices. The frequency of its use in agriculture is used to track production procedures and to evaluate plant-related diseases, water/soil pollutants, and insect pest invasions [104].

Nanomaterials for Remediation of Contaminated Soils

Removing poisonous heavy metals from polluted water in wastewater treatment has become a significant problem for the agricultural sector. These metals pose a significant threat to human wellbeing since they can bioaccumulate in the food chain [105]. NMs can be applied in abiotic remediation and bioremediation of contaminated soils. The sorption–desorption reactions with soil constituents determine the toxicity and activity of these soil pollutants. Consequently, suitable alterations in contaminated soils can impact the toxicity and activity of soil contaminants [106]. These alterations can inactivate components that bind the soil contaminants, thereby reducing their activity and bioavailability [107]. For instance, nano zero-valent iron enhanced the soil washing performance of metal-polluted soil by adsorbing the metal and isolating it by magnetic separation [108]. In remediation experiments, adsorbent recovery offers a chance for contaminant recovery (either for their utilization as a resource or their ex situ treatment) and reuse. Furthermore, reusing nanomaterials in additional treatment cycles is gaining attention because it can minimize the cost of remediation. Therefore, the development of stable nanomaterials that could still retain their efficacy for numerous treatment cycles is highly challenging [109]. Nanomaterials are also used in advanced oxidation processes (AOPs), which use various oxidants to mineralize and catalyze organic contaminants. Iron nanoparticles can also be used, along with chelating agents. For instance, iron nanoparticles were used along with different chelating agents in Fenton oxidation to remediate pyrene-polluted soil [110]. The results showed varying oxidation efficiencies, dependent on the nature of chelating agents and were optimal for sodium pyrophosphate. Notably, depending on iron contents and soil nature, chemical oxidation can be effectively catalyzed by endogenous iron [111].

On the other hand, bioremediation is a natural, on-site, cheap, eco-friendly, and versatile technique for decontaminating harmful pollutants [112]. Nanomaterials are now integrated with bioremediation to proffer solutions to overcome associated limitations with this novel technology. Nanobioremediation holds excellent promise for decontaminating polluted soils with both inorganic and organic contaminants. Highly reactive NMs can be used in the dehalogenation or dechlorination process for remediating persistent organic pollutants, followed by bioremediation for more effectiveness [113]. NMs are also helpful in remediating soil organic pollutants by enhancing the availability of these contaminants to the biological agents used in bioremediation. Adherence of contaminants to nanomaterials could result in accumulation in plants with tiny nanoparticles and altered selective membrane permeability due to phytotoxic nanomaterials, which may also encourage organic contaminant uptake [105]. Nanobioremediation can also help improve the efficacy of phytoremediating heavy metals in polluted soils. The effect of titanium oxide nanoparticles on the accumulation of cadmium by soybean plants showed enhanced cadmium uptake and reduced toxicity of the same in the plants by preventing oxidative damage, which is due to scavenging free radicals produced as a result of cadmium contamination [114]. Nano-hydroxyapatite and nano-carbon black significantly minimized lead phytotoxicity to ryegrass while increasing its potential for phytoextraction [115]. It is worthy of note that various nanomaterials have different effects on the uptake of heavy metal or its toxicity in plants [105].

Nanomaterials as Plant Elicitors

Plant elicitors can be described as activated specialized metabolites important for plants' adaptation to environmental stress [116]. They can be biotic (insects, microorganisms, etc.) or abiotic (salinity, wounds, light, etc.) elicitors. Recently, it was reported that many nanomaterials could act as elicitors, enhancing the plant defense mechanism against abiotic (salinity, temperature, etc.) stressors and biotic stressors such as viruses and bacteria [117]. For example, iron oxide and gold nanoparticles reportedly promote tobacco resistance to mosaic virus disease [118,119]. Metal, metallic oxides, and carbon-based nanomaterials have shown potential in various experimental conditions. For instance, the medicinal qualities of *Calendula officinalis* improved when a combination of silver nanoparticles and methyl jasmonate was applied. These compounds increased the saponin content compared to the control treatment, while all treatments showed a reduction in anthocyanin and flavonoid content [120]. Additionally, several metallic alloys, such as silver, gold, and copper, were used in treating *Silybum marianum* at different ratios. Improved germination frequency and root and shoot development were observed in the treatments, depending on the nanoparticle [121]. In addition, the callus of *Stevia rebaudiana* treated with zinc oxide nanoparticles reportedly showed improved flavonoid and phenolic content and generation of reactive oxygen species, thereby boosting antioxidant responses [122].

Furthermore, carbon nanomaterials have shown high efficiency against abiotic stresses, enhancing crop performance. For instance, incubating carbon nanodots with *Oryza sativa* showed mitigation of salt stress with the assumption that the tolerance is caused by the nanodot's reactive oxygen species scavenging properties, which lowers the seedlings' exposure to these radicals and minimizes oxidative damage [123]. Other carbon-based nanomaterials, such as single-walled carbon nano-horns, have been used in testing different crops. The results revealed that carbon nano-horns could enhance the germination of selected plants and improve organ growth in crops such as soybean and tomato [53].

3.3.3. Nanotechnology for Crop Improvement

Nanobiotechnology

In recent years, crop improvement has been the most pressing need for plant breeders and farmers to meet the booming demand for products derived from plants [124]. An essential biotechnological tool, next-generation sequencing (NGS), is being promoted to support crop development initiatives. Genome-assisted breeding is an application of the NGS process, an integrated method for locating and choosing genetic variants. Currently, nanoparticles are employed to generate efficient carriers for gene transformation, and nanobiotechnology tools are expected to develop new strategies for replacing the genetic material of one species with another [125]. Nanotechnology application has been initiated in the era of genetic engineering, with the targeted delivery of CRISPR/Cas9 single guide RNA (sgRNA) innovation [57], and this procedure has been effectively utilized in genome alteration in plants [126].

Nanoparticle technology, being a useful biotechnological tool, has been applied to improve some plant species' agronomic parameters. Mukarram et al. [127] evaluated the optimum concentration of SiNPs and their potential advantageous effects on lemon grass agronomy, focusing on induced physiological responses along the way. Lemon grass is an industrially beneficial plant with medicinal properties and is exploited for its essential oil production [127]. The result showed that Si-NPs up-regulate gaseous exchange, photosynthesis, antioxidants, and nitrogen metabolism. A general trend of crop enhancement was triggered in lemongrass after the foliar application of SiNPs. In addition, some crop plant pest species, such as planthoppers, have developed high resistance to conventional pesticides and insecticides. Hence, there is an urgent need to develop green, economical, sustainable control and preventive technology for these pests. Cheng et al. [128] reported the uptake and translocation of SiO_2 nanomaterials, with the help of fluorescence technology, from root to shoot. SiO_2 nanomaterials induced the resistance of rice plants to planthopper, primarily via physical defense.

Nanobiofortification

Nutrient deficiency in food crops, commonly in developing countries, severely compromises human health. Yet, measures taken during the COVID-19 pandemic further reduced the production and availability of food crops, resulting in increased hunger and malnutrition. Biofortification of food crops with essential micronutrients is vital in improving food quality, improving its availability and promoting human health. Nano-biofortification of food crops is a budding biotechnological area that exploits the chemistry of plants and nanomaterials for controlled and sustainable nutrient delivery. These nano-nutrients can potentially mitigate humanity's struggle against diseases such as COVID-19. In addition, the promising use of nano-nutrients has distinguished opportunities to treat COVID-19 as anti-COVID-19 nanoparticles or nano-medicine [129]. Nano-biofortification can be achieved through agronomic nano-biofortification, selective breeding, and genetic manipulation with nanomaterials. Agronomic nano-biofortification may be applied through fertilizing food crops with nano-nutrients via direct physical application to the soil or indirectly through the foliar spray, seed priming, or seedling immersion into nano-fertilizer solutions [130]. Selective plant breeding practices such as plant-growth-promoting microbes can enhance micronutrient availability in soil and bioavailability in food crops by producing chelating agents such as siderophores [131]. Nano-biofortification can also be achieved using advanced tools such as genetic engineering to produce transgenic crops through direct gene transfer to modify target genes to precisely create desired genotypes. This strategy can optimize the accumulation of nano-nutrients in edible tissues without adversely impacting other physiological and developmental features of food crops [130].

Nanotechnology in Plant Stress Management

Agricultural productivity faces various environmental factors that can lead to plant stress conditions [132]. Stressed conditions can arise from biotic (insects, microorganisms, animals, competitive species) or abiotic (heat, drought, salinity, heavy metals, flood) stresses. Hence, there is a need to enable accelerated adaptation of plants in coping and thriving under various stress factors. However, plants can accomplish this coping mechanism through diverse strategies such as the regulation of hormones, stress gene expression, toxic metal uptake regulation, activation of plant enzymes, and avoiding drought stress by shortening the plant life cycle [133]. Advances in agricultural nanotechnology suggest that nanoparticles can improve crop production under stressed conditions. Table 2 presents recent experiments applying nanomaterials in combating various environmental stressors.

While salt stress limits crop production in cultivated lands globally, the application of silicon oxide has reportedly shown increased plant fresh weight and dry weight and improved seed growth and chlorophyll content in squash plants and tomatoes under salt stress [141]. Similarly, foliar application of iron sulfate nanoparticles showed that sunflower cultivars positively tolerated salt stress. The results showed improved shoot dry weight, leaf area, net carbon dioxide assimilation rate and chlorophyll content and increased iron content [142]. UV-B stress in plants can also be effectively managed using silicon

NPs [143]. Furthermore, nanoparticles, via foliar application, are reported to be effective in the detoxification of rice plants by regulating cadmium accumulation [144]. Banerjee et al. [145] examined the ameliorative efficacy of SiNPs in molecular damage and yield loss in IR-64 plants irrigated with fluoride-polluted water throughout their life cycle. The results showed improved overall growth of fluoride-stressed plants (Figure 4). By effectively reducing fluoride uptake and bioaccumulation in mature tissues, nano-Si-priming of IR-64 seedlings produced significant levels of fluoride tolerance. The agronomic characteristics and yield of SiNP-treated and stressed seedlings were significantly enhanced, and fluoride bioaccumulation decreased in the grains. Other studies also showed the efficacy of silicon nanoparticles against lead and cadmium accumulation in wheat and rice [146,147].

Table 2. Application of nanomaterials in combating environmental stressors.

Nanomaterial	Mode of Application	Findings	References
Chitosan/Silver/Manganese-Magnesium ferrite Nanocomposite	Foliar and Root Application	Improved the growth and biomass of cabbage under cadmium stress	[134]
Copper Oxide (CuO)	Root Application	Inhibited arsenic bioaccumulation in rice plant	[135]
Selenium Oxide (SeO) and Zinc Oxide (ZnO)	Seed Priming	Elevated seed germination and early seedling growth upon salt stress	[136]
Cerium Oxide (CeO)	Foliar Application	Alleviated salinity stress by improving growth parameters and antioxidant defense system	[137]
Gold Nanoparticles (Au)	Foliar Application	Improved salt tolerance in wheat plant	[138]
Titanium Oxide (TiO_2)	Seedlings	Improved the adaptability of plants under UV-B stress	[139]
Biosynthesized Ferric Oxide (Fe_2O_2)	Foliar Application	Optimum survival rate in wheat plants exposed to drought stress	[140]

Figure 4. Comparison between a stressed and nanoparticle-managed plant [145].

Furthermore, nanomaterials have shown the potential to regulate stress gene expression. For instance, a microarray analysis after the application of silver nanoparticles in Arabidopsis showed that some genes were up-regulated or down-regulated [148]. Some up-regulated genes were associated with their response to metals and oxidative stress. In contrast, the down-regulated genes were related to their response to pathogens and hormonal stimuli, including systemic acquired resistance and gene regulated by auxin, which is involved in organ size and growth [148].

4. Challenges of Nanotoxicity

Recent advances in technology have postulated that nanomaterials have the potential to revolutionize nutrition, crop production, food packaging, and food availability. Issues associated with latent toxicological profiles of nanomaterials are poorly understood, and hence, so too is the resistance to global acceptance of its applications. The toxicity of NMs is dependent on several factors, such as particle size, surface-area-to-mass ratio, structure, and solubility [149]. Due to their small size and high surface-area-to-mass ratio, NPs can enter cells and cellular organelles, interfering with their normal biological functioning. Their large surface area may boost the production of reactive oxygen species such as superoxide anion and hydroxyl radicals, resulting in oxidative stress. For instance, positively and negatively charged nano-diamonds of different ranges were tested on the intestines of *Daphnia magna*. The results showed that the nano-diamonds with the higher range (15 nm) induced more reactive oxygen species than the lower range (5 nm) [150]. Continued ROS production in a cell may also lead to gene deletions or mutations that could cause cell carcinogenicity, which promotes the proliferation of tumors and cancer [149].

In nanotoxicology, in vitro and in vivo assessment are vital tools in toxicity testing. In vitro assessments have lowered the number of animal studies required for toxicity testing, reducing ethical concerns. Genotoxic and cytotoxic effects of nanomaterials have been examined in primary human epidermal cells using the comet assay [151]. Results showed increased DNA damage in cells exposed to ZnO NPs for 6 h, compared to the control group. The in vitro micronucleus assay and the mycoplasma backward mutation assay have also been used to investigate the cytotoxic effects of AgNPs on strains of *Salmonella typhimurium* [152]. Their reports showed no mutations in any of the test strains. This could be a result of low concentrations of NPs used. In vivo experiments, which usually use rats or mice as model organisms, provide data that enable reports that are not accessible from in vitro studies. An in vivo study was conducted on experimental mouse models to highlight the probable toxicological impact of NPs on their brains. Upon intravenous injection, silica-coated magnetic nanoparticles were found in the animal's brain [153]. A recent in vivo study reported the cytotoxicity of Se nanorods and spherical NPs in female mice. Their findings revealed an accumulation of NPs in the kidney, resulting in oxidative stress and tissue damage [154]. The toxicity of nanomaterials cannot be overlooked as it can potentially impact all organisms, and ultimately humans.

5. Issues of Risk and Regulatory Policies on Nano Crop Farming

5.1. Risk Assessment and Management in the Nano-Agricultural Sector

Assessing the risk associated with using nano-enabled products on human and environmental health is an uphill task because of the varying facets of application, such as different nano-materials used, different nano-formulations, and several exposure dosages and concentrations [155]. Hence, the properties of the active ingredient, concentrations, and the nano-substance should be the focus in identifying associated hazards and risks. The application of engineered nanoparticles as biofertilizers, pesticides, or bioremediation should be monitored and viewed according to their lifespan and persistence in the environment, from field application to residue disposal [156]. However, concerns regarding the fate and transportation of these nanomaterials into the environment are yet to be addressed (Figure 5). These issues certainly need to be considered because nanomaterials' interaction with environmental factors may alter their physicochemical characterization and, inherently, the toxicological profile [155,157]. This reiterates the necessity and importance of characterization—evaluating the composition, chemical form, shape, size, surface area, and functionalization—following the synthesis of nanomaterials. As much as those engaged in synthesizing nanomaterials and their incorporation into agricultural products receive the greatest exposure, contamination of water systems and residual presence in food products may typify other sources of exposure for the general populace. Risk assessment enables tangible forecasts of identified and given risks, allowing their adequate management [158].

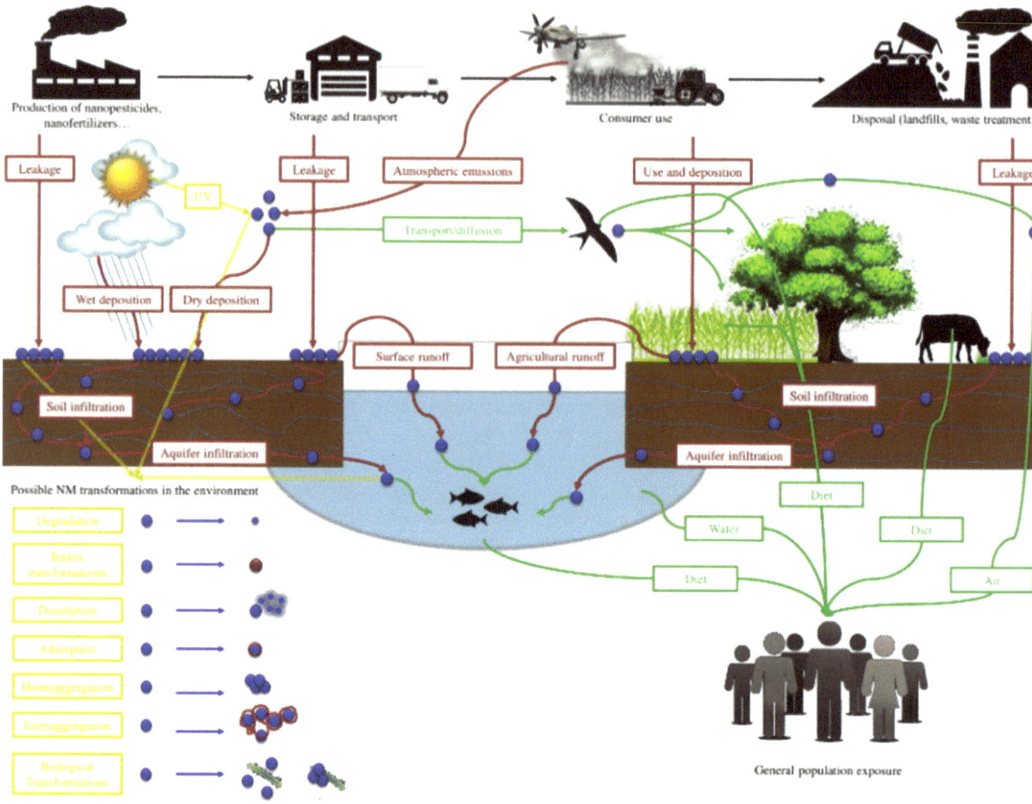

Figure 5. The life cycle of nano-enabled products in agriculture [155].

However, there are still doubts and misgivings regarding exposures and risks in the emerging nanotechnological applications in crop farming, making it crucial to embrace an innovative management strategy before the intentional and direct application of nanomaterials in the environment. This implies that dynamic, continuously evaluated, improved, and verified strategies will be employed in nano-risk management [159]. A management plan that focuses on elimination strategies or alternative exposures should be adopted to address nano-related risks in agricultural practices. In addition, risk reduction via the utilization of administrative structures at all stages of the life cycle of nanomaterials is down to personal protective equipment [155]. Furthermore, the principles of well-established green chemistry can be adopted to produce innovative and safe nanomaterials, thus avoiding reliance on procedures that might lead to environmental contaminants and health risks for the populace [160]. The decision to protect human health and the environment can have an impact and reduce cost when applied early in creating the design and development of a product. This concept has been embraced by green engineering [161]. Hence, the advent of green nanotechnology focuses on the full life cycle that can better equip users to reduce the generation of new hazards through inadvertent consequences.

5.2. Regulatory Aspects and Policies

The potential of nanotechnology applications in various areas for human well-being and development could make life better and easier [162]. This novel and state-of-the-art technology, however, has some demerits. There are existing concerns regarding the possible toxicological impacts of engineered nanomaterials on the environment and, more importantly, public health [163,164]. In as much as the use of nanomaterials is still emerging, nanotechnology could serve to improve and transform many technologies in addition to several industrial sectors. However, the absence of adequate disposal techniques has led to a constant increase in the level of nanomaterials in the environment. Ecological authorities and scientists need to theorize about the influence of nanomaterials, proffering strategies and regulations for the utilization and disposal of nanomaterials to avoid complications associated with their use. Several studies have also been carried out on the exposure limits of nanomaterials [165,166]. In addition, in the USA, manufacturers are mandated to make detailed information on nanomaterials available to the United States Environmental Protection Agency (USEPA) for assessment to ensure that these products do not result in any danger and/or environmental hazards. Presently, the European Union (EU) legislation distinctly reiterates that nanomaterials are to be included in various regulations related to products and consumers [58]. In addition, voluntary structures for data submission on nanomaterials in European nations are serving to collect data regarding awareness and toxicity levels in production and market distribution [167]. Therefore, carrying out an appropriate life-cycle evaluation, in addition to the risk assessment evaluation, is vital for nanomaterials before large-scale application [168].

6. Conclusion and Future Prospects

The COVID-19 pandemic negatively impacted every facet of life, including food supply, safety, and security. Despite various measures put in place to curb the spread of the disease, societies improvised and developed progressive ways to continue living. Although the virus spread is slowing down, the ripple effects of the stringent measures taken to curb its spread have continued, especially in areas of crop production, supply chains, and the availability of affordable food. Efforts in many countries to increase and sustain food demand and supply, food safety, food security, and trade have been inadequate.

Nanotechnology, which had been identified for useful applications in all fields of life, has continued to present high potential in the fight against diseases [169] and also to improve crop production. The usefulness of nanotechnology as a tool for modern agricultural development and crop farming is a promising driver of the economy in coming years for sustainable and viable agriculture. While nanotechnological applications are currently being utilized in medicine to combat COVID-19, there is still a lack of consensus regarding their acceptance in crop production due to the dearth of safety standards that can monitor nanomaterials. In as much as the potential of nanotechnology to improve and increase crop production exists, field applications that can engineer increased crop production are slow due to regulatory and policy-making issues [32]. In the European Union, the regulatory authority regularly scrutinizes the usage of nanomaterials in the agri-food sector, addressing safety standards and risk management issues [170]. For instance, the European Commission Food Law Regulation ensures that nano-based foods have pre-market approval and provides guidelines for the production and distribution of food. In addition, the international standard organization has certain standards regarding the manufacture and toxicity assessment of nano-based products, which are followed by most countries worldwide. With the pressure on regulating agencies by non-state actors and the growing nano-agri sector, new regulatory approaches such as self-regulation, co-regulation, and meta-regulation are receiving wider recognition [32]. There is a need to collaborate with private agencies, industries, academic and research institutions, stakeholders, government bodies, and the public to weigh up the positive and negative effects of nanotechnology and its products in crop production.

However, future research could be focused on but not limited to the following:

(i) Collaborative efforts of researchers to strengthen controlled green synthesis of nanomaterials, which will be vital for the development of stable, cheap, efficient, multi-functional, and eco-friendly nanomaterials;

(ii) Further research on the synthesis, characterization, and application of bio-synthesized nanomaterials for their wider field application in plants of high economic importance;

(iii) The need to understand nanoparticles' mechanism of interaction with plant systems, their behavior, and persistence in agricultural soils;
(iv) Exploration of more field applications, useful for large-scale implementation of nano-based techniques.

Agriculture is one of the major sources of food for humans, yet this sector was one of the most impacted by the COVID-19 pandemic, and as such the assistance of prominent technology such as nanotechnology could have latent and notable advantages. With the increasing global population and climate change that has affected land use, COVID-19 further aggravated food insecurity, malnutrition and hunger because it triggered stringent measures that affected crop production and supply. Nanotechnology has found many applications in agricultural activities, including the replacement and/or more effective use of agrochemicals, stress management agents, or even environmental remediation agents [12]. Modern nanotechnological applications have the potential to solve many problems related to conventional agricultural practices such as poor control of micronutrient release, poor amount of mineral micronutrients available to plants, high rate of chemical and fertilizer run-offs, poor nutrient-use efficiency of plants, and so on [12]. Since there had been shortage of food supply and increased demand, nanotechnological applications are emerging in food preservation and packaging to maintain the freshness of the limited supply. Although the fate, behavior, and impact of nanomaterials on the environment are yet to be firmly understood and ascertained by environmental scientists and researchers, the emergence of artificial intelligence and machine learning being integrated with nano-precision farming is also progressing to monitor the safety of nanomaterials in food products, even though it is still in its infancy [34]. Nanomaterial application may enhance agronomic parameters and yield of crops, but various plant species may respond differently [12]. Hence, before commercial use, rigorous investigations into the assessment, evaluation, and optimization of the nanomaterials for different plant species are required.

Since the interaction of nanoparticles with the plant body is strongly affected by plant physiology, the results from species to species is significantly different [171]. In addition, farmers are unwilling to utilize this technology as an alternative strategy in the fields because of the low-cost benefit economics of nanomaterials production [155]. Due to high-cost input in some laboratory experiments that have been restricted to few plants in growth chambers or hydroponics, the interesting findings are seldom reproduced in the field. In conclusion, the importance and beneficial applications of nanotechnology have been stated, allaying consumers' concerns regarding the safety of this progressing technology. Hence, adequate understanding, more funding, and wider field applications of nanotechnological advances are necessary to enhance its development.

Author Contributions: Conceptualization, F.O.A., O.I.A. and G.A.D.; Writing–original draft preparation, F.O.A., O.I.A. and G.A.D.; Writing–review and editing, F.O.A., O.I.A., E.F.A. and G.A.D.; Visualization, F.O.A. and O.I.A.; Supervision, O.I.A., E.F.A. and G.A.D. All authors have read and agreed to the published version of the manuscript.

Funding: This research received no external funding.

Institutional Review Board Statement: Not applicable.

Informed Consent Statement: Not applicable.

Data Availability Statement: Data that support these findings available within this article.

Conflicts of Interest: The authors declare no conflict of interest.

References

1. Su, S.; Wong, G.; Shi, W.; Liu, J.; Lai, A.C.K.; Zhou, J.; Liu, W.; Bi, Y.; Gao, G.F. Epidemiology, genetic recombination, and pathogenesis of coronaviruses. *Trends Microbiol.* **2016**, *6*, 490–502. [CrossRef] [PubMed]
2. Wang, N.; Shi, X.; Jiang, L.; Zhang, S.; Wang, D.; Tong, P.; Guo, D.; Fu, L.; Cui, Y.; Liu, X.; et al. Structure of MERS-CoV spike receptor-binding domain complexed with human receptor DPP4. *Cell Res.* **2013**, *23*, 986–993. [CrossRef]
3. Tabish, S.A. COVID-19 Pandemic: The crisis and the longer-term perspectives. *J. Cardiol.* **2020**, *13*, 41–44. [CrossRef]

4. Du, T.A. Outbreak of a novel coronavirus. *Nat. Rev. Microbiol.* **2020**, *18*, 123.
5. World Health Organization (WHO). WHO Announces COVID-19 Outbreak a Pandemic. Europe: World Health Organization. 2020. Available online: http://www.euro.who.int/en/health-topics/health-emergencies/coronavirus-covid-19/news/news/2020/3/who-announces-covid-19-outbreak-a-pandemic (accessed on 16 September 2022).
6. World Health Organization (WHO). Coronavirus Disease (COVID-19) Outbreak Situation. Geneva: World Health Organization. 2020. Available online: https://www.who.int/emergencies/diseases/novel-coronavirus-2019 (accessed on 16 September 2022).
7. Aba, S.C.; Baiyeri, K.P. Application of biological and digital technologies in resolving the negative effects of COVID-19 pandemic on crop production in Nigeria. *J. Trop. Agric. Food Environ. Ext.* **2021**, *20*, 46–51. [CrossRef]
8. Mirian, O.; Danjuma, Y.; Amaonyeze, N. Impact of corona virus disease-2019 (COVID-19) pandemic on social lives and interactions of Nigerian citizens. *Arch. Med.* **2021**, *13*, 15.
9. Andam, K.; Edeh, H.; Oboh, V.; Pauw, K.; Thurlow, J. Impacts of COVID-19 on food systems and poverty in Nigeria. *Adv. Food Secur. Sustain.* **2020**, *5*, 5–145.
10. FAO; IFAD; UNICEF; WFP; WHO. *The State of Food Security and Nutrition in the World 2020. Transforming Food Systems for Affordable Healthy Diets*; FAO: Rome, Italy, 2020. [CrossRef]
11. Ashraf, S.A.; Siddiqui, A.J.; Elkhalifa, A.E.O.; Khan, M.I.; Patel, M.; Alreshidi, M.; Moin, A.; Singh, R.; Snoussi, M.; Adnan, M. Innovations in nanoscience for the sustainable development of food and agriculture with implications on health and environment. *Sci. Total Environ.* **2021**, *768*, 144990. [CrossRef]
12. Usman, M.; Farooq, M.; Wakeel, A.; Nawaz, A.; Cheema, S.A.; Rehman, H.; Ashraf, I.; Sanaullah, M. Nanotechnology in agriculture: Current status, challenges and future opportunities. *Sci. Total Environ.* **2020**, *721*, 137778. [CrossRef]
13. Hamad, H.T.; Al-Sharify, Z.T.; Al-Najjar, S.Z.; Gadooa, Z.A. A review on nanotechnology and its applications on fluid flow in agriculture and water recourses. *Mater. Sci. Eng.* **2020**, *870*, 012038. [CrossRef]
14. Paliwal, P.; Sargolzaei, S.; Bhardwaj, S.K.; Bhardwaj, V.; Dixit, C.; Kaushik, A. Grand Challenges in Bio-Nanotechnology to Manage the COVID-19 Pandemic. *Front. Nanotechnol.* **2020**, *2*, 2673–3013. [CrossRef]
15. Hofmann, T.; Lowry, G.V.; Ghoshal, S.; Tufenkji, N.; Brambilla, D.; Dutcher, J.R.; Gilbertson, L.M.; Giraldo, J.P.; Kinsella, J.M.; Landry, M.P.; et al. Technology readiness and overcoming barriers to sustainably implement nanotechnology-enabled plant agriculture. *Nat. Food* **2020**, *1*, 416–425. [CrossRef]
16. He, X.; Deng, H.; Hwang, H. The current application of nanotechnology in food and agriculture. *J. Food Drug Anal.* **2019**, *27*, 1–21. [CrossRef] [PubMed]
17. Bai, R.G.; Sabouni, R.; Husseini, G. Green nanotechnology—A road map to safer nanomaterials. In *Applications of Nanomaterials*; Elsevier: Amsterdam, The Netherlands, 2018; pp. 133–159. [CrossRef]
18. Saratale, R.G.; Karuppusamy, I.; Saratale, G.D.; Pugazhendhi, A.; Kumar, G.; Park, Y.; Ghodake, G.S.; Bharagava, R.N.; Banu, J.R.; Shin, H.S. A comprehensive review on green nanomaterials using biological systems: Recent perception and their future applications. *Colloids Surf. B Biointerfaces* **2018**, *170*, 20–35. [CrossRef] [PubMed]
19. Campos, J.; Estefânia, V.R.; Proença, P.L.F.; Oliveira, J.L.; Pereira, A.E.S.; de Morais Ribeiro, L.N.; Fernandes, F.O.; Gonçalves, K.C.; Polanczyk, R.A.; Pasquoto-Stigliani, T.; et al. Carvacrol and linalool co-loaded in β-cyclodextrin-grafted chitosan nanoparticles as sustainable biopesticide aiming pest control. *Sci. Rep. Nat.* **2018**, *8*, 1–14. [CrossRef] [PubMed]
20. Oliveira, J.L.; de Campos, E.V.R.; Pereira, A.E.S.; Pasquoto, T.; Lima, R.; Grillo, R.; de Andrade, D.J.; de Santos, F.A.; dos Fraceto, L.F. Zein nanoparticles as eco-friendly carrier systems for botanical repellents aiming sustainable agriculture. *J. Agric. Food Chem.* **2018**, *66*, 1330–1340. [CrossRef]
21. Pascoli, M.; Lopes-Oliveira, P.J.; Fraceto, L.F.; Seabra, A.B.; Oliveira, H.C. State of the art of polymeric nanoparticles as carrier systems with agricultural applications: A mini review. *Energy Ecol. Environ.* **2018**, *3*, 137–148. [CrossRef]
22. Shang, Y.; Hasan, M.d.K.; Ahammed, G.J.; Li, M.; Yin, H.; Zhou, J. Applications of Nanotechnology in Plant Growth and Crop Protection: A Review. *Molecules* **2019**, *24*, 14. [CrossRef]
23. Harismah, K.; Mirzaei, M. COVID-19: A serious warning for emergency health innovation. *Adv. J. Sci. Eng.* **2020**, *1*, 32–33.
24. Ozkendir, O.M.; Askar, M.; Kocer, N.E. Influence of the Epidemic COVID-19: An Outlook on Health, Business and Scientific Studies. *Lab-Silico* **2020**, *1*, 26–30.
25. Sportelli, M.C.; Izzi, M.; Kukushkina, E.A.; Hossain, S.I.; Picca, R.A.; Ditaranto, N.; Cioffi, N. Can nanotechnology and materials science help the fight against SARS-CoV-2. *Nanomaterials* **2020**, *10*, 802. [CrossRef] [PubMed]
26. HLPE. *High Level Panel of Experts on Food Security and Nutrition. Impacts of COVID-19 on Food Security and Nutrition: Developing Effective Policy Responses to Address the Hunger and Malnutrition Pandemic*; HLPE: Rome, Italy, 2020. [CrossRef]
27. Bochtis, D.; Benos, L.; Lampridi, M.; Marinoudi, V.; Pearson, S.; Sorensen, C.G. Agricultural workforce crisis in light of the COVID-19 pandemic. *Sustainability* **2020**, *12*, 8212. [CrossRef]
28. Balana, B.; Oyeyemi, M.; Ogunniyi, A.; Fasoranti, A.; Edeh, H.; Aiki, J.; Andam, K.S. The effects of COVID-19 policies on livelihoods and food security of smallholder farm households in Nigeria: Descriptive results from a phone survey. *Int. Food Policy Res. Inst. (IFPRI) Discuss. Pap.* **2020**, *01979*, 34.
29. Ilesanmi, F.; Ilesanmi, O.; Afolabi, A. The effects of the COVID-19 pandemic on food losses in the agricultural value chains in Africa: The Nigerian case study. *Public Health Pract.* **2021**, *2*, 100087. [CrossRef]

30. Javed, R.; Bilal, M.; Ali, J.S.; Khan, S.; Cheema, M. Nanotechnology: A Tool for the Development of Sustainable Agroindustry. In *Agricultural and Environmental Nanotechnology*; Interdisciplinary Biotechnological Advances; Fernandez-Luqueno, F., Patra, J.K., Eds.; Springer: Singapore, 2023. [CrossRef]
31. WHO/ILO Policy Brief. *Preventing and Mitigating COVID-19 at Work*; WHO: Geneva, Switzerland, 2021.
32. Saritha, G.N.G.; Anju, T.; Kumar, A. Nanotechnology-Big impact: How nanotechnology is changing the future of agriculture? *J. Agric. Food Res.* **2022**, *10*, 100457. [CrossRef]
33. Singh, D.P.; Packirisamy, G. Applications of nanotechnology to combat the problems associated with modern food. *J. Sci. Food Agric.* **2022**, *103*, 479–487. [CrossRef]
34. United States Department of Agriculture—National Institute of Food and Agriculture (USDA-NIFA). Nanotechnology in Agriculture and Food Systems | National Institute of Food and Agriculture. 2022. Available online: https://www.usda.gov/ (accessed on 27 January 2023).
35. Chaudhary, V.; Royal, A.; Chavali, M.; Yadav, S.K. Advancements in research and development to combat COVID-19 using nanotechnology. *Nanotechnol. Environ. Eng.* **2021**, *6*, 1–15. [CrossRef]
36. Elkodous, M.A.; El-Sayyad, G.S.; Nasser, H.A.; Elshamy, A.A.; Morsi, M.; Abdelrahman, I.Y.; Kodous, A.S.; Mosallam, F.M.; Gobara, M.; El-Batal, A.I. Therapeutic and diagnostic potential of nanomaterials for enhanced biomedical applications. *Colloids Surf. B Biointerfaces* **2019**, *180*, 411–428. [CrossRef]
37. Akinsiku, A.A.; Ajanaku, K.O.; Dare, E.O. Green synthesis of pseudo-cubic Ag/Ni bimetallic nanoparticles using *Senna occidentalis* leaf extract. *J. Phys. Conf. Ser.* **2019**, *1299*, 012133. [CrossRef]
38. Alayande, S.O.; Akinsiku, A.A.; Akinsipo, O.B.; Ogunjinmi, E.O.; Dare, E.O. Green synthesized silver nanoparticles and their therapeutic applications. In *Comprehensive Analytical Chemistry*; Elsevier: Amsterdam, The Netherlands, 2021; Volume 94, pp. 585–611. ISSN 0166–0526X. [CrossRef]
39. Hossain, Z.; Yasmeen, F.; Komatsu, S. Nanoparticles: Synthesis, morphophysical effects and proteomic responses of crop plants. *Int. J. Mol. Sci.* **2020**, *21*, 3056. [CrossRef]
40. Ferreira, C.D.; Nunes, I.L. Oil nanoencapsulation: Development, application, and incorporation into the food market. *Nanoscale Res. Lett.* **2019**, *14*, 9. [CrossRef]
41. Jafari, S.M. *An Overview of Nanoencapsulation Techniques and Their Classification—Nanoencapsulation Technologies for the Food and Nutraceutical Industries Amsterdam*; Academic Press: Cambridge, MA, USA, 2017; pp. 1–34.
42. Suganya, V.; Anuradha, V. Microencapsulation and nanoencapsulation: A review. *Int. J. Pharm. Clin. Res.* **2017**, *9*, 233–239. [CrossRef]
43. Patra, J.K.; Das, G.; Fraceto, L.F.; Campos, E.V.R.; Rodriguez-Torres, M.D.P.; Acosta-Torres, L.S.; Diaz-Torres, L.A.; Grillo, R.; Swamy, M.K.; Sharma, S.; et al. Nano based drug delivery systems: Recent developments and future prospects. *J. Nanobiotechnol.* **2018**, *16*, 71. [CrossRef]
44. Wani, T.A.; Masoodi, F.A.; NabiBaba, W.; Ahmad, M.; Rahmanian, N.; Jafari, S.M. Chapter 11—Nanoencapsulation of Agrochemicals, Fertilizers, and Pesticides for Improved Plant Production. *Adv. Phytonanotechnol.* **2019**, 279–298.
45. Robledo, N.; Loópez, L.; Bunger, A.; Tapia, C.; Abugoch, L. Effects of antimicrobial edible coating of thymol nanoemulsion/quinoa protein/ chitosan on the safety, sensorial properties, and quality of refrigerated strawberries (*Fragaria ananassa*) under commercial storage environment. *Food Bioprocess. Technol.* **2018**, *11*, 1566–1574. [CrossRef]
46. Che Marzuki, N.H.; Wahab, R.A.; Abdul Hamid, M. An overview of nanoemulsion: Concepts of development and cosmeceutical applications. *Biotechnol. Biotechnol. Equip.* **2019**, *33*, 779–797. [CrossRef]
47. Feng, J.; Zhang, Q.; Liu, Q.; Zhu, Z.; McClements, D.J.; Jafari, S.M. *Chapter 12—Application of Nanoemulsions in Formulation of Pesticides*; Nanoemulsions, Jafari, S.M., McClements, D.J., Eds.; Academic Press: Cambridge, MA, USA, 2018; pp. 379–413.
48. Ko, F.K.; Wan, Y. *Introduction to Nanofiber Materials*; Cambridge University Press: Cambridge, UK, 2014; pp. 1–9.
49. Xu, N.; Li, Z.; Huangfu, X.; Cheng, X.; Christodoulatos, C.; Qian, J.; Chen, M.; Chen, J.; Su, C.; Wang, D. Facilitated transport of nTiO_2-kaolin aggregates by bacteria and phosphate in water-saturated quartz sand. *Sci. Total Environ.* **2020**, *713*, 136589. [CrossRef]
50. Anzar, N.; Hasan, R.; Tyagi, M.; Yadav, N.; Narang, J. Carbon nanotube-A review on Synthesis, Properties and plethora of applications in the field of biomedical science. *Sens. Int.* **2020**, *1*, 100003. [CrossRef]
51. Lahiani, M.; Nima, Z.; Villagarcia, H.; Biris, A.S.; Khodakovskaya, M. Assessment of effects of the long-term exposure of agricultural crops to carbon nanotubes. *J. Agric. Food Chem.* **2018**, *66*, 6654–6662. [CrossRef]
52. Patel, D.K.; Kim, H.B.; Dutta, S.D.; Ganguly, K.; Lim, K.T. Carbon nanotubes-based nanomaterials and their agricultural and biotechnological applications. *Materials* **2020**, *13*, 1679. [CrossRef]
53. Lahiani, M.H.; Chen, J.; Irin, F.; Puretzky, A.A.; Green, M.J.; Khodakovskaya, M.V. Interaction of carbon nanohorns with plants: Uptake and biological effects. *Carbon* **2015**, *81*, 607–619. [CrossRef]
54. Sanvicens, N.; Pastells, C.; Pascual, N.; Marco, M.P. Nanoparticle-based biosensors for detection of pathogenic bacteria. *TrAC Trends Anal. Chem.* **2009**, *28*, 1243–1252. [CrossRef]
55. Singh, A.; Tiwari, S.; Pandney, J.; Lata, C.; Singh, I.K. Role of nanoparticles in crop improvement and abiotic stress management. *J. Biotechnol.* **2021**, *337*, 57–70. [CrossRef]
56. Tirani, M.M.; Haghjou, M.M.; Ismaili, A. Hydroponic grown tobacco plants respond to zinc oxide nanoparticles and bulk exposures by morphological, physiological and anatomical adjustments. *Funct. Plant Biol.* **2019**, *46*, 360–375. [CrossRef]

57. Acharya, A.; Pal, P.K. Agriculture nanotechnology: Translating research outcome to field applications by influencing environmental sustainability. *NanoImpact* **2020**, *19*, 100232. [CrossRef]
58. Duhan, J.S.; Kumar, R.; Kumar, N.; Kaur, P.; Nehra, K.; Duhan, S. Nanotechnology: The new perspective in precision agriculture. *Biotechnol. Rep.* **2017**, *125*, 11–23. [CrossRef]
59. Zhang, P.; Guo, Z.; Ullah, S.; Melagraki, G.; Afantitis, A.; Lynch, I. Nanotechnology and artificial intelligence to enable sustainable and precision agriculture. *Nat. Plants* **2021**, *7*, 864–876. [CrossRef]
60. Day, W. Engineering precision into variable biological systems. *Ann. Appl. Biol.* **2005**, *146*, 155–162. [CrossRef]
61. Dong, J.; Liu, X.; Chen, Y.; Yang, W.; Du, X. User-safe and efficient chitosan-gated porous carbon nanopesticides and nanoherbicides. *J. Colloid Interface Sci.* **2021**, *594*, 20–34. [CrossRef]
62. Haq, I.U.; Ijaz, S. Use of Metallic Nanoparticles and Nanoformulations as Nanofungicides for Sustainable Disease Management in Plants. In *Nanobiotechnology in Bioformulations*; Prasad, R., Kumar, V., Kumar, M., Choudhary, D., Eds.; Springer: Cham, Switzerland, 2019; pp. 289–316.
63. Shahzad, K.; Manzoor, F. Nanoformulations and their mode of action in insects: A review of biological interactions. *Drug Chem. Toxicol.* **2019**, *44*, 1–11. [CrossRef]
64. Zheng, Q.; Wang, R.; Qin, D.; Yang, L.; Lin, S.; Cheng, D.; Huang, S.; Zhang, Z. Insecticidal efficacy and mechanism of nanoparticles synthesized from chitosan and carboxymethyl chitosan against *Solenopsis invicta* (*Hymenoptera*: Formicidae). *Carbohydr. Polym.* **2021**, *260*, 117839. [CrossRef]
65. Jameel, M.; Shoeb, M.; Khan, M.T.; Ullah, R.; Mobin, M.; Farooqi, M.K.; Adnan, S.M. Enhanced Insecticidal Activity of Thiamethoxam by Zinc Oxide Nanoparticles: A Novel Nanotechnology Approach for Pest Control. *ACS Omega* **2020**, *5*, 1607–1615. [CrossRef] [PubMed]
66. Ayoub, H.A.; Khairy, M.; Rashwan, F.A.; Abdel-Hafez, H.F. Synthesis and characterization of silica nanostructures for cotton leaf worm control. *J. Nanostructured Chem.* **2017**, *7*, 91–100. [CrossRef]
67. Elnaggar, M.; Abdelsalam, N.; Fouda, M.; Mackled, M.; Al-Jaddadi, M.; Ali, H.; Siddiqui, M.H.; Kandil, E. Soil Application of Nano Silica on Maize Yield and Its Insecticidal Activity Against Some Stored Insects After the Post-Harvest. *Nanomaterials* **2020**, *10*, 739. [CrossRef] [PubMed]
68. Wang, Z.; Zhu, W.; Chen, F.; Yue, L.; Ding, Y.; Xu, H.; Rasmann, S.; Xiao, Z. Nanosilicon enhances maize resistance against oriental armyworm (*Mythimna separata*) by activating the biosynthesis of chemical defenses. *Sci. Total Environ.* **2021**, *778*, 146378. [CrossRef] [PubMed]
69. El-Ashry, R.M.; El-Saddony, M.T.; El-Sobki, A.E.A.; El-Tahan, A.M.; Al-Otaibi, S.; El-Shehawi, A.M.; Saad, A.M.; Elshaer, N. Biological Silicon Nanoparticles maximize the efficiencies of nematicides against biotic stress induced by *Meloidogyne incognita* in eggplant. *Saudi J. Biol. Sci.* **2022**, *29*, 920–932. [CrossRef] [PubMed]
70. Pereira, A.E.S.; Grillo, R.; Mello, N.F.S.; Rosa, A.H.; Fraceto, L.F. Application of poly(epsilon-caprolactone) nanoparticles containing atrazine herbicide as an alternative technique to control weeds and reduce damage to the environment. *J. Hazard. Mater.* **2014**, *268*, 207–215. [CrossRef] [PubMed]
71. Grillo, R.; Pereira, A.E.; Nishisaka, C.S.; de Lima, R.; Oehlke, K.; Greiner, R.; Fraceto, L.F. Chitosan/tripolyphosphate nanoparticles loaded with paraquat herbicide: An environmentally safer alternative for weed control. *J. Hazard. Mater.* **2014**, *8*, 163–171. [CrossRef]
72. de Sousa, B.T.; Pereira, A.E.S.; Fraceto, L.F.; Oliveira, H.C.; Dalazen, G. Post-emergence herbicidal activity of nanoatrazine against Alternanthera tenella Colla plants compared to other weed species. *Heliyon* **2022**, *8*, e09902. [CrossRef]
73. Ahmad, H.; Venugopal, K.; Rajagopal, K.; De Britto, S.; Nandini, B.; Pushpalatha HGKonappa, N.; Udayashankar, A.C.; Geetha, N.; Jogaiah, S. Green Synthesis and Characterization of Zinc Oxide Nanoparticles Using Eucalyptus globules and Their Fungicidal Ability Against Pathogenic Fungi of Apple Orchards. *Biomolecules* **2020**, *10*, 425. [CrossRef]
74. Chen, J.; Wu, L.; Lu, M.; Lu, S.; Li, Z.; Ding, W. Comparative study on the fungicidal activity of metallic MgO nanoparticles and macroscale MgO against Soil-borne fungal phytopathogens. *Front. Microbiol.* **2020**, *11*, 365. [CrossRef] [PubMed]
75. Irshad, M.A.; Nawaz, R.; Zia ur Rehman, M.; Imran, M.; Ahmad, J.; Ahmad, S.; Inam, A.; Razzaq, A.; Rizwan, M.; Ali, S. Synthesis and characterization of titanium dioxide nanoparticles by chemical and green methods and their antifungal activities against wheat rust. *Chemosphere* **2020**, *258*, 127352. [CrossRef] [PubMed]
76. Oussou-Azo, A.; Nakama, T.; Nakamura, M.; Futagami, T.; Vestergaard, M. Antifungal Potential of Nanostructured Crystalline Copper and Its Oxide Forms. *Nanomaterials* **2020**, *10*, 1003. [CrossRef] [PubMed]
77. Jahangirian, H.; Rafiee-Moghaddam, R.; Jahangirian, N.; Nikpey, B.; Jahangirian, S.; Bassous, N.; Saleh, B.; Kalantari, K.; Webster, T.J. Green Synthesis of Zeolite/$Fe(_2)O(_3)$ Nanocomposites: Toxicity & Cell Proliferation Assays and Application as a Smart Iron Nanofertilizer. *Int. J. Nanomed.* **2020**, *15*, 1005–1020.
78. Cota-Ruiz, K.; Ye, Y.; Valdes, C.; Deng, C.; Wang, Y.; Hernández-Viezcas, J.A.; Duarte-Gardea, M.; Gardea-Torresdey, J.L. Copper nanowires as nanofertilizers for alfalfa plants: Understanding nano-bio systems interactions from microbial genomics, plant molecular responses and spectroscopic studies. *Sci. Total Environ.* **2020**, *742*, 140572. [CrossRef]
79. Naseem, F.; Zhi, Y.; Farrukh, M.A.; Hussain, F.; Yin, Z. Mesoporous ZnAl2SiO nanofertilizers enable high yield of *Oryza sativa* L. *Sci. Rep.* **2020**, *10*, 10841. [CrossRef]
80. Pérez-de-Luque, A.; Rubiales, D. Nanotechnology for parasitic plant control. *Pest Manag. Sci.* **2009**, *65*, 540–545. [CrossRef]

81. Grillo, R.; dos Santos, N.Z.P.; Maruyama, C.R.; Rosa, A.H.; de Lima, R.; Fraceto, L.F. Poly(γ-caprolactone) nanocapsules as carrier systems for herbicides: Physico-chemical characterization and genotoxicity evaluation. *J. Hazard. Mater.* **2012**, *231–232*, 1–9. [CrossRef]
82. Rani, P.U.; Madhusudhanamurthy, J.; Sreedhar, B. Dynamic adsorption of α-pinene and linalool on silica nanoparticles for enhanced anti-feedant activity against agricultural pests. *J. Pestic. Sci.* **2014**, *87*, 191–200. [CrossRef]
83. Abigail, M.E.A.; Melvin, S.S.; Chidambaram, R. Application of rice husk nano-sorbents containing, 2,4-dichlorophenoxyacetic acid herbicide to control weeds and reduce leaching from soil. *J. Taiwan Inst. Chem. Eng.* **2016**, *63*, 318–326. [CrossRef]
84. Sharma, P.; Sharma, A.; Sharma, M.; Bhalla, N.; Estrela, P.; Jain, A.; Thakur, P.; Thakur, A. Nanomaterial Fungicides: In Vitro and In Vivo Antimycotic Activity of Cobalt and Nickel Nanoferrites on Phytopathogenic Fungi Global Challenges. *Glob. Chall.* **2017**, *1*, 1770071. [CrossRef]
85. Pariona, N.; Mtz-Enriquez, A.I.; Sánchez-Rangel, D.; Carrión, G.; Paraguay-Delgad, F.; Rosas-Saito, G. Green-synthesized copper nanoparticles as a potential antifungal against plant pathogens. *RSC Adv.* **2019**, *9*, 18835. [CrossRef] [PubMed]
86. Le, V.T.; Bach, L.G.; Pham, T.T.; Le, N.T.T.; Ngoc, U.T.P.; Tran, D.H.N.; Nguyen, D.H. Synthesis and antifungal activity of chitosan-silver nanocomposite synergize fungicide against *Phytophthora Capsici*. *J. Macromol. Sci. Part A* **2019**, *56*, 522–528. [CrossRef]
87. Giannousi, K.; Avramidis, I.; Dendrinou-Samara, C. Synthesis, characterization and evaluation of copper based nanoparticles as agrochemicals against *Phytophthora infestans*. *RCS Adv.* **2013**, *3*, 21743–21752. [CrossRef]
88. Salama, D.M.; El-Aziz, M.E.A.; Rizk, F.A.; Elwahed, M.S.A. Applications of nanotechnology on vegetable crops. *Chemosphere* **2021**, *266*, 129026. [CrossRef] [PubMed]
89. Askary, M.; Amirjani, M.R.; Saberi, T. Comparison of the effects of nano-iron fertilizer with iron-chelate on growth parameters and some biochemical properties of *Catharanthus roseus*. *J. Plant Nutr.* **2017**, *40*, 974–982. [CrossRef]
90. Bastani, S.; Hajiboland, R.; Khatamian, M.; Saket-Oskoui, M. Nano iron (Fe) complex is an effective source of Fe for tobacco plants grown under low Fe supply. *J. Soil Sci. Plant Nutr.* **2018**, *18*, 524–541. [CrossRef]
91. Elrys, A.S.; Abdo, A.I.E.; Desoky, E.S.M. Potato tubers contamination with nitrate under the influence of nitrogen fertilizers and spray with molybdenum and salicylic acid. *Environ. Sci. Pollut. Res.* **2018**, *25*, 7076–7089. [CrossRef]
92. Abbasifar, A.; ValizadehKaji, B.; Iravani, M.A. Effect of green synthesized molybdenum nanoparticles on nitrate accumulation and nitrate reductase activity in spinach. *J. Plant Nutr.* **2020**, *43*, 13–27. [CrossRef]
93. Osman, S.A.; Salama, D.M.; Abd El-Aziz, M.E.; Shaaban, E.A.; AbdElwahede, M.S. The influence of MoO3-NPs on agro-morphological criteria, genomic stability of DNA, biochemical assay, and production of common dry bean (*Phaseolus vulgaris* L.). *Plant Physiol. Biochem.* **2020**, *151*, 77–87. [CrossRef]
94. Al-Ubaidy, R.M.; Mohammed, M.M.; Al-Zaidy, A.K. Influence of chemical fertilizers and foliar spraying with humic acid in growth and yield of red cabbage. *Biochem. Cell. Arch.* **2019**, *19*, 1215–1219.
95. Abdulhameed, M.F.; Taha, A.A.; Ismail, R.A. Improvement of cabbage growth and yield by nanofertilizers and nanoparticles. *Environ. Nanotechnol. Monit. Manag.* **2021**, *15*, 100437. [CrossRef]
96. Abdel-Aziz, H.M.M.; Hasaneen, M.N.A.; Omer, A.M. Nano chitosan-NPK fertilizer enhances the growth and productivity of wheat plants grown in sandy soil. *Span. J. Agric. Res.* **2016**, *14*, 17. [CrossRef]
97. Mastronardi, E.; Tsae, P.; Zhang, X.; Monreal, C.; DeRosa, M.C. Strategic role of nanotechnology in fertilizers: Potential and limitations. In *Nanotechnologies in Food and Agriculture*; Rai, M., Ribeiro, C., Mattoso, L., Duran, N., Eds.; Springer: Cham, Switzerland, 2015; pp. 25–67.
98. Monreal, C.M.; DeRosa, M.; Mallubhotla, S.C.; Bindraban, P.S.; Dimkpa, C. Nanotechnologies for increasing the crop use efficiency of fertilizer-micronutrients. *Biol. Fertil. Soils* **2016**, *52*, 423–437. [CrossRef]
99. Arduini, F.; Cinti, S.; Scognamiglio, V.; Moscone, D. Nanomaterials in electrochemical biosensors for pesticide detection: Advances and challenges in food analysis. *Microchim. Acta* **2016**, *183*, 2063–2083. [CrossRef]
100. Antonacci, A.; Arduini, F.; Moscone, D.; Palleschi, G.; Scognamiglio, V. Nanostructured (bio)sensors for smart agriculture. *TrAC Trends Anal. Chem.* **2018**, *98*, 95–103. [CrossRef]
101. Tarafdar, J.C.; Sharma, S.; Raliya, R. Nanotechnology: Interdisciplinary science of applications. *Afr. J. Biotechnol.* **2013**, *12*, 219–226.
102. Walia, S.; Acharya, A. Fluorescent cadmium sulfide nanoparticles for selective and sensitive detection of toxic pesticides in aqueous medium. *J. Nanoparticle Res.* **2014**, *16*, 2778. [CrossRef]
103. Dar, A.I.; Walia, S.; Acharya, A. Citric acid-coated gold nanoparticles for visual colorimetric recognition of pesticide dimethoate. *J. Nanoparticle Res.* **2016**, *18*, 233. [CrossRef]
104. Hu, W.; Wan, L.; Jian, Y.; Ren, C.; Jin, K.; Su, X.; Bai, X.; Haick, H.; Yao, M.; Wu, W. Electronic noses: From advanced materials to sensors aided with data processing. *Adv. Mater. Technol.* **2019**, *4*, 1800488. [CrossRef]
105. Gong, X.; Huang, D.; Liu, Y.; Peng, Z.; Zeng, G.; Xu, P.; Cheng, M.; Wang, R.; Wan, J. Remediation of contaminated soils by biotechnology with nanomaterials: Biobehavior, applications, and perspectives. *Crit. Rev. Biotechnol.* **2018**, *38*, 455–468. [CrossRef] [PubMed]
106. Hamid, Y.; Tang, L.; Hussain, B.; Usman, M.; Gurajala, H.K.; Rashid, M.S.; He, Z.; Yang, X. Efficiency of lime, biochar, Fe containing biochar and composite amendments for Cd and Pb immobilization in a co-contaminated alluvial soil. *Environ. Pollut.* **2020**, *257*, 113609. [CrossRef] [PubMed]

107. Robinson, B.H.; Bañuelos, G.; Conesa, H.M.; Evangelou, M.W.H.; Schulin, R. The phytomanagement of trace elements in soil. *Crit. Rev. Plant Sci.* **2009**, *28*, 240–266. [CrossRef]
108. Boente, C.; Sierra, C.; Martínez-Blanco, D.; Menéndez-Aguado, J.M.; Gallego, J.R. Nanoscale zero-valent iron-assisted soil washing for the removal of potentially toxic elements. *J. Hazard. Mater.* **2018**, *350*, 55–65. [CrossRef]
109. Kumar, L.; Ragunathan, V.; Chugh, M.; Bharadvaja, N. Nanomaterials for remediation of contaminants: A review. *Environ. Chem. Lett.* **2021**, *19*, 3139–3163. [CrossRef]
110. Jorfi, S.; Rezaee, A.; Moheb-ali, G.A.; Jaafarzadeh, N.A. Pyrene removal from contaminated soils by modified Fenton oxidation using iron nano particles. *J. Environ. Health Sci. Eng.* **2013**, *11*, 17. [CrossRef]
111. Santos, A.; Firak, D.S.; Emmel, A.; Siedlecki, K.; Lopes, A.; Peralta-Zamora, P. Evaluation of the Fenton process effectiveness in the remediation of soils contaminated by gasoline: Effect of soil physicochemical properties. *Chemosphere* **2018**, *207*, 154–161. [CrossRef]
112. Barbato, R.A.; Reynolds, C.M. 22- Bioremediation of contaminated soils. In *Principles and Application of Soil Microbiology*, 3rd ed.; Elsevier: Amsterdam, The Netherlands, 2021; pp. 607–631.
113. Le, T.T.; Nguyen, K.H.; Jeon, J.R.; Francis, A.J.; Chang, Y.S. Nano/bio treatment of polychlorinated biphenyls with evaluation of comparative toxicity. *J. Hazard. Mater.* **2015**, *287*, 335–341. [CrossRef]
114. Singh, J.; Lee, B.K. Influence of nano-TiO_2 particles on the bioaccumulation of Cd in soybean plants (*Glycine max*): A possible mechanism for the removal of Cd from the contaminated soil. *J. Environ. Manag.* **2016**, *170*, 88–96. [CrossRef]
115. Liang S-x Jin, Y.; Liu, W.; Li, X.; Shen, S.G.; Ding, L. Feasibility of Pb phytoextraction using nano-materials assisted ryegrass: Results of a one-year field-scale experiment. *J. Environ. Manag.* **2017**, *190*, 170–175. [CrossRef]
116. Anjum, S.; Anjum, I.; Hano, C.; Kousar, S. Advances in nanomaterials as novel elicitors of pharmacologically active plant specialized metabolites: Current status and future outlooks. *RSC Adv.* **2019**, *9*, 40404–40423. [CrossRef]
117. Paramo, L.A.; Feregrino-Perez, A.A.; Guevara, R.; Mendoza, S.; Esquivel, K. Nanoparticles in agroindustry: Applications, toxicity, challenges and trends. *Nanomaterials* **2020**, *10*, 1654. [CrossRef] [PubMed]
118. Kongala, S.I.; Nadendla, S.R.; Mamidala, P. Internalization and induction of defense responses in tobacco by harpinPss conjugated gold nanoparticles as a foliar spray. *Colloid Interface Sci. Commun.* **2021**, *43*, 100438. [CrossRef]
119. Cai, L.; Cai, L.; Jia, H.; Liu, C.; Wang, D.; Suna, X. Foliar exposure of Fe_3O_4 nanoparticles on *Nicotiana benthamina*—Evidence for nanoparticles uptake, plant growth promoter and defense response elicitor against plant virus. *J. Hazard. Mater.* **2020**, *393*, 122415. [CrossRef]
120. Ghanati, F.; Bakhtiarian, S. Effect of Methyl Jasmonate and Silver Nanoparticles on Production of Secondary Metabolites by *Calendula officinalis* L (*Asteraceae*). *Trop. J. Pharm. Res.* **2014**, *13*, 1783–1789. [CrossRef]
121. Salman-Khan, M.; Zaka, M.; Haider Abbasi, B.; Rahman, L.; Shah, A. Seed germination and biochemical profile of *Silybum marianum* exposed to monometallic and bimetallic alloy nanoparticles. *IET Nanobiotechnology* **2016**, *10*, 359–366. [CrossRef] [PubMed]
122. Javed, R.; Yucesan, B.; Zia, M.; Gurel, E. Elicitation of Secondary Metabolites in Callus Cultures of *Stevia rebaudiana* Bertoni Grown Under ZnO and CuO Nanoparticles Stress. *Sugar Technol.* **2018**, *20*, 194–201. [CrossRef]
123. Li, Y.; Gao, J.; Xu, X.; Wu, Y.; Zhuang, J.; Zhang, X.; Zhang, H.; Lei, B.; Zheng, M.; Liu, Y.; et al. Carbon Dots as a Protective Agent Alleviating Abiotic Stress on Rice (*Oryza sativa* L.) through Promoting Nutrition Assimilation and the Defense System. *ACS Appl. Mater. Interfaces* **2020**, *12*, 33575–33585. [CrossRef]
124. Santana, I.; Wu, H.; Hu, P.; Giraldo, J.U. Targeted delivery of nanomaterials with chemical cargoes in plants enabled by a bio-recognition motif. *Nat. Commun.* **2020**, *11*, 2045. [CrossRef]
125. Demirer, G.S.; Zhang, H.; Matos, J.L.; Goh, N.S.; Cunningham, F.J.; Sung, Y.; Chang, R.; Aditham, A.J.; Chio, L.; Cho, M.; et al. High aspect ratio nanomaterials enable delivery of functional genetic material without DNA integration in mature plants. *Nat. Nanotechnol.* **2019**, *14*, 456–464. [CrossRef]
126. Miller, J.B.; Zhang, S.; Kos, P.; Xiong, H.; Zhou, K.; Perelman, S.S.; Zhu, H.; Siegwart, D.J. Non-viral CRISPR/Cas gene editing in vitro and in vivo enabled by synthetic nanoparticle co-delivery of Cas9 mRNA and sgRNA. *Angew. Chem. Int. Ed.* **2017**, *56*, 1059–1063. [CrossRef] [PubMed]
127. Mukarram, M.; Khan, M.M.A.; Corpas, F.J. Silicon nanoparticles elicit an increase in lemongrass (*Cymbopogon flexuosus* (Steud.) Wats) agronomic parameters with a higher essential oil yield. *J. Hazard. Mater.* **2021**, *412*, 125254. [CrossRef] [PubMed]
128. Cheng, B.; Chen, F.; Wang, C.; Liu, X.; Yue, L.; Cao, X.; Wang, Z.; Xing, B. The molecular mechanisms of silica nanomaterials enhancing the rice (*Oryza sativa* L.) resistance to planthoppers. *Sci. Total Environ.* **2021**, *767*, 144967. [CrossRef] [PubMed]
129. Gatadi, S.; Madhavi, Y.V.; Nanduri, S. Nanoparticle drug conjugates treating microbial and viral infections: A review. *J. Mol. Struct.* **2021**, *1228*, 129750. [CrossRef]
130. Kapoor, P.; Dhaka, R.K.; Sihag, P.; Mehla, S.; Sagwal, V.; Singh, Y.; Langaya, S.; Balyan, P.; Singh, K.P.; Xing, B.; et al. Nanotechnology-enabled biofortification strategies for micronutrients enrichment of food crops: Current understanding and future scope. *NanoImpact* **2022**, *26*, 100407. [CrossRef]
131. Khan, A.; Singh, J.; Upadhayay, V.K.; Singh, A.V.; Shah, S. Microbial biofortification: A Green technology through plant growth promoting microorganisms. In *Sustainable Green Technologies for Environmental Management*; Springer: New York, NY, USA, 2019; pp. 255–269.
132. Roychoudhury, A. Silicon-nanoparticles in crop improvement and agriculture. *Int. J. Adv. Res. Biotechnol. Nanotechnol.* **2020**, *3*, 54.

133. Ahmadian, K.; Jalilian, J.; Pirzad, A. Nano-fertilizers improved drought tolerance in wheat under deficit irrigation. *Agric. Water Manag.* **2021**, *244*, 106544. [CrossRef]
134. Abdel-Maksoud, M.I.A.; Bekhit, M.; El-Sherif, D.M.; Sofy, A.R.; Sofy, M. Gamma radiation-induced synthesis of a novel chitosan/silver/Mn-Mg ferrite nanocomposite and its impact on cadmium accumulation and translocation in *brassica* plant growth. *Int. J. Biol. Macromol.* **2022**, *194*, 306–316. [CrossRef]
135. Liu, J.; Li, J.; Wolfe, K.; Perrotta, B.; Cobb, G.P. Mobility of arsenic in the growth media of rice plants (*Oryza sativa* subsp. japonica. 'Koshihikari') with exposure to copper oxide 3 nanoparticles in a life-cycle greenhouse study. *Sci. Total Environ.* **2021**, *774*, 145620. [CrossRef]
136. El-Badri, A.M.; Batool, M.; Wang, C.; Hashem, A.M.; Tabl, K.M.; Nishawy, E.; Kuai, J.; Zhou, G.; Wang, B. Selenium and zinc oxide nanoparticles modulate the molecular and morpho-physiological process during seed germination of *Brassica napus* under salt stress. *Ecotoxicol. Environ. Saf.* **2021**, *225*, 112695. [CrossRef]
137. Mohammadi, M.H.Z.; Panahirad, S.; Navai, A.; Bahrami, M.K.; Kulak, M.; Gohari, G. Cerium oxide nanoparticles improve growth parameters and antioxidant defense system in Moldavian Balm (*Dracocephalum moldavica*) under salinity stress. *Plant Stress* **2021**, *1*, 100006. [CrossRef]
138. Wahid, I.; Rani, P.; Kumari, S.; Ahmad, R.; Hussain, S.J.; Alamri, S.; Tripathy, N.; Khan, M.I.R. Biosynthesized gold nanoparticles maintained nitrogen metabolism, nitric oxide synthesis, ions balance and stabilizes the defense systems to improve salt stress tolerance in wheat. *Chemosphere* **2022**, *287*, 132142. [CrossRef] [PubMed]
139. Wang, J.; Li, M.; Feng, J.; Yan, X.; Chen, H.; Han, R. Effects of TiO_2-NPs pretreatment on UV-B stress tolerance in *Arabidopsis thaliana*. *Chemosphere* **2021**, *281*, 130109. [CrossRef]
140. Noor, R.; Yasmin, H.; Ilyas, N.; Nosheen, A.; Hassan, M.N.; Mumtaz, S.; Khan, N.; Ahmad, A.; Ahmad, P. Comparative analysis of iron oxide nanoparticles synthesized from ginger (*Zingiber officinale*) and cumin seeds (*Cuminum cyminum*) to induce resistance in wheat against drought stress. *Chemosphere* **2022**, *292*, 133201. [CrossRef] [PubMed]
141. Siddiqui, M.H.; Al-Whaibi, M.H. Role of nano-SiO_2 in germination of tomato (*Lycopersicum esculentum* seeds Mill.). *Saudi J. Biol. Sci.* **2014**, *21*, 13–17. [CrossRef] [PubMed]
142. Torabian, S.; Zahedi, M.; Khoshgoftar, A.H. Effects of foliar spray of nano-particles of $FeSO_4$ on the growth and ion content of sunflower under saline condition. *J. Plant Nutr.* **2017**, *40*, 615–623. [CrossRef]
143. Tripathi, D.K.; Singh, S.; Singh, V.P.; Prasad, S.M.; Dubey, N.K.; Chauhan, D.K. Silicon nanoparticles more effectively alleviated UV-B stress than silicon in wheat (*Triticum aestivum*) seedlings. *Plant Physiol. Biochem.* **2017**, *110*, 70–81. [CrossRef]
144. Wang, S.H.; Wang, F.Y.; Gao, S.C. Foliar application with nano-silicon alleviates Cd toxicity in rice seedlings. *Environ. Sci. Pollut. Res.* **2015**, *22*, 2837–2845. [CrossRef]
145. Banerjee, A.; Singh, A.; Sudarshan, M.; Roychoudhury, A. Silicon nanoparticle-pulsing mitigates fluoride stress in rice by finetuning the ionomic and metabolomics balance and refining agronomic traits. *Chemosphere* **2021**, *262*, 127826. [CrossRef]
146. Ali, S.; Rizwan, M.; Hussain, A.; ur Rehman, M.Z.; Ali, B.; Yousaf, B.; Wijaya, L.; Alyemeni, M.N.; Ahmad, P. Silicon nanoparticles enhanced the growth and reduced the cadmium accumulation in grains of wheat (*Triticum aestivum* L.). *Plant Physiol. Biochem.* **2019**, *140*, 1–8. [CrossRef]
147. Hussain, B.; Lin, Q.; Hamid, Y.; Sanaullah, M.; Di, L.; Khan, M.B.; Yang, X. Foliage application of selenium and silicon nanoparticles alleviates Cd and Pb toxicity in rice (*Oryza sativa* L.). *Sci. Total Environ.* **2020**, *712*, 136497. [CrossRef]
148. Banerjee, J.; Kole, C. Plant Nanotechnology: An overview on concepts, strategies, and tools. In *Plant Nanotechnology*; Kole, C., Kumar, D., Khodakovskaya, M., Eds.; Springer: Cham, Switzerland, 2016; pp. 1–14.
149. Singh, S.; Jaiswal, V.; Singh, J.K.; Semwal, R.; Raina, D. Nanoparticle formulations: A smart era of advanced treatment with nanotoxicological imprints on the human body. *Chem. -Biol. Interact.* **2023**, *373*, 110355. [CrossRef]
150. Domínguez, G.A.; Torelli, M.D.; Buchman, J.T.; Haynes, C.L.; Hamers, R.J.; Klaper, R.D. Size dependent oxidative stress response of the gut of Daphnia magna to functionalized nanodiamond particles. *Environ. Res.* **2018**, *167*, 267–275. [CrossRef] [PubMed]
151. Sharma, V.; Singh, S.K.; Anderson, D.; Tobin, D.J.; Dhawan, A. Zinc oxide nanoparticle induced genotoxicity in primary human epidermal keratinocytes. *J. Nanosci. Nanotechnol.* **2011**, *11*, 3782–3788. [CrossRef] [PubMed]
152. Li, Y.; Chen, D.H.; Yan, J.; Chen, Y.; Mittelstaedt, R.A.; Zhang, Y.; Biris, A.S.; Heflich, R.H.; Chen, T. Genotoxicity of silver nanoparticles evaluated using the ames test and in vitro micronucleus assay. *Mutat. Res.* **2012**, *745*, 4–10. [CrossRef] [PubMed]
153. Kim, J.S.; Yoon, T.J.; Yu, K.N.; Kim, B.G.; Park, S.J.; Kim, H.W.; Lee, J.K.; Cho, M.H. Toxicity and tissue distribution of magnetic nanoparticles in mice. *Toxicol. Sci.* **2006**, *89*, 338–347. [CrossRef]
154. Stepankova, H.; Michalkova, H.; Splichal, Z.; Richtera, L.; Svec, P.; Vaculovic, T.; Pribyl, J.; Kormunda, M.; Rex, S.; Adam, V.; et al. Unveiling the nanotoxicological aspects of Se nanomaterials differing in size and morphology. *Bioact. Mater.* **2023**, *20*, 489–500. [CrossRef]
155. Iavicoli, I.; Leso, V.; Beezhold, D.H.; Shvedova, A.A. Nanotechnology in agriculture: Opportunities, toxicological implications and occupational risks. *Toxicol. Appl. Pharamacology* **2017**, *329*, 96–111. [CrossRef]
156. Shatkin, J.A.; Kim, B. Cellulose nanomaterials: Life cycle risk assessment and environmental health and safety roadmap. *Environ. Sci. Nano* **2015**, *2*, 497–499. [CrossRef]
157. Handy, R.D.; Shaw, B.J. Toxic effects of nanoparticles and nanomaterials: Implications for public health, risk assessment and the public perception of nanotechnology. *Health Risk Soc.* **2007**, *9*, 125–144. [CrossRef]

158. Savolainen, K.; Alenius, H.; Norppa, H.; Pylkkänen, L.; Tuomi, T.; Kasper, G. Risk assessment of engineered nanomaterials and nanotechnologies—A review. *Toxicology* **2010**, *269*, 92–104. [CrossRef] [PubMed]
159. Iavicoli, I.; Leso, V.; Ricciardi, W.; Hodson, L.L.; Hoover, M.D. Opportunities and challenges of nanotechnology in the green economy. *Environ. Health* **2014**, *7*, 78. [CrossRef] [PubMed]
160. Bergeson, L.L. Sustainable nanomaterials: Emerging governance systems. *ACS Sustain. Chem. Eng.* **2013**, *1*, 724–730. [CrossRef]
161. Pandiyan, G.K.; Prabaharan, T. Implementation of nanotechnology in fuel cells. *Mater. Today Proc.* **2020**, *1*, 368. [CrossRef]
162. Trump, B.D.; Hristozov, D.; Malloy, T.; Linkov, I. Risk associated with engineered nanomaterials: Different tools for different ways to govern. *Nano Today* **2018**, *21*, 9–13. [CrossRef]
163. Kühnel, D.; Nickel, C.; Hellack, B.; van der Zalm, E.; Kussatz, C.; Herrchen, M.; Meisterjahn, B.; Hund-Rinke, K. Closing gaps for environmental risk screening of engineered nanomaterials. *NanoImpact* **2019**, *15*, 100173. [CrossRef]
164. Rodríguez-Ibarra, C.; Déciga-Alcaraz, A.; Ispanixtlahuatl-Meráz, O.; Medina-Reyes, E.I.; Delgado-Buenrostro, N.L.; Chirino, Y.I. International landscape of limits and recommendations for occupational exposure to engineered nanomaterials. *Toxicol. Lett.* **2020**, *322*, 111–119. [CrossRef]
165. Johnston, L.J.; Gonzalez-Rojano, N.; Wilkinson, K.J.; Xing, B. Key challenges for evaluation of the safety of engineered nanomaterials. *NanoImpact* **2020**, *18*, 100219. [CrossRef]
166. Hermann, A.; Diesner, M.O.; Abel, J.; Hawthorne, C.; Greßmann, A. *Assessment of Impacts of a European Register of Products Containing Nanomaterials*; Federal Environment Agency: Dessau-Roßlau, Germany, 2014; p. 142.
167. Salieri, B.; Turner, D.A.; Nowack, B.; Hischier, R. Life cycle assessment of manufactured nanomaterials: Where are we? *NanoImpact* **2018**, *10*, 108–120. [CrossRef]
168. Tripathi, M.; Yadav, S.N.; Prasad, N.; Kumar, A. Nanotechnology: An Aid to Human Welfare in COVID-19 Pandemic Era. *Virol. Immunol. J.* **2020**, *4*, 000252.
169. Mitter, N.; Hussey, K. Moving policy and regulation forward for nanotechnology applications in agriculture. *Nat. Nanotechnol.* **2019**, *14*, 508–510. [CrossRef] [PubMed]
170. Kah, M.; Tufenkji, N.; White, J.C. Nano-enabled strategies to enhance crop nutrition and protection. *Nat. Nanotechnol.* **2019**, *14*, 532–540. [CrossRef] [PubMed]
171. Siddiqui, M.H.; Al-Whaibi, M.H.; Firoz, M.; Al-Khaishany, M.Y. Role of nanoparticles in plants. In *Nanotechnology and Plant Sciences*; Springer: Berlin/Heidelberg, Germany, 2015; pp. 19–35.

Disclaimer/Publisher's Note: The statements, opinions and data contained in all publications are solely those of the individual author(s) and contributor(s) and not of MDPI and/or the editor(s). MDPI and/or the editor(s) disclaim responsibility for any injury to people or property resulting from any ideas, methods, instructions or products referred to in the content.

water

MDPI

Article

Investigating the Potential of Cosmic-Ray Neutron Sensing for Estimating Soil Water Content in Farmland and Mountainous Areas

Yifei Jiang [1,2], Kefan Xuan [1,2], Chen Gao [1,2], Yiren Liu [1,2], Yuan Zhao [3], Haodong Deng [4], Xiaopeng Li [1,*] and Jianli Liu [1,5,*]

1 Institute of Soil Science, Chinese Academy of Sciences, Nanjing 210008, China; jyf@issas.ac.cn (Y.J.); kfxuan@issas.ac.cn (K.X.); gaochen@issas.ac.cn (C.G.); liuyiren@issas.ac.cn (Y.L.)
2 University of Chinese Academy of Sciences, Beijing 100049, China
3 College of Environmental & Resource Sciences, Shanxi University, Taiyuan 030006, China; zhaoyuan383@163.com
4 College of Hydrology and Water Resources, Hohai University, Nanjing 211100, China; ddenghaodong1997@163.com
5 University of Chinese Academy of Sciences, Nanjing 211135, China
* Correspondence: lixp@issas.ac.cn (X.L.); jlliu@issas.ac.cn (J.L.)

Abstract: The conventional methods of estimating soil water content (SWC) are mainly based on in situ measurements at sampling points and remote sensing measurements over an entire region. In view of these methods, cosmic-ray neutron sensing (CRNS) has received increasing attention in recent years as a mesoscale, noncontact SWC estimation technology that can provide more accurate and timely estimates of SWC over a larger area. In this study, we estimated SWC using both CRNS and soil-mounted detectors in farmland and mountainous areas, and evaluated the accuracy of the estimations at two experimental sites. Ultra-rapid adaptable neutron-only simulation (URANOS) was used to simulate the detection radius and depth of the two experimental sites and to obtain the spatial weights of the CRNS footprint. The results show that the theoretical range of detection was reduced in farmland compared to mountainous areas during the experimental period, suggesting that farmland retained more SWC even with less precipitation. Spatial weights were simulated to calculate the SWC of sampling points, and the weighted and averaged SWC were then correlated with CRNS. The weighting calculation improves the accuracy of CRNS estimations, with a determination coefficient (R^2) of 0.645 and a root mean square error (RMSE) of 0.046 $cm^3 \cdot cm^{-3}$ for farmland, and reproduces the daily dynamics of SWC. The R^2 and RMSE in mountainous areas are 0.773 and 0.049 $cm^3 \cdot cm^{-3}$, respectively, and the estimation accuracy of CRNS cannot be improved by the weighting calculation. The estimation accuracy of CRNS is acceptable in both regions, but the mountainous terrain obstructs neutron transmission, causing a deviation between the actual and theoretical neutron footprints in mountainous areas. Thus, the accuracy of SWC estimation is limited in mountainous terrain. In conclusion, this study demonstrates that CRNS is suitable for use in farmland and mountainous areas and that further attention should be given to the effects of topography and vegetation when it is applied in mountainous environments.

Keywords: cosmic-ray neutron sensing; soil water content estimation; farmland; mountainous terrain

Citation: Jiang, Y.; Xuan, K.; Gao, C.; Liu, Y.; Zhao, Y.; Deng, H.; Li, X.; Liu, J. Investigating the Potential of Cosmic-Ray Neutron Sensing for Estimating Soil Water Content in Farmland and Mountainous Areas. *Water* **2023**, *15*, 1500. https://doi.org/10.3390/w15081500

Academic Editors: Redmond R. Shamshiri, Muhammad Imran, Muhammad Sultan and Yuguang Zhou

Received: 20 March 2023
Revised: 6 April 2023
Accepted: 9 April 2023
Published: 12 April 2023

Copyright: © 2023 by the authors. Licensee MDPI, Basel, Switzerland. This article is an open access article distributed under the terms and conditions of the Creative Commons Attribution (CC BY) license (https://creativecommons.org/licenses/by/4.0/).

1. Introduction

Soil water content (SWC) is an important factor in research in agriculture, forestry, hydrology, and other fields. Many methods are available for determining SWC, including time domain reflectometry (TDR) and frequency-domain reflectometry (FDR) [1,2] for point-scale SWC determination, and satellite-based remote sensing [3,4] for regional or global SWC estimation. The measurement scales of these methods are either a few centimetres or a few kilometres. Cosmic-ray neutron sensing (CRNS) has shown promise for filling the gap between the point scale and remote sensing scale in recent studies [5,6].

The use of cosmic-ray neutron sensing to determine SWC was first proposed by Zreda et al. [7] during preliminary research on cosmic-ray neutron transport [8,9]. The theory of CRNS can be described as follows: fast neutrons with medium energy near the land surface are created by cosmic rays interacting with the atmosphere. Fast neutrons experience significant energy loss and are converted into thermal neutrons when reacting with hydrogen nuclei. Therefore, the intensity of fast neutrons is inversely correlated to the hydrogen atom intensity near the ground surface. The fast neutron intensity detected by CRNP (cosmic ray neutron probe) can indicate the hydrogen content near the ground surface. The hydrogen content near the ground surface is the main component of soil water content, which can be calculated from the measured fast neutron density.

As a passive detection method, the detection area of CRNS is determined by the neutron footprint, which is defined as the area around the probe from which 86% of counted neutrons arise. The maximum footprint of CRNS at sea level is 335 m in radius according to the calculation of Zreda et al. [7] using Monte Carlo N-Particle transport code (MCNP), and the measurement depth ranges from 0.76 m in dry soils to 0.12 m in wet soils. MCNP simulations explain the theoretical footprint and measurement depth of CRNS, but the neutron footprint in the horizontal direction is correlated with neutron intensity, which is affected by atmospheric pressure, altitude, and other hydrogen sources [10]. A deviation occurs between the theoretical neutron footprint and the actual measured footprint when using CRNS under different ecological conditions [11]. Köhli et al. [12] considered the influencing factors and improved the footprint simulations via the Monte Carlo code ultra rapid adaptable neutron-only simulation (URANOS). These simulations indicate that the revised footprint radius is 240 m at sea level, and the measurement depth ranges from 0.15 m to 0.83 m under different soil moisture conditions. The simulation results of URANOS also indicate that the signal strength per radial distance is highly nonlinear and can be described as a weighting function.

The revised footprint simulated with URANOS has proven compatible with field studies of CRNS and has been applied in several studies [5,13–15]. However, the application of CRNS in different scenarios still requires specific calculations. The most suitable research sites for CRNS are flat terrains with relatively dry climates, which provide stable signals for CRNP. Therefore, CRNS is more widely used in farmland and agro-pastoral areas [16–18]. Forest ecosystems provide a wide variety of hydrogen pools, creating a worst-case scenario [19]. Heidbüchel et al. [20] suggested that several factors, such as the presence of a litter layer and spatially-heterogeneous soil moisture conditions within the sensor footprint, could cause inaccurate estimations of the footprint when applying CRNS to forests. Currently, research on the applicability of CRNS in humid forest conditions is limited.

In flat terrain conditions, CRNS has been used to measure field-scale root zone soil moisture [21], aboveground water equivalent [22], the retrieval of soil ice content [23], etc. However, open questions remain regarding how to account for spatial heterogeneity. Previous studies have suggested that neutrons can travel tens of centimetres in soil with the respective moisture conditions, beyond which distance they will be absorbed [24]. This finding implies that a portion of the neutrons are absorbed by the soil when they traverse intricate terrain. The simulation of the CRNS detection range is based on the assumption that the underlying surface is flat in the horizontal direction. The discrepancy between the simulated and actual detection ranges of CRNS may lead to uncertainty in the estimation. In contrast, the accuracy of CRNS estimation in complex mountainous terrain needs further study.

Therefore, we have conducted the present experiment to achieve three objectives: (i) investigate the potential of CRNS for estimating SWC in farmland and mountainous areas, (ii) examine the characteristics of CRNS footprints in farmland and mountainous areas, and (iii) explore the influence of terrain and vegetation factors on the accuracy of CRNS estimation.

2. Experimental Site and Instrumentation

To study the applicability of CRNS under different ecological conditions, two experimental sites were established to represent the typical environmental conditions of the Huang-Huai-Hai Plain in northern China and the Taihu Basin in southeastern China. The aim of this research is to evaluate the effectiveness of CRNS in estimating SWC in flat farmland areas with a warm-temperate, semihumid monsoon climate, and steep headwater hillslopes in a subtropical humid climate.

2.1. Cosmic-Ray Neutron Sensing Probe and Meteorological Equipment

We developed a cosmic-ray neutron sensing probe (CRNP) for this study using a proportional counter tube filled with BF_3, which can detect neutrons in the fast energy range. The fast neutron is absorbed while passing through the tube and induces a pulse of electrical current that is sent to the pulse module. Then a counting module (CR300, Campbell Scientific, Logan, UT, USA) records an electrical pulse signal that is proportional to the neutron density (Figures 1 and 2).

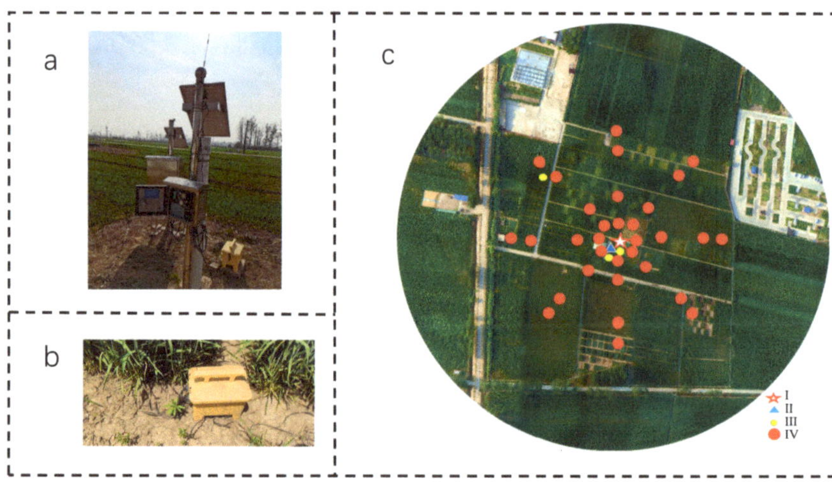

Figure 1. Location of the cosmic-ray neutron sensing probe and the experimental sampling design at Fengqiu station. (**a**) Cosmic-ray neutron sensing probe. (**b**) Frequency domain reflection (FDR) auto soil moisture monitor. (**c**) Bird's-eye view of the study site, I: cosmic-ray neutron sensing probe; II: meteorological instrument; III: FDR auto soil moisture monitors; IV: soil sampling points.

Meteorological observation equipment (H21-USB, Bourne, MA, USA) was installed near the CRNP to record atmospheric pressure, atmospheric temperature, and air humidity data. At the same time, local precipitation data were recorded by the equipment, and all meteorological data were recoded once an hour (Figures 1 and 2).

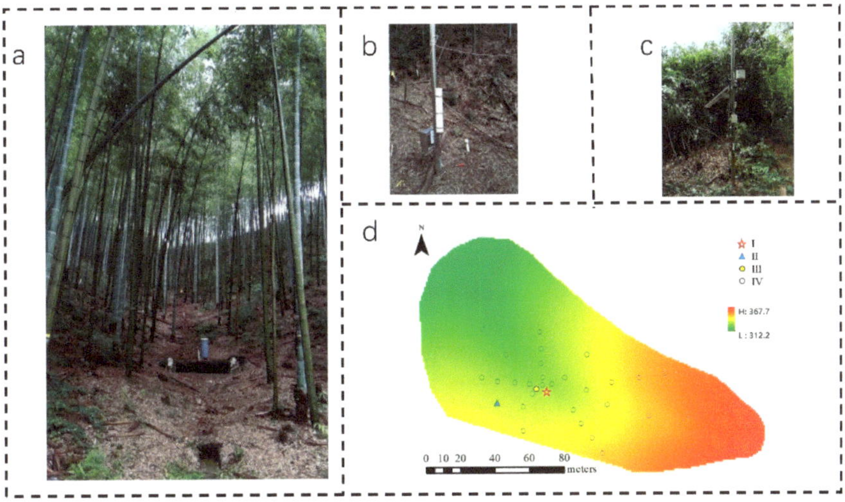

Figure 2. Location of the cosmic-ray neutron sensing probe and the experimental sampling design at Hemuqiao station. (**a**) View of the study site. (**b**) CRNP system. (**c**) Meteorological instrument. (**d**) Digital elevation map and sampling points of the study site, I: CRNP; II: Meteorological instrument; III: FDR auto soil moisture monitor; IV: Soil sampling points.

2.2. Study Site 1

The study was conducted at the Fengqiu State Key Agro-Ecological Experimental Station, Fengqiu Country, Henan Province, China (114°34′ E, 35°01′ N). The elevation at the Fengqiu station is 67 m above the mean sea level. The site has a typical monsoon climate at medium latitudes with annual temperatures averaging 13.9 °C and annual precipitation averaging 605 mm, 50~90% of which occurs from May to October. The field has been cultivated with winter wheat and summer maize, two crops in a 1-year rotation, for at least 50 years.

For this study site, the CRNP was installed in the middle of the Fengqiu Station (Figure 1). Soil samples were taken from 32 points at radial distances of 1, 25, 50, 100, and 125 m from the CRNP in eight directions. Samples were taken using a soil corer 80 cm in length and 5 cm in diameter, which enables samples to be divided into 8 layers with 10 cm-long soil columns.

To determine the continuous change in soil moisture, three FDR auto soil moisture monitors with a determination frequency of once per hour were installed at distances of 1, 25, and 100 m from the CRNS probe. The FDR monitors were calibrated by the oven-drying method.

2.3. Study Site 2

A headwater hillslope with an area of 0.42 ha was selected for the study, which was conducted at the Hemuqiao experimental station in southeast China (119°47′ E, 30°34′ N). The elevation in Hemuqiao ranges from 312 to 367 m above mean sea level. The study site has hydrologic characteristics typical of the humid hilly areas of southeast China, with an average annual precipitation of approximately 1580 mm. The site's average annual temperature and evapotranspiration are 14.6 °C and 805 mm, respectively. The vegetation type is moso bamboo (Phyllostachys edulis), which covers all areas of the hillslopes. The shape of the hillslope is similar to a saddle, with a valley in the middle and two hillsides on both sides. The slope of the hill can reach 30.5°, and the soil depth ranges from 13 cm to 171 cm, with uneven distribution.

At this study site, the CRNP was installed in the middle of the hillslope, and soil samples were taken from 31 points in eight directions of the CRNP (Figure 2). Samples were taken using a soil corer 30 cm in length and 5 cm in diameter, which enables the sample be divided into 3 layers with 10 cm-long soil columns. At the first sampling, soil moisture was measured by a TDR monitor, and soil samples were simultaneously collected. Then the soil columns were oven-dried to determine their SWC and calibrate the TDR. Due to topographical factors, the TDR soil moisture monitor was used in subsequent soil moisture sampling of this study site. Meanwhile, an FDR auto soil moisture monitor was installed near the CRNP to determine the rate of continuous change in soil moisture at study site 2.

3. Methodology

3.1. Calibration

Cosmic ray neutron sensing probes (CRNPs) are particle detectors that measure the neutron intensity near the land surface [25]. The CRNP neutron counts were affected by atmospheric pressure, air humidity, and incoming neutron flux.

Atmospheric pressure determines the probability that fast neutrons will collide with particles while moving through the atmosphere and produce energy decay. Therefore, atmospheric pressure affects the density of fast neutrons near the land surface. The pressure correction factor f_p is given by [25]:

$$f_p = \exp\frac{P - P_0}{L} \tag{1}$$

where L is the mass attenuation length for high-energy neutrons (g·cm^{-2}), which was determined to be 138 g·cm^{-2} in this study; P is the pressure at the specific site; and P_0 is an arbitrary reference pressure (the long-term average pressure at the specific site, KPa).

Air humidity indicates the atmospheric water vapour content, which affects the density of fast neutrons near the land surface. The atmospheric water vapour correction factor f_w is defined as [26]:

$$f_w = 1 + 0.0054(\rho - \rho_0) \tag{2}$$

where ρ is the absolute humidity at the time of measurement (g·m^{-3}) and ρ_0 is the absolute humidity at the reference time (g·m^{-3}).

Fast neutron intensity near the land surface is sensitive to temporal changes in incoming neutron flux, which should be corrected with the factor f_i and can be expressed as [25]:

$$f_i = \frac{I_m}{I_0} \tag{3}$$

where I_m is the measured neutron intensity of the cosmic-ray neutron monitors, which are designed to detect high-energy secondary neutrons. I_0 is a specified baseline reference intensity at a given time. The neutron monitor at Yang-Ba-Jing, China was used in our study.

The CRNP neutron counts were then corrected with the three factors given above in the form of:

$$N = N_m \times f_p \times f_w / f_i \tag{4}$$

where N is the calibrated CRNP neutron count and N_m is the measured neutron count from CRNP.

3.2. Estimation of SWC

The function for converting neutron count into SWC was improved by Desilets et al. [11] and is expressed as:

$$\theta = \frac{\rho_{bd}}{\rho_w}\left(\frac{a_0}{N/N_0 - a_1} - a_2\right) \tag{5}$$

where θ is the volumetric water content (cm^3·cm^{-3}), ρ_{bd} is the soil bulk density, ρ_w is the water density, N is the neutron count calibrated with Equation (4), N_0 is the counting rate over dry soil under the same reference conditions, and a_i is the fitting parameter. The parameters were defined as a_0 = 0.0808, a_1 = 0.372, and a_2 = 0.115 for water contents higher than 0.02 kg kg^{-1}.

3.3. Horizontal Weight

Currently, the most reliable function for describing the spatial weighting of CRNS is the weighting function obtained from the neutron transport distribution fitted by Monte Carlo simulations. The Monte Carlo-based URANOS also provides a simpler method to obtain neutron transport distributions without fitting the weighting function.

In this study, URANOS was used to obtain the neutron transport distribution curves at the two sites. Then, the distribution curves were integrated to determine the contribution weight of the CRNS soil sampling points. This method simplifies the step of fitting the neutron transport distribution curve into the weighting function and avoids overfitting errors. URANOS accounts for factors affecting neutron transport, such as air humidity and soil bulk density, resulting in simulation weights that more closely approximate true weights.

3.4. Vertical Weight

The penetration depths of the detected neutrons for different distances from the CRNP proved to be different. Therefore, the conventional vertical weighting function, which used a linear relation from Franz et al. [24], needed to be improved. In this work, we used the revised vertical weighting function given by Schrön et al. [13] and expressed it as:

$$W = exp\frac{-2d}{D} \tag{6}$$

where W is the weight at different depths, d (cm). D denotes the effective penetration depth, defined as the depth within which 86% of neutrons probed the soil. The value of D for different distances can be calculated by Schrön et al. [13].

3.5. Estimation of Measurement Accuracy

The measurement accuracy of CRNS was quantified using the linear correlation between the SWC and CRNS determination results. Calculation indexes include the coefficient of determination (R^2) and root mean square error (RMSE). The R^2 and RMSE are given by:

$$R^2 = \frac{\sum_{i=1}^{n}\left(x_i - \bar{x}\right)\left(y_i - \bar{y}\right)^2}{\sum_{i=1}^{n}\left(x_i - \bar{x}\right)^2 \sum_{i=1}^{n}\left(y_i - \bar{y}\right)^2} \tag{7}$$

$$\text{RMSE} = \sqrt{\frac{1}{n}\sum_{i=1}^{n}(x - y)^2} \tag{8}$$

where n is the number of data points and x and y are the SWCs measured by the point samples and CRNS, respectively.

4. Results and Discussion

4.1. Footprint Estimation at Different Study Sites

Footprint and weighting functions show good applicability in the farmlands of Fengqiu due to their flat topography. However, the CRNP at the Hemuqiao station had an unascertained footprint because of its complex terrain. Therefore, we conducted multiple simulations under different air humidity and soil moisture conditions to determine the variation in the neutron footprint under different ecological conditions.

The intensity of cosmic-ray neutrons depends on the hydrogen content above the ground. In our simulations, the maximum detection radius and depth were obtained in the driest condition of the study sites. In contrast, the minimum detection radius and depth were obtained in the wettest condition of the study sites. The maximum and minimum detection depths in Fengqiu were 31.2 cm and 16.9 cm, respectively, both of which were shallower than those of Hemuqiao (see Table 1). In Franz et al.'s research [24], the detection depth was primarily determined by soil hydrogen sources, which included surface water, SWC, and lattice water, indicating that the soil hydrogen source reserve is higher in Fengqiu than in Hemuqiao.

Table 1. The footprint of the CRNS simulated by URANOS at different study sites. The theoretical values of the detection radius and detection depth were obtained from simulation results under extremely dry and extremely humid conditions.

Study Site	Maxim Radius	Minimum Radius	Maxim Depth	Minimum Depth
Fengqiu	139 m	127 m	31.2 cm	16.9 cm
Hemuqiao	153 m	125 m	45.9 cm	19.1 cm
Theoretically	218 m	123 m	65.9 cm	13.8 cm

Our simulation indicated that the detection radius was more significantly affected by air humidity than by SWC (see Appendix A). However, no obvious difference in detection depth was found under different air humidity conditions (see Appendix B). We assumed that the detection depth at a specific point was affected only by its SWC. The two study sites had different maximum and minimum detection depths due to the difference in soil water conditions. Fengqiu is an irrigated agricultural field that provides a vast quantity of hydrogen, including crops and soil water. In contrast, the terrain of Hemuqiao is a mountain slope, where the soil cannot retain water for a long time due to the steep gradient and uneven distribution of soil depth (see Section 2.3). Therefore, the maximum and minimum detection depths of Hemuqiao were larger than those of Fengqiu.

According to previous studies, the effective detection radius [7,24,26,27] is influenced by a combination of factors, including SWC, atmospheric pressure, water vapour, and aboveground biomass. As acknowledged by previous studies, the horizontal footprint is determined mainly by atmospheric density and humidity [25]. Hemuqiao has a higher altitude than Fengqiu, and its soil hydrogen inventory is less than that at Fengqiu during the driest period, resulting in a maximum detection radius of 153 m, which was larger than the detection region of Fengqiu. However, the minimum detection radii of the two study sites were comparable due to their complex meteorological and vegetation conditions. First, Hemuqiao had a higher average annual precipitation than Fengqiu. Second, a litter layer and moso bamboo had an interception effect on precipitation, which resulted in higher air humidity at Hemuqiao during the precipitation period. The combination of these factors resulted in similar minimum detection radii at Hemuqiao and Fengqiu, which were 125 m and 127 m, respectively.

4.2. SWC Variation and Seasonal Characteristics at Different STUDY Sites

The time series of SWC at the two sites are shown in Figure 3. The soil water measured by FDR was calibrated with the oven drying method; hence, it could represent the actual SWC time series (blue dotted line). In this research, SWC measured by the CRNS had a similar trend to the SWC time series during the growing season. This variation pattern was observed at both study sites, indicating that CRNS can characterize the time series of SWC. SWC measured by the CRNS was inversely proportional to the neutron intensity. The detected neutrons interacted with the hydrogen in the footprint and carried information about its hydrogen inventory [7]. The SWC ranged from 0.10 $cm^3 \cdot cm^{-3}$ to 0.40 $cm^3 \cdot cm^{-3}$ at the Fengqiu station and from 0.05 $cm^3 \cdot cm^{-3}$ to 0.30 $cm^3 \cdot cm^{-3}$ at the Hemuqiao station (see Figure 3). As expected, the range of SWC and the footprint simulations in Section 4.1 consistently showed that Fengqiu stored a larger amount of SWC during the experimental periods. In general, in this study, Fengqiu provided more hydrogen with higher humidity than Hemuqiao despite the occurrence of more frequent precipitation and greater total precipitation at the Hemuqiao station. Differences in SWC storage between Fengqiu and Hemuqiao are most likely predominantly influenced by topography. Thin soil thickness and a large terrain gradient characterize Hemuqiao [28], while, in Fengqiu, flat farmland, which retains more soil water than the land in Hemuqiao, can provide abundant hydrogen resources.

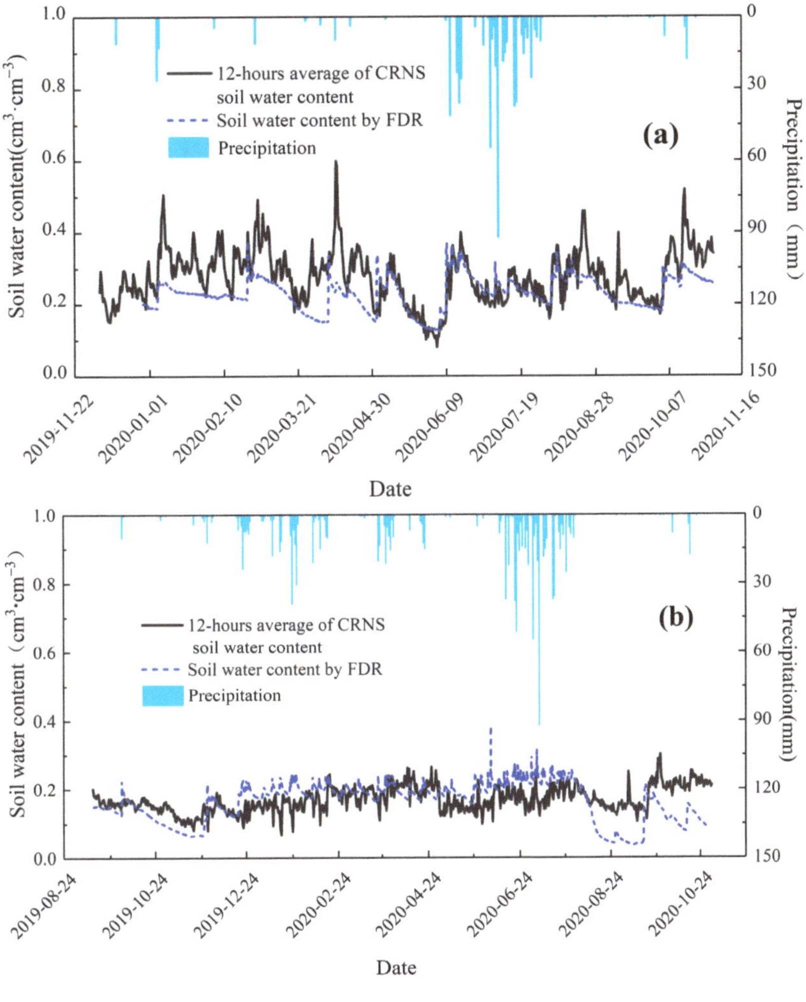

Figure 3. Comparison of SWC measured by the CRNS and the FDR at the Fengqiu station (**a**) and the Hemuqiao station (**b**). N_0 corresponds to the neutron count rate on 4 June 2020, at the Fengqiu station, while N_0 corresponds to the neutron count rate on 18 November 2019, at the Hemuqiao station.

Seasonal differences in SWC time series measured by the CRNS were observed at the two study sites. For most of our study period at the Fengqiu station, SWC measured by CRNS was slightly higher than that in the SWC time series. The diversity was even more obvious in the arid climatic conditions that dominate from January to April, when precipitation is relatively rare. SWC measured by CRNS was also sensitive to other factors, such as vegetation growth and atmospheric pressure [16,27]. The climatic conditions and crop growth may have decreased the neutron count accuracy and precision. The difference in SWC time series suggested that the CRNS captured the changes in both SWC and other hydrogen sources. However, the period from August to November at the Hemuqiao station is arid because of the low precipitation (see Figure 3b), and SWC measured by CRNS decreased as SWC decreased. The time series of SWC measured by CRNS was less volatile and lower than the SWC measured at the Hemuqiao station, except during the arid periods. The SWC measured by the CRNS relied on the neutron count rates of the CRNP. Neutrons are transported in the soil over distances of a few tens of centimetres [11] and, therefore, were blocked by ridges on the hillslope. Particular mountain terrain caused the radius of the footprint received by the CRNP at the Hemuqiao station to be smaller than that of the URANOS simulation. The narrow neutron fluctuation range measured by CRNS led to the narrow time series fluctuation range in Hemuqiao.

The time series of daily precipitation during the study period is also shown in Figure 3. Distinct seasonality in precipitation was observed at both study sites due to the monsoon climate. The total precipitation during the study period for Fengqiu and Hemuqiao was 647.4 mm and 1155.2 mm, respectively. The rare precipitation and higher SWC of Fengqiu also indicated that farmland has a higher capacity to store water than mountainous regions. Transient changes in SWC generated by precipitation were sensitively observed by CRNS at the Fengqiu station (see Figure 3a). The CRNS accurately captured each precipitation event, although some precipitation events did not have a significant effect on the SWC. The response characteristics to precipitation indicated that CRNS was more sensitive to changes in SWC than other soil-mounted detectors [18,29].

4.3. Spatial Characteristics of SWC at Different Study Sites

In this study, the data from the sampling points were calibrated by the weighting function (Sections 3.3 and 3.4) for comparison with CRNS. In addition, direct averaging of sampling points was considered and used for estimating the SWC (see Figure 4b,d). With direct averaging of sampling points, the R^2 values were 0.487 and 0.798 for Fengqiu and Hemuqiao, respectively. The RMSE values were 0.054 and 0.049 $cm^3 \cdot cm^{-3}$ for Fengqiu and Hemuqiao, respectively. The weighting calculation improved the accuracy of CRNS estimation in Fengqiu compared to the average calculation, indicating that the weighting simulation was consistent with the spatial characteristics of the CRNS footprint. Previous studies reported that the neutron density detected by CRNP is heterogeneously distributed within the CRNS footprint [30–32]. Despite this result, the weighting calculation did not significantly improve the CRNS estimation accuracy of Hemuqiao.

The linear positive correlation between the SWC measured by CRNS and the SWC at the sampling points indicated that CRNS is capable of accurately estimating the SWC. The estimation accuracy was determined by the coefficient of determination (R^2) and the RMSE. In the context of 6 samplings, the R^2 for Fengqiu was 0.645, while the RMSE was 0.049 $cm^3 \cdot cm^{-3}$; in the context of 7 samplings, the R^2 and RMSE for Hemuqiao were 0.773 and 0.049 $cm^3 \cdot cm^{-3}$, respectively (see Figure 4a,c). The estimation accuracy of CRNS increased as the sampling frequency increases, indicating that CRNS was well correlated with SWC [33,34]. Although the R^2 for Fengqiu was lower than that of Hemuqiao, the linear fitting curve for Fengqiu was closer to the 1:1 line than that of Hemuqiao, suggesting that CRNS estimated a more accurate SWC at the Fengqiu station.

It is important to note that topographical differences between the two experimental sites affected the neutron footprint. The distance that neutrons travelled before detection was reduced by the obstruction of the mountain, according to Kohli et al. and Franz et al. [12,24]. The CRNP was installed in the middle of the hillslope at the Hemuqiao station, which was in the saddle of the valley. Therefore, the flanking hillsides blocked part of the neutron footprint, narrowing the actual detection footprint of the CRNS to the region inside the valley. The obstruction of neutron transmission received by CRNS in the horizontal direction could lead to an unacceptable level of precision for SWC estimation in mountainous terrain [35].

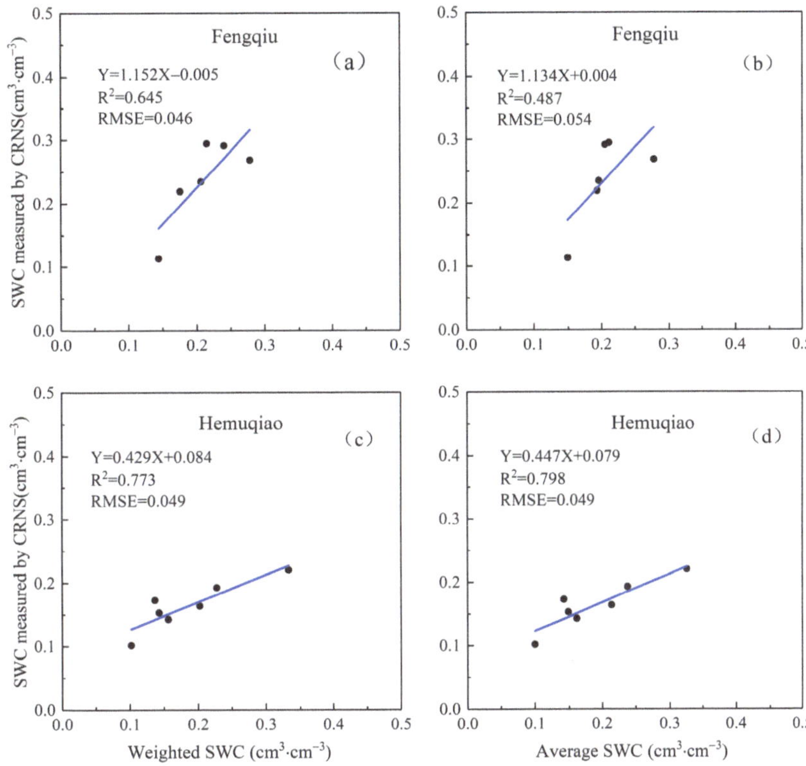

Figure 4. The SWC correlation between CRNS and point sampling. (**a**) The SWC correlation between CRNS and weighted point sampling in Fengqiu. (**b**) The SWC correlation between CRNS and averaged point sampling in Fengqiu. (**c**) The SWC correlation between CRNS and weighted point sampling in Hemuqiao. (**d**) The SWC correlation between CRNS and averaged point sampling in Hemuqiao.

The time series variation of CRNS is an accurate indicator of the accuracy of CRNS estimates. Significant differences between the two experimental sites in the correlation daily variations of CRNS and FDR were observed in Figure 5. At Fengqiu, the determination coefficient between CRNS and FDR was 0.390 with an RMSE of 0.072 $cm^3 \cdot cm^{-3}$. However, no significant correlation between the daily variation in CRNS and FDR was observed at Hemuqiao. The time series of SWC measured by FDR represented a continuous change in SWC at the experimental site. Therefore, the determination coefficient of 0.059 indicated that CRNS could not accurately represent the diurnal variation in SWC at Hemuqiao. CRNS counts depend on the proximity of neutron density to the surface, and thus fluctuate within a certain range. Increasing the sampling interval can improve statistical stability, as previously reported in several studies [21,33]. This study assumed that the daily or 12 h variation in the CRNS could adequately describe the temporal variation in SWC without increasing the error due to long sampling intervals.

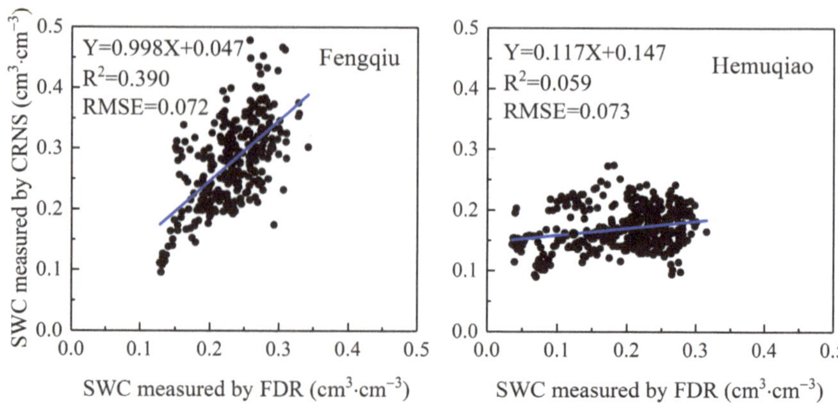

Figure 5. Correlation between the daily variations in SWC measured by CRNS and FDR.

The vegetation factor at Hemuqiao was also nonnegligible for the accuracy of CRNS estimates. The pole height of the moso bamboo is over 20 m. Most of its canopy extends above the CRNP, which is the largest difference between forest vegetation and farm vegetation. Fortunately, the hydrogen content of moso bamboo is assumed to be constant, and only the leaves change seasonally. Research has shown that, during the wet season, the hydrogen storage in tree canopies and litter layers can reach 0.067 $cm^3 \cdot cm^{-3}$ of soil volumetric water content [20]. However, it is not possible to determine the dynamic changes in bamboo forest contributions to the hydrogen pool during wet conditions with high interception storage. Therefore, we recommend that calibration for forest vegetation be considered in further studies to obtain more accurate estimates.

5. Conclusions

This study applied CRNS to estimate the SWC at two different experimental sites: a typical monsoonal farmland and a humid mountainous region. The URANOS simulations of the footprint and weighting functions show that the detection radius of CRNS ranges between 125 m and 153 m for different humidity conditions in mountainous regions, and the detection depth of CRNS ranges between 19.1 cm and 45.9 cm for different humidity conditions. In farmland, the detection radius of CRNS ranges from 127 m to 139 m, and the detection depth ranges from 16.9 cm to 31.2 cm depending on different humidity conditions. In general, farmland has a smaller CRNS footprint than mountainous regions due to its flat topography and thicker soils, which can retain more hydrogen resources.

The SWC measured by CRNS is generally slightly higher than that measured by the soil-mounted detectors, which is likely due to the influence of hydrogen sources other than the soil. Furthermore, CRNS is more sensitive to changes in SWC than other soil-mounted detectors, suggesting that correction of nonsoil hydrogen sources should be considered when monitoring SWC with CRNS.

The comparison between the CRNS and point sampling estimates show that the estimation accuracy varies between the two different experimental sites, but the difference is not statistically significant. The use of weighting calculation in farmland significantly improves the estimation accuracy of CRNS, indicating the necessity of using spatial weighting simulations for CRNS in order to obtain more accurate results. However, the accuracy of SWC estimation using CRNS in mountainous areas cannot be further improved by using weighting calculations. Previous research has indicated that the use of CRNS is suitable for estimating the SWC of farmland; however, there are many uncertainties when applying CRNS to estimate SWC in mountainous areas due to the terrain obstructing neutron transmission and tall vegetation introducing errors into the CRNS estimation results. Future studies should pay greater attention to the effects of topography and vegetation when CRNS is applied in mountainous environments.

Author Contributions: Conceptualization, Y.J., X.L. and J.L.; methodology, Y.J., X.L. and J.L.; software, Y.J. and K.X.; validation, Y.J. and C.G.; investigation, Y.J., Y.Z. and Y.L.; resources, Y.J., H.D. and K.X.; data curation, Y.J.; writing—original draft preparation, Y.J.; writing—review and editing, Y.J. and J.L.; supervision, J.L. and X.L.; funding acquisition, J.L. and X.L. All authors have read and agreed to the published version of the manuscript.

Funding: This work was supported by the National Key Research and Development Program of China (Funding number, 2022YFD1500502), the Strategic Priority Research Program of the Chinese Academy of Sciences (grant No. XDA28010401), and the National Natural Science Foundation of China (grant Nos. 42177302 and 41877021).

Data Availability Statement: The data presented in this study are available on request from the corresponding author.

Acknowledgments: We would like to express our appreciation to the Fengqiu State Key Argo-Ecological Experimental Station and Hemuqiao Experimental Station for providing the experimental site support for this research.

Conflicts of Interest: The authors declare no conflict of interest.

Appendix A. Detection Radius (m) of Simulations at Different SWCs ($cm^3 \cdot cm^{-3}$) and air Humidities ($g \cdot cm^{-2}$)

Table A1. Detection radius obtained by simulations under different SWCs ($cm^3 \cdot cm^{-3}$) and air humidities ($g \cdot cm^{-2}$). The pattern of variations in the detection radius values indicates the impact of SWC and air humidity on detection radius.

SWC	0.10	0.15	0.20	0.25	0.30	0.35	0.40	0.45
Radius with the air humidity of 10	187	178	172	166	161	157	153	150
Air humidity	5	10	15	20	25	30	33	
Radius with SWC of 0.15	184	178	173	168	164	160	158	

Note: Multiple regression analysis showed that both air humidity and SWC had a significant impact on the detection radius ($p < 0.001$), and the adjusted R^2 was 0.976.

Appendix B. Detection Depth (cm) of Simulations at Different SWCs ($cm^3 \cdot cm^{-3}$) and air Humidities ($g \cdot cm^{-2}$)

Table A2. Detection depth obtained by simulations under different SWCs ($cm^3 \cdot cm^{-3}$) and air humidities ($g \cdot cm^{-2}$). The pattern of variations in the detection depth values indicates the impact of SWC and air humidity on detection depth.

SWC	0.10	0.15	0.20	0.25	0.30	0.35	0.40	0.45
Depth with the air humidity of 10	35.3	29.2	24.8	21.6	19.1	16.9	15.2	13.7
Air humidity	5	10	15	20	25	30	33	
Depth with SWC of 0.15	29.1	29.2	29.2	29.2	29.3	29.4	29.4	

Note: Multiple regression analysis was used to investigate the effects of air humidity and SWC on the detection depth, and the adjusted R^2 was 0.954. SWC had a significant impact on the detection depth ($p < 0.001$), but there was no significant impact observed on air humidity to the detection depth ($p = 0.919$).

References

1. Friel, R.; Or, D. Frequency analysis of time-domain reflectometry (TDR) with application to dielectric spectroscopy of soil constituents. *Geophysics* **1999**, *64*, 707–718. [CrossRef]
2. Huisman, J.A.; Bouten, W.; Vrugt, J.A.; Ferre, P.A. Accuracy of frequency domain analysis scenarios for the determination of complex dielectric permittivity. *Water Resour. Res.* **2004**, *40*, W02401. [CrossRef]
3. Vreugdenhil, M.; Dorigo, W.; Broer, M.; Haas, P.; Eder, A.; Hogan, P.; Bloeschl, G.; Wagner, W. Towards a High-Density Soil Moisture Network for the Validation of Smap in Petzenkirchen, Austria. In Proceedings of the 2013 IEEE International Geoscience and Remote Sensing Symposium (Igarss), Melbourne, Australia, 21–26 July 2013; pp. 1865–1868.

4. Yang, Z.W.; Crow, W.; Hu, L.; Di, L.P.; Mueller, R. Smap Data for Cropland Soil Moisture Assessment—A Case Study. In Proceedings of the 2017 Ieee International Geoscience and Remote Sensing Symposium (Igarss), Fort Worth, TX, USA, 23–28 July 2017; pp. 1996–1999.
5. Li, D.; Schrön, M.; Köhli, M.; Bogena, H.; Weimar, J.; Jiménez Bello, M.A.; Han, X.; Martínez Gimeno, M.A.; Zacharias, S.; Vereecken, H.; et al. Can Drip Irrigation be Scheduled with Cosmic-Ray Neutron Sensing? *Vadose Zone J.* **2019**, *18*, 190053. [CrossRef]
6. Mujumdar, M.; Goswami, M.M.; Morrison, R.; Evans, J.G.; Ganeshi, N.; Sabade, S.S.; Krishnan, R.; Patil, S.N. A study of field-scale soil moisture variability using the COsmic-ray Soil Moisture Observing System (COSMOS) at IITM Pune site. *J. Hydrol.* **2021**, *597*, 126102. [CrossRef]
7. Zreda, M.; Desilets, D.; Ferré, T.P.A.; Scott, R.L. Measuring soil moisture content non-invasively at intermediate spatial scale using cosmic-ray neutrons. *Geophys. Res. Lett.* **2008**, *35*, L21402. [CrossRef]
8. Hendrick, L.D.; Edge, R.D. Cosmic-Ray Neutrons near Earth. *Phys Rev.* **1966**, *145*, 1023–1025. [CrossRef]
9. Kodama, M.; Kudo, S.; Kosuge, T. Application of Atmospheric Neutrons to Soil-Moisture Measurement. *Soil Sci.* **1985**, *140*, 237–242. [CrossRef]
10. Sato, T.; Niita, K. Analytical functions to predict cosmic-ray neutron spectra in the atmosphere. *Radiat Res.* **2006**, *166*, 544–555. [CrossRef]
11. Desilets, D.; Zreda, M.; Ferré, T.P.A. Nature's neutron probe: Land surface hydrology at an elusive scale with cosmic rays. *Water Resour. Res.* **2010**, *46*, W11505. [CrossRef]
12. Köhli, M.; Schrön, M.; Zreda, M.; Schmidt, U.; Dietrich, P.; Zacharias, S. Footprint characteristics revised for field-scale soil moisture monitoring with cosmic-ray neutrons. *Water Resour. Res.* **2015**, *51*, 5772–5790. [CrossRef]
13. Schrön, M.; Kohli, M.; Scheiffele, L.; Iwema, J.; Bogena, H.R.; Lv, L.; Martini, E.; Baroni, G.; Rosolem, R.; Weimar, J.; et al. Improving calibration and validation of cosmic-ray neutron sensors in the light of spatial sensitivity. *Hydrol. Earth Syst. Sci.* **2017**, *21*, 5009–5030. [CrossRef]
14. Schrön, M.; Rosolem, R.; Köhli, M.; Piussi, L.; Schröter, I.; Iwema, J.; Kögler, S.; Oswald, S.E.; Wollschläger, U.; Samaniego, L.; et al. Cosmic-ray Neutron Rover Surveys of Field Soil Moisture and the Influence of Roads. *Water Resour. Res.* **2018**, *54*, 6441–6459. [CrossRef]
15. Badiee, A.; Wallbank, J.R.; Fentanes, J.P.; Trill, E.; Scarlet, P.; Zhu, Y.C.; Cielniak, G.; Cooper, H.; Blake, J.R.; Evans, J.G.; et al. Using Additional Moderator to Control the Footprint of a COSMOS Rover for Soil Moisture Measurement. *Water Resour. Res.* **2021**, *57*, e2020WR028478. [CrossRef]
16. Baroni, G.; Oswald, S.E. A scaling approach for the assessment of biomass changes and rainfall interception using cosmic-ray neutron sensing. *J. Hydrol.* **2015**, *525*, 264–276. [CrossRef]
17. Li, X.; Xu, X.; Li, X.; Xu, C.; Wang, K. Field scale soil water prediction based on areal soil moisture measurements using cosmic-ray neutron sensing in a karst landscape. *J. Hydrol.* **2022**, *605*, 127395. [CrossRef]
18. Tan, X.; Zhang, L.; He, C.; Zhu, Y.; Han, Z.; Li, X. Applicability of cosmic-ray neutron sensor for measuring soil moisture at the agricultural-pastoral ecotone in northwest China. *Sci. China Earth Sci.* **2020**, *63*, 1730–1744. [CrossRef]
19. Bogena, H.R.; Huisman, J.A.; Baatz, R.; Franssen, H.J.H.; Vereecken, H. Accuracy of the cosmic-ray soil water content probe in humid forest ecosystems: The worst case scenario. *Water Resour. Res.* **2013**, *49*, 5778–5791. [CrossRef]
20. Heidbüchel, I.; Güntner, A.; Blume, T. Use of cosmic-ray neutron sensors for soil moisture monitoring in forests. *Hydrol. Earth Syst. Sci.* **2016**, *20*, 1269–1288. [CrossRef]
21. Peterson, A.M.; Helgason, W.D.; Ireson, A.M. Estimating field-scale root zone soil moisture using the cosmic-ray neutron probe. *Hydrol. Earth Syst. Sci.* **2016**, *20*, 1373–1385. [CrossRef]
22. Tian, Z.C.; Li, Z.Z.; Liu, G.; Li, B.G.; Ren, T.S. Soil water content determination with cosmic-ray neutron sensor: Correcting aboveground hydrogen effects with thermal/fast neutron ratio. *J. Hydrol.* **2016**, *540*, 923–933. [CrossRef]
23. Mwangi, S.; Zeng, Y.; Montzka, C.; Yu, L.; Su, Z. Assimilation of Cosmic-ray Neutron Counts for the Estimation of Soil Ice Content on the Eastern Tibetan Plateau. *J. Geophys. Res. Atmos.* **2020**, *125*, e2019JD031529. [CrossRef]
24. Franz, T.E.; Zreda, M.; Ferre, T.P.A.; Rosolem, R.; Zweck, C.; Stillman, S.; Zeng, X.; Shuttleworth, W.J. Measurement depth of the cosmic ray soil moisture probe affected by hydrogen from various sources. *Water Resour. Res.* **2012**, *48*, W08515. [CrossRef]
25. Zreda, M.; Shuttleworth, W.J.; Zeng, X.; Zweck, C.; Desilets, D.; Franz, T.; Rosolem, R. COSMOS: The COsmic-ray Soil Moisture Observing System. *Hydrol. Earth Syst. Sci.* **2012**, *16*, 4079–4099. [CrossRef]
26. Rosolem, R.; Shuttleworth, W.J.; Zreda, M.; Franz, T.E.; Zeng, X.; Kurc, S.A. The Effect of Atmospheric Water Vapor on Neutron Count in the Cosmic-Ray Soil Moisture Observing System. *J. Hydrometeorol.* **2013**, *14*, 1659–1671. [CrossRef]
27. Iwema, J.; Schrön, M.; Koltermann Da Silva, J.; Schweiser De Paiva Lopes, R.; Rosolem, R. Accuracy and precision of the cosmic-ray neutron sensor for soil moisture estimation at humid environments. *Hydrol. Process.* **2013**, *35*, e14419. [CrossRef]
28. Liu, J.; Han, X.; Chen, X.; He, R.; Wu, P. Prediction of soil thicknesses in a headwater hillslope with constrained sampling data. *Catena* **2019**, *177*, 101–113. [CrossRef]
29. Dimitrova-Petrova, K.; Geris, J.; Wilkinson, M.E.; Rosolem, R.; Verrot, L.; Lilly, A.; Soulsby, C. Opportunities and challenges in using catchment-scale storage estimates from cosmic ray neutron sensors for rainfall-runoff modelling. *J. Hydrol.* **2020**, *586*, 124878. [CrossRef]

30. Desilets, D.; Zreda, M. Spatial and temporal distribution of secondary cosmic-ray nucleon intensities and applications to in situ cosmogenic dating. *Earth Planet. Sci. Lett.* **2003**, *206*, 21–42. [CrossRef]
31. Shuttleworth, J.; Rosolem, R.; Zreda, M.; Franz, T.E. The COsmic-ray Soil Moisture Interaction Code (COSMIC) for use in data assimilation. *Hydrol. Earth Syst. Sci.* **2013**, *17*, 3205–3217. [CrossRef]
32. Kohli, M.; Schron, M.; Schmidt, U. Response functions for detectors in cosmic ray neutron sensing. *Nucl. Instrum. Meth. A* **2018**, *902*, 184–189. [CrossRef]
33. Rivera Villarreyes, C.A.; Baroni, G.; Oswald, S.E. Integral quantification of seasonal soil moisture changes in farmland by cosmic-ray neutrons. *Hydrol. Earth Syst. Sci.* **2011**, *15*, 3843–3859. [CrossRef]
34. Zreda, M.; Shuttleworth, W.J.; Zweck, C.; Zeng, X.; Ferre, T. The Cosmic-ray Soil Moisture Observing System (COSMOS): A non-invasive, intermediate scale soil moisture measurement network. In Proceedings of the BHS Third International Symposium, Managing Consequences of a Changing Global Environment EGU General Assembly Conference Abstracts, Newcastle, UK, 19–23 July 2010.
35. Francke, T.; Heistermann, M.; Köhli, M.; Budach, C.; Schrön, M.; Oswald, S.E. Assessing the feasibility of a directional cosmic-ray neutron sensing sensor for estimating soil moisture. *Geosci. Instrum. Methods Data Syst.* **2022**, *11*, 75–92. [CrossRef]

Disclaimer/Publisher's Note: The statements, opinions and data contained in all publications are solely those of the individual author(s) and contributor(s) and not of MDPI and/or the editor(s). MDPI and/or the editor(s) disclaim responsibility for any injury to people or property resulting from any ideas, methods, instructions or products referred to in the content.

AgriEngineering

Review

The Significance and Ethics of Digital Livestock Farming

Suresh Neethirajan

Farmworx Research Institute, Van der Waals Straat, 6705 Wageningen, The Netherlands; sneethir@gmail.com

Abstract: The emergence of precision and digital livestock farming presents an opportunity for sustainable animal farming practices that enhance animal welfare and health. However, this transformation of modern animal farming through digital technology has several implications for the technological, social, economic, and environmental aspects of farming. It is crucial to analyze the ethical considerations associated with the digitalization of modern animal farming, particularly in the context of human–animal relationships and potential objectification. This analysis can help develop frameworks for improving animal welfare and promoting sustainability in animal farming. One of the primary ethical concerns of digital livestock farming is the potential for a digital divide between farmers who have access to advanced technologies and those who do not. This could lead to a disparity in animal welfare and health outcomes for different groups of animals. Additionally, the use of artificial intelligence in digital livestock farming may lead to a loss of personal connection between farmers and animals, which could impact the animal's well-being. Another ethical concern of digital livestock farming is the potential for the objectification of animals as mere data points. The use of sensors and other monitoring technologies can provide valuable data on animal health and behavior, but it is important to remember that animals are sentient beings with complex emotional and social needs. The use of digital technologies should not lead to neglect of animal welfare or a lack of human responsibility toward animals. Furthermore, social context becomes essential while integrating technologies in livestock farming to overcome ethics. By considering the cultural and societal norms of different communities, we can ensure that the use of digital technologies does not undermine these values. To address these ethical challenges, the development of standards and codes of conduct for the adoption and implementation of digital livestock farming tools and platforms can help ensure that animal welfare and sustainability are prioritized. This can help alleviate the privacy concerns of stakeholders and improve sustainability in animal farming practices. Additionally, the use of virtual and augmented reality technologies can provide a way to enhance human–animal interactions and provide more personalized care to animals, further promoting animal welfare.

Keywords: precision livestock farming; digital livestock farming; smart farming; societal impacts; data ownership; open access; sustainability; animal ethics; digital divide; digital agriculture

Citation: Neethirajan, S. The Significance and Ethics of Digital Livestock Farming. *AgriEngineering* **2023**, *5*, 488–505. https://doi.org/10.3390/agriengineering5010032

Academic Editors: Muhammad Sultan, Yuguang Zhou, Redmond R. Shamshiri and Muhammad Imran

Received: 23 December 2022
Revised: 22 February 2023
Accepted: 24 February 2023
Published: 26 February 2023

Copyright: © 2023 by the author. Licensee MDPI, Basel, Switzerland. This article is an open access article distributed under the terms and conditions of the Creative Commons Attribution (CC BY) license (https://creativecommons.org/licenses/by/4.0/).

1. Introduction

The increasing demand for animal products has led to significant changes in animal farming practices, with a shift toward more intensive and industrialized methods of animal production. As a result, animal welfare and sustainability have become important concerns for consumers, policymakers, and stakeholders in the animal agriculture industry. This has led to the development of new technologies and farming practices, such as precision and digital livestock farming, which offer potential benefits for animal welfare and sustainability.

Technologies can be considered a double-edged sword—something that is clearly exemplified in livestock farming. For instance, deepfake technologies have been drawing attention lately. Such technologies are used mostly in malicious applications, but they can also be integrated with virtual reality to enhance animal welfare. To illustrate, suppose that a calf is suffering emotionally owing to the absence of a mother that has been sent for slaughter. The technologies in question can be harnessed to create a digital simulation of the mother cow that may provide comfort to the calf and enhance its emotional well-being [1].

Indeed, there are problems that are presently considered unsolvable in the livestock farming sector but that can, in fact, be solved or might become solvable using digitization approaches, which make possible innovations such as the following:

- Transitioning from subjective judgments to objective methods of measurement;
- Moving from one-dimensional to multidimensional approaches in solving problems from a holistic standpoint;
- Becoming able to solve problems by moving from a coarse-grained to a fine-grained level of analysis in order to understand, design and develop solutions;
- Moving away from reactive to predictive approaches.

Two such digitized approaches in livestock farming are precision livestock farming and digital livestock farming, the latter of which has emerged from the former. Precision livestock farming and digital livestock farming are two examples of digitized approaches that offer benefits such as real-time monitoring, predictive analytics, and automated decision-making. The integration of digital technologies into animal farming practices raises several ethical concerns, particularly with regard to human–animal relationships and objectification. There is a need to analyze the ethical considerations associated with digital livestock farming and develop frameworks for improving animal welfare and sustainability.

The research question that inspired this opinion review is, "What are the ethical considerations associated with digital livestock farming, and how can they be addressed to improve animal welfare and sustainability?" The review aims to provide a critical analysis of the ethical challenges of digital livestock farming technologies and their potential impact on animal welfare and sustainability. It also offers recommendations for ethical considerations, including the development of standards and codes of conduct for the adoption and implementation of digital livestock farming tools and platforms. The review is intended to inform policymakers, animal agriculture industry stakeholders, and consumers about the potential benefits and ethical challenges of digital livestock farming and to encourage discussion and action to promote animal welfare and sustainability in animal farming practices.

2. Precision Livestock Farming, Digital Livestock Farming and Smart Livestock Farming

To begin, let us explore the core notions that shall occupy the stakeholders, namely precision livestock farming, digital livestock farming and, in addition smart livestock farming.

Precision livestock farming (PLF) terminology has been used for over 10 years. PLF may be defined as "the management of livestock production using the principles and technology of process engineering" [2]. Its purpose is to "improve the efficiency of production while increasing animal and human welfare, via applying advanced information and communication technologies (ICT), targeted resource use and precise control of the production process" [3].

More recently, PLF has been defined as farm management based on continuous automatic real-time monitoring and control of production/reproduction, animal health and welfare, and the environmental impact of livestock production [4–6]. However, as we shall see presently, this is a definition that reflects the ways in which PLF has been moving toward digital livestock farming since it emphasizes real-time monitoring and control, something that has been characteristic of the digital farming transformation. Accordingly, the previous definition offered above is more appropriate for our purposes of observing the distinction between precision livestock farming and digital livestock farming.

Despite the recent prevalence of PLF, we are now slowly moving from PLF to digital livestock farming (DLF) and, relatedly, smart farming (SF). DLF may be defined as digital agriculture or farming, applied specifically to livestock farming. Digital agriculture, in turn, is defined as follows [7]:

We define digital agriculture as the use of elaborate and detailed digital information in guiding decisions along the agri-food value chain. This includes the use of high-volume, variable source data ("big data") that creates critical insights, modeling, actionable analytics and automation for precision livestock farming. Importantly, it is not restricted to farm animal production alone but also can span all or part of the value chain; the transformational aspect of digitization offers a bridge between the consumer and animal caretaker or the farmer.

There is a marked difference in precision and accuracy with respect to each type of farming. PLF is mostly an incremental way of finding solutions, while DLF features a higher degree of integration and encompasses sustainability [8] and predictive possibilities that account for time-scale phenomena [9]. The primary focus of PLF is on post-data collected measurements for supporting decision-making and directing farming practices. By contrast, DLF is about real-time and predictive capabilities, which are made possible by self-learning artificial intelligence algorithms [10]. Digitalization is about creating actionable intelligence via data and creating meaningful value for all of the stakeholders in livestock farming, all while ensuring the well-being of animals.

Thus, DLF incorporates and integrates precision and smart animal farming concepts. Additionally, the emphasis of DLF is not anymore on precision, which was already achieved at the stage of PLF. Instead, now that precision has been achieved, the emphasis is on the integration of precise data into digital systems. Thus, we see that DLF goes beyond PLF.

One useful way of thinking about PLF is in terms of a multi-layered structure. First, we may speak in terms of input systems, such as wearable devices, sensors, recording equipment, etc. These input devices generate data, which constitutes the second layer. On the third layer, we find data management and storage. This is followed by a further layer which includes processing, interpretation and, finally, feedback.

To clarify, however, one should bear in mind that, in certain specific cases, some digital animal farming technologies can be considered more as an incremental acquisition of understanding through observation. An example of this would be mobile, on-farm real-time detection tools that measure the health of farm animals in the form of data for commercial use and to address animal caretaker concerns.

There are some useful points of contrast that can be appealed to in order to elaborate on the differences between PLF and DLF:

In a conventional system such as precision livestock farming (PLF), sensors (wearable and minimally invasive or non-invasive) and video monitoring technologies collect animal measurements and send this data to a computer [11]. Algorithms process this data and create insights or enable farmers to make decisions. This data can be compared to a target value or a model, for example, in monitoring temperature. If the temperature is between two given values, it is within an acceptable range, but if it falls outside that range, there is something amiss that may require some decision-making. The underlying condition behind the anomalous temperature could be variations or fluctuating hormonal levels inside an animal. Here, the interaction between various physiological parameters has to be considered and evaluated. Thus, conventional PLF basically collects data based on health- or welfare-indicating parameters after the occurrence of an event. By contrast, with digital animal farming, we have the capability to move from reactive to predictive measurements; even before disease onset, the data analysis can tell when it will occur [10,12].

It is also important to point out that the transition from PLF to DLF can be analyzed under a two-stage approach. The initial stage is characterized by an emphasis on digitalization. The second stage, in turn, involves a more precise distinction between digital livestock farming and smart livestock farming (SLF), where both of these can be regarded as successors to PLF. Smart farming is a development that focuses on the usage of information and communication technology (ICT) in the cyber-physical livestock farm management cycle [13]. Now, many of the ethical and social implications of both farming systems will be the same. However, SLF also has specific implications of its own. Some work needs to be done to give more weight to SLF compared to DLF in discussions of new trends in farming. Another important question to explore here is whether SLF complements DLF.

Digital and/or precision livestock farming is an emerging area of research, and the technologies are currently being developed rapidly without much consideration of the implications of ethics in relation to the animals and the bonding with the animal caretakers and other social elements. There is a need for understanding the elements of ethics from the social context and from the animals' individual needs perspective.

3. The Ethics of DLF

Ethics is a crucial dimension of analysis to consider when discussing DLF [14]. In this regard, there are multiple issues to bear in mind, which bring to the fore the ethical complexities created by DLF. Here, we discuss several of those issues.

3.1. Animal Contentment

Animal contentment is closely associated with productivity. As one may say, "A happy cow is a productive cow" [15]. It is the fundamental responsibility of farmers and farming industries to ensure that animals live fully in their short life span (such as by exhibiting natural behavior, demonstrating playfulness and enjoying a quality of life). DLF incorporates this concern in the development of solutions since it can be used to promote animal welfare [16]. Of course, sometimes society can be considered venal, as the main focus is on maximizing profits and efficiency, not on animal welfare. As researchers, however, we know that it is not sustainable to consider profit alone as a goal in livestock farming, and DLF serves as a powerful tool in this ethical pursuit. Here are some ways in which DLF can help achieve this goal:

Precision management: DLF can provide real-time data on animal health, behavior, and performance. This information can be used to identify potential health problems or changes in behavior that may indicate stress, pain, or discomfort. With this information, farmers can intervene early to prevent or treat health issues, resulting in better animal welfare and productivity.

Environmental monitoring: DLF can also provide real-time data on environmental conditions, such as temperature, humidity, and air quality, which can affect animal health and welfare. By monitoring these factors, farmers can make adjustments to improve animal comfort and reduce the risk of disease.

Efficient resource management: DLF can help farmers manage resources more efficiently, including feed, water, and energy. By optimizing the use of these resources, farmers can reduce waste and minimize environmental impact while also improving profitability.

Traceability and transparency: DLF can provide a detailed record of each animal's life cycle, including its origin, breeding, feed, and medication history. This information can be used to ensure the safety and quality of animal products and to provide consumers with transparency about where their food comes from and how it was produced. In addition, transparency in the use of DLF and the sharing of data with consumers would increase accountability and drive positive change in the industry.

Animal Welfare: To ensure animal welfare in DLF, it is important to prioritize the use of digital tools that facilitate real-time monitoring and early detection of animal health issues. This would allow for timely interventions and treatment, minimizing animal suffering. DLF can certainly help farmers optimize animal welfare, reduce environmental impact, and improve profitability. However, it is important to remember that DLF is a tool, and it should be used in conjunction with other ethical and sustainable farming practices to achieve the best possible outcomes for animals, the environment, and society as a whole.

3.2. Disruption and Critical Analysis

DLF has the potential to be disruptive in management [17]. Now, being disruptive does not necessarily mean that technology is negative. For example, by alleviating physiological uncertainty in animals differently from how it was intended to be alleviated, predictive digital farming tools can be self-fulfilling and provide their own raison d'être. Nonetheless, normative questions and a constant reflection on reality are necessary for developing solutions via DLF. We should learn from our mistakes, for example, even when poorly supported predictions end up being true. That is, we do not learn from mistakes if we cannot recognize mistakes as such. Proper use of digital technology has to consider and look at ways to realize and take stock of our mistakes and learn from them.

To avoid the self-fulfilling prophecies of predictive digital farming tools and the misuse of technology, it is essential to subject all digital tools to rigorous ethical scrutiny before they are deployed. This should involve a multidisciplinary approach that includes experts in animal welfare, ethics, and data privacy. The use of ethical principles such as the precautionary principle and the principle of non-maleficence could help to guide decision-making.

3.3. Digital Divide and Obsolescence

"Digital divide" is another concept that we have to consider in connection to ethics [18,19]. Not everyone is capable of using DLF technologies. The digital divide between educated and not technically savvy users will lead to a technological rift. Animals' inability to access digital tools in farms may lead to detrimental isolation from social interactions and may affect the nature and quality of interactions between farm animals and caretakers. A related issue is that of obsolescence. It has been estimated that engineering knowledge becomes obsolete, on average, every 5.2 years [20]. Technology obsolescence is also a prevalent issue. This may deepen the technological divide between those who are able to keep up with the pace of technological change and those who are not.

To address the digital divide in the livestock farming industry, stakeholders should prioritize providing access to digital tools and training for all farmers. This could involve the development of user-friendly digital tools and training programs that cater to individuals with varying levels of technical proficiency. To mitigate obsolescence, the development of digital tools should be agile and adaptable, with regular updates and support provided to users.

3.4. Social Considerations

A further crucial issue to consider is the social dimension of DLF. Of particular note here is the question of social attachment. Some groups with a more individualistic approach may see social attachment as less beneficial and perhaps even as a burden. Nonetheless, a concern for well-being in relation to technology should take into account that users and farm animals are diverse and be inclusive [21]. How to fight digital discrimination in technological adoption and build inclusion should be considered part of the development of digital tools for livestock farming. This can be done by integrating values into the design of those tools. Indeed, technology is not just an artifact, it is a tool, and it can also be an institution and an intermediary. Design is part, as well, of social, political, and moral values, which may lead to some fundamental challenges that involve resolving potential ethical conflicts.

To ensure social inclusion in DLF, it is important to involve all stakeholders in the design and deployment of digital tools. This would enable the integration of diverse perspectives and values and could help to mitigate digital discrimination. Additionally, the use of participatory design approaches could help to ensure that digital tools are developed with the needs and preferences of all users in mind.

3.5. Greenhouse Gas Emissions

Digital-tool design and deployment in modern animal farming should also consider the individual carbon footprint associated with animals, processes and facilities, as well as other greenhouse emissions like methane. Greenhouse emissions are of peak importance in considering the environmental implications of farming. This is a particularly clear example, among many others, of how digital farming requires analyzing whether the digital tools deployed improve society in addition to enhancing food productivity. Doing so should consider various platforms and applications of those tools and the context in which they are being applied. DLF offers various ways of monitoring and perhaps reducing emissions, both methane [22,23]. To reduce greenhouse gas emissions in DLF, stakeholders should prioritize the use of digital tools that help to minimize the carbon footprint of livestock farming. This could involve the use of precision feeding systems that reduce feed waste and associated methane emissions, as well as the use of renewable energy sources to power digital tools and other farm processes.

3.6. Artificial Intelligence and Public Policy

The uptake of artificial intelligence-enabled digital tools for livestock farming via public–private partnerships and arrangements is becoming a main policy tool. The public sector is not very approachable for citizens when considering AI algorithms, which therefore become black boxes, raising questions of transparency [24]. Relating this issue to freedom of information, we may note that proprietary information is completely different and varies between the public and private sectors. The public sector emphasizes transparency, legitimacy and shared benefit for all, while the private sector focuses on efficiency and profitability.

With that being said, we should bear in mind that, as is often the case, "The medium is the message". Digital tools can be perceived in both a favorable and less favorable light. It is important to consider how to create avenues so that digital tools shape our ideas about modern animal farming in a responsible manner.

To ensure transparency and ethical use of AI in DLF, it is essential to involve all stakeholders in the development of public policy around the use of AI in the industry. This could involve the establishment of independent bodies that oversee the use of AI in livestock farming, ensuring that AI systems are transparent, explainable, and accountable. Additionally, public–private partnerships should prioritize shared benefits and the interests of all stakeholders, including farmers, consumers, and animals.

3.7. Human–Animal Relationships

The gap between the farm animal's and the farmer's intellect can possibly be bridged using the magic of visualization platforms (Figure 1) via digital technologies. Due to the complexity of the vocalization of farm animals and the verbal depth and breadth involved, visualization tools can offer avenues for better clarity of understanding [25–27]. This can lead to a more humane form of farming, which is of vital importance, given that alienation and dehumanization are some of the chief current concerns around farming. Additionally, healthy human–animal relationships are central to farming productivity [28]. The integration of digital technologies in livestock farming challenges traditional human–animal relationships by changing the way humans interact with animals. With real-time monitoring, automated decision-making, and predictive analytics, the role of the farmer shifts from that of a caretaker to that of a supervisor, with less direct involvement in animal care. This raises questions about the value of human–animal relationships and whether the use of digital technologies leads to the objectification of animals.

Moreover, digital technologies can lead to the reduction of human–animal interaction, which can impact animal welfare negatively. For instance, if animals are housed in automated facilities, they may not receive the necessary physical contact and socialization required for their well-being, leading to decreased welfare. Therefore, it is important to ensure that the use of digital technologies does not compromise human–animal relationships and that animals continue to receive proper care and socialization.

3.8. Privacy and Data Protection

The integration of digital technologies in livestock farming requires the collection, storage, and analysis of large amounts of data. This data includes sensitive information such as animal health, behavior, and genetics, which could be used to infer private information about the farmers and the farm's business operations. Moreover, if this data falls into the wrong hands, it could lead to issues such as cyber-attacks, fraud, or theft, compromising the privacy and safety of farmers and animals.

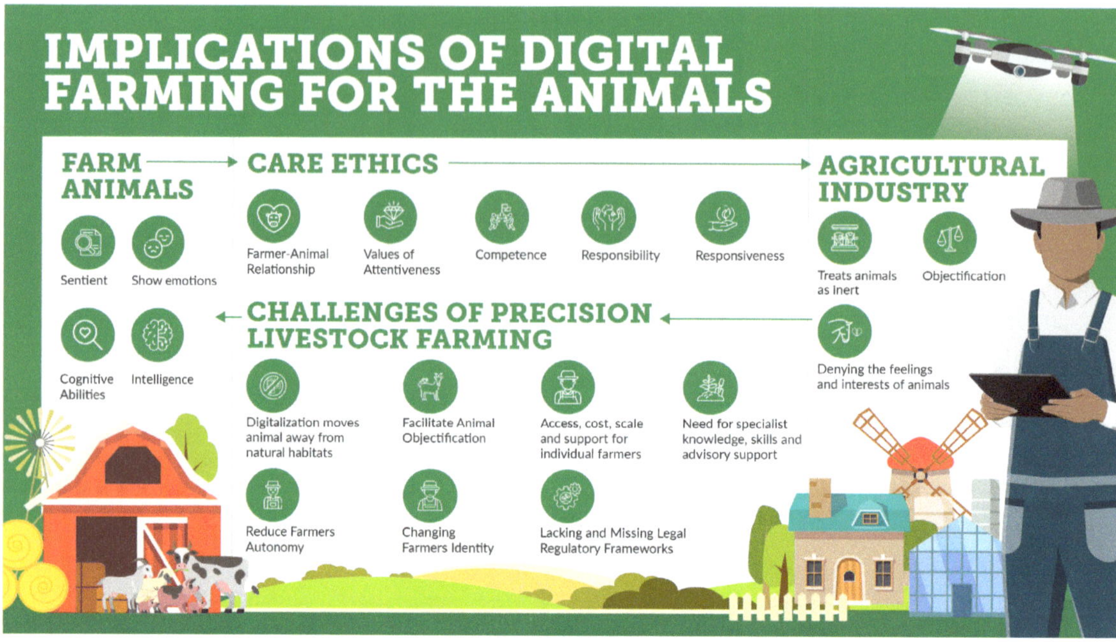

Figure 1. Implications of digital livestock farming for the animals.

Therefore, it is essential to have adequate data protection and security measures in place to ensure that data is not misused or exploited. This includes secure storage, data anonymization, and secure data transfer protocols. Farmers must be transparent about data collection and usage and obtain informed consent from stakeholders before collecting data.

3.9. Bias and Discrimination

Digital technologies in livestock farming rely on algorithms to analyze data and make decisions. However, these algorithms may have inherent biases, leading to discrimination and unfair treatment of animals. For instance, algorithms may not consider the diversity of animal breeds or account for individual differences in behavior, leading to inaccurate predictions or decisions.

Therefore, it is essential to ensure that algorithms used in digital livestock farming are fair, transparent, and unbiased. This includes regular testing, validation, and audit of the algorithms to prevent unintended discrimination or biases. The development of ethical standards and guidelines for algorithm development and use is also essential to prevent potential discrimination and ensure fairness.

3.10. Environmental Impact

Digital livestock farming has the potential to improve environmental sustainability by reducing waste, optimizing resource use, and minimizing the environmental impact of livestock production. However, the use of digital technologies in animal farming can also lead to increased energy consumption, e-waste, and carbon emissions from the production, maintenance, and disposal of digital devices and infrastructure.

Therefore, it is crucial to consider the environmental impact of digital livestock farming and adopt sustainable practices to mitigate its negative effects. This includes using renewable energy sources, recycling and disposing of electronic waste appropriately, and reducing unnecessary energy consumption.

3.11. Creating New Values

Current values do not suffice to solve the ethical conundrums of DLF, which calls for the creation of new values. For example, consider the issue of how digital tools bring about value changes. Farmers can conceptualize digital tools differently based on the context in which they are applied. They may see them as tools that extend their agency or as an imposition [29]. It is important, therefore, to engage in careful deliberations to interpret the significance of technology adoption among farmers as it unfolds. This may lead to the creation of new values and perspectives that provide solutions to current problems.

Ethical issues surrounding DLF require a multidisciplinary approach that integrates diverse perspectives and values. By involving all stakeholders in the development and deployment of digital tools, we can ensure that DLF promotes animal welfare, reduces environmental impact, and meets the needs of all users.

3.12. The Prospect of Monopolies

Another possible concern is that a single multinational company may take control of the digital technology market for agriculture, creating a monopoly—something that would harm the freedom of small-scale farmers [30]. This could be a real danger because a monopoly would have the power to determine all sorts of crucial factors, such as the inner workings of programs, digital environments and algorithms, assumptions, modeling techniques, etc., thereby compromising objectivity. Furthermore, there needs to be accountability with respect to algorithms and modeling, something that monopolies threaten. For example, a model may prioritize the achievement of certain breeding goals at the expense of animal well-being. This calls for a proper balancing of values that requires inclusive solutions and scrutiny.

The issue described here becomes more pressing, considering how a priority for farmers clearly is to have access to economical systems that integrate many different parameters and assumptions into their models. This is due to the fact that margins are very low in the farming sector. Hence, livestock farming scientists have a responsibility to balance a wide range of variables that go into modeling, a process requiring oversight and openness.

3.13. A Civic Approach

When technologies are being developed for livestock farming, a true civic approach to science is needed that allows us to incorporate viewpoints from the public at large, farmers and ethicists [19]. We cannot predict the future; rather, we are making it, so we must all partake in it as members of civil society. Of course, we hope that whatever emerges from new technologies such as DLF will be positive, but the results cannot be fully predicted at the beginning of the whole enterprise. If ethicists are involved in the very beginning of such new research, they could provide some valuable foresight on the consequences of technological development. This is the case even when technologies are developed with good intentions since, to emphasize, the issue is that of unexpected consequences. Accordingly, at the earliest stage, during the technology's development, ethics have to be considered and integrated into the design and planning processes.

3.14. Social Issues

In relation to issues pertaining to the social dimension of DLF, we should mention how digital animal farming tools can be adopted and how their development proceeds, depending on how we think about them. A broader social perspective of technological development is essential (Table 1). One example is the DeLaval brushes and other comfort technologies available in dairy farms. There is evidence suggesting that such technologies enhance cow productivity [31]. However, we may go beyond that issue and focus on the social significance of the technologies. What does it mean in terms of societal trends when such devices are developed for the comfort of animals in ways analogous to human luxuries? Is this simply a matter of pursuing productivity?

Table 1. Overview of the Ethical implications of Digital Livestock Farming per level.

Level	Potential Improvements	Concerns and Knowledge Gaps
Animal	▪ Automatic and continuous sampling of objective parameters to improve health and welfare; ▪ Potential for increased autonomy, such as choosing when to be milked.	▪ This could lead to increased objectification of animals, being just another part of the system and through digitalization of the habitat; ▪ Decreases human–animal interactions; ▪ Accuracy, reliability, and efficiency of the machinery needs to be validated; ▪ Algorithms and target values must be carefully considered to take into account individual differences; ▪ A holistic, multi-modal approach has to be taken to assess each platform per species, e.g., through a bioethical analysis.
Farmer	▪ Limits repetitive, time- and labor-consuming tasks done by the farmer; ▪ Increases productivity while minimizing environmental impact leading to potentially more income after initial investment; ▪ Automatic and constant monitoring of the entire farm, even remotely.	▪ Need for a new specialist, digital skills and advisory support; ▪ Reduces farmer's autonomy as they depend more on technology; ▪ Changes the job and will lead to job losses; ▪ Interpretation of complex systems can increase mental workload; ▪ Lack of trust between farmers and tech companies; ▪ Lack of a framework regarding data ownership, usage, storage and misuse.
Society and the Planet	▪ Addresses the increasing customer demand for better animal welfare; ▪ Decreases farm animal and human health issues by reducing disease and the need for antibiotics; ▪ It can aid in reducing pollution, emissions of greenhouse gases and energy consumption.	▪ Provides more transparency toward the customer due to increased traceability; ▪ Lack of public acceptance, but society is also relatively unaware of the advantages and disadvantages of PLF; ▪ Other attitudes are still unknown, such as those of veterinarians and stakeholders; ▪ The added value of PLF has not been quantified yet, and there is a need for Life Cycle Assessment.

3.15. The Limits of Biology and the Case of Dead-On-Arrival

Digital technologies enable the development of phenotyping and breeding of pigs that will have good heart conditions despite their body weight. Digital technologies can be used to derive digital biomarkers and use such biomarkers to select desirable heart traits through breeding in pigs. Digital biomarkers are digital measurements of phenotypes that are categorized according to the intended use of the collected data [32]. Biomarkers may allow for selective breeding since they facilitate the observation of phenotypes that are partly determined by genes that can be selected.

A related issue is that of heat stress. Pigs may often die during transportation from commercial farmers to slaughterhouses due to heat stress [33]. One possible way to stop this may be to use biomarkers in order to constantly monitor pigs and take preventative or palliative measures, if necessary, to reduce the incidence of dead-on-arrival cases. Another possibility is to use the data from biomarkers to breed more heat-resistant pigs. This is of particular importance given how global warming is expected to affect livestock well-being, which calls for a search for ways to adapt [34].

Similarly, digital biomarkers may provide a solution in the realm of antibiotic-related challenges. Often, farmers are unable to identify diseases in livestock before they arise due to limited capacities for phenotype observation and analysis. This means that farmers are often forced to apply antibiotics to livestock, which may have a negative effect on sales since consumers may be reluctant to consume meat from animals that have received antibiotics. One solution to this would be to use biomarkers in order to identify diseases preventively and avoid the need to apply antibiotics as a curative measure.

A further example that highlights the potential uses of digital biomarkers concerns so-called "exploding chickens" or broiler chickens—that is, chickens that are raised to reach their maximum weight as early as possible for commercial purposes. Broiler chickens raise significant concerns about animal welfare. One common issue that affects the well-being of those chickens is respiratory problems. Biomarking can help identify such issues in order to potentially address them since certain respiratory diseases are correlated with detectable abnormal sounds in broiler chickens [35].

Relatedly, the use of biomarkers for selective breeding can lead to potential conflicts between the constant drive to push the boundaries of biological limits and ethical qualms. One may see this in the case of broiler chickens. Apart from the problems arising from their living conditions and the physiological consequences of forcing chickens to grow so quickly, we may identify a certain unease that arises from the idea of engineering life in order to attain the limits of what is possible in terms of productivity. This may lead to concerns that life is being excessively commoditized and reduced to a mere asset whose productivity must be maximized without any regard for what is natural or for the integrity of farm animals.

Inevitably, this leads to a conundrum that can only be solved through a delicate balancing act. Livestock productivity is of paramount importance in order to secure consumer demands, guarantee profitability and guarantee global food security [36]. It is no longer possible to raise animals according to traditional methods that used to require animals to grow for extensive periods of time before being ready for consumption. The demands of productivity must, therefore, be reconciled with animal welfare concerns about the implications of the "race to the limit" in extracting economic value from livestock. One possible solution may be to develop a framework that will allow us to determine how much is too much optimization, making sure to leave some room for farm animals to enjoy some minimal levels of quality of life and something approaching, to the extent possible, a completely natural life cycle.

3.16. Ethical Approaches

The ethical dimension of DLF can be explored by looking at four strategies (Figure 2) in understanding the impact of DLF and the way that it accounts for animal well-being.

Strategy 1: Sorry. If the animal becomes ill, we cannot help it, and we cannot cure it. The only option is for it to be culled. An example is diagnosing a seriously ill animal, such as a pig, as being antibiotic resistant. Most of the time, the animal, or the particular group of pigs that it belongs to, is culled, as it would not be economical for the farmer to treat them and also to avoid the possibility of contagion by other animal units within the farm. "Sorry" may be considered the least ethical strategy, as it involves the culling of animals that are deemed too costly to treat or too contagious to keep alive. While culling may be necessary in some cases, it is important to ensure that it is done in a humane and respectful manner and that alternatives to culling are explored where possible.

This solution involves culling animals that cannot be helped or cured when they become ill. While this approach may be necessary in some cases, it can raise ethical concerns if animals are being culled unnecessarily or if the reasons for culling are based purely on economic considerations rather than animal welfare. One way to address this issue is to improve disease prevention measures through the use of digital tools and monitoring systems, such as real-time monitoring of animal health and behavior. By identifying and addressing potential health issues early on, farmers can reduce the need for culling and improve overall animal welfare.

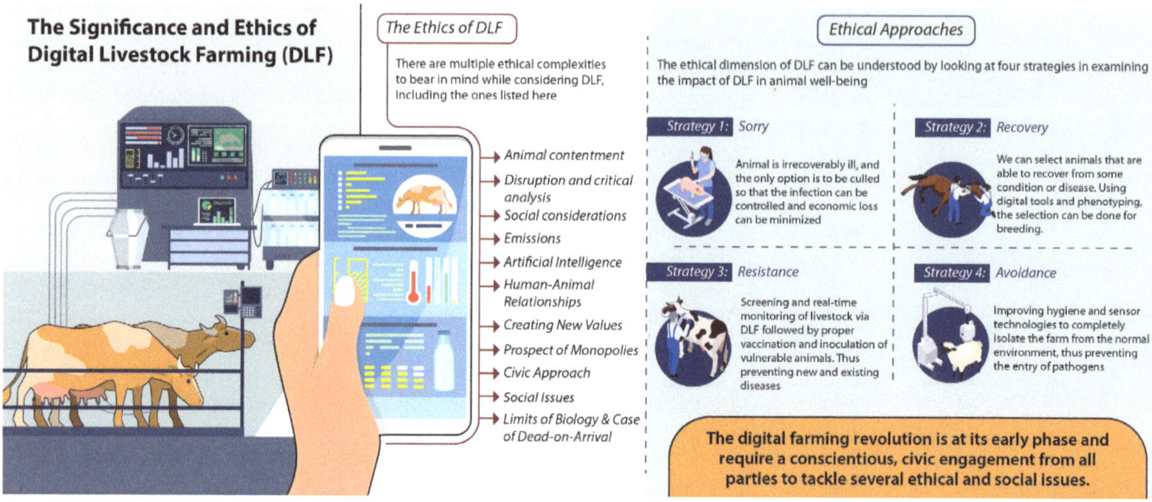

Figure 2. Roadmap to the digital transformation of modern livestock farming while integrating social and ethical awareness. Four strategies for the exploration of the ethical dimensions of digital livestock farming.

Strategy 2: Recovery. Farmers hope that the animal can recover. Here, we can select animals that are able to recover from some condition or disease. "Recovery" is a more optimistic strategy that focuses on breeding animals that are more resilient to disease and better able to recover from illness. This strategy emphasizes the importance of selecting desirable traits and genetics and may involve the use of digital phenotyping to identify animals that are better able to recover from certain conditions.

This solution involves selecting animals that are able to recover from certain conditions or diseases using digital phenotyping and breeding. By selecting desirable traits, such as good heart health or heat resistance, farmers can breed healthier animals and reduce the risk of disease or mortality. However, this approach can raise concerns about the ethical implications of genetic modification and selective breeding. It is important to consider the potential risks and benefits of these practices and ensure that animals are not subjected to unnecessary harm or suffering in the process.

Strategy 3: Resistance. This strategy emphasizes the importance of proactive measures to prevent illness and disease rather than simply reacting to outbreaks as they occur. DLF can be used to screen for potential diseases and to monitor animals in real-time, allowing farmers to intervene early and prevent the spread of disease.

This solution involves using inoculations or other forms of protection to prevent illness in animals, as well as screening and real-time monitoring using digital platforms to identify and prevent diseases. This approach can be effective in reducing disease and mortality rates, but it can also raise ethical concerns about the use of antibiotics and other treatments that may have negative effects on animal health or the environment. It is important to balance the benefits of disease prevention with the potential risks and ensure that animals are not being subjected to unnecessary harm or suffering.

Strategy 4: Avoidance. Here, the farm animal is not concerned with the system. By manipulating the environment, disease-causing pathogens are prevented from entering, thereby protecting the animal. In this way, the animal may be shielded from becoming ill, such as by improving hygiene or by sensor technologies that block certain factors and players in the environment. Rather than making the animal fit for the environment, by using digital tools, we adapt the environment to the needs of the animal. "Avoidance" is a strategy that focuses on adapting the environment to meet the needs of the animal. This strategy emphasizes the importance of providing a healthy and hygienic environment for animals and may involve the use of sensor technologies to monitor and control environmental factors that can affect animal well-being.

This solution involves manipulating the environment to prevent disease-causing pathogens from entering, using digital tools such as sensor technologies and improved hygiene practices. This approach can be effective in reducing disease and mortality rates, but it can also raise ethical concerns about the impact of these interventions on animal behavior and well-being. It is important to ensure that animals are not being subjected to unnecessary stress or discomfort as a result of these interventions and that their natural behaviors and needs are being respected.

The key to addressing ethical problems in digital livestock farming is to balance the need for productivity and profitability with the need to ensure that animals are being treated humanely and that their welfare is being prioritized. By using a combination of these four solutions, and carefully considering the potential risks and benefits of each approach, it is possible to develop more ethical and sustainable practices in digital livestock farming.

4. Technological Change and Innovation

Aspects of DLF having to do with technological change and innovation shall be discussed. There are currently some interesting novel tools and developments happening worldwide under the DLF umbrella. However, we are at the tip of the iceberg, or at best, we have just started scratching the surface. DLF will be here for at least a couple of decades, with many significant developments still lying ahead of us.

4.1. Market Growth

Generally speaking, based on market research analysis, we can say that the precision agriculture market is projected to reach US $52.3 billion by 2030 at an annual growth rate of 10.67% [37]. This is a large number, but it includes both the crop and livestock farming sectors. Focusing on livestock, veterinary animals and farm animals alone, growth would be around US $19.37 billion by 2030 [38]. The whole DLF market can be categorized into multiple segments (animal monitoring technologies, animal biotechnologies, sensors/wearables, data (without the hardware component), etc.). We have to look carefully at what context the technologies are being applied in while considering these numbers. Obviously, North America will be the dominant player in the DLF area, followed by Europe, while the core and seminal R&D work will be led by EU countries.

4.2. Adoption

Based on surveys done by research groups, we can confidently say that farmers are more than willing, and indeed happy, to embrace DLF technologies, as it helps them to cut time in farm management and provides convenience, such as remote management and predictive capabilities [39]. Nonetheless, there are some reservations in farmers' minds. One of the major concerns we hear from them is the narrow profit margins and questions about who "owns" the data collected via these tools [40].

When mobile phones were introduced 30 years ago, they were very heavy, large brick-like objects. People had concerns about their potential to cause job loss, cancer, brain damage, and many other issues. And yet, everyone now lives with a mobile phone for multiple hours a day. Despite this, not much has resulted in terms of disruption regarding health (generally speaking). The common public has adapted, embraced and learned to live with mobile phone technologies. Extrapolating on this, we may perhaps conclude that digital livestock farming technology will not necessarily lead to bad outcomes down the road.

4.3. Changes in Worker Profiles

We live in a new world, a digital world. This is a new situation where it is not enough merely to regard farmers as laborers who may have extensive experience with phenotyping and may immediately detect issues affecting livestock, together with a pertinent solution.

Rather, we must go beyond this paradigm of experience-based competency that may be overly reliant on subjectivity. Digitization can introduce a supplemental element of objectivity in this respect. This will also reduce the work burden on farmers, who will profit from the ability of digital technologies and platforms to assist in decision-making and animal management. If something breaks, the system will try to fix it on its own, but with the intervention and approval of the human expert or the farmer, creating a dual and complementary system. This may also lead to a new professional class of "digital experts" who may fix things akin to how a plumber or electrician may do so when a farmer is not able to resolve an issue on their own.

With respect to more specific developments in DLF, below are a couple of noteworthy developments that are influencing each other:

4.4. Digital Twins

A digital twin can be defined as "a virtual representation of a physical asset enabled through data and simulators for real-time prediction, optimization, monitoring, controlling, and improved decision-making" [41]. In this way, the physical animal may be seen as the source of the data for its virtual twin.

In the past, farming was entirely based on monitoring physical animals, but now we may also use their digital twins as proxies that can be used to monitor them without physical contact or direct observation. For example, this could replace the need for blood samples, visually measuring and observing the animals instead and integrating the sampled data into their virtual twins. This represents a gradual shift of balance, in which the "virtual animal" becomes increasingly important and influences the physical animal, as the former can be used to make predictions about the latter.

4.5. Artificial Intelligence

Artificial intelligence can automate the development of analytical models via algorithms that basically "learn" from data. The algorithms are then able to update the models continuously and instantaneously, paving the path for real-time predictions. This offering of real-time modeling and periodical predictions makes AI powerful. However, the models are restricted to forecasting the effects of data that is similar to that obtained in the past. AI-enabled models tend to produce inaccurate results upon extrapolation beyond observed bounds.

Biased data or missing data in modeling, with or without an AI-based and data-driven method, can provide flawed models while exploring the future based only on the past. The over-dependence on data-driven models for decision-making based on past lessons and past experiences in modern animal farming is risky and makes the model fragile. This can be a major shortcoming during pandemics or other major incidents in animal farming. Data-driven models can indeed be worse when developing simulations or prediction platforms if they are exposed to previously unseen events. However, AI algorithms can overcome this hurdle by enabling the prediction of situations and incidents that have never occurred before by running hypothetical scenarios outside of conventional bounds. This can be done via enhanced theory and causal hypotheses about the farming system by taking a systematic approach, thoroughly understanding the processes involved, and implementing high-fidelity methodologies.

Generative adversarial networks and self-learning approaches come in handy to enable accurate predictions about various states/phases of livestock farming activity, which can then match the predicted output by simulated trajectories. Even a huge volume of "outside-the-known-parameter" data can be generated via simulation to train the model to such an extent that it becomes able to provide critical insights regarding unknown future events [10].

To be able to create sensible predictions, AI-enabled simulations have to precisely reflect the processes involved in the systems to which they are being applied. In terms of hybrid modeling (consider a gray box model), mixing various modeling paradigms is possible via AI approaches, which allows the animal scientist to create a "bouquet" (combining multiple flowers from various plants) or a "cocktail" into a powerful platform.

Agent-based simulation models can account for deviations from optimization, environmental exploitation, or lapses of rationality. The AI models can then predict how the system can reconfigure itself via "adaptation" in the context of animal farming. This specific capability of AI will allow animal farming to stay ahead of human responses and interventions. Methods and technologies to build very powerful simulations are available already. The trend now is not to rely merely on collecting data from the real world but to venture into a "metaverse" or "virtual world" and create and develop various "what-if scenarios", converting the virtual digital world into trillions of terabytes of data at a meager cost.

Resource optimization can be taken to the maximum level via digital/smart farming approaches. There is a remote possibility of "desire discrepancy" due to digitalization, which may influence the next generation of consumers to refrain from eating artificially "selected" meat in the future. For example, it is known that consumers are often reluctant to adopt new technology-based food, in great measure, on how the technology is presented [42]. Due to an evolving consciousness caused by digitalization among the Gen Z and Gen Alpha cohorts, the emphasis on the transparency of meat production would matter in making choices about food. However, these are issues that seem eminently addressable by finding approaches to engage productively with consumer attitudes.

5. Responsible Research and Innovation

Responsible Research and Innovation (RRI) is an approach to innovation that seeks to integrate ethical, social, and environmental considerations [43] into the research and innovation process. It aims to ensure that the development and use of new technologies align with the values and needs of society.

Digital livestock farming is an innovative way of managing livestock that uses digital technologies such as big data, sensors, and the Internet of Things (IoT) to improve efficiency and productivity. The use of sensors and big data enables farmers to collect and analyze data on various aspects of animal health, behavior, and performance, helping them to make informed decisions about the management of their animals.

Farmers' perceptions play a crucial role [44] in the adoption of innovation in agriculture. Policymakers and researchers need to understand farmers' views and concerns in order to develop policies and technologies that are acceptable and useful to farmers. This requires engaging with farmers and involving them in the innovation process. By understanding farmers' views and involving them in the innovation process, we can ensure that the benefits of these technologies are shared equitably and that the livestock industry becomes more sustainable and resilient.

In a digitally focused era, RRI places emphasis on engaging stakeholders such as farmers, animal caretakers, researchers, veterinarians, animal scientists and policymakers in the design and implementation of research activities in a bid to ensure that the Digital Livestock Farming technologies are ethical, safe, and beneficial to society.

5.1. Ethics, Law, and Governance

From an ethical standpoint, digital livestock farming raises issues of animal rights and welfare, as well as data privacy and security. From a legal perspective, the use of digital tools for livestock management is subject to existing legislation, as well as new laws such as GDPR. Finally, from a governance perspective, there is a need to ensure that digital livestock farming is regulated in a way that is fair and effective.

5.2. Digital & Precision Livestock Big Data

One of the key aspects of digital livestock farming is the generation of large amounts of data. This data can be used to gain insights into livestock behavior, health, and performance. It can also be used to develop digital & precision livestock farming systems, which employ data-driven decision-making to optimize animal husbandry and production. Big data refers to the massive amounts of data generated by various sources, including sensors and IoT devices. The ethical dimensions of digital livestock farming and the use of big data include privacy concerns, data ownership, and the potential for unequal access to these technologies. It is important to ensure that the benefits of these technologies are shared equitably and that farmers have control over the data they generate. The use of big data in digital & precision livestock farming raises important ethical and legal considerations, namely, the protection of data privacy, data security, and transparency. Furthermore, there is a need to ensure that data-driven decision-making is fair and equitable and that any decisions made with the use of big data do not discriminate against certain groups.

In order for digital livestock farming to revolutionize the way farmers raise and manage animals and for this to happen responsibly and ethically, it is important to consider the ethical, legal and governance challenges associated with this type of farming. The use of big data in precision livestock farming must be carefully regulated.

5.3. *Micro-Innovations*

Micro-innovations refer to small improvements or adaptations [45] made to existing technologies or the development of new ones. In the context of digital livestock farming, these innovations can include the use of sensors, big data, and artificial intelligence (AI) to optimize the management of livestock and improve the sustainability of animal agriculture. The ethical and social implications of digital livestock farming are significant and need to be considered in order to ensure the responsible and sustainable development of these technologies. For example, privacy concerns and data ownership issues need to be addressed to ensure that farmers retain control over the data they generate and that their rights and interests are protected.

Responsible technological innovation is crucial in ensuring that the benefits of digital livestock farming are realized while avoiding any unintended consequences that might negatively impact the resilience [46] of the livestock industry. The role of technology as a mediator of resilience-enhancing social behavior should not be underestimated, as it can help animal farmers to adopt more sustainable practices and overcome uncertainties and challenges they face.

Micro-innovations in digital livestock farming have the potential to improve the efficiency and sustainability of animal agriculture, but it is important to approach these innovations responsibly and with consideration for the ethical and social implications involved. By taking a responsible and holistic approach to technological innovation, we can ensure that the benefits of these technologies are shared equitably and that the livestock industry becomes more resilient and sustainable.

6. Summary

The ethics of digital livestock farming is an emerging field that focuses on the ethical implications of using digital technologies in animal agriculture. As the use of digital technologies in agriculture continues to grow, there are several key areas where future research in the ethics of digital livestock farming is likely to focus:

Animal welfare: One of the most important ethical considerations in digital livestock farming is the impact of digital technologies on animal welfare. Future research is likely to focus on developing and implementing digital technologies that can improve animal welfare, such as sensors and other monitoring devices that can detect signs of distress or illness in animals and trigger appropriate interventions.

Data privacy and ownership: Another key ethical concern in digital livestock farming is the collection, storage, and use of data generated by digital technologies. Future research will need to address issues of data privacy and ownership to ensure that farmers and other stakeholders have control over the data generated by these technologies and that it is used in an ethical and responsible manner.

Environmental sustainability: Digital livestock farming has the potential to improve the environmental sustainability of animal agriculture by reducing waste, improving efficiency, and decreasing the environmental impact of livestock production. Future research will need to explore the environmental benefits and costs of different digital technologies and their impact on the overall sustainability of animal agriculture.

Labor and social justice: Digital livestock farming is likely to have a significant impact on the labor market in agriculture, and there are concerns about the impact of these technologies on farmworkers and their communities. Future research will need to address issues of labor and social justice in the context of digital livestock farming, including the potential for automation to displace human workers and the need to ensure fair labor practices and working conditions.

Ethical governance: Finally, future research in the ethics of digital livestock farming will need to address issues of ethical governance, including the development of standards and guidelines for the responsible use of digital technologies in animal agriculture and the establishment of regulatory frameworks that can ensure that these technologies are used in an ethical and responsible manner.

To achieve a more ethical approach to DLF, there needs to be a shift in the goals and values of livestock farming, moving beyond profit as the sole objective. The use of DLF can aid in this ethical pursuit by providing tools and techniques to improve animal welfare, reduce environmental impacts, and promote sustainable practices.

Research on the ethical implications of DLF is a growing field, and many organizations and researchers are actively engaged in studying the impact of DLF on animal welfare, environmental sustainability, and ethical considerations. For example, organizations such as the World Wildlife Fund, the Humane Society, and Compassion in World Farming are actively involved in researching and advocating for ethical practices in livestock farming, including the use of DLF.

Furthermore, academic research in fields such as animal science, agricultural engineering, and computer science is also exploring the potential benefits and ethical implications of DLF. For instance, researchers are exploring the use of digital biomarkers to monitor animal health and welfare, the use of precision feeding systems to reduce feed waste and improve nutrient utilization, and the use of predictive analytics to improve herd management and reduce environmental impacts.

Achieving ethical DLF requires a multidisciplinary approach involving collaboration between stakeholders such as farmers, researchers, industry organizations, and animal welfare advocates. It also requires a commitment to ongoing research and evaluation of the impact of DLF on animal welfare, environmental sustainability, and ethical considerations to ensure that technological advancements are used in a responsible and sustainable manner.

The digital farming revolution is just beginning, and it calls for a conscientious civic engagement of all the parties involved, including farmers, scientists, ethicists, and consumers. In this way, DLF is a prime example of the complex interconnections that make us all co-actors in the construction of a common future.

Funding: This research received no external funding.

Institutional Review Board Statement: Not applicable.

Informed Consent Statement: Not applicable.

Data Availability Statement: Not applicable.

Acknowledgments: The author thanks Bart Gremmen, Ethics in Life Sciences of the Wageningen University & Research, for his valuable input and for sharing critical insights in the development of this article.

Conflicts of Interest: The author declares no conflict of interest.

References

1. Neethirajan, S. Is Seeing Still Believing? Leveraging Deepfake Technology for Livestock Farming. *Front. Vet. Sci.* **2021**, *8*, 740253. [CrossRef] [PubMed]
2. Wathes, C.M.; Kristensen, H.H.; Aerts, J.-M.; Berckmans, D. Is precision livestock farming an engineer's daydream or nightmare, an animal's friend or foe, and a farmer's panacea or pitfall? *Comput. Electron. Agric.* **2008**, *64*, 2–10. [CrossRef]
3. Banhazi, T.; Lehr, H.; Black, J.; Crabtree, H.; Schofield, P.; Tscharke, M.; Berckmans, D. Precision Livestock Farming: An international review of scientific and commercial aspects. *Int. J. Agric. Biol. Eng.* **2012**, *5*, 1–9.
4. Berckmans, D. Precision livestock farming technologies for welfare management in intensive livestock systems. *Rev. Sci. Tech. L'oie* **2014**, *33*, 189–196. [CrossRef] [PubMed]
5. Berckmans, D. General introduction to precision livestock farming. *Anim. Front.* **2017**, *7*, 6–11. [CrossRef]
6. Aquilani, C.; Confessore, A.; Bozzi, R.; Sirtori, F.; Pugliese, C. Review: Precision Livestock Farming technologies in pasture-based livestock systems. *Animal* **2021**, *16*, 100429. [CrossRef] [PubMed]
7. Shepherd, M.; Turner, J.A.; Small, B.; Wheeler, D. Priorities for science to overcome hurdles thwarting the full promise of the 'digital agriculture' revolution. *J. Sci. Food Agric.* **2020**, *100*, 5083–5092. [CrossRef]
8. Basso, B.; Antle, J. Digital agriculture to design sustainable agricultural systems. *Nat. Sustain.* **2020**, *3*, 254–256. [CrossRef]
9. Ozdogan, B.; Gacar, A.; Aktas, H. Digital Agriculture Practices in The Context of Agriculture 4.0. *J. Econ. Financ. Account.* **2017**, *4*, 186–193. [CrossRef]
10. Morota, G.; Ventura, R.V.; Silva, F.F.; Koyama, M.; Fernando, S.C. Big data analytics and precision animal agriculture symposium: Machine learning and data mining advance predictive big data analysis in precision animal agriculture. *J. Anim. Sci.* **2018**, *96*, 1540–1550. [CrossRef]
11. Ivanov, S.; Bhargava, K.; Donnelly, W. Precision Farming: Sensor Analytics. *IEEE Intell. Syst.* **2015**, *30*, 76–80. [CrossRef]
12. Koltes, J.E.; Cole, J.B.; Clemmens, R.; Dilger, R.N.; Kramer, L.M.; Lunney, J.K.; McCue, M.E.; McKay, S.D.; Mateescu, R.G.; Murdoch, B.M.; et al. A Vision for Development and Utilization of High-Throughput Phenotyping and Big Data Analytics in Livestock. *Front. Genet.* **2019**, *10*, 1197. [CrossRef] [PubMed]
13. Wolfert, S.; Ge, L.; Verdouw, C.; Bogaardt, M.-J. Big Data in Smart Farming—A Review. *Agric. Syst.* **2017**, *153*, 69–80. [CrossRef]
14. Van der Burg, S.; Bogaardt, M.-J.; Sjaak, W. Ethics of smart farming: Current questions and directions for responsible innovation towards the future. *NJAS—Wagening. J. Life Sci.* **2019**, *90–91*, 100289. [CrossRef]
15. Llonch, P.; Mainau, E.; Ipharraguerre, I.R.; Bargo, F.; Tedó, G.; Blanch, M.; Manteca, X. Chicken or the Egg: The Reciprocal Association Between Feeding Behavior and Animal Welfare and Their Impact on Productivity in Dairy Cows. *Front. Veter- Sci.* **2018**, *5*, 305. [CrossRef]
16. Schillings, J.; Bennett, R.; Rose, D.C. Exploring the Potential of Precision Livestock Farming Technologies to Help Address Farm Animal Welfare. *Front. Anim. Sci.* **2021**, *2*, 1–17. [CrossRef]
17. Yadav, S.; Kaushik, A.; Sharma, M.; Sharma, S. Disruptive Technologies in Smart Farming: An Expanded View with Sentiment Analysis. *Agriengineering* **2022**, *4*, 424–460. [CrossRef]
18. Rotz, S.; Duncan, E.; Small, M.; Botschner, J.; Dara, R.; Mosby, I.; Reed, M.; Fraser, E.D.G. The Politics of Digital Agricultural Technologies: A Preliminary Review. *Sociol. Rural.* **2019**, *59*, 203–229. [CrossRef]
19. Rotz, S.; Gravely, E.; Mosby, I.; Duncan, E.; Finnis, E.; Horgan, M.; LeBlanc, J.; Martin, R.; Neufeld, H.T.; Nixon, A.; et al. Automated pastures and the digital divide: How agricultural technologies are shaping labour and rural communities. *J. Rural. Stud.* **2019**, *68*, 112–122. [CrossRef]
20. Ignatova, E. Professional Standard as an Incentive for the Development of Polytechnic Education. *Adv. Econ. Bus. Manag. Res.* **2020**, *128*, 1833–1837. [CrossRef]
21. Hackfort, S. Patterns of Inequalities in Digital Agriculture: A Systematic Literature Review. *Sustainability* **2021**, *13*, 12345. [CrossRef]
22. Muñoz-Tamayo, R.; Agudelo, J.F.R.; Dewhurst, R.J.; Miller, G.; Vernon, T.; Kettle, H. A parsimonious software sensor for estimating the individual dynamic pattern of methane emissions from cattle. *Animal* **2019**, *13*, 1180–1187. [CrossRef] [PubMed]
23. Ma, S.; Li, J.; Wei, W. The carbon emission reduction effect of digital agriculture in China. *Environ. Sci. Pollut. Res.* **2022**, *29*, 1–18. [CrossRef]
24. Wischmeyer, T. Artificial Intelligence and Transparency: Opening the Black Box. *Regul. Artif. Intell.* **2019**, 75–101. [CrossRef]
25. Du, X.; Lao, F.; Teng, G. A Sound Source Localisation Analytical Method for Monitoring the Abnormal Night Vocalisations of Poultry. *Sensors* **2018**, *18*, 2906. [CrossRef]
26. Du, X.; Carpentier, L.; Teng, G.; Liu, M.; Wang, C.; Norton, T. Assessment of Laying Hens' Thermal Comfort Using Sound Technology. *Sensors* **2020**, *20*, 473. [CrossRef] [PubMed]
27. Friel, M.; Kunc, H.P.; Griffin, K.; Asher, L.; Collins, L.M. Positive and negative contexts predict duration of pig vocalisations. *Sci. Rep.* **2019**, *9*, 1–7. [CrossRef]
28. Mota-Rojas, D.; Broom, D.M.; Orihuela, A.; Velarde, A.; Napolitano, F.; Alonso-Spilsbury, M. Effects of human-animal relationship on animal productivity and welfare. *J. Anim. Behav. Biometeorol.* **2020**, *8*, 196–205. [CrossRef]
29. Wang, Y.-N.; Jin, L.; Mao, H. Farmer Cooperatives' Intention to Adopt Agricultural Information Technology—Mediating Effects of Attitude. *Inf. Syst. Front.* **2019**, *21*, 565–580. [CrossRef]

30. Syukta, M. Big Data in Agriculture: Property Rights, Privacy and Competition in Ag Data Services. *Int. Food Agribus. Manag. Rev.* **2016**, *19*, 57–74.
31. Keeling, L.; De Oliveira, D.; Rustas, B.-O. *Precision Dairy Farming Use of Mechanical Rotating Brushes in Dairy Cows—A Potential Proxy for Performance and Welfare?* Wageningen Academic Publishers: Leeuwarden, The Netherlands, 2016; pp. 343–347.
32. Neethirajan, S.; Kemp, B. Digital Phenotyping in Livestock Farming. *Animals* **2021**, *11*, 2009. [CrossRef] [PubMed]
33. Averós, X.; Knowles, T.G.; Brown, S.N.; Warriss, P.D.; Gosálvez, L.F. Factors affecting the mortality of weaned piglets during commercial transport between farms. *Vet. Rec.* **2010**, *167*, 815–819. [CrossRef] [PubMed]
34. Bernabucci, U. Climate change: Impact on livestock and how can we adapt. *Anim. Front.* **2019**, *9*, 3–5. [CrossRef] [PubMed]
35. Liu, L.; Li, B.; Zhao, R.; Yao, W.; Shen, M.; Yang, J. A Novel Method for Broiler Abnormal Sound Detection Using WMFCC and HMM. *J. Sens.* **2020**, *2020*, 2985478. [CrossRef]
36. Baldi, A.; Gottardo, D. Livestock production to feed the planet. Animal Protein: A Forecast of Global Demand Over the Next Years. *Relations* **2017**, *5*, 65–71. [CrossRef]
37. Precedence Research. *Digital Agriculture Market, s.l.: Precedence Research*; Precedence: Ottawa, ON, Canada, 2020.
38. Grand View Research. *Livestock Monitoring Market Size Worth $19.37 Billion by 2030, s.l.: Grand View Research*; Grand View Research: San Francisco, CA, USA, 2020.
39. Blasch, J.; van der Kroon, B.; van Beukering, P.; Munster, R.; Fabiani, S.; Nino, P.; Vanino, S. Farmer preferences for adopting precision farming technologies: A case study from Italy. *Eur. Rev. Agric. Econ.* **2020**, *49*, 33–81. [CrossRef]
40. Papst, F.; Saukh, O.; Römer, K.; Grandl, F.; Jakovljevic, I.; Steininger, F.; Mayerhofer, M.; Duda, J.; Egger-Danner, C. Embracing Opportunities of Livestock Big Data Integration with Privacy Constraints. In Proceedings of the 9th International Conference on the Internet of Things, Bilbao, Spain, 22–25 October 2019; Volume 27, pp. 1–4. [CrossRef]
41. Rasheed, A.; San, O.; Kvamsdal, A.T. *Digital Twin: Values, Challenges and Enablers*; IEEE Access: New York City, NY, USA, 2020; Volume 8, pp. 21980–22012. [CrossRef]
42. Siegrist, M.; Hartmann, C. Consumer acceptance of novel food technologies. *Nat. Food* **2020**, *1*, 343–350. [CrossRef]
43. Burget, M.; Bardone, E.; Pedaste, M. Definitions and conceptual dimensions of responsible research and innovation: A literature review. *Sci. Eng. Ethics* **2017**, *23*, 1–19. [CrossRef]
44. Vecchio, Y.; Di Pasquale, J.; Del Giudice, T.; Pauselli, G.; Masi, M.; Adinolfi, F. Precision farming: What do Italian farmers really think? An application of the Q methodology. *Agric. Syst.* **2022**, *201*, 103466. [CrossRef]
45. Charatsari, C.; Lioutas, E.D.; De Rosa, M.; Vecchio, Y. Technological Innovation and Agrifood Systems Resilience: The Potential and Perils of Three Different Strategies. *Front. Sustain. Food Syst.* **2022**, *6*, 872706. [CrossRef]
46. Owen, R.; Macnaghten, P.; Stilgoe, J. *Responsible Research and Innovation: From Science in Society to Science for Society, with Society*; Routledge Taylor and Francis Group: New York, NY, USA, 2017; pp. 117–126. [CrossRef]

Disclaimer/Publisher's Note: The statements, opinions and data contained in all publications are solely those of the individual author(s) and contributor(s) and not of MDPI and/or the editor(s). MDPI and/or the editor(s) disclaim responsibility for any injury to people or property resulting from any ideas, methods, instructions or products referred to in the content.

sustainability

Article

Determining the Influence of a Magnetic Field on the Vibration and Fuel Consumption of a Heavy Diesel Engine

Yousef Darvishi [1], Seyed Reza Hassan-Beygi [2,*], Jafar Massah [2], Marek Gancarz [3,4,*], Arkadiusz Bieszczad [3] and Hamed Karami [5]

1 Department of Biosystems Engineering, University of Tehran, Tehran 3391653755, Iran
2 Department Agro-Technology, College of Abouraihan, University of Tehran, Tehran 3391653755, Iran
3 Faculty of Production and Power Engineering, University of Agriculture in Kraków, Balicka 116 B, 30-149 Kraków, Poland
4 Institute of Agrophysics Polish Academy of Sciences, Doświadczalna 4, 20-290 Lublin, Poland
5 Department of Petroleum Engineering, College of Engineering, Knowledge University, Erbil 44001, Iraq
* Correspondence: rhbeigi@ut.ac.ir (S.R.H.-B.); m.gancarz@urk.edu.pl (M.G.)

Abstract: Most of the fuels used in internal combustion engines are liquid fuels. The magnetic behavior of fuel leads to a change in the interaction of hydrocarbon and oxygen molecules. This study aimed to evaluate the fuel consumption and engine vibration (time domain) of the Perkins A63544 diesel engine using magnetized fuel. The vibration of an internal combustion engine can cause failure in engine components and discomfort and injury to users. Engine vibration behavior changes due to changes in fuel types and engine combustion. Therefore, in this study, the vibration behavior of the tractor engine (Perkins model, four-stroke, direct injection diesel) was evaluated in stationary mode at different engine speeds due to changes in fuel types. Three accelerometers (CTC AC102 model) were used to measure the vibration acceleration. The fuels used included diesel as a normal control and fuels that had been subjected to magnetic field intensities of 1000, 2000, 3000, and 4000 gauss. The longitudinal, vertical, and lateral vibration signals with 5 levels of engine speed were measured. The results illustrated that the vibration root mean square (RMS) values were essentially ($p < 0.01$) affected by the engine speed, fuel type, and their interactions. It was found that for the 4000-gauss magnetized fuel, the average vibration acceleration using the five velocity settings reduced by 15%, 15.30%, 12.40%, 12.35%, and 15.38% compared to the respective control fuels. The results showed that engine fuel consumption and specific fuel consumption decreased by 2.3% using the 4000-gauss magnetized fuel compared with the normal control fuel.

Keywords: vibration; specific fuel consumption; magnetized fuel; RMS

Citation: Darvishi, Y.; Hassan-Beygi, S.R.; Massah, J.; Gancarz, M.; Bieszczad, A.; Karami, H. Determining the Influence of a Magnetic Field on the Vibration and Fuel Consumption of a Heavy Diesel Engine. *Sustainability* **2023**, *15*, 4088. https://doi.org/10.3390/su15054088

Academic Editors: Muhammad Sultan, Yuguang Zhou, Redmond R. Shamshiri and Muhammad Imran

Received: 27 December 2022
Revised: 12 February 2023
Accepted: 16 February 2023
Published: 23 February 2023

Copyright: © 2023 by the authors. Licensee MDPI, Basel, Switzerland. This article is an open access article distributed under the terms and conditions of the Creative Commons Attribution (CC BY) license (https://creativecommons.org/licenses/by/4.0/).

1. Introduction

In recent years, many approaches have been proposed to diminish fuel consumption and emissions from internal combustion engines. Emissions such as NOx and CO produced by burning fossil fuels are one of the main causes of pollution in the atmosphere. Accumulation of CO_2 and other greenhouse gases in the atmosphere is causing changes in the Earth's climate and seems to have dire consequences for humans and other organisms [1–8]. In another survey concerning the magnetic field, a two-stroke engine was utilized in the fuel line; fuel consumption diminished by about 9–14%, and CO, as well as HC pollutants, dropped by 30 and 40%, respectively [9,10]. Generally, fuels used in internal combustion engines are a combination of several molecules. These molecules are made of atoms (electrons, protons, and neutrons). Positive and negative charges in these fuels can be observed in magnetic movements. By applying a strong magnetic field, the internal levels of molecules are changed. However, because these molecules have not yet been released, they cannot react with oxygen molecules actively. Fuel molecules or chains of hydrocarbons must be ionized to be released. Ionization phenomena and the release of molecules are achieved by applying a magnetic field [11]. Using a magnetic field, the internal energy of fuel can be increased, which means some alteration in the molecular surface of fuel leads to a rise in the internal energy of the fuel, its molecular segregation, and more reactivity, thereby improving combustion [12]. In a research study, petroleum, diesel, and gasoline fuels were situated in magnetic fields of varying intensity and duration. The results verified a 10% reduction in gasoline viscosity and a 4% fall in diesel viscosity [13]. In another study, the mechanical characteristics of the rotating system of two double disks were examined. The Hall current effect, Arrhenius activation energy, and chemical reaction entropy generation were investigated. The results showed that the time-independent three-dimensional flow of Maxwell nanofluid was induced using two parallel rotating disks because of the Hall current effect [14]. Moreover, the entropy generation in unsteady magneto-bioconvection Casson nanofluid flow in the presence of Cattaneo–Christov heat and mass flux theory was investigated. The Buongiorno nanofluid model was used to consider the influences of Brownian motion, viscous dissipation, and thermophoresis [15]. Torsional vibrations (acceleration) in engines exhibit a phenomenon that is not typical of vibrations. These effects are rooted in the geometry of the reciprocating mechanism and emerge through dynamic analysis. Apart from the fact that the system changes by crankshaft rotation, the outcomes show that torsional vibration is influenced by friction between the piston and the cylinder [16]. In another research study, the vibration of the two-wheel tractor was investigated during transportation. It was shown that when the rotational speed of the engine increases, the RMS average acceleration of vibration generally rises in all situations and gears, and the highest increase in RMS acceleration value is in light gears [17]. The present research aimed to look at the impact of magnetized fuel on the vibration function and fuel consumption of diesel engines of MF-399 tractors. In this survey, the vibration of the MF-399 tractor engine at five levels of engine rotational speed under load was measured using magnetized fuel, and the values of acceleration RMS and vibration signals were recorded.

2. Materials and Methods

2.1. Diesel Engine

In this survey, an MF399 tractor engine made by Tractorsazi Co., Tabriz, Iran, was utilized. Table 1 shows the technical characteristics of the engine during all experiments. Because the tractor was used in a stationary state, the tractor engine was the main source of vibrations. The moving parts of an engine are the crankshaft, connecting rods, and pistons. Vibration in mutual motors is caused by changes in the gas weight interior of the cylinder and substituting inertia strengths focused on diverse motor parts.

Table 1. Technical details of an MF 399 tractor engine.

Model Type	Perkins A63544
Builder	Iran tractor co.
Combustion type	Direct injection
Cylinders	6
Cylinder stroke	128 mm
Bore of cylinders	98.5 mm
Volume of cylinders	5.8 liter
Topmost power	2300 rpm, 110 hp
Topmost torque	1300 rpm, 375 N m

2.2. The Magnetizer Machine

To magnetize diesel fuel, an alternative magnetic field manufacturer with a maximum capacity of 6000 gauss was made by the agricultural and technical group of Abouraihan-Tehran University. As can be seen in Figure 1, this device contains the following main parts:

Figure 1. Magnetic fuel conditioner device. (1) Fuel tank. (2) The route of fuel rotation. (3) Power supply. (4) Cooling system.

2.3. *The Vibration-Measuring Device*

In this research, the vibration of the MF-399 tractor engine was estimated both in stationary mode and under full load on an asphalt road in an open space. The research was conducted in the renewable energy laboratory of Tarbiat-Modares University (TMU). The experiments were carried out at five different intensities of the magnetic field (0 to 4000 gauss) and at five different levels of engine speeds. The measurement of the vibration was a 5 (field intensity) × 5 (engine speeds) experiment on a randomized design and five replications. The total number of experiments is listed as:

$$\text{The number of experiments} = (5)\ \text{fuel} \times (5)\ \text{engine speeds} \times (5)\ \text{iterations}$$

Because stationary conditions can be utilized as the basis of measurement to compare vibration in another condition, the experiments in this study included stationary mode. Examination of variance was conducted using the Multi-Factorial Adjusted Show in the SAS 9.3 Measurable Computer program (SAS Institute, Cary, NC, USA). The results of these tests provide recognizable proof of the contrasts in vibrations in response to the fuel and engine speeds. Extra investigations were carried out to independently determine the interaction between the fuel and engine speeds for each treatment. The RMS of speeding up was factually analyzed employing an adjusted two-factorial examination of variance (analysis of variance, ANOVA). In each experiment, the number of five-time domain signals (repeat) with duration of 4 s was recorded with a sampling rate of 120,000.

2.4. *The Accelerometers*

The acceleration sensor type CTC AC102 was employed to record acceleration signals, and its specifications are presented in Table 2. The output data of the acceleration sensor were converted to digital data using an electronic analog-to-digital converter, and then the digital signals were recorded on a hard drive. The acceleration sensors and vibration measurement setup are illustrated in Figure 2.

Table 2. Specifications of Model AC102 accelerometer manufactured by CTC (USA).

Characteristic	Range
Sensitivity	100 mV/g
Frequency response	0.4 to 14000 Hz
Dynamic range	±50 g, at peak
Supply voltage	17–31 VDC
Temperature range	−50–120 °C
Natural resonant frequency	26,000 Hz

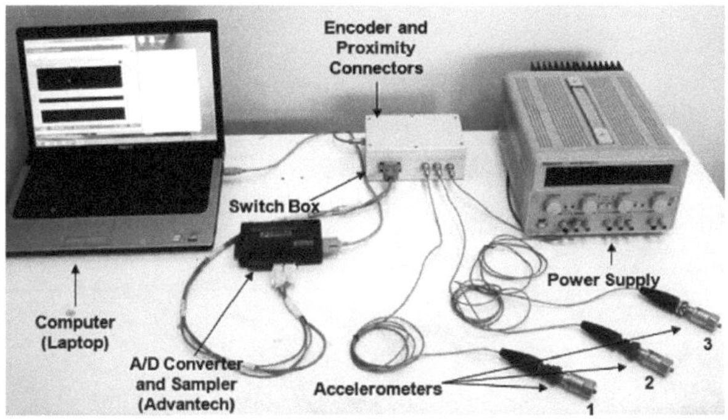

Figure 2. The setup used for the engine vibration measurement.

2.5. Specific Fuel Consumption

Specific fuel consumption is characterized as the mass of the fuel expended to create 1 kWh of genuine engine work. The specific fuel specific fuel consumption may be a comparative variable that shows how an engine changes fuel vitality to real work. In this regard, fuel consumption is accessible in terms of ml/min. Engine-specific fuel consumption was calculated from Equation (1) [18]:

$$\text{SFC} = \frac{60\rho \times \text{FC}}{P} \tag{1}$$

All the tests were repeated three times, and the mean values were reported. The information was measurably analyzed utilizing the two factors completely randomized design in order to study the impacts of engine speeds and fuel types on debilitated outflows and fuel consumption. Furthermore, Duncan's multiple range tests were used to compare the means. It should be noted that the engine was warmed up for about 10 min before measuring the vibration signals for the control sample (0 gauss). For the other experimental runs, the engine was fueled by the magnetized samples for about 10 min before measuring the vibration signals. As a result, for each sample, the engine was stable and the new sample fuel was placed in the fuel circuit.

3. Results and Discussion

3.1. Fuel Specifications

The specifications of pure diesel fuel (0 gauss refers to the condition before magnetic treatment was applied) and diesel fuel magnetized within the permissible range of each item measured in the biodiesel laboratory at TMU are shown in Table 3. As shown, the values obtained for each fuel, including the density, viscosity, and flash point of all fuels, were in the standard range. The 4000-gauss magnetic field intensity provides the lowest value of specific gravity, kinematic viscosity, and flashpoint. With increasing magnetic field intensity, the fuel density reduced, so the lowest density was for the magnetic field of 4000 G. The hypothesis that a magnetic field diminishes the viscosity of many fluids has been reported in the literature [19]. Moreover, a recent report indicated that a magnetic field causes changes in oil viscosity. The results of experiments have shown that the significant reduction in the viscosity of paraffin and crude oil is caused by a combination of long chains [20]. Due to the greater surface area of the combustion chamber, smaller droplets evaporate faster, which causes faster and more complete combustion. In addition, in injection systems (gasoline and diesel), finer fuel droplets increase the combustion speed, and consequently, combustion is more complete [21].

Table 3. Fuel details.

Permissible Range	Accuracy	4000 G	3000 G	2000 G	1000 G	0 G	Fuel Specifications
820–850	±1 kg/m^3	813.5	815	816	824	826.1	Specific gravity (kg/m^3)
1.9–6	±0.1 cST	2.43	2.44	2.45	2.51	2.54	Kinematic viscosity at 40 °C (CST)
Max 130	±1 °C	60.1	62	62.1	64	65.4	Flashpoint (°C)

3.2. Time-Domain Signals

Figure 3 depicts a section of the time domain signal in three directions—lateral, longitudinal, and vertical—at a speed of 1700 rpm and diesel fuel (0 gauss) for the engine under full load. The combustion cycles are shown in this figure. As can be observed, the vibration acceleration in the lateral direction exceeds that in the vertical direction, and the vertical vibration acceleration surpasses that in the longitudinal direction. Therefore, the time domain acceleration signals significantly depend on the direction of measurement. As it is clear the maximum and minimum values of the acceleration signals of vibration recorded at the lateral and longitudinal directions are attributed to the six-cylinder engine being well balanced in the vertical direction than the two other directions. Moreover, the experimental results showed that the time domain acceleration signal values of vibration decrease by applying the magnetic field intensity. Therefore, the value of the time domain acceleration signals using 1000 Gauss field intensity is significantly less than the value of time domain acceleration signals related to fuel without a magnetic field. The reason why acceleration is higher in the lateral direction is the good dynamic balance of the six-cylinder engine, which causes the vibration to reduce on the vertical side than on the lateral one, as well as the blast force produced by the engine in the lateral direction [22].

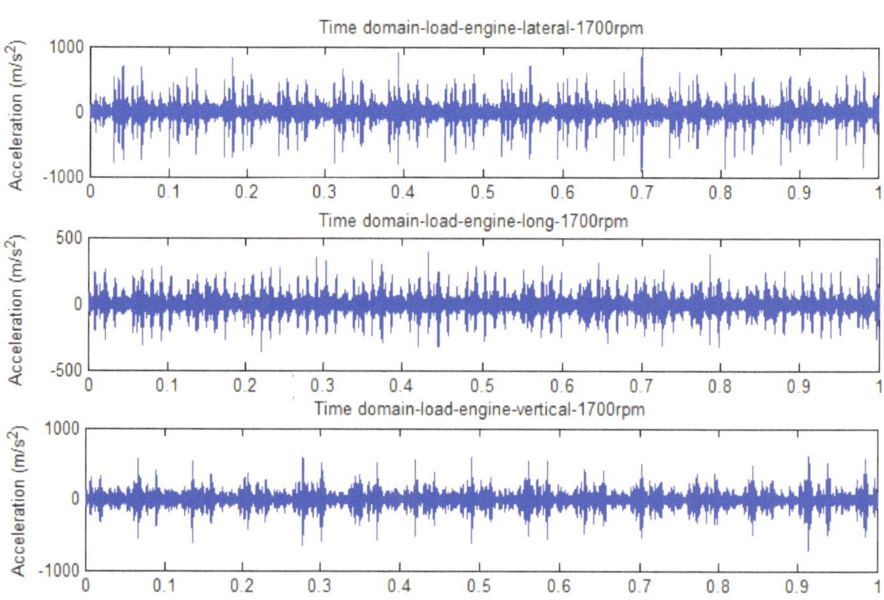

(a)

Figure 3. *Cont.*

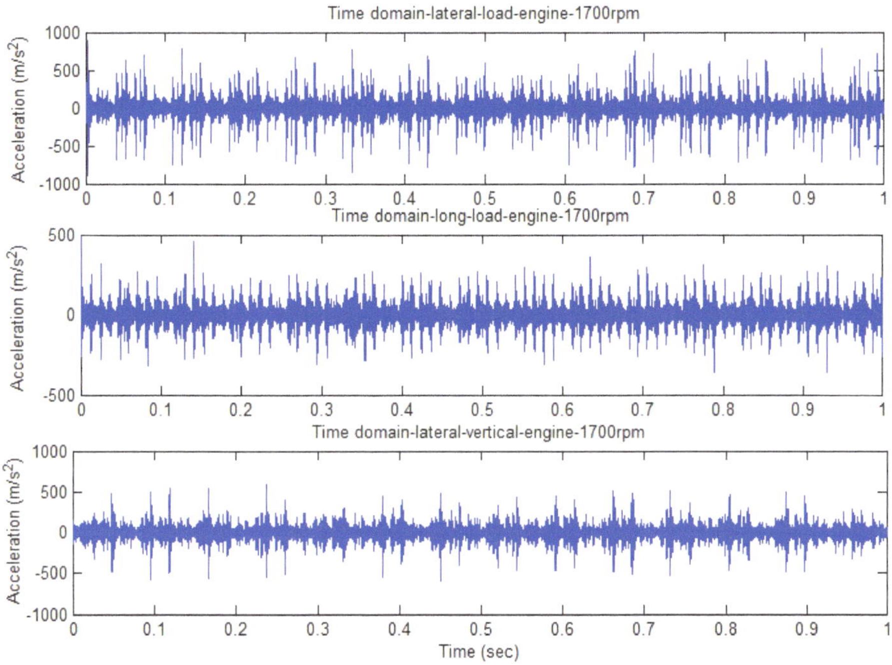

(**b**)

Figure 3. Vibration acceleration signals in the time domain at 1700 rpm engine speed in the lateral, longitudinal, and vertical directions. (**a**) Without a magnetic field. (**b**) Diesel fuel under a 1000-gauss field.

3.3. Evaluation of RMS of Vibration Acceleration Signals

For evaluation of the engine's vibration at different speeds and field intensities applied to fuel, the acceleration RMS was used for each direction of measurement. Five replications against each treatment were considered. Acceleration RMS was calculated using Equation (2):

$$a\text{RMS} = \sqrt{\frac{1}{N}\sum_{k=1}^{N} ak^2} \tag{2}$$

where a_{RMS} is the root mean square value of the acceleration signal, (m/s^2), a_k is the acceleration amplitude at time of k within the time domain signal, and N, is the entire range of samples (N = 120,000) for 4 s [22].

3.4. Effects of Variable Levels on Vibration

A comparison between the test results for the interaction effects of the direction of measurement and fuel type on RMS is given in Table 4. Duncan's multiple range test was used to compare treatment means. Means were significantly different when $p < 0.01$. For the engine under full load, the RMS significantly affected engine speed (ES) for the longitudinal (x), vertical (z), and lateral (y) axes. The RMS of acceleration increases significantly when the engine speed increases from 1900 to 2100 rpm; this is true for each ES (engine speed) and all axes ($p < 0.05$; Table 3). A similar approach was detailed for the Mitsubishi power tiller [13,16]. As is evident from this table, in the lateral direction, the RMS acceleration values at every level of the fuel are significantly greater than the horizontal and vertical lengths. The reason why acceleration is higher in the lateral direction is the good dynamic balance of the six-cylinder engine, which causes vibration to reduce on the vertical side than on the lateral, and also the existence of blast forces in rotary engines [18]. In the longitudinal direction, there are no significant differences between RMS values of acceleration for pure diesel fuel at 1000-gauss, and RMS acceleration at 4000 gauss reduced significantly. This could be due to the low intensity of the 1000-gauss field influencing the molecular structure of the fuel. By increasing the field intensity to 4000-gauss, the fuel's molecular structure probably alters, so that the resulting combustion is more uniform. It was clarified that under a fairly powerful magnetic field, hydrocarbon molecules turn into a de-cluster mode with a smaller form, which makes it possible to find more contact space to react with oxygen, and this situation results in a more thorough combustion and fuel consumption decrease [20,23,24]. As regards RMS values of acceleration in the vertical direction, there are significant differences between pure diesel fuel and other fuels, and RMS acceleration reduces significantly at 4000-gauss.

Table 4. RMS acceleration and mean square (MS) comparison derived from a completely randomized design (CRD) and interaction effects of directions and engine speeds.

Applied Field Intensity (FI)	4000 G	3000 G	2000 G	1000 G	0 G
Longitudinal (X)	60.816 ^{o}	68.816 ^{n}	71.816 ^{m}	78.816 ^{l}	79.816 ^{l}
Lateral (Y)	97.797 ^{f}	98.79 ^{e}	104.729 ^{c}	119.400 ^{a}	120.853 ^{a}
Vertical (Z)	81.040 ^{r}	84.016 ^{i}	91.484 ^{h}	95.167 ^{g}	96.484 ^{g,f}

The means with the same letters showed no significant differences ($p > 0.05$).

In the longitudinal direction, there is no significant difference between the RMS value of (0 G) fuel acceleration and (1000 G), and with increasing field strength, the acceleration of these fuels slowly decreases. In the vertical direction, too, no significant difference is observed among the RMS values of pure diesel and 1000 G, but compared to other fuels, the RMS value for 2000 G fuel significantly drops and the smallest RMS acceleration value is in the vertical direction. The highest RMS value of acceleration in 1000 G and 0 G fuels is observed in the lateral direction, and the minimum value is in 4000 G fuel in the longitudinal direction. Overall, the RMS of vibration acceleration recorded in the lateral direction is the greatest (average 107.52 m/s^2) and the smallest value is in the longitudinal direction (average 71.23 m/s^2).

3.5. The Interaction Effects of Field Intensity and Engine Speed

Table 5 illustrates the evaluation of means for different ES and FS (magnetic field strength). As can be seen, the vibration dramatically rises at 2100 rpm, and the rate of the increase is substantially fast. To discover why vibration increases at some engine speeds, the output torque curve of the engine can be used [25]. The average of increased vibrations for all field intensities from 1700 to 2100 rpm is 2.02, 3.43, 7.65, and 8.57 m/s^2, respectively. Figure 4 illustrates the output torque and power curves performance for Perkins engine model A63544. There are several important points in this diagram. The maximum torque is 431 Nm at 1200 rpm. The maximum power is 82 kW at 2200 rpm. The topmost vibration always occurs at an engine speed where the ES, torque, or power is the greatest. Furthermore, in this engine, the product of power and torque (P-T) reaches the highest engine speed of 2000 rpm. The reason for this claim is this tractor's power take-off [26]. Moreover, Table 5 shows that proportional to the field intensity increase from 0 to 4000 gauss, the average acceleration of vibration reduces by 15%, 15.30%, 12.40%, 12.35%, and 15.38%, respectively (for five ES). This indicates an average 29% reduction in vibration acceleration. The reason why the value of vibration acceleration falls is probably that as the magnetic field increases, the power required to overcome the hydrocarbon clusters of the fuel structure is provided by the engine, as a result of which the fuel burns more constantly and less vibration is produced in the internal parts of the engine [27–31]. Using a magnetic field, the internal energy of fuel can be increased, which means some alteration in the molecular surface of fuel leads to a rise in the internal energy of the fuel, its molecular segregation, and more reactivity, thereby improving combustion [32–34].

Table 5. RMS acceleration and mean square (MS) comparison derived from the completely randomized design (CRD) for each engine speed.

FS	ES (rpm)				
	1700	1800	1900	2000	2100
0 G	87.439 [e,f]	91.929 [e]	98.451 [c]	108.598 [b]	118.606 [a]
1000 G	85.111 [f]	87.554 [e,f]	95.505 [d]	107.324 [b]	114.969 [a]
2000 G	81.474 [j,k]	83.561 [j,k]	91.177 [j]	102.875 [g,h]	110.325 [b]
3000 G	79.259 [j]	82.923 [j,i]	90.768 [h]	100.629 [h]	108.795 [g]
4000 G	73.487 [l]	77.867 [k,l]	86.219 [kl]	95.177 [j]	100.356 [h]

The means with the same letters showed no significant differences ($p > 0.05$). Subset for alpha = 0.05.

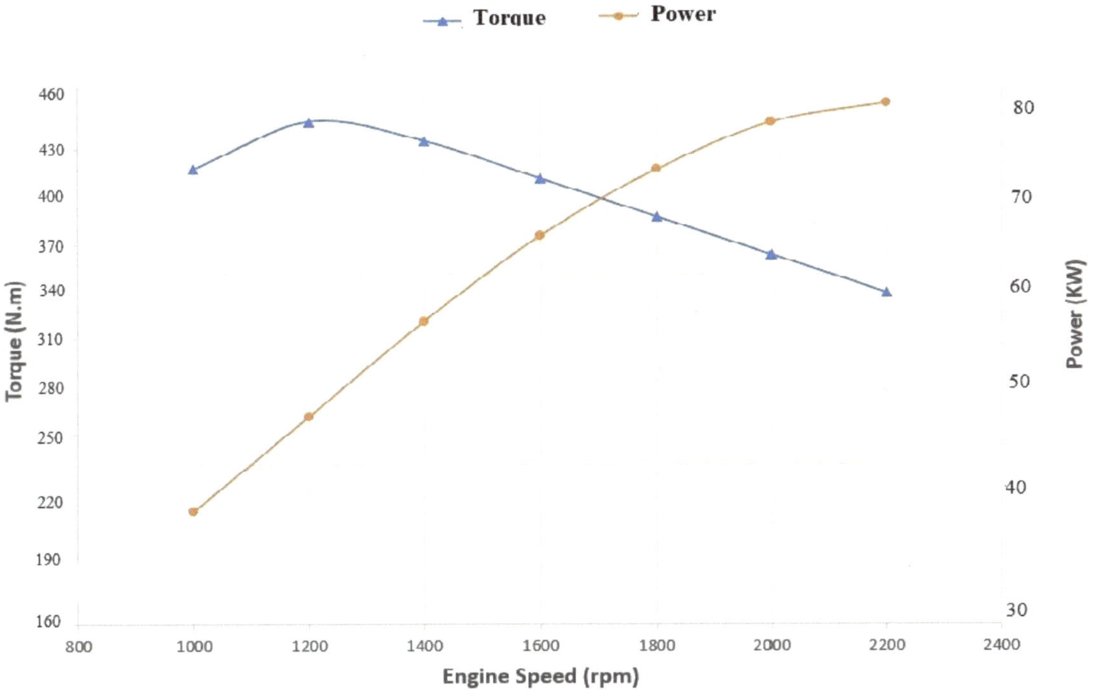

Figure 4. The engine torque power variations in terms of the engine speed.

3.6. Specific Fuel Consumption

Changes in the specific fuel consumption as a function of engine speed are shown in Figure 5. As can be observed in this figure, the diagram of all fuels shows the same trend, while the diagram of 4000 gauss has lower specific fuel consumption.

The maximum amount of specific fuel consumption reduction compared to control fuel was 2.3% for 4000 gauss. The coefficient of determination (R^2) related to the regression equations was fitted and shown on the engine specific fuel consumption data as a function of engine speed for each magnetic field density. Duncan's multiple range tests were used to compare whether the mean values of the engine vibration and specific fuel consumption varied significantly when the engine speed and magnetic field density changed. The results of Duncan's multi-range test are shown in Table 6. Letters (a, b, c, etc.) were utilized to find whether mean values had notable differences at a 0.5% probability level or not using Duncan's multiple range tests. The same letters indicate that there was no remarkable difference at the 0.5% probability level. Moreover, other surveys conducted by scientists approved the results of this study. For instance, in an investigation, the experimental results showed that the fuel consumption of the diesel engine was less when the engine had a magnetic field than that without it at a higher load. The results showed that consumption of the fuel with a magnetic field was less. The reduction in fuel consumption was about 8% [35].

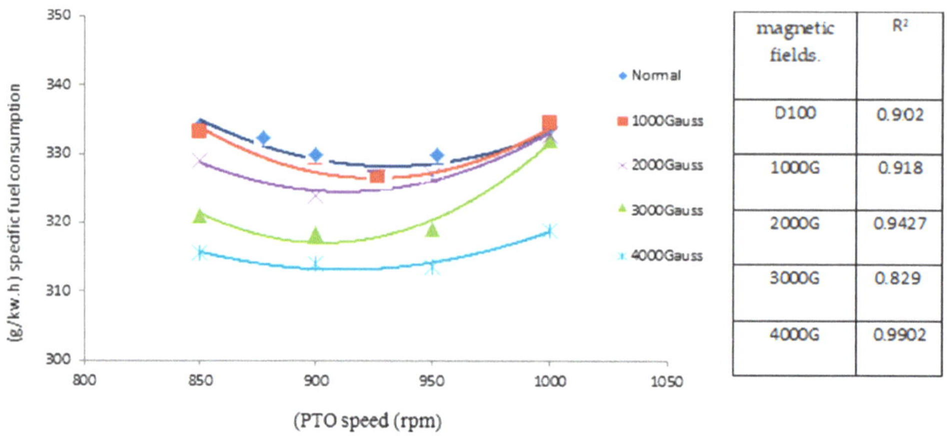

Figure 5. Engine speed and engine specific fuel consumption for different magnetic fields.

Table 6. The magnetic field and SFC.

Magnetic Field (Gauss)	Specific Fuel Consumption (g/kW h)
0 G	330.45 [a]
1000	330.26 [a]
2000	329.296 [a]
3000	326.95 [a,b]
4000	322.6 [b]

The means with the same letters showed no significant differences ($p > 0.05$). Subset for alpha = 0.05.

Table 7 shows Duncan's multi-range test results comparing the average specific fuel consumption and engine speed. As shown, the mean values of specific fuel consumption at speeds of 1700 to 1800 rpm are not significant but significantly decrease at a speed of 1900 rpm. Maximum average specific fuel consumption was seen at a speed of 1700 rpm and 334.86 g/kWh, and minimum average specific fuel consumption was achieved at a speed of 950 rpm and 319 g/kWh.

Table 7. Engine speed and SFC.

Engine Speed (rpm)	Specific Fuel Consumption
1700	334.86 [a]
1800	331 [a]
1900	319 [c]
2000	326 [b]

The means with similar letters showed no significant differences ($p > 0.05$). Subset for alpha = 0.05.

4. Conclusions

In this study, the fuels used included diesel fuel as standard control and fuels that have been subjected to magnetic field intensity at 1000, 2000, 3000, and 4000 gauss. The levels of the magnetic field were selected based on the literature survey as well as the magnetic field ability of the used unit. Some experiments were conducted in this research to evaluate the effects of magnetic fuel on engine parameters such as engine vibration. Effects of magnetic fields with different intensities were studied on vibration signals of a six-cylinder internal combustion engine. Vibration signals in the longitudinal, vertical, and lateral with 5 levels of engine speed were measured.

From the discussions, it can be concluded that as engine speed increases from 1700 to 2100 rpm, the RMS vibration acceleration average generally increases in all fuels and directions, and this research clarified that by applying a magnetic field to diesel fuel, the MF399 tractor engine vibration significantly decreases provided that the magnetic field density is sufficient.

Moreover, the time domain acceleration signals were significantly dependent on the measurement direction. furthermore, the experimental results showed that the time domain acceleration signal values of vibration decrease by applying the magnetic field intensity. Therefore, the value of the time-domain acceleration signals using a magnetic field intensity was significantly less than the level of acceleration signals related to fuel without a magnetic field.

Along with the engine speed increase from 1700 to 2100 rpm, the RMS average vibration acceleration generally increases in all fuels.

The experimental results showed that the values of RMS vibration acceleration decreases when the magnetic field strength is increased. Therefore, the minimum value of RMS was obtained using a field intensity of 4000 gauss. Moreover, specific fuel consumption declined by about 3/2%.

5. Patents

Patent No. 84531, Title: Magnetic diesel fuel using a magnetizer, Iran Industrial Patent Office.

Author Contributions: Conceptualization, Y.D., S.R.H.-B., and J.M.; methodology, Y.D.; software, Y.D.; validation, S.R.H.-B., H.K., and M.G.; formal analysis, M.G.; investigation, Y.D.; resources, H.K.; data curation, Y.D.; writing—original draft preparation, Y.D.; writing—review and editing, Y.D., M.G. and A.B.; visualization, Y.D.; supervision, S.R.H.-B.; project administration, Y.D.; funding acquisition, S.R.H.-B. All authors have read and agreed to the published version of the manuscript.

Funding: This research received no external funding.

Informed Consent Statement: Not applicable.

Data Availability Statement: Data sharing is not applicable to this article.

Conflicts of Interest: The authors declare no conflict of interest.

Abbreviations

a_{RMS}	Root mean square of acceleration
a_k	Acceleration amplitude at time of k within the time domain signal
FC	Fuel consumption
P	Power
SFC	Special fuel consumption
N	The entire range of samples
ρ	Volumetric mass

References

1. Darvishi, S.; Hassan-Beygi, S.R.; Ghobadian, B.; Massah, J. Magnetizing the perkins A63544 engine fuel and its vibratory assessment. *J. Engine Res.* **2015**, *38*, 17–26.
2. Demirbas, A.H.; Demirbas, I. Importance of rural bioenergy for developing countries. *Energy Convers. Manag.* **2007**, *48*, 2386–2398. [CrossRef]
3. Darvishi, Y.; Hassan-Beygi, S.R.; Zarafshan, P.; Hooshyari, K.; Malaga-Toboła, U.; Gancarz, M. Numerical Modeling and Evaluation of PEM Used for Fuel Cell Vehicles. *Materials* **2021**, *14*, 7907. [CrossRef] [PubMed]

4. Karami, H.; Rasekh, M.; Darvishi, Y. Effect of temperature and air velocity on drying kinetics and organo essential oil extraction efficiency in a hybrid dryer. *Innov. Food Technol.* **2017**, *5*, 65–75.
5. Karami, H.; Rasekh, M.; Darvishi, Y.; Khaledi, R. Effect of drying temperature and air velocity on the essential oil content of Mentha aquatica L. *J. Essent. Oil Bear. Plants* **2017**, *20*, 1131–1136. [CrossRef]
6. Kurji, H.J.; Imran, M.S. Magnetic field effect on compression ignition engine performance. *ARPN J. Eng. Appl. Sci.* **2018**, *13*, 3943–3949.
7. Sahoo, R.R.; Jain, A. Experimental analysis of nanofuel additives with magnetic fuel conditioning for diesel engine performance and emissions. *Fuel* **2019**, *236*, 365–372. [CrossRef]
8. Sidheshware, R.K.; Ganesan, S.; Bhojwani, V.K. Enhancement of internal combustion engine efficiency by magnetizing fuel in flow line for better charge combustion. *Heat Transf. Res.* **2020**, *51*, 419–431. [CrossRef]
9. Faris, A.S.; Al-Naseri, S.K.; Jamal, N.; Isse, R.; Abed, M.; Fouad, Z.; Kazim, A.; Reheem, N.; Chaloob, A.; Mohammad, H. Effects of magnetic field on fuel consumption and exhaust emissions in two-stroke engine. *Energy Procedia* **2012**, *18*, 327–338. [CrossRef]
10. Thiyagarajan, S.; Sonthalia, A.; Edwin Geo, V.; Ashok, B.; Nanthagopal, K.; Karthickeyan, V.; Dhinesh, B. Effect of electromagnet-based fuel-reforming system on high-viscous and low-viscous biofuel fueled in heavy-duty CI engine. *J. Therm. Anal. Calorim.* **2019**, *138*, 633–644. [CrossRef]
11. Govindasamy, P.; Dhandapani, S. Effects of EGR & magnetic fuel treatment system on engine emission characteristics in a bio fuel engine. In Proceedings of the International Conference on Mechanical Engineering, Dhaka, Bangladesh, 26–28 December 2009; pp. 26–28.
12. Tao, R.; Xu, X. Viscosity reduction in liquid suspensions by electric or magnetic fields. *Int. J. Mod. Phys. B* **2005**, *19*, 1283–1289. [CrossRef]
13. Marsili, A.; Ragni, L.; Santoro, G.; Servadio, P.; Vassalini, G. PM—Power and machinery: Innovative systems to reduce vibrations on agricultural tractors: Comparative analysis of acceleration transmitted through the driving seat. *Biosyst. Eng.* **2002**, *81*, 35–47. [CrossRef]
14. Khan, N.S.; Shah, Q.; Sohail, A.; Kumam, P.; Thounthong, P.; Muhammad, T. Mechanical aspects of Maxwell nanofluid in dynamic system with irreversible analysis. *ZAMM-J. Appl. Math. Mech. Z. Angew. Math. Mech.* **2021**, *101*, e202000212. [CrossRef]
15. Khan, N.S.; Humphries, U.W.; Kumam, W.; Kumam, P.; Muhammad, T. Assessment of irreversibility optimization in Casson nanofluid flow with leading edge accretion or ablation. *ZAMM J. Appl. Math. Mech. Z. Angew. Math. Mech.* **2022**, *102*, e202000207. [CrossRef]
16. Ahmadian, H.; Hassan-Beygi, S.R.; Ghobadian, B. Power tiller vibration acceleration envelope curves on transportation mode. *J. Vibroeng.* **2013**, *15*, 1431–1441.
17. Rocha, N.; González, C.; Marques, L.C.d.C.; Vaitsman, D.S. A preliminary study on the magnetic treatment of fluids. *Pet. Sci. Technol.* **2000**, *18*, 33–50. [CrossRef]
18. Taghizadeh-Alisaraei, A.; Ghobadian, B.; Tavakoli-Hashjin, T.; Mohtasebi, S.S. Vibration analysis of a diesel engine using biodiesel and petrodiesel fuel blends. *Fuel* **2012**, *102*, 414–422. [CrossRef]
19. Okoronkwo, C.; Nwachukwu, C.; Ngozi–Olehi, L.; Igbokwe, J. The effect of electromagnetic flux density on the ionization and the combustion of fuel (an economy design project). *Am. J. Sci. Ind. Res.* **2010**, *1*, 527–531. [CrossRef]
20. Saksono, N. Magnetizing kerosene for increasing combustion efficiency. *J. Teknol. Ed.* **2005**, *2*, 155–162.
21. Dhandapani, S. Theoretical and Experimental Investigation of Catalytically Activated Lean Burnt Combustion. Ph.D. Thesis, Indian Institute of Technology, Chennai, India, 1991.
22. Taghizadeh, A.A.; Tavakoli, H.T.; Ghobadian, B. An Analysis and Evaluation of Vibrations of Power Tiller in the Stationary Conditions. *Iran. J. Biosyst. Eng.* **2010**, *41*, 279677.
23. Faraji, H.; Heidarbeigi, K.; Samadi, S. Effect of magnetized ethanol-gasoline blends on vibration and sound emissions of a single cylinder SI engine. *Res. Sq.* 2020. [CrossRef]
24. Kita, R.J.; Kulish, P.A. Electromagnetic Device for the Magnetic Treatment of Fuel. U.S. Patents US5829420A, 3 November 1998.
25. Langer, T.H.; Ebbesen, M.K.; Kordestani, A. Experimental analysis of occupational whole-body vibration exposure of agricultural tractor with large square baler. *Int. J. Ind. Ergon.* **2015**, *47*, 79–83. [CrossRef]
26. Zhang, J.; Yao, H.; Chen, L.; Zheng, E.; Zhu, Y.; Xue, J. Vibration characteristics analysis and suspension parameter optimization of tractor/implement system with front axle suspension under ploughing operation condition. *J. Terramech.* **2022**, *102*, 49–64. [CrossRef]
27. Pera, I. Magnetizing Hydrocarbon Fuels and Other Fluids. U.S. Patent 4716024A, 29 December 1987.
28. Munpollasri, S.; Poo-arporn, Y.; Donphai, W.; Sirijaraensre, J.; Sangthong, W.; Kiatphuengporn, S.; Jantaratana, P.; Witoon, T.; Chareonpanich, M. How magnetic field affects catalytic CO_2 hydrogenation over Fe-Cu/MCM-41: In situ active metal phase—Reactivity observation during activation and reaction. *Chem. Eng. J.* **2022**, *441*, 135952. [CrossRef]
29. Jiancun, G.; Xigang, Y.; Shoutao, H.; Le, W.; Zijin, H.; Xu, S.; Ruxia, L. Effects of Magnetic Fields on Combustion and Explosion. *Chem. Technol. Fuels Oils* **2022**, *58*, 379–390. [CrossRef]
30. Liu, X.; Xie, N.; Xue, J.; Li, M.; Zheng, C.; Zhang, J.; Qin, Y.; Yin, Y.; Dekel, D.R.; Guiver, M.D. Magnetic-field-oriented mixed-valence-stabilized ferrocenium anion-exchange membranes for fuel cells. *Nat. Energy* **2022**, *7*, 329–339. [CrossRef]

31. Niaki, S.R.A.; Zadeh, F.G.; Niaki, S.B.A.; Mouallem, J.; Mahdavi, S. Experimental investigation of effects of magnetic field on performance, combustion, and emission characteristics of a spark ignition engine. *Environ. Prog. Sustain. Energy* **2020**, *39*, e13317. [CrossRef]
32. Attar, A.; Arulprakasajothi, M.; Vasulkar, D.; Gorde, N.; Kharat, S.; Kulkarni, S. Investigation of impact of the magnetic field through halbach array on Hydrocarbon fuel. *Int. J. Ambient. Energy* **2022**, *43*, 2124–2129. [CrossRef]
33. Sidheshware, R.K.; Ganesan, S.; Bhojwani, V. Experimental investigation on the viscosity and specific volume of gasoline fuel under the magnetisation process. *Int. J. Ambient. Energy* **2022**, *43*, 486–491. [CrossRef]
34. Zhou, S.; Gao, J.; Luo, Z.; Hu, S.; Wang, L.; Wang, T. Role of ferromagnetic metal velvet and DC magnetic field on the explosion of a C3H8/air mixture-effect on reaction mechanism. *Energy* **2022**, *239*, 122218. [CrossRef]
35. Patel, P.M.; Rathod, G.P.; Patel, T.M. Effect of magnetic field on performance and emission of single cylinder four stroke diesel engine. *IOSR J. Eng.* **2014**, *4*, 28–34. [CrossRef]

Disclaimer/Publisher's Note: The statements, opinions and data contained in all publications are solely those of the individual author(s) and contributor(s) and not of MDPI and/or the editor(s). MDPI and/or the editor(s) disclaim responsibility for any injury to people or property resulting from any ideas, methods, instructions or products referred to in the content.

AgriEngineering

Article

The Next Generation of Cotton Defoliation Sprayer

Jyoti Neupane [1], Joe Mari Maja [1,*], Gilbert Miller [2], Michael Marshall [3], Matthew Cutulle [3], Jeremy Greene [3], Jun Luo [4] and Edward Barnes [5]

1 Department of Agricultural Sciences, Clemson University, 240 McAdams Hall, Clemson, SC 29634, USA
2 Edisto Research and Education Center, Clemson University, 64 Research Road, Blackville, SC 29817, USA
3 Department of Plant and Environmental Sciences, Clemson University, 171 Poole Agricultural Center, Clemson, SC 29634, USA
4 School of Mathematical and Statistical Sciences, Clemson University, Martin O-15, Clemson, SC 29634, USA
5 Cotton Incorporated, 6399 Weston Parkway, Cary, NC 27513, USA
* Correspondence: jmaja@clemson.edu

Abstract: Chemical spraying is one of the most important and frequently performed intercultural agricultural operations. It is imperative to select the appropriate spraying technology as a selection of ineffective one leads to the wastage of a considerable volume of applied chemicals to the non-target area. Many precision technologies have been developed in the past few decades, such as image processing based on real-time variable-rate chemical spraying systems, autonomous chemical sprayers using machine vision and nozzle control, and use of unmanned aerial and ground vehicles. Cotton defoliation is a natural physiological process, but untimely and inadequate leaf defoliation by natural process hinders the mechanical cotton harvest. Induced defoliation is practiced by applying defoliants to address the issue with the natural process of defoliation. This paper covers spraying technologies in agriculture, cotton plants, cotton defoliation, new defoliant spraying systems, and the recent field test. The new spraying system attached to an autonomous mobile robot aims to improve the delivery of defoliant chemicals by adding a spray unit on the side of the plant. Preliminary results of the water-sensitive paper test at the field showed adequate penetration with low flow rates. This is a huge development as there is a huge potential to save on the cost of applying defoliant chemicals.

Keywords: cotton production; defoliation chemical; intelligent spray; robotics

Citation: Neupane, J.; Maja, J.M.; Miller, G.; Marshall, M.; Cutulle, M.; Greene, J.; Luo, J.; Barnes, E. The Next Generation of Cotton Defoliation Sprayer. *AgriEngineering* **2023**, *5*, 441–459. https://doi.org/10.3390/agriengineering5010029

Academic Editors: Muhammad Sultan, Yuguang Zhou, Redmond R. Shamshiri and Muhammad Imran

Received: 25 January 2023
Revised: 14 February 2023
Accepted: 20 February 2023
Published: 22 February 2023

Copyright: © 2023 by the authors. Licensee MDPI, Basel, Switzerland. This article is an open access article distributed under the terms and conditions of the Creative Commons Attribution (CC BY) license (https://creativecommons.org/licenses/by/4.0/).

1. Introduction

Chemical spraying is one of the most important and frequently performed intercultural operations in agriculture [1]. Based on the target, chemical pesticides are classified as herbicides, insecticides, and fungicides to kill weeds, insects, and pathogens, respectively [2]. Similarly, some chemicals are used to accelerate the harvesting process, such as harvest aids and defoliant [3]. Even though using agrochemicals can effectively enhance crop production, it significantly impacts production cost, human health, and the environment [4]. Pesticide resistance is another major problem due to the overuse of chemicals [5]. During the green revolution, more agrochemicals were used to achieve targeted production, which created environmental degradation, human health effects, and high yields [6,7].

Moreover, pesticides can remain active in soil and water for an extended period, which causes soil degradation, eutrophication, algal bloom, and decline in the fisheries [6]. Thus, it is vital to consider spraying methods, plant characteristics, the nature of chemicals, and weather conditions while applying agrochemicals to achieve an efficient and less harmful effect. According to a Food and Agriculture (FAO) report, 2.7 million tons of pesticides (active ingredients), equivalent to USD 41.1 billion, was used for agriculture purpose worldwide. The United States of America rank first by consuming nearly 0.5 million tons of pesticides then followed by Brazil and China. Previously, China used to be top consumer; however, they significantly cutoff the use of pesticides [8].

1.1. Spraying Technologies

It is imperative to select the appropriate spraying technology, as a selection of an ineffective one leads to the wastage of a considerable volume of applied chemicals in the non-target area. Advancements in new technologies in precision agriculture, such as the use of sensors, artificial intelligence, and unmanned vehicles contributed to the growth of agriculture.

Introduction of the different sprayer technology, including tunnel sprayer technology, has been proven most effective for the orchards and vineyards, which enclose the target spray mixture to reduce airborne drift and soil contamination. Moreover, it is based on the air circulation and liquid recycling principle as mentioned by Ade et al. where they found 20~50% of applied chemical liquid recycled with the application of tunnel sprayer on a vineyard [9]. The above statement is supported by Jamar et al. as their study also found on an average 30% recycling of spraying volume which contributed to more environmental sustainability compared with traditional machines. Tunnel sprayer found to suitable for dwarf trees using traditional hydraulic nozzles [10].

A tower sprayer that sprays chemicals horizontally with airflow direction is suitable for taller plants. Tower air conveyor sprayers were found to have greater advantages on the distribution of air and pesticides through the trees, which help to minimize the loss of pesticides during spraying [11] but the airspeed, airflow, and power should consider properly as these factors significantly affects the penetrations of spray using tower sprayers in orchard [12]. Cannon air blast sprayers with cylindrical outlets create high air velocity jets to break spray mixture into finer droplets and have good canopy penetration, especially in the orchards such as blueberry which spray across the tops of several rows of blueberry resulting in high field capacity and minimum mechanical damage to developing fruits [13]. Similarly, the air-assisted spraying technology is used in cereal production to control insect pests [14].

Ultra-low volume (ULV) sprayers are very effective for controlling pests and insects in cotton plants [15]. Reduced fluid application rate, drift, and wastage are the primary objectives of ULV. Likewise, electrostatic techniques are emerging technology in agriculture. The concept behind it is the retention of particle charge, optimizing the deposition field. Some of the examples of this concept are air-assisted electrostatic spray, aerodynamic electrostatic spray, postharvest electrostatic sprays, and electrostatic pollination [16].

Conventional tractor-mounted boom sprayers spray on the upper sides, which is ineffective for controlling sucking insects as those insects' shelter on the bottom side of the leaves. The agrochemical application of conventional sprayers result in the wastage of applied chemicals, as it applies the agrochemicals at a uniform rate, resulting in off-target losses, economic losses, and environmental hazards. With advancements in technology, new concepts of variable rate of spraying technologies have been promising to apply the chemicals exactly to the required location.

Many precision technologies have been developed in the past few decades, such as an image processing based on real-time variable-rate chemical spraying system [17] autonomous chemical sprayer using machine vision and nozzle control [18], use of unmanned aerial vehicle [19,20], and electrical robots [21].

Esau et al. reported that a machine vision smart sprayer for spot-application of agrochemical in wild blueberry resulted in an 11.6% saving of fungicide application along with a significant increase in the health of plant and harvestable yield [22]. Llorens et al. and Asaei et al. reported saving of more than 50% spraying volume with machine vision variable rate application as compared to constant rate conventional sprayer in orchards [23,24]. Through the use of machine learning, high spectral images have been increasing their ability to achieve precise and accurate action. Similarly, the use of unmanned vehicles is also increasing for multiple purpose. In agriculture, the application of unmanned aerial vehicle (UAV) is not only limited to spraying, but it is widely being use for mapping, planting, crop monitoring, and harvesting (Figure 1). Despite of many advantages it has some significant limitations such as battery and flight time [25]. In cotton, researchers are using unmanned aerial vehicles to spray chemicals, predict application volume using remote sensing images [26].

Figure 1. Multiple application of UAV in Agriculture.

Moreover, the use of unmanned ground vehicles is also increasing in agriculture. Mobile robots which mainly composed of two system: hydraulic subsystem with liquid tank and electrical system with pump, pressure regulator, electric flux regulator, flow rate meter, and electric on/off valves [21]. The robots are sometime referring as 'electrical robots' when they are powered by electrical system. There is huge scope and opportunity of robotics in agriculture. Like UAV, the mobile robots also have multiple application such as scouting, spraying, weeding, and harvesting (Figure 2) [27].

Figure 2. Mobile robots' different applications in Agriculture.

1.2. Cotton

Cotton (*Gossypium hirsutum*), a member of the *Malvaceae* family, is a tropical shrub [28]. Cotton is a diversified and historical plant, which is justified by its fire-adaptive herbaceous perennial nature in northwest Australia to small trees in southwest Mexico which can cope with the dry season by dropping its leaves, and the evidence suggest its emergence roughly before 12.5 million years [29]. There are 49 different species in the Gossypium genus. It exhibits tremendous morphological and ecological diversity, as well as a long history of chromosomal evolution. Human cultivation and modification of several species, including G. arboretum, G. herbaceum, G. barbadense, and G. hirsutum has led to significant agronomic advancement and germplasm dispersion.

G. arboretum is cultivated in some cotton markets leading countries such as India and Pakistan [30]. Likewise, G. herbaceum is cultivated in Africa and Asia on a small scale, whereas recently, the cultivation of tetraploid cultivars such as G. barbadenseis has been increasing in several regions of Asia and America due to its long, strong, and fine fibers [31]. Despite the number of advantages, its production is limited to 10% of total world production due to its low yield. Most of the world's cotton is dominated by the modern cultivar of G. hirsutum (i.e., upland cotton) [29].

Cotton, which is considered 'White gold', before the industrial revolution, the market was only restricted to domestic within a small region. Later, after the establishment of many textile companies and intercontinental markets, cotton became one of the major crops in the world. The U.S. has been one of the most important cotton exporters in the world since the beginning of the 1790s. The availability of enormous land and labor, and climatic suitability for cotton production in the U.S., especially in the upper southern part of the county, such as the Carolinas and Georgia, contributed to becoming the top cotton exporter in the world. The cotton industry primarily defines the U.S. position in the world economy. The United States is the world's third largest cotton producer after China and India [29].

Cotton is a multipurpose crop mainly grown for fiber and seeds. Although it is widely known for its natural fiber, its uses are not limited to the textile industry. Several activities are involved between cotton cultivation and cotton consumption, such as cultivation, harvesting, transportation, spinning, and production of another by-product. It is also believed to have some medicinal uses. In ancient times, people used its young shoot, leaves, and flowers to cure diseases such as asthma, convulsion, and skin diseases [32]. Cotton seed products include hulls, kernels, and linters. Hulls mainly provide the raw materials for rubbers and plastics, and the kernel provides raw materials for soap, glycerin, refined oil, animal feed, and fertilizers, whereas linters provide materials for the textile industry [33]. Because of its diversified uses, the cotton plant significantly influences the world economy even though it is not a food crop.

1.3. Physiology and Growth Pattern

An understanding of the growth habit of cotton is very crucial for its better management. It has indeterminate growth habits. Cotton is a perennial plant and favorable condition such as high rainfall, nitrogen fertilization, and fertile soil encourages high vegetative growth as compared to reproductive growth. From the economic perspective, reproductive growth is very important so application of growth regulator to achieve desirable vegetative and reproductive growth is beneficial [34]. The maximum growth of cotton occurs between 40 and 80 days after emergence (DAE). Likewise, canopy coverage also follows a similar pattern, such as height, in relation to days after emergence [35]. This result is also supported by Sun et al. work on cotton growth analysis using LiDAR [36].

At the vegetative stage, most of the carbohydrates produced are used for root and leaf development, but once the plant reaches the reproductive stage, the transport of carbohydrates shifts to developing fruits [37]. The cotton plant has two different kinds of branches: fruiting (sympodia) and vegetative (monopodia). The 'stop-and-go' and the zigzag growth habit of fruiting branches set it apart from vegetative branches, which have a straight growth habit. Cotton leaves' photosynthetic capacity varies with age, reaching a maximum value about 20 days of leaf age. Low fertility and water stress cause premature aging of the cotton leaf canopy, which further reduces the photosynthetic capacity and subsequently reduces the yield of the crop [37].

Generally, two different colors of flower are seen in cotton, i.e., white and pink. The white flower stage, or the first day the flower is open, is when pollination takes place. After pollination, the flower changes color to pink. As being indeterminate plant, the flowering occurs both horizontally (in same node) and vertically (in different node) simultaneously. The horizontal flowering interval is 5–6 days, whereas the vertical flowering interval is 2–3 days. Approximately three weeks after fertilization, fiber development starts and continues until it reaches the full staple length. When fiber reaches its maximum stable length, it begins maturing, or thickening, which happens because of the deposition of additional layers of cellulose [38]. The economic part of cotton is its fiber. Historically, growers used to harvest cotton boll manually which very tedious work and inconvenient for commercial production. Later, the concept of defoliation came to make harvesting more efficient. Along with defoliant chemical, boll opener and growth inhibitors are also used to facilitate mechanical harvesting, maintain quality, encourage boll opening, and control the regrowth of the cotton plant.

1.4. Cotton Defoliation

Defoliation is the shedding of leaves by natural or artificial phenomena. Cotton defoliation is a natural physiological process, but untimely and inadequate leaf defoliation by natural process hinders the mechanical cotton harvest [39]. Therefore, induced defoliation is practiced by applying defoliants to facilitate mechanical harvesting [40]. Cotton defoliation is a major factor influencing mechanical harvesting, fiber quality, and the costs of cotton cultivation [41].

The importance of defoliation in cotton has been realized with increased mechanical/autonomous harvesting [42]. Several factors, such as plant condition, atmospheric temperature, and moisture during defoliant's application influence the success of cotton defoliation [43]. The defoliation practice accelerated boll opening and thus higher yield [44]. Historically, calcium cyanamide dust was first used as defoliants to promote aeration and thus reduce the boll rots in 1938 [45]. Chemical application is important from a production perspective and the most dangerous agricultural operation [46]. Therefore, we should apply chemicals precisely to minimize costs and their adverse impacts on environments and living beings.

Generally, chemical defoliants are applied through ground-based or aerial-based vehicles such as tractor-mounted boom sprayer, unmanned aerial spraying (drone spraying), and unmanned ground sprayer (robots spraying). Each system has its own merits and demerits and selection of these system depend on available technology, economic budget, cultivated area. Pivot attached sprayer system and conventional tractor-mounted sprayer resulted in better defoliation than chemigation [47]. However, UVA spraying system found to have significantly higher pesticides utilization rate, agronomic properties, yield, and fiber quality of cotton as compared to tractor-mounted boom sprayer [48]. Conventional spraying systems resulted in higher losses of chemical due to off target application. Thus, to tackle this problem many advance technologies such as pulse width modulation technology [49], LiDAR-guided system [50], unmanned aerial vehicles [51], unmanned ground vehicles such as mobile robots [52], and variable rate of application techniques has been developed to minimize the off-target loss and maximize the efficiency of chemical spraying. For cotton defoliation sprayers, US farmers usually use a boom sprayer. The same boom sprayer is used for pesticide and fertilizer applications. A study compared an air-sleeve sprayer with a hydraulic nozzle sprayer for droplet coverage, insect control, and cotton defoliation. The air-sleeve sprayer was found to be more efficient than the hydraulic nozzle sprayer based on droplet coverage and defoliation [53]. As large machinery such as tractors cause soil compaction, mechanical crop damage, and yield losses [54], the use of unmanned aerial vehicles (UAV) is increasing in cotton production, mainly for spraying of harvest aids. Good droplet coverage and leaf defoliant retention are crucial for proper defoliation. The application of vegetable oil adjuvant and harvest adds (defoliants and boll opener) was reported to increase defoliation and boll opening rates [55] significantly. A study was conducted to compare the efficacy of UAV spraying and traditional ground-based spraying in cotton. Based on the final cotton yield, the UAV sprayings were more efficient than traditional ground-based spraying [54,56]. There are few reasons why aerial applications used for defoliations are not very common in the US. This is due to the Federal Aviation Authority (FAA) policy on using spray drone where one needs to file waivers and the steps of getting one takes some time and effort for the user to have the drone spraying on the field especially as these defoliants are chemicals. Other issues are off-target of the products [57]. For ground-based, it is apparent that manual spraying is still being used especially in India. The most popular sprayer unit is the mobile backpack (MBP), an air-assisted sprayer [58].

1.5. Pulse Width Modulation (PWM) Technology for Spraying

The basic concept behind the pulse width modulation is waveforms or the switching frequency, which exhibits the varying duty cycles of the power switches, and it is the concept of electronic power conversion [59]. The application of PWM has been increasing in agriculture mainly to improve the efficiency of pesticide application with higher precision [60]. Moreover, the nozzle flow rate can be controlled by using PWM, which can significantly reduce pesticide consumption, off-target drift losses, environmental risk, and production cost [61]. Although lower PWM results in a lower flow rate and minimizes the chemical use, we need to ensure the desired droplet size, distribution, and coverage. Only minimizing chemical use by ignoring these parameters should not be the primary goal. Therefore, selecting the right PWM, nozzle type, and pressure are critical for proper spraying. A study on the droplet size and nozzle tip pressure on a PWM sprayer found that the use of at least 40% duty cycle with a minimum pressure of 276 kpa and non-venturi type of nozzles to optimize and homogenize the spray droplet size across spray application [62].

Similarly, a study was conducted to investigate the on/off latency in PWM nozzles and determine the effect of active nozzles on spray fan pattern latency and flow characteristics to simulate dynamic spray coverage. A 20 ms delay in nozzle pressure development during each cycle was reported, irrespective of the number of nozzles activated. Moreover, the spray coverage was found within $\pm 10\%$ of the target rate for each PWM duty cycle most of the time [63]. The application of PWM can be found in various crops such as apples [49], vines [64], and Palmer amaranth [65]. However, minimal studies about the application of PWM have been found for row crops such as cotton.

Boatwright et al. reported that intelligent spraying technology have higher efficiency in controlling disease and pests with reduced spray volume and drift in peach orchards compared to conventional air blast systems [66]. Likewise, Chen et al. used drones equipped with sensors and multispectral cameras to generate a defoliation prescription map using RGB and multispectral images [56]. The spectral indices and cotton defoliation rate has strong correlation. Thus, their work highlighted the application of UAV remote sensing for cotton defoliation. Based on the review mentioned above, the large number of works has been conducted on conventional boom sprayer and UAV spraying; however, very limited study has been conducted on application of unmanned ground vehicle together with pulse width modulation technology for cotton defoliation.

1.6. Defoliation Time

The timing of the defoliant application plays a significant role in cotton yield because premature defoliation reduces yield and late defoliation can deteriorate fiber quality due to bad weather [34]. Gormus et al. concluded that early application of defoliants significantly reduces seed cotton yield, boll number per plant, micronaire, and fiber length [67]. Various methods have been used to determine the time of defoliants application on cotton. Traditionally, counting nodes above the cracked boll (NACB), and calculating the percentage of the open boll methods were used which are more subjective and may differ according to growers' perception and observation. Generally, when 50–60% boll is open, and NACB is less than or equal to four, is considered safe for defoliation [34].

Early application of defoliant (prior to 60% open boll) found to decrease yield and fiber quality thus Snipes and Baskin suggested to apply defoliant when 60% of boll are open [68]. However, it is important to consider variety of cultivated cotton as cultivar effect can be significant and the timing can be different for different cultivar [69].

However, there are no clear-cut guidelines for the perfect timing of defoliants application. Some literature mentioned COTMAN concept to determine the timing of defoliation. COTMAN is a cotton management software program developed on concept of cotton plant growth and development. At first the concept of COTMAN was used to measure node above white flower (NAWF) for management of insect. Later, this concept adopted for defoliation. When NAWF is equal to 5 it is considered as physiological cutout [70].

According to COTMAN program when the NAWF = 5 and HU = 472 consider as appropriate time for defoliation. The amount of heat accumulated during a 24 h time when the average ambient temperature is greater than 15.5 degree (base threshold temperature) is define as heat unit (HU) in cotton. However, Bynum and Cothren findings contradict COTMAN's recommendation based on their experiment [71]. They consider NAWF (=3, 4, 5) and HU's (361, 417, 472, 528, and 583) as indicator for defoliation. A total of 29% more lint yield was observed when plant was defoliated at 60 percent of open boll timing than compared to COTMAN recommendation (i.e., NAWF = 5 and 472 HU). Similarly, Clay et al. studied the effectiveness of defoliation at various heat unit accumulations: 630 HU, 730 HU, 830 HU, 930 HU, 1030 HU, 1130 HU, and 1330 HU and its impact on lint yield and fiber quality [72]. The highest defoliation was found at 830 HU. Moreover, lint yield and gin turnout were found to be highest in early defoliation but fiber qualities except the fiber strength were insignificant to the defoliation timing. The HU accumulation can differ from one location to another based on their weather condition and generalizing single research findings to the varied location might not be a good idea. Therefore, future multilocation research should be conducted to determine defoliation timing considering different methods.

1.7. Defoliant and Factors Affecting Defoliation

The defoliants and harvest aids are categorized into two groups based on their nature of action: herbicidal and hormonal. Herbicidal defoliants injure the plant, causing it to produce ethylene in response which promotes leaf abscission [73]. Some examples of herbicidal defoliants are Carfentrazone-ethyl, Thidiazuron, diuron, and Tribufos. In the case of the herbicidal defoliant, for a successful leaf drop, each leaf should encounter the chemical, as chemical penetration and evenness of the spray play a significant role in defoliation [74]. In contrast, hormonal defoliants enhance ethylene production and inhibit auxin transport in the plant, encouraging leaf abscission [73]. The hormonal defoliants are generally more sensitive to temperature and crop conditions than herbicidal defoliants [75]. It is very important to select appropriate chemical for efficient result. Application of harvest aids such as Thidiazuron and ethephon were found to have a significant effect on cotton defoliation and boll opening [76].

Several factors affect the efficiency of chemical defoliants such as plant characteristics, applied nutrients, weather conditions, types of chemicals, and spraying methods. Wang et al. found to have a significant effect of nitrogen and harvest aid on defoliation efficiency and cotton yield [76]. With rising N levels, the defoliation and boll opening percentages were significantly reduced and 150 kg N/ha together with 900 + 3000 g ai ha^{-1} of Thidiazuron + Ethephon was recommended for good cotton yield and efficient defoliation [76]. Weather factors that influence harvest-aid performance are temperature, relative humidity, seasonal rainfall, and the occurrence of precipitation shortly following application. High seasonal night temperatures promote crop maturity and susceptibility to defoliation. Harvest aid work best when nighttime temperatures are 15.5 degrees Celsius or higher [34]. In general, herbicidal type defoliants such as tribufos (S, S, S, -tributyl phosphonodithioate) have lower minimum temperatures (12.7–15.6 °C) than hormonal types (15.6–18.3 °C) such as ethephon and thidiazuron (N-phenyl-N′-1,2,3- thiadiazol-5-ylurea) [76]. Defoliation occurs twice as quickly at 35 degrees Celsius than it does at 25 degrees Celsius because the rate of activity doubles with a 10-degree Celsius increase in temperature [76]. Defoliation was 17% with no adjuvant at five days after treatment (DAT) when day/night temperatures 29.5/21 degree Celsius, 37% with crop oil concentrate added, 40% with ammonium sulfate, and 75% with both adjuvants combined whereas there was less than 10% leaf drop at 21/12.7-degree Celsius day/night temperature under all treatments at 5 DAT [77].

In general, excessive watering before and during defoliation can result in more vegetative growth of plants. Whereas extremely dry condition during defoliants application may hinder the defoliants activity. Terminating the irrigation at least 24 days before the defoliation resulted in good defoliation with single application of defoliant [78].

Plant characteristic such as leaf area index was found to have a significant effect on droplet deposition and defoliation rate [79], but the distribution of droplet depends on several factors which affects the defoliation. Plant characteristics (e.g., height, leaf area, and density) and spraying technology (spraying volume, nozzle types, pressure, and vehicle speed) influence the droplet coverage and distribution in the plant canopy [80].

Droplet deposition found significantly affected by droplet size, spray volume, and flight height of UAV. The droplet deposition in the lower part of the canopy gradually increases with the increase in the droplet size and volume of spray. Deposition is influenced by an interaction between flight height and droplet size. when the flight height is 1 m, no variation in the deposition among the droplet size (100, 150, and 200 m) was found. However, the average deposition of coarse droplet found greater when the flight height was 2 or 3 m [81]. The upper leaves intercept droplets and reduce the number of droplets available to deposit onto the lower parts of the leaf canopy [82]. Zhu et al. reported that spray deposition decreased dramatically from the top to the bottom of the canopy throughout the growing season [83]. This statement is also supported by Meng et al., who found the lowest droplet coverage on the bottom half of the cotton [84].

The main objective of this work is to evaluate the new spraying system developed by studying the droplet characteristics and defoliation at the different treatment levels and how does this affect the yield.

2. Materials and Methods

2.1. Study Site

The research was conducted at the Edisto Research and Education Center in Blackville, SC, USA. Two locations at the research farm were selected (Field 1: 33.34736,−81.31925; Field 2: 33.35398,−81.31024). Both locations utilized a completely randomized experimental design, with 4 replications in Field 1 and 2 replications in Field 2. In Field 1, cotton was planted in 6 rows. Each row was divided into 2 groups (P1 and P2), where each group was further divided into 3 smaller rows with a length of 9.8 m (32 ft.) (See Figure 3).

Figure 3. Cotton planted in Field 1 with 6 rows and 2 groups.

In Field 2, cotton was planted in late May of 2022 and has a smaller size as compared to Field 1 (Figure 4). Six rows were selected with length ranging from 21 m to 27 m. Cotton on both fields were planted by skipping one row (skip row) to facilitate the movement of the mobile robot.

Figure 4. Cotton planted in Field 2.

2.2. Autonomous Mobile Platform and Sprayer Unit

This work utilized the mobile platform (Husky A200, Clearpathrobotics, ON, CA) from other research work at the sensor and automation laboratory (see Figure 5). The mobile platform is lightweight and powerful enough to pull or carry payloads up to 75 kg [85]. It is equipped with an inertial measuring unit (UM7, CH Robotics, Victoria, Australia), global positioning system (Swiftnav, Swift Navigation, CA, USA), motors, encoders, and laser scanner (UST-10LX, Hokuyo, Osaka, Japan) for its navigation.

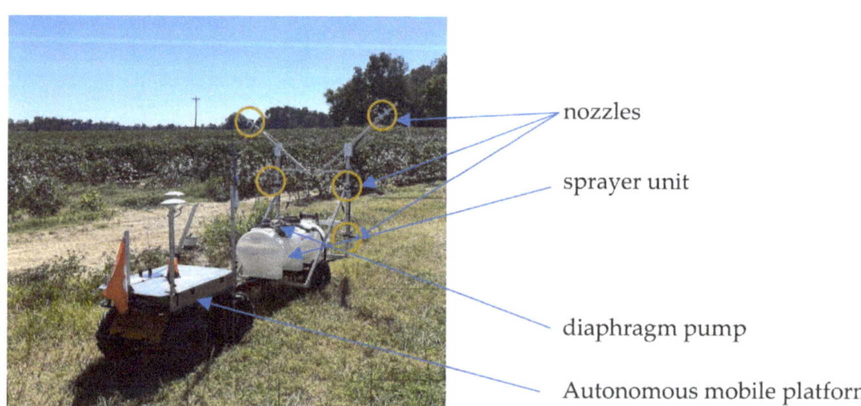

Figure 5. Autonomous mobile platform with the pull behind sprayer.

The sprayer unit (Model #1598042, County Line, Austin, TX, USA) is a 94 L 2-nozzle trailer sprayer with a built-in 12V diaphragm 9.5 L/min pump. The rated pressure of the pump based on the specification was 482 KPa. The sprayer was retrofitted with 6 nozzles (Model #625147-001, Capstan Ag Systems Inc, Topeka, KS, USA), and all valves, O-rings, flynut, and other sprayer parts were provided by Wilger Inc. (Wilger Industries, SK, Canada). Three nozzles were situated on each side, with nozzles located at 38 cm, 84 cm, and 145 cm from the ground as shown on Figure 5. The nozzle from the bottom to the top were designated as low, mid, and top nozzles, respectively. An aluminum extrusion was used to hold the nozzles and the top nozzle were position on an angle of 40 degrees. The extrusion holding the top nozzle can be adjusted by three hex screws. This is intended for crop height changes during the field test. The distance of the bottom and middle nozzles was based on the spread of the tip used while the top was based on the height of the crop.

Each nozzle was retrofitted with a spray tip (ER110-06, Wilger Industries, SK, Canada). The ER series spray tip is a conventional flat fan nozzle with relative fine spray with consistent pattern. A link to the UGV and sprayer can be found in the Supplementary Materials that show the UGV and sprayer unit performance on the field.

2.3. Sprayer Controller

The controller developed specifically for this project was an ARM Cortex-M4-based microcontroller (MK66FX1M0VMD18, NXP, Eindhoven, The Netherlands) with wireless transceiver (Telemetry Radio V3, Holybro, Hong Kong, China), microSD card socket, and global positioning system (GPS) module (MTK3339, GlobalTop Technology Inc., Kaohsiung, Taiwan). The ARM Cortex-M4 comes with a 256 Kb Static Random-Access Memory (SRAM), 1280 Kb of Flash RAM, 4 Kb of EEPROM, 6 UART, 3 SPI, 4 i2c, 2 USB controllers, and 1 Ethernet port. It also has 100 programmable GPIO pins with 25 16-bit Timer and 4 32-bit Timer. The sprayer controller has a separate external power for the pump (as shown in Figure 6—pump power) and can be controlled remotely. The pump power is a separate power specific only for the diaphragm pump to minimize issues with high amp power requirement. Although this functionality was not used for this current field test. The top board as shown in Figure 6 is the spray unit controller while the bottom board is the ARM cortex controller.

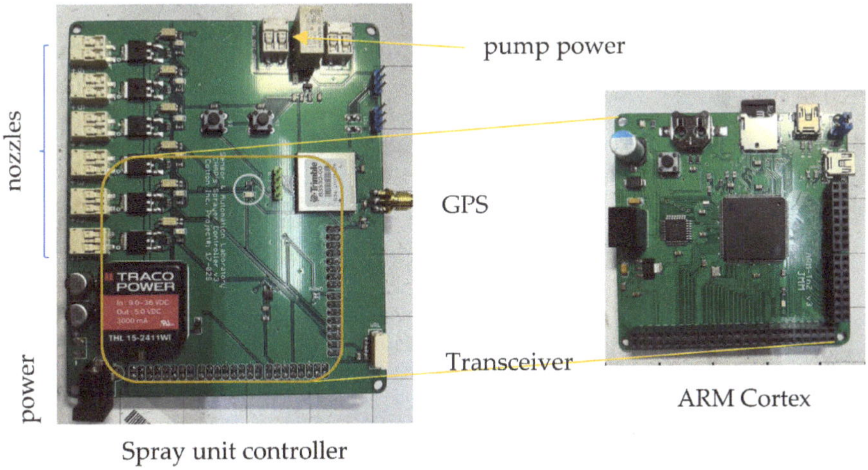

Figure 6. Sprayer controller developed for this project.

The sprayer controller was configured to use 10 Hz for the pulse width modulation frequency due to the specification requirement for the Capstan Ag coil assembly. Six GPIO pins were also configured to generate the pulse and each of the assigned pins can be configured from 0 to 100% duty cycle. Note that the duty cycle settings must be configured to the tip use and pressure of the spraying system. In this work, each nozzle was tested with the different duty cycles and determined the correct volume for the duty cycles—20%, 40%, 60%, 80%, and 100%, as shown in Table 1. Only three duty cycles (20%, 40%, and 60%) were used as a treatment for the field test.

Table 1. Laboratory test on duty cycle vs. Volume rate.

Duty Cycle (%)	Volume/Sec (mL/s)
20	13.0
40	22.8
60	31.9
80	42.7
100	49.5

The firmware code that runs on the board regularly monitor for incoming character for a new update on the duty cycle for each of the nozzle. At the same time, the controller also transmits the GPS, voltage of the board, and the different duty cycle assigned to each of the nozzle. The whole sprayer controller was housed in a waterproof enclosure with transparent lid at the back of the sprayer unit as shown in Figure 7.

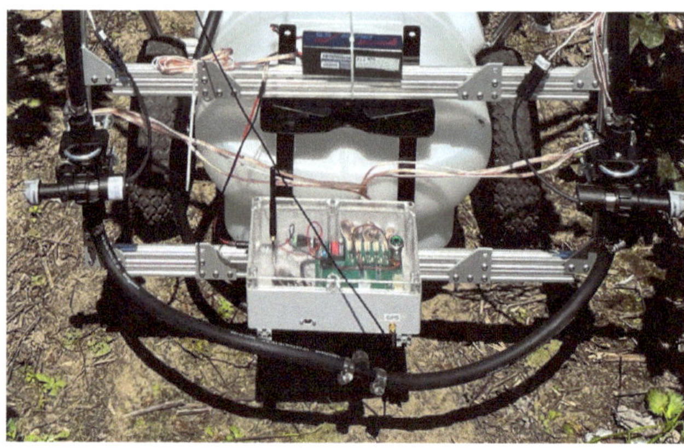

Figure 7. Placement of the sprayer unit enclosure.

2.4. Cotton Cultivars and Defoliant Chemicals

Delta pine cultivars (DP 2038B3XF and D10 DP 2055) were planted in the first and second fields, respectively. The first field was planted on 12 May 2022, while the second field was planted on 25 May 2022. The management practices used for both fields were according to the South Carolina cotton growers' guide. A mixture of three different chemicals was used for the spraying; the information is shown in Table 2.

Table 2. Chemical formulations, active ingredients, and application rates used for this research.

Product Formulation	Active Ingredient	Rate	Remarks
Folex 6 EC	Tribufos	454 g/38 L	Cotton defoliant
Free fall SC	Thidiazuron	91 g/38 L	Cotton defoliant
Super boll	Ethephon	907 g/38 L	Plant regulator

2.5. Data Collection and Analysis

A total of 80 plants from Field 1 and 40 plants from Field 2 were randomly selected and tagged with red thread before the treatment application. Three different duty cycles were used for the field test as treatment: 20% (13.0 mL/s), 40% (22.8 mL/s), and 60% (31.9% mL/s). In this field test, control represented the plants where defoliation was conducted using a conventional tractor-mounted sprayer. The data on plant height, node count, and total boll was collected. The number of leaves on the same tagged plants was counted at the 0th, 4th, 8th, 12th, 16th, and 20th days after treatment application (DAT). The defoliation rate was calculated using Equation (1):

$$Defoliation\ Rate = \frac{lfc_n - lfc_{n+1}}{lfc_n} \times 100 \tag{1}$$

where *lfc* is leaf count, and *n* is the days when the count was made. Note *n* is 0, 4, 8, 12, 16, and 20.

Similarly, a water-sensitive paper (WSP) was used to study the droplet characteristics. Ten plants were randomly selected per treatment, and WSP were placed at three different canopy heights, i.e., lower, middle, and upper canopy as shown in Figure 8. The paper was collected after drying and kept in a yellow envelope to prevent moisture contamination. The WSP papers were then scanned using a WSP scanner (DropScope, SprayX, São Carlos, Brazil), providing different droplet characteristics. Likewise, cotton bolls were harvested manually on a standard sample length of 3 m row to study the yield after ginning, and the sample cotton fiber was sent to the Cotton Incorporated Laboratory for quality analysis. One-way ANOVA and Tukey test were conducted to study treatment effects and mean separation.

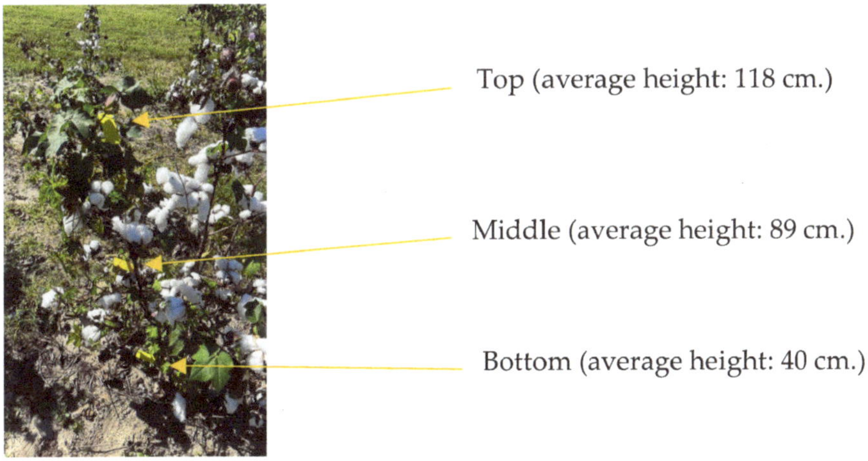

Figure 8. WSP placement during the field sprayer test.

3. Results and Discussion

3.1. Defoliation Rate

Overall, the control has higher mean defoliation (Figure 9). Here control represents the current practice in which spraying was conducted using the conventional tractor-mounted sprayer. There is no significant difference in defoliation across the three treatment levels, meaning a 20% duty cycle does the same results as a 60% duty cycle. Even though, the volume output is significantly higher at 60% compared to the 20% duty cycle. Note that there were three nozzles that sprayed one side of the cotton plants, which means that we do not have any issues with penetration as compared to the tractor-mounted spray where the nozzles were pointing to the top of the cotton canopy. Our spray systems successfully delivered the defoliants to the three levels of the cotton plants. The results could be attributed that even with less volume of defoliants (20%) is enough to defoliate the plants if the chemicals were sprayed from top to bottom. This is a very promising result as there are no study yet on cotton defoliation that sprayed the side of the cotton. This means that farmers can reduce the use of defoliants and focus on the delivery of the chemicals.

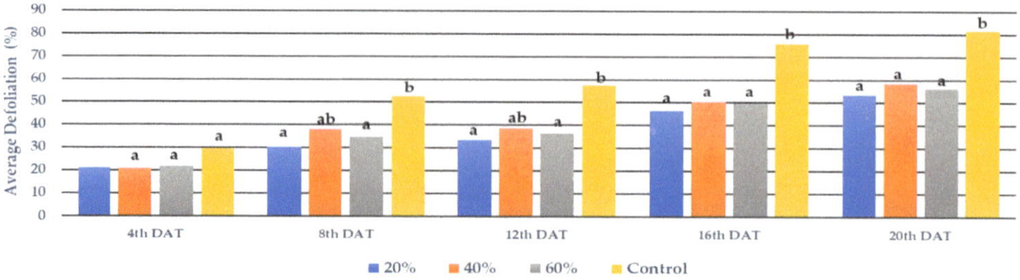

Figure 9. Average defoliation percentage on different days after defoliant application in Field 1.

Moreover, during the spraying on the research plot with the new sprayer, there was an unexpected light rainfall just after the defoliant application. In contrast, the rain stopped when the control plot was sprayed with the tractor-mounted spray. This could be the reason behind the high defoliation rate on control treatment. In addition, two consecutive freezing nights one week after the spraying could have affected the defoliation on the three treatments. Freezing night temperatures can kill the leaves before the development of the abscission layer between the petiole and stem.

The control has higher mean defoliation in Field 2, as shown in Figure 10. The results were similar to Field 1 except on the 16th day, where all treatments, including the control, had statistically similar defoliation. The defoliation is significantly higher at 20% on the 4th, 8th, and 12th day compared to the 60% duty cycle. In contrast, there is no significant difference in defoliation rate across the treatments on the 16th and 20th day.

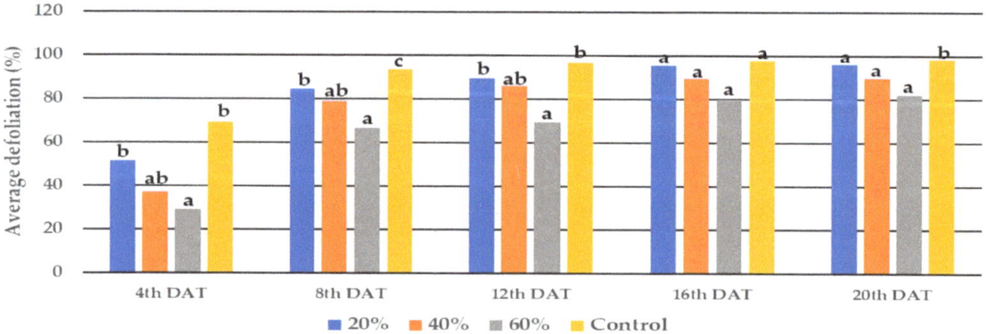

Figure 10. Average defoliation percentage on different days after defoliant application (Field 2).

3.2. *Droplet Characteristics*

The ANOVA analysis for droplet deposition, density, and droplet drift potential (see Tables 3–5) showed no significant treatment effect. Meaning using 20%~60%duty cycles have the same effect on droplet characteristics.

Table 3. ANOVA analysis for droplet deposition (* $p \leq 0.05$, ** $p \leq 0.01$).

Source	Degree of Freedom	Sum of Squares	Mean of Squares	F-Value	*p*-Value
Treatment (duty cycle)	2	121.3	60.66	2.58	0.08269
Block (Canopy height)	2	328.8	164.39	6.99	0.00167 **
Interaction	4	277.4	69.36	2.95	0.02567 *
Residuals	72	1692.1	23.5		

Table 4. ANOVA analysis for droplet density (** $p \leq 0.01$).

Source	Degree of Freedom	Sum of Squares	Mean of Squares	F-Value	*p*-Value
Treatment (duty cycle)	2	60,365	30,183	2.557	0.08457
Block (Canopy height)	2	146,508	73,254	6.206	0.0032 **
Interaction	4	46,643	11,661	0.988	0.41978
Residuals	72	849933	11,805		

Table 5. ANOVA analysis for drift potential (** $p \leq 0.01$).

Source	Degree of Freedom	Sum of Squares	Mean of Squares	F-Value	*p*-Value
Treatment (duty cycle)	2	0.00464	0.02318	1.634	0.2022
Block (Canopy height)	2	0.1918	0.09589	6.762	0.00204 **
Interaction	4	0.1102	0.02754	1.942	0.11262
Residuals	72	1.0209	0.01418		

3.3. Cotton Yield

Figure 11 presents the average cotton boll number and weight in 10-foot cotton rows per treatment from both fields. No statistical comparisons are present in this figure, and all treatments weigh more than the control in both fields. As mentioned in the methodology section, the treatment plot has skipped rows; in contrast, control rows do not have skip rows. With the increase in row spacing, the competition of plants for nutrients and sunlight will decrease, which could cause a higher yield on the treatment side. Similarly, Field 1 has a higher yield in every treatment than Field 2, which could be due to differences in soil nutrient level, planting date, and soil moisture.

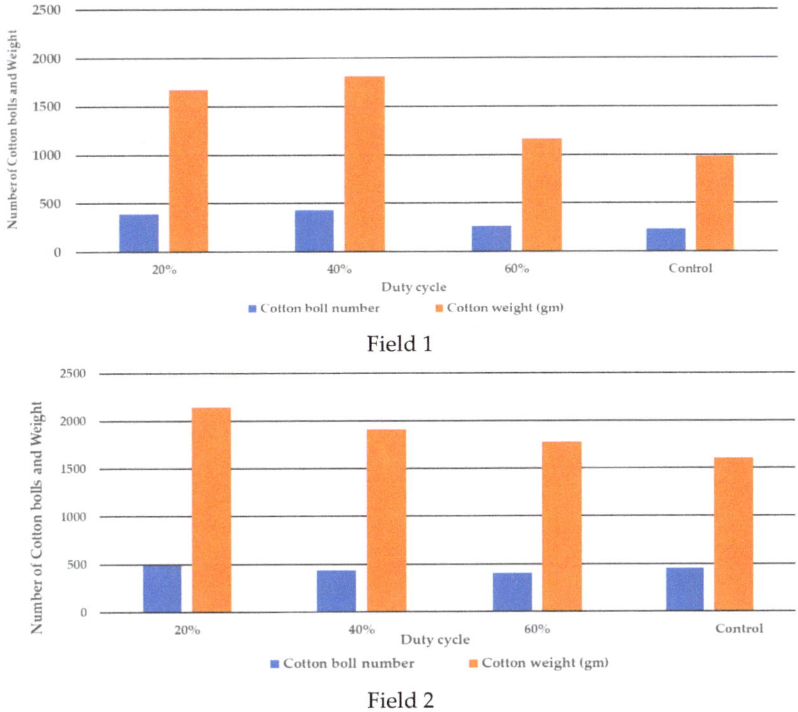

Figure 11. Average cotton boll and weight data for both fields across treatments.

4. Discussions

The results for Figure 9 showed that the duty cycles and defoliation rate combination of using UGV and specialized spraying system has a huge potential to apply precision agriculture technology. The results showed that chemical penetration is important as compared to volume.

5. Conclusions

The combination of using UGV and a specialized spraying system has a huge potential to apply precision agriculture technology. This preliminary work showed an unexpected result. Although the assumptions that more chemicals should result in a higher defoliation rate, results on these experiments showed that higher duty cycles (high volumes) do not result in higher defoliation rates. For example, in Field 1, the treatments and control have insignificant effects on the 4th day, but showed that on the 16th to 20th days, the control showed a higher defoliation rate than the treatments. Note that the 40% treatment is not statistically different from the rest of the treatment and control on the 8th and 12th day. In contrast, in Field 2, the 20% duty cycle has a higher defoliation rate on the 4th, 8th, and 12th days compared to other treatments, although not significant. The results showed that chemical penetration is more important as compared to volume, as shown in the results of Figures 9 and 10. Overall, these preliminary results for both fields showed that defoliation can be achieved even with a smaller amount of defoliant chemicals, as long as it is delivered properly to the cotton plants. Delivery of the defoliant chemicals is more important than volume.

Droplet characteristics (droplet distribution, density, and potential droplet drift) were insignificant across the treatments. Therefore, using only a 20% duty cycle is enough to defoliate based on the result of the field experiment. Higher spacing in treatment rows provides a better opportunity for nutrients and sunlight and thus results in higher yields than control rows. More experiments are needed to validate these results further and although this work is focused on cotton, the technology developed can also be applied to other agricultural crops.

Supplementary Materials: The following supporting information can be viewed at: Video S1: UGV Cotton Defoliation Sprayer during field test (https://youtu.be/i8SamBbH3N4 accessed on 24 January 2023).

Author Contributions: Conceptualization, J.M.M., J.N. and G.M.; methodology, J.M.M., J.N., J.L., G.M., M.M. and M.C.; software, J.N. and J.M.M.; validation, J.N. and J.L.; formal analysis, J.N. and J.L.; investigation, J.N. and J.M.M.; resources, J.M.M., G.M., M.M., J.G. and E.B.; data curation, J.N.; writing—original draft preparation, J.N. and J.M.M.; writing—review and editing, J.N. and J.M.M.; visualization, J.M.M.; supervision, J.M.M., G.M. and M.M.; project administration, J.M.M.; funding acquisition, J.M.M. and E.B. All authors have read and agreed to the published version of the manuscript.

Funding: This work was partially supported by a grant from Cotton Inc. Project No. 17-209 and is based on work supported by NIFA/USDA under project number SC-1700611. The authors express their gratitude to Clemson University Professional Internship and Co-op Program (UPIC). Student engagement is one of Clemson University's areas of investment. To meet this goal Clemson developed an on-campus internship and co-op program (UPIC) in 2012 to offer students the opportunity to work closely with a member or members of Clemson's faculty or administration in an on-campus or Clemson affiliated position.

Data Availability Statement: Not applicable.

Acknowledgments: The authors would like to thank Wilger Inc. for providing sprayer materials and the Edisto-REC Shop (Tim Still and Kim Still) for the help on the spraying equipment. Thanks to Jerry Han for developing the sprayer dashboard software to wirelessly control the sprayer.

Conflicts of Interest: The authors declare no conflict of interest.

References

1. Ahmad, F.; Sultanm, M. *Technology in Agriculture*; IntechOpen: London, UK, 2021.
2. Yadav, I.C.; Devi, N.L. Pesticides classification and its impact on human and environment. *Environ. Sci. Eng.* **2017**, *6*, 140–158.
3. Jin, D.; Wang, X.; Xu, Y.; Gui, H.; Zhang, H.; Dong, Q.; Sikder, R.K.; Yang, G.; Song, M. Chemical defoliant promotes leaf abscission by altering ros metabolism and photosynthetic efficiency in *Gossypium hirsutum*. *Int. J. Mol. Sci.* **2020**, *21*, 2738. [CrossRef] [PubMed]
4. Grisso, R.D.; Dickey, E.C.; Schulze, L.D. The Cost of Misapplication of Herbicides. *Appl. Eng. Agric.* **1989**, *5*, 344–347. [CrossRef]
5. Hawkins, N.J.; Bass, C.; Dixon, A.; Neve, P. The evolutionary origins of pesticide resistance: The evolutionary origins of pesticide resistance. *Biol. Rev.* **2019**, *94*, 135–155. [CrossRef]
6. Wilson, C. Environmental and human costs of commercial agricultural production in South Asia. *Int. J. Soc. Econ.* **2000**, *27*, 816–846. [CrossRef]
7. Bjørling-Poulsen, M.; Andersen, H.R.; Grandjean, P. Potential developmental neurotoxicity of pesticides used in Europe. *Environ. Health* **2008**, *7*, 50. [CrossRef]
8. FAO. *Pesticides Use, Pesticides Trade and Pesticides Indicators—Global, Regional and Country Trends, 1990–2020*; FAOSTAT Analytical Briefs 2022, no. 46; FAO: Rome, Italy, 2022. [CrossRef]
9. Ade, G.; Molari, G.; Rondelli, V. Recycling tunnel sprayer for pesticide dose adjustment to the crop environment. *Trans. ASABE* **2007**, *50*, 409–413. [CrossRef]
10. Jamar, L.; Mostade, O.; Huyghebaert, B.; Pigeon, O.; Lateur, M. Comparative performance of recycling tunnel and conventional sprayers using standard and drift-mitigating nozzles in dwarf apple orchards. *Crop Prot.* **2010**, *29*, 561–566. [CrossRef]
11. Cunningham, G.P.; Harden, J. Air-tower sprayers increase spray application efficiency in mature citrus trees. *Aust. J. Exp. Agric.* **1998**, *38*, 871–877. [CrossRef]
12. Panneton, B.; Lacasse, B.; Thériault, R. Penetration of spray in apple trees as a function of airspeed, airflow, and power for tower sprayers. *Can. Biosyst. Eng.* **2005**, *47*, 2–13.
13. Hanson, E.; Hancock, J.; Ramsdell, D.C.; Schilder, A.; VanEe, G.; Ledebuhr, R. Sprayer type and pruning affect the incidence of blueberry fruit rots. *HortScience* **2000**, *35*, 235–238. [CrossRef]
14. Gimenes, M.; Raetano, C.; Dal Pogetto, M.; Prado, E.; de Souza Christovam, R.; Rezende, D.; de Almeida Costa, S. Air-assistance in spray booms which have different spray volumes and nozzle types for chemically controlling *Spodoptera frugiperda* on corn. *J. Plant Prot. Res.* **2012**, *52*, 247–253. [CrossRef]
15. Cauquil, J. Cotton-pest control: A review of the introduction of ultra-low-volume (ULV) spraying in sub-Saharan French-speaking Africa. *Crop Prot.* **1987**, *6*, 38–42. [CrossRef]
16. Edward Law, S. Agricultural electrostatic spray application: A review of significant research and development during the 20th century. *J. Electrost.* **2001**, *51–52*, 25–42. [CrossRef]
17. Tewari, V.K.; Pareek, C.M.; Lal, G.; Dhruw, L.K.; Singh, N. Image processing based real-time variable-rate chemical spraying system for disease control in paddy crop. *Artif. Intell. Agric.* **2020**, *4*, 21–30. [CrossRef]
18. Terra, F.P.; Nascimento, G.H.; Duarte, G.A.; Drews-Jr, P.L.J. Autonomous agricultural sprayer using machine vision and nozzle control. *J. Intell. Robot. Syst.* **2021**, *102*, 38. [CrossRef]
19. Faiçal, B.S.; Costa, F.G.; Pessin, G.; Ueyama, J.; Freitas, H.; Colombo, A.; Fini, P.H.; Villas, L.; Osório, F.S.; Vargas, P.A.; et al. The use of unmanned aerial vehicles and wireless sensor networks for spraying pesticides. *J. Syst. Archit.* **2014**, *60*, 393–404. [CrossRef]
20. Faiçal, B.S.; Freitas, H.; Gomes, P.H.; Mano, L.Y.; Pessin, G.; de Carvalho, A.C.P.L.F.; Krishnamachari, B.; Ueyama, J. An adaptive approach for UAV-based pesticide spraying in dynamic environments. *Comput. Electron. Agric.* **2017**, *138*, 210–223. [CrossRef]
21. Cantelli, L.; Bonaccorso, F.; Longo, D.; Melita, C.D.; Schillaci, G.; Muscato, G. A small versatile electrical robot for autonomous spraying in agriculture. *AgriEngineering* **2019**, *1*, 391–402. [CrossRef]
22. Esau, T.; Zaman, Q.; Groulx, D.; Farooque, A.; Schumann, A.; Chang, Y. Machine vision smart sprayer for spot-application of agrochemical in wild blueberry fields. *Precis. Agric.* **2018**, *19*, 770–788. [CrossRef]
23. Llorens, J.; Gil, E.; Llop, J.; Escolà, A. Variable rate dosing in precision viticulture: Use of electronic devices to improve application efficiency. *Crop Prot.* **2010**, *29*, 239–248. [CrossRef]
24. Asaei, H.; Jafari, A.; Loghavi, M. Site-specific orchard sprayer equipped with machine vision for chemical usage management. *Comput. Electron. Agric.* **2019**, *162*, 431–439. [CrossRef]
25. Kim, J.; Kim, S.; Ju, C.; Son, H.I. Unmanned Aerial Vehicles in Agriculture: A Review of Perspective of Platform, Control, and Applications. *IEEE Access* **2019**, *7*, 105100–105115. [CrossRef]
26. Chen, P.; Xu, W.; Zhan, Y.; Wang, G.; Yang, W.; Lan, Y. Determining application volume of unmanned aerial spraying systems for cotton defoliation using remote sensing images. *Comput. Electron. Agric.* **2022**, *196*, 106912. [CrossRef]
27. Barnes, E.; Morgan, G.; Hake, K.; Devine, J.; Kurtz, R.; Ibendahl, G.; Sharda, A.; Rains, G.; Snider, J.; Maja, J.M.; et al. Opportunities for Robotic Systems and Automation in Cotton Production. *AgriEngineering* **2021**, *3*, 339–362. [CrossRef]
28. Maiti, R.; Kumari, C.A.; Huda, A.K.S.; Mandal, D.; Begum, S. *Advances in Cotton Science: Botany, Production, and Crop Improvement*; CRC Press: Boca Raton, FL, USA, 2020.

29. Stewart, J.M.; Oosterhuis, D.; Heitholt, J.J.; Mauney, J.R. *Physiology of Cotton*; Springer Science & Business Media: Berlin/Heidelberg, Germany, 2009.
30. Rönnbäck, K.; Theodoridis, D. Cotton cultivation under colonial rule in India in the nineteenth century from a comparative perspective. *Econ. Hist. Rev.* **2022**, *75*, 374–395. [CrossRef]
31. Shaheen, T.; Tabbasam, N.; Iqbal, M.A.; Ashraf, M.; Zafar, Y.; Paterson, A.H. Cotton genetic resources. A review. *Agron. Sustain. Dev.* **2012**, *32*, 419–432.
32. Wegier, A.; Alavez, V.; Piñero, D. Cotton: Traditional and modern uses. In *Ethnobotany of Mexico: Interactions of People and Plants in Mesoamerica*; Lira, R., Casas, A., Blancas, J., Eds.; Springer: New York, NY, USA, 2016; pp. 439–456. [CrossRef]
33. Singh, B.P. Overview of industrial crops. In *Industrial Crops and Uses*; CAB International: Wallingford, UK, 2010. [CrossRef]
34. Jones, M.; Farmaha, B.; Greene, J.; Marshall, M.; Mueller, J. *South Carolina Cotton Growers Guide*; Clemson University: Clemson, SC, USA, 2019.
35. Chu, T.; Chen, R.; Landivar, J.A.; Maeda, M.M.; Yang, C.; Starek, M.J. Cotton growth modeling and assessment using unmanned aircraft system visual-band imagery. *J. Appl. Remote Sens.* **2016**, *10*, 036018. [CrossRef]
36. Sun, S.; Li, C.; Paterson, A.H.; Jiang, Y.; Xu, R.; Robertson, J.S.; Snider, J.L.; Chee, P.W. In-field high throughput phenotyping and cotton plant growth analysis using lidar. *Front. Plant Sci.* **2018**, *9*, 16. [CrossRef]
37. Ritchie, G.; Bednarz, C.W.; Jost, P.H.; Brown, S.M. *Cotton Growth and Development*; University of Georgia: Athens, GA, USA, 2004.
38. Wright, D.L.; Sprenkel, R.K.; Marois, J.J.; Plant, C. *Cotton Growth and Development*; University of Florida, Agronomy Department, Florida Cooperative Extension Services, IFAS Extension, Publication SSAGR-238: Gainsville, FL, USA, 2005.
39. Xin, F.; Zhao, J.; Zhou, Y.; Wang, G.; Han, X.; Fu, W.; Deng, J.; Lan, Y. Effects of dosage and spraying volume on cotton defoliants efficacy: A case study based on application of unmanned aerial vehicles. *Agronomy* **2018**, *8*, 85. [CrossRef]
40. Meng, L.; Zhang, L.; Qi, H.; Du, M.; Zuo, Y.; Zhang, M.; Tian, X.; Li, Z. Optimizing the application of a novel harvest aid to improve the quality of mechanically harvested cotton in the North China Plain. *J. Integr. Agric.* **2021**, *20*, 2892–2899. [CrossRef]
41. Addicott, F.T.; Lynch, R.S. Defoliation and desiccation: Harvest-aid practices. *Adv. Agron.* **1957**, *9*, 67–93. [CrossRef]
42. Karademir, E.; Karademir, C. *Cotton Production under Abiotic Stress*; Iksad Publications: Ankara, Turkey, 2021. Available online: https://www.google.com/url?sa=t&rct=j&q=&esrc=s&source=web&cd=&ved=2ahUKEwiLtPXO8OL8AhWHD1kFHTEwC-gQFnoECA8QAQ&url=https%3A%2F%2Fiksadyayinevi.com%2Fwp-content%2Fuploads%2F2021%2F12%2FCOTTON-PRODUCTION-UNDER-ABIOTIC-STRESS.pdf&usg=AOvVaw13l7ZwnBpKaBmE4an88BQ5 (accessed on 22 November 2022).
43. Snipes, C.E.; Cathey, G.W. Evaluation of defoliant mixtures in cotton. *Field Crops Res.* **1992**, *28*, 327–334. [CrossRef]
44. Meena, R.A.; Monga, D.; Ratna, S. Effect of defoliation on maturity behaviour and seed cotton yield in cotton. *J. Cotton Res. Dev.* **2016**, *30*, 63–65.
45. Eaton, F.M. Physiology of the cotton plant. *Annu. Rev. Plant. Physiol.* **1955**, *6*, 299–328. [CrossRef]
46. Abbas, I.; Liu, J.; Faheem, M.; Noor, R.S.; Shaikh, S.A.; Solangi, K.A.; Raza, S.M. Different real-time sensor technologies for the application of variable-rate spraying in agriculture. *Sens. Actuators A Phys.* **2020**, *316*, 112265. [CrossRef]
47. Sumner, H.R.; Dowler, C.C.; Garvey, P.M. Application of agrichemicals by chemigation, pivot-attached sprayer systems, and conventional sprayers. *Appl. Eng. Agric.* **2000**, *16*, 103. [CrossRef]
48. Dou, Z.; Fang, Z.; Han, X.; Liu, Y.; Duan, L.; Zeeshan, M.; Arshad, M. Comparison of the Effects of Chemical Topping Agent Sprayed by a UAV and a Boom Sprayer on Cotton Growth. *Agronomy* **2022**, *12*, 1625. [CrossRef]
49. Salcedo, R.; Zhu, H.; Zhang, Z.; Wei, Z.; Chen, L.; Ozkan, E.; Falchieri, D. Foliar deposition and coverage on young apple trees with PWM-controlled spray systems. *Comput. Electron. Agric.* **2020**, *178*, 105794. [CrossRef]
50. Mahmud, M.S.; Zahid, A.; He, L.; Choi, D.; Krawczyk, G.; Zhu, H.; Heinemann, P. Development of a LiDAR-guided section-based tree canopy density measurement system for precision spray applications. *Comput. Electron. Agric.* **2021**, *182*, 106053. [CrossRef]
51. Xue, X.; Lan, Y.; Sun, Z.; Chang, C.; Hoffmann, W.C. Develop an unmanned aerial vehicle based automatic aerial spraying system. *Comput. Electron. Agric.* **2016**, *128*, 58–66. [CrossRef]
52. Mahmud, M.S.A.; Abidin, M.S.Z.; Emmanuel, A.A.; Hasan, H.S. Robotics and automation in agriculture: Present and future applications. *Appl. Model. Simul.* **2020**, *4*, 130–140. Available online: http://arqiipubl.com/ojs/index.php/AMS_Journal/article/view/130 (accessed on 8 February 2022).
53. Manor, G.; Hofner, A.; Or, R.; Phishler, G.; Epstein, Y.; Nakash, T.; Jacobi, M. Air Stream Facilitated Application of Cotton Foliage Treatments. *Trans. ASAE* **1989**, *32*, 0037–0038. [CrossRef]
54. Cavalaris, C.; Karamoutis, C.; Markinos, A. Efficacy of cotton harvest aids applications with unmanned aerial vehicles (UAV) and ground-based field sprayers–A case study comparison. *Smart Agric. Technol.* **2022**, *2*, 100047. [CrossRef]
55. Xiao, Q.; Xin, F.; Lou, Z.; Zhou, T.; Wang, G.; Han, X.; Fu, W. Effect of aviation spray adjuvants on defoliant droplet deposition and cotton defoliation efficacy sprayed by unmanned aerial vehicles. *Agronomy* **2019**, *9*, 217. [CrossRef]
56. Chen, P.; Xu, W.; Zhan, Y.; Yang, W.; Wang, J.; Lan, Y. Evaluation of Cotton Defoliation Rate and Establishment of Spray Prescription Map Using Remote Sensing Imagery. *Remote Sens.* **2022**, *14*, 4206. [CrossRef]

57. Leon, R.G.; Wright, D.L.; Brecke, B.J. 2020 Cotton Defoliation and Harvest Aid Guide, UF-IFAS Extension 2020. Available online: https://edis.ifas.ufl.edu/publication/AG188 (accessed on 2 February 2023).
58. Singh, M.; Sharma, K.; Mishra, P.K.; Dixit, A.K.; Patel, B. Defoliation Effect on Cotton Crop by using Different Sprayers. *Agric. Eng. Today* **2017**, *41*, 31–34.
59. Holtz, J. Pulse width modulation for electronic power conversion. *Proc. IEEE* **1994**, *82*, 1194–1214. [CrossRef]
60. Zhu, H.; Lan, Y.; Wu, W.; Hoffmann, W.C.; Huang, Y.; Xue, X.; Fritz, B. Development of a PWM precision spraying controller for unmanned aerial vehicles. *J. Bionic Eng.* **2010**, *7*, 276–283. [CrossRef]
61. Salcedo, R.; Zhu, H.; Ozkan, E.; Falchieri, D.; Zhang, Z.; Wei, Z. Reducing ground and airborne drift losses in young apple orchards with PWM-controlled spray systems. *Comput. Electron. Agric.* **2021**, *189*, 106389. [CrossRef]
62. Butts, T.R.; Butts, L.E.; Luck, J.D.; Fritz, B.K.; Hoffmann, W.C.; Kruger, G.R. Droplet size and nozzle tip pressure from a pulse-width modulation sprayer. *Biosyst. Eng.* **2019**, *178*, 52–69. [CrossRef]
63. Mangus, D.L.; Sharda, A.; Engelhardt, A.; Flippo, D.; Strasser, R.; Luck, J.D.; Griffin, T. Analyzing the nozzle spray fan pattern of an agricultural sprayer using pulse width modulation technology to generate an on-ground coverage map. *Trans. ASABE* **2017**, *60*, 315–325.
64. Grella, M.; Gioelli, F.; Marucco, P.; Zwertvaegher, I.; Mozzanini, E.; Mylonas, N.; Nuyttens, D.; Balsari, P. Field assessment of a pulse width modulation (Pwm) spray system applying different spray volumes: Duty cycle and forward speed effects on vines spray coverage. *Precis. Agric.* **2022**, *23*, 219–252. [CrossRef]
65. Womac, A.R.; Melnichenko, G.; Steckel, L.; Montgomery, G.; Hayes, R.M. Spray tip effect on glufosinate canopy deposits in Palmer amaranth (*Amaranthus palmeri*) for pulse-width modulation versus air-induction technologies. *Trans. ASABE* **2016**, *59*, 1597–1608.
66. Boatwright, H.; Zhu, H.; Clark, A.; Schnabel, G. Evaluation of the intelligent sprayer system in peach production. *Plant Dis.* **2020**, *104*, 3207–3212. [CrossRef]
67. Gormus, O.; El Sabagh, A.; Kurt, F. Impact of defoliation timings and leaf pubescence on yield and fiber quality of cotton. *J. Agric. Sci. Technol.* **2017**, *19*, 903–915.
68. Snipes, C.E.; Baskin, C.C. Influence of early defoliation on cotton yield, seed quality, and fiber properties. *Field Crops Res.* **1994**, *37*, 137–143. [CrossRef]
69. Faircloth, J.C.; Edmisten, K.L.; Wells, R.; Stewart, A.M. The influence of defoliation timing on yields and quality of two cotton cultivars. *Crop Sci.* **2004**, *44*, 165–172. [CrossRef]
70. Bourland, F.M.; Tugwell, N.P.; Oosterhuis, D.M.; Cochran, M.J. *Initial Development of COTMAN Program*; COTMAN Crop Management System; Arkansas Agricultural Experiment Station: Fayetteville, AR, USA, 2008; pp. 15–19.
71. Bynum, J.B.; Cothren, J.T. Indicators of last effective boll population and harvest aid timing in cotton. *Agron. J.* **2008**, *100*, 1106–1111. [CrossRef]
72. Clay, P.A.; Young, K.M.; Taylor, E.R. *Effect of Heat Unit Accumulation on Cotton Defoliation, Lint Yield and Fiber Quality*; University of Arizona: Tucson, AZ, USA, 2006. Available online: https://repository.arizona.edu/handle/10150/198201 (accessed on 24 January 2023).
73. Gwathmey, C.; Craig, C., Jr. Defoliants for cotton. In *Encyclopedia of Pest Management*; CRC Press: Boca Raton, FL, USA, 2007; pp. 135–137.
74. Weicai, Q.; Xinyu, X.; Longfei, C.; Qingqing, Z.; Zhufeng, X.; Feilong, C. Optimization and test for spraying parameters of cotton defoliant sprayer. *Int. J. Agric. Biol. Eng.* **2016**, *9*, 63–72.
75. Logan, J.; Gwathmey, C.O. Effects of weather on cotton responses to harvest-aid chemicals. *J. Cotton Sci.* **2002**, *6*, 1–12.
76. Wang, H.; Gao, K.; Fang, S.; Zhou, Z. Cotton yield and defoliation efficiency in response to nitrogen and harvest aids. *Agron. J.* **2019**, *111*, 250–256. [CrossRef]
77. Hake, S.J.; Kerby, T.A.; Hake, K.D. *Cotton Production Manual*; University of California, Agriculture and Natural Resources: Berkeley, CA, USA, 1996.
78. Wills, G.D.; Snipes, C.E. *Influence of Various Compounds and Temperature on Activity of Dropp Defoliant on Cotton*; Office of Agricultural Communications (Publications Section), Division of Agriculture, Forestry, and Veterinary Medicine, Mississippi State University: Starkville, MS, USA, 1996.
79. Nelson, J.M.; Bartels, P.G.; Hart, G. Effect of Plant Water Status on Defoliation and Yield of Upland Cotton for Short-Season Production. In *Cotton: A College of Agriculture Report*; College of Agriculture, University of Arizon: Tucson, AZ, USA, 1992.
80. Liao, J.; Zang, Y.; Luo, X.; Zhou, Z.; Zang, Y.; Wang, P.; Hewitt, A.J. The relations of leaf area index with the spray quality and efficacy of cotton defoliant spraying using unmanned aerial systems. *Comput. Electron. Agric.* **2020**, *169*, 105228. [CrossRef]
81. De Lima, I.D.S., Jr.; Degrande, P.E.; de Souza, C.; Viegas Neto, A.L. Droplet distribution as a function of cotton interrow spacing and angles of sprayer displacement. *Eng. Agrícola* **2018**, *38*, 927–933. [CrossRef]
82. Chen, P.; Ouyang, F.; Wang, G.; Qi, H.; Xu, W.; Yang, W.; Zhang, Y.; Lan, Y. Droplet distributions in cotton harvest aid applications vary with the interactions among the unmanned aerial vehicle spraying parameters. *Ind. Crops Prod.* **2021**, *163*, 113324. [CrossRef]
83. Zhu, H.; Dorner, J.W.; Rowland, D.L.; Derksen, R.C.; Ozkan, H.E. Spray penetration into peanut canopies with hydraulic nozzle tips. *Biosyst. Eng.* **2004**, *87*, 275–283. [CrossRef]

84. Meng, Y.; Song, J.; Lan, Y.; Mei, G.; Liang, Z.; Han, Y. Harvest aids efficacy applied by unmanned aerial vehicles on cotton crop. *Ind. Crops Prod.* **2019**, *140*, 111645. [CrossRef]
85. Maja, J.M.; Polak, M.; Burce, M.E.; Barnes, E. CHAP: Cotton-Harvesting Autonomous Platform. *AgriEngineering* **2021**, *3*, 199–217. [CrossRef]

Disclaimer/Publisher's Note: The statements, opinions and data contained in all publications are solely those of the individual author(s) and contributor(s) and not of MDPI and/or the editor(s). MDPI and/or the editor(s) disclaim responsibility for any injury to people or property resulting from any ideas, methods, instructions or products referred to in the content.

agriculture

Article

Engineering Design, Kinematic and Dynamic Analysis of High Lugs Rigid Driving Wheel, a Traction Device for Conventional Agricultural Wheeled Tractors

Hafiz Md-Tahir [1,2], Jumin Zhang [1], Yong Zhou [1], Muhammad Sultan [2], Fiaz Ahmad [2], Jun Du [1], Amman Ullah [2], Zawar Hussain [2] and Junfang Xia [1,*]

[1] Key Laboratory of Agricultural Equipment in Mid-Lower Yangtze River, Ministry of Agriculture, College of Engineering, Huazhong Agricultural University, No. 1, Shizishan Street, Hongshan District, Wuhan 430070, China
[2] Department of Agricultural Engineering, Bahauddin Zakariya University, Multan 60800, Pakistan
* Correspondence: xjf@mail.hzau.edu.cn; Tel.: +86-027-87282120

Abstract: Traction failure on loose terrain is common in conventional agricultural wheeled tractors due to poor traction ability and lower power transfer efficiency of drive wheels, which leads to excessive energy consumption and soil compaction in agriculture. To overcome the problem, this paper presents a new design of a rigid lugged wheel for use in field tillage operations. This wheel was designed according to field operational requirements and also provided with rubber pads for smooth on-road drives. Kinematic and dynamic analysis of new wheel designs were carried out to study how they move, how they interact with the soil, and how they generate drive force in loose terrain soil. The relationship of wheel lug motion trajectories, displacement, and velocity of the wheel relative to field conditions, different travel reduction rates, and lug penetration/wheel sinkage were analyzed. Wheel-terrain interaction and shear stress-shear displacement relationships when the wheel is driven in soft, deformable terrain were studied using classic soil mechanics principles. It is found that the component of thrust in the direction of driving, i.e., driving force, is ranged between 81.52% and 86.17%, while the vertical component is reported to be less than 30% and further decreases to 9%, which is the compaction avoiding factor. The relationships, thus developed, of wheel parameters, soil stress and thrust characteristics, and wheel drive force were derived and revealed that the traction performance, power transfer efficiency, and trafficability of tractors in loose terrain can be improved by using the newly proposed wheel. A finite element method was used to analyze the designed wheel model for structural stability and optimization. The theoretical analysis results of the new drive wheel are convincing, so further tests and field operation research are recommended for sustainable adoption.

Keywords: intensive tillage; high traction demand; loose field soil; traction failure; rigid lugged wheel; kinematic and dynamic analysis; traction performance

Citation: Md-Tahir, H.; Zhang, J.; Zhou, Y.; Sultan, M.; Ahmad, F.; Du, J.; Ullah, A.; Hussain, Z.; Xia, J. Engineering Design, Kinematic and Dynamic Analysis of High Lugs Rigid Driving Wheel, a Traction Device for Conventional Agricultural Wheeled Tractors. *Agriculture* **2023**, *13*, 493. https://doi.org/10.3390/agriculture13020493

Academic Editor: Jin He

Received: 30 December 2022
Revised: 9 February 2023
Accepted: 13 February 2023
Published: 19 February 2023

Copyright: © 2023 by the authors. Licensee MDPI, Basel, Switzerland. This article is an open access article distributed under the terms and conditions of the Creative Commons Attribution (CC BY) license (https://creativecommons.org/licenses/by/4.0/).

1. Introduction

Conventional agricultural tractors are prominent off-road ground vehicles and the main power source in field tillage operations for crop cultivation. Tillage and ploughing are the major cultivation operations, so the traction demand for tillage is increasing with the development of mechanized agriculture [1]. Consequently, the size of tractors has increased in recent years, which requires maintaining the power-to-weight ratio of tractors for efficient operations, soil management, and energy conservation [2]. Weight control for all agricultural vehicles, particularly tractors, is strongly recommended for arable soil management and for reducing the energy demand of crop production [1,3–5].

Tractor traction or drawbar force development through wheel-soil interaction is a complex process that adversely affects the terrain, soil, and energy demand of the tractor [1]. Researchers around the world are working to find solutions for efficient traction development to prevent soil compaction and promote energy conservation. All the attempts that were made for the purpose are adhesion-based and focused on increasing the wheel-terrain contact area and normal axle loads at contact points [3,6–8]. All-wheel drives, dual and multiple wheels, ballasting techniques, inflation pressure adjustments, and split and continuous type rubber tracks are examples of practiced methods. Due to their size, weight, and resistance to movement, these methods cannot be used in agricultural field operations [4].

Lugged-type rigid wheels have some advantages over tire wheels for traction intensification on loose deformable terrains. Many forms of lugged wheels are in use in agriculture, especially in low-land farming and paddy fields, on conventional tractors and boat-type tractors that have limited applicability in some specific conditions [9]. Some new types of moveable lugged wheels and transformable walking wheels were developed for ground vehicles that are specific to small machines and not applicable to conventional agricultural tractors [10,11]. To further enhance the applicability and advantage of lug-type rigid wheels, it is required to improve their structure and design characteristics, with a special focus on agricultural field conditions.

With regard to the agricultural environment and field conditions, energy crises, agricultural sustainability issues, soil degradation caused by mechanical compaction issues (mainly by machine weight), and environmental quality problems, it is of great importance to design highly efficient traction wheels for agricultural tractors [9,11–13]. For designing and studying agricultural vehicles, especially agricultural wheeled tractors, it is important to consider machine weight, field mobility, traction ability (pull ability), soil trafficability, and soil behavior under applied wheel forces. It is more effective to use terramechanics principles, soil mechanics theories, and vehicle engineering principles to make the operations energy-efficient, economical, and environment-friendly [14–17].

A new design of the traction device "rigid lugged wheel" presented in this paper was designed for the high traction demand of intensive tillage operations with a similar mounting mechanism to tire wheels [4,18]. It is easily replaceable with tire wheels and also provides rubber pad accessories to use for on-road driving. Rubber pad assemblies can be mounted and dismounted to meet field and road drive requirements. It is used in the field without rubber pads, where sharp edge high lugs (150 mm) penetrate to soil and apply exertion pressure to the terrain soil, causing thrust in response to the lug working surface. The horizontal component of this thrust force is the driving force of the wheel that moves the tractor in the forward direction and also provides drawbar force to draw the tillage implements/ploughs through the soil. The lugged wheel is intended to generate traction force through insertion pressure rather than shearing pressure, and the majority of vertical pressure is applied in the lateral direction. The thrust or drive force of a rigid lugged wheel is independent of normal axle load and only depends on the soil thrust (soil strength) and the area of the working faceplate of the lug. Theoretical analysis is conducted for the newly designed rigid lugged wheel that provides in-depth theoretical modeling to assess and confirm its working principles. It will elaborate on how this wheel would help to maximize the traction ability of agricultural wheeled tractors and how it interacts with the terrain. The FEM analysis will help optimize the model structure of the wheel for lighter weight and longer life for feasible adoption.

2. Design and Development

2.1. Design Concept

A wheel prototype with a new design structure for a high-lug rigid wheel was proposed. An easily mountable and replaceable lugged wheel, which is a novel mechanism for agricultural wheeled tractors, ensures full-time four-wheel drive in complex off-road terrain and diversified agricultural environments. Additionally, it is convenient to control the traction mechanism of wheeled tractors. As the rigid lugged wheel was designed to drive tractors on soft soil surfaces, due to the penetration of lugs into the soil, it may face some resistance force. The rigidly lugged wheel derives its driving force from the lugs' penetration into the soil, resulting in a pre-tillage effect on the terrain soil beneath the tracks.

When the Agricultural wheeled tractor is driven on soft terrain, the wide surface rim of the rigid lugged wheel contacts the soft surface to support most of the weight and avoid soil compaction. The trafficability of Agricultural wheeled tractors can be improved due to the larger contact area of the newly designed rigid lugged wheel as rear drive wheels. When Agricultural wheeled tractors are driven on hard terrain or paved roads, the road pad accessory components are mounted to the wheel and the rear wheels are adjusted to contact the ground for a smoother drive.

2.2. Design Requirements

The rigid lugged wheel as a driving wheel for the agricultural tractor can adjust its status based on the driving area and the different characteristics of the agricultural field soil. On soft terrain, due to the height of the lugs, the wheel will improve the tractive force performance, field operation energy consumption, and trafficability of the field soil. The rigid lugged wheel's main requirement is field operations and traction performance on soft, relatively weaker soils with less trafficability.

The height of the lugs of the rigid lugged wheel is relative to the normal tillage depth of 120 to 180 mm, so the lugs will get supported by the subsurface, and with a higher contact area, the ground pressure on soft soil will get reduced, which will help to avoid soil compaction. The working surface or working faceplate of the lug is the object of study in the design process. The driving force and traction ability of the wheel depend on the working surface's contact with the soil; as much as the lugs penetrate the soil, they deliver the driving force through soil contact.

The design parameters of the rigid lugged wheel should be determined in such a way that it can produce maximum driving force with optimum energy consumption and minimum damage to the working environment. Rim width, lug length, lug height, and lug angle, all should be in optimum working performance position. The rim setting should be adjusted to withstand wet soil working and avoid soil sticking and wheel blocking during wet soil field operations.

Based on the above considerations, three significant problems must be solved. First, a mechanism should be designed to transform the wheel into a lightweight and highly efficient to generate driving force, whereas by implementing tracks the weight of the tractor and internal motion resistance increased by a significant amount. Second, the power transmission or power transfer mechanism to pull the tillage implement must be solved. Third, these two mechanisms can be integrated to reduce the amount of energy and space required and also avoid soil compaction in agriculture.

2.3. Structure and Working Principle

The model of the proposed rigid lugged wheel was developed in Pro-E, and MATLAB was employed for the solution of the design equation and design calculations. This computerized model of a rigid lugged wheel is held to make further design calculations, justifications and confirmations, kinematic and dynamic analysis, and drive force calculations. Pro-E software is an element-based tool that first develops the parts individually and then unites them by joining the parts in patterns to make the full assembly. The preliminary design of the rigid lugged wheel is presented in Figure 1, where: 1. The rim of the wheel, 2. Lug (150 mm) and lug working faceplate (the main research object), 3. Rubber pad (for on-road drive) and fixing assembly, 4. Wheel axis, 5. Lug angle ($\alpha = 30°$), 6. The direction of motion or rolling of the wheel.

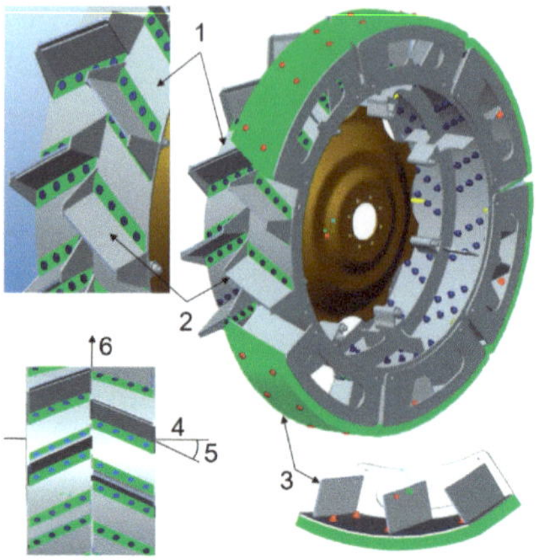

Figure 1. An illustration of the complete modeled structure of the rigid lugged wheel.

2.3.1. Wheel Structure

The whole wheel structure of the rigid lugged wheel consists of three main functional components. (1) Rim, including internal strengthening supports and mounting accessories. (2) Lugs that are arranged on the rim in herringbone arrangement at a 30° mounting angle with the axis of the wheel and lug height of 150 mm were set for optimum traction and driveability performance. (3) Road pad/rubber pad assemblies with internal supports and outer rubber pad that can be mounted to the wheel for on-road drive.

The main function and purpose of a rigid lugged wheel are field drive and adequate traction force provision for high-draught implements in diverse agricultural field conditions. Therefore, the road pad assemblies are only accessory items for minor road drives, while the rim and lugs are two major components for the functional requirements. The main design considerations were given to the two components, and some details of the designs are as follows:

The rim of the rigid lugged wheel is designed according to mounting provisions available for the tire wheel, as the same is used for rigid lugged wheel mounting, and according to available space and the drive mechanism. The width of the rigid lugged wheel rim was kept as wide as possible according to the drive criteria of the tractor and optimum working performance. Rim setting was adjusted to withstand all terrain/soil conditions and all working environments, such as dry and hard-untilled surfaces, soft loose tilled soil, and wet or saturated soil working conditions, and to avoid soil sticking and wheel blocking during wet soil field operations.

The main research object of this study was the lug of the rigid lugged wheel, so the design parameters of the lug are of much importance. On the lug, the working faceplate was the more specific object to design, arrange, and analyze for optimum wheel performance. The driving force of the lug is dependent on working surface parameters such as length and width, the mounting angle of the lug with the wheel axis, the angle of the lug's working faceplate according to the reference axis, and its interaction with soil (soil penetration). Therefore, consideration was given to lug design, and analysis was performed for lug design and lug arrangements.

2.3.2. Design of Lug Working Faceplate

Figure 2a shows the fixed coordinate system "xyz", where: the x-axis is the iron wheel axis, the y-axis points to the tractor's forward direction, and the z-axis is vertically upward. In the "$x_o y_o z_o$" coordinate system, that is obtained by rotating "xyz" about 30 degrees in the positive direction (counter-clockwise) around the z-axis.

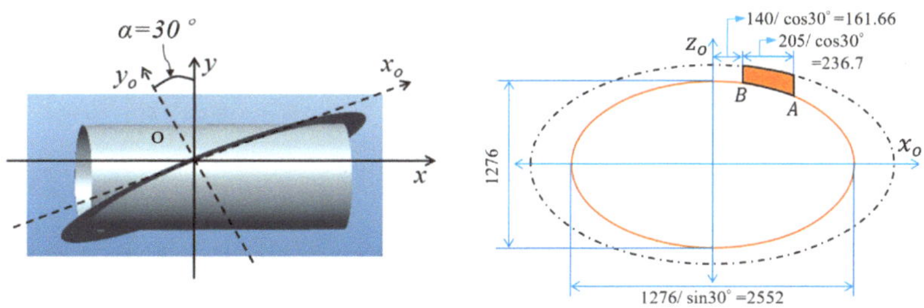

Figure 2. Lug face plate design. (**a**) Design model illustration, (**b**) Ellipses of lug working faceplate.

In this way, the z-axis remains the same, and the transformation relationship between the coordinate systems "xyz" and "$x_oy_oz_o$" is as follows:

$$\begin{cases} x = x_o \cos\alpha - y_o \sin\alpha \\ y = y_o \cos\alpha + x_o \sin\alpha \\ z = z_o \end{cases} \quad (1)$$

In the "$x_oy_oz_o$" coordinate system, the small ellipse at the lug base of the iron wheel (as shown in Figure 2b) corresponds to the following equation:

$$\frac{{x_o}^2 \sin^2\alpha}{R^2} + \frac{{z_o}^2}{R^2} = 1 \quad (2)$$

where: R = 1276/2 = 638 mm is the radius of the iron wheel rim; α = 30°; the small elliptic equation (2), now written as a parametric equation:

$$\begin{cases} x_o = \frac{R}{\sin\alpha} \cdot \sin\theta \\ y_o = 0 \\ z_o = R.\cos\theta \end{cases} \quad (3)$$

In Equation (3), θ is a small ellipse parameter, and the equation of a small ellipse in the xyz coordinate system is:

$$\begin{bmatrix} x \\ y \\ z \end{bmatrix} = A \begin{bmatrix} \frac{R}{\sin\alpha} \cdot \sin\theta \\ 0 \\ R\cos\theta \end{bmatrix}$$

By putting it in Equation (1), it results in:

$$\begin{bmatrix} x \\ y \\ z \end{bmatrix} = A \begin{bmatrix} \frac{R}{\tan\alpha} \cdot \sin\theta \\ R\sin\theta \\ R\cos\theta \end{bmatrix} \quad (4)$$

The range of x in the equation (according to Figure 2b) is $345 \geq x \geq 140$.

2.3.3. Mounting Angle of the Lug Working Faceplate

As shown in Figure 3a,b, line 1: iron wheel axis; plane 4: is a plane perpendicular to line 1 wheel axis; point p1: is the intersection of plane 4 and line 1 wheel axis; point p2: is the intersection of plane 4 and smaller elliptic curve; line 2: the intersection of point p1 and point p2. line 3: intersect line 2 at point p2 and go along the wheel lug surface and plane 4; angle gamma (γ): mounting angle of iron wheel lug faceplate; that is the angle between line 2 and line 3 at vertex point p2.

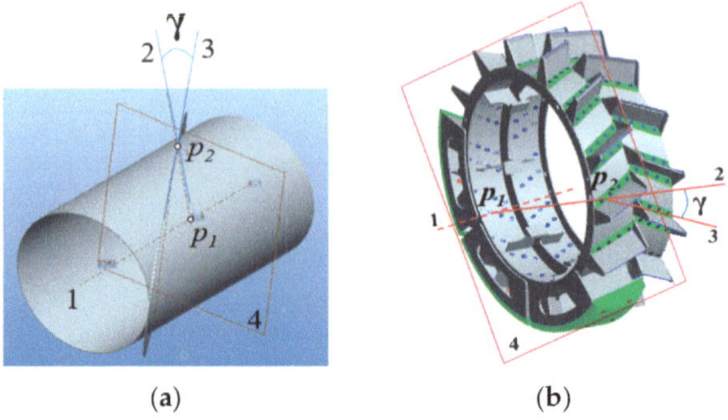

Figure 3. Schematic explanation diagram of working faceplate mounting angle (**a**) Design model, (**b**) Prototype "labeling is explained in paragraphs where cited".

In Figure 3, the obliquely upward direction vector of line 2 obtained from Equation (4) is as follows:

$$\vec{n}_2 = R\sin\theta.\ \vec{j} + R\cos\theta.\vec{k} \tag{5}$$

In the coordinate system xyz, line 3 is parallel to the z-axis, so the angle between line 2 and line 3 is the angle between the vector and the z-axis.

$$\cos\theta = \cos\gamma, \tag{6}$$

where: $\theta = \gamma$.

That is, the installation angle of the lug is equal to the parameter theta (θ) in the small elliptic parametric equation.

From Equation (4):

$$x = \frac{R}{\sin\alpha}\,.\sin\theta \;\Rightarrow\; \theta = \gamma = sin^{-1}\left(\frac{x\tan\alpha}{R}\right) \tag{7}$$

In the above equation, the value range of x is [140,345]; α = 30°; R = 638 mm;

The width of the lug along the axis of the iron wheel is 205 mm (see Figure 2b). The variable x in Equation (7) is replaced by the lug width parameter w (range: 0 to 205):

$$\theta = \gamma = sin^{-1}\left(\frac{(w+140)\tan\alpha}{R}\right) \tag{8}$$

According to Equation (8), the relationship curve between the installation angle of the lug and the width "w" along the axis of the wheel can be drawn, as shown in Figure 4.

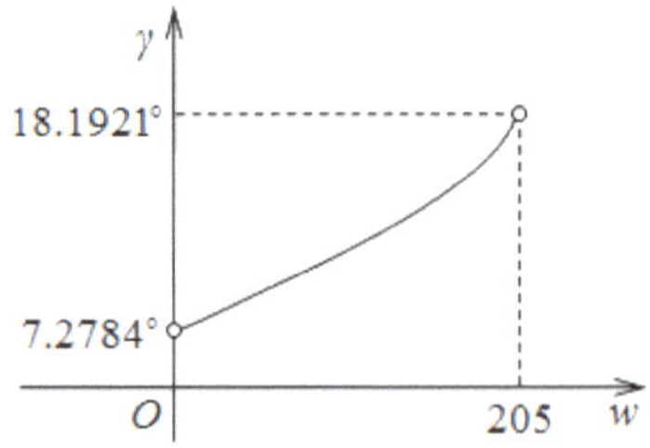

Figure 4. The relationship between the installation angle of the lug and its axial width.

To verify the correctness of Equation (8), the values of the installation angle "γ" at different values of w are obtained by a ProE three-dimensional modeling test, and the results are compared with those calculated by Equation (8). The results are shown in Table 1.

Comparisons between measured values of the three-dimensional model of installation angle gamma and calculated values of theoretical Equation (8) at different values of w.

Table 1. Measured and calculated values of the installation angle (γ) of the working faceplate in the 3D model.

	w (mm)	0	41	82	123	164	205
γ°	Model measurements	7.27844	9.4272	11.5894	13.7684	15.9680	18.1921
	Theoretical calculations	7.278357	9.427167	11.589370	13.768447	15.968037	18.192075

Because of the ProE 5.0 version of the three-dimensional software, its measurement accuracy is four digits after the decimal point. From Table 1, we can see that Equation (8) for calculating the installation angle γ is correct.

3. Kinematic Analysis

3.1. Driving Wheel Rotational and Translational Motion Relationship

It is assumed that when the tractor is driven on hard non-deformable or soft deformable terrain, the vehicle body maintains its movement in the horizontal plane. The vertical position is not affected by the rotation of the driving wheel, which is the rigid lugged wheel. Therefore, the driving wheel axis maintains a horizontal plane of movement. The radius R_o represents the rolling radius of the wheel on hard un-deformable terrain; on soft, deformable terrain, when lugs penetrate the soil and the rim is not submerged, the rolling radius of the wheel is R, which is the radius of the rim. If the rim is submerged, R_1 represents the rolling radius of the wheel.

In Figure 5, the point "O" represents the wheel center or axle point of the rear axle of the tractor that moves along the horizontal direction with velocity "v" when the driving wheel rotates with an angular velocity "ω". This is pure rotational movement along imaginary plane 1, which is the ground surface. A perpendicular line passes through the axle point O, and the point of intersection with plane 1 is the instantaneous axis of rotation.

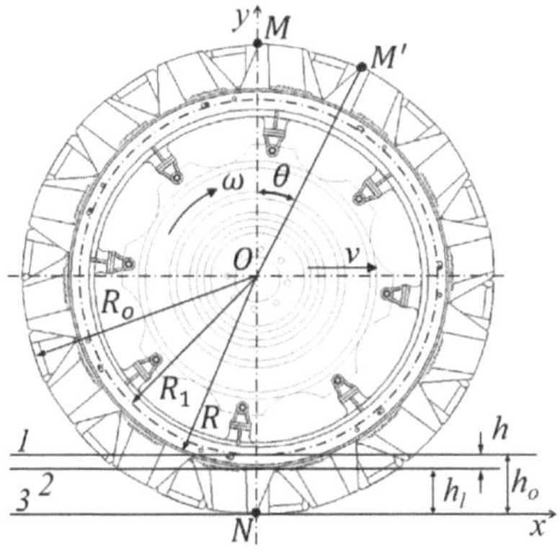

Figure 5. The rolling condition diagram of the rigid lugged wheel.

To produce the driving force through wheel-terrain interaction to move the tractor on the soft, deformable soil, the wheel rotates under the action of applied torque and normal load on the wheel. In this process, some slippage happens between the wheel and the terrain due to the shearing and sliding of surface soil. Ideally, the driving wheel rotates without slipping or sliding.

Figure 5 shows that the actual rolling motion of the driving wheel is in accordance with plan 3. The velocity v_t is the theoretical velocity of the point "O", and "R_o" is the maximum rolling radius, with "ω" as the angular velocity. Equation (9) can be used to calculate the theoretical velocity.

$$v_t = R_o.\omega \tag{9}$$

During the driving of a tractor for practical applications such as soil tillage, a certain amount of slip is always produced by the driving wheel, even without towed load conditions. Therefore, the actual velocity "v" of the tractor is less than the theoretical velocity "v_t". Equation (10) is used to calculate the actual velocity.

$$v = R_1.\omega \tag{10}$$

where, R_1 is the equivalent radius or actual rolling radius of the wheel in operational conditions. The loss of velocity caused by the slipping or shearing of the driving wheel over the terrain surface during the driving process is called travel reduction T_r and can be calculated by the following equation:

$$T_r = \frac{v_t - v}{v_t} = 1 - \frac{v}{v_t} \tag{11}$$

The rolling radius of the driving wheel varies with field operational conditions, terrain properties, wheel structural parameters, applied normal and drawbar loads, and several other factors. Therefore, in practice, it is inconvenient to use the actual travel reduction for theoretical computations and analysis. Therefore, the maximum radius R_o of the rigid lugged wheel is used for the calculation of travel reduction by using Equation (12).

$$T_r = 1 - \frac{R_1}{R_o} \tag{12}$$

The rolling radius of the tractor wheel changes with the applied drawbar or traction load, which can be determined during the field-driving test. If the tractor operated at wheel rpm n under a certain applied load and traveled a certain distance s, the rotating radius of the wheel could be calculated by the following equation:

$$R_1 = \frac{s}{2\pi n} \tag{13}$$

The tractor wheel radius, R_1 may vary according to the terrain and operational conditions; it is used for kinematic and dynamic analysis and determination and calculation of tractor traction performance parameters in field operations with the applied drawbar.

3.2. Motion Trajectory of the Lug Tip of the Rigid Lugged Wheel

While the driving wheel "rigid lugged wheel" rotates on the loose terrain/ field soil surface, the lugs of the wheel continuously pass through the soil. This process forms the cut marks in the soil that reflect the interaction between the wheel and the terrain soil. Therefore, it is important to analyze the wheel-soil interaction relative to the motion of the wheel to determine the stresses of the wheel lug on soil and the soil cutting and shearing process. The theoretical shape and size of the lug cut hole are determined by the motion trajectory, and they change with the structural parameters of the wheel lug, slip rate, i.e., travel reduction of the driving wheel, and height of lug penetration/wheel submergence.

When the rigid lugged wheel is driven on a flat, non-deformable surface, the terrain surface contact points are located on the tips of the lugs, which are the highest points on the wheel's outer radius along the vertical axis. Therefore, kinematic analysis of the rigid lugged wheel is equivalent to finding the rotational and translational conditions of the contact points. A diagram representing the rotational conditions of the rigid lugged wheel with axes and plans of motion is shown in Figure 5, where point O represents the center of the wheel or wheel rotation axis that is at a height from the plan of motion (line 3) equivalent to the radius R_o. A point M located on the tip of the lug in rest condition at time $t = 0$ has coordinates $X_M = 0$, which means the horizontal displacement is 0, and $Y_M = 2R$, which is the height of point M from the ground surface equal to the wheel diameter.

Assuming the wheel begins to rotate and moves with a uniform angular velocity ω and a linear velocity v. After a time t, the point M turns into M'. By subtending an angle θ (Figure 5), the coordinates of the point M' with respect to the ground-surface contact point reference frame can be determined by the following equations:

$$X_M = R_o\,\theta + R_o \sin(\theta) \tag{14}$$

$$Y_M = R_o + R_o \cos(\theta) \tag{15}$$

The above equations are for ideal motion without slipping on a hard plain surface. However, in real conditions, in which wheels slip and slide on the terrain surface, the slip factor should be included, and the maximum radius, R_o, should be replaced with the equivalent radius, R_1, that is defined in Equation (13). Therefore, the equations to determine the position of the point are as follows when $(\theta = \omega t)$:

$$X_M = (1 - T_r)\,R_1\,\omega t + R_1 \sin(\omega t) \tag{16}$$

$$Y_M = R_1 + R_1 \cos(\omega t) \tag{17}$$

From the above equations, it is found that the motion trajectory of point M is associated with the rolling radius of the rigid lugged wheel and the wheel travel reduction rate. The motion trails of point M with different travel reduction rates are shown in Figure 6, and with different lug penetration or wheel submergence in Figure 7. The vertical axis (y-axis) represents the vertical position of the point, and the horizontal axis (x-axis) represents the horizontal position of the point at any instant of time while the wheel is rolling. The curves 1, 2, 3, and 4 in Figure 6 represent the trajectories of point M with a travel reduction rate of 0.0, 0.1, 0.25, and 0.4, respectively. Curve 1 is the trajectory of a point with ideal motion without slip, which means no travel reduction, and this pattern of motion is called cycloidal motion. Curves 2, 3, and 4 show the reduction in horizontal distance traveled per rotation of the wheel to calculate the travel reduction rate. As the travel reduction rate increases due to slipping and shearing between the surfaces, it produces more soil breakage and damage and reduces the distance traveled. Curves 1, 2, 3, and 4 in Figure 7 represent the trajectories of point M with lug penetration and wheel submergence of 0, 75, 150, and 165 mm, respectively. The effect of lug penetration and wheel submergence on rolling motion and horizontal distance traveled is obvious.

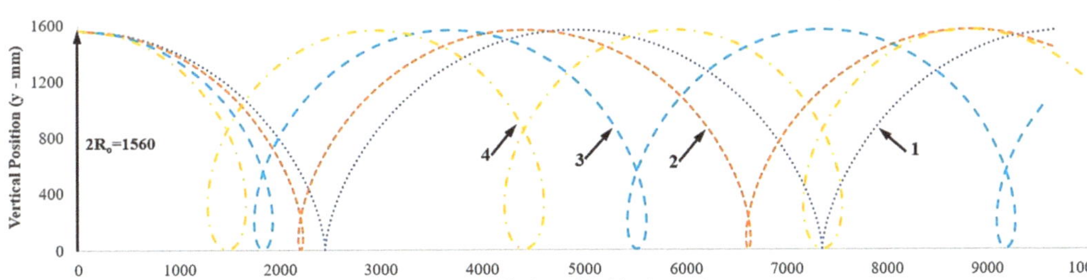

Figure 6. Motion trail or trajectories of point M at different travel reduction rates, Corves 1, 2, 3 and 4 represent travel reduction rate of 0.0, 0.1, 0.25 and 0.4 respectively.

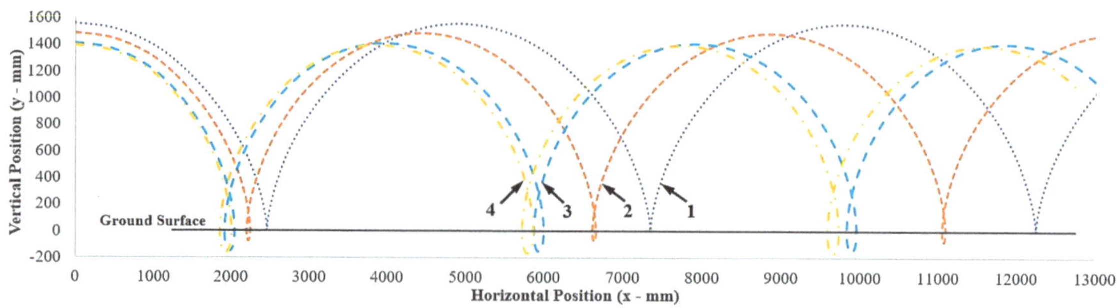

Figure 7. Motion trail or trajectories of point M at different lug penetration and wheel submergence, Corves 1, 2, 3 and 4 represent submergence of 0, 75, 150 and 165 respectively.

The instantaneous velocity v_M of point M at any instant of the time can be determined by the following equation for the center of the wheel reference frame when the wheel is rotating with uniform angular velocity ω and translating with a uniform linear velocity v:

$$v_M = R_1\ \omega \sin(\omega t) + R_1\ \omega\ cos(\omega t) \tag{18}$$

When the rigid lugged wheel undergoes pure rotation at a certain velocity, the relationship between the vertical displacement and velocity of the edge points of the wheel lugs is shown in Figure 8a. The vertical axes (y-axis) denote the vertical displacement and velocity of the point. The horizontal axis (x-axis) denotes the horizontal displacement. Curves 1 and 2 represent the vertical displacement and velocity of the point M respectively. When the rigid lugged wheel goes through pure rotation at a slip rate of 0.4 on the soft deformable terrain surface, the relationship between the vertical displacement and velocity of the edge points of the wheel lugs is shown in Figure 8b. Curves 1 and 2 represent the vertical displacement and velocity, respectively.

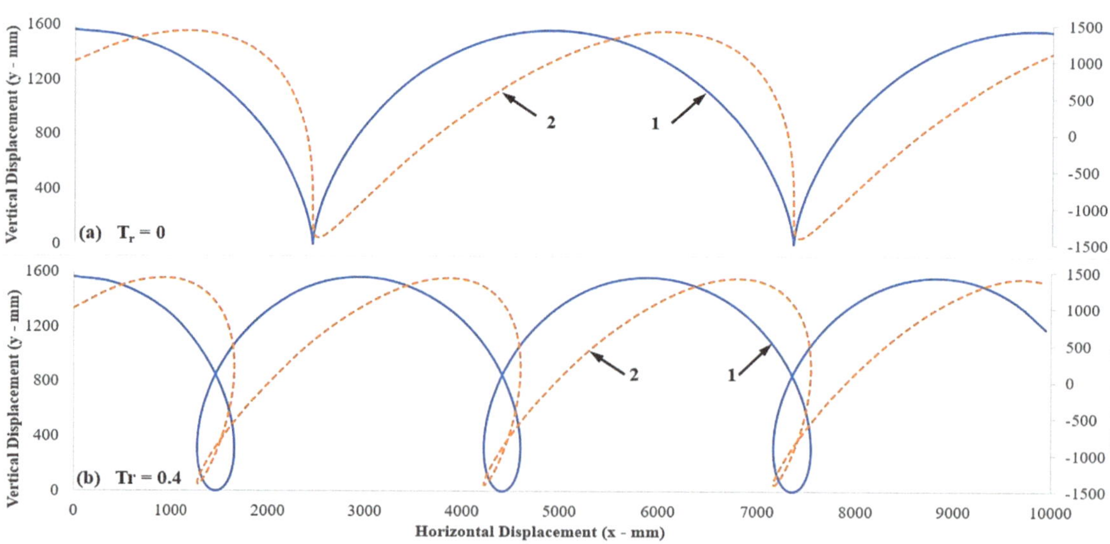

Figure 8. Relationships of vertical displacement and velocity of lug edge point M at different travel reduction rates (**a**) 0, and (**b**) 0.4, and corves 1 and 2, represent vertical displacement and velocity respectively.

A comparative analysis of Figure 8a,b indicates that when the drive wheel slips, the displacement and velocity of the wheel lug tips (contact points) are repeated in the space. Thus, the vertical displacement and velocity of the wheel occur at lug tips or edge points, and the motion condition that is produced is more complex than pure rotation.

The impacts of the soil on the wheel during the wheel-soil interaction process produce power losses. The motion or rolling resistance is produced by the wheel penetration/sinkage and the soil reaction force, which is generated as the wheel lug penetrates and breaks/excavates the soil. The wheel lug impacting the soil backward also produces power losses, but it is necessary to obtain a wheel thrust or drive force from the soil.

4. Dynamic Analysis

4.1. Determination of the Lugged Wheel Driving Force

4.1.1. The Normal Vector of the Lug Working Surface

In the $x_o y_o z_o$ coordinate system, the normal vector of the working surface of the lug is ($\vec{n}_o = \vec{j}$), by combining Equation (1), the normal vector of the lug working surface in the xyz coordinate system is:

$$\vec{n} = A.\,\vec{n}_o\,,\ \text{where}: \begin{bmatrix} n_x \\ n_y \\ n_z \end{bmatrix} = \begin{bmatrix} -\sin\alpha \\ \cos\alpha \\ 0 \end{bmatrix} \tag{19}$$

In the process of cultivation or tillage, the tractor's forward speed is v_m, along the positive direction of the y-axis. The angular speed of the iron wheel is ω, along the negative direction of the x-axis. Then, there is a rotation and translation transformation relationship between the coordinate system $x_1y_1z_1$ fixed on the iron wheel at the time 't' and the fixed coordinate system xyz at the beginning, and the rotation transformation relationship is shown in Figure 9.

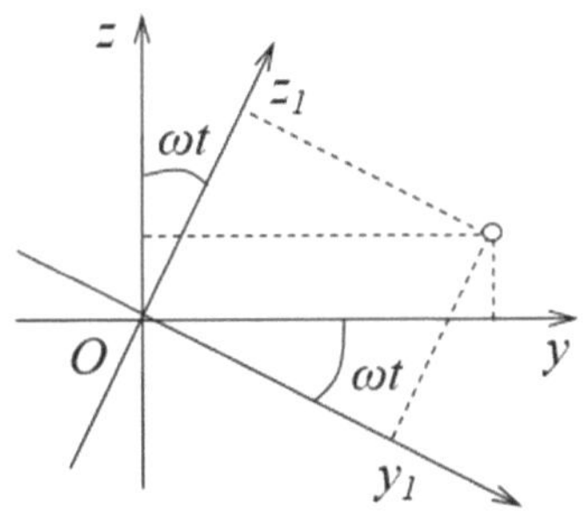

Figure 9. Coordinate rotation transformation in the tillage process.

From Figure 9, following equation is derived:

$$\begin{cases} x_1 = x \\ y_1 = y\cos\omega t - z\sin\omega t \\ z_1 = z\cos\omega t + y\sin\omega t \end{cases} \tag{20}$$

Assume:

$$\begin{bmatrix} x_1 \\ y_1 \\ z_1 \end{bmatrix} = B \begin{bmatrix} x \\ y \\ z \end{bmatrix} \equiv \begin{bmatrix} x \\ y \\ z \end{bmatrix} = B^{-1} \begin{bmatrix} x_1 \\ y_1 \\ z_1 \end{bmatrix},$$

where

$$B = \begin{bmatrix} 1 & 0 & 0 \\ 0 & \cos\omega t & -\sin\omega t \\ 0 & \sin\omega t & \cos\omega t \end{bmatrix}$$

and

$$B^{-1} = \begin{bmatrix} 1 & 0 & 0 \\ 0 & \cos\omega t & \sin\omega t \\ 0 & -\sin\omega t & \cos\omega t \end{bmatrix}$$

Considering the rotation and translation relationship between the coordinate systems $x_1y_1z_1$ and xyz, there is:

$$\begin{bmatrix} x \\ y \\ z \end{bmatrix} = B \begin{bmatrix} x_1 \\ y_1 \\ z_1 \end{bmatrix} + \begin{bmatrix} 0 \\ v_m t \\ 0 \end{bmatrix} \tag{21}$$

By adding Equation (19) to Equation (20), the expression of the normal direction vector of the lug working surface in the fixed coordinate system xyz is obtained as follows:

$$\vec{n}_t = \begin{bmatrix} n_{tx} \\ n_{ty} \\ n_{tz} \end{bmatrix} = B^{-1}\,.\,\vec{n} = \begin{bmatrix} -\sin\alpha \\ \cos\alpha\,.\cos\omega t \\ -\cos\alpha\,.\,\sin\omega t \end{bmatrix} \tag{22}$$

4.1.2. Driving Force Analysis of the Lug

Most of the upland farming areas in China cultivated two crops a year, so farmers tilled the soil twice a year with rotary or vertical tillage implements. Considering the physical characteristics of farmland soil, we can characterize it as an isotropic and completely plastic material. This study assumes that the depth of surface soil compacted by the iron wheel due to submergence of the iron wheel rim of radius R = 638 mm during cultivation is h = 15 mm, as shown in Figure 10. According to this figure:

$$R\,(1-\cos\beta) = h \; and \; \beta = \cos^{-1}\left(\frac{R-h}{R}\right) \tag{23}$$

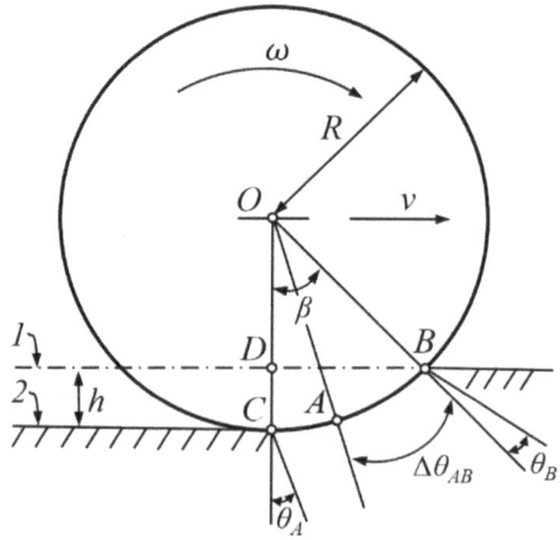

Figure 10. Schematic of the rigid lugged wheel in working condition: 1—Front untilled soil surface line; 2—Back compressed soil surface line; h: compressed depth of cultivated soil = 15 mm.

By the above equation, the value of β is:

$$\beta = 12.4488^\circ = 0.217273\; rad.$$

As the width of the iron wheel B = 440 mm, the wheel soil contact area of the iron wheel rim is:

$$S_o = (R \times \beta) \times B, \tag{24}$$

It results in: S_o = 60992.75144 mm^2.

Equation (4) was used to determine the corresponding position of two points A and B in Figure 2b, where x_A = 345 and x_B = 140 were on the lug in the fixed coordinate system *xyz*:

$$\begin{bmatrix} x_A \\ y_A \\ z_A \end{bmatrix} = \begin{bmatrix} \frac{R}{\tan\alpha} \cdot \sin\theta_A \\ R\sin\theta_A \\ R\cos\theta_A \end{bmatrix} \text{ and } \begin{bmatrix} x_B \\ y_B \\ z_B \end{bmatrix} = \begin{bmatrix} \frac{R}{\tan\alpha} \cdot \sin\theta_B \\ R\sin\theta_B \\ R\cos\theta_B \end{bmatrix}.$$

By taking x_A = 345 and x_B = 140 into the above equation, it results in θ_A = 18.19° and θ_B = 7.28°, respectively, and by putting them in Equation (21), the trajectory equations of points A and B are obtained in return as:

$$\begin{bmatrix} x_{At} \\ y_{At} \\ z_{At} \end{bmatrix} = B^{-1} \begin{bmatrix} x_A \\ y_A \\ z_A \end{bmatrix} + \begin{bmatrix} 0 \\ v_m t \\ 0 \end{bmatrix} = \begin{bmatrix} 345 \\ R\sin\theta_A\cos\omega t + R\cos\theta_A\sin\omega t + v_m t \\ -R\sin\theta_A\sin\omega t + R\cos\theta_A\cos\omega t \end{bmatrix}$$

That resulted in the following equation:

$$\begin{bmatrix} x_{At} \\ y_{At} \\ z_{At} \end{bmatrix} = \begin{bmatrix} 345 \\ R\sin(\omega t + \theta_A) + v_m t \\ R\cos(\omega t + \theta_A) \end{bmatrix} \tag{25}$$

and

$$\begin{bmatrix} x_{Bt} \\ y_{Bt} \\ z_{Bt} \end{bmatrix} = B^{-1} \begin{bmatrix} x_B \\ y_B \\ z_B \end{bmatrix} + \begin{bmatrix} 0 \\ v_m t \\ 0 \end{bmatrix} = \begin{bmatrix} 140 \\ R\ \sin\theta_B\ \cos\omega t + R\ \cos\theta_B\ \sin\omega t + v_m t \\ -R\ \sin\theta_B\ \sin\omega t + R\ \cos\theta_B\ \cos\omega t \end{bmatrix}$$

$$\begin{bmatrix} x_{Bt} \\ y_{Bt} \\ z_{Bt} \end{bmatrix} = \begin{bmatrix} 140 \\ R\sin(\theta_B + \omega t) + v_m t \\ R\cos(\theta_B + \omega t) \end{bmatrix} \tag{26}$$

Based on ProE three-dimensional software modeling and measurement, the area of the lug working surface shown in Figure 2b, is S = 34,343.2 mm^2. The force or soil thrust on the lug's working surface is related to the reaction force of the lug on the soil. Regardless of the friction force between the soil and the lug surface, it is assumed that the normal pressure of the soil on the driving or working surface of the lug is p_l, the unit of which is N·m^{-2}. By using Equation (22), the positive pressure of soil, i.e., soil thrust, on the working surface of the lug is as follows:

$$\vec{F} = -\begin{bmatrix} n_{tx} \\ n_{ty} \\ n_{tz} \end{bmatrix} .S.p_l = \begin{bmatrix} S.p_l.\sin\alpha \\ -S.p_l.\cos\alpha\cos\omega t \\ S.p_l.\cos\alpha\sin\omega t \end{bmatrix} \tag{27}$$

In the equation:

S = area of lug working surface (34,343.2 mm^2) it varies with the height of penetration of the lug, p_l to be determined represents the compressive or normal stress of soil on the lug working surface, ω =1.28 rad s^{-1}, and 2.37 s $\geq$ t $\geq$ 2.18 s.

Equation (27) shows that the axial component of the soil thrust acting on the lug, i.e., the x-axis component, is a constant, i.e.,

$$F_x = S.p_l.\sin\alpha = 0.5Sp_l$$

Therefore, the axial component accounts for 50% of the total force.

The component of soil thrust acting on the lug working surface that helps to move the whole machine in the forward direction is the y-axis component, which is to say, the driving force provided by the interaction of the wheel lug and soil is:

$$\begin{cases} F_y = -S \,.\, p_l \,.\, \cos\alpha \,.\, \cos\omega t \\ F_y = (0.9413 \sim 0.9950).\, S.\, p_l.\, \cos\alpha \\ F_y = (0.8152 \sim 0.8617) S.\, p_l \end{cases} \tag{28}$$

It can be seen that the lug's driving force is proportional to the angle α the smaller the angle, the greater the driving force. In addition, the driving force of the lug is related to the relative alignment of the working surface in the soil. In the beginning, the driving force is the smallest when the lug starts to penetrate and gradually increases until it is completely submerged in the soil. The fluctuation range of the driving force is 81.52% to 86.17%.

The vertical component of soil thrust acting on the lug is as follows:

$$\begin{cases} F_z = S \,.\, p_l \,.\, \cos\alpha \,.\, \sin\omega t \\ F_z = (0.3375 \sim 0.1001).\, S.\, p_l.\, \cos\alpha \\ F_z = (0.2923 \sim 0.0867).\, S.\, p_l \end{cases} \tag{29}$$

It can be seen that the vertical force decreases from 29.23% to 8.67%. The smaller vertical force helps to reduce the compaction effect of the whole machine on the soil [19,20].

4.1.3. Soil Thrust-Driving Force Relationship Analysis of Lugged Wheel

Predicting the normal pressure p of soil applied to the wheel contact surface or $(p_l = p)$ applied to the lug working surface is critical for analyzing the tractor drive wheel's traction performance. The principles used to predict the soil mechanical behavior and response under wheel loading are called the pressure-sinkage relationship and shear failure criteria. The pressure-sinkage relationship suggested by scientists who worked on soil mechanics is presented in Equation (30) [21].

$$p = kz^n \tag{30}$$

This pressure sinkage relationship states that the average pressure applied to a plate on a uniform terrain is dependent on an exponential function of sinkage 'z', in which the exponent 'n' represents a soil property reliant on the size of the wheel and 'k' is the proportionality constant. A more explanatory and descriptive form of the pressure sinkage relationship of vehicle wheels is presented in Equation (31), which is a more widely relied-upon expression and is used in the practice of terramechanics and vehicle traction mechanics [21].

$$p = \left(\frac{k_1}{b} + k_2\right) z^n \tag{31}$$

The proportionality constant 'k' has been revised to justify the dimensions of the contact area. The letter b represents the wheel width, and k_1 and k_2 remain parameters of soil sinkage. The soil parameters n, k_1, and k_2 are empirical and have no intrinsic physical meaning. The parameters can be determined by pressure-sinkage data from field experiments. Although the pressure sinkage relationship is a semi-empirical expression, it is most important in terramechanics and tractor traction mechanics as the shear strength of the soil is mainly related to normal load (ground pressure). Additionally, the soil thrust developed under the drive wheel or wheel drive force and motion resistance is relative to sinkage (i.e., the relationship between pressure and sinkage).

The shear stress-displacement relationship, or share failure expression for maximum share stress and applied normal pressure, can be expressed by the Mohr-Coulomb equation that adequately describes the relation in many cases [17].

$$\tau_{max} = (c + p\ tan\phi) \tag{32}$$

To describe the shared stress-displacement relationship for a deformable terrain, an expression was suggested by [22], in which the researcher [23] proposed a modification for loose plastic soil that is presented in the following equation and widely adopted in practice.

$$\frac{\tau}{\tau_{max}} = \left(1 - e^{-\frac{j}{K}}\right) \tag{33}$$

Therefore, Equation (32) has the following form:

$$\tau = (c + p\ tan\phi)\left(1 - e^{-\frac{j}{K}}\right) \tag{34}$$

In the above expression, τ is the shear stress, j is the soil shearing displacement, K is the shear modulus, which can be determined by fitting the shear stress-shear displacement equation [23] to each data set, c is the soil cohesion, and ϕ is internal friction angle, which can be determined by the Mohr-Coulomb failure criteria.

The share displacement that is developed along the contact patch of the rigid wheel can be determined by analysis of the slip velocity v_j on the wheel-terrain interface, as shown in Figure 11a [17,24,25]. The tangential component of the absolute velocity v of a certain point on the circumference of the wheel relative to the terrain is described as slip velocity v_j [17]. The magnitude of the slip velocity v_j of a point on the rigid wheel circumference defined by the angle β can, therefore, be expressed as below [24].

$$v_j = r\omega\{1 - (1 - T_r)\cos\beta\} \tag{35}$$

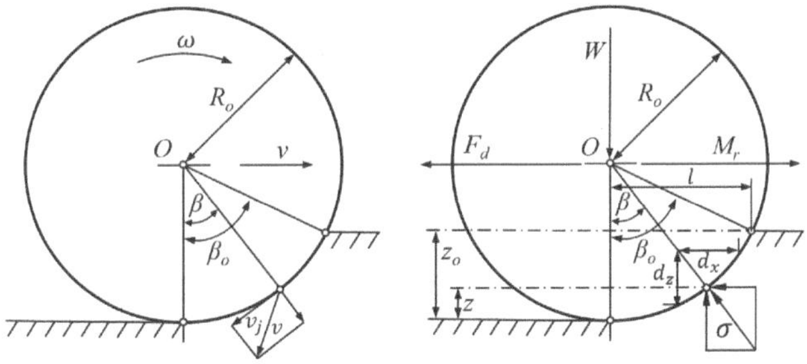

Figure 11. The schematic force diagram of shear-displacement under the wheel terrain interface (**a**) slip velocity v and (**b**) with applied radial pressure σ.

The parameter j, which is the shared displacement of a point along the contact surface for rigid wheels and loose soil, can be determined by the following equation:

$$\begin{gathered} j = \int_0^t v_j\, dt = \int_0^{\beta_o} r\omega[1 - (1 - T_r)\cos\beta]\frac{d\beta}{\omega} \\ j = r(\beta_o - \beta) - (1 - T_r)(\sin\beta_o - \sin\beta) \end{gathered} \tag{36}$$

According to soil shear theory and wheel-terrain interaction, the rigid lugged wheel has two main soil reaction forces along the y-axis: (1) F_s due to the shear stress τ between the wheel bottom surface and soil surface, and (2) F_l due to the normal terrain pressure p on the lug working surface.

The thrust-slip relationship for a rigid wheel, Equation (37), can be determined by using the shear stress-shear displacement relationship, Equation (34), by considering uniformly distributed normal pressure and the flat wheel-terrain contact patch of contact length l and contact area A [17].

$$F = (Ac + p\tan\phi)\left[1 - \frac{K}{T_r\, l}\left(1 - e^{-T_r\, l/K}\right)\right] \tag{37}$$

where c, ϕ, and K are the cohesion, angle of shearing resistance or internal friction, and shear deformation parameters of the terrain material, "the surface soil", respectively. T_r is the travel reduction due to the wheel slip, and p is the normal pressure on the wheel-terrain contact surface that is equal to the normal load of the wheel on the contact surface.

As the front contact angle of the rigid lugged wheel is determined (i.e., $0 - \beta_o$) (Figure 11) [17], the rigid lugged wheel thrust force or drive force, F_s, can be determined by integrating the stresses over the wheel-terrain contact surface.

$$F = F_s = bR_o\left[\int_0^{\beta_o} \tau(\beta)\cos\beta\, d\beta\right] = bR_o\left[\int_0^{\beta_o} (c + p\ tan\phi)\left(1 - e^{-j/K}\right) cos\ \beta\ d\beta\right]$$

$$F = F_s = bR_o\left[\int_0^{\beta_o}\left(c + \left(\frac{K_1}{b} + K_2\right)R_o{}^n(\cos\beta - \cos\beta_o)^n tan\phi\right)\left(1 - e^{-(R_o(\beta_o-\beta)-(1-T_r)(\sin\beta_o - \sin\beta))/K}\right)cos\beta d\beta\right] \quad (38)$$

From the above equation, it is found that the driving force of a rigid lugged wheel is related directly to wheel width, radius, and sinkage and inversely to travel reduction rate.

When the tractor is operating on soft terrain and the majority of the rigid wheel's lugs are submerged in the terrain, the terrain material/soil fills the spaces between the lugs. In this case, the rigid lugged wheel acts like a smooth rigid wheel (Figure 11), so Equation (38), can be used for the thrust/drive force analysis of a rigid lugged wheel. However, over medium to hard terrain, the entire wheel load may be supported on the tips or faces of lugs, and there is minimum or no contact of the carcass (rim surface) with the terrain. In this case, the thrust-sinkage analysis, already done for the single lug, Equation (28), can be used by integrating the stresses over the whole contact area of lugs or the number of lugs by considering ($\omega t = \beta$) as follows:

$$F = F_l = -S\cos\alpha\int_0^{\beta_o} p_l.\cos\beta d\beta = -S\cos\alpha\left[\int_0^{\beta_o}\left(\frac{K_1}{b} + K_2\right)R_o{}^n(\cos\beta - \cos\beta_o)^n\cos\beta d\beta\right] \quad (39)$$

The net driving force, F_d, produced by wheel-terrain interaction can be found by:

$$F_d = F - M_r \quad (40)$$

Rolling resistance or motion resistance M_r to the wheel of a moving tractor on deformable terrain could be produced due to the penetration of lugs and sinkage of the wheel. The power consumed in the motion resistance of the tire wheel is equal to the work conducted on soil compaction and rut formation, but in the case of high lugs rigid wheel, work is expended on the breaking and loosening of soil by the penetration of sharp-edged lugs.

$$M_r = b\int_0^{\beta_o} \sigma\ R\ Sin\beta\ d\beta \quad (41)$$

The radial pressure σ applied on the wheel due to wheel-terrain interaction (Figure 11b) is equal to the normal pressure p below a sinkage plate at the same sinkage depth z in a pressure sinkage test; therefore, the following expression is developed.

$$\sigma R.sin\beta\ d\beta = p\ dz \quad (42)$$

By combining the pressure-sinkage relationship with the above equation and including the effect of travel reduction rate, the motion resistance of a rigid wheel can be determined as follows:

$$M_r = \frac{b}{(1-T_r)}\int_0^{z_o}\left(\frac{k_1}{b} + k_2\right)z_o{}^n\ dz = \frac{b}{(1-T_r)}\left[\left(\frac{k_1}{b} + k_2\right)\frac{(R_o(\cos\beta - \cos\beta_o))^{n+1}}{n+1}\right] \quad (43)$$

where z_o represents the wheel's maximum sinkage, as shown in Figure 11b. The above analysis shows that the motion resistance of a rigid lugged wheel depends on the wheel width, wheel radius, and sinkage depth; a higher travel reduction rate also causes higher motion resistance.

From the dynamic analysis of the rigid lugged wheel, it is found that the forces on the wheel include shear and normal stresses on the wheel rim and normal pressure on the wheel lug working surface. The normal pressure on the lug working surface results in a higher driving force from the rigid lugged wheel that is independent of the wheel axle's normal load. The motion resistance of a rigid lugged wheel could be higher due to the penetration of lugs into the soil, but this will help to break and loosen the soil and get compensated by the lower soil tillage draft. In this way, the rigid lugged wheel will help to consume tractor power efficiently and avoid soil compaction in soil tillage operations.

5. Model Evaluation (FEM Analysis) for Structural Stability and Optimization

The designed model of the rigid lugged wheel was analyzed in the ANSYS workbench platform for the structural optimization of a new wheel. The behavior of the structural components under applied forces on the designed model was analyzed on assessing stresses, strains, and deformations resulting from the analysis of applied loads. The purpose of this simulation analysis was to check the material suitability and design parameters to select the best-fit material and optimum parameters, e.g., the thickness of the steel sheet used for the wheel structure with the minimum wheel structure weight.

5.1. Material Properties (Engineering Data)

The material was selected based on design criteria, working conditions, and operational requirements. It was a Manganese Iron alloy called high Carbon Manganese steel (Steel No. 15 Mn) sheets that contain Carbon (0.12–0.18), Silicon (0.17–0.37), and Manganese (0.70–1.00) percent as chemical mass fractions. The engineering properties data of the material are as follows: the density of 15Mn steel is 7.8×10^3 kg m^{-3}, the modulus of elasticity is 200–220 G Pa, and the Poisson's ratio is 0.25–0.30. In the field, tractor wheels are constantly stressed and, in extreme cases, face high-intensity impact loads. The material used for the fabrication of the newly designed modern type rigid lugged wheel was chosen considering all possible loading conditions so that it can safely withstand applied stresses and all types of loads.

5.2. Static Structural Analysis

The structure of the rigid lugged wheel under static loads was analyzed for structural stability and optimization using ANSYS Workbench mechanical simulation software. ProE's (Creo Elements) 2D and 3D engineering design platform was used to generate 3D wheel model geometry, and the entire wheel structure assembly was saved in "igs" file format, which is compatible with and importable into ANSYS Workbench. In ANSYS R17 Workbench, the static structural project solver was used for analysis, engineering data was modified according to test specifications and analysis requirements, and the '.igs' geometry file was imported into the project schematic. After making some corrections and modifications, the final structure model was prepared, and the mesh was applied in Figure 12a,b. The meshing, loads, supports, and displacement setup were defined, and resulting behavior parameters, e.g., stress-strain and deformations were selected. The project was analyzed by the problem solver option in the design modeler of ANSYS. Results were generated in graphical form (color schematic), as presented in Figure 12c–h which explained the effect of applied loads on the wheel body.

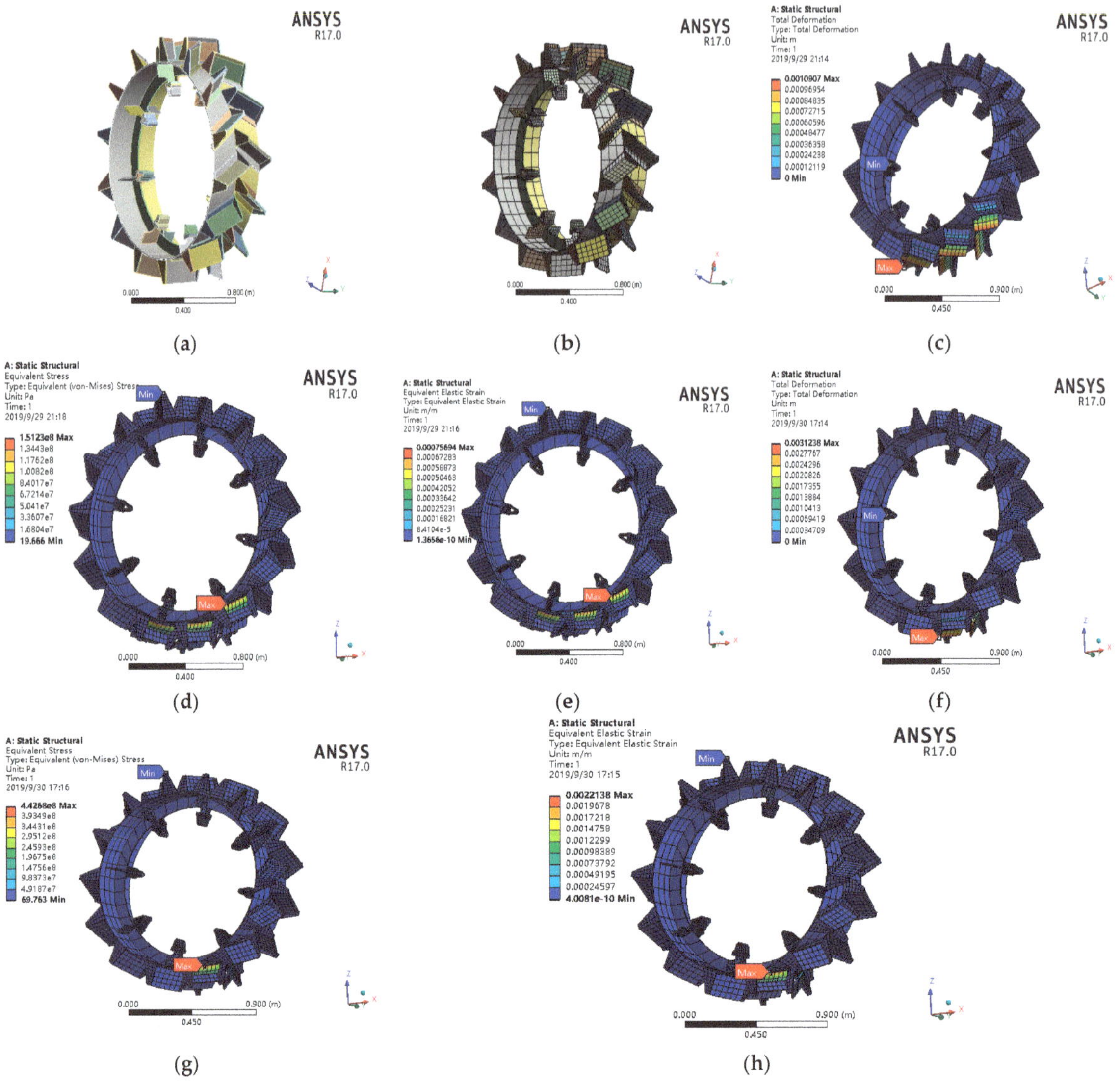

Figure 12. Static structural simulation analysis process in ANSYS under applied loads and results of the analysis in different surface soil conditions.

The reaction force, or soil thrust, developed on the wheel against the normal weight of the tractor and applied torque that acted on the lug working surface as the lugs of the wheel were in direct contact with the soil. The normal component of reaction forces to the lug-working surface provides the drive and traction force to the tractor. Therefore, the total force applied to the lugs would be equal to the maximum drawbar force of the tractor. To avoid the failure risk of the wheel structure in adverse field conditions, 20% of the total force was included as a safety factor.

In terms of the contact surface with the wheel, two different field conditions were considered. The first one was soft, easily penetrable soil where the whole lug can penetrate easily; in this case, a total of six lugs were considered to be in contact with the soil. In another case, the soil was hard enough that it allowed very little penetration, and only two lugs were in contact with the soil surface. The total force was divided accordingly in both cases; when force was divided among six lugs, there was less force per lug, so the stresses and deformations were less (Figure 12c–e) compared to another case where total force was applied only on two lugs, where stresses and deformations were higher (Figure 12f–h).

In both cases, the overall effects of loads and resulting stresses and deformations were very low and there was no situation was observed that affect the structural stability. Therefore, it is recommendable to suggest a reduced thickness of the steel sheet that might help to reduce the weight of the wheel, and ultimately, the total weight of the tractor helps to get optimum traction with minimum damage to the soil and the environment.

6. Wheel Prototype Development and Driving Tests

The designed rigid lugged wheel model was converted to a prototype after the structure stability analysis in the ANSYS workbench and verification of the designed parameters. Two rigid lugged wheel full-body structures were fabricated and developed according to the size and operational requirements of the specified tractor for this study. At a local industrial manufacturing factory, wheels were developed using a specific material (15Mn steel).

On-road drive and field tests were carried out to verify the design parameters, theoretical analysis, and traction performance (Figure 13), which represented the significant improvement in the wheeled tractor's traction performance with rigid lugged wheels as drive wheels compared to conventional tire wheels. Traction and wheel-soil interaction performance of a rigid lugged wheel is already reported [2,4], and further studies and analyses are under process to verify the effectiveness of the rigid lugged wheel for off-road drive operations of the conventional wheeled tractor.

(a) (b) (c)

Figure 13. Rigid lugged wheel on-road drive tests, (**a**) concrete, (**b**) asphalt, and (**c**) field drive and traction test.

7. Conclusions

Traction demand from conventional agricultural wheeled tractors has increased in recent years for cultivation operations according to the requirement of intensive traction, improved trafficability, and drivability without increasing machine weight to avoid soil compaction and excessive energy consumption. A new design of a high lug rigid wheel is developed and tested as a traction device. The main component of the research object is the design of the lug working faceplate. The working principle of the rigid lugged wheel is theoretically confirmed by kinematic and dynamic analysis. The relationships and analysis are elucidated for wheel design parameters, terrain soil characteristics, rolling motion, and drive force generation behavior. A computer simulation of the designed model by the finite element method is presented that verifies the structure's stability and provides a reference for optimization. The complete wheel structure is developed and tested on field conditions while the theoretical analysis/modeling and design parameters optimization are in process. The results of theoretical analysis and field tests are consistent, indicating that the rigid lugged wheel is well suited to soft deformable terrain, with improved traction and drivability. Theoretical analysis is performed for the newly designed rigid lugged wheel that provides in-depth theoretical modeling to assess and confirm its working principles. Insertion pressure, rather than shearing pressure, generates driving power in a rigid lugged wheel with a sharp edge and flat contact surface. In this way, it can produce higher traction, give A pre-tillage effect to the top soil, and reduces the tillage draught requirement. As confirmed by theoretical analysis, the component of thrust along the direction of driving ranged from 81.52% to 86.17%, which is actually the driving force and relates to lug faceplate alignment and lug penetration depth. Moreover, the vertical component of thrust reported was less than 30% and decreases further to less than 9% as the wheel displaces in the soil, which will avoid soil compaction under the wheel. To test lugged wheels' field performance and conduct a comparative analysis with tire wheels, traction, and wheel-soil interaction performance are already been reported [2,4], which provide the validation base for these theoretical results. Further research on the rigid lugged wheel is required to improve the wheel structure parameters and adaptability in various adverse field conditions. The results of this study could provide a trusted technical reference for the future design and construction of off-road agricultural vehicles, agricultural tractors, and particularly traction devices/drive wheels for field operations.

8. Patents

High-power tractor drive wheel. Zhang Jumin, Xia Junfang, Zhou Yong, and Hafiz Md-Tahir. Huazhong Agricultural University. Patent No. ZL 2015 10922181.8. (2019). State Intellectual Property Administration Office of China, PR China.

Author Contributions: Conceptualization, H.M.-T. and J.Z.; methodology, H.M.-T.; software, H.M.-T. and J.Z.; validation, H.M.-T. and J.Z.; formal analysis, H.M.-T., M.S. and F.A.; investigation, H.M.-T.; resources, J.X., J.Z., Y.Z. and J.D.; data curation, H.M.-T.; writing—original draft preparation, H.M.-T.; writing—review and editing, H.M.-T., A.U. and Z.H.; visualization, H.M.-T., J.Z. and J.X.; supervision, J.X., J.Z. and Y.Z.; project administration, J.X., J.Z., Y.Z. and J.D.; funding acquisition, J.X. and J.Z. All authors have read and agreed to the published version of the manuscript.

Funding: Financially supported by: 1. National key technology research and development projects, National Natural Science Foundation of P.R. China, (Grant No. 32271994), entitled "Study on mechanism of detachment and consumption reduction in the soil-straw-machine interface for rice straw returning to soil in up-land paddy rotation farming". 2. Ministry of Science and Technology of P. R. China, (Grant No. 2017YFD0301300) entitled "Key technology research on intelligent production of diversified cereal crops in double cropping system".

Institutional Review Board Statement: Not applicable.

Data Availability Statement: The data presented in this study are available on request from the corresponding author. The data are not publicly available due to institutional privacy.

Acknowledgments: This study was conducted as part of the Ph.D. study of H.M.-T., which was supported by the China Government Scholarship through the China Scholarship Council, which is acknowledged. The authors would like to thank research group members and Engineering College Workshop staff for assistance in the rigid lugged wheel development and execution of the field experiments.

Conflicts of Interest: The authors declare no conflict of interest.

References

1. Battiato, A.; Diserens, E. Tractor traction performance simulation on differently textured soils and validation: A basic study to make traction and energy requirements accessible to the practice. *Soil Tillage Res.* **2017**, *166*, 18–32. [CrossRef]
2. Md-Tahir, H.; Zhang, J.; Xia, J.; Zhou, Y.; Zhou, H.; Du, J.; Sultan, M.; Mamona, H. Experimental Investigation of Traction Power Transfer Indices of Farm-Tractors for Efficient Energy Utilization in Soil Tillage and Cultivation Operations. *Agronomy* **2021**, *11*, 168. [CrossRef]
3. Arvidsson, J.; Westlin, H.; Keller, T.; Gilbertsson, M. Rubber track systems for conventional tractors–Effects on soil compaction and traction. *Soil Tillage Res.* **2011**, *117*, 103–109. [CrossRef]

4. Md-Tahir, H.; Zhang, J.; Xia, J.; Zhang, C.; Zhou, H.; Zhu, Y. Rigid lugged wheel for conventional agricultural wheeled tractors—Optimising traction performance and wheel–soil interaction in field operations. *Biosyst. Eng.* **2019**, *188*, 14–23. [CrossRef]
5. Hamza, M.; Anderson, W. Soil compaction in cropping systems: A review of the nature, causes and possible solutions. *Soil Tillage Res.* **2005**, *82*, 121–145. [CrossRef]
6. Battiato, A.; Diserens, E. Influence of tyre inflation pressure and wheel load on the traction performance of a 65 kW MFWD tractor on a cohesive soil. *J. Agric. Sci.* **2013**, *5*, 197.
7. Molari, G.; Bellentani, L.; Guarnieri, A.; Walker, M.; Sedoni, E. Performance of an agricultural tractor fitted with rubber tracks. *Biosyst. Eng.* **2012**, *111*, 57–63. [CrossRef]
8. Rasool, S.; Raheman, H. Improving the tractive performance of walking tractors using rubber tracks. *Biosyst. Eng.* **2018**, *167*, 51–62. [CrossRef]
9. Keen, A.; Hall, N.; Soni, P.; Gholkar, M.D.; Cooper, S.; Ferdous, J. A review of the tractive performance of wheeled tractors and soil management in lowland intensive rice production. *J. Terramechanics* **2013**, *50*, 45–62. [CrossRef]
10. Hermawan, W.; Yamazaki, M.; Oida, A.J. Theoretical analysis of soil reaction on a lug of the movable lug cage wheel. *J. Terramechanics* **2000**, *37*, 65–86. [CrossRef]
11. Xie, X.; Gao, F.; Huang, C.; Zeng, W. Design and development of a new transformable wheel used in amphibious all-terrain vehicles (A-ATV). *J. Terramechanics* **2017**, *69*, 45–61. [CrossRef]
12. Raper, R. Agricultural traffic impacts on soil. *J. Terramechanics* **2005**, *42*, 259–280. [CrossRef]
13. Srivastava, A.K.; Goering, C.E.; Rohrbach, R.P.; Buckmaster, D.R. *Engineering Principles of Agricultural Machines*, 2nd ed.; American Society of Agricultural and Biological Engineers: St. Joseph, MI, USA, 2006.
14. Bekker, M. Evolution of approach to off-road locomotion. *J. Terramechanics* **1967**, *4*, 49–58. [CrossRef]
15. Okello, A. A review of soil strength measurement techniques for prediction of terrain vehicle performance. *J. Agric. Eng. Res.* **1991**, *50*, 129–155. [CrossRef]
16. Söhne, W. Terramechanics and its influence on the concepts of tractors, tractor power development, and energy consumption. *J. Terramechanics* **1976**, *13*, 27–43. [CrossRef]
17. Wong, J.Y. *Terramechanics and Off-Road Vehicle Engineering: Terrain Behaviour, Off-Road Vehicle Performance and Design*, 2nd ed.; Butterworth-Heinemann is an imprint of Elsevier: Amsterdam, The Netherlands, 2010.
18. Zhang, J.; Xia, J.; Zhou, Y.; Md-Tahir, H. High Power Tractor Drive Wheel. China. Patent No. ZL 2015 1 0922181.8. CNIPA: China National Intellectual Property Administration (CNIPA), 2019.
19. Patel, S.; Mani, I. Effect of multiple passes of tractor with varying normal load on subsoil compaction. *J. Terramechanics* **2011**, *48*, 277–284. [CrossRef]
20. Taghavifar, H.; Mardani, A. Effect of velocity, wheel load and multipass on soil compaction. *J. Saudi Soc. Agric. Sci.* **2014**, *13*, 57–66. [CrossRef]
21. Wong, J.Y. *Theory of Ground Vehicles*; John Wiley & Sons. Inc.: New York, NY, USA, 2001.
22. Bekker, M.G. *Theory of Land Locomotion*; The University of Michigan Press: Ann Arbor, MI, USA, 1956.
23. Janosi, Z.; Hanamoto, B. The analytical determination of drawbar pull as a function of slip for tracked vehicles in deformable soils. In Proceedings of the ISTVS 1st Intrnational Conference on Mechanics of Soil-Vehicle Systems, Edizioni Minerva Tecnica, Torino, Italy, 1961.
24. Wong, J.-Y.; Reece, A. Prediction of rigid wheel performance based on the analysis of soil-wheel stresses: Part II. Performance of towed rigid wheels. *J. Terramechanics* **1967**, *4*, 7–25. [CrossRef]
25. Wong, J.; Preston-Thomas, J. On the characterization of the shear stress-displacement relationship of terrain. *J. Terramechanics* **1983**, *19*, 225–234. [CrossRef]

Disclaimer/Publisher's Note: The statements, opinions and data contained in all publications are solely those of the individual author(s) and contributor(s) and not of MDPI and/or the editor(s). MDPI and/or the editor(s) disclaim responsibility for any injury to people or property resulting from any ideas, methods, instructions or products referred to in the content.

energies

MDPI

Article

Improvement of Piglet Rearing's Energy Efficiency and Sustainability Using Air-to-Air Heat Exchangers—A Two-Year Case Study

Hauke F. Deeken *, Alexandra Lengling, Manuel S. Krommweh † and Wolfgang Büscher

Institute of Agricultural Engineering, University of Bonn, Nußallee 5, 53115 Bonn, Germany
* Correspondence: deeken@uni-bonn.de; Tel.: +49-228-73-68734
† Current address: Federal Office for Agriculture and Food, Deichmanns Aue 29, 53179 Bonn, Germany.

Abstract: Pig farming in mechanically ventilated barns requires much electricity for ventilation or exhaust air purification. Furthermore, thermal energy is needed to fulfill the animals' temperature requirements, especially in piglet rearing. Electrical and thermal energy input leads to CO_2 emissions and operating costs. Up to 90% of heat losses are due to the exhausted air. Heat exchangers can recover thermal energy from the warm exhaust air and transfer it to the cold fresh air. This study aimed to investigate energy consumption, efficiency, CO_2 emissions, and energy costs when using heat exchangers in a German piglet rearing barn under practical conditions in combination with exhaust air purification. The following parameters were obtained for a two-year period: air temperatures, air flow rates, and electricity and liquefied natural gas consumption; the latter were used to calculate CO_2 emissions and energy costs. In total, 576,042 $kWh_{el,th}$ and 616,893 $kWh_{el,th}$ (years 1 and 2) of energy were provided, including 290,414 kWh_{th} and 317,913 kWh_{th} of thermal energy recovered. Using heat exchangers reduced CO_2 emissions by up to 37.5% and energy costs by up to 19.7% per year. The study shows that piglet rearing can increase its ecological and environmental sustainability by using heat recovery.

Keywords: air purification; bio scrubber; climate change; CO_2 emissions; emissions; heat recovery; HVAC; livestock husbandry; pig farming; resource-saving

Citation: Deeken, H.F.; Lengling, A.; Krommweh, M.S.; Büscher, W. Improvement of Piglet Rearing's Energy Efficiency and Sustainability Using Air-to-Air Heat Exchangers—A Two-Year Case Study. *Energies* **2023**, *16*, 1799. https://doi.org/10.3390/en16041799

Academic Editors: Muhammad Sultan, Yuguang Zhou, Redmond R. Shamshiri and Muhammad Imran

Received: 24 January 2023
Revised: 8 February 2023
Accepted: 9 February 2023
Published: 11 February 2023

Copyright: © 2023 by the authors. Licensee MDPI, Basel, Switzerland. This article is an open access article distributed under the terms and conditions of the Creative Commons Attribution (CC BY) license (https://creativecommons.org/licenses/by/4.0/).

1. Introduction

In order to reduce the effects of global climate change, the increase in the average global temperature must be limited to 1.5 °C compared with the pre-industrial level. As part of the United Nations Framework Convention on Climate Change, member states have committed to meet national and international climate protection targets (Federal Climate Protection Act). For example, Germany must reduce greenhouse gas (GHG) emissions (referring to CO_2 equivalents, CO_2eq, including carbon dioxide, nitrous oxide, and methane emissions with various global warming potentials) by 88% by 2040 compared with 1990 levels [1]. In detail, the major proportion of CO_2 emissions arises from fossil fuel combustion processes [2], providing energy for mobility and transport, or thermal and electrical energy. Similar information can be applied currently to many other states, and the importance of GHG mitigation extends to all states considering global climate change.

Agricultural animal husbandry contributes directly and indirectly to climate change with significant GHG and ammonia (NH_3) emissions [3]. Emission mitigation measurements are, therefore, increasingly the subject of research and development; using proven mitigation technologies is becoming more important or even mandatory for (German and Dutch) farmers [4–6]. GHG emissions are an essential indicator for deriving the carbon footprint of animal husbandry and the corresponding provision of animal-based food. Sustainability is composed of environmental (e.g., climate change impact of GHG emissions), economic (e.g., rentability of animal husbandry), and social aspects (e.g., air quality for animals and workers within the barn building). The sustainability of animal-based food production plays an increasing role, considering the growing global population and the availability of plant-based food or meat substitutes.

Pig farms have a high demand for electrical energy, e.g., to operate the ventilation and feeding systems [7–12]. An exhaust air purification system—used for mitigating ammonia, particulate matter, and odor emissions—increases this energy demand. It is therefore associated with acquisition costs and increased operating costs for the farmer [13–15]. The demand for exhaust air purification systems is increasing in Germany [4] and other central European states (e.g., The Netherlands and Denmark) or regions with high livestock intensities [5,6]. The provision of the electrical energy used is often associated with further climate-relevant emissions; for example, the German electricity mix was responsible for 0.427 kg CO_2eq kWh_{el}^{-1} in 2020 [16]. However, the GHG emission intensity of electricity generation varies widely in the various countries, depending on the sources of electricity provision [17]. Consequently, the operation of air purification systems involves a trade-off between reduced emissions of environmentally relevant substances (e.g., NH_3 and particulate matter) and increased climate-relevant emissions (e.g., CO_2). In the long term, using regenerative sources for electricity or low-emission on-site energy sources can make sense.

In addition to electrical energy, piglet rearing barns have a high demand for thermal energy to meet the piglets' temperature requirements [18,19]. However, some parameters of exhaust air purification depend on the exhaust air temperature, e.g., the biological ammonia separation efficiency decreases with increasing exhaust air temperatures [20,21]. Reducing exhaust air temperature using heat recovery (HR) before air purification could positively affect the purification parameters. However, this topic can be considered an open question and needs research.

Continuous air exchange is essential for maintaining an appropriate animal housing climate, which is vital for animal welfare and development. Especially in winter, it is often difficult to keep the barn climate stable: cold outside air only allows a reduced air supply into the barn without exposing the animals to draughts and temperature stress, and control settings of the heating system can lead to air temperature fluctuations [22]. At the same time, reduced ventilation increases harmful gas concentrations like ammonia in the barn. The incoming and barn air heating is associated with additional thermal energy demand, e.g., by combusting fossil fuels. This combustion is associated with increased operating costs and additional CO_2 emissions [e.g., 0.237 kg CO_2 kWh_{th}^{-1} combusting liquefied natural Gas (LNG)] primarily in the winter period [23].

Heat recovery systems transfer thermal energy from a warmer to a colder medium. The recovered thermal energy can therefore be reused [24]. In animal barns, 70–90% of thermal energy losses are caused by the warm exhaust air leaving the barn [25,26]. A heat exchanger can transfer parts of this energy to the cold fresh air. The preheated supply air helps to achieve a more stable barn climate and reduce fossil fuel combustion, especially in the cold months. At the same time, heat recovery can lead to savings in fuel costs and CO_2 emissions [27,28].

There are several technical principles of heat recovery systems, such as air-to-air heat exchangers [22,29,30], water-to-air-based systems [31–34], or heat pumps [31,35–37] in the field of mechanically ventilated barn building construction. In addition, geothermal energy (e.g., underfloor air circulation [38–40]) and solar thermal energy [40–42] can play an essential role in barn air conditioning. However, these technologies are neither new in the field of heating, ventilation, and air conditioning (HVAC) or in industrial processes [43–45] nor in (German) agricultural animal husbandry; some scientific studies date back several decades [46–48]. Furthermore, demand in agriculture has not been constant in recent decades and has been partly dependent on the development of energy prices [29].

Bokkers et al. (2010) [27] published a study describing the effects of heat exchangers in practical use in broiler husbandry and farmers' experiences before and after their installation. This survey complements scientific studies on the performance parameters of heat exchangers, e.g., [22,28,30,33], and practical guides on how to use heat exchangers in practice, e.g., [49]. Besides, heat recovery can be used to mitigate ammonia concentrations and emissions in hen housing using manure belt drying [30]. Although, some of these studies have short investigation periods (e.g., one month [22,30]) or inaccuracies in the calculation methods (e.g., unclear additional electricity needed to operate the heat exchangers [28,29]). Moreover, the advancing climate change, fossil fuel shortages, partly increasing energy costs, and the social focus on more sustainability and environmental responsibility has led to a change in mentality. Heat recovery can improve the sustainability of processes and should also be increasingly focused on animal husbandry in mechanically ventilated barns. Furthermore, it should be recognized and used in practice as the best available technique depending on production systems and climate regions [6].

Therefore, the object of this study was to investigate the energy consumption and efficiency of a piglet rearing barn using an air-to-air heat exchanger and an exhaust air purification system in a long-term case study. For this purpose, the thermal energy recovered was quantified, and various performance parameters of the heat exchangers were determined. At the same time, the amount of CO_2 emission mitigation and energy cost savings were quantified for two cases: the barn with heat exchangers (reality) and the barn without heat exchangers (hypothetical, calculated case). The study aimed to demonstrate the potential of using a heat recovery system to make pig farming more sustainable.

In detail, the results are farm-specific as a single case study, so generalizability to global animal husbandry is only possible with caution and limitations. However, the presentation of the findings is intended to ensure a high grade of transferability. Furthermore, due to the long-term measurements of two years, the results offer detailed insights concerning the influence of different operating situations. By presenting ecological and economic savings potentials, the farmers' and stakeholders' interest in investing in energy-efficient, resource-saving, and emission-mitigating technologies is to be increased. We hypothesize that those technologies and their combination in an efficient way can contribute to more sustainable agriculture in the future.

2. Materials and Methods

2.1. Barn Building, Ventilation System, and Technical Equipment

The study was conducted under practical conditions in a piglet rearing barn in Germany (Lower Saxony), which had outside dimensions of 6.923 × 4.324 × 292.5 cm [811.4 cm height including the roof, Length × Width × Height (L × W × H); for abbreviations, please see the Nomenclature in the Supplementary Material, Section S1]. Data recording took place over two years, from December 2019 to December 2021 (730 days total). The barn building of the farm included 4140 animal places (AP) in piglet rearing, 128 AP for gilt rearing, and 110 AP in the breeding center. Since the animals were of different weights, the total mass of the animals in the barn could be calculated using livestock units (LU, 1 LU = 500 kg animal life weight). The nominal livestock unit (based on AP per animal compartments) was 124.2 LU for piglet rearing, 15.4 LU for gilt rearing, and 33.0 LU for the breeding center. However, the barn building was not fully occupied during the trial period. The following mean values were considered for year 1 (17 December 2019–15 December 2020) 98.4 LU in piglet rearing, 12.1 LU in gilt rearing, and 31.8 LU in the breeding center; and for year 2 (16 December 2020–15 December 2021) 101.0 LU in piglet rearing, 11.3 LU in gilt rearing and 32.2 LU in the breeding center.

The barn was mechanically ventilated with a central underfloor extraction system (Figure 1), leading to a central exhaust air collection duct. Furthermore, the barn was equipped with six exhaust fans with a diameter of 990 mm each (A3G990-AZ02-35, ebm-papst Mulfingen GmbH & Co. KG, Mulfingen, Germany). From the central exhaust air collection duct, air could be supplied to two air-to-air heat exchangers using two of the six exhaust fans. These two fans were operated continuously and identically throughout the year. At higher ventilation rates, the excess exhaust air could bypass the heat exchangers in two ways: (1) the exhaust air was conveyed by the two fans mentioned above, but the air did not flow through the heat exchanger heat transfer plates using bypass air shutters inside the heat exchanger shell, or (2) the four other exhaust fans conveyed the exhaust air in a regular manner. The air was then supplied to a following exhaust air purification system before leaving the barn.

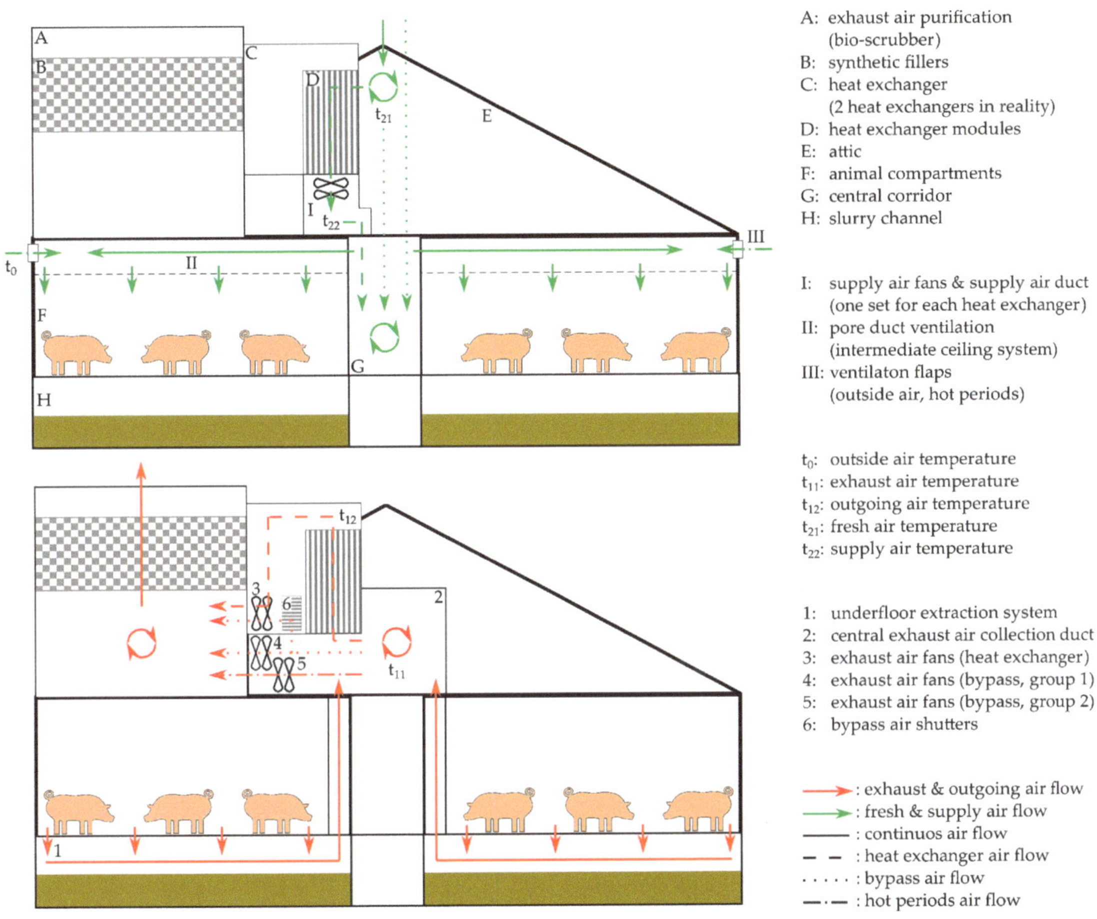

Figure 1. Schematic sketch of the barn building and the parts in the heat recovery, ventilation, and exhaust air purification systems. Moreover, the air flow pattern is shown for the supply (upper sketch) and exhaust (lower sketch) air flow considering the varying air flow management and the corresponding measurement points for the temperature loggers during the trials.

The supply air system consisted of a pore duct ventilation with an intermediate ceiling system. Fresh air could be led in different ways to the central corridor from where a mixed supply air was guided to the barn compartments and then through the pore duct to the animals. In winter, the fresh air from the attic was passed through the two heat exchangers before being guided into the supply air duct and the central corridor. During warmer temperatures, the fresh air from the attic bypassed the heat exchangers directly to the central corridor, or outside air could enter the central corridor via air shutters. In addition, during hot periods, outside air could enter the barn compartments' pore ducts directly through additional air shutters below the eaves. If the fresh air was conveyed through the heat exchangers, this was done by additional fans (one fan per heat exchanger, and these two fans were operated identically and in parallel throughout the year; Model FC080-6DQ.6K.A7_XBS, Ziehl-Abegg SE, Künzelsau, Germany).

The ventilation system was controlled automatically by a digital controller depending on the temperatures of outside air, supply air, room air in the animal compartments, the number of animals housed, and the minimum air flow rates for each animal compartment. The control system also regulated the management of flow paths (e.g., whether the exhaust air bypassed the heat exchangers or outside air was directed into the central corridor) and the air flow rates conveyed along the different paths. The control settings were developed by the ventilation system manufacturer (hdt Anlagenbau GmbH, Diepholz, Germany) to meet legal requirements and ensure good air quality in the animal compartments. The controller of the ventilation system recorded all operating parameters at intervals of about 30 s; adjustments to the system control were made at irregular intervals of a few minutes.

The heat exchangers were two recuperative air-to-air heat exchangers (Product type WT-BTK 200) from the manufacturer hdt Anlagenbau GmbH with a size of 332 $\times$ 114.6 $\times$ 442 cm (L $\times$ W $\times$ H) each. Each heat exchanger had an exchange surface of 266.64 m^2 made out of high heat acrylonitrile butadiene styrene (ABS-HH) exchange plates (thickness 0.5 mm). Each plate had a nominal thermal transmittance (U-value) of $\approx$ 5.79 W $(K\ m^2)^{-1}$ (thermal conductivity of ABS-HH 0.180–0.250 W $(K\ m)^{-1}$; for more details, see Table S1). Recuperative systems are characterized by continuous thermal flow according to the physical heat transfer process without storage processes [24]. Inside the exchangers, fresh and exhaust air were passed by each other in opposite directions (counter-flow) in a vortex flow process so the thermal energy could be transferred from the warm exhaust air to the cooler fresh air. However, thermoformed plastic plates separated the two air flows to prevent pollution and ensure the hygienic quality of the fresh air. The transfer area was increased through the helical surfaces of the plastic heat transfer plates used in the exchangers, and the turbulent air flow improved heat transfer. Each heat exchanger was designed for a maximum air volume flow rate of approximately 16,800 $m^3\ h^{-1}$; the sum of both heat exchangers corresponded to about one-sixth of the maximum air volume flow rate for ventilating the barn building in warm summer periods. As described above, both heat exchangers were continuously operated identically, i.e., the same exhaust and supply air volume flow rates were conveyed through both heat exchangers at all times (see Section 2.3).

The barn was equipped with LNG radiant heaters for heating the compartments. In each piglet pen (approximately 300 $\times$ 500 cm), a part of the pen (approximately 150 $\times$ 200 cm) were heated during the first 7 to 14 days of the rearing period (timed control). After this period, the radiant heaters were cleaned, disinfected, and used in the next compartment.

The exhaust air purification system installed in the barn was a bio-scrubber from the manufacturer Devrie Technical Solutions B.V. (Vriezenveen, The Netherlands), certified by the German Agricultural Society [50]. By sprinkling the synthetic fillers with slightly acidic or alkaline scrubbing water (target pH 6.5–6.8), a habitat was provided for microorganisms that decomposed the ammonia and odor molecules contained in the exhaust air, thus purifying the air. In addition, fine dust and bioaerosols were filtered out of the air. The separation efficiency for the exhaust air purification system used in the barn was at least 70% for ammonia and dust, following the specifications of [4], and achieved a prescribed maximum odor concentration of 300 odor units per cubic meter of exhaust air during certification.

2.2. Measurement of Air Temperature

Temperature and relative humidity (the latter was not further considered in this study) were measured continuously for the two-year trial period in measurement intervals of 15 min. The following measurement points apply continuously throughout the two-year trial period (Figure 1): outside air (t_0, below the eaves of the shady north side of the barn), and inside the barn, the exhaust air (t_{11}), outgoing air (t_{12}), fresh air (t_{21}), and supply air (t_{22}). The terminology of the air masses partly contradicts the nomenclature of typical HVAC applications but adopts terms used in agricultural practice and former reports [33]. In detail, the air entering the heat exchanger on the exhaust side corresponds to extract air in HVAC terminology (in this study, exhaust air, since it is the air that is exhausted from the animal compartments), and the air leaving the heat exchanger corresponds to exhaust air in HVAC terminology (in this study, outgoing air, since it is the air that is discharged to the purification system). Fresh, supply, and outgoing air parameters were measured for one heat exchanger; the same values were assumed for the second since both heat exchangers were used in parallel throughout the trial. Data were recorded using three different types of data loggers (Testo 174 T, Testo SE & Co. KGaA, Titisee-Neustadt, Germany, and Tinytag Plus 2 TGP-4500 and TGP-4505, Gemini Data Loggers Ltd., Chichester, United Kingdom) and additionally by implemented sensors of the manufacturer of the heat exchangers (WTR 190-A1-1A2/PT1000, promesstec GmbH, Schüttorf, Germany) at each measuring point. For each point, the mean values of all sensors were used for the subsequent calculations. All sensors were checked before and during the measurement period (eight times during the two-year trials). For more details (e.g., sensor type, measuring range, accuracy) see Table S2.

The temperature variation (difference between the daily minimum and maximum temperature, $\Delta t = t_{max} - t_{min}$) is an important parameter to calculate the rate of amplitude damping of buildings or HVAC systems [51,52]. In the context of the conducted study, amplitude damping refers to the ability of the heat exchangers to reduce the supply air temperature variation Δt_{22} compared with outside air temperature variation Δt_0:

$$\text{amplitude damping}_{HR} = \left(\frac{\Delta t_{22}}{\Delta t_0} - 1\right) * 100 = \left(\frac{t_{22,max} - t_{22,min}}{t_{0,max} - t_{0,min}} - 1\right) * 100 \quad (1)$$

where:

- amplitude damping_{HR} = amplitude damping due to fresh air conditioning via heat recovery of the heat exchanger in %;
- Δt_{22} = daily temperature variation of the supply air in K;
- Δt_0 = daily temperature variation of the outside air in K;
- $t_{22,max}$ = daily maximum supply air temperature in °C;
- $t_{22,min}$ = daily minimum supply air temperature in °C;
- $t_{0,max}$ = daily maximum outside air temperature in °C;
- $t_{0,min}$ = daily minimum outside air temperature in °C.

2.3. Calculation of the Air Flow Rates

The thermal power of the heat exchangers was calculated based on the air mass flow rate of the fresh air conveyed through the heat exchanger heat transfer plates (see Section 2.4). However, in these trials, measuring the air volume flow rate directly (e.g., using measuring fans) was impossible due to the constructional and technical implementation of the heat exchangers into the barn building (see Section 3.6).

Instead, all fans' air volume flow rates were calculated based on the fan speed and the differential pressures (upstream and downstream of the fans). The derivation of the two variables was also carried out for each 15-min measurement interval. The supplementary material file gives a detailed explanation of this procedure (see Section S3, Figures S1 and S2).

Briefly, the fan speeds were calculated using the fans' analog control [V] after the corresponding correlations had been derived in preliminary tests. For this, the speeds were recorded using an optical tachometer (model K06538, Voltcraft, Hirschau, Germany) and reflective markers (supply air fans from Ziehl-Abegg SE, Künzelsau, Germany) or using the BUS connection and software supplied by the manufacturer (exhaust air fans from ebm-papst Mulfingen GmbH & Co. KG, Mulfingen, Germany). Differential pressures within the ventilation system were measured using digital differential pressure transmitters [twice the model DE2752M042CK00MWU3231 (measuring range 0–100 Pa) and twice the model DE2752M042CK00MWU3233 (measuring ranges 0–200 Pa and 0–300 Pa), FISCHER Mess- und Regeltechnik GmbH, Bad Salzuflen, Germany] at several places: the central exhaust air collection duct, one heat exchanger (the heat exchanger with the temperature sensors), pressure chamber beneath the synthetic fillers of the bio-scrubber, and in the supply air duct right in front of and behind the supply air fans.

The fans were arranged in several operating groups, controlled separately by the digital controller of the ventilation system (see Section 2.1). The supply air fans convey air through the heat exchanger as long as thermal energy is needed to raise the room air temperatures in the animal compartments. If the supply air fans have the highest electrical power input, i.e., the maximum fan speed, and the exhaust bypass air shutters are closed, this is called full workload of the heat exchanger (in the corresponding measuring interval). If the thermal energy demand decreases, the supply air flow rate is reduced, and the exhaust bypass air shutters open (=partial workload operation) until the heat exchanger stops providing thermal energy (heat exchanger off). The supply air fans were operated at $t_{21} \leq 30.5$ °C, however 99.0% of the heat exchanger operation intervals were observed for $t_{21} \leq 23.0$ °C [minimum of 6597 (m^3 supply air) h^{-1} total capacity for $t_{21} = 23.2$ °C, maximum 30,864 (m^3 supply air) h^{-1} for $t_{21} = 16.4$ °C].

For the user, the information concerning the ventilation system's performance is given as a percentage of the performance of the entire ventilation system [%] (=air flow rate of the exhaust air fans). The extreme values of the system control were 19.8% at $t_{21} = -1.9$ °C [16,354 (m^3 exhaust air) h^{-1} total capacity] as well as 96.5% at $t_0 = 30.8$ °C [223,889 (m^3 exhaust air) h^{-1} total capacity]. The exhaust bypass air shutters of the heat exchangers were opened for varying fresh air temperatures depending on the room air temperatures (operating range $t_{21} \leq 12.2$ °C). The four remaining exhaust fans for bypass operation were controlled uniformly until 8 April 2020 (operating range $t_0 \geq 11.7$ °C), after which they were divided into two groups to provide more precise control of the barn ventilation (group A: operating range $t_0 \geq -8.4$ °C, group B: operating range $t_0 \geq 12.9$ °C).

2.4. Calculation of the Thermal Power and the Effectiveness

The thermal power $\dot{Q}_{HR}$ of the heat exchanger (the one with the temperature and differential pressure sensors) was calculated using the following equation for each 15-min measurement interval [53]:

$$\dot{Q}_{HR,th} = \frac{\dot{m}_2 * c_{pl} * (t_{22} - t_{21})}{1000} \quad (2)$$

where:

- $\dot{Q}_{HR,th}$ = thermal power of the heat recovery system in kW;
- $\dot{m}_2$ = supply air mass flow rate in kg h^{-1}, calculated based on the air volume flow rate and the corresponding air density at varying air temperatures ($\dot{m}_2 = \dot{V}_2\ (-0.00482\ t_{22} + 1.274)$; where $\dot{V}_2$ = air volume flow rate of the supply air fan in $m^3\ h^{-1}$ [38,54]);
- c_{pl} = specific heat capacity of dry air in Wh_{th} (kg K)$^{-1}$ (c_{pl} = 1.005 kJ_{th} (kg K)$^{-1}$ = 0.28 Wh_{th} (kg K)$^{-1}$);
- t_{21} = fresh air temperature in °C;
- t_{22} = supply air temperature in °C.

The calculated thermal power was adopted for the second heat exchanger since both were operated identically (assuming identical temperatures and air flow rates). In the further course of the article, reference is always made to the summed thermal power of both heat exchangers. The amount of thermal energy recovered [kWh_{th}] in a given period corresponds to $Q_{HR,th}$. Due to the calculation of the air volume flow rates (Section 2.3) and the possible inaccuracy (see Section 3.6), an uncertainty analysis was carried out concerning the calculation of the thermal power (see Section S4 and Figure S3 [34,55–58]). This possible uncertainty consequently also influenced other performance parameters (e.g., $COP_{heating}$ and $PF_{heating}$, see Sections 2.5 and 2.7).

Another critical performance parameter of heat exchangers is the dimensionless effectiveness of sensible heat transfer. This parameter describes the ratio of the actual thermal power and the maximum possible thermal power of the heat recovery system for each measurement interval. The latter is determined by (a) the temperature difference between the exhaust and fresh air, which is usually the largest temperature difference of the air masses, and (b) the minimum supply or exhaust air mass flow rate and the specific heat capacity, which determines how much air emits or absorbs thermal energy. The product of (a) and (b) limits the amount of thermal energy transferable from the warmer to the colder air flow for each measurement interval. The effectiveness calculation can differ in the various studies in this field of research (see Section 3.2). However, in this study, it was calculated as follows [46,47,59]:

$$\varepsilon = \frac{\dot{Q}_{HR,th}}{\dot{Q}_{max,th}} = \frac{\dot{m}_2 * c_{pl} * (t_{22} - t_{21})}{\dot{m}_{min} * c_{pl} * (t_{11} - t_{21})} \tag{3}$$

where:

- ε = effectiveness of the sensible heat recovery system (heat exchangers);
- $\dot{Q}_{max,th}$ = maximum possible thermal power of the heat recovery system in kW;
- $\dot{m}_{min}$ = the minimum of exhaust air $\dot{m}_1$ or supply air mass flow rate $\dot{m}_2$ for each measurement interval in kg h^{-1};
- t_{11} = exhaust air temperature in °C.

2.5. Calculation of Energy Input, Coefficient of Performance, and Performance Factor

Electricity consumption of the investigated barn was recorded by individual electronic electricity meters (model DSZ15D-3x80A, Eltako GmbH, Fellbach, Germany) for each electric consumer, e.g., ventilation system, air purification system, feeding system, air compressor, and social rooms, over the whole period. The consumption of LNG was recorded using the ordered quantities and the gas tank level.

Moreover, the fans' electricity and the barn building's LNG consumption were calculated for the 15-min measurement intervals. For this purpose, the electrical power consumption [W] of the fans was also measured in the preliminary tests (see Section 2.3, Sections S3 and S4, especially Figure S1) depending on their analog control [V]. Thus, based on the control software's values, the eight fans' power consumption could be calculated for each interval. The documented gas consumption was related to the recorded outside air temperatures, and the corresponding correlation was used to calculate the daily consumption [(kg LNG) d^{-1}] of the barn building, which was transferred to the measurement intervals (see Section S5, Figure S4). These values were subsequently compared with the data from the actual electricity meters and purchased LNG quantities and showed only minor differences (+2.3% for LNG consumption, −2.1% for electrical energy consumption).

The coefficient of performance ($COP_{heating}$) represents the ratio of the thermal power [kW_{th}] (see Section 2.4) transferred by a heat recovery system to the required electrical power [kW_{el}] to operate a heat recovery system. It is an indicator of the energy efficiency of heat recovery units and was calculated using the following general equation for each 15-min interval [53,60]:

$$COP_{heating} = \frac{\dot{Q}_{HR,th}}{P_{el}} \tag{4}$$

where:

- $COP_{heating}$ = coefficient of performance during heating;
- P_{el} = electrical power of various consumers in kW.

This parameter is essential for evaluating heat recovery systems and is mainly used for heat pumps or cooling units. Several former articles in the context of heat recovery in animal barns used COP for heat recovery systems, including recuperative systems, e.g., [33–36]; thus, we adopt the term. An index has been added to $COP_{heating}$ to clarify that energy efficiency is considered and calculated only during heating periods.

The concept of $COP_{heating}$ for heat exchangers is debatable: heat exchangers recover thermal energy via passive heat transfer. The heat exchangers used in this study did not absorb electrical energy; consequently, quotient formation as in $COP_{heating}$ was, strictly speaking, not possible. However, they were part of the barn building and its ventilation system, and their presence and use resulted in increased electrical power consumption of the ventilation fans due to the rising air flow resistance (see Section 2.7). Consequently, the ventilation system had an increased electrical energy consumption in order to be able to recover thermal energy. This input is compared with the recovered thermal energy. This ratio enables comparisons between heat exchangers and other heat recovery systems, such as heat pumps.

Moreover, the $COP_{heating}$ was calculated by the thermal and electrical power [kW], which means that it was calculated for one operating moment, i.e., for each measurement interval. In the further article, when talking about the quotient of thermal energy recovered [kWh_{th}] and electrical energy consumption over a certain period [kWh_{el}], the term Performance factor ($PF_{heating}$) is used.

Since these are practice measurements in a barn building with several energy consumers within the air ventilation and treatment systems, different $PF_{heating}$ values (or corresponding $COPs_{heating}$ for each measurement interval) were calculated. These were the $PF_{V,heating}$ that related to the electrical power of the fans (ventilation system, V, $Q_{V, el}$) and the $PF_{VA,heating}$ that related to the electrical power of the fans (ventilation system, V, $Q_{V, el}$) and the exhaust air purification system (A, $Q_{A, el}$; summed to, $Q_{V+A, el}$) with the uniform thermal energy recovery of the heat exchangers ($Q_{HR,th}$). For the single measurement intervals, $COP_{V,heating}$ and $COP_{VA,heating}$ were calculated.

2.6. *Calculation of Substituted CO_2 Emissions and Heating Costs*

Based on the energy consumption and the recovered thermal energy, the resulting CO_2 emissions, costs, and the corresponding reductions using heat recovery were calculated for the barn building and the various consumers.

2.6.1. Calculation of Substituted CO_2 Emissions

To determine the CO_2 emissions, the factors $m_{CO2, el}$ = 0.427 kg CO_2 kWh_{el}^{-1} for electrical energy (electricity supplied in the German power grid, [16]) and $m_{CO2,th}$ = 0.237 kg CO_2 kWh_{th}^{-1} for thermal energy (combustion of LNG) were assumed [61–63]. Within this article, a simplified calculation of CO_2 emissions was made, with the barn building serving as the system boundary. This means that the emission factor of the thermal energy $m_{CO2,th}$ only considered the CO_2 emissions caused by the actual combustion, and CO_2eq emissions in the course of LNG extraction, transport, and supply were not considered. For electrical energy, the emission factor $m_{CO2,el}$ was simplified as pure CO_2 emissions, even though other GHG emissions may have occurred during electricity generation.

For example, the following equation is applied for the CO_2 emissions resulting from the operation of the exhaust air purification system (similar calculations apply for the other consumers):

$$CO_2\ Emissions_A = Q_{A,el} * m_{CO2,\ el} \quad (5)$$

where:

- CO_2 $Emissions_A$ = CO_2 emissions based on the electrical energy consumption of the exhaust air purification system (A) for a given period in kg CO_2;
- $Q_{A,el}$ = Electrical energy consumption of the exhaust air purification system (A) for a given period in kWh_{el};
- $m_{CO2, el}$ = CO_2 emission factor for electric energy [electricity supplied in the German power grid in (kg CO_2) kWh_{el}^{-1}].

2.6.2. Initial Capital Cost and Fixed Costs

The initial capital cost used to calculate the payback period was based on the barn-specific costs supplied by the farmer and the offer sheets of the manufacturers. The initial capital cost included the costs for the technology and material itself, assembly, and German taxes (value-added tax 19%).

Subsequently, initial capital costs were used to calculate the annual fixed costs. For this, a depreciation of 15 years was assumed. The interest rate (0.5%) and insurance costs (EUR 300 a^{-1} for Case 1, and EUR 100 a^{-1} for Case 2, see Section 2.7) were added.

2.6.3. Energy Costs

For the calculation of the saved energy costs, the monthly prices for electrical energy C_{el} (mean EUR 0.2697 ± 0.0030 kWh_{el}^{-1}) during the trial period were used [64,65]. The thermal energy costs C_{th} (in this trial, LNG costs; mean EUR 0.0658 ± 0.0169 kWh_{th}^{-1}) were calculated on a monthly base via weekly online price comparisons for the German energy market.

For example, the following equation is applied for the energy costs resulting from the operation of the ventilation system (similar calculations apply for the other consumers):

$$\text{Energy}\ \cos \text{ts}_V = Q_V * C_{el} \tag{6}$$

where:

- $\text{Energy}\ \cos \text{ts}_V$ = energy costs based on the electrical energy consumption of the ventilation system (V) for a given period in EUR;
- $Q_{V,el}$ = electrical energy consumption of the ventilation system (V) for a given period in kWh_{el};
- C_{el} = energy costs for electrical energy (e.g., electricity of the German power grid) in EUR kWh_{el}^{-1}.

2.6.4. Operating and Maintenance Costs and Annual Variable Costs

The operating and maintenance costs include water and labor costs for cleaning and maintenance (EUR 1200 a^{-1} for Case 1, and EUR 1150 a^{-1} for Case 2, see Section 2.7) and the repair costs (2% of the initial capital cost). These costs were added to the energy costs to calculate the annual variable costs.

2.6.5. Payback Period

The payback period is required for economic savings, primarily due to fuel cost savings using heat recovery to equal the initial capital cost. To calculate this, the initial capital costs were divided by the annual cost savings (difference between total annual costs of Case 2 and Case 1, see Section 2.7). This was done for the initial capital cost of the heat recovery system in Case 1 and the total initial capital cost difference between Case 2 and Case 1.

2.7. Comparison between the Actual Barn Situation (with Heat Recovery) and the Hypothetical Situation (without Heat Recovery)

The procedure described above was used to determine the energy efficiency of the actual barn using a heat recovery system. As described above, a heat exchanger does not consume electrical energy itself; however, the use of heat exchangers causes higher electrical energy consumption of the fans due to higher flow resistance of the conveyed air flow [27,28]. Therefore, this additional electrical energy input can be opposed to the recovered thermal energy.

In order to be able to evaluate the energy efficiency of the heat recovery system, the energy consumption of the actual barn situation with a heat recovery system (Case 1) was compared with a hypothetical case without a heat recovery system (Case 2, see Section S6 for more information). Case 2 should be an exact copy of Case 1, i.e., the same air flow rates, temperatures, and thermal energy input were assumed. For this hypothetical case (without a heat recovery system), the energy consumptions for each 15-min measurement interval were calculated. For both cases, the overall animal husbandry and management should be consistent, including the thermal energy input to ensure consistent room air temperatures in the animal compartments (inspired by [30]). Therefore, the recovered thermal energy in Case 1 was substituted by LNG combustion in hypothetical Case 2. Consequently, the demand for LNG increased in Case 2.

The demand for electrical energy for Case 2 was calculated as follows:

1. The energy consumption of the supply air fans (see Section 2.1) was not considered further since supply air fans are commonly unnecessary in ventilation systems without heat exchangers.
2. For the two exhaust air fans behind the heat exchangers, the calculation was done in several steps:
    a. It was assumed that the same fans in each case convey the identical air volume flow rates.
    b. Since the heat exchangers are associated with additional pressure losses, the corresponding differential pressure was used, which would prevail if no heat exchangers were installed (see Figure S5). This is the differential pressure between the central exhaust air collection duct (including adjustment of the pressure conditions in this duct if the supply air fans were switched off and did not "force" any air into the barn) and the measuring point behind the fans (before the exhaust air purification). These values were recorded using digital differential pressure transmitters (see Section 2.3).
    c. The correlation equations (see Section 2.3 and Figure S2) were used to calculate the fan speed required at the assumed differential pressures to deliver the actual (unchanged) air volume flow rates. This corresponded to the operating parameters of the fans in Case 2 (without heat recovery).
    d. It was derived which analog control and electric power input of the fans would be necessary under these circumstances.
3. A comparable method was used for the four additional exhaust air fans (bypass), adjusting the hypothetical pressure conditions in the central exhaust air collection duct.

The difference in electricity consumption between Case 1 and 2 will be referred to as the electrical energy consumption for the heat recovery and was calculated as follows:

$$Q_{HR,\ el} = Q_{V,\ Case\ 1,el} - Q_{V,\ Case\ 2,el} \tag{7}$$

where:

- $Q_{HR,\ el}$ = electrical energy consumption to operate a heat recovery system (two heat exchangers) in the barn building's ventilation system for a given period in $kWh_{el;}$
- $Q_{V,\ Case\ 1,el}$ = electrical energy consumption of the ventilation system (V) in Case 1 (with heat recovery system) for a given period in $kWh_{el;}$
- $Q_{V,\ Case\ 2,el}$ = electrical energy consumption of the ventilation system (V) in Case 2 (without heat recovery system) for a given period in $kWh_{el.}$

To determine the energy efficiency of the heat recovery, the value $PF_{HR,heating}$ was calculated: This is derived from the recovered thermal energy $Q_{HR,th}$ and electrical energy consumption for the heat recovery $Q_{HR,\ el}$ to operate the heat recovery system (see Section 2.5, Equation (4)). The uncertainty of $COP_{heating}$ calculation for $COP_{HR,heating}$ $\omega_{COP_{HR,heating}}$ was also calculated (see Section S4).

Subsequently, the total quantities of energy (electrical and thermal) were calculated, which were input into the animal husbandry, i.e., into the barn building, through electricity consumption or liquid gas combustion. This was calculated for the two cases as follows:

$$Q_{total} = Q_{total,\ el} + Q_{total,th} \tag{8}$$

$$Q_{Case1} = Q_{total,\ el} + Q_{LNG,Case1,th} \quad (9)$$

$$Q_{Case2} = Q_{total,\ el} - Q_{HR,\ el} + Q_{LNG,Case1,th} + Q_{extra\ LNG,Case2,th} \quad (10)$$

where:

- Q_{total} = energy consumption to operate the barn building for a given period in $kWh_{el,th;}$
- $Q_{total,el}$ = electrical energy consumption to operate the barn building for a given period in $kWh_{el;}$
- $Q_{total,th}$ = thermal energy consumption to operate the barn building for a given period in $kWh_{th;}$
- Q_{Case1} = energy consumption to operate the barn building in Case 1 (with heat recovery) for a given period in $kWh_{el,th;}$
- $Q_{LNG,\ Case1,th}$ = thermal energy provided due to combustion of LNG in Case 1 for a given period in $kWh_{th;}$
- Q_{Case2} = energy consumption to operate the barn building in Case 2 (without heat recovery) for a given period in $kWh_{el,th;}$
- $Q_{extra\ LNG,\ Case2,th}$ = $Q_{HR,th}$ = additional thermal energy provided due to combustion of LNG in Case 2 for a given period in kWh_{th}; this corresponds to the recovered thermal energy in Case 1 to ensure equal indoor air temperatures in the barn building.

Finally, the energy input to the barn building was compared between Cases 1 and 2. In short, in Case 2, the electrical energy consumption to operate a heat recovery system $Q_{HR,\ el}$ was subtracted, and the recovered thermal energy in Case 1 $Q_{HR,th}$ was replaced by additional thermal energy provided due to the combustion of LNG $Q_{extra\ LNG,\ Case2,th}$. In addition, the energy quantities of Equations (8)–(10) were multiplied by the respective specific emission factors and energy costs for electricity and LNG combustion (see Section 2.6) to determine the savings in CO_2 emissions and energy operating costs.

2.8. Measurement Data Analysis

In general, missing or invalid data points (for the 15-min measurement intervals) were not considered for the derivation of correlations or calculations (valid data points 93.3%; 65,421 intervals of the total of 70,084 intervals). However, the cumulated monthly values were calculated using interpolated values for the missing data points based on the corresponding monthly mean values.

Furthermore, only trial days were considered for calculating temperature amplitude damping, which provided more than 22 h of the day valid data.

Data analysis and visualization were done using Microsoft Office Excel 2019 (Microsoft Corporation, Redmond, Washington, DC, USA). Data were analyzed descriptively by sample size, mean values, standard deviation, minimum and maximum.

3. Results and Discussion

3.1. Operating Parameters and Workload of the Heat Exchangers

Figure 2 shows the heat exchangers' monthly workload over the trial period of two years. The figure shows that the workload from October to March was almost constant at 100%. From April to September 2020, the heat exchangers' workload ranged from 45% to approximately 90%, with the lowest operating workload in August. In the second year, the heat exchangers' workload showed a similar range, with the lowest workload in June. In his studies on an air-to-air heat exchanger, Rösmann [28] was also able to achieve an annual workload of 88.4%, with minimum values in summer of 51.2%. The results presented here are, therefore, comparable. Since the efficiency of a heat exchanger increases with the increasing temperature difference between supply and exhaust air [66], it could be assumed that the heat exchanger generates no benefit in summer. The results refute this assumption and show that the heat exchangers also operate during summer.

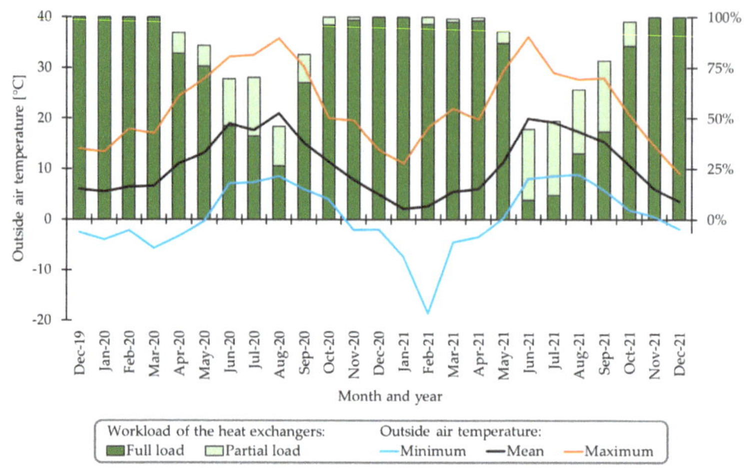

Figure 2. Workload of the heat exchangers (full or partial load) and outside air temperatures t_0 (minimum, maximum, and monthly mean values) on a monthly base over the two-year trial period.

Figure 3 shows the mean workload of the heat exchangers in August 2020. For this purpose, the mean values of each hour of the day were calculated over the 31 days. The heat exchangers reached a workload of 50 to 80% during the night, despite relatively high outside air temperatures of 16–18 °C. The workload dropped to a minimum of 10% at a maximum daily mean outside air temperature of 26 °C and rose to 50% in the evening at an outside air temperature of 20 °C. Looking at the selected courses, it is again clear that a benefit is generated from the heat exchangers, even in warm summer, due to the thermal power during the cooler night hours.

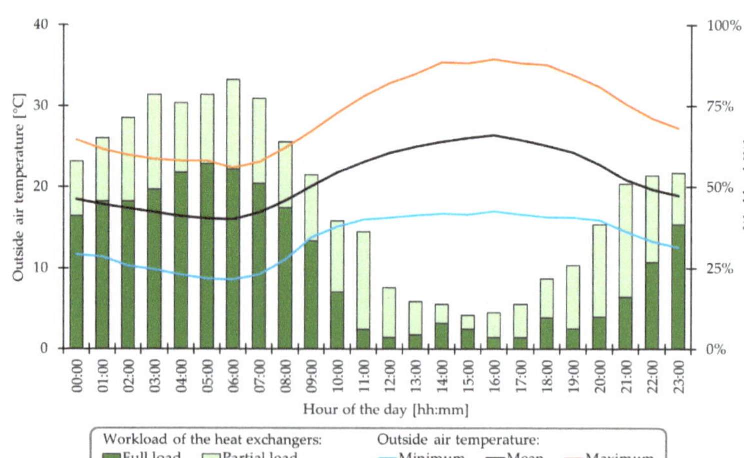

Figure 3. Workload of the heat exchangers (full or partial load) and outside air temperatures t_0 (minimum, maximum and mean values) in August 2020 concerning the mean values for the hours of the day (over the 31 days of the month).

Figure 4 shows the temperature profile [°C] for a single day for outside, fresh, supply, outgoing, and exhaust air (4 April 2020). This day was a typical example of the temperature courses during the spring season and the corresponding heat exchanger operation, i.e., heating during the night and turning off in the afternoon. A closer look at the temperature courses indicates that the fresh air was already warmer than the outside air. This preheating within the attic was already reported by [38] and can influence the thermal power of heat recovery systems.

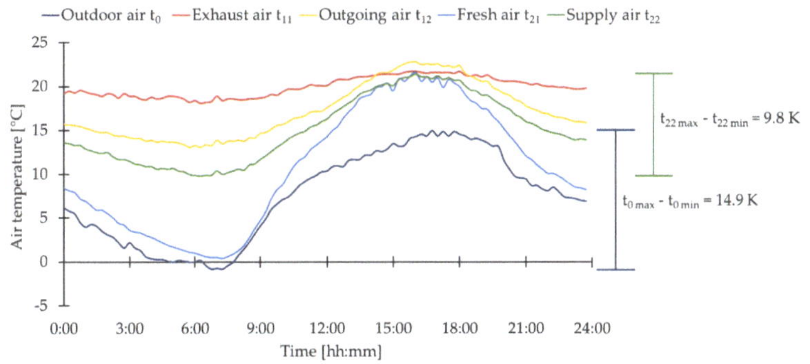

Figure 4. Course of the outside air, exhaust air, outgoing air, fresh air, and supply air temperatures [°C] during 4 April 2020 and the temperature variations Δt_0 and Δt_{22} [K] for the outside and supply air on this day.

The temperature variation of the outside air was 14.9 K (minimum of −0.8 °C and a maximum of 14.1 °C). The temperature variation of the supply air was 9.8 K (minimum of 9.8 °C and a maximum of 19.6 °C). Consequently, the heat exchangers could dampen the temperature amplitude by about −34%. Significant temperature variations occurred, especially in spring and autumn, when temperatures were already in a double-digit range during the day but could drop into negative temperature ranges at night. However, noticeable outside air temperature variations could also occur during the summer periods (Figure 3), with corresponding heat exchanger workloads during the colder hours. Furthermore, the amplitude damping described was typical in the trial period. The mean damping for each day within the two-year trial was −34.7 ± 12.8%. The more detailed evaluation shows that the percentage damping depended neither on the daily mean outside air temperature nor the daily outside air temperature variation (Figure S6). This means the relative amplitude damping was consistent throughout the year, but the absolute differences between outside and supply air temperature variations changed with the amount of outside air temperature variations during the seasons.

Significant temperature variations over a day can increase heating costs and thus CO_2 emissions [67]. Furthermore, significant temperature variations must be avoided in terms of animal welfare because the animals can quickly reach the limits of their adaptability and be exposed to temperature stress [68,69]. The results presented here align with other studies' results, demonstrating lower supply air temperature variations and amplitude damping using heat exchangers or other heat recovery systems [22,29,38]. The decreasing temperature difference between incoming supply and compartments' room air reduces the risk of draughts and promotes more homogenous temporal and spatial air temperatures and air flow [22]. In poultry production, this has led to improved animal distribution in the barn building [22,27].

Furthermore, Han et al. (2013) [22] reported that average air flow rates were about 10–20% higher in a broiler barn building with heat exchangers compared with a comparable building without heat exchangers. Goselink et al. (2019) [30] reported an increase of 12%. This is especially important in cold situations: preheating the fresh air facilitates higher air flow rates for an improved discharge of harmful gases, e.g., CO_2 or NH_3. For poultry production, higher air flow rates can positively influence litter quality and animal health [22,27]. In addition, the practice survey showed that heat exchangers could tend ($p = 0.07$) to improve the daily weight gain of broilers [27]. There is still a need for research on piglet rearing concerning possible benefits.

Within this investigated period, it was once more shown that the temperature difference between supply and fresh air depended on the fresh air temperatures. The correlation ($t_{22} = -0.3897 \times t_{21} + 8.9326$; $R^2 = 0.9534$) was high, and the maximum temperature difference was 14.0 K on 13 February 2021 (second year of trial; outside air temperature −17.9 °C, fresh air temperature −11.0 °C, supply air temperature 2.9 °C, supply air volume flow rate 22.790 $m^3\ h^{-1}$, thermal power 113.8 kW_{th}).

3.2. Thermal Power, Recovered Thermal Energy, and the Effectiveness

Table 1 shows the minimum, mean and maximum monthly temperature profiles [°C] of outside, fresh, and supply air for the first year of the trial period and the thermal power [kW_{th}].

Table 1. Outside, fresh and supply air temperatures, and the thermal power of the heat exchangers $\dot{Q}_{HR,th}$ (minimum, maximum, and monthly mean values) on a monthly base for the first trial year (17 December 2019–15 December 2020).

Year 1 2019/2020	Outside Air t_0 [A] [°C]			Fresh Air t_{21} [B] [°C]			Supply Air t_{22} [B] [°C]			Thermal Power $\dot{Q}_{HR,th}$ [B] [kW_{th}]		
	Min.	**Mean**	**Max.**	**Min.**	**Mean**	**Max.**	**Min.**	**Mean**	**Max.**	**Min.**	**Mean**	**Max.**
December *	−2.5	6.0	14.0	−0.9	7.1	14.9	8.0	13.0	17.3	17.6	49.1	80.8
January	−4.0	5.5	13.4	−3.2	6.4	14.5	6.4	12.8	17.7	23.8	54.2	84.9
February	−2.1	6.5	17.9	−0.3	7.3	17.8	8.8	13.3	19.6	15.1	51.0	82.8
March	−5.6	6.6	17.0	−4.1	8.3	20.3	6.4	13.7	21.1	2.6	48.2	98.9
April	−3.3	11.0	24.4	−2.4	11.3	22.2	8.0	15.3	22.6	0.5	34.9	98.1
May	−0.3	13.1	27.8	1.7	12.9	23.0	9.7	16.6	23.3	0.3	31.2	72.1
June	7.2	19.0	32.2	8.0	17.2	24.1	13.8	19.9	24.4	0.3	23.1	53.9
July	7.3	17.7	32.6	8.5	16.7	23.9	14.3	19.5	24.0	0.3	23.7	55.7
August	8.6	21.0	35.8	9.2	16.9	23.9	15.3	20.0	24.4	0.3	25.0	55.3
September	6.0	15.1	30.3	7.5	14.5	25.2	13.9	18.1	25.6	0.3	31.1	62.2
October	4.0	11.4	20.0	5.0	12.2	21.9	11.5	16.1	22.1	0.6	33.4	64.2
November	−2.0	7.9	19.6	−0.7	9.1	20.3	8.6	14.2	20.8	1.9	41.1	76.6
December **	0.9	4.5	9.4	2.5	5.7	10.1	10.6	12.5	15.1	34.1	56.3	78.6
Year 1 total	−5.6	11.6	35.8	−4.1	10.8	25.2	6.4	15.5	25.6	0.3	39.6	98.9

[A] Values whether the heat exchangers were operating or not (based on 15-min measurement intervals).
[B] Values during heating of the heat exchangers (based on 15-min measurement intervals).
* 17–31 December 2019.
** 1–15 December 2020.

In the first year of the investigations (2020), the fresh air temperature was on average 10.8 °C while the supply air temperature was on average 15.5 °C. Thus, the fresh air was heated by an average of 4.7 K when it passed through the heat exchangers, considering a mean thermal power of 39.6 kW_{th}. The maximum thermal power was 98.9 kW_{th} at a fresh air temperature of −4.1 °C and an attained supply air temperature of 6.4 °C (supply air volume flow rate of 17.272 $m^3\ h^{-1}$ per heat exchanger). The temperature difference was thus 10.8 K. On the other hand, the lowest thermal power was achieved at a fresh air temperature of 24.2 °C and a supply air temperature of 24.3 °C. Comparable values were also obtained in the second year of the investigation, with a maximum thermal power of 115.6 kW_{th}. As already shown by former studies [28,33], the thermal power increases with decreasing outside air temperature. The higher the temperature difference between fresh and exhaust air, the more heat that can be transferred. During the heat exchangers' heating, the exhaust air showed constant temperature levels of 21.5 ± 2.3 °C for year 1 and 20.9 ± 2.4 °C for year 2.

Figure 5 shows the correlation between the thermal power [kW_{th}] and the fresh air temperature [°C]. The highest thermal power could be achieved at low fresh air temperatures following a linear correlation.

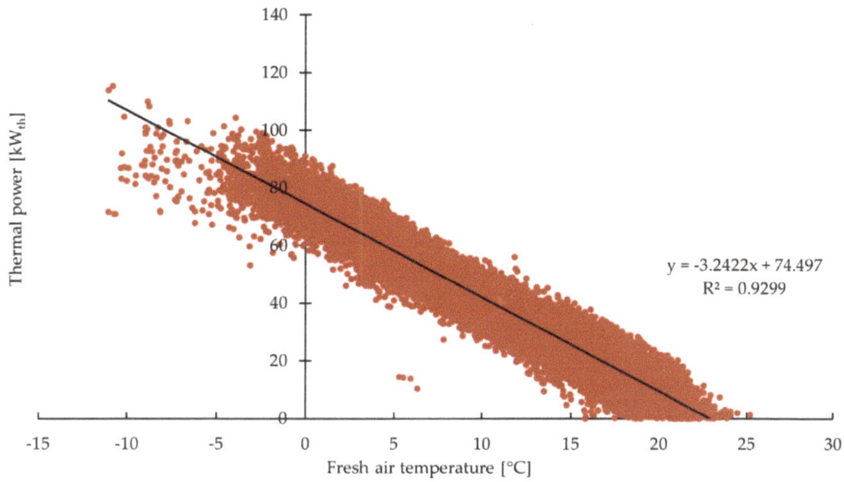

Figure 5. Correlation of the thermal power $\dot{Q}_{HR,th}$ [kW_{th}] of the two heat exchangers and the fresh air temperature [°C] when heat exchangers heated the incoming fresh air.

The effectiveness ε of the heat exchangers is presented in Figure 6. The mean value of the parameter during the two-year trial was 0.49 ± 0.13. It is shown that the values were at a constant level and decreased considerably at higher fresh air temperatures, thus with decreasing temperature differences between fresh and exhaust air. While 86.6% of the data were in the range $0.4 \leq \varepsilon \leq 0.6$, 0.6% of the values were $\varepsilon > 1$ and outside the actual definition range of the parameter. A detailed examination (Figure S7) shows that these values mainly occurred at comparatively low exhaust air mass flows, and effectiveness increased with decreasing air flow rates. Due to varying fan control in short intervals, the ratio of exhaust and supply air mass flow rates drifted widely, allowing a calculated effectiveness of $\varepsilon > 1$. Since these irregularities of the fan control occurred in short periods within the two-year trial and were resolved after detection, these values are considered outliers, which can occur in the context of investigation under field conditions.

The effectiveness investigated in this study was considerably higher than that of earlier barn building studies [46,47]. Moreover, various publications state different equations for calculating heat exchangers' effectiveness; some studies consider the exhaust or supply air mass flow rates [46,47], and some do not [29,30,41,70]. Therefore, comparing those publications and effectiveness values is only of limited informative value. However, the investigated heat exchangers' effectiveness seems to be high in the context of barn building ventilation.

3.3. Energy Input, Coefficient of Performance, and Performance Factor

The $PF_{heating}$ described the energy efficiency of the heat recovery and was determined for various reference variables. Table 2 gives the annual electrical energy consumption for both trial years of all relevant consumers, the thermal energy supply, and the $PFs_{heating}$ for different variables. There was no noticeable difference in electricity consumption between the two trial years. This is because both years were conducted under the same conditions with minimal modifications during the test period. Electricity consumption was higher in summer than in winter in both years due to increased ventilation rates resulting from high temperatures.

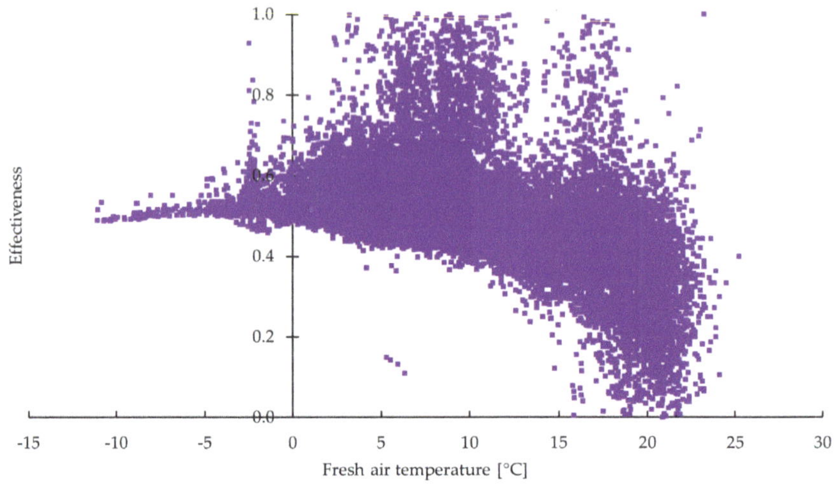

Figure 6. Dependency of the effectiveness (ε) of the two heat exchangers and the fresh air temperature [°C] when heat exchangers heated the incoming fresh air. This figure shows 99.4% of the values given, 0.6% (360 of 55,082 measurement intervals) are considered irregular outliers outside the nominal range of effectiveness (0–1).

Table 2. Cumulated electrical energy consumption [kWh_{el}], thermal energy supply [kWh_{th}], and the $PFs_{heating}$ for trial years 1 and 2.

	Year 1 *	Year 2 **
Cumulated electrical energy consumption [kWh_{el}]		
Ventilation system [A] $Q_{V,Case1,el}$	66,066	67,954
… of this: electrical energy consumption for the heat recovery $Q_{HR,el}$ [A]	41,888 (63.4%)	40,514 (59.6%)
Air purification system [A] $Q_{A,el}$	25,179	25,176
Others [B]	39,904	44,590
SUM $Q_{Case1,el}$	131,149	138,260
Cumulated thermal energy supply [kWh_{th}]		
LNG combustion [A] $Q_{LNG,Case1,th}$	154,479	160,720
Heat recovery [A] $Q_{HR,th}$	290,414	317,913
SUM $Q_{Case1,th}$	444,893	478,633
… of this: heat recovery	65.3%	66.4%
$PF_{HR,heating}$	6.9	7.8
$PF_{V,heating}$	5.4	5.9
$PF_{VA,heating}$	3.8	4.2

[A] Values are based on calculated data.
[B] Values are based on electricity meters.
* 17 December 2019–15 December 2020.
** 16 December 2020–15 December 2021.

In detail, the cumulated thermal energy supply $Q_{Case1,th}$ accounted for the larger share of total energy demand, with 77% and 78% for the two years, respectively (Table 2). The electrical energy consumption $Q_{Case1,el}$ accounted for the remaining 22–23%, with ventilation $Q_{V,Case1,el}$ taking the most significant proportion at 11%. The data collected are comparable to those previously described in the literature [7,71].

For the ventilation system, 59.6–63.4% was attributed as electrical energy consumption for the heat recovery $Q_{HR,el}$. This means that only 36.6–40.4% of the corresponding electrical energy for ventilation would be necessary if the ventilation system was operated without heat exchangers. This is because using the heat exchangers causes an increased air flow resistance, which the fans must overcome. Additional electrical energy is required for this [24]. The values align with the former literature [22].

However, when considering the thermal energy input to operate the barn building $Q_{Case1,th}$, 65–66% was provided by heat recovery $Q_{HR,th}$, and only 34–35% was provided by the combustion of LNG $Q_{LNG,Case1,th}$ (for more detailed information, see Figure S8). The electrical energy consumption for the heat recovery $Q_{HR,el}$ was 40,514–41,888 kWh_{el}. The total thermal energy recovered $Q_{HR,th}$ using the heat exchangers was 290,414–317,913 $kWh_{th}\ a^{-1}$. This means that the total energy consumption to operate the barn building in Case 1 Q_{Case1} was 285,628 $kWh_{el,th}$ in reality with heat exchangers (year 2: 298,980 $kWh_{el,th}$) and would be in Case 2 Q_{Case2} 534,154 $kWh_{el,th}$ if operated without heat exchangers (year 2: 576,379 $kWh_{el,th}$). Therefore, heat exchangers reduced the energy demand of these piglet rearing buildings by −47% (year 2: −48%, see Section 3.4). The energy demand of Case 2 (without heat exchangers) is comparable to former literature concerning German pig farms [7,18] in terms of energy demand per animal place (AP, concerning 4140 piglet rearing places) and year [129–139 $kWh_{el,th}$ (AP a)$^{-1}$]. This value decreases noticeably in Case 1 (with heat exchangers) to 69–72 $kWh_{el,th}$ (AP a)$^{-1}$. The data presented demonstrate the great potential to reduce energy consumption through the use of heat exchangers and thus act in an economically and ecologically sustainable manner. In addition, e.g., [22,27,28,30,33,36,38] could already show in their investigations that reductions are possible by heat recovery technologies in barn buildings.

Figure 7 shows the various $PFs_{heating}$ as a function of the fresh air temperature (see Sections 2.5 and 2.7). In general, it can be said that the lower the fresh air temperature, the higher the $PF_{heating}$. As the fresh air temperature increased and the thermal power decreased, the $COP_{heating}$ followed a negative, linear correlation. In detail, the general linear correlation was less appropriate for $COP_{HR,heating}$ with fresh air temperatures above 15 °C. The increasing values can be explained based on the opening of the bypass air shutters and the corresponding effects on the differential pressures. In this case, the electrical energy consumption for the heat recovery $Q_{HR,el}$ decreased noticeably while the thermal power followed the linear correlation, i.e., the $COP_{heating}$ increased.

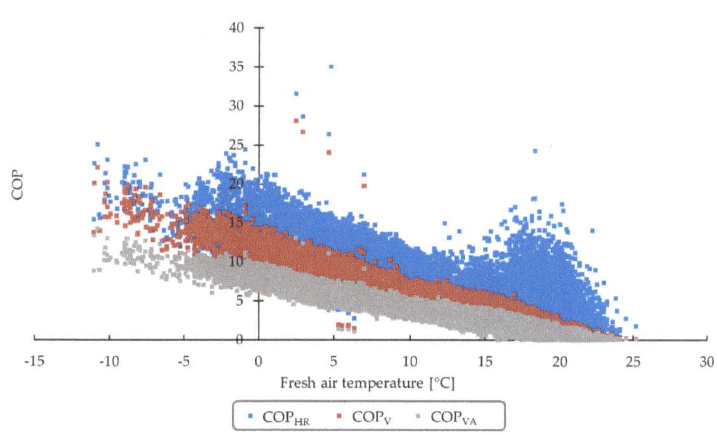

Figure 7. Correlation of the different $COPs_{heating}$ of the two heat exchangers and the fresh air temperature [°C] when heat exchangers heated the incoming fresh air. $COP_{HR,heating}$ considers the thermal power $\dot{Q}_{HR,th}$ and the electrical power input for the heat recovery $\dot{Q}_{HR,el}$; $COP_{V,heating}$ considers the thermal power $\dot{Q}_{HR,th}$ and the electrical power input of the ventilation system $\dot{Q}_{V,el}$ in Case 1; $COP_{VA,heating}$ considers the thermal power $\dot{Q}_{HR,th}$ and the electrical power input of the ventilation system in Case 1 plus the air purification system $\dot{Q}_{V+A,el}$. For $COP_{HR,heating}$, 99.18% of the measurement intervals show an improvement in energy efficiency ($COP_{HR,heating}$ > 1).

The $PF_{HR,heating}$ averaged 6.9–7.8 per year and reached a $COP_{HR,heating}$ maximum of 25.1 at a fresh temperature of −10.8 °C (excluding outliers). Even in the warm summer months, the $PF_{HR,heating}$ was between 2.5 and 3.5 (June–August), so an energy benefit was also available here (see Figure 8). [33] determined comparable $COP_{heating}$ values between 7.1 and 11.5 during tests with an exchange scrubber in winter. Lower values were reported by [27,34]. Another study could determine higher $COP_{heating}$ or $PF_{heating}$ values by using a heat exchanger [28]. However, in that study, no detailed information is given about the distinction between electrical energy consumption for the barn ventilation and the electrical energy consumption for operating the heat recovery $Q_{HR,el}$. The study presented here is, to the author's knowledge, the first that presents such a detailed approach to investigate this topic and to obtain more precise information about the energy efficiency of heat exchangers in their practical use in barn buildings.

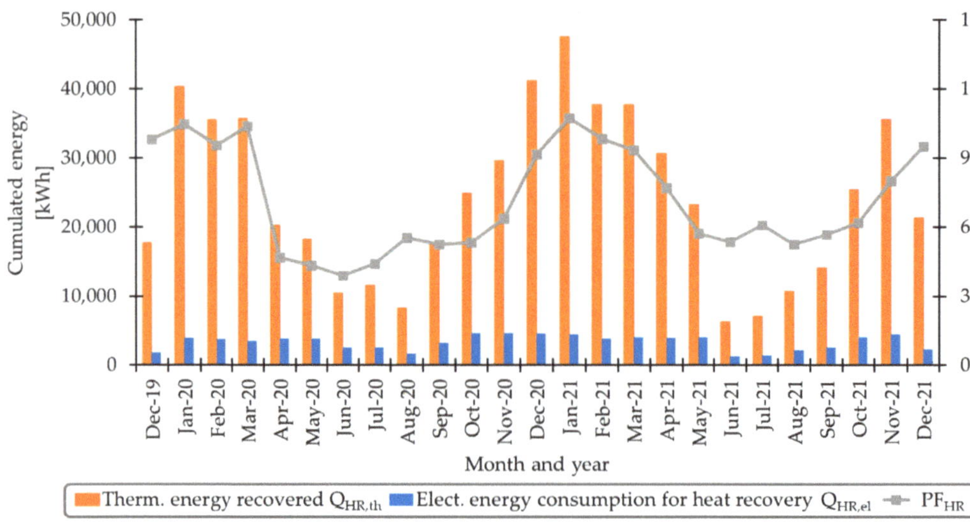

Figure 8. Course of the cumulated thermal energy recovered $Q_{HR,th}$ [kWh_{th}], the cumulated electrical energy consumption for the heat recovery $Q_{HR,el}$ [kWh_{el}], and the corresponding $PF_{HR,heating}$ of the two heat exchangers over the various trial months (cumulated or mean values, respectively).

Moreover, the $COP_{heating}$ and $PF_{heating}$ of the investigated air-to-air heat exchanger were, in most parts, higher than the $COPs_{heating}$ or $PFs_{heating}$ of other heat recovery technologies mentioned in the various cited studies concerning barn buildings. Morshed et al. (2018) [72], for example, reported $PF_{heating}$ values of 1.5 on a daily average when using an earth-to-air heat exchanger. Studies investigating heat pumps were able to determine values between 2.5 and 4.1 [35,36]. Nevertheless, the comparison based on the $COP_{heating}$ or $PF_{heating}$ between different studies is only possible to a limited extent since factors such as study design, seasonal influences (especially outside and fresh air temperatures), and technical equipment affect the trial conditions and the calculated $COPs_{heating}$ or $PFs_{heating}$, respectively.

3.4. Substituted CO_2 Emissions and Energy Costs

Table 3 shows the total amount of consumed energy to operate the barn, the proportions of recovered thermal energy included therein, and the resulting calculated CO_2 and cost savings for Case 1 and 2 for both trial years. For year 1, the total energy consumption corresponded to values of 95.42 $kWh_{el,th}$ $(m^2\ a)^{-1}$ for the barn building for Case 1 and 178.44 $kWh_{el,th}$ $(m^2\ a)^{-1}$ for Case 2, respectively; for year 2, 99.88 $kWh_{el,th}$ $(m^2\ a)^{-1}$ for Case 1 and 192.54 $kWh_{el,th}$ $(m^2\ a)^{-1}$ for Case 2, respectively.

In both years, the total energy consumption for operating the barn building was reduced by nearly half through heat exchangers (year 1: 46.5%; year 2: 48.1%). The energy savings correspond to reduced connected CO_2 emissions of 35.5% (year 1) and 37.5% (year 2). In total, 50,950 kg and 58,056 kg CO_2 per year were mitigated, respectively. The recovered thermal energy resulted in cost savings of EUR 4540 and EUR 12,028 per year for the investigated barn; this corresponds to savings of EUR 1.10 and 2.91 $(AP\ a)^{-1}$.

Table 3. Total energy consumption operating the piglet rearing barn during the trial period in Case 1 (with heat recovery) and Case 2 (without heat recovery), and the corresponding savings of energy consumption, CO_2 emissions, and energy costs using the heat recovery system.

	Unit	Year 1 *	Year 2 **
Electrical energy consumption to operate the barn building $Q_{total,el}$ (A)	$kWh_{el}\ a^{-1}$	131,149	138,260
… of this: electrical energy consumption for the heat recovery $Q_{HR,el}$ (B)	$kWh_{el}\ a^{-1}$	41,888	40,514
Thermal energy input to operate the barn building $Q_{total,el}$ (C)	$kWh_{th}\ a^{-1}$	444,893	478,633
… of this: thermal energy recovered by the heat recovery system $Q_{HR,th}$ (D)	$kWh_{th}\ a^{-1}$	290,414	317,913
Case 1: Energy consumption to operate the barn building (with heat recovery) Q_{Case1} (A + C − D)	$kWh_{el,th}\ a^{-1}$	285,628	298,980
Case 2: Energy consumption to operate the barn building (without heat recovery) Q_{Case2} (A − B + C)	$kWh_{el,th}\ a^{-1}$	534,154	576,379
Saved energy		−46.5%	−48.1%
Total	$kWh_{el,th}\ a^{-1}$	248,526	277,400
… per animal place [A]	$kWh_{el,th}\ (AP\ a)^{-1}$	60	67
… per nominal LU [B]	$kWh_{el,th}\ (LU\ a)^{-1}$	1440	1608
… per real LU [C]	$kWh_{el,th}\ (LU\ a)^{-1}$	1773	1919
Saved CO_2 emissions		−35.5%	−37.5%
Total	$kg\ a^{-1}$	50,950	58,056
… per animal place [A]	$kg\ (AP\ a)^{-1}$	12.3	14.0
… per nominal LU [B]	$kg\ (LU\ a)^{-1}$	295.3	336.4
… per actual LU [C]	$kg\ (LU\ a)^{-1}$	363.5	414.2

Table 3. *Cont.*

	Unit	Year 1 *	Year 2 **
Saved energy costs		−9.5%	−19.7%
Total	EUR a^{-1}	4540	12,028
... per animal place [A]	EUR (AP a)$^{-1}$	1.1	2.9
... per nominal LU [B]	EUR (LU a)$^{-1}$	26.3	69.7
... per actual LU [C]	EUR (LU a)$^{-1}$	32.4	85.8

[A] Considering the 4140 piglet rearing places within the barn.
[B] Considering the 4140 piglet rearing places (0.03 LU AP^{-1}), 128 gilt rearing places (0.12 LU AP^{-1}), and 110 breeding center pig places (0.30 LU AP^{-1}) within the barn.
[C] Considering the actual animals held in the barn building. The following mean values were considered: for year 1 * 98.4 LU in piglet rearing, 12.1 LU in gilt rearing and 31.8 LU in the breeding center; for year 2 ** 101.0 LU in piglet rearing, 11.3 LU in gilt rearing and 32.2 LU in the breeding center.
* 17 December 2019–15 December 2020.
** 16 December 2020–15 December 2021.
LU = Livestock Unit (1 LU = 500 kg animal life weight), AP = animal place.

The results show that heat exchangers can significantly increase energy efficiency and economic and ecological sustainability. Despite the higher demand for electrical energy due to heat recovery, cost savings can still be achieved. In this case study, the saving of energy costs followed a high correlation depending on the fresh air temperature (savings per hour [EUR h^{-1}] = $0.2244 \times t_{21} - 4.3377$; $R^2 = 0.7138$). For the two-year trial, the comparison between specific heating costs using LNG (6.58 ± 1.69 ct $kWh_{th}{}^{-1}$) and using the heat exchangers (concerning LNG substitution and electrical energy consumption for the heat recovery $Q_{HR,el}$, and electricity price 26.97 ± 0.30 ct $kWh_{el}{}^{-1}$) underlined this fact (Figure S9). The specific heating costs of the heat exchangers were 5.02 ± 19.38 ct $kWh_{th}{}^{-1}$ and 87.2% of the time below the actual German LNG costs (based on the 15-minute intervals when the heat exchanger was heating).

This leads to a payback period after a few years, depending on the individual initial capital costs. In detail, the savings of the first trial year were considered in a barn-specific calculation: The installation of the heat recovery system cost around EUR 66,400 with a payback period of 13.0 years. In Case 2 (without heat recovery), however, there were additional costs for installing an LNG heating system with a higher heating capacity than in Case 1. The bottom line is that Case 1 would be only EUR 8300 more expensive than Case 2. The payback period for these additional costs is 1.6 years. These calculations include the maintenance costs for the heat exchangers (see Section 2.6 and Table S3).

Other studies have also demonstrated energy-efficient technologies' economic and ecological benefits. In comparison with the former literature in this field of research, the study presented here is based on a larger scale (heat recovery for the whole barn building and not only for one animal compartment, e.g., [36]), for a more extended measurement period (two years and not only trials during the winter months, e.g., [22,30,33]), and more detailed (calculation of the electrical energy consumption for the heat recovery $Q_{HR,el}$ and not only an estimation of a specific ratio of the total electricity consumption of the fans, e.g., [28,29]).

Nevertheless, the investigated system's application and other heat recovery systems can be extended to other areas of farm animal husbandry. Several studies have, for example, already dealt with the use of heat recovery systems in poultry husbandry and have also been able to demonstrate increases in sustainability [22,27,30]. Depending on the husbandry stage and form, the barn ventilation systems are fundamentally similar so that heat recovery systems can be usefully integrated into the housing concept for both animal species. Some challenges of animal husbandry can be improved in this way, whereby the animal species-specific prerequisites must be researched and considered (see Section 3.1). However, the use of such systems remains limited to force-ventilated systems, so it is not easy to implement them, for example, in dairy farming, which is primarily free-ventilated. Nevertheless, heat recovery systems can be considered a sustainable technology in pig and poultry production which politics should fund in the future.

Concerning climate change and the legal requirements for mitigating emissions [1], the benefits shown here should be utilized more effectively. Ecologically, using a heat exchanger also brings advantages since fossil fuels could be saved, and thus resulting CO_2 emissions could be noticeably reduced. As mentioned above, the CO_2 savings depend on the fresh air temperature too (savings per hour $[(\text{kg CO}_2)\ \text{h}^{-1}] = 0.7492 \times t_{21} - 16.293$; $R^2 = 0.9275$; Figure S10).

3.5. Relevance for Modern, Global Pig Husbandry

Figure 9 shows the annual savings per animal place using heat recovery [EUR (AP year)$^{-1}$] depending on the prices for electricity and liquefied natural gas [ct $\text{kWh}_{el,th}{}^{-1}$] for different scenarios. The economic benefit is affected by the ratio of electricity and LNG prices. The heat recovery systems offer the highest profit at high heating and low electricity costs.

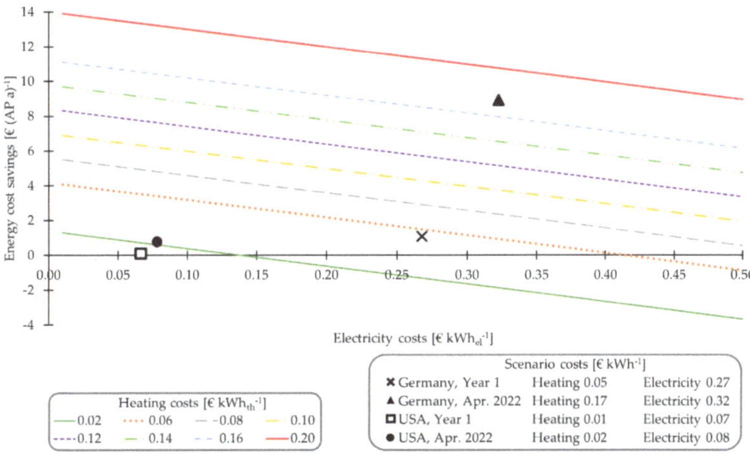

Figure 9. Annual energy cost savings [EUR (AP a)$^{-1}$] for various scenarios concerning electricity and LNG costs [EUR $\text{kWh}_{el,th}{}^{-1}$]. The annual reduction quantities consider the 4140 animal places (AP) for piglet rearing in the barn. The calculations are based on the electrical energy consumption for the heat recovery $Q_{HR,el}$ (41,888 kWh_{el}) and heat recovered $Q_{HR,th}$ (290,414 kWh_{th}) of trial year 1 (17 December 2019–15 December 2020). GER = Germany, USA = United States of America.

In April 2022, with a LNG price of 17.37 ct $\text{kWh}_{th}{}^{-1}$ and electricity costs of 32.27 ct $\text{kWh}_{el}{}^{-1}$, savings of EUR 36,940.49 could be achieved in Germany. In contrast, the given scenario of current energy costs in the US show (savings of EUR 3219.60 a^{-1} in the scenario USA April 2022) that the savings can differ noticeably between the various countries. Since the (German) prices for fuels are fluctuating and following an increasing trend [64], it can be assumed that the use of heat exchangers can also be described as profitable if the development of energy prices remains steady. Crises and advancing climate change can influence this trend.

Figure 10 demonstrates the possible reductions in CO_2 emissions in several different application scenarios; depicted are the five scenarios of the world's largest pork producers [73]. The magnitude of the reductions strongly depends on the energy supply's individual emission factors and the $PF_{HR,heating}$. Fossil fuels show varying CO_2 emissions during their combustion; of the standard heating fuels, the gases show lower emissions in this regard [61]. In addition, the CO_2 emissions of the individual country-specific electricity supply vary: Brazil (electricity generation primarily using hydropower) and France (nuclear power plants) show significantly lower emission factors than China (coal-fired power plants), for example [17]. Consequently, in the extreme situations shown (Brazil—heating oil vs. China—natural gas), CO_2 emission reduction per animal place and year could be around twice as high. Therefore, the potential for CO_2 emission reductions in China, the world's largest pork producer [73], is lower than in other countries. Nevertheless, the positive values show noticeable improvements in the carbon footprint of piglet rearing. This could be further improved with a large-scale or farm-specific conversion of power generation to low-emission processes. Furthermore, the comparison between the economic (Figure 9) and environmental (Figure 10) benefits shows that, while the calculated $PF_{HR,heating}$ does not provide financial benefits in some pricing scenarios, it benefits the carbon footprint in all scenarios presented here. However, possible future pricing of CO_2 emissions or corresponding reductions, e.g., taxation, certificate trading, and higher sales revenues, could also further improve the economic situation.

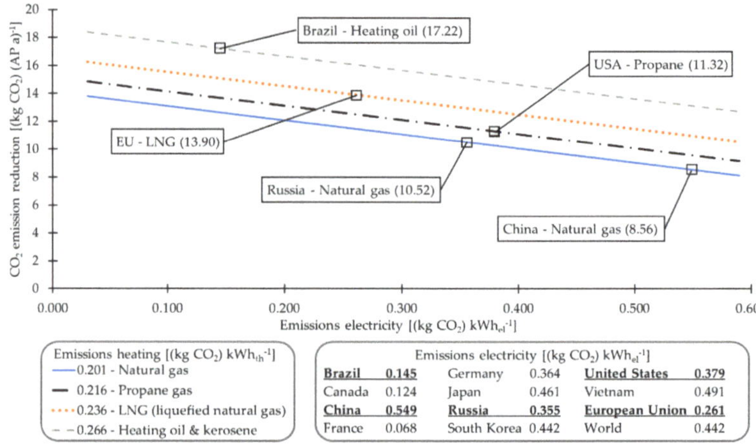

Figure 10. Annual CO_2 emission reduction [(kg CO_2) (AP a)$^{-1}$] for various scenarios concerning emission factors for electricity and heat input [(kg CO_2) $kWh_{el,th}{}^{-1}$]. The annual reduction quantities consider the 4140 animal places (AP) for piglet rearing in the barn. The calculations are based on the electrical energy consumption for the heat recovery $Q_{HR,el}$ (41,888 kWh_{el}) and heat recovered $Q_{HR,th}$ (290,414 kWh_{th}) of trial year 1 (17 December 2019–15 December 2020). Emission factors for fossil fuels are given by [27,61] and for the countries' or regions' electricity production, respectively, for the year 2021 by [17] (the factor for Germany differs compared with the other calculations in this article due to various literature values and reference years). EU = European Union.

The possibility of combining several energy-efficient technologies should not be disregarded in future animal husbandry. Due to their large roof areas, animal barns especially offer much potential for photovoltaic systems. By supplying self-generated solar energy to their network, farmers could partially cover their own electricity needs, e.g., for the use of heat recovery systems. Some studies have already investigated the advantages of combining different renewable energy technologies and photovoltaic systems in animal barns [74,75]. The sustainability of livestock production benefits economically and environmentally due to low CO_2 emissions from on-farm photovoltaic electricity generation [0.048 (kg CO_2) $kWh_{el}{}^{-1}$ [17]]. Another possibility for independent electricity generation is the use of wind energy [0.011 (kg CO_2) $kWh_{el}{}^{-1}$ [17]]. Additionally, using energy storage media could increase this benefit.

Exhaust air purification systems are listed as the best available technique and effective for emission mitigation in European Countries [5,6,76–78], especially in countries with large-scale operations or high farm animal density. Furthermore, these systems are or will become mandatory for some large farms, e.g., [4]. The additional operating costs can be reduced or compensated by combining them with a heat exchanger. Both systems require or benefit from central supply and exhaust air conduction. Furthermore, heat exchangers have been used to mitigate NH_3 formation in a laying hen barn, using the heated supply air for manure belt drying. Focusing on heat recovery, varying combinations are possible, e.g., a staged and independent sequence of systems as described in this study or integrated combinations as described by Krommweh et al. (2021) [32,33]. The high $COP_{heating}$ values, in this case, $COP_{VA,heating}$ or the values of [32,33], further show an improvement in the energy efficiency for the combination of systems. Should the retrofitting of an exhaust air purification system be necessary, a simultaneous installation of a heat recovery system seems to make sense because of, e.g., simultaneous reconstruction of the flow paths, saving of construction costs, or compensation of the initial and operating costs by the heating cost reduction.

Consequently, there should be significant synergy effects if heat recovery, i.e., CO_2 emission mitigation, and exhaust air purification systems, i.e., NH_3, particulate matter, and odor emission mitigation, are combined with on-site energy provision systems, i.e., CO_2 emission mitigation. Additional studies are necessary to investigate the consequences of this system combination concerning its sustainability.

3.6. Limitations of the Study and Examination of the Methodological Procedure

Heat recovery systems are well-known and studied in many application areas. It is also known that these systems can positively influence energy efficiency and sustainability. This study, therefore, does not provide any new findings in this regard. Instead, this case study is intended to provide further numerical data to complement the small selection of scientific studies in animal barn applications. The main focus here was that the two-year trial—to the authors' knowledge, the most extended study period to date in this specific field of research—reflects many of the possible operating conditions and provides further planning data for future barn construction. This is especially true for barn buildings under comparable environmental conditions, e.g., in moderate climate zones, but the authors are sure that the given correlations and findings apply to stakeholders in other regions, too.

The trial described here took place in a conventional piglet rearing barn in an ongoing practical operation. However, measurements in practice are limited in their continuity and accuracy compared with test bench measurements. Adjustments in the operation and systems management or fluctuating animal numbers influenced the control and performance parameters of the ventilation and heat recovery systems. In addition, adaptations for measuring were only possible to a limited extent. Measuring fans are frequently used technologies to determine air flow rates, especially in the field of emission measurements of barn buildings [58,79,80]. Another applied technology is the Fan Assessment Numeration System (FANS) for in situ measurements of the actual air flow rates to determine emission mass flow rates [81–84]. However, the dimensioning of the heat exchanger's outer shell and the resulting flow paths were not large or long enough to install these measurement technologies.

Calvet et al. [58] state that the procedure chosen here, i.e., air flow rate calculation using manufacturer's protocol, is the most inaccurate of the listed measurement procedures for determining the air flow rates in barn buildings, with a measurement inaccuracy of approx. 20%. Furthermore, fouling on the fans can cause the actual air flow rate to vary considerably from the manufacturer's protocols [58,82,83,85]. Previous studies have called for increased focus on uncertainty analysis in studies such as this [34,55–57], which is particularly important for the inaccuracy mentioned above. The uncertainty analysis (Section S4) indicated a mean value for the uncertainty of the thermal power of $\omega_{\dot{Q}_{HR}} = 9.20 \pm 3.34\ kW_{th}$ or 26.3% ± 32.8%, respectively, and $\omega_{COP_{HR,heating}}$ of 1.93 ± 1.51 or 30.6% ± 48.3%, respectively, for $COP_{HR,heating}$. The overall uncertainties for the two-year trial were $\omega_{Q_{HR}} = 21.0\%$ and $\omega_{COP_{HR,heating}} = 25.5\%$, respectively. The methodical procedure explains the extreme outliers (see Section S4).

This issue can be seen as a big concern in this case study. However, the presented study showed essential findings, e.g., physical correlations, in practical barn building applications. This study advanced a similar methodology from [30] by combining the manufacturer's protocol information with in situ differential pressure measurements and preliminary tests. With this approach, comparable measurements can be practiced in many barn buildings of varying designs. Consequently, e.g., barn design comparisons are possible with little effort for retrofitting measurement technology and lead to valid findings under consideration of possible inaccuracies or second-order errors [58]. Furthermore, derived air flow rates align in magnitude with the planning values concerning barn ventilation [53,86,87]. The calculated exhaust air flow rates exceed the planning values of DIN 18910 [53] by +24.7% (at outside air temperatures of −5.2 to −4.8 °C) or +77.6% (at outside air temperatures of 29.8 to 30.2 °C). Earlier studies report an increase in air flow rates of up to 20% for heat recovery in broiler barn buildings [22,27,30], which agrees with the magnitude of the study presented here. The higher values at high outdoor temperatures are due to the farmer's specific wishes in planning the barn, e.g., increased heat load removal.

The results presented for comparing the barn building with and without heat exchangers show noticeable savings to increase sustainability (see Section 3.4). For this, barn management in Case 2 was considered identical to Case 1 (Section 2.7). The savings described here exceed the empirical data collected in surveys by Bokkers et al. (2010) [27]. Consequently, it is possible that barn buildings are not identically operated if they are operated with or without heat exchangers, i.e., reduced airflow rates without heat exchangers to reduce heat losses. Therefore, the comparison presented here shows the potential for maximum savings instead of practical savings.

Further research should focus on open research questions considering improvements in energy efficiency and sustainability, changes in animal behavior, and benefits for animal welfare and development for varying animal species in long-term trials.

4. Conclusions

The results of this case study indicate that the investigated heat exchangers achieve high workloads throughout the year (up to 100% in 14 of the total 24 trial months) and acceptable performance in the warmer summer (minimum 45%). An energetic benefit is not limited to the cold season but can be generated throughout the year.

Furthermore, using heat exchangers can improve the sustainability of pig farming, especially piglet production, in all three relevant aspects:

- Ecological, e.g., reduction of energy input, mitigation of CO_2 emissions Due to the amount of recovered thermal energy (290,414 kWh_{th} in year one and 317,913 kWh_{th} in year two), the total energy consumption to operate the barn building could be significantly decreased (−46.5% in year 1 and −48.1% in year 2), and costs could be saved. The important $PF_{HR,heating}$ indicated a mean value of 6.9 and 7.8 for the two trial years, a noticeable energy efficiency improvement. At the same time, CO_2 emissions could be reduced by up to 37.5% by substituting fossil fuel combustion. The transfer to international application scenarios shows that piglet rearing's carbon footprint can be improved substantially for all scenarios, i.e., for the five largest pork-producing countries.
- Social, e.g., barn air quality, animal welfare, occupational safety The heat recovery reduced supply air temperature variation and led to a temperature amplitude damping by an average of −34.7%, especially in the transitional seasons. The literature review shows this can lead to higher air flow rates, more even ventilation of the barn building, and a more constant indoor climate. This can be exceptionally favorable regarding indoor air quality (e.g., improved NH_3 exhaustion) and, thus, animal health, animal welfare, and occupational safety for the farm staff.
- Economic, e.g., reduction of fuel and energy costs In this study, EUR 2.90 (animal place year)$^{-1}$ were saved (up to EUR 12,028 per year for the entire barn building). This led to a payback period of the additional installation costs of the heat recovery system

of 1.6 years. The transfer to international application scenarios shows that savings depend noticeably on farm-specific electricity and heating costs.

Especially in combination with exhaust air purification systems, the use of heat exchangers seems to be reasonable: ecologically relevant emissions are reduced using exhaust air purification systems, e.g., NH_3, particulate matter, odor; in parallel, environmental emissions can be reduced using heat recovery systems, e.g., CO_2 emissions. The heat recovery can compensate for increased operating costs due to the exhaust air purification system. Thus, the study can help increase farmers' willingness to invest in energy-efficient technologies and promote sustainable future pig farming.

Supplementary Materials: The following supporting information can be downloaded at: https://www.mdpi.com/article/10.3390/en16041799/s1, Section S1. Nomenclature. Section S2. Heat Exchanger & Sensors. Table S1: Technical specifications and parameters of each air-to-air-heat exchanger used in the presented trial. Table S2: Technical specifications and parameters of the temperature sensors used in the presented trial. Section S3. Characteristics of the fans and calculation of the air flow rates for the example of the exhaust air fans (A3G990-AZ02-35, ebm-papst Mulfingen GmbH & Co. KG, Mulfingen, Germany). Figure S1: Fan speed [n min^{-1}] and electric power input [W] of ebm exhaust air fans in the investigated barn building depending on the analog control [V] of the fans. Figure S2: Air volume flow rate [m^3 h^{-1}] of the ebm exhaust air fans depending on the fan speed [n min^{-1}] for exemplary pressure differences [Pa]. Section S4. Uncertainty analysis for the calculation of the thermal power. Figure S3: Magnitude of uncertainty $\omega_{\dot{Q}_{HR}}$ on absolute [kW_{th}] or relative [%] base for the calculated thermal power depending on the thermal power [kW_{th}]. A2 shows 7378 of the 55,435 values (13.30%), and R2 shows 54,894 values (99.02%). Section S5. Calculation of the daily LNG consumption. Figure S4: Daily LNG consumption [L d^{-1}] depending on the daily mean outside air temperature [°C]. Section S6. Calculation of the electrical energy consumption of the exhaust air fans in Case 2 (fan type: A3G990-AZ02-35, ebm-papst Mulfingen GmbH & Co. KG, Mulfingen, Germany). Figure S5: Pressure difference $\Delta Pa_{1\text{-}2}$ [Pa] ($\Delta Pa_{1\text{-}2} = Pa_1 - Pa_2$) within the central exhaust air collection duct due to the absence of supply air ventilators (Pa_2, Case 2, without heat recovery and supply air fans) compared with the actual ventilation system operation (Pa_1, Case 1, with heat recovery and supply air fans) depending on the performance of the entire ventilation system [%]. These pressure differences $\Delta Pa_{1\text{-}2}$ were derived by the data of the 15-minute measurement intervals during the ongoing trial (differential pressure in the central exhaust air collection duct with turned on/off supply air fan; n = 2366 intervals for the mean value calculations of all integer performance values). Section S7. Results and Discussion—Supplementary Figures & Tables. Figure S6: Correlation of amplitude damping [%] and daily supply air temperature variation [K] depending on the daily mean outside air temperature [°C] or daily outside air temperature variation [K]. Temperature variation: difference between the (daily) minimum and maximum temperature. In the context of the conducted study, amplitude damping refers to the ability of the heat exchangers to reduce the supply air temperature variation Δt_{22} compared with outside air temperature variation Δt_0. The values represent 635 days within the two-year measurement period (see Section 2.8). Figure S7: Correlation of the actual thermal power $\dot{Q}_{HR,th}$ and the maximum possible thermal power $\dot{Q}_{max,th}$ [kW_{th}] of the two heat exchangers and the fresh air temperature t_{21} [°C] in the upper diagram, or air mass flow rate of the exhaust air fans (heat exchanger) [kg h^{-1}] in the lower diagram, respectively, when heat exchangers heated the incoming fresh air. Figure S8: Correlation of the contribution [%] of recovered heat $Q_{HR,th}$ using the heat exchangers from the total thermal energy supply in the barn building $Q_{total,th}$ (heat recovery $Q_{HR,th}$ plus LNG combustion $Q_{LNG,Case1,th}$) and the monthly mean outside air temperature [°C]. Values are based on the monthly cumulated sums over the two-year trial. Figure S9: Correlation of the specific heating costs [ct $kWh_{th}{}^{-1}$] of the two heat exchangers and the fresh air temperature [°C] when heat exchangers heated the incoming fresh air. The calculated values are based on the substituted LNG costs (6.58 ± 1.69 ct $kWh_{th}{}^{-1}$) and the extra electrical energy costs (26.97 ± 0.30 ct $kWh_{el}{}^{-1}$) using the heat recovery. This figure shows an excerpt (98.66% of total values); some single outliers show values of up to 3660.92 ct $kWh_{th}{}^{-1}$. However, 87.18% of all values are below EUR 0.0658 $kWh_{th}{}^{-1}$, and thus the specific heating costs of heat recovery are lower than the average LNG costs. LNG = liquefied natural gas. Figure S10: Correlation of the CO_2 emissions saved [kg h^{-1}] using the two heat exchangers and the fresh air temperature [°C] when heat exchangers heated the incoming fresh air. The calculated values are based on the substituted CO_2

emissions based on LNG combustion (0.237 kg kWh_{th}^{-1}) and the CO_2 emissions based on electrical energy consumption (0.427 kg kWh_{el}^{-1}) using heat recovery. Table S3: Simplified economic analysis and return of investment calculation of the two heat exchangers based on the results in year 1 under the experimental conditions.

Author Contributions: Conceptualization: H.F.D., A.L., M.S.K. and W.B.; methodology: H.F.D., M.S.K. and W.B.; software: H.F.D. and A.L.; validation: H.F.D., A.L. and M.S.K.; formal analysis: H.F.D., A.L. and M.S.K.; investigation: H.F.D.; resources: W.B.; data curation: H.F.D. and A.L.; writing—original draft preparation: H.F.D. and A.L.; writing—review and editing: H.F.D., A.L., M.S.K. and W.B.; visualization: H.F.D. and A.L.; supervision: W.B.; project administration: W.B.; funding acquisition: W.B. All authors have read and agreed to the published version of the manuscript.

Funding: This research was funded by The Ministry for Environment, Agriculture, Conservation and Consumer Protection of the State of North Rhine-Westphalia, grant number: 41.2019.01. The APC was funded by the Open Access Publication Fund of the University of Bonn.

Data Availability Statement: The datasets used and analyzed in this study are available from the corresponding author on reasonable request.

Acknowledgments: The authors would like to thank Heinrich Dönselmann-Theile, Daniel Dönselmann-Theile and Christopher Dönselmann-Theile and the whole company staff of hdt Anlagenbau GmbH (Diepholz, Germany) for their technical support and help during the entire research project and trials. Furthermore, we thank Bernd Förthmann and his family (Bahrenborstel, Germany) for providing the barn building, background information and optimal research conditions during our trials. Finally, we would like to thank our colleagues of the scientific workshop (Institute of Agricultural Engineering, Rheinische Friedrich-Wilhelms-Universität Bonn, Germany) for supporting the investigations.

Conflicts of Interest: The authors declare no conflict of interest. The funders had no role in the design of the study; in the collection, analyses, or interpretation of data; in the writing of the manuscript, or in the decision to publish the results.

References

1. Bundes-Klimaschutzgesetz vom 12. Dezember 2019 (BGBl. I S. 2513), das durch Artikel 1 des Gesetzes vom 18. August 2021 (BGBl. I S. 3905) Geändert Worden Ist. Available online: https://www.bgbl.de/xaver/bgbl/start.xav?start=//*%5B@attr_id=%27bgbl119s2513.pdf%27%5D#__bgbl__%2F%2F*%5B%40attr_id%3D%27bgbl119s2513.pdf%27%5D__1673355701326 (accessed on 10 January 2023).
2. Umweltbundesamt. Treibhausgas-Emissionen in Deutschland. 2022. Available online: https://www.umweltbundesamt.de/daten/klima/treibhausgas-emissionen-in-deutschland#emissionsentwicklung (accessed on 5 October 2022).
3. Rösemann, C.; Haenel, H.-D.; Dämmgen, U.; Döring, U.; Wulf, S.; Eurich-Menden, B.; Freibauer, A.; Döhler, H.; Schreiner, C.; Osterburg, B.; et al. *Calculations of Gaseous and Particulate Emissions from German Agriculture*; Thünen-Report No. 67; Johann Heinrich von Thünen-Institut: Braunschweig, Germany, 2019.
4. Erste Allgemeine Verwaltungsvorschrift zum Bundes-Immissionsschutzgesetz (Technische Anleitung zur Reinhaltung der Luft—TA Luft) vom 18. August 2021. Available online: https://www.verwaltungsvorschriften-im-internet.de/bsvwvbund_18082021_IGI25025005.htm (accessed on 10 January 2023).
5. Melse, R.W.; Ogink, N.W.M.; Rulkens, W.H. Overview of European and Netherlands' regulations on airborne emissions from intensive livestock production with a focus on the application of air scrubbers. *Biosyst. Eng.* **2009**, *104*, 289–298. [CrossRef]
6. Delgado Sancho, L.; Georgitzikis, K.; Giner Santonja, G.; Montobbio, P.; Roudier, S.; Scalet, B.M. *Best Available Techniques (BAT) Reference Document for the Intensive Rearing of Poultry or Pigs: Industrial Emissions Directive 2010/75/EU (Integrated Pollution Prevention and Control)*; Publications Office of the European Union: Luxembourg, 2017; ISBN 978-92-79-70214-3.
7. Eckel, H.; Büscher, W.; Feller, B.; Fritzsche, S.; Gaio, C.; Kämper, H.; Neiber, J. *Energiebedarf in der Schweine- und Hühnerhaltung*; Kuratorium für Technik und Bauwesen in der Landwirtschaft: Darmstadt, Germany, 2014; ISBN 9783945088012.
8. Todde, G.; Murgia, L.; Caria, M.; Pazzona, A. A Comprehensive Energy Analysis and Related Carbon Footprint of Dairy Farms, Part 1: Direct Energy Requirements. *Energies* **2018**, *11*, 451. [CrossRef]
9. Ramirez, C.; Patel, M.; Blok, K. How much energy to process one pound of meat? A comparison of energy use and specific energy consumption in the meat industry of four European countries. *Energy* **2006**, *31*, 2047–2063. [CrossRef]
10. Blancard, S.; Martin, E. Energy efficiency measurement in agriculture with imprecise energy content information. *Energy Policy* **2014**, *66*, 198–208. [CrossRef]
11. Bos, J.F.F.P.; Haan, J.d.; Sukkel, W.; Schils, R.L.M. Energy use and greenhouse gas emissions in organic and conventional farming systems in the Netherlands. *NJAS Wagening. J. Life Sci.* **2014**, *68*, 61–70. [CrossRef]

12. Nguyen, T.L.T.; Hermansen, J.E.; Mogensen, L. Fossil energy and GHG saving potentials of pig farming in the EU. *Energy Policy* **2010**, *38*, 2561–2571. [CrossRef]
13. Hahne, J.; Arends, F.; Beverborg, R.; Niehoff, A.-L.; Bönsch, S.; Hortmann-Scholten, A. *Aktuelle Entwicklung Kosten-Nutzenanalyse und Vollzugsempfehlungen für den Einsatz von Abluftreinigungsanlagen in der Tierhaltung*; Umweltbundesamt: Dessau-Roßlau, Germany, 2015.
14. Wagner, S.; Angenendt, E.; Beletskaya, O.; Zeddies, J. Costs and benefits of ammonia and particulate matter abatement in German agriculture including interactions with greenhouse gas emissions. *Agric. Syst.* **2015**, *141*, 58–68. [CrossRef]
15. Grimm, E. (Ed.) *Exhaust Air Treatment Systems for Animal Housing Facilities. Techniques—Performance—Costs*; KTBL: Darmstadt, Germany, 2008; ISBN 9783939371601.
16. Icha, P.; Kuhs, G. *Entwicklung der Spezifischen Kohlendioxid-Emissionen des Deutschen Strommix in den Jahren 1990–2019*; Umweltbundesamt: Dessau-Roßlau, Germany, 2020. Available online: https://www.umweltbundesamt.de/sites/default/files/medien/1410/publikationen/2020-04-01_climate-change_13-2020_strommix_2020_fin.pdf (accessed on 5 October 2022).
17. Zieliński, M.; Fletcher, J.; Ewen, M.; Fulghum, N.; Tunbridge, P.; Jones, D.; Lolla, A.; Candlin, A.; Worthington, B.; Moore, C.; et al. Global Electricity Review 2022. 2022. Available online: https://ember-climate.org/insights/research/global-electricity-review-2022/#supporting-material-downloads (accessed on 19 December 2022).
18. Seifert, C.; Wietzke, D.; Fritzsche, S. Energy for heating and ventilation in pig production on farms. *Landtechnik* **2009**, *64*, 423–425.
19. Tierschutz-Nutztierhaltungsverordnung in der Fassung der Bekanntmachung vom 22. August 2006 (BGBl. I S. 2043), die zuletzt durch Artikel 1a der Verordnung vom 29. Januar 2021 (BGBl. I S. 146) Geändert Worden Ist. Available online: https://www.gesetze-im-internet.de/tierschnutztv/BJNR275800001.html (accessed on 10 January 2023).
20. Melse, R.W.; van Wagenberg, A.V.; Mosquera, J. Size Reduction of Ammonia Scrubbers for Pig and Poultry Houses: Use of Conditional Bypass Vent at High Air Loading Rates. *Biosyst. Eng.* **2006**, *95*, 69–82. [CrossRef]
21. Melse, R.W.; Ploegaert, J.P.M.; Ogink, N.W.M. Biotrickling filter for the treatment of exhaust air from a pig rearing building: Ammonia removal performance and its fluctuations. *Biosyst. Eng.* **2012**, *113*, 242–252. [CrossRef]
22. Han, H.; Kim, K.; Jang, K.-J.; Han, G.-S.; Lee, I.B. Energy Consumption and Indoor Environment of Broiler Houses with Energy Recovery Ventilators. *Appl. Eng. Agric.* **2013**, *29*, 751–759. [CrossRef]
23. Jungbluth, T.; Büscher, W.; Krause, M. *Technik Tierhaltung*, 2nd ed.; Ulmer (UTB): Stuttgart, Germany, 2017; ISBN 9783825242435.
24. Heinrich, G.; Franzke, U. *Wärmerückgewinnung in Lüftungstechnischen Anlagen*; C. F. Müller GmbH: Heidelberg, Germany, 1993; ISBN 3-7880-7498-1.
25. Lindley, J.A.; Whitaker, J.H. *Agricultural Buildings and Structures*, Rev. ed.; American Society of Agricultural Engineers: St. Joseph, MI, USA, 1996; ISBN 0929355733.
26. Averberg, U.; Block, K.; Brüggemann, C.; Brügger, E.; Cielejewski, H.; Feldmann, R.; Feller, B.; Fübbeker, A.; Gers-Grapperhaus, C.; Hartmann, H.-B.; et al. *Energieeffizienzverbesserung in der Landwirtschaft*; Verband der Landwirtschaftskammern e.V.: Berlin, Germany, 2009. Available online: https://www.landwirtschaftskammer.de/landwirtschaft/technik/pdf/energieeffizienzverbesserung.pdf (accessed on 10 January 2023).
27. Bokkers, E.A.M.; van Zanten, H.H.E.; van den Brand, H. Field study on effects of a heat exchanger on broiler performance, energy use, and calculated carbon dioxide emission at commercial broiler farms, and the experiences of farmers using a heat exchanger. *Poult. Sci.* **2010**, *89*, 2743–2750. [CrossRef]
28. Rösmann, P. Einsatz von Regenerativen Energiequellen zum Heizen und Kühlen von zwangsbelüfteten Tierställen. Ph.D. Thesis, Rheinische Friedrich-Wilhelms-Universität Bonn, Bonn, Germany, 2012.
29. Rösmann, P.; Büscher, W. Rating of an air-to-air heat exchanger in practice. *Landtechnik* **2010**, *65*, 418–420.
30. Goselink, Y.S.M.; Ramirez, B.C. Characterization of an Air-to-Air Heat Exchanger for Manure Belt Drying Ventilation in an Aviary Laying Hen House. *J. Appl. Poult. Res.* **2019**, *28*, 1359–1369. [CrossRef]
31. Inno + BV. The Triple EEE System. Available online: https://inno-plussystems.com/en/products/triple-eee-heat-recovery-system-for-stables/ (accessed on 10 November 2022).
32. Krommweh, M.S.; Büscher, W. Heating performance of a laboratory pilot-plant combining heat exchanger and air scrubber for animal houses. *Sci. Rep.* **2021**, *11*, 6872. [CrossRef]
33. Krommweh, M.S.; Deeken, H.F.; Licharz, H.; Büscher, W. Heating Performance and Ammonia Removal of a Single-Stage Bioscrubber Pilot Plant with Integrated Heat Exchanger under Field Conditions. *Energies* **2021**, *14*, 6484. [CrossRef]
34. Shah, S.B.; Lentz, Z.A.; van Heugten, E.; Currin, R.D.; Singletary, I. Tempering ventilation air in a swine finishing barn with a low-cost earth-to-water heat exchanger. *J. Renew. Sustain. Energy* **2017**, *9*, 23901. [CrossRef]
35. Licharz, H.; Rösmann, P.; Krommweh, M.S.; Mostafa, E.; Büscher, W. Energy Efficiency of a Heat Pump System: Case Study in Two Pig Houses. *Energies* **2020**, *13*, 662. [CrossRef]
36. Jeong, M.G.; Rathnayake, D.; Mun, H.S.; Dilawar, M.A.; Park, K.W.; Lee, S.R.; Yang, C.J. Effect of a Sustainable Air Heat Pump System on Energy Efficiency, Housing Environment, and Productivity Traits in a Pig Farm. *Sustainability* **2020**, *12*, 9772. [CrossRef]
37. Choi, H.C.; Salim, H.M.; Akter, N.; Na, J.C.; Kang, H.K.; Kim, M.J.; Kim, D.W.; Bang, H.T.; Chae, H.S.; Suh, O.S. Effect of heating system using a geothermal heat pump on the production performance and housing environment of broiler chickens. *Poult. Sci.* **2012**, *91*, 275–281. [CrossRef]
38. Krommweh, M.S.; Rösmann, P.; Büscher, W. Investigation of heating and cooling potential of a modular housing system for fattening pigs with integrated geothermal heat exchanger. *Biosyst. Eng.* **2014**, *121*, 118–129. [CrossRef]

39. Deglin, D.; van Caenegem, L.; Dehon, P. Subsoil Heat Exchangers for the Air Conditioning of Livestock Buildings. *J. Agric. Eng. Res.* **1999**, *73*, 179–188. [CrossRef]
40. Islam, M.M.; Mun, H.-S.; Bostami, A.B.M.R.; Park, K.-J.; Yang, C.-J. Combined active solar and geothermal heating: A renewable and environmentally friendly energy source in pig houses. *Environ. Prog. Sustain. Energy* **2016**, *35*, 1156–1165. [CrossRef]
41. Poole, M.R.; Shah, S.B.; Grimes, J.L.; Boyette, M.D.; Stikeleather, L.F. Evaluation of a novel, low-cost plastic solar air heater for turkey brooding. *Energy Sustain. Dev.* **2018**, *45*, 1–10. [CrossRef]
42. Cordeau, S.; Barrington, S. Performance of unglazed solar ventilation air pre-heaters for broiler barns. *Sol. Energy* **2011**, *85*, 1418–1429. [CrossRef]
43. Cuce, P.M.; Riffat, S. A comprehensive review of heat recovery systems for building applications. *Renew. Sustain. Energy Rev.* **2015**, *47*, 665–682. [CrossRef]
44. Mardiana-Idayu, A.; Riffat, S.B. Review on heat recovery technologies for building applications. *Renew. Sustain. Energy Rev.* **2012**, *16*, 1241–1255. [CrossRef]
45. Dai, B.; Liu, C.; Liu, S.; Wang, D.; Wang, Q.; Zou, T.; Zhou, X. Life cycle techno-enviro-economic assessment of dual-temperature evaporation transcritical CO_2 high-temperature heat pump systems for industrial waste heat recovery. *Appl. Therm. Eng.* **2023**, *219*, 119570. [CrossRef]
46. McGinnis, D.S.; Ogilvie, J.R.; Pattie, D.R.; Blenkhorn, K.W.; Turnbull, J.E. Shell-and-Tube Heat Exchanger for Swine Buildings. *Can. Agric. Eng.* **1983**, *25*, 69–74. Available online: https://library.csbe-scgab.ca/docs/journal/25/25_1_69_ocr.pdf (accessed on 10 January 2023).
47. McGinnis, D.S. Computer model of an air-to-air, shell-and-tube heat exchanger. *Can. Agric. Eng.* **1984**, *26*, 151–162. Available online: https://library.csbe-scgab.ca/docs/journal/26/26_2_151_ocr.pdf (accessed on 10 January 2023).
48. Barber, E.M.; Classen, H.L.; Thacker, P.A. Energy use in the production and housing of poultry and swine—An overview. *Can. J. Anim. Sci.* **1989**, *69*, 7–21. [CrossRef]
49. Dyck, J. Ventilation Heat Exchangers: Save Energy, Save Money and Reduce Greenhouse Gases in Livestock Barns: FACTSHEET 19-013, AGDEX 717. Available online: https://www.publications.gov.on.ca/030012 (accessed on 10 January 2023).
50. DLG e.V. *Abluftreinigungsanlage „Biologic Clean Air Kombiwäscher BCA 70/90": Devriecom b. v.: DLG-Prüfbericht 5879*; DLG e.V. Testzentrum Technik und Betriebsmittel: Groß-Umstadt, Germany, 2009. Available online: https://www.huber-kontech.ch/dynasite.cfm?cmd=cdownload_download_download_download&id=157&skipfurl=1 (accessed on 10 January 2023).
51. Moro, J.L. *Baukonstruktion—Vom Prinzip zum Detail: Band 4 Prinzipien*, 1st ed.; Springer: Berlin/Heidelberg, Germany, 2020; ISBN 9783662547366.
52. Asan, H. Numerical computation of time lags and decrement factors for different building materials. *Build. Environ.* **2006**, *41*, 615–620. [CrossRef]
53. *DIN EN 13053*; Ventilation for Buildings—Air Handling Units—Rating and Performance for Units, Components and Sections; German Version EN 13053:2019. Beuth Verlag: Berlin, Germany, 2020.
54. *DIN 18910*; Thermal Insulation for Closed Livestock Buildings—Thermal Insulation and Ventilation—Principles for Planning and Design for Closed Ventilated Livestock Buildings. Beuth Verlag: Berlin, Germany, 2017.
55. Spitler, J.; Gehlin, S. Measured Performance of a Mixed-Use Commercial-Building Ground Source Heat Pump System in Sweden. *Energies* **2019**, *12*, 2020. [CrossRef]
56. Gehlin, S.D.; Spitler, J.; Berglöf, K.; Mazzotti Pallard, W.; Witte, H. *IEA HPT Annex 52—Long-Term Performance Monitoring of GSHP Systems for Commercial, Institutional and Multi-Family Buildings*; Heat Pump Centre: Boras, Sweden, 2021. [CrossRef]
57. Henry, Z.A.; Zoerb, G.C.; Birth, G.S. *Instrumentation and Measurement for Environmental Sciences*, 3rd ed.; American Society of Agricultural Engineers: St. Joseph, MI, USA, 1991; ISBN 0929355164.
58. Calvet, S.; Gates, R.S.; Zhang, G.; Estellés, F.; Ogink, N.W.; Pedersen, S.; Berckmans, D. Measuring gas emissions from livestock. buildings: A review on uncertainty analysis and error sources. *Biosyst. Eng.* **2013**, *116*, 221–231. [CrossRef]
59. Mehendale, S.S. Single-Phase Heat Exchangers. In *Handbook of Thermal Science and Engineering*; Springer: Berlin/Heidelberg, Germany, 2018; pp. 1447–1472. [CrossRef]
60. *VDI 3803-5*; Air-Conditioning, System Requirements—Heat Recovery Systems (VDI Ventilation Code of Practice). Beuth Verlag: Berlin, Germany, 2013.
61. Juhrich, K. *CO_2-Emissionsfaktoren für Fossile Brennstoffe*; Umweltbundesamt: Dessau-Roßlau, Germany, 2016. Available online: https://www.umweltbundesamt.de/sites/default/files/medien/1968/publikationen/co2-emissionsfaktoren_fur_fossile_brennstoffe_korrektur.pdf (accessed on 5 October 2022).
62. IPCC. Chapter 2 Stationary Combustion. In *2006 IPCC Guidelines for National Greenhouse Gas Inventories*; Intergovernmental Panel on Climate Change: Genf, Switzerland, 2006. Available online: https://www.ipcc-nggip.iges.or.jp/public/2006gl/pdf/2_Volume2/V2_2_Ch2_Stationary_Combustion.pdf (accessed on 5 October 2022).
63. Deutscher Bundestag. *Sachstand, CO_2—Bilanzen*; Einzelfragen zu Energieträgern, Insbesondere Flüssiggas; German Bundestag: Berlin, Germany, 2018. Available online: https://www.bundestag.de/resource/blob/550728/61522d07688eb301e4edf6b8d2e68f41/wd-8-003-18-pdf-data.pdf (accessed on 5 October 2022).
64. DESTATIS. Daten zur Energiepreisentwicklung—Lange Reihen bis Juni 2022. Statistisches Bundesamt. 2022. Available online: https://www.destatis.de/DE/Themen/Wirtschaft/Preise/Publikationen/Energiepreise/energiepreisentwicklung-pdf-5619001.html (accessed on 5 October 2022).

65. DESTATIS. Strompreise für Haushalte: Deutschland, Halbjahre, Jahresverbrauchsklassen, Preisarten. Statistisches Bundesamt. 2022. Available online: https://www-genesis.destatis.de/genesis/online?operation=abruftabelleBearbeiten&levelindex=0&levelid=1659707542422&auswahloperation=abruftabelleAuspraegungAuswaehlen&auswahlverzeichnis=ordnungsstruktur&auswahlziel=werteabruf&code=61243-0001&auswahltext=&werteabruf=Werteabruf#abreadcrumb (accessed on 5 October 2022).
66. Bonkoß, K.; Neiber, J.; Neser, S. *Energieeinsparung in der Schweinehaltung*, 1st ed.; Bayerische Landesanstalt für Landwirtschaft: Freising-Weihenstephan, Germany, 2012.
67. Büscher, W.; Franke, G.; Haidn, B.; Müller, H.J.; Niethammer, F.; Leuschner, P. *DLG-Arbeitsunterlage. Lüftung von Schweineställen. 3 überarbeitete Fassung*; Deutsche Landwirtschafts-Gesellschaft e.V.: Frankfurt, Germany, 2004.
68. Bianca, W. The significance of meteorology in animal production. *Int. J. Biometeor.* **1976**, *20*, 139–156. [CrossRef]
69. Mikovits, C.; Zollitsch, W.; Hörtenhuber, S.J.; Baumgartner, J.; Niebuhr, K.; Piringer, M.; Anders, I.; Andre, K.; Hennig-Pauka, I.; Schönhart, M.; et al. Impacts of global warming on confined livestock systems for growing-fattening pigs: Simulation of heat stress for 1981 to 2017 in Central Europe. *Int. J. Biometeorol.* **2019**, *63*, 221–230. [CrossRef]
70. Diener, R.G.; Avery, J.F.; Moseley, J.L.; McNeer, M.L. Analysis for an Underground Heat Exchanger for Poultry Buildings. *J. Agric. Engng. Res.* **1986**, *34*, 187–198. [CrossRef]
71. Köhler, A.; Grahn, D.; Rebbe, F.; Berthold, J.; Vollmer, M. Energieeffizienz in der Abluftreinigung (Schweinehaltung): Schriftenreihe, Heft 19/2014. 2014. Available online: https://publikationen.sachsen.de/bdb/artikel/22353 (accessed on 10 January 2023).
72. Morshed, W.; Leso, L.; Conti, L.; Rossi, G.; Simonini, S.; Barbari, M. Cooling performance of earth-to-air heat exchangers applied to a poultry barn in semi-desert areas of South Iraq. *Int. J. Agric. Biol. Eng.* **2018**, *11*, 47–53. [CrossRef]
73. US Department of Agriculture; USDA Foreign Agricultural Service. Global Pork Production in 2021 and 2022, by Country (in 1,000 Metric Tons), Statista. 2022. Available online: https://www.statista.com/statistics/273232/net-pork-production-worldwide-by-country/ (accessed on 19 December 2022).
74. Yüksel, A.N.; Türkboylan, E.Y. Using the Photovoltaic Cells for Ventilation and Cooling of the Animal Barns. In Proceedings of the 1st International Congress on Agricultural Structures and Irrigation, Antalya, Turkey, 26–28 September 2018; p. 49, ISBN 978-605-81136-0-2.
75. Buragohain, S.; Mohanty, K.; Mahanta, P. Hybridization of solar photovoltaic and biogas system: Experimental, economic and environmental analysis. *Sustain. Energy Technol. Assess.* **2021**, *45*, 101050. [CrossRef]
76. Bourguignon, D. *Air Quality: Pollution Sources and Impacts, EU Legislation and International Agreements: Study*; Directorate-General for Parliamentary Research Services European Parliament: Brussels, Belgium, 2018; ISBN 978-92-846-3477-4.
77. United Nations Economic Commission for Europe. United Nations Economic Commission for Europe Framework Code for Good Agricultural Practice for Reducing Ammonia Emissions, ECE/EB.AIR/129. 2015. Available online: https://unece.org/fileadmin/DAM/env/documents/2015/AIR/EB/ECE_EB.AIR_129_ENG.pdf (accessed on 10 January 2023).
78. United Nations Economic Commission for Europe. Guidance Document on Preventing and Abating Ammonia Emissions from Agricultural Sources, ECE/EB.AIR/120. 2014. Available online: http://www.unece.org/fileadmin/DAM/env/documents/2012/EB/ECE_EB.AIR_120_ENG.pdf (accessed on 10 January 2023).
79. Lengling, A.; Reckels, B.; Schwennen, C.; Hölscher, R.; Waldmann, K.H.; Visscher, C.; Büscher, W. Validation of a new resource-efficient feeding system for fattening pigs using increased crude fiber concentrations in diets: Feed intake and ammonia emissions. *Animals* **2020**, *10*, 497. [CrossRef]
80. Overmeyer, V.; Trimborn, M.; Clemens, J.; Hölscher, R.; Büscher, W. Acidification of slurry to reduce ammonia and methane emissions: Deployment of a retrofittable system in fattening pig barns. *J. Environ. Manag.* **2023**, *331*, 117263. [CrossRef]
81. Gates, R.S.; Casey, K.D.; Xin, H.; Wheeler, E.F.; Simmons, J.D. Fan Assessment Numeration System (FANS) design and calibration specifications. *Trans. ASAE* **2004**, *47*, 1709–1715. [CrossRef]
82. Casey, K.D.; Ford, S.E.; McClure, J.W.; Zhang, Y.; Gates, R.S. Determining Fan Performance Using FANS: An Investigation of Performance Impacts. *Appl. Eng. Agric.* **2007**, *23*, 333–338. [CrossRef]
83. Casey, K.D.; Gates, R.S.; Wheeler, E.F.; Xin, H.; Liang, Y.; Pescatore, A.J.; Ford, M.J. On-Farm Ventilation Fan Performance Evaluations and Implications. *J. Appl. Poult. Res.* **2008**, *17*, 283–295. [CrossRef]
84. Sama, M.P.; Day, G.B.; Pepple, L.M.; Gates, R.S. Fourth-Generation Fan Assessment Numeration System (FANS) Design and Performance Specifications. *Trans. ASABE* **2017**, *60*, 507–516. [CrossRef]
85. Simmons, J.D.; Lott, B.D. Reduction of poultry ventilation fan output due to shutters. *Appl. Eng. Agric.* **1997**, *13*, 671–673. [CrossRef]
86. Pedersen, S.; Sällvik, K. *Heat and Moisture Production at Animal and House Levels*; 4th Report of Working Group on Climatization of Animal Houses; Research Centre Bygholm, Danish Institute of Agricultural Sciences: Horsens, Denmark, 2002; ISBN 87-88976-60-2.
87. Pedersen, S.; Blanes-Vidal, V.; Joergensen, H.; Chwalibog, A.; Haeussermann, A.; Heetkamp, M.J.W.; Aarnink, A.J.A. Carbon Dioxide Production in Animal Houses: A Literature Review; X BC 08 008. 2008. Available online: https://cigrjournal.org/index.php/Ejounral/article/view/1205 (accessed on 19 December 2022).

Disclaimer/Publisher's Note: The statements, opinions and data contained in all publications are solely those of the individual author(s) and contributor(s) and not of MDPI and/or the editor(s). MDPI and/or the editor(s) disclaim responsibility for any injury to people or property resulting from any ideas, methods, instructions or products referred to in the content.

AgriEngineering

Article

UAV-Based Wireless Data Collection from Underground Sensor Nodes for Precision Agriculture

Lucas Holtorf [1], Igor Titov [1], Frank Daschner [2] and Martina Gerken [1,*]

[1] Integrated Systems and Photonics, Faculty of Engineering, Kiel University, 24118 Kiel, Germany
[2] Microwave Engineering, Faculty of Engineering, Kiel University, 24118 Kiel, Germany
* Correspondence: mge@tf.uni-kiel.de

Abstract: In precision agriculture, information technology is used to improve farm management practices. Thereby, productivity can be increased and challenges with overfertilization and water consumption can be addressed. This requires low-power and wireless underground sensor nodes for monitoring the physical, chemical and biological soil parameters at the position of the plant roots. Three ESP32-based nodes with these capabilities have been designed to measure soil moisture and temperature. A system has been developed to collect the measurement data from the sensor nodes with a drone and forward the data to a ground station, using the LoRa transmission standard. In the investigations of the deployed system, an increase in the communication range between the sensor node and the ground station, from 300 m to 1000 m by using a drone, was demonstrated. Further, the decrease in the signal strength with the increasing sensor node depth and flight height of the drone was characterized. The maximum readout distance of 550 m between the sensor node and drone was determined. From this, it was estimated that the system enables the readout of the sensor nodes distributed over an area of 470 hectares. Additionally, analysis showed that the antenna orientation at the sensor node and the drone influenced the signal strength distribution around the node due to the antenna radiation pattern. The reproducibility of the LoRa signal strength measurements was demonstrated to support the validity of the results presented. It is concluded that the system design is suitable for collecting the data of distributed sensor nodes in agriculture.

Keywords: precision agriculture; unmanned aerial vehicle (UAV); drone; LoRa; Internet of Things (IoT); Internet of Underground Things (IoUT)

Citation: Holtorf, L.; Titov, I.; Daschner, F.; Gerken, M. UAV-Based Wireless Data Collection from Underground Sensor Nodes for Precision Agriculture. *AgriEngineering* **2023**, *5*, 338–354. https://doi.org/10.3390/agriengineering5010022

Academic Editors: Muhammad Sultan, Yuguang Zhou, Redmond R. Shamshiri and Muhammad Imran

Received: 27 December 2022
Revised: 7 February 2023
Accepted: 8 February 2023
Published: 9 February 2023

Copyright: © 2023 by the authors. Licensee MDPI, Basel, Switzerland. This article is an open access article distributed under the terms and conditions of the Creative Commons Attribution (CC BY) license (https://creativecommons.org/licenses/by/4.0/).

1. Introduction

The world's growing population and limited natural resources increase the demand for precision agriculture. The goal is to improve farm management by using information technology in order to increase crop quality and profitability [1]. Additionally, the problems that arise with today's high yield agriculture can also be addressed. Overfertilization results in health hazards, pollution, reduced product quality and a loss of soil fertility [2]. The extensive, uncontrolled irrigation of cropland leads to the deterioration of water sources, salt build-up on fields and soil erosion [3].

The development of precise and cost effective sensing solutions is recognized as one of the key challenges for increasing the adoption of precision agriculture [1]. Monitoring soil parameters, such as moisture, temperature, pH, nitrate or nitrite, is crucial for efficient plant growth and root development [4]. This allows for the precise control of the nutrient content to enhance productivity [5]. Weather, management practices, different soil types and local topology can cause significant variations in the parameters of interest, even within a single agricultural field and over short timespans [6]. Traditional practices, such as in-situ measurements conducted by an appropriate operator or offsite laboratory analysis of soil samples, are not suitable for the spatial and temporal measurement densities required. Thus, the use of distributed, independently operating sensor nodes with in-situ measurement capabilities is desired [7].

As the manual operation of wired sensor nodes deployed on large agricultural fields is both impractical and expensive, low-power and long-range data transmission standards, such as Long Range (LoRa), have been established for the transmission of sensor data [8,9]. The sensor nodes should not obstruct the normal work in the field, which introduces the need for wireless underground sensor nodes. Burying sensor nodes in the soil helps to protect them from extreme weather conditions and farm equipment [7]. A drawback is that the achievable communication distance is significantly reduced, because the soil introduces variable high path losses [10]. Researchers at the University of Applied Sciences in Osnabrück have proposed a system for data readout from a sensor buried 60 cm in the ground at a receiving station 350 m away using LoRa communication [8].

Typically, the distances in agriculture are significantly larger than 350 m. Thus, a drone can be a solution for the data collection of sensor nodes distributed in large areas without the need to install additional infrastructure. The deployment of unmanned aerial vehicles (UAVs) in the context of precision agriculture is becoming a valuable tool for farmers and researchers. Barbedo provides a comprehensive review on the use of UAVs for monitoring and assessing plant stresses [11]. Barnetson et al. measured plant pasture biomass and quality with a UAV [12]. Bukowiecki et al. used an UAV with an integrated camera for the estimation of the green area index (GAI) of winter wheat to calibrate the Sentinel-2 data for crop monitoring and yield prediction [13]. Fan et al. discuss plant classification based on drone data [14]. Quino et al. introduced drones in combination with RFID tags for taking a plant inventory [15]. The numerical fluid dynamics simulation by Marturano et al. investigates the air flow around a drone, which is relevant for the placement of sensors on drones [16].

The potential of combing drones with LoRa communication for data collection has recently been proposed in the literature. Behjati et al. demonstrated a system that used drones with an attached LoRaWAN® gateway to collect the data from water quality sensors and livestock monitoring equipment on a farm [17]. The authors investigated the packet loss performance of the LoRa connection, taking into account the different spreading factors and drone speeds. Additionally, they optimized the flight path to maximize the flight range of the fixed-wing vertical takeoff drone. The forwarding of the sensor data from the gateway to a server was conducted using LTE. Park et al. demonstrated a system for collecting environmental sensor data on a tree farm via LoRa using a drone with an attached LoRaWAN® gateway [18]. They successfully applied a proposed client-server communication scheme, but encountered some issues with the gateway in low temperature conditions and limitations of the flight time of the drone due to the additional weight.

Other groups have worked on different subproblems of the approach to collect sensor data with drones. Caruso et al. derived an analytical model for data acquisition with a drone and LoRa from a large regular grid of sensor nodes, as would be found in agriculture [19]. The model allowed for the determination of the optimal spacing of the sensor nodes, the time a sensor needs to be in range of a sensor node and the required velocity of the drone to optimize the probability of successful data collection.

Furthermore, Zhang et al. improved the UAV data collection efficiency of distributed sensor nodes [20]. They developed an algorithm to organize sensor nodes that are close to each other into clusters, with a cluster head that transmits the sensor data of the whole cluster to the drone. A genetic algorithm was then used to minimize the flight path of the UAV between the clusters. Additionally, an adaptive scheme for changing the transmission data rate based on the received signal strength and the signal to noise ratio was reported.

Pan et al. proposed a dynamic UAV speed control scheme in order to adapt to different sensor device densities [20,21]. Based on an analytical model of the connection situation, the number of successful sensor node connections per second was maximized.

Zorbas and O'Flynn worked on the problem of LoRa transmission packet collision when many sensor nodes are located in a specific area. Their protocol allowed for the data collection of 80 nodes in a 1500 m × 1500 m area without packet collisions [22].

In this paper, a system was investigated to collect the data from buried sensor nodes and forward that data to a ground station by using the LoRa transmission standard and a UAV. When operating such a system, a challenge is that it has to be designed in such a way that the data collection is secured while the minimizing battery consumption of the sensor nodes and taking into account the limited flight time of the drone. The necessary optimization of the readout scheme requires knowledge about the different influences on the LoRa transmission quality when the underground sensor node communicates with an aboveground UAV.

To the best of the authors' knowledge, there have not been many published experimental demonstrations on this topic. A similar setup using the NB-IoT standard has been analyzed in simulations by Castellanos et al. The authors demonstrated that an UAV could collect the data within 50 s from 2000 sensors buried in a grid with 10 m spacing under a 20 hectare potato field. Additionally, the impact of soil moisture on the number of successfully served sensor nodes due to changing damping has been shown [23].

Experimental work has recently been published by two other groups. Cariou et al. buried a sensor node at a depth of 15 cm and demonstrated the data collection capability of their system for flight heights of 20 m to 60 m and distances up to 150 m [24]. Hossain et al. developed a similar system and conducted measurements of the received signal strength of the LoRa packages close to the point where the node was buried [25]. A detailed comparison between the system architectures and methods used and the approach described here is given in Section 4.

The general setup (see Figure 1a) consists of a sensor node buried in a field that transmits a series of measurements upon the request of the repeater drone, using the LoRa transmission standard. The repeater drone receives the data and forwards it immediately to the user ground station. Low power sensor nodes were implemented and a protocol for their energy efficient operation was developed. In several experiments, different aspects of the system described were analyzed:

- Dependency of received LoRa signal strength on flight height of the drone and on the burial depth of the sensor node.
- Demonstration of the range extension capability.
- Determination of the maximum distance between the sensor node and drone with the setup.
- Investigation of the effect of the antenna placement inside the sensor node on the received LoRa signal strength.
- Investigation of the repeatability of the signal strength over distance measurements.

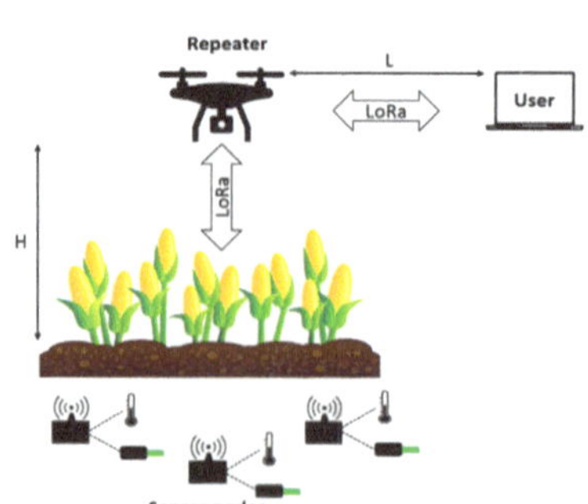

(a)

(b)

Figure 1. *Cont.*

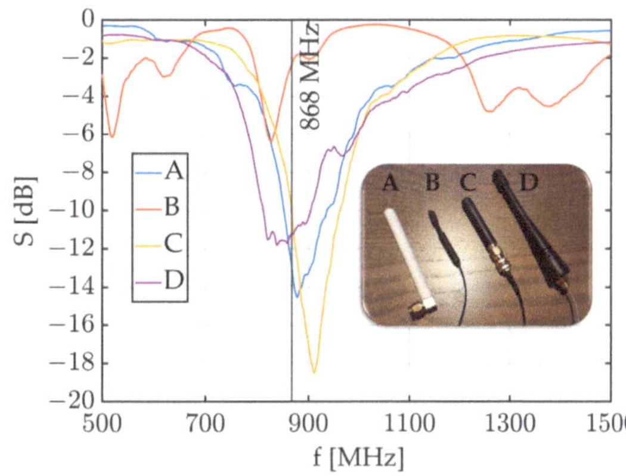

(**c**) (**d**)

Figure 1. (**a**) Schematic of the investigated communication setup. (**b**) Picture of a sensor node deployed in a maize field before closing the hole with soil. (**c**) Picture of the drone with attached component housing. (**d**) Characterization of different antennas in the range around the 868 MHz LoRa frequency with the scattering parameter S.

This paper is organized as follows. Section 2 presents the materials and methods. Section 3 contains the experimental results. The system design and the results are compared side-by-side with those of the relevant literature in Section 4 (discussion). Finally, the conclusions are presented in Section 5. Some of these results have recently been published as a conference contribution [26].

2. Materials and Methods

2.1. Materials

Three test sensor nodes, based on the ESP32 boards with a LoRa communication chip operating at 868 MHz, were used (WiFi LoRa 32 V2, Heltec Automation Technology, Chengdu, Sichuan, China). Figure 1d shows the characterization of the different antennas for the LoRa communication. The incorporated antenna from Heltec (Antenna "B") showed an unsatisfactory performance at the desired communication frequency of 868 MHz. For the further experiments, we used the antennas marked as "A" and "C", due to their superior performance around 868 MHz and small antenna size. The spreading factor (SF) was set to 10 to balance between the required power and time required for data transmission and the achievable range. The bandwidth (BW) was set to 125 kHz and the maximum available output power of 20 dBm was used. Those parameters where chosen which favored longer communication ranges over data rates and transmission power consumption.

A sensor node operates as follows to minimize power consumption. The ESP32 stays in deep sleep mode and wakes up at regular intervals for 5 s to obtain the data from the connected sensors and to listen for data requests from the drone. When it receives a request, it transmits the data stored since the last request and then goes back to sleep. The Heltec board has a rather high deep sleep current, of approximately 1 mA [27]. Therefore, the overall power consumption is determined by the deep sleep power, by the power required while waiting for a request from the drone and by the ratio between the time spent in deep sleep mode vs. time spent waiting for a request. It was decided to favor the transmission range over transmission power consumption because it was assumed that the amount of data collected and the frequency of drone readouts would be low. In the experiments, a wakeup interval of 5 s was chosen to acquire more data about the transmission quality. This interval would need to be increased when operating the system for longer timespans.

Each sensor node is powered by a 3.6 V lithium thionyl chloride (Li/SOCL2) battery (SL-2780, Tadiran Batteries, Kiryat Ekron, Israel), due to the high energy capacity of 19 Ah and long lifetime (up to 10 years). In the experiments, a capacitive moisture and temperature sensor (Adafruit STEMMA Soil Sensor, Adafruit Industries LLC, New York, NY, USA) was connected. An IP68 enclosure protects the components from the ingress of water and dust.

The drone used to investigate the LoRa repeater function was the DJI Phantom 3 Advanced (DJI Technology, Shenzhen, China) (Figure 1c). DJI states a maximum flight time of approximately 23 min [28], but in the experiments, flight times of only approximately 15 min were achieved. A Heltec WiFi LoRa 32 V2 board with antenna "A" was attached to provide LoRa connectivity. It was powered by a 1100 mAh LiPo battery pack. The position of the drone was tracked by the onboard inertial measurement unit (IMU) and an additional NEO-6M GPS module connected to the Heltec board.

In order to acquire the data over LoRa and record it, a ground station consisting of a readout laptop and connected Arduino Uno with a Dragino LoRa Shield (SX1276 Shield-868, Dragino Technology, Shenzhen, China), was used. This module is also based on Semtech's SX1276 LoRa transceiver chip. The received LoRa packages were loaded onto the laptop over a USB serial interface and saved to a CSV file with a Python script. The post-collection analysis of the data was also conducted in Python.

The SX1276 chip provides information about the received signal strength indication (RSSI). This value can be used to evaluate the signal attenuation and was the main parameter used for interpreting the following experiments. Therefore, the RSSI of both LoRa communication paths (sensor to drone and drone to ground station) was acquired. The RSSI of the transmission between the sensor node and the drone was transmitted to the ground station, together with the GPS data from the NEO-6M module and the sensor data.

2.2. Methods

In order to analyze the different aspects of the usage of a drone as a repeater for LoRa communication, a series of experiments were conducted on the Experimental Farm Hohenschulen of Kiel University.

2.2.1. Investigation of Flight Height and Burial Depth Dependency

In the first series of experiments, the LoRa signal attenuation of the communication between the sensor node and drone was investigated, with the drone hovering at an increasing flight level H directly above the sensor node (Figure 2). There were two measurement series with sensor node depth D of 30 cm and 60 cm. The RSSI was recorded every 5 m until a 40 m height level was reached.

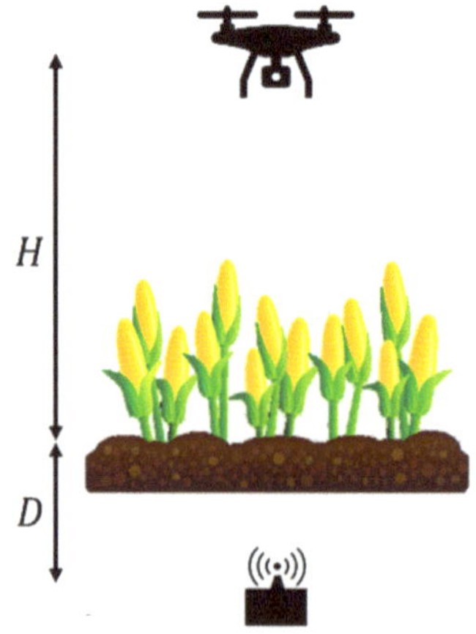

Figure 2. Schematic of the experimental setup used for investigating the dependency of the received signal strength at the drone on flight height H and burial depth D.

2.2.2. Demonstration of the Range Extension Capability

In a subsequent experiment, the communication range extension capability of the described system was investigated. Hence, the communication range L was compared for two different situations: direct sensor node to ground station communication (S1) and intermediate signal repetition at the drone (S2) (Figure 3).

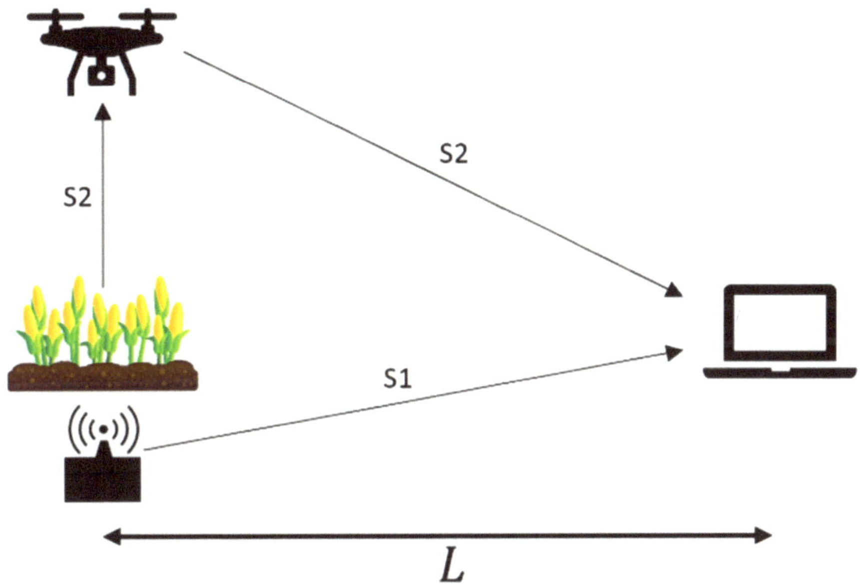

Figure 3. Schematic of the experimental setup used for demonstrating the extended range of the data collection system design. S1 shows the direct sensor node to ground station communication path and S2 the path with intermediate signal repetition at the drone.

Firstly, the system was configured so that the sensor node, buried at a depth of 30 cm, communicated directly with the ground station, bypassing the drone. The distance between the sensor node and ground station was increased while measuring the RSSI and the GPS position. Secondly, the experiment was conducted by using the drone as a repeater. The drone hovered 30 m above the sensor node and then, again, recorded the RSSI and GPS position every 50 m while the ground station was moved away.

2.2.3. Investigation of the Communication Range between Sensor Node and Drone

When applying the described system in a real-world scenario, a drone would not hover above each sensor node but would conduct the data collection from a certain distance. In order to maximize the area in which the sensor nodes can be deployed and read out, the range R for the communication between a sensor node and the drone has to be determined (Figure 4). For this purpose, a sensor node was buried at a depth of 30 cm and the communication was started with the drone hovering above the sensor node. The drone was then flown away at a constant height of 30 m while recording both its GPS position and the RSSI of the packages received at the drone from the sensor node.

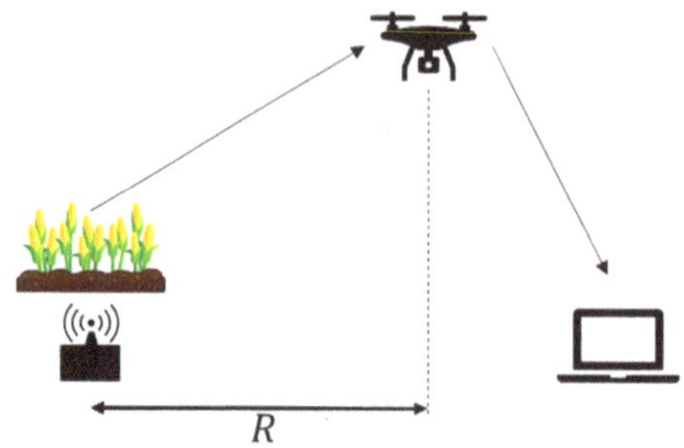

Figure 4. Schematic of the experimental setup used for the determination of the communication range between sensor node and drone.

2.2.4. Investigation of the Antenna Radiation Pattern

Commercially available antennas for LoRa communication are usually helical antennas. The radiation pattern has a toroidal shape and is symmetric around the z-axis of the antenna (Figure 5b). When using compact enclosures for the sensor nodes, the antennas must be oriented horizontally (Figure 5a). Consequently, the radiation pattern above the ground is not symmetric around the point where the drone is buried.

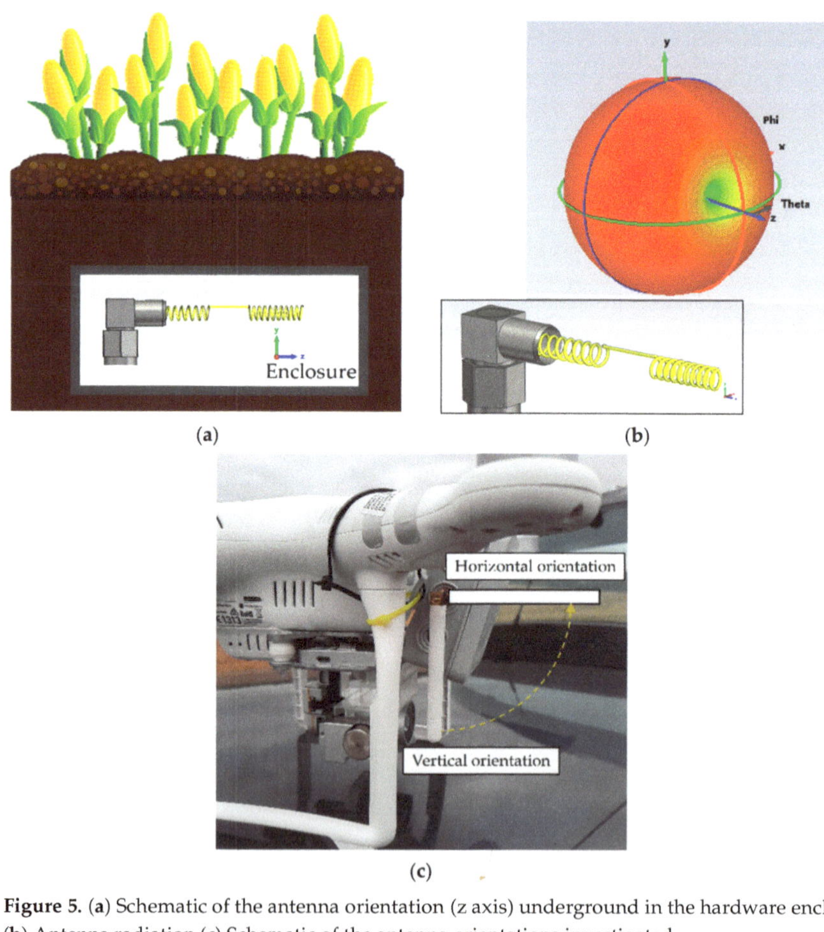

(a) (b) (c)

Figure 5. (**a**) Schematic of the antenna orientation (z axis) underground in the hardware enclosure. (**b**) Antenna radiation (**c**) Schematic of the antenna orientations investigated.

How this affects the RSSI of the data packages sent from the node to the drone was investigated. To achieve this, the sensor node was buried in a depth of 30 cm, with the antenna z-axis oriented in a known direction in the ground plane. Then, the drone was flown at a constant height of 30 m and with varying positions around the point at which the sensor node was buried. During flight, the sensor data was continuously read out and the GPS coordinates of the drone were acquired each time a package from the sensor node arrived. This enabled the correlation of the RSSI with the position of the drone. The antenna on the drone was oriented with the helix z-axis pointing downwards (Figure 5c) and, therefore, its radiation pattern is symmetric, minimizing distortions caused by the drone orientation.

2.2.5. Investigation of Reproducibility

The interpretation of the described experiments requires the reproducibility of the measurement data. In order to verify this, a sensor node was buried at a depth of 30 cm and three flights were carried out with the same path, directly passing above the sensor node at a constant height of 25 m (Figure 6). Each flight was started directly above the sensor node and then the drone was flown away to a defined distance (flight phase P1). Afterwards, it was steered back along the same path, passing over the sensor node, and then flown away again from the sensor node to a turning point (P2). From that point, the drone was flown back to the sensor node (P3). During flight, the data packages were continuously read out from the sensor node. Additionally, the chosen trajectory allows the analysis of how the RSSI changes when flying over the sensor node during data collection.

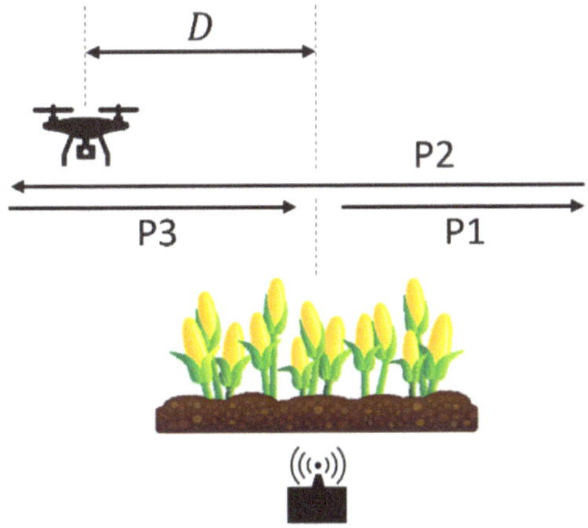

Figure 6. Schematic of the experimental setup and the different flight phases.

3. Results

3.1. *Investigation of Flight Height and Burial Depth Dependency*

The RSSI for a sensor node depth *D* of 30 cm was −63 dBm at a drone height of 5 m and −77 dBm at a height of 40 m, as shown in Figure 7. For a sensor node depth *D* of 60 cm the values were −64 dBm and −81 dBm, respectively, at drone heights of 5 m and 40 m. During the experiments, a loss of LoRa packages was observed at an RSSI of approximately −120 dBm. From this data, a repeater flight level of 130 m was estimated as possible for a sensor node buried at 60 cm depth. In a second set of experiments, it was found that the attenuation for flying the drone at a constant height to different distances *L* was in a similar range, with ~0.3 dB/m. This experiment was conducted on 23 November 2020.

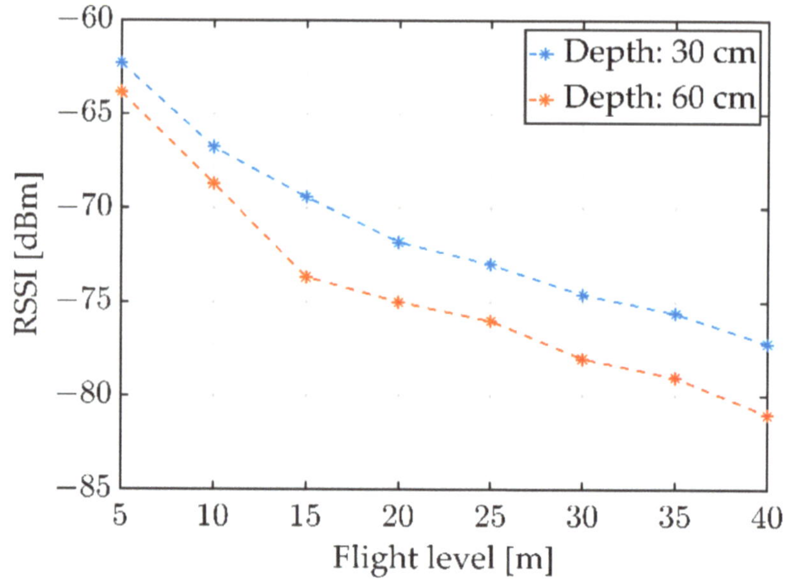

Figure 7. Comparison of the relationship between flight level and RSSI for the two sensor node depths of 30 cm and 60 cm.

3.2. Demonstration of the Range Extension Capability

Figure 8a shows a satellite image of the paths away from the sensor node for both parts of the experiment. The positions at which the last packages with sensor data were received are marked. The graph depicting the RSSI over distance (Figure 8b) shows that the LoRa packages were received up to L = 316 m (blue) for the case without using a drone. When using the drone to receive and transmit the data, the last package was received at a distance of L = 1000 m (orange). Thus, the communication range is enhanced by at least three times with the drone as a repeater. The height profile of the path between the sensor node and ground station (Figure 8c) particularly shows the advantage of the system. It can be seen that, for direct communication between node and ground station, the signal is lost after crossing the top of a hill, as the line of sight (LoS) is lost. By using the drone, a LoS can be ensured even in landscape with hills and other obstacles that might obstruct the signal. This experiment was conducted on 6 August 2021.

3.3. Investigation of the Communication Range between Sensor Node and Drone

The RSSI measurement results of the experiment can be seen in Figure 9a. The RSSI decreases with the increasing distance until the last data package is received at a distance of R = 550 m. Consequently, the drone can be used as a data collector for sensor nodes within a R_{max} = 550 m range.

From this data, it is possible to estimate the maximum area A_{max} on which the deployed sensor nodes can be read (Figure 9b). Assuming an average flying velocity of 18 km/h and a flight time of 15 min, a maximum flight distance of the drone S_{max} = 4500 m can be calculated. This allows for the determination of the length L, considering the given flightpath with minimal overlap (Figure 9b):

$$L = \frac{1}{2}(S_{max} - 4R_{max}) = 1150 \text{ m}. \tag{1}$$

Following this, the maximum area

$$A_{max} = 4R_{max}L + (4+\pi)R_{max}^2 = 4.69 \text{ km}^2 \tag{2}$$

is calculated. Thus, utilizing a drone for sensor readout enables an area of approximately 470 hectares to be scanned within only 15 min. The data can be transmitted online to the user within a certain range (in this setup approximately 1 km) or by storage and the subsequent transmission of the data packages upon return to the base. This experiment was conducted on the 31 May 2022.

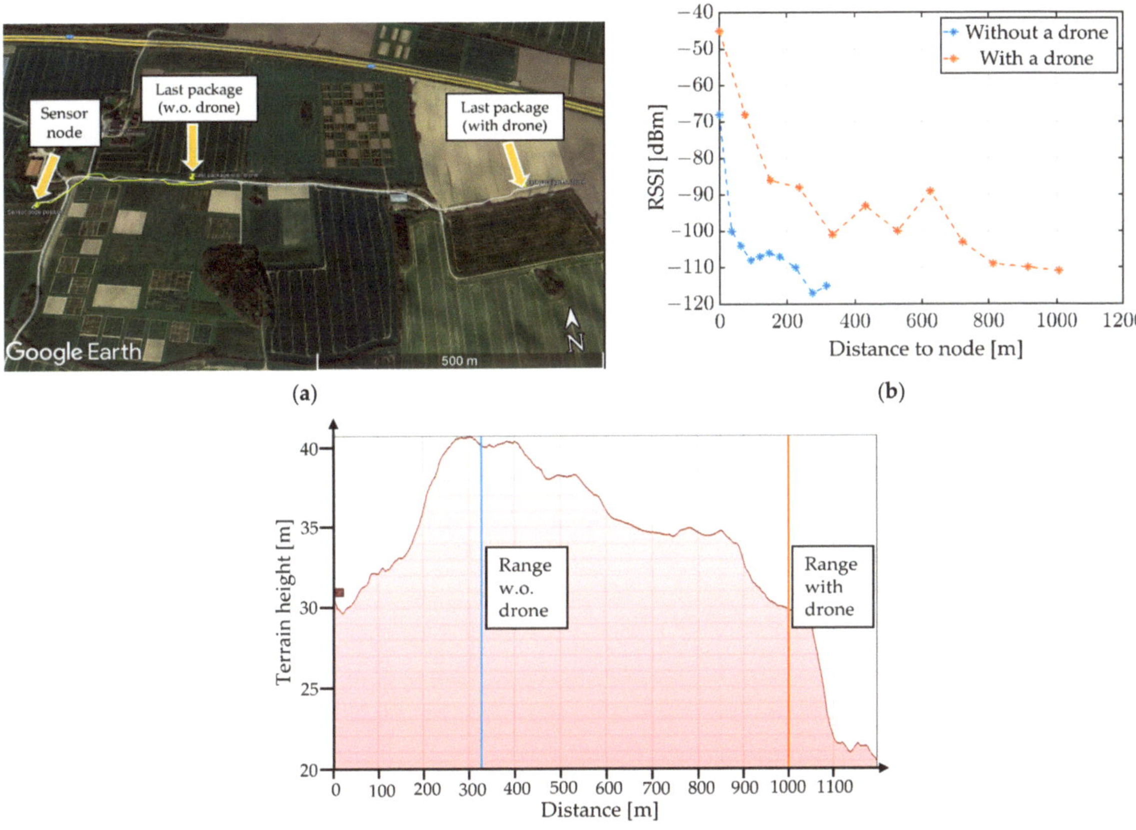

Figure 8. (**a**) Satellite view of the travelled path away from the sensor node (Map data: Google, Landsat/Copernicus) (**b**) Comparison of the RSSI of the packages received by the ground station with and without a drone as repeater. Sensor node depth is 30 cm. Drone is hovered 30 m above the sensor node as a repeater. (**c**) Height profile of the traveled path (Map data: Google, Landsat/Copernicus).

3.4. Investigation of the Antenna Radiation Pattern

The results of this experiment are visualized as point clouds (Figure 10). Each colored dot corresponds to the position of the drone when it received a package with the given RSSI. Figure 10a shows the point cloud for the vertical antenna orientation. The RSSI of the packages is higher at the perpendicular point to the antenna orientation than the inline point. This supports the hypothesis that the antenna orientation of the sensor node influences the RSSI of the transmitted packages. This fact can be important when optimizing flight paths for data collection, as the RSSI of data transmission not only depends on the distance between the sensor node and drone, but also on the sensor node orientation.

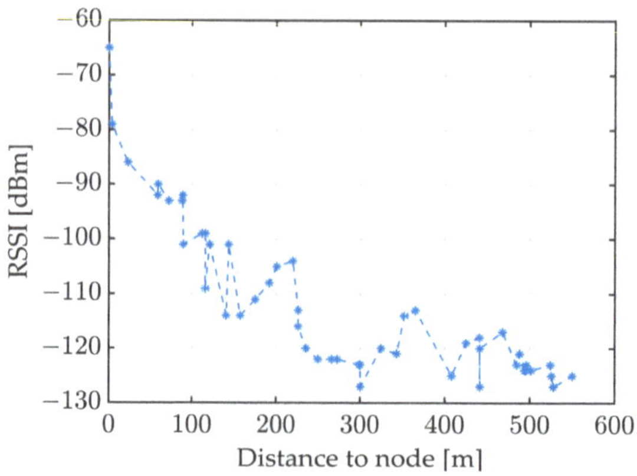

(**a**)

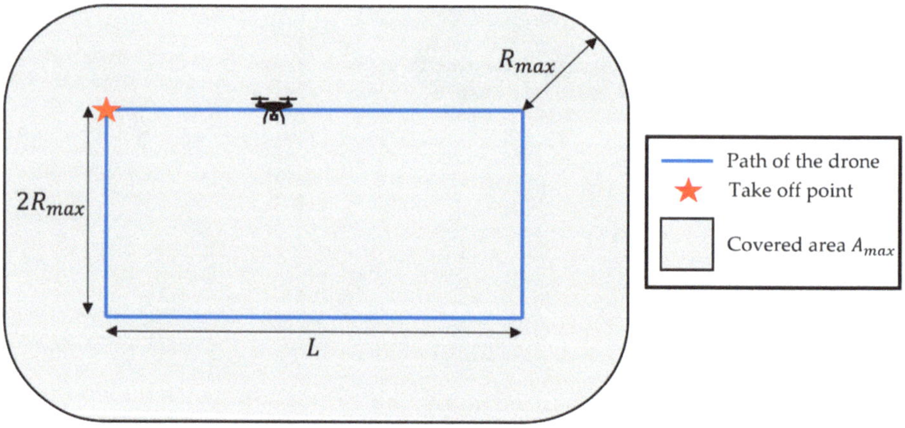

(**b**)

Figure 9. (**a**) Reduction in the RSSI with the increasing distance between the sensor node and drone. The sensor node depth was 30 cm. The drone had a constant flight height of 30 m. (**b**) Schematic of the assumed flight configuration for calculating the maximum covered area A_{max} = 470 ha.

In addition, the antenna orientation on the drone was changed from a vertical to a horizontal orientation (Figure 10b). This creates an additional dependency of the RSSI on the yaw angle of the drone, as the symmetry of the radiation pattern of the drone's antenna is lost. Therefore, the RSSI is less predictable; therefore, for practical applications, a vertical orientation of the antenna is to be preferred over a horizontal orientation. This experiment was conducted on 7 July 2022.

3.5. Investigation of Reproducibility

Firstly, the change in the RSSI over the pathway of a single flight was evaluated (see schematic in Figure 6). When flying away in phase P1, the RSSI decreased with distance D from the sensor node and increased again when turning around to fly phase P2 (Figure 11). After passing over the sensor node, the RSSI decreased, then in P3, it increased again until arriving back over the sensor node. The main observation is that the RSSI is lower when flying away from the sensor node than when flying towards it. It is believed that the reason for this is probably due to the radiation pattern of the antenna. For this experiment, the antenna was mounted with the z-axis facing downwards; thus, the radiation pattern (Figure 5b) is symmetric around the yaw axis of the drone, but not around the pitch axis. When flying towards the sensor node, the node lies closer to the xy-plane of the antenna radiation pattern, where the maximum energy of the incoming signal is received by the drone's antenna. When flying away, the signal from the sensor node comes from an angle, where less energy of the incoming signal is converted from electromagnetic waves to an electric signal by the antenna, therefore the RSSI is lower.

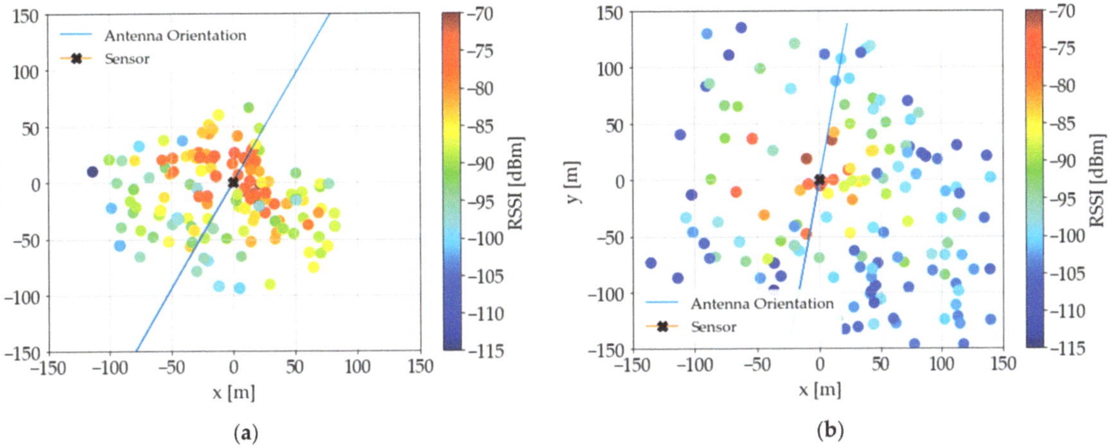

Figure 10. (**a**) Visualization of the experimental results for vertical antenna orientation on the drone as a point cloud. Each point corresponds to the position relative to the sensor node that a package was received and is colored to show the RSSI value of the transmission. Additionally, a line is placed along the orientation of the antenna of the sensor node. Sensor node depth was 30 cm. Drone had a constant flight height of 30 m. (**b**) Experimental results for horizontal antenna orientation on the drone.

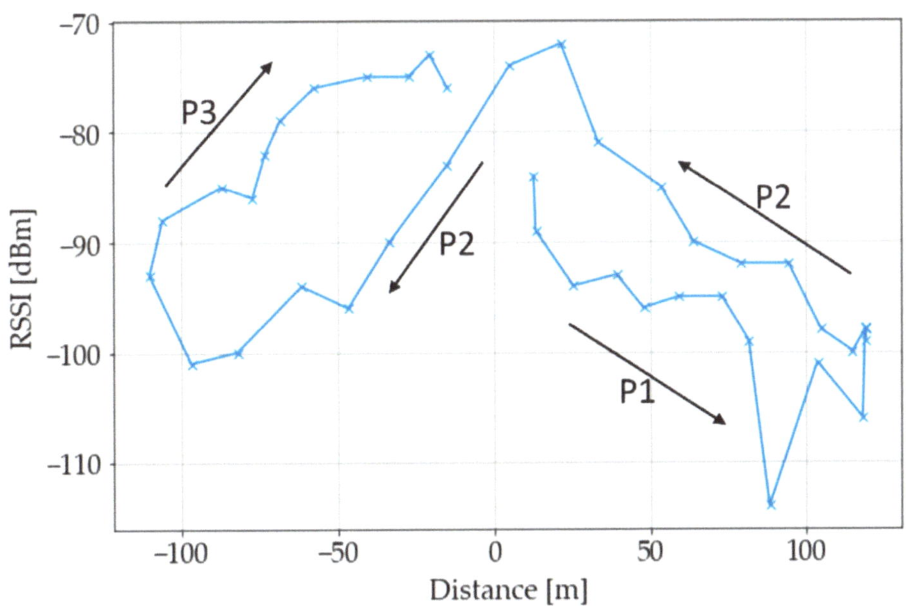

Figure 11. RSSI over distance D diagram of the third overflight to explain the behavior.

Secondly, the change in the RSSI over distance was compared between the three flights. During P1, the RSSI values differ, at a maximum of 10 dB for similar distances from the sensor node, and the qualitative development of the RSSI is comparable (Figure 12a). The same is true for the second flight phase (Figure 12b), where the maximum difference is also 10 dBm, but for most distances the difference is smaller, at approximately 5 dB. For the third flight phase, the differences are mostly 4 dB, and when close to the sensor node, the difference goes up to approximately 10 dB (Figure 12c).

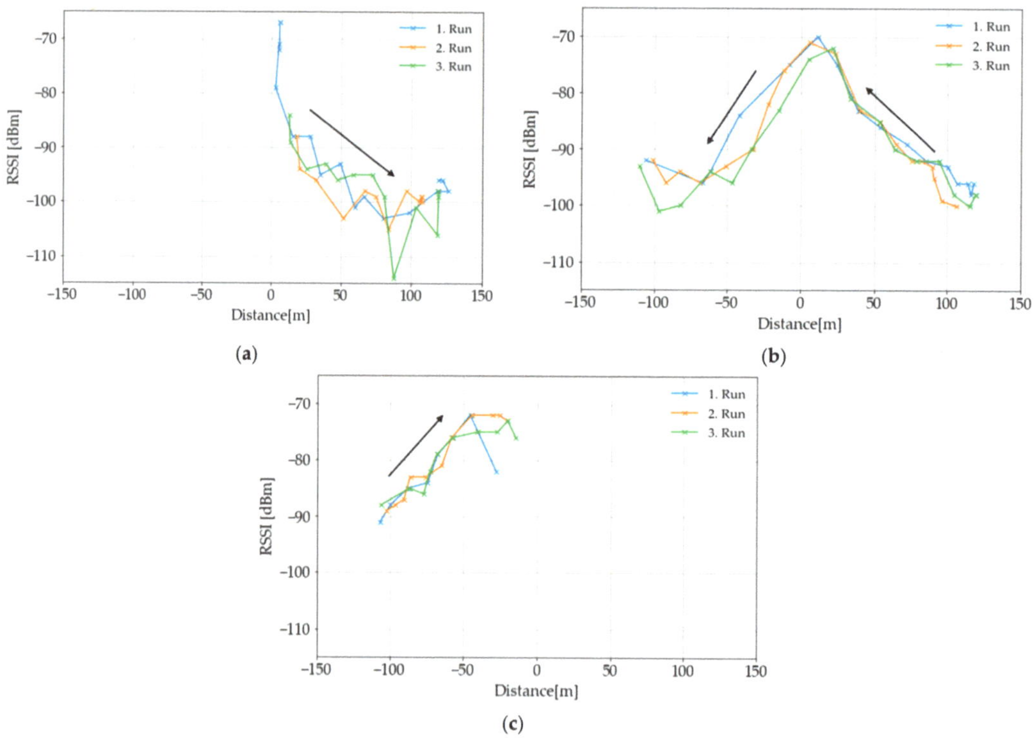

Figure 12. RSSI over distance diagram of the three flights, divided into the different flight phases for better clarity. The arrows show the direction of time evolution. Sensor node depth is 30 cm. Drone has a constant flight height of 25 m. (**a**) Phase 1. (**b**) Phase 2. (**c**) Phase 3.

The differences can be caused by errors in the reproduction of the desired flight path. These errors are most significant close to the sensor node because, here, small deviations in the flight path can lead to comparatively high deviations in the RSSI value. Additionally, the GPS positioning is also only accurate to a few meters, particularly the NEO6M GPS module as it only uses between six and eight satellites. The RSSI also depends on the pitch angle of the drone, as described before. The pitch angle also determines the speed of the drone. Therefore, speed differences between the flights can cause differences in the RSSI measurements. The large RSSI differences at the turning points of the drone are likely to be explained by this, because the speed and orientation changes at these points are difficult to reproduce between different flights. This experiment was conducted on the 29 August 2022.

4. Discussion

In this section, the system setup, methods and results are compared to the work of Cariou et al. [24] and Hossain et al. [25], because they share the same approach of collecting data from underground sensor nodes by using an UAV and the LoRa transmission technology.

The system setup of Cariou et al. consisted of a sensor node that was designed to be buried in the ground for several months. It awakened regularly from the sleep mode to transmit data from a soil moisture sensor, on request, to a collector node attached to a drone via LoRa. This transmission used an SF of 7, a BW of 125 kHz and a transmission power (TxP) of 14 dBm. The collector node saved the data frames on an SD card and transmitted them over ZigBee to a network gateway when in close proximity.

The setup of Hossain et al. used a sensor node consisting of a Raspberry Pi for control, an RFM95 for LoRa connectivity and a soil moisture sensor. The sensor node was powered from above ground by a 12 V car battery and was not optimized for low power operation. Data packages were sent continuously to a LoRaWAN gateway attached to the drone, configured with SF 7, BW 125 kHz and a TxP of 20 dBm. The gateway also stored the data until the drone landed.

The SF, BW, and TxP for the LoRa transmissions of both groups were configured for higher data rates and lower transmission ranges between sensor node and drone. SF 10, BW 125 kHz, and TxP 20 dBm were used, which has to be considered when comparing the RSSI values and achieved transmission ranges. Another difference was the way the data from the drone was further processed after being received. Both Cariou et al. and Hossain et al. used an approach in which the data was collected and was only accessible to the user at the end of the flight. The data packages were transmitted online to the user with LoRa. This allowed for the diagnoses of problems during flight and access the data without delay. However, this also reduced the data rate further as the drone LoRa module could not collect new data packages from other nodes while retransmitting old packages to the ground station. Another approach was the direct upload of the data to the cloud via a 4G internet connection, as demonstrated in [17,29]. This obviously requires the local availability of a 4G connection.

In their experiments, Cariou et al. first measured the RSSI in a range of distances (0 to 150 m) from the UAV to the buried sensor node (15 cm deep) and for heights of 20/40/60 m. Their results are comparable to the results in Sections 3.1 and 3.3, which considered a 30 m flight height and 30 cm node depth. Their data showed that for high enough lateral distances, the RSSI is higher for higher flight heights. Secondly, they conducted an overflight experiment over the buried sensor node at different UAV speeds to show that the system was able to collect 100 packages from the node, independent of the speed. Their results showed RSSI measurements that are comparable to the results in Section 3.5, but the authors did not further investigate or explain this behavior. The authors further optimized the path for the data collection of 25 buried sensor nodes.

Hossain et al. measured the RSSI with a receiver gateway attached to a wooden pole with increasing height (0.3 to 2 m) and increasing lateral distance (0 to 5 m) relative to the sensor node (30 cm node depth), both for cross- and co-polarization between the sensor node and gateway antenna. The results were also compared with numerical simulations. Furthermore, the authors measured the packet loss for distances between the node and drone of up to 82 m, again comparing cross- and co-polarization. Additionally, they conducted trajectory flights at four heights, from 0.91 m to 3.66 m around the sensor node, covering a rectangle spanning −24 to 24 m and −27.4 m to 24 m. During flight, they again measured the RSSI for cross- and co-polarization. This experiment, although methodically similar to Section 2.2.4, considered a smaller lateral range scale and smaller flight heights. From the results, the authors concluded that the RSSI was largely independent of the polarization. The RSSI measurements are mostly complementary to the work of Cariou et al. [24] and the work described here as they focused on a scale closer to the buried sensor node.

Overall, this work shows similarities to both these other works when the methods and the results are considered. A common point of interest is the RSSI distribution around the sensor node, as described before, but different length scales were investigated. Furthermore, an attempt was made to estimate the maximum coverable area for the system setup. A particular new emphasis of this work is the impact of the orientation of the antenna at the sensor node and at the drone. Additionally, experiments on using a LoRa link from drone to ground station are presented.

Comparing the chosen system setups and methods, it can be observed that there are different assumptions about the sensor node density and the overall amount of data that needs to be collected. Cariou et al. and Hossain et al. considered smaller length scales in their experiments and configured the LoRa communication more for higher data rates than for communication distances. This is suitable for higher sensor node densities or if large amounts of data need to be transmitted at each connection. The system described here was designed more towards covering a greater area with lower sensor node density and lower amounts of data for each transmission.

5. Conclusions

In conclusion, a setup for communication with an underground network of LoRa sensor nodes has been demonstrated using a drone to collect the data. A much larger area can be covered than previously, without the need for the installation of additional ground-based hardware. The maximum communication ranges between an underground sensor node and drone and the dependency of the received signal strength on the flight height and burial depth for the setup was experimentally determined. The evaluation of the antenna orientation of the sensor node and the overflight experiment hinted at other possibilities for drone flight path optimization. In addition, the reproducibility of the RSSI measurements for different distances between the sensor node and drone was also demonstrated.

The long-term goal is a fully autonomous data collecting system for several sensor nodes based on a drone. The sensor nodes were operated in the soil and demonstrated successful connection for up to 49 days. Currently, sensors for physical and chemical soil parameters are integrated into the sensor nodes for the relevant data collection. For this purpose, a microfluidic unit for the extraction of soil water has been developed [30], as well as a compact sensor chip based on organic optoelectronic devices for nitrite and nitrate determination [31]. To enable the data collection for a larger number of sensor nodes, a more sophisticated communication protocol must be implemented to avoid package collisions. Additionally, adapting the spreading factor and the bandwidth according to the respective distances and sensor node densities is important to maximize the area coverage and collection performance of the system. This can be further improved by optimizing the drone path between sensor nodes. Regarding the system setup, it is still to be determined whether LoRa is the optimal way to transmit the data from the drone to the ground station. Possibly, a transmission directly over LTE to a cloud is advantageous to point-to-point LoRa, particularly for situations where high data rates are required. Another important aspect is the optimization of the sensor node antenna to the ground-to-air transmission situation.

Author Contributions: Conceptualization, L.H., I.T., F.D. and M.G.; methodology, L.H., I.T., F.D. and M.G.; software, L.H.; validation, L.H. and I.T.; formal analysis, L.H., I.T. and M.G.; investigation, L.H. and I.T.; resources, M.G.; data curation, L.H. and I.T.; writing—original draft preparation, L.H., I.T. and M.G.; writing—review and editing, L.H., I.T., F.D. and M.G.; visualization, L.H., I.T. and F.D.; supervision, I.T. and M.G.; project administration, M.G.; funding acquisition, M.G. All authors have read and agreed to the published version of the manuscript.

Funding: This project received funding from the European Regional Development Fund (EFRE) by the European Union (OPTOCHIP, LPW-E/1.2.2/1303).

Data Availability Statement: Data underlying the results presented in this paper are not publicly available at this time but may be obtained from the authors upon reasonable request.

Acknowledgments: The authors thank the Institute of Crop Science and Plant Breeding for providing the drone and for farm access at the Experimental Farm Hohenschulen.

Conflicts of Interest: The authors declare no conflict of interest.

References

1. Robert, P.C. Precision agriculture: A challenge for crop nutrition management. *Plant Soil* **2002**, *247*, 143–149. [CrossRef]
2. Shiva, V. *The Violence of the Green Revolution: Third World Agriculture, Ecology and Politics*; Zed Books: London, UK, 1991; ISBN 9780862329655.
3. Hillel, D.; Braimoh, A.K.; Vlek, P.L.G. *Soil Degradation Under Irrigation*; Springer: Dordrecht, The Netherlands, 2008.
4. Bardhan, K.; York, L.M.; Hasanuzzaman, M.; Parekh, V.; Jena, S.; Pandya, M.N. Can smart nutrient applications optimize the plant's hidden half to improve drought resistance? *Physiol. Plant.* **2021**, *172*, 1007–1015. [CrossRef]
5. Forde, B.; Lorenzo, H. The nutritional control of root development. *Plant Soil* **2002**, *232*, 51–68. [CrossRef]
6. Gebbers, R.; Adamchuk, V.I. Precision agriculture and food security. *Science* **2010**, *327*, 828–831. [CrossRef] [PubMed]
7. Vuran, M.C.; Salam, A.; Wong, R.; Irmak, S. Internet of underground things: Sensing and communications on the field for precision agriculture. In Proceedings of the 2018 IEEE 4th World Forum on Internet of Things (WF-IoT), Singapore, 5–8 February 2018. [CrossRef]
8. Grunwald, A.; Schaarschmidt, M.; Westerkamp, C. LoRaWAN in a rural context: Use cases and opportunities for agricultural businesses. In Proceedings of the Mobile Communication-Technologies and Applications; 24. ITG-Symposium, Osnabrueck, Germany, 15–16 May 2019; pp. 1–6.
9. Salam, A.; Vuran, M.C.; Irmak, S. Di-Sense: In situ real-time permittivity estimation and soil moisture sensing using wireless underground communications. *Comput. Netw.* **2019**, *151*, 31–41. [CrossRef]
10. Wohwe Sambo, D.; Forster, A.; Yenke, B.O.; Sarr, I.; Gueye, B.; Dayang, P. Wireless Underground Sensor Networks Path Loss Model for Precision Agriculture (WUSN-PLM). *IEEE Sens. J.* **2020**, *20*, 5298–5313. [CrossRef]
11. Barbedo, J. A Review on the Use of Unmanned Aerial Vehicles and Imaging Sensors for Monitoring and Assessing Plant Stresses. *Drones* **2019**, *3*, 40. [CrossRef]
12. Barnetson, J.; Phinn, S.; Scarth, P. Estimating Plant Pasture Biomass and Quality from UAV Imaging across Queensland's Rangelands. *AgriEngineering* **2020**, *2*, 523–543. [CrossRef]
13. Bukowiecki, J.; Rose, T.; Kage, H. Sentinel-2 Data for Precision Agriculture?-A UAV-Based Assessment. *Sensors* **2021**, *21*, 2861. [CrossRef]
14. Fan, D.; Su, X.; Weng, B.; Wang, T.; Yang, F. Research Progress on Remote Sensing Classification Methods for Farmland Vegetation. *AgriEngineering* **2021**, *3*, 971–989. [CrossRef]
15. Quino, J.; Maja, J.M.; Robbins, J.; Fernandez, R.T.; Owen, J.S.; Chappell, M. RFID and Drones: The Next Generation of Plant Inventory. *AgriEngineering* **2021**, *3*, 168–181. [CrossRef]
16. Marturano, F.; Martellucci, L.; Chierici, A.; Malizia, A.; Di Giovanni, D.; d'Errico, F.; Gaudio, P.; Ciparisse, J.-F. Numerical Fluid Dynamics Simulation for Drones' Chemical Detection. *Drones* **2021**, *5*, 69. [CrossRef]
17. Behjati, M.; Mohd Noh, A.B.; Alobaidy, H.A.H.; Zulkifley, M.A.; Nordin, R.; Abdullah, N.F. LoRa Communications as an Enabler for Internet of Drones towards Large-Scale Livestock Monitoring in Rural Farms. *Sensors* **2021**, *21*, 5044. [CrossRef]
18. Park, S.; Yun, S.; Kim, H.; Kwon, R.; Ganser, J.; Anthony, S. Forestry Monitoring System Using LoRa and Drone. In Proceedings of the 8th International Conference on Web Intelligence, Mining and Semantics, Novi Sad, Serbia, 25–27 June 2018. [CrossRef]
19. Caruso, A.; Chessa, S.; Escolar, S.; Barba, J.; Lopez, J.C. Collection of Data With Drones in Precision Agriculture: Analytical Model and LoRa Case Study. *IEEE Internet Things J.* **2021**, *8*, 16692–16704. [CrossRef]
20. Zhang, Z.; Zhou, C.; Sheng, L.; Cao, S. Optimization Schemes for UAV Data Collection with LoRa 2.4 GHz Technology in Remote Areas without Infrastructure. *Drones* **2022**, *6*, 173. [CrossRef]
21. Pan, Q.; Wen, X.; Lu, Z.; Li, L.; Jing, W. Dynamic Speed Control of Unmanned Aerial Vehicles for Data Collection under Internet of Things. *Sensors* **2018**, *18*, 3951. [CrossRef] [PubMed]
22. Zorbas, D.; O'Flynn, B. Collision-Free Sensor Data Collection using LoRaWAN and Drones. In Proceedings of the 2018 Global Information Infrastructure and Networking Symposium (GIIS), Thessaloniki, Greece, 23–25 October 2018; IEEE: Piscataway, NJ, USA, 2018.
23. Castellanos, G.; Deruyck, M.; Martens, L.; Joseph, W. System Assessment of WUSN Using NB-IoT UAV-Aided Networks in Potato Crops. *IEEE Access* **2020**, *8*, 56823–56836. [CrossRef]
24. Cariou, C.; Moiroux-Arvis, L.; Pinet, F.; Chanet, J.-P. Data Collection from Buried Sensor Nodes by Means of an Unmanned Aerial Vehicle. *Sensors* **2022**, *22*, 5926. [CrossRef]
25. Hossain, F.F.; Messenger, R.; Captain, G.L.; Ekin, S.; Jacob, J.D.; Taghvaeian, S.; O'Hara, J.F. Soil Moisture Monitoring Through UAS-Assisted Internet of Things LoRaWAN Wireless Underground Sensors. *IEEE Access* **2022**, *10*, 102107–102118. [CrossRef]
26. Titov, I.; Holtorf, L.; Daschner, F.; Gerken, M. Drone as LoRa® Repeater for Readout of Low-power Sensor Nodes in Precision Agriculture. In Proceedings of the EASS 2022; 11th GMM-Symposium, Erfurt, Germany, 5–6 July 2022; pp. 1–3.
27. Heltec Automation Technology Co., Ltd. WiFi LoRa 32 (V2): LoRa Node Development Kit. Available online: https://resource.heltec.cn/download/Manual%20Old/WiFi%20Lora32Manual.pdf (accessed on 15 January 2023).
28. DJI Technology Co., Ltd. Phantom 3 Advanced: User Manual V1.8. Available online: https://dl.djicdn.com/downloads/phantom_3/en/Phantom_3_Advanced_User_Manual_en_v1.8_160719.pdf (accessed on 15 January 2023).
29. Almalki, F.A.; Soufiene, B.O.; Alsamhi, S.H.; Sakli, H. A Low-Cost Platform for Environmental Smart Farming Monitoring System Based on IoT and UAVs. *Sustainability* **2021**, *13*, 5908. [CrossRef]

30. Böckmann, S.; Titov, I.; Gerken, M. Extraction of Soil Solution into a Microfluidic Chip. *AgriEngineering* **2021**, *3*, 783–796. [CrossRef]
31. Titov, I.; Köpke, M.; Gerken, M. Monolithic Integrated OLED–OPD Unit for Point-of-Need Nitrate Sensing. In *Optical Sensors and Sensing Congress 2022 (AIS, LACSEA, Sensors, ES), Proceedings of the Optical Sensors and Sensing Congress 2022 (AIS, LACSEA, Sensors, ES), Vancouver, BC, Canada, 11 July 2022*; Optica Publishing Group, Ed.; Optica Publishing Group: Washington, DC, USA, 2022.

Disclaimer/Publisher's Note: The statements, opinions and data contained in all publications are solely those of the individual author(s) and contributor(s) and not of MDPI and/or the editor(s). MDPI and/or the editor(s) disclaim responsibility for any injury to people or property resulting from any ideas, methods, instructions or products referred to in the content.

AgriEngineering

Article

Operational, Economic, and Environmental Assessment of an Agricultural Robot in Seeding and Weeding Operations

Mahdi Vahdanjoo [1,*], René Gislum [1] and Claus Aage Grøn Sørensen [2]

[1] Department of Agroecology, Aarhus University, Forsøgsvej 1, 4200 Slagelse, Denmark
[2] Department of Electrical and Computer Engineering, Aarhus University, Finlandsgade 22, 8200 Aarhus, Denmark
* Correspondence: mahdi.vahdanjoo@agro.au.dk

Abstract: The development of robotic-based agricultural machinery systems has significantly increased in recent years. Many autonomous systems have not yet been measured based on sustainability and economic performances, even though automation is regarded as an opportunity to increase safety, dependability, productivity, and efficiency. The operational aspect, economic viability, and environmental impact of replacing conventional machinery with robotized alternatives are the primary focus of this study. The robot considered in this research is designed for extensive fieldwork, where PTO and external hydraulics are required. This robot is equipped with two 75 (hp) Kubota diesel engines with a total engine gross power of up to 144 (hp). Both robotic system and conventional machinery were described, and different scenarios were used to examine various operational and environmental indicators, as well as individual cost elements, considering various field sizes and working widths of implements used in seeding and weeding operations. The findings demonstrate that the robotic system outperforms conventional machinery in terms of operational efficiency by as much as 9%. However, the effective field capacity comparison reveals that the conventional system has a field capacity that is up to 3.6 times greater than that of the robotic system. Additionally, the total cost per hour of the robotic system is up to 57% lower than that of the conventional system. The robotic system can save up to 63.3% of fuel during operation, resulting in the same percentage reduction in CO_2 emissions as the conventional system, according to a comparison of fuel consumption.

Keywords: economic assessment; environmental impact; operational management; precision agriculture; robotic system; sustainability

Citation: Vahdanjoo, M.; Gislum, R.; Sørensen, C.A.G. Operational, Economic, and Environmental Assessment of an Agricultural Robot in Seeding and Weeding Operations. *AgriEngineering* **2023**, *5*, 299–324. https://doi.org/10.3390/agriengineering5010020

Academic Editors: Muhammad Sultan, Yuguang Zhou, Redmond R. Shamshiri and Muhammad Imran

Received: 11 November 2022
Revised: 25 January 2023
Accepted: 27 January 2023
Published: 1 February 2023

Copyright: © 2023 by the authors. Licensee MDPI, Basel, Switzerland. This article is an open access article distributed under the terms and conditions of the Creative Commons Attribution (CC BY) license (https://creativecommons.org/licenses/by/4.0/).

1. Introduction

In recent years, the introduction of agricultural machinery systems based on robotic applications has increased significantly [1–3]. The main reason behind the automation of agricultural processes is the need for more efficiency. This can be achieved by reducing the operational time, reducing the energy required to perform machine operations, and increasing the crop yield. Moreover, recent advances in mechanical engineering, sensory perception, computing, and human interfacing are laying the foundation for the introduction and application of autonomous robotic systems.

Automation is seen as a chance to boost productivity, safety, reliability, and efficiency. However, the sustainable environmental and economic performances of many autonomous systems have yet to be measured [4]. There have been few studies on the economic and operational feasibility of using autonomous machines in agriculture. However, the economic, operational, and environmental aspects of robotic systems have not been the subject of comprehensive assessments [5]. Earlier studies such as Goense (2003) [6] indicate the positive feasibility of autonomous vehicles when utilized for up to 23 (h/day) by setting up scenarios for balanced cropping practices and cropping rotations for grains.

Have (2004) [7] showed the effects of automation on machinery sizes and costs for soil tillage and crop seeding operations including all machinery costs as well as timeliness costs. Results showed 20% higher investment, 80% lower labour requirement, and double the working hours, as well as showing that changing to autonomous vehicles would lower the machinery sizes and investment to 50–60% and the overall costs to 65%. Sørensen et al. (2005) [8] assessed the feasibility of a plant-nursing robot for weeding operations and found that profitability gains ranging from 20 to 50% are achievable through targeted applications. Specific studies for robotic seeding in sugar beets indicate a 7–10% cost reduction [9]. Lightweight robotic systems are expected to reduce soil compaction and its adverse effects on soil properties such as field readiness and soil workability which can cause poor drainage and increase surface runoff [10]. As a conclusion, indications are that applying an autonomous system would reduce the size, initial cost of machinery, and operation cost [11]. There are different types of agricultural robots that can automate seeding or weeding operations. Dahlia 4.3 is a lightweight and fully automatic robot that uses solar power as source of energy [12]. Farming-GT is another agricultural robot that is fully electrical and lightweight [13]. Farm Droid FD20 is another fully automatic and lightweight robot that is solar-powered [14]. Naio-OZ is another fully electrical and lightweight robot that can automate sowing and weeding operations [15]. For this research, the Agrointelli-Robotti [16] was used, which is a powerful field robot that uses diesel as a source of energy to accomplish power-intensive operations.

Seeding as the basic operation in crop production has a major effect on the productivity and environmental footprint of agriculture. Recent studies have shown a growing interest in the application of robotic systems in precision seeding operations. Improper tillage and seeding practices can cause extensive yield losses [9]. Bhimanpallewar & Narasingarao (2020) [17] presented an automatic seeding and fertilizer micro-dosing robot for farmers to perform precision farming. The designed robot can carry out the seeding and micro-dose fertilizing operations based on the type of seeds and plants, including quantifying the number of seeds and the amount of applied fertilizer. Moreover, the performance of this robot in terms of working time and energy consumption was analyzed in each soil profile for each type of mentioned seeds. However, no comprehensive evaluation of capacity and efficiency, and comparison with traditional methods were carried out. Neha S. Naik et al. (2016) [18] proposed a prototype of an autonomous agriculture robot that is specifically designed for seed operation. The designed robot is a four-wheeled vehicle that can perform efficient seed sowing at optimal depth and distances between crops and their rows. However, no analysis of different performance parameters for the designed robot was carried out. Sunitha et al. (2017) [19] designed an agricultural robot for plowing and seed-sowing operations in the field however, the detailed performance of the robot was not assessed.

There are three approaches to weeding: mechanical weeding, chemical weeding, or a combination of them. In mechanical weeding, the weed plants are removed by uprooting, cutting, and/or flaming. In chemical weeding, the weeds are removed by spraying herbicides on the plants. Recent studies have shown that weeding robots have received significant attention from researchers. Gonzalez-de-Soto et al. (2016) [20] presented an agricultural robotic solution for precise spraying. The designed system consists of an autonomous mobile robot based on a modified commercial tractor (for example equipped with high-tech perception and actuation systems), a real-time machine vision system that can detect weeds, and a rapid-response spraying system. Laboratory characterization and field tests revealed that the proposed system was reliable and can treat approximately 99.5% of the detected weeds while achieving significant herbicide savings. However, the author did not evaluate all the performance parameters of this robot from an agronomic, operational, and sustainability point of view, as well as not comparing conventional methods. Berenstein & Edan (2017) [21], presented a human–robot collaboration for site-specific spraying. The article provides details on the robotic platform design, the human–robot collaboration framework, and the tools used for the robotic sprayer to collaborate with a remote human operator for target detection and spraying. The field experiment proved the feasibility of human–robot collaboration for spraying specific targets. Moreover, the proposed system can help to remove humans from hazardous pesticide environments as well as to reduce the use of pesticides by up to 50% by minimizing the quantity of sprayed material while maintaining the crop yield. In the presented work, only a limited performance evaluation was carried out.

This paper proposes a targeted methodology for the evaluation of operational, economic, and environmental aspects of a robotic system in arable farming. This methodology is based on analyzing the basic unit tasks and processes for estimating the operational performance of the robotic system and traditional agricultural machinery systems.

In the following sections, a conventional machinery system and a robotic system are described and analyzed through varying scenarios, by considering different field sizes, and different working widths for the implements in seeding and weeding operations, followed by a breakdown of all costs and environmental impact assessment. The hypothesis is that this methodology can assess the operational aspect, economic feasibility, and environmental impact of replacing conventional practices with a respective robotized alternative.

2. Materials and Methods

2.1. The Robot

The analyzed robot (Figure 1) performs site-specific agricultural operations including seedbed preparation, seeding, hoeing, weeding, harrowing, soil sampling, spraying, and mowing. Version 150D of this robot is designed for extensive fieldwork, where PTO and external hydraulics are required for example in seeding operations. The mentioned version of the robot is equipped with two 75 (hp) Kubota diesel engines (total engines gross power up to 144 (hp)). One engine is dedicated to propulsion, the other runs the PTO and external hydraulic system. The forward speed of the robot in autonomous mode is up to 5 (km/h) and the high speed in manual mode is up to 10 (km/h). The approximate weight of the robot is 3100 (kg), and it has 2-wheel steering (able to do zero turns). The robot navigates precisely with Real-Time kinematic (RTK) GPS technology that has an accuracy of 2 (cm). The logging data collected from the control center of the robot were analyzed to measure the in-field operations of the robot [16].

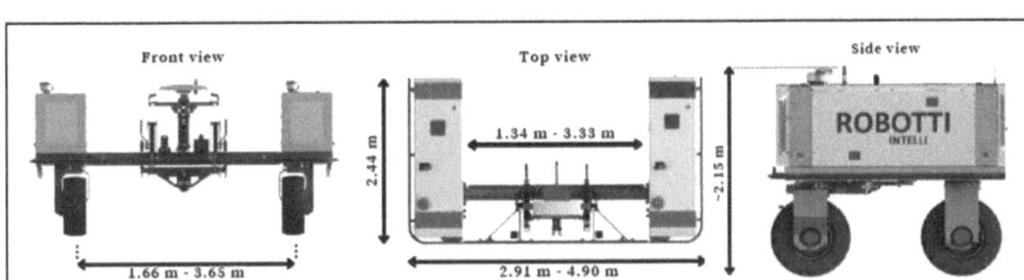

Figure 1. The robotic system.

2.2. *Monitoring, Measuring, and Analyzing of Robot's In-Field Operations*

By using the GPS antenna and sensors, it is possible to measure the in-field performance of the robot. The motions of the robot are decomposed into different task (time/distance) elements. The generated data can be aggregated into useful information for determining the fieldwork patterns, turning types, and evaluation of performance parameters such as field capacity, field efficiency, and so on. These data can be used for analyzing the current coverage method to identify the potential efficiency of the applied robotic system. This includes a comparison of robot vs. conventional method in terms of performance parameters. Figure 2 is the plotted GPS coordinates related to the location of the robot. During the operation, the GPS antenna of the robot sends the coordinates of the robot's location and based on the logging data from the robot (e.g., the mode of the robot:

manual driving, automatic and working, automatic and not working) we can plot with different colors to distinguish different task elements of the robot.

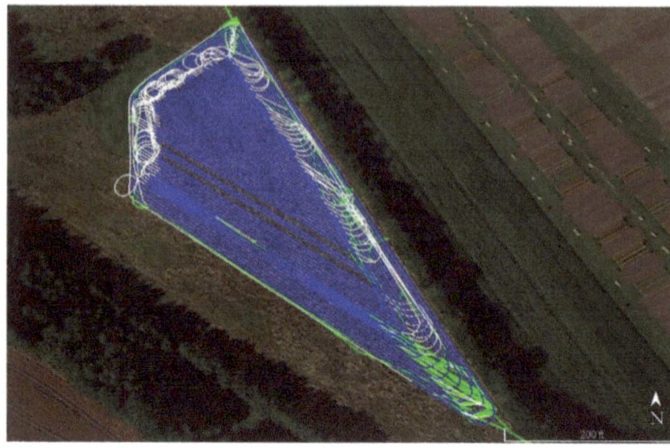

Figure 2. An example of analysis for the robot's in-field operation. "Manual driving": When the robot drives manually by the robot's operator (Green color). "Automatic and working": When the robot automatically performs the operation in the field (Blue color). "Automatic and not working": When the robot automatically moves but not performing the operation for example when it turns in the headland part (White color).

2.3. *Operational Factors*

In this study, specific operational indicators were considered, including field efficiency, effective field capacity, theoretical field capacity, energy consumption rate, and power requirements of the machine. These factors are important for evaluating the capacity and efficiency of an agricultural machine [22].

2.3.1. Field Efficiency

It is a comparison of the actual amount of work done by a machine compared to what it would do without any loss of time or capacity.

$$Field\ Efficiency\ (\%) = \frac{100 \times T_p}{T_t} \quad (1)$$

where T_p = productive time (actual working) and, T_t = total time (considering the static and idle time).

2.3.2. Effective Field Capacity

The effective field capacity is determined by considering the working speed, the working width of the implement, and the machine's field efficiency.

$$Effective\ Field\ Capacity\ (ha/h) = \frac{S \times W \times Field\ Efficiency}{C} \quad (2)$$

where S = working speed (km/h), W = working width of implement (m), and a constant $C = 10$.

2.3.3. Theoretical Field Capacity

The theoretical field capacity is determined by noting the theoretical work speed and the machine's theoretical working width.

$$Theoretical\ Field\ Capacity\ (ha/h) = V_t \times W_t \times K \quad (3)$$

where V_t = theoretical working speed, W_t = theoretical working width, and a constant $K = 0.1$.

2.4. Economic Factors

The total cost of an agricultural machine can be calculated as the summation of two main parts, the ownership cost, and the operational cost [3].

2.4.1. Ownership Costs

The costs of ownership include depreciation, interest cost, taxes, insurance, and housing facilities [3]. Depreciation is a type of expense that results from the obsolescence and age of a machine [23]. The degree of mechanical wear may cause a reduction in the initial value of a particular machine. Sometimes a major design change or emerging new technology can make an older machine suddenly obsolete, causing a sharp drop in its residual value [23]. However, the economic life and total hours of usage are usually considered as the main factors for calculating the residual value of a machine. To estimate the annual depreciation, the machine's age, and the salvage value at the end of the economic life must be specified. The economic life of a machine is the number of years for which costs should be estimated. The salvage value (SV) is assumed to be 10% of the machine's purchase price (PP) [3]

$$c_d = \frac{PP - SV}{Age} \tag{4}$$

where PP = purchase price, SV = Salvage value, and Age = economic life.

If the operator borrows money to purchase a machine, the lender sets the interest rate to charge. However, if the farmer uses his own capital, the rate will depend on the opportunity cost for that capital elsewhere in the farm business. If only part of the money is borrowed, an average of the two rates should be used. The average annual interest charge is computed by subtracting the trade-in or salvage value from the purchase price, multiplying this difference by the rate of interest, and dividing by 2 [23]. In this research, it is assumed that the interest rate i is equal to 9 percent.

$$Interest = \frac{PP - SV}{2} \times i \tag{5}$$

where i = interest rate.

The cost of taxes and insurance is usually much lower than depreciation and interest, but it should be considered. In this research, a cost estimate equal to 1.0 percent of the purchase price is considered as the cost of taxes and insurance [24].

Providing shelter, tools, and maintenance equipment leads to fewer repairs in the field and less damage to mechanical parts. In this research, an estimated charge of 1.0 percent of the purchase price is considered for housing costs [24]. The estimated costs of depreciation, interest, taxes, insurance, and housing are added together to find the total ownership cost.

2.4.2. Operational Costs

The operating cost of a system is the sum of maintenance costs, energy costs (i.e., fuel or electrical energy consumption), lubrication, labor costs, and farm-to-field transportation costs. For the calculation of the repair and maintenance cost for the conventional machinery system, the ASABE standards estimation process is applied [25]

$$C_{rm} = RF1 \times \left(\frac{h}{1000}\right)^{RF2} \times PP \tag{6}$$

where $RF1$ and $RF2$ are the repair and maintenance factors and h, expressed in hours, stands for the accumulated working hours of the machinery.

The performance rate of agricultural machines depends on the speed that can be achieved and the optimal use of time [25]. Field speeds may be limited by heavy yields, rough soil, and adequacy of operator control [25]. Small or irregularly shaped fields, heavy yields, and high-capacity machines may cause a substantial reduction in field efficiency [25]. Typical speeds and field efficiency for seeding operation are presented in Table 1.

Table 1. Operational data and repair factors for conventional machine types [25].

Machine	Field Efficiency		Field Speed		Estimated Life	Total Life R&M Cost	Repair Factors	
	Range %	Typical %	Range (km/h)	Typical (km/h)	H	% of List Price	RF1	RF2
Grain drill	55–80	70	6.5–11.0	8.0	1500	75	0.32	2.1
Tractors (4-wheel drive & crawler)					16,000	80	0.003	2.0

For the case of conventional machinery, the prediction of the energy cost, that is the engine fuel consumption during the operation, was based on the specific volumetric fuel consumption approach as it is described in Table 2.

Table 2. Average fuel consumption for different modes of tractor operation [26].

Average Fuel Consumption (John Deere 7250 R)			
Working width sowing machine (m)	3	4	6
Power demand (kW)	90	105	155
Mode	**Consumption (L/h)**		
Working mode (sowing)	13.82	15.64	23.08
Transportation mode	9.61	11.21	16.55
Static mode	4.30	4.30	4.30

In the standard ASABE and "Farm power and machinery management" [27], the required energy for pulling a Rod weeder is in the range of 1.5–4.8 (kW h/ha). Therefore, based on the "Nebraska OECD Tractor Test 2085–Summary 934, John Deere 7250r Diesel, E23 Transmission" [28], the amount of fuel consumption for the tractor in the weeding operation was considered 9.43 (L/h).

According to the standard ASAE D497.4 FEB03 [29], the amount of oil consumption in (L/h) can be estimated for tractor and robot based on their maximum PTOP. For the PTOP between 134–200, the amount of oil consumption for a diesel engine is 0.111 (L/h) and for PTOP over 200, the amount of oil consumption is equal to 0.135 (L/h).

In a robotic system, although the system performs autonomously, manual labor may require supervising the so-called residual tasks (such as system advancements, reconfigurations, and potential safety monitoring) [3]. In this research, for the labor cost of the robotic system, half of the hourly wage of the conventional system was considered. Finally, we could also consider the transportation cost for the robotic system in the case where the distance to the field is significant.

2.5. Environmental Factors (CO_2 Emission)

The amount of emission CO_2 gas during the seeding and weeding operation can be estimated based on the amount of consumed fuel by the tractor or robot. According to other studies [30–32], consuming 100 L of diesel fuel emits 275–376 (kg) of CO_2. Therefore, in this study, it is assumed that the consumption of 1 L of diesel fuel during the operation produces 2.75 (kg) of CO_2 gas related to the greenhouse effect.

2.6. *Comparing the Performance Parameters of a Robotic System with Conventional Agricultural Machinery*

2.6.1. Comparing the Draft Force Required to Pull an Implementation

The objective of this test was to find out how much energy it takes for an implement to be pulled by the robot. The results of this test were used to compare the energy requirements of the robot with the conventional machinery. An eight-directional string gauge device demonstrated in Figure 3 was mounted on the robot, and thereafter a Kongskilder seeder was mounted onto the string gauge device.

Figure 3. Mounted string gauge on the robot.

Two different implementations were mounted on the robot in this pulling test. Figures 4 and 5 demonstrate the applied seeder and weeder machines. The specification of these machines is listed as follow:

Figure 4. Grain drill seeder, 3 (m)/25 rows, 5 (km/h), 4 (cm) depth. The seeder was filled with weights in the center to represent the weight of the seed. Total weight 210 (kg) (4 × 30 (kg) + 2 × 45 (kg)).

Figure 5. Mechanical weeder (Rod weeder), 3 (m), 5 (km/h), 2 (cm) depth.

The test field was in Denmark and the robot drove the planned route automatically. Only four tracks were considered for this test. The length of the tracks is T1 = 131 (m), T2 = 141 (m), T3 = 170 (m), and T4 = 189 (m). Figure 6 demonstrates the test field.

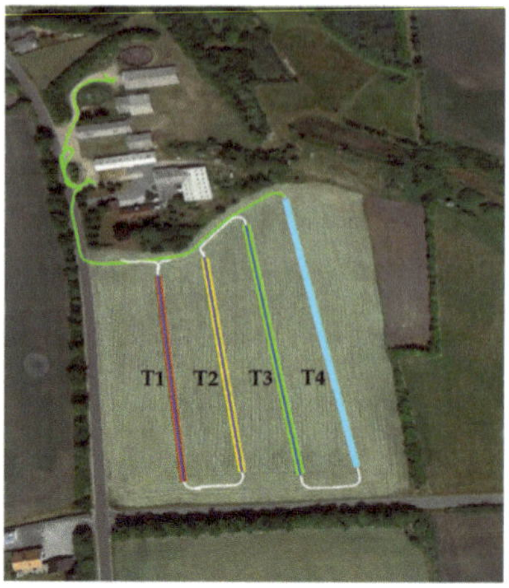

Figure 6. The sample field was used for the draft force test. T1, . . . , T4 represent the field work tracks.

The draft force (N) required to pull the implementation (in the horizontal direction) can be estimated based on Equation (7). The soil and machine parameters for the weeder and seeder were presented in Table 3.

$$Draft\ (N) = F_i\left[A + (B \times S) + \left(C \times S^2\right)\right]WT \quad (7)$$

F_i = dimensionless soil texture adjustment parameter; i = 1 for fine, 2 for medium, and 3 for coarse-textured soils; A, B, and C = machine-specific parameters; S = field speed (km/h); W = machine width (m); T = tillage depth (cm) for major tools. The value of T for minor tillage tools and seeding implements is equal to 1 (dimensionless).

Table 3. Machine and soil parameters for rod weeder and grain drill [25].

Implement	Width	Machine Parameters			Soil Parameters			Range ±%
		A	B	C	F_1	F_2	F_3	
				Major Tillage Tools				
Rod Weeder	m	210	10.7	0.0	1.0	0.85	0.65	25
				Minor Tillage Tools				
Grain Drill	rows	720	0.0	0.0	1.0	0.92	0.79	35

2.6.2. Case Study Description

A case study is presented to show how to apply the methodological approach in this study. The purpose of the paper is to evaluate the operational, economical, and environmental aspects of the robotic system and compare them with conventional machinery. Different scenarios were defined based on the size of the fields (less than 1 (ha), between 1 to 10 (ha), and more than 10 (ha)), type of operations (seeding and weeding), and different working widths for the conventional machinery.

Table 4 shows how to calculate the ownership and operational costs for the robotic and conventional systems in seeding operations. As an example, the purchase price of the robot with seeding attachment sets to EUR 144,500, then the salvage value is 10% of the purchase price which is equal to EUR 14,450. The economic life of the robot is assumed to be 6000 h, equal to 10 years. The average interest rate is assumed to be equal to 9%. Depreciation can be evaluated based on the formula (4). The amount of interest can be calculated by the formula (5). The cost of insurance and taxes as well as the cost of housing assumed to be 1% of the purchase price. Then the ownership cost is equal to the summation of depreciation, interest, insurance and taxes, and housing costs. When the field is too far, a trailer is going to use for the transportation of the robot to the field. In this case study, the distance to the fields is short and the transportation cost was not considered.

Table 4. The calculation of ownership costs for robotic and conventional machinery in seeding operation.

		Robot-Working Width 3 [m]	Tractor-Working Width 3 [m]	Tractor-Working Width 4 [m]	Tractor-Working Width 6 [m]
Ownership costs parameters	Purchase price (EUR)	144,500	269,200	273,200	280,700
	Salvage value (EUR)	14,450	26,920	27,320	28,070
	Economical life (years)	10	15	15	15
	Average interest rate (%)	9	9	9	9
	Depreciation (EUR/year)	13,005	16,152	16,392	16,842
	Interest (EUR/year)	5852	10,903	11,065	11,368
	Insurance and Taxes (EUR/year)	1445	2692	2732	2807
	Housing (EUR/year)	1445	2692	2732	2807
Ownership costs (EUR/year)		21,747	32,439	32,921	33,824
Economical life (h)		6000	16,000	16,000	16,000
Ownership costs (EUR/h)		3.62	2.03	2.06	2.11

In the case of conventional machinery, a John Deere tractor model (7R250) was considered as an example to show how to calculate the total cost for conventional machinery. For example, in the case of "tractor-working width 3 [m]" by setting the purchase price of the tractor with seeding attachment equal to 269,200 (EUR), then the salvage value is 10% of that, which is equal to EUR 26,920. The economic life of the tractor is assumed to be 16,000 h, equal to 15 years (each year is 1067 (h)) [33]. The average interest rate is assumed to be equal to 9%. The depreciation value and the amount of interest can be calculated by Equations (4) and (5) respectively. The cost of insurance and taxes as well as the cost of housing assumed to be 1% of the purchase price. Then the ownership cost is equal to the summation of depreciation, interest, insurance and taxes, and housing costs.

Table 5 shows how to calculate the ownership and operational costs for the robotic and conventional systems in weeding operations. The same procedure as above was applied by considering the price of weeding attachments in each case.

Table 5. The calculation of ownership costs for robotic and conventional machinery in weeding operation.

		Robot-Working Width 2.4 [m]	Tractor-Working Width 2.4 [m]	Tractor-Working Width 4 [m]	Tractor-Working Width 6 [m]
Ownership costs parameters	Purchase price (EUR)	144,200	268,900	275,700	284,600
	Salvage value (EUR)	14,420	26,890	27,570	28,460
	Economical life (years)	10	15	15	15
	Average interest rate (%)	9	9	9	9
	Depreciation (EUR/year)	12,978	16,134	16,542	17,076
	Interest (EUR/year)	5840	10,890	11,166	11,526
	Insurance & Taxes (EUR/year)	1442	2689	2757	2846
	Housing (EUR/year)	1442	2689	2757	2846
Ownership costs (EUR/year)		21,702	32,402	33,222	34,294
Economical life (h)		6000	16,000	16,000	16,000
Ownership costs (EUR/h)		3.62	2.03	2.08	2.14

3. Results and Discussion

3.1. The Result of Draft Force Test

The average of the horizontal force during the seeding operation was plotted in Figure 7 and the required energy was calculated as follows:

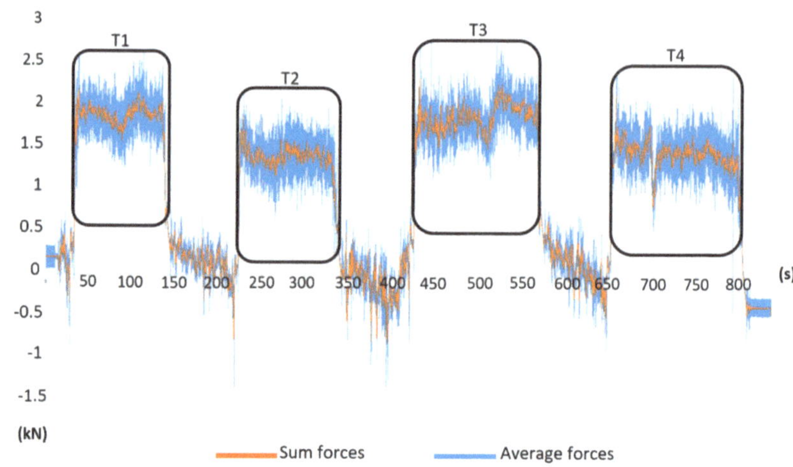

Figure 7. Plotted average horizontal force during the seeding operation. Vertical axis represents the forces (kN) and horizontal axis shows the time (s).

The required average power for covering each track can be calculated based on this formula, Power = Force × Distance/Time, therefore, for T1 the amount of power is equal to 2.223 (kW). To calculate the amount of (kW h), the following formula was used: kW hr = Power (kW) × Time (h). For example, in Seeding/T1, the (kW h) can be calculated as follow: kW hr = 2.223 × (109/3600) = 0.067 (kW h). The amount of covered area for each track can be calculated as follow: Work area = working width × track length. For example, in Seeding/T1: Work area = 3 (m) × 131 (m) = 393 (m^2) = 0.0393 (ha) (The working width of the implementations was 3 m). Finally, the amount of kW h/ha for Seeding/T1 can be calculated as follows: kW h/ha = 0.067 (kW h) × 1/0.0393(ha) = 1.7 (kW h/ha). Table 6 shows the result of the draft force test in the seeding operation.

Table 6. Calculation of required energy for pulling the seeder.

Calculation of Required Energy for Pulling the Seeder					
	Seeding/T1	**Seeding/T2**	**Seeding/T3**	**Seeding/T4**	**Average**
The average horizontal force (kN)	1.85	1.39	1.87	1.44	1.64
Travel time (s)	109	114	143	150	129
Traveled distance (track length) (m)	131	141	170	189	157.75
kW h	0.067	0.055	0.088	0.076	0.072
kW h/ha	1.70	1.30	1.73	1.33	1.52

Figure 7 shows the average horizontal force during the seeding operation. T_i, $i = 1, \ldots 4$ represents the track (working area) ID. The plotted forces in the boxes related to the horizontal force during the automatic performance of robot. The other horizontal forces related to the driving or turning in the headland part.

The same calculation as the seeder was considered for the mechanical weeding operation. The average horizontal force during the weeding operation was plotted in Figure 8 and the results were summarized in Table 7.

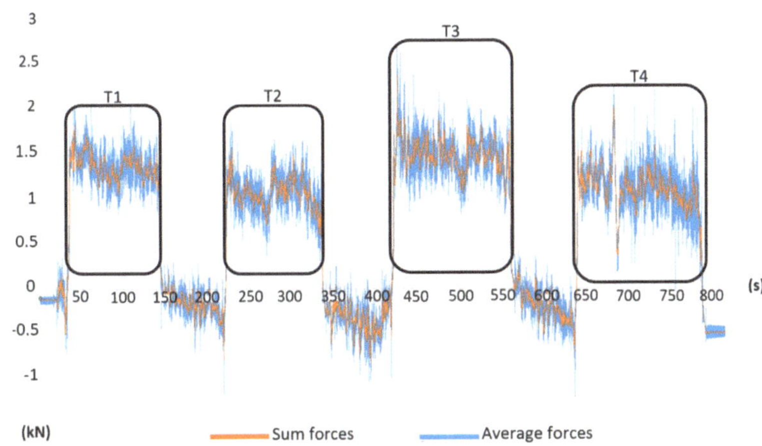

Figure 8. Plotted average of horizontal force (kN) during the weeding operation. Vertical axis represents the forces (kN) and horizontal axis shows the time (s).

Table 7. The results of the calculation of required energy for the weeding operation.

Calculation of Required Energy for Pulling the Weeder					
	Weeder/T1	**Weeder/T2**	**Weeder/T3**	**Weeder/T4**	**Average**
The average horizontal force (kN)	1.38	1.07	1.50	1.12	1.27
Travel time (s)	106	115	139	148	127
Traveled distance (track length) (m)	131	141	170	189	157.75
kW h	0.050	0.042	0.071	0.059	0.056
kW h/ha	1.28	0.99	1.39	1.04	1.18

Like Figure 7, Figure 8 also shows the average horizontal force during the weeding operation.

Based on the standard ASABE [25,27], we can see that for example for seeder (Grain drill) the required (kW h/ha) for pulling the implement is in this range (1.1–3.9) and for weeder (Rod weeder) is (1.5–4.8). The comparison shows that the required energy to pull the implementation with the robot is in the lower range than the required energy from the tractor.

3.2. Seeding Operation

In the first scenario, a test field (Figure 9) with a size of 0.56 (ha) (less than 1 hectare) was selected for the seeding operation. The driving was connected to route planning and the operation was executed automatically. The duration of operation was 67 min, the amount of automatic working time was 36 min, and the rest was automatic turning time.

Table 8 shows the result of in-field analysis for the robotic system and conventional machinery. The field efficiency and effective field capacity can be calculated based on formulas 1 and 2 respectively. The amount of fuel consumption for the tractor in different modes can be evaluated based on the data from Table 2. The amount of oil consumption for the robot can be calculated by assuming the consumption rate equal to 0.111 (L/h) and for the tractor, it can be estimated by assuming the consumption rate equal to 0.135 (L/h) [27]. The energy cost can be calculated by multiplying the fuel consumption by the fuel cost. The lubrication cost can be calculated in the same way as the energy cost. The price of diesel and lubricant is assumed to be EUR 2 and EUR 0.3 respectively. The labour cost for the robot is assumed to be EUR 14.8/h and for the tractor assumed to be double this amount and equal to EUR 28.6/h. The repair and maintenance cost for the robot is equal to EUR 7009 per year and by dividing it by 600 h, it is equal to EUR 11.7/h. The repair and maintenance cost for the tractor can be calculated based on the information from Table 1.

Then the operational cost is the summation of all the previous costs (energy, lubrication, labour, repair, and maintenance). In the end, if the ownership cost is divided by 6000 h for the robot and 16,000 h for the tractor, then it is going to have the same unit as operational cost (EUR/h) and the summation of operational and ownership costs will show the total cost. The amount of CO_2 emission can be estimated by multiplying 2.75 (kg CO_2) by the amount of fuel consumption.

Figure 9. Seeding operation with the robot in a small triangle field.

Table 8. The results of in-field operation for the first scenario for the robot and the tractor with different working widths.

	Robot-Working Width 3 [m]	Tractor-Working Width 3 [m]	Tractor-Working Width 4 [m]	Tractor-Working Width 6 [m]
Field working time [min]	36	15.4	11.8	8
Non-working time [min]	31	13.7	10.6	7.12
Static time [min]	0	0.95	0.34	0.34
Total time [min]	67	30.05	22.74	15.46
Field efficiency [%]	53.7	51.2	51.9	51.7
Effective field capacity [ha/h]	0.81	1.36	1.84	2.92
Fuel consumption [L]	3.5	5.82	5.12	5.06
Oil consumption [L]	0.12	0.07	0.05	0.03
Energy cost [EUR]	7	11.64	10.24	10.12
Lubrication cost [EUR]	0.04	0.02	0.015	0.01
Labour cost [EUR]	16.53	14.82	11.22	7.63
Repair and Maintenance [EUR]	13.1	6.76	5.2	3.62
Operational costs [EUR]	36.67	33.24	26.68	21.38
Operational costs [EUR/h]	32.84	66.37	70.4	83
Ownership costs [EUR/h]	3.62	2.03	2.06	2.11
Total cost per hour [EUR/h]	36.46	68.4	72.46	85.11
Total cost [EUR]	40.71	34.26	27.46	21.93
CO_2 emission [kg]	9.63	16	14.08	13.92

Figures 10 and 11 demonstrate the comparison of operational factors between the robot and the tractor. The results show that the robot has slightly better field efficiency than the tractor. However, the tractor has a higher effective field capacity than the robot.

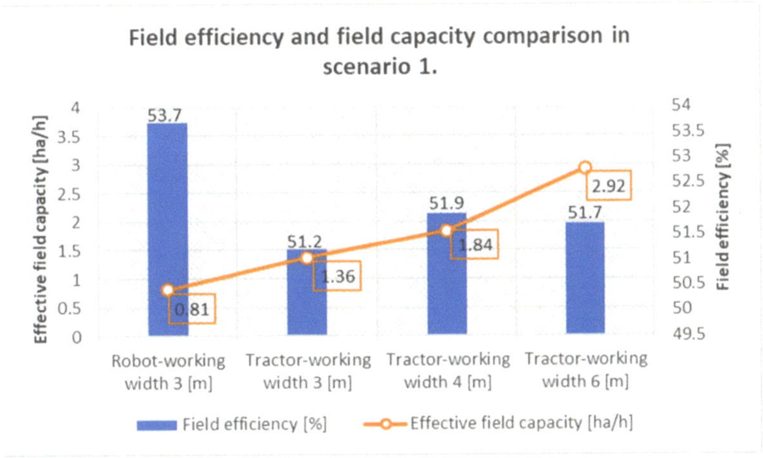

Figure 10. Comparison of field efficiency and field capacity for robot and tractor in scenario 1.

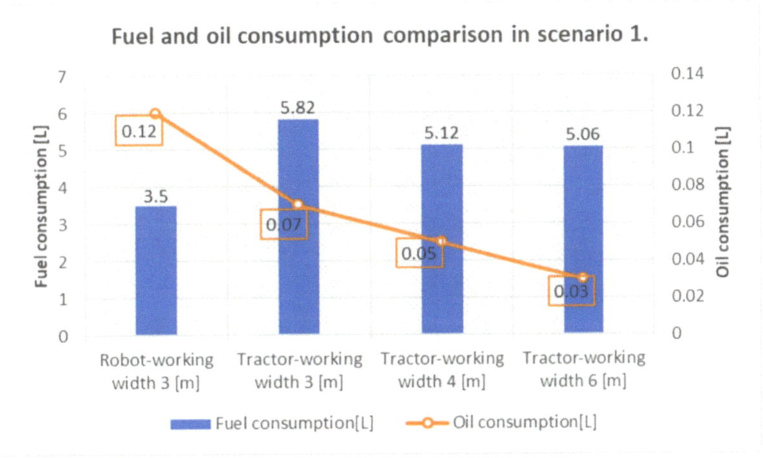

Figure 11. Comparison of fuel and oil consumption for robot and tractor in scenario 1.

Figure 11 shows that the robot has less fuel consumption than the tractor and it can save up to 39.8% of fuel till the end of the operation. The results also show that the robot has more oil consumption than the tractor.

Based on Figure 12, the comparison of operational costs shows that the robot has up to 60% less operational costs per hour than the tractor. A comparison of total costs shows that the robotic solution has up to 57% less total cost per hour than the tractor. Considering the duration of this operation, we can calculate the total cost of the operation. The results show that the tractor with 6-m working width has up to 46% less total cost than the robot. Moreover, the comparison of CO_2 emission reveals that the robot emits up to 39.8% less CO_2 than the tractor in this operation.

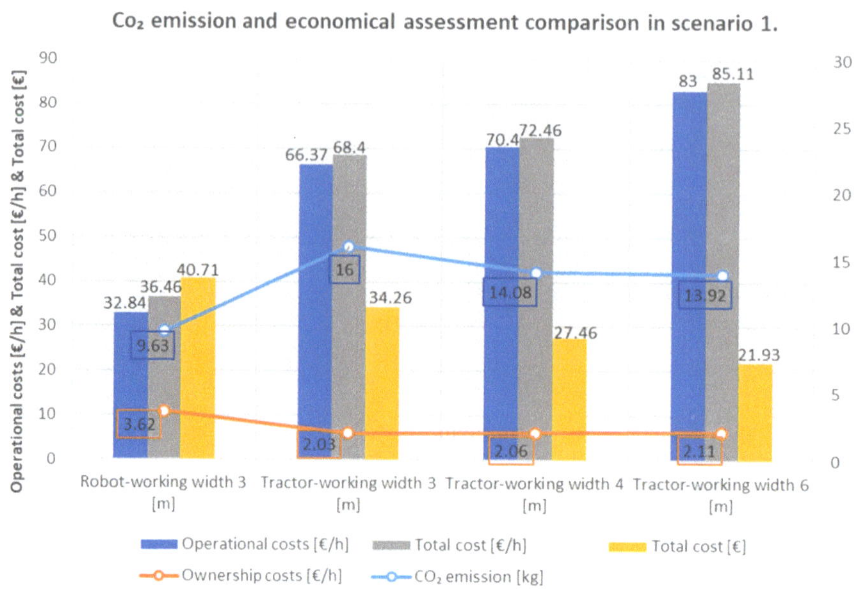

Figure 12. The comparison of operational and total cost and CO_2 emission.

In the second scenario, a sample field (Figure 13) with a size of fewer than 10 hectares was selected for seeding operation with the robot. The robot followed the initial plan generated by the route planning and the whole operation was accomplished automatically.

Figure 13. A sample field with a size equal to 7.5 (ha) was used for the seeding operation with the robot.

Table 9 shows the result of in-field operation for the robot and the tractor with different sizes of the implementation. The procedure for calculating the operational factors was the same as described in Table 8.

Table 9. The results of in-field operation for the second scenario for the robot and the tractor with different working widths.

	Robot-Working Width 3 [m]	Tractor-Working Width 3 [m]	Tractor-Working Width 4 [m]	Tractor-Working Width 6 [m]
Field working time [min]	167	167.2	126.1	83.6
Non-working time [min]	70	80.3	68.9	43.9
Static time [min]	9	12.75	4.5	4.5
Total time [min]	246	260.25	199.5	132
Field efficiency [%]	67.9	64.2	63.2	63
Effective field capacity [ha/h]	1.63	1.73	2.27	3.37
Fuel consumption [L]	43.5	52.28	46.06	44.6
Oil consumption [L]	0.46	0.59	0.45	0.3
Energy cost [EUR]	87	104.56	92.12	89.2
Lubrication cost [EUR]	0.138	0.177	0.135	0.09
Labour cost [EUR]	60.68	128.4	98.42	65.12
Repair and Maintenance [EUR]	47.97	58.56	45.55	30.89
Operational costs [EUR]	195.79	291.7	236.23	185.3
Operational costs [EUR/h]	47.75	67.25	71.05	84.23
Ownership costs [EUR/h]	3.62	2.03	2.06	2.11
Total cost per hour [EUR/h]	51.37	69.28	73.11	86.34
Total cost [EUR]	210.62	300.5	243.09	189.95
CO_2 emission [kg]	119.63	143.77	126.67	122.65

Figure 14 shows the comparison of field efficiency and field capacity between the robot and tractor in scenario 2. Based on the results, the robot has up to 7% better field efficiency than the tractor in this sample field. The comparison of effective field capacity shows that tractor has a higher field capacity than the robot.

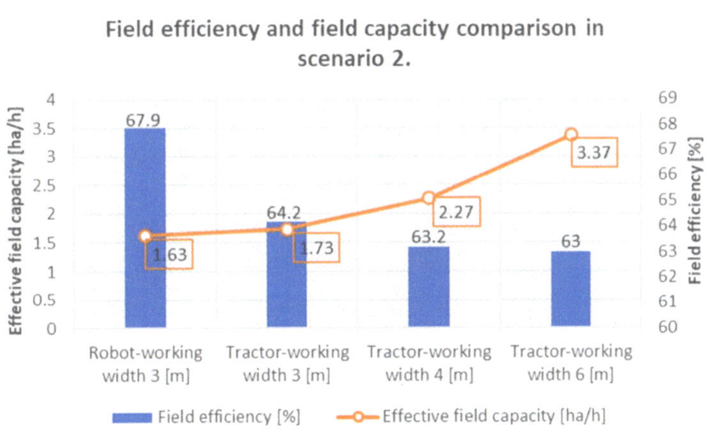

Figure 14. The comparison of field efficiency and field capacity for the robot and tractor in the second scenario.

According to Figure 15, the comparison of fuel consumption shows that the robot has up to 16.8% less fuel consumption than the tractor. Regarding oil consumption, the robot consumes less oil than the tractor with the same size of implementation. However, the tractor with 6 m size implementation, has less oil consumption than the robot.

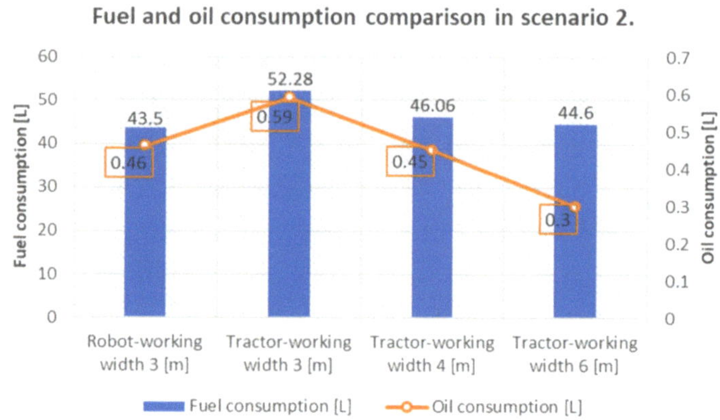

Figure 15. The comparison of fuel and oil consumption between robot and tractor in the second scenario.

Figure 16 demonstrates that in this sample field, the robot has up to 43.3% less operational costs per hour than the tractor with a different size of implementation. The comparison of total cost shows that the robot has up to 40% less total cost per hour than the tractor. The comparison of the total cost of operation shows that the robot has up to 30% less cost than the tractor with 3 and 4 m of working width. The total cost of operation for the tractor with 6-m working width is 10% less than the robot. Moreover, the comparison of CO_2 emission shows that the robot has up to 16.8% less CO_2 emission than the tractor in this operation.

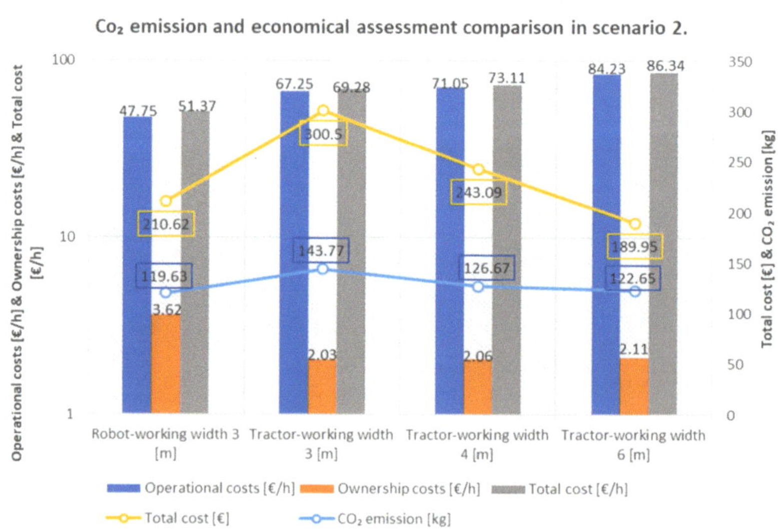

Figure 16. The comparison of economical assessment and CO_2 emission between robot and tractor in the second scenario.

In the third scenario, a sample field (Figure 17) with a size of more than 10 hectares was selected for seeding operation with the robot. The robot followed the initial plan generated by the route planning and all the operation was accomplished automatically.

Figure 17. The sample field in the third scenario with a size equal to 21 ha.

Table 10 represents the result of in-field operation for the robot and the tractor with different sizes of implementation.

Table 10. The results of in-field operation for the third scenario for the robot and the tractor with different working widths.

	Robot-Working Width 3 [m]	Tractor-Working Width 3 [m]	Tractor-Working Width 4 [m]	Tractor-Working Width 6 [m]
Field working time [min]	514.1	479.5	399.3	267.4
Non-working time [min]	89.7	117.4	101	69
Static time [min]	118.2	96.1	70.4	42
Total time [min]	722	693	571	378
Field efficiency [%]	71.2	69.2	69.9	70.7
Effective field capacity [ha/h]	1.71	1.87	2.52	3.82
Fuel consumption [L]	121.1	136.14	128	124.9
Oil consumption [L]	1.34	1.56	1.28	0.85
Energy cost [EUR]	242.2	272.28	256	249.8
Lubrication cost [EUR]	0.4	0.47	0.38	0.255
Labour cost [EUR]	178.1	341.9	281.7	186.48
Repair and Maintenance [EUR]	140.79	155.93	130.38	88.45
Operational costs [EUR]	561.5	770.6	668.5	525
Operational costs [EUR/h]	46.66	66.72	70.24	83.33
Ownership costs [EUR/h]	3.62	2.03	2.06	2.11
Total cost per hour [EUR/h]	50.28	68.75	72.3	85.44
Total cost [EUR]	605.04	794.06	688.06	538.27
CO_2 emission [kg]	333.03	374.38	352	343.48

Based on Figure 18, the comparison of field efficiencies shows that the robot has up to 3% better field efficiency than the tractor. The tractor has higher effective field capacity than the robot.

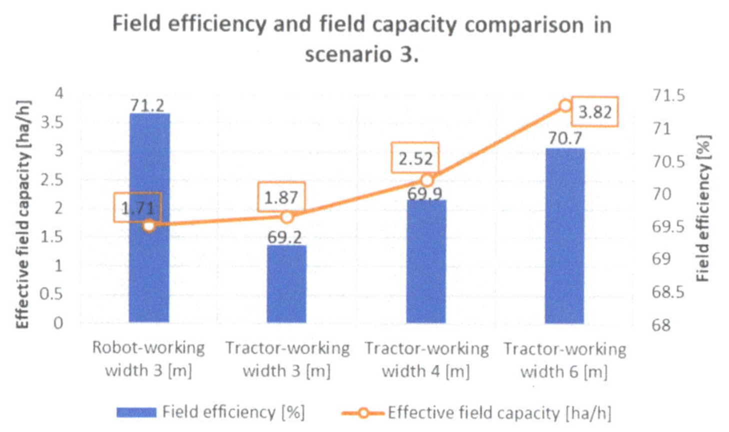

Figure 18. The comparison of field efficiency and field capacity for robot and tractor in scenario 3.

According to Figure 19, the robot has up to 11% less fuel consumption than the tractor. Regarding oil consumption, the robot consumes less oil than the tractor with the same size of implementation but, the tractor with 6 m size of implementation, shows less oil consumption than the robot.

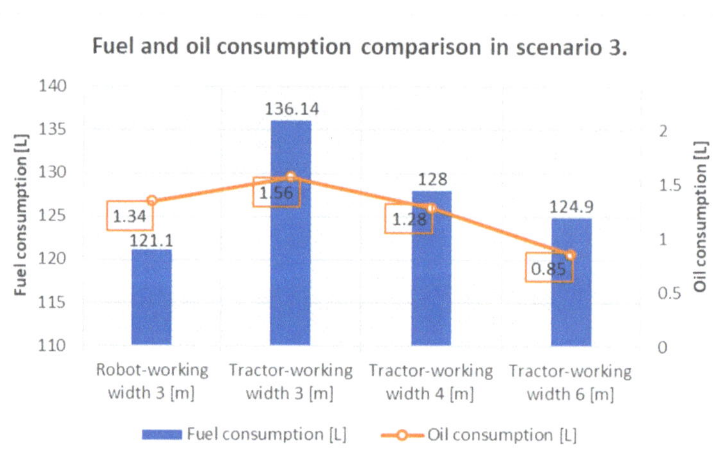

Figure 19. The comparison of fuel and oil consumption between robot and tractor in scenario 3.

Figure 20 shows that the robot has up to 44% less operational costs per hour than the tractor. The comparison of total cost shows that the robot has up to 41% less total cost per hour than the tractor. The comparison of the total cost of the operation shows that the robot has up to 28% less cost than the tractor with a 3–4 m working width. The tractor with 6 m working width has 11% less cost than the robot. Moreover, the comparison of CO_2 emission shows that the robot has up to 11% less CO_2 emission than the tractor.

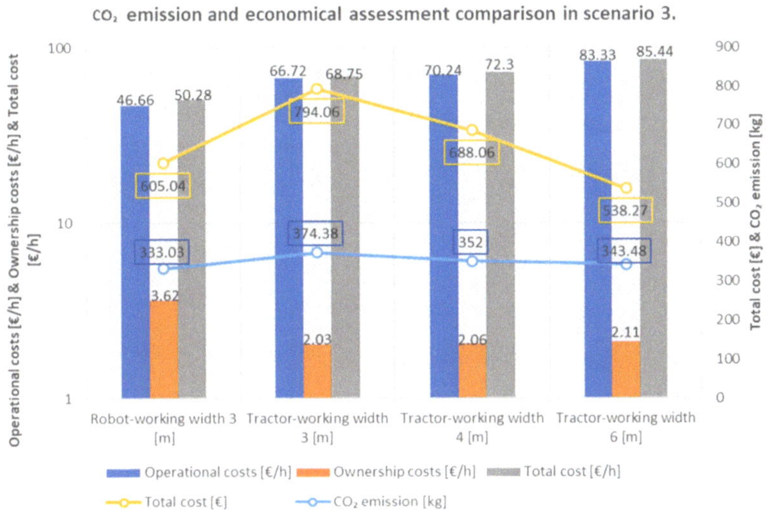

Figure 20. The comparison of economical assessment and CO_2 emission between robot and tractor in the third scenario.

3.3. Weeding Operation

In the fourth scenario, the sample field in the first scenario (Figure 21) with a size equal to 0.56 (ha) was selected for the weeding operation. The driving was connected to route planning and the operation was accomplished automatically and the duration of the operation was almost 76 min.

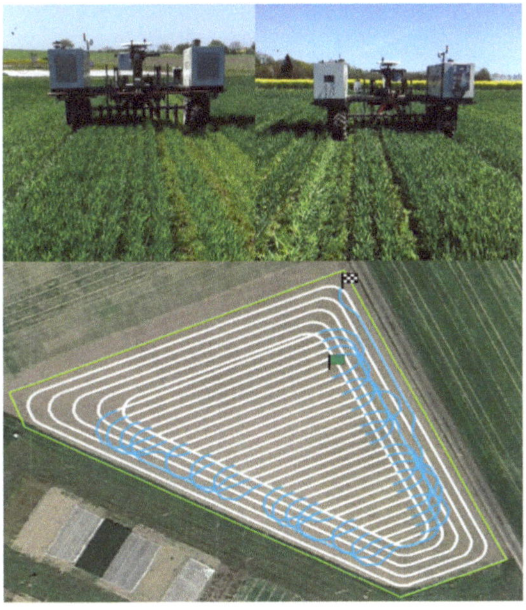

Figure 21. Weeding operation with the robot in a triangle shape field.

Table 11 represents the result of in-field operation for the robot and the tractor with different sizes of implementation. In this operation, the working speed of the robot was set to 3.5 (km/h) and the travel speed of the robot was set to 5 (km/h). The working speed of the tractor was 5 (km/h) with the travel speed equal to 7 (km/h).

Table 11. The result of in-field operation for the robot and the tractor with different sizes of implementation in weeding operation.

	Robot-Working Width 2.4 [m]	Tractor-Working Width 2.4 [m]	Tractor-Working Width 4 [m]	Tractor-Working Width 6 [m]
Field working time [min]	43.23	26.5	14.8	9.98
Non-working time [min]	31.32	14.3	9.2	4.6
Static time [min]	1	6	4.65	4.65
Total time [min]	75.55	46.8	28.65	19.23
Field efficiency [%]	57.2	56.6	51.7	51.9
Effective field capacity [ha/h]	0.48	0.67	1.03	1.56
Fuel consumption [L]	2.7	7.36	4.5	3.02
Oil consumption [L]	0.14	0.11	0.06	0.04
Energy cost [EUR]	5.4	14.72	9	6.04
Lubrication cost [EUR]	0.04	0.03	0.02	0.01
Labour cost [EUR]	18.64	23.1	14.13	9.5
Repair and Maintenance [EUR]	14.73	10.5	6.6	4.6
Operational costs [EUR]	38.81	48.35	29.75	20.15
Operational costs [EUR/h]	30.82	62	62.3	62.9
Ownership costs [EUR/h]	3.62	2.03	2.08	2.14
Total cost per hour [EUR/h]	34.44	64.03	64.38	65.04
Total cost [EUR]	43.37	49.94	30.74	20.85
CO_2 emission [kg]	7.43	20.24	12.37	8.31

Figure 22 represents the comparison of operational factors between the robot and the tractor. The comparison of field efficiency shows that the robot has up to 9.6% better field efficiency than the tractor. However, the tractor has a higher effective field capacity than the robot.

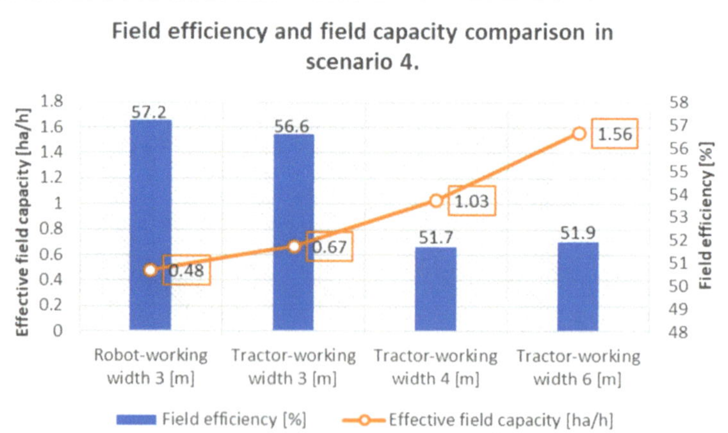

Figure 22. The comparison of field efficiency and field capacity between the robot and tractor in scenario 4.

Figure 23 shows the comparison of fuel and oil consumption between robot and tractor. The comparison shows that the robot has up to 63.3% less consumption than the tractor with different working widths of implementation. Moreover, the robot consumes more oil than the tractor in this case.

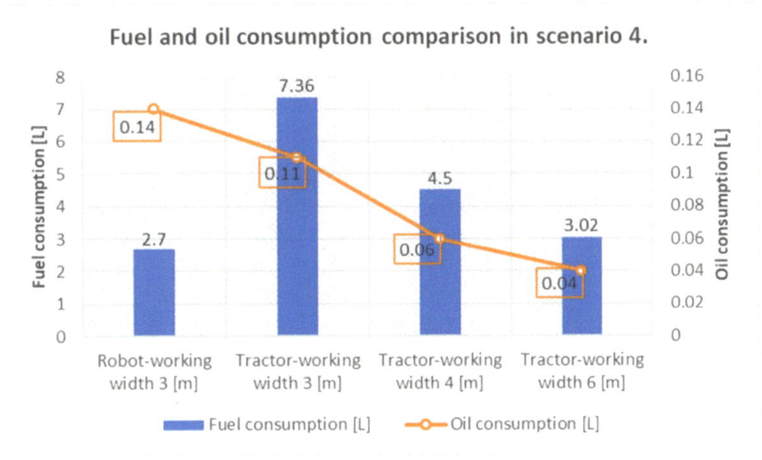

Figure 23. The comparison of fuel and oil consumption for the robot and the tractor in weeding operation.

Based on Figure 24, the comparison of operational costs shows that the robot has up to 51% less operational cost per hour than the tractor. A comparison of total costs per hour shows that the robot has up to 47% less total cost per hour than the tractor. The results shows that the tractor with 6 m working width has the lowest total cost. Moreover, the comparison of CO_2 emission shows that the robot has up to 63% less CO_2 emission than the tractor in weeding operation.

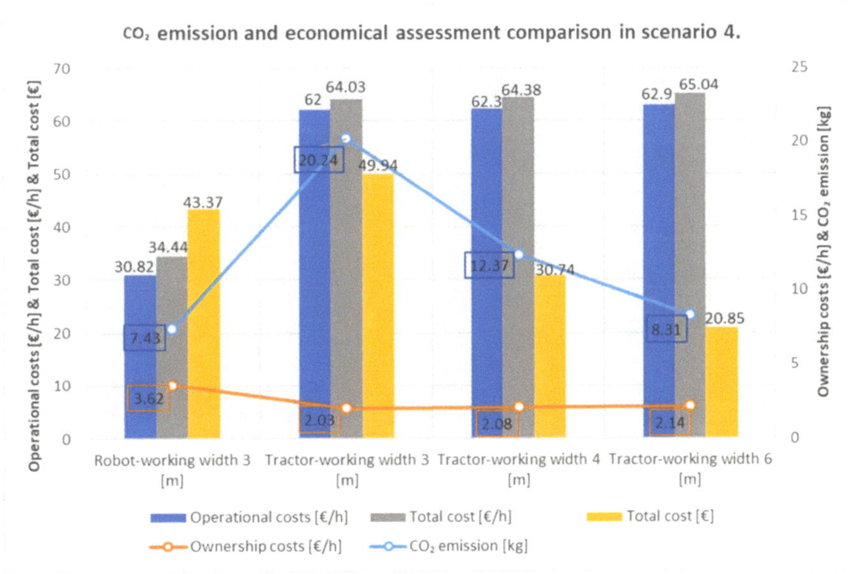

Figure 24. The comparison of economical assessment and CO_2 emission between robot and tractor.

The results show that in all the scenarios, the robot has less effective field capacity than the tractor since the tractor has a higher working speed and higher working width. The field speed of the tractor in seeding operation is up to 11 (km/h) and for the robot, it is up to 8 (km/h). Therefore, the tractor can accomplish the operation faster than the robot. The theoretical field capacity of this robot is 2.4 (ha/h), however, the working speed of the robot is varying based on the type of operation and the condition of the soil.

The shape of the agricultural fields is another important factor that can affect the operation of the robot. The robot has better field efficiency in the fields with long working areas (tracks) such as the sample fields in scenarios 2 and 3.

The size of the field is another factor that can affect the operation of the robot. The results show that by increasing the size of the field the efficiency of the robot is going to increase. One advantage of robots is that they can work continuously for a longer time and this benefit of robots can better show in a large field where the duration of the operation is longer. In the third scenario where the operation takes a longer time, we consider 40 min as lunchbreak for the operator of the tractor. Table 12 shows the result of this change in the operational and economic factors.

Table 12. The results of in-field operation in the third scenario for the robot and the tractor by considering 40 min break time for the tractor.

	Robot-Working Width 3 [m]	Tractor-Working Width 3 [m]	Tractor-Working Width 4 [m]	Tractor-Working Width 6 [m]
Field working time [min]	514.1	479.5	399.3	267.4
Non-working time [min]	89.7	117.4	101	69
Static time [min]	118.2	96.1	70.4	42
Idle time [min]	-	40	40	40
Total time [min]	722	733	611	418
Field efficiency [%]	71.2	65.4	65.3	64
Effective field capacity [ha/h]	1.71	1.77	2.35	3.46
Fuel consumption [L]	121.1	136.14	128	124.9
Oil consumption [L]	1.34	1.56	1.28	0.85
Energy cost [EUR]	242.2	272.28	256	249.8
Lubrication cost [EUR]	0.4	0.47	0.38	0.255
Labour cost [EUR]	178.1	349.4	291.2	199.25
Repair and Maintenance EUR]	140.79	164.93	139.5	97.8
Operational costs [EUR]	561.5	787.1	687.1	547.11
Operational costs [EUR/h]	46.66	64.43	67.47	78.53
Ownership costs [EUR/h]	3.62	2.03	2.06	2.11
Total cost per hour [EUR/h]	50.28	66.46	69.53	80.64
Total cost [EUR]	604.2	811.92	708.05	561.8
CO_2 emission [kg]	333.03	374.38	352	343.48

The results show that the field efficiency and field capacity for tractor with working widths 3, 4, and 6 m were reduced by 5%, 6%, and 9% respectively. The labour cost, repair and maintenance cost, operational cost, and the total cost of the seeding operation were increased by up to 9%.

The weight of agricultural machinery as an important factor can affect the amount of compaction on the soil. The weight of this robot is 3100 (kg), and the tractor (John Deere 7R250) is 12,058 (kg). The comparison shows that the weight of the tractor is almost four times higher than the robot. The higher weight of implementation can also put more stress on the topsoil layer which can cause more soil compaction [34,35].

One of the benefits of agricultural robots is that they can operate instead of a human in a difficult situation and accomplish repetitive tasks. For example, when a lot of hand weeding is required, this job can put too much effort and stress on workers. Applying robotic solutions in these types of field tasks can reduce the total operational time as well as being able to reduce the workload of workers and increase the leisure time on family farms [8,36].

About the price of the tractor, there are some factors that can affect the price of a tractor such as engine power, size, brand, and so on. For this research, we needed a tractor with at least 150 kW power to be able to carry out all the operations. There is a wide range of prices for this type of tractor in the market. To fairly show the effect of the tractor's price on the ownership costs, we consider a range of EUR 110,000–270,000 for the purchase price of the tractor. Table 13 shows the calculation of ownership costs for a cheaper tractor in both seeding and weeding operations.

Table 13. The calculation of ownership costs for the tractor in seeding and weeding operations considering the lowest purchase price for the tractor from the mentioned range.

	Seeding Operation			
		Tractor (3 m)	**Tractor (4 m)**	**Tractor (6 m)**
Ownership costs parameters	Purchase price (EUR)	114,500	118,500	126,000
	Salvage value (EUR)	11,450	11,850	12,600
	Economical life (years)	15	15	15
	Average interest rate (%)	9	9	9
	Depreciation (EUR/year)	6870	7110	7560
	Interest (EUR/year)	4637	4799	5103
	Insurance & Taxes (EUR/year)	1145	1185	1260
	Housing (EUR/year)	1145	1185	1260
Ownership costs (EUR/year)		13,797	14,279	15,183
Economical life (h)		16,000	16,000	16,000
Ownership costs (€/h)		0.86	0.89	0.95
	Weeding operation			
		Tractor (2.4 m)	Tractor (4 m)	Tractor (6 m)
Ownership costs Parameters	Purchase price (EUR)	114,200	121,000	129,900
	Salvage value (EUR)	11,420	12,100	12,990
	Economical life (years)	15	15	15
	Average interest rate (%)	9	9	9
	Depreciation (EUR/year)	6852	7260	7794
	Interest (EUR/year)	4625	4900	5261
	Insurance & Taxes (EUR/year)	1142	1210	1299
	Housing (EUR/year)	1142	1210	1299
Ownership costs (EUR/year)		13,761	14,580	15,653
Economical life (h)		16,000	16,000	16,000
Ownership costs (EUR/h)		0.86	0.91	0.98

The results show that considering a cheap price tractor can reduce the ownership costs in both seeding and weeding operations by up to 57%. However, the ownership costs take a small part of the total cost of the operation.

Agricultural robots can affect the labor market as well. For example, they can take the place of low-skilled labour in the field to save on wage and transaction costs [2] or may also allow replacing skilled labor with unskilled labor because of new automation technology in steering systems which enables driving tractor autonomously without need to a skillful operator. Moreover, agricultural robots require robot software and sensor experts which underlines the demand for more high-skilled labor due to digitalization in agriculture [37].

Considering all the positive aspects of the robotic solutions, there are some concerns about these autonomous systems such as unresolved liability issues, market maturity and availability of robots, increased dependence on the manufacturer, lack of compatibility with other technology, high level of specialization of technology, the safety of humans and animals and ethical concerns [38].

4. Conclusions

In this paper, the operational, economic, and environmental analysis of an agricultural robotic system was presented and some of the important operational indicators as well as individual cost elements with some environmental factors were provided. Additionally, a case study of a robotic system performing seeding and weeding operations in three different fields was presented and the calculated operational, economic, and environmental factors were compared to the factors of the respective conventional system. Based on the results, the robotic system has 3–9.6% better operational efficiency than conventional machinery. However, the comparison of effective field capacity shows that the conventional system has a 2–3.6 times bigger field capacity than the robotic system. The difference in the field capacity is due to the larger working width of the tractor's implementations and the higher operating speed of the tractor. Moreover, the total cost per hour of the operation executed by a robotic system in all the scenarios ranged between 40–57% less than the total cost per hour of the operation executed by conventional agricultural machinery. The tractor with 6-m working width has up to 46% less total cost than the robot due to the higher field capacity. The comparison of fuel consumption between the robotic system and the conventional one shows that the robotic system can save 11–63.3% of diesel during the entire operation. The labor cost highly affects the cost of the robotic system due to the longest duration of the operations and removing human workers from the robotic system can make it more economic, however, completely removing humans from the robotic system is not currently possible. Furthermore, the comparison of the amount of emission CO_2 gas during the operation between robotic and conventional systems shows that the robotic system can emit 11–63.3% less CO_2 gas into the environment. Finally, a comparison of the weight between the robotic and conventional systems shows that the robotic system is four times lighter than the conventional system and it has fewer effects on the soil structures in comparison with the heavy conventional machinery. In a future study, the social effects of these new robotic systems on the farmers and workers in the agricultural field are going to be considered.

Author Contributions: Conceptualization, M.V., R.G. and C.A.G.S.; methodology, M.V. and R.G.; software, M.V.; validation, M.V., R.G. and C.A.G.S.; formal analysis, M.V.; investigation, M.V.; resources, R.G. and C.A.G.S.; data curation, M.V.; writing—original draft preparation, M.V.; writing—review and editing, M.V., R.G. and C.A.G.S.; visualization, M.V.; supervision, R.G. and C.A.G.S.; project administration, C.A.G.S.; funding acquisition C.A.G.S. All authors have read and agreed to the published version of the manuscript.

Funding: This research was funded by Innovation Fund Denmark, Grant 9092-00007B AgroRobottiFleet.

Conflicts of Interest: Authors declare no conflict of interest.

References

1. Moysiadis, V.; Tsolakis, N.; Katikaridis, D.; Sørensen, C.; Pearson, S.; Bochtis, D. Mobile robotics in agricultural operations: A narrative review on planning aspects. *Appl. Sci.* **2020**, *10*, 3453. [CrossRef]
2. Marinoudi, V.; Sørensen, C.; Pearson, S.; Bochtis, D. Robotics and labour in agriculture. A context consideration. *Biosyst. Eng.* **2019**, *184*, 111–121. [CrossRef]
3. Lampridi, M.; Kateris, D.; Vasileiadis, G.; Marinoudi, V.; Pearson, S.; Sørensen, C. A case-based economic assessment of robotics employment in precision arable farming. *Agronomy* **2019**, *9*, 175. [CrossRef]
4. Saidani, M.; Pan, E.; Kim, H.; Greenlee, A.; Wattonville, J.; Yannou, B.; Leroy, Y.; Cluzel, F. Assessing the environmental and economic sustainability of autonomous systems: A case study in the agricultural industry. *Procedia CIRP* **2020**, *90*, 209–214; ISSN 2212-8271. [CrossRef]

5. Duckett, T.; Pearson, S.; Blackmore, S.; Grieve, B. *Agricultural Robotics: The Future of Robotic Agriculture. UK-RAS White Papers*; UK-RAS: Leeds, UK, 2018; ISSN 2398-4414. [CrossRef]
6. Goense, D. The economics of autonomous vehicles. In Proceedings of the VDI-MEG Conference on Agricultural Engineering, VDI-Tatung Landtechnic, Hannover, Germany, 7–8 November 2003.
7. Have, H. Effects of automation on sizes and costs of tractor and machinery. In Proceedings of the European Society of Agricultural Engineers, Paper 285, Leuven, Belgium, 12–16 September 2004.
8. Sørensen, C.G.; Madsen, N.A.; Jacobsen, B.H. Organic farming scenarios: Operational analysis and costs of implementing innovative technologies. *Biosyst. Eng.* **2005**, *91*, 127–137. [CrossRef]
9. Pedersen, S.M.; Fountas, S.; Sørensen, C.G.; Van Evert, F.K.; Blackmore, B.S. Robotic seeding: Economic perspectives. In *Precision Agriculture: Technology and Economic Perspectives. Progress in Precision Agriculture*; Pedersen, S., Lind, K., Eds.; Springer: Cham, Switzerland, 2017. [CrossRef]
10. Shaheb, M.R.; Venkatesh, R.; Shearer, S.A. A review on the effect of soil compaction and its management for sustainable crop production. *J. Biosyst. Eng.* **2021**, *46*, 417–439. [CrossRef]
11. Pedersen, S.M.; Fountas, S.; Have, H.; Blackmore, B.S. Agricultural Robots—System Analysis and economic feasibility. *Precis. Agric.* **2006**, *7*, 295–308. [CrossRef]
12. Robotic Solutions for Sustainable Farming. Available online: https://dahliarobotics.com (accessed on 23 January 2023).
13. Powerful, Energy-Efficient and Autonomous. Available online: https://farming-revolution.com (accessed on 23 January 2023).
14. Product. Available online: https://farmdroid.dk/en/product (accessed on 23 January 2023).
15. Autonomous Oz Weeding Robot. Available online: https://www.naio-technologies.com/en/oz (accessed on 23 January 2023).
16. Robotti 150D, Agrointelli. Available online: https://agrointelli.com/robotti/150D/ (accessed on 14 December 2022).
17. Bhimanpallewar, R.; Narasingarao, M. AgriRobot: Implementation and evaluation of an automatic robot for seeding and fertilizer microdosing in precision agriculture. *Int. J. Agric. Resour. Gov. Ecol.* **2020**, *16*, 33. [CrossRef]
18. Naik, N.S.; Shete, V.V.; Danve, S.R. Precision agriculture robot for seeding function. In Proceedings of the International Conference on Inventive Computation Technologies (ICICT), Coimbatore, India, 26–27 August 2016; p. 13. [CrossRef]
19. Sunitha, K.A.; Suraj, G.S.G.S.; Sowrya, C.H.P.N.; Sriram, G.A.; Shreyas, D.; Srinivas, T. Agricultural robot designed for seeding mechanism. In *IOP Conference Series: Materials Science and Engineering*; IOP Publishing: Bristol, UK, 2017; Volume 197.
20. Gonzalez-de-Soto, M.; Emmi, L.; Perez-Ruiz, M.; Aguera, J.; Gonzalez-de-Santos, P. Autonomous systems for precise spraying: Evaluation of a robotized patch sprayer. *Biosyst. Eng.* **2016**, *146*, 165–182. [CrossRef]
21. Berenstein, R.; Edan, Y. Human-robot collaborative site-specific sprayer. *J. Field Robot.* **2017**, *34*, 1519–1530. [CrossRef]
22. Oduma, O.; Oluka, S.; Nwakuba, N.; Ntunde, D. Agricultural field machinery selection and utilization for improved farm operations in South-East Nigeria: A review. *Poljopr. Teh.* **2019**, *44*, 44–58. [CrossRef]
23. Pflueger, B. How to Calculate Machinery Ownership and Operating Costs. SDSU Extension Circulars 2015. Paper 485. Available online: http://openprairie.sdstate.edu/extension_circ/485 (accessed on 30 January 2023).
24. Edwards, W. Estimating Farm Machinery Costs: Ag Decision Maker, Lowa State University Extension and Outreach. Available online: https://www.extension.iastate.edu/agdm/crops/html/a3-29.html (accessed on 31 October 2022).
25. American Society of Agricultural and Biological Engineers (ASABE). *Agricultural Machinery Management—ASAE EP496.3 FEB2006 (R2015)*; ASABE: St. Joseph, MI, USA, 2006.
26. Tihanov, G.; Ivanov, N. Fuel consumption of a machine-tractor unit in direct sowing of wheat. *Agric. Sci. Technol.* **2021**, *13*, 40–42. [CrossRef]
27. Hunt, D.; Wilson, D. *Farm Power and Machinery Management*, 11th ed.; Waveland Press: Long Grove, IL, USA, 2015; (ED-TECH).
28. NTTL. Nebraska Tractor Test 2085 for John Deere 7250R.Lincoln, NE: Nebraska Tractor Test Laboratory 2014. Available online: http://tractortestlab.unl.edu (accessed on 30 January 2023).
29. ANSI ASABE D497-4-2003 Scribd. Available online: https://www.scribd.com/document/345945527/ANSI-ASABE-D497-4-2003 (accessed on 1 November 2022).
30. Manzone, M.; Calvo, A. Energy and CO_2 analysis of poplar and maize crops for biomass production in north Italy. *Renew Energy* **2016**, *86*, 675–681. [CrossRef]
31. Soane, B.D.; Ball, B.C.; Arvidsson, J.; Basch, G.; Moreno, F.; Roger-Estrade, J. No-till in northern, western and south-western Europe: A review of problems and opportunities for crop production and the environment. *Soil Tillage Res.* **2012**, *118*, 66–87. [CrossRef]
32. Filipovic, D.; Kosutic, S.; Gospodaric, Z.; Zimmer, R.; Banaj, D. The possibilities of fuel savings and the reduction of CO_2 emissions in the soil tillage in Croatia. *Agric. Ecosyst. Environ.* **2006**, *115*, 290–294. [CrossRef]
33. American Society of Agricultural and Biological Engineers (ASABE). *Agricultural Machinery Management Data—ASAE D497.5 FEB2006*; ASABE: St. Joseph, MI, USA, 2006.
34. Augustin, K.; Kuhwald, M.; Brunotte, J.; Duttmann, R. Wheel load and wheel pass frequency as indicators for soil compaction risk: A four-year analysis of traffic intensity at Field Scale. *Geosciences* **2020**, *10*, 292. [CrossRef]
35. Pulido-Moncada, M.; Munkholm, L.J.; Schjønning, P. Wheel load, repeated wheeling, and traction effects on subsoil compaction in Northern Europe. *Soil Tillage Res.* **2019**, *186*, 300–309. [CrossRef]

36. Redhead, F.; Snow, S.; Vyas, D.; Bawden, O.; Russell, R.; Perez, T.; Brereton, M. Bringing the Farmer Perspective to Agricultural Robots. In *Proceedings of the 33rd Annual ACM Conference Extended Abstracts on Human Factors in Computing Systems 2015, Seoul, Republic of Korea, 18–23 April 2015*; ACM: New York, NY, USA, 2015; pp. 1067–1072. [CrossRef]
37. Rotz, S.; Gravely, E.; Mosby, I.; Duncan, E.; Finnis, E.; Horgan, M.; LeBlanc, J.; Martin, R.; Neufeld, H.T.; Nixon, A.; et al. Automated pastures and the digital divide: How agricultural technologies are shaping labour and rural communities. *J. Rural. Stud.* **2019**, *68*, 112–122. [CrossRef]
38. Spykman, O.; Gabriel, A.; Ptacek, M.; Gandorfer, M. Farmers' perspectives on field crop robots—Evidence from Bavaria, Germany. *Comput. Electron. Agric.* **2021**, *186*, 106176. [CrossRef]

Disclaimer/Publisher's Note: The statements, opinions and data contained in all publications are solely those of the individual author(s) and contributor(s) and not of MDPI and/or the editor(s). MDPI and/or the editor(s) disclaim responsibility for any injury to people or property resulting from any ideas, methods, instructions or products referred to in the content.

MDPI AG
Grosspeteranlage 5
4052 Basel
Switzerland
Tel.: +41 61 683 77 34

MDPI Books Editorial Office
E-mail: books@mdpi.com
www.mdpi.com/books

Disclaimer/Publisher's Note: The title and front matter of this reprint are at the discretion of the Topic Editors. The publisher is not responsible for their content or any associated concerns. The statements, opinions and data contained in all individual articles are solely those of the individual Editors and contributors and not of MDPI. MDPI disclaims responsibility for any injury to people or property resulting from any ideas, methods, instructions or products referred to in the content.

www.ingramcontent.com/pod-product-compliance
Lightning Source LLC
Chambersburg PA
CBHW042014260926
51850CB00020B/3406
* 9 7 8 3 7 2 5 8 2 7 1 1 4 *